S0-CYD-656

STRUCTURED CATALYSTS AND REACTORS

CHEMICAL INDUSTRIES

A Series of Reference Books and Textbooks

Consulting Editor

HEINZ HEINEMANN
Berkeley, California

1. *Fluid Catalytic Cracking with Zeolite Catalysts,* Paul B. Venuto and E. Thomas Habib, Jr.
2. *Ethylene: Keystone to the Petrochemical Industry,* Ludwig Kniel, Olaf Winter, and Karl Stork
3. *The Chemistry and Technology of Petroleum,* James G. Speight
4. *The Desulfurization of Heavy Oils and Residua,* James G. Speight
5. *Catalysis of Organic Reactions,* edited by William R. Moser
6. *Acetylene-Based Chemicals from Coal and Other Natural Resources,* Robert J. Tedeschi
7. *Chemically Resistant Masonry,* Walter Lee Sheppard, Jr.
8. *Compressors and Expanders: Selection and Application for the Process Industry,* Heinz P. Bloch, Joseph A. Cameron, Frank M. Danowski, Jr., Ralph James, Jr., Judson S. Swearingen, and Marilyn E. Weightman
9. *Metering Pumps: Selection and Application,* James P. Poynton
10. *Hydrocarbons from Methanol,* Clarence D. Chang
11. *Form Flotation: Theory and Applications,* Ann N. Clarke and David J. Wilson
12. *The Chemistry and Technology of Coal,* James G. Speight
13. *Pneumatic and Hydraulic Conveying of Solids,* O. A. Williams
14. *Catalyst Manufacture: Laboratory and Commercial Preparations,* Alvin B. Stiles
15. *Characterization of Heterogeneous Catalysts,* edited by Francis Delannay
16. *BASIC Programs for Chemical Engineering Design,* James H. Weber
17. *Catalyst Poisoning,* L. Louis Hegedus and Robert W. McCabe
18. *Catalysis of Organic Reactions,* edited by John R. Kosak
19. *Adsorption Technology: A Step-by-Step Approach to Process Evaluation and Application,* edited by Frank L. Slejko
20. *Deactivation and Poisoning of Catalysts,* edited by Jacques Oudar and Henry Wise
21. *Catalysis and Surface Science: Developments in Chemicals from Methanol, Hydrotreating of Hydrocarbons, Catalyst Preparation, Monomers and Polymers, Photocatalysis and Photovoltaics,* edited by Heinz Heinemann and Gabor A. Somorjai
22. *Catalysis of Organic Reactions,* edited by Robert L. Augustine
23. *Modern Control Techniques for the Processing Industries,* T. H. Tsai, J. W. Lane, and C. S. Lin

24. *Temperature-Programmed Reduction for Solid Materials Characterization,* Alan Jones and Brian McNichol
25. *Catalytic Cracking: Catalysts, Chemistry, and Kinetics,* Bohdan W. Wojciechowski and Avelino Corma
26. *Chemical Reaction and Reactor Engineering,* edited by J. J. Carberry and A. Varma
27. *Filtration: Principles and Practices: Second Edition,* edited by Michael J. Matteson and Clyde Orr
28. *Corrosion Mechanisms,* edited by Florian Mansfeld
29. *Catalysis and Surface Properties of Liquid Metals and Alloys,* Yoshisada Ogino
30. *Catalyst Deactivation,* edited by Eugene E. Petersen and Alexis T. Bell
31. *Hydrogen Effects in Catalysis: Fundamentals and Practical Applications,* edited by Zoltán Paál and P. G. Menon
32. *Flow Management for Engineers and Scientists,* Nicholas P. Cheremisinoff and Paul N. Cheremisinoff
33. *Catalysis of Organic Reactions,* edited by Paul N. Rylander, Harold Greenfield, and Robert L. Augustine
34. *Powder and Bulk Solids Handling Processes: Instrumentation and Control,* Koichi Iinoya, Hiroaki Masuda, and Kinnosuke Watanabe
35. *Reverse Osmosis Technology: Applications for High-Purity-Water Production,* edited by Bipin S. Parekh
36. *Shape Selective Catalysis in Industrial Applications,* N. Y. Chen, William E. Garwood, and Frank G. Dwyer
37. *Alpha Olefins Applications Handbook,* edited by George R. Lappin and Joseph L. Sauer
38. *Process Modeling and Control in Chemical Industries,* edited by Kaddour Najim
39. *Clathrate Hydrates of Natural Gases,* E. Dendy Sloan, Jr.
40. *Catalysis of Organic Reactions,* edited by Dale W. Blackburn
41. *Fuel Science and Technology Handbook,* edited by James G. Speight
42. *Octane-Enhancing Zeolitic FCC Catalysts,* Julius Scherzer
43. Oxygen in Catalysis, Adam Bielański and Jerzy Haber
44. *The Chemistry and Technology of Petroleum: Second Edition, Revised and Expanded,* James G. Speight
45. *Industrial Drying Equipment: Selection and Application,* C. M. van't Land
46. *Novel Production Methods for Ethylene, Light Hydrocarbons, and Aromatics,* edited by Lyle F. Albright, Billy L. Crynes, and Siegfried Nowak
47. *Catalysis of Organic Reactions*, edited by William E. Pascoe
48. *Synthetic Lubricants and High-Performance Functional Fluids*, edited by Ronald L. Shubkin
49. *Acetic Acid and Its Derivatives*, edited by Victor H. Agreda and Joseph R. Zoeller
50. *Properties and Applications of Perovskite-Type Oxides*, edited by L. G. Tejuca and J. L. G. Fierro
51. *Computer-Aided Design of Catalysts*, edited by E. Robert Becker and Carmo J. Pereira
52. *Models for Thermodynamic and Phase Equilibria Calculations*, edited by Stanley I. Sandler

53. *Catalysis of Organic Reactions*, edited by John R. Kosak and Thomas A. Johnson
54. *Composition and Analysis of Heavy Petroleum Fractions*, Klaus H. Altgelt and Mieczyslaw M. Boduszynski
55. *NMR Techniques in Catalysis,* edited by Alexis T. Bell and Alexander Pines
56. *Upgrading Petroleum Residues and Heavy Oils*, Murray R. Gray
57. *Methanol Production and Use*, edited by Wu-Hsun Cheng and Harold H. Kung
58. *Catalytic Hydroprocessing of Petroleum and Distillates*, edited by Michael C. Oballah and Stuart S. Shih
59. *The Chemistry and Technology of Coal: Second Edition, Revised and Expanded,* James G. Speight
60. *Lubricant Base Oil and Wax Processing*, Avilino Sequeira, Jr.
61. *Catalytic Naphtha Reforming: Science and Technology*, edited by George J. Antos, Abdullah M. Aitani, and José M. Parera
62. *Catalysis of Organic Reactions*, edited by Mike G. Scaros and Michael L. Prunier
63. *Catalyst Manufacture,* Alvin B. Stiles and Theodore A. Koch
64. *Handbook of Grignard Reagents,* edited by Gary S. Silverman and Philip E. Rakita
65. *Shape Selective Catalysis in Industrial Applications: Second Edition, Revised and Expanded*, N. Y. Chen, William E. Garwood, and Francis G. Dwyer
66. *Hydrocracking Science and Technology*, Julius Scherzer and A. J. Gruia
67. *Hydrotreating Technology for Pollution Control: Catalysts, Catalysis, and Processes*, edited by Mario L. Occelli and Russell Chianelli
68. *Catalysis of Organic Reactions*, edited by Russell E. Malz, Jr.
69. *Synthesis of Porous Materials: Zeolites, Clays, and Nanostructures,* edited by Mario L. Occelli and Henri Kessler
70. *Methane and Its Derivatives*, Sunggyu Lee
71. *Structured Catalysts and Reactors*, edited by Andrzej Cybulski and Jacob A. Moulijn
72. *Industrial Gases in Petrochemical Processing*, Harold Gunardson
73. *Clathrate Hydrates of Natural Gases: Second Edition, Revised and Expanded*, E. Dendy Sloan, Jr.

ADDITIONAL VOLUMES IN PREPARATION

Fluid Cracking Catalysts, edited by Mario L. Occelli and Paul O'Connor

STRUCTURED CATALYSTS AND REACTORS

edited by
Andrzej Cybulski
Institute of Physical Chemistry
Polish Academy of Sciences
Warsaw, Poland

Jacob A. Moulijn
Delft University of Technology
Delft, The Netherlands

Marcel Dekker, Inc. New York • Basel • Hong Kong

Library of Congress Cataloging-in-Publication Data

Structured catalysts and reactors / edited by Andrzej Cybulski, Jacob
A. Moulijn.
p. cm. — (Chemical industries ; v. 71)
Includes bibliographical references and index.
ISBN: 0-8247-9921-6
1. Catalysts. 2. Chemical reactors. I. Cybulski, Andrzej.
II. Moulijn, Jacob A. III. Series.
TP156.C35S77 1998
660′.2995—dc21 97-33104
CIP

The publisher offers discounts on this book when ordered in bulk quantities. For more information, write to Special Sales/Professional Marketing at the address below.

This book is printed on acid-free paper.

Marcel Dekker, Inc.
270 Madison Avenue, New York, New York 10016
http://www.dekker.com

Current printing (last digit):
10 9 8 7 6 5 4 3 2 1

PRINTED IN THE UNITED STATES OF AMERICA

Preface

Heterogeneous catalytic processes are among the main ways to decrease the consumption of raw materials in chemical industries and to decrease the emission of pollutants of all kinds to the environment via an increase in process selectivity. Selectivity can be improved by the modification of catalyst composition and surface structure and/or by the modification of pellet dimensions, shape, and texture, i.e., pore size distribution, pore shape, length, and cross-sectional surface area (distribution). Until recently, the limiting factor in the latter modifications has been the particles' size, to which the length of diffusion in pores is related. The size should not be too small because of the significantly higher pressure drop for such small particles. Shell catalysts, which contain catalytic species concentrated near the outer particle's surface, are a remedy for improving selectivity, keeping pressure drop at a reasonable level. Pressure drop can be the limiting factor even for such catalysts, e.g., when large quantities of raw materials must be processed or when the higher pressure drop results in the significantly higher consumption of raw materials. For instance, converting huge amounts of natural gas in remote areas would require equipment characterized by low pressure drop. Otherwise the cost of processing would be too high to make the process economical. Too high a pressure drop in catalytic car mufflers would result in an increase in fuel consumption by several percent. This would mean a several-percent-higher consumption of crude oil for transportation. An inherent feature of conventional packed-bed reactors is their random and structural maldistributions. A structural maldistribution in fixed-bed reactors originates from the looser packing of particles near the reactor walls. This results in a tendency to bypass the core of the bed, even if the initial distribution of fluid(s) is uniform. The uniformly distributed liquid tends to flow to the walls, and this can drastically alter its residence time from the design value. Random maldistributions result in: (1) a nonuniform access of reactants to the catalytic surface, worsening the overall process performance, and (2) unexpected hot spots and thermal runaways of exothermic reactions (mainly in three-phase reactions).

Structured catalysts (reactors) are promising as far as the elimination of these drawbacks of fixed beds is concerned. Three basic kinds of structured catalysts can be distinguished:

1. *Monolithic catalysts (honeycomb catalysts),* in the form of continuous unitary structures that contain small passages. The catalytically active material is present on or inside the walls of these passages. In the former case, a ceramic or metallic support is coated with a layer of material in which active ingredients are dispersed.

2. *Membrane catalysts* are structures with permeable walls between passages. The membrane walls exhibit selectivity in transport rates for the various components present. A slow radial mass transport can occur, driven by diffusion or solution/diffusion mechanisms in the permeable walls.
3. *Arranged catalysts.* Particulate catalysts arranged in *arrays* belong to this class of structured catalysts. Another group of arranged catalysts are *structural catalysts*, derived from structural packings for distillation and absorption columns and static mixers. These are structures consisting of superimposed sheets, possibly corrugated before stacking. The sheets are covered by an appropriate catalyst support in which active ingredients are incorporated. The structure is an open cross-flow structure characterized by intensive radial mixing.

Usually, structured catalysts are structures of large void fraction ranging from 0.7 to more than 0.9, compared to 0.5 in fixed beds. The path the fluids follow in structured reactors is much less twisty (e.g., straight channels in monoliths) than that in conventional reactors. Finally, structured reactors are operated in a different hydrodynamic regime. For single-phase flow the regime is laminar, and the eddies characteristic of packed beds are absent. For multiphase systems various regimes exist, but here also the eddies typical of packed beds are absent. For these reasons, the pressure drop in structured catalysts must be significantly lower than that for the randomly packed bed of particles. Indeed, pressure drop in monolithic reactors is up to two orders of magnitude lower than that in packed-bed reactors.

Catalytic species are incorporated either into a very thin layer of a porous catalytic support deposited on the structured elements or into the thin elements themselves. The short diffusion distance inside the thin layer of the structured catalysts results in higher catalyst utilization and can contribute to an improvement of selectivity for processes controlled by mass transfer within the catalytic layer. Contrary to conventional fixed-bed reactors, the thickness of the catalytic layer in monolithic reactors can be significantly reduced with no penalty paid for the increase in pressure drop. Membrane catalysts provide a unique opportunity to supply reactants to the reaction mixture gradually along the reaction route or to withdraw products from the reaction mixture as they are formed. The former mode of carrying out complex reactions might be very effective in controlling undesired reactions whose rates are strongly dependent on the concentration of the added reactant. The latter mode might result in higher conversions for reversible reactions which are damped by products. The use of catalytic membranes operated in any of these modes can also contribute to significant improvement in selectivity.

The regular structure of the arranged catalysts prevents the formation of the random maldistributions characteristic of beds of randomly packed particles. This reduces the probability of the occurrence of hot spots resulting from flow maldistributions. Scale-up of monolithic and membrane reactors can be expected to be straightforward, since the conditions within the individual channels are scale invariant.

Finally, structured catalysts and reactors constitute a significant contribution to the search for better catalytic processes via improving mass transfer in the catalytic layer and thus improving activity and selectivity (also using membrane catalysts), decreasing operation costs through lowering pressure drop, and eliminating maldistributions.

Structured catalysts, mainly monolithic ones, are now used predominantly in environmental applications, first of all in the cleaning of automotive exhaust gases. Monolithic reactors have become the most commonly used sort of chemical reactors: several hundred million small monolithic reactors are moving with our cars! Monolithic cleaners of flue gases are now a standard unit. Monolithic catalysts are also close to commercialization in the combustion of fuels for gas turbines, boilers, heaters, etc. The catalytic combustion reduces NO_x formation, and the use of low-pressure-drop catalysts makes the process more economical. Some special features of monolithic catalysts make the burning of LHV fuels in monolithic units much easier than in particulate catalysts. There are some characteristics that make structured catalysts also of interest for three-phase reactions. Several three-phase processes are in the development stage. One, the hydrogenation step in the production of H_2O_2 using the alkylanthraquinone process, has already reached full scale, with several plants in operation.

Interest in structured catalysts is steadily rising due to the already proven, and potential, advantages of the catalysts. Some review articles regarding different aspects of structured catalysts have been published in the last decade [see A. Cybulski and J. A. Moulijn, *Catal. Rev.-Sci. Eng. 36*(1):179–270 (1994); H. P. Hsieh, *Catal. Rev.-Sci. Eng. 33*(1&2):1–70 (1991); S. Irandoust and B. Andersson, *Catal. Rev.-Sci. Eng. 30*(3):341–392 (1988); L. D. Pfefferle and W. C. Pfefferle, *Catal. Rev.-Sci. Eng. 29*(2–3):219–267 (1987); and G. Saracco and V. Specchia, *Catal. Rev.-Sci. Eng. 36*(1):305–384 (1994)]. These articles do not cover the whole area of structured catalysts and reactors. Moreover, the science and applications of structured catalysts and reactors are developing very fast. The time has come to devote an entire book to structured catalysts and reactors. In this compilation an attempt is made to give detailed information on all structures known to date and on all aspects of structured catalysts and reactors containing them: catalyst preparation and characterization, catalysts and process development, modeling and optimization, and finally reactor design and operation. As such, the book is dedicated to all readers who are involved in the development of catalytic processes, from R&D to process engineering. A very important area of structuring in catalysis is that directed at a catalytic surface, microstructure, and structuring the shapes and size of the catalytic bodies. This area is essentially covered by publications concerning more fundamental approaches to heterogeneous catalysis. A lot of the relevant information is scale independent and, as a consequence, is not unique to structured catalytic reactors. Therefore these activities are described only briefly in the book.

The book starts with an overview on structured catalysts (Chapter 1). The rest of the book is divided into four parts. The first three parts deal with structures differing from each other significantly in conditions for mass transfer in the reaction zone. The fourth part is dedicated to catalyst design and preparation.

Part I deals with monolithic catalysts. Chapters 2 and 3 deal with configurations, microstructure, physical properties, and the manufacture of ceramic and metallic monoliths. Monolithic catalysts for cleaning the exhaust gas from gasoline-fed engines are dealt with in Chapter 4, including fundamentals and exploitation experience. Chapter 5 is devoted to commercial and developmental catalysts for protecting the environment. Fundamentals and applications of monoliths for selective catalytic reduction are given. Unconventional reactors used in this field (reverse-flow reactors, rotating monoliths) are also discussed. Materials, activity, and stability of catalysts for catalytic combustion and practical aspects of applications of monolithic catalysts in this area are discussed in Chapter 6. The use of monolithic catalysts for noncombustion and nonenvironmental purposes is discussed in Chapter 7. Chapter 8 is devoted to the modeling of monolithic catalysts for two-phase processes (gaseous reactants/solid catalyst). Chapters 9–11 deal with three-phase monolithic processes. Both catalytic and engineering aspects of these processes are discussed.

Arranged catalysts allowing for convective mass transfer over the cross section of the reactor are discussed in Part II. Conventional particulate catalysts arranged in arrays are dealt with in Chapters 12 and 13. Current and potential applications of ordered structures of different kinds (parallel-passage and lateral-flow reactors, bead-string reactor) are mentioned. Chapter 14 is devoted to structural catalysts derived from static mixers and column packings, typified by low pressure drop. Configurations, methods of incorporation of catalytically active components into the structure, commercial and developmental catalysts, flow characteristics, mixing, mass and heat transfer, current and potential applications in two- and three-phase processes, reactor configurations, and design procedures are discussed. Chapter 15 deals with catalytic filters in general.

Part III of the book provides information about the structured catalysts of monolithic type with permeable walls, i.e., catalytically active membranes. Catalytic membranes create a unique opportunity to couple processes opposite in character (e.g., hydrogenation/dehydrogenation, endothermic/exothermic) via the combination of reaction and separation. Catalytic membranes can allow for the easy control of reactant addition or product withdrawal along the reaction route. Chapter 16 deals with membrane reactors with metallic walls permeable to some gases. The properties of metallic membranes, permeation mechanisms in metallic membranes, the preparation of membranes, commercial membranes, modeling and design, engineering and operating considerations, and finally current and potential applications of metallic membranes are discussed. Chapter 17 presents inorganic membrane reactors—materials, membrane microstructures, commercial membranes, cross-flow reactors, modeling and design, engineering and operating issues, current and potential applications. Chapter 18 is dedicated to the special sort of catalytic filters used for cleaning exhausts from diesel engines. Recent developments in the field of advanced membranes, in the form of zeolitic membranes, are discussed in Chapter 19. Cross-flow reactors are the subject of Chapter 20.

The last part of the book (Part IV) discusses techniques for incorporating catalytic species into the structured catalytic support (Chapter 21) and the computer-aided design of catalysts aiming at an optimal porous structure for the catalytic support (Chapter 22).

The amount of detail in this book varies, depending on whether the catalyst/reactor is in the developmental stage or has been commercialized. The know-how gained in process development has commercial value, and this usually inhibits the presentation of the details of the process/reactor/catalyst. Consequently, well-established processes/reactors/catalysts are described more generally. Projects at an earlier stage presented in this book are being developed at universities, which usually reveal more details. Each chapter was designed as a whole that can be read without reference to the others. Therefore repetitions and overlapping between the chapters of this book are unavoidable.

The authors of individual chapters are top specialists in their areas. They comprise an international group of scientists and practitioners (Great Britain, Italy, The Netherlands, Poland, Russia, Sweden, Switzerland, and the U.S.A.) from universities and companies that are advanced in the technology of structured catalysts. The editors express their gratitude to all of the contributors for sharing their experience. The editors also appreciate the administrative assistance of Mrs. Qwen Klis and the help of Ms. Annelies van Diepen in the preparation of the subject index.

Andrzej Cybulski
Jacob A. Moulijn

Contents

Contributors

Bengt Andersson Department of Chemical Reaction Engineering, Chalmers University of Technology Göteborg, Sweden

Oliver Bailer Department of Technology & Development, Catalyst Technology, Sulzer Chemtech Ltd., Winterthur, Switzerland

Alessandra Beretta Dipartimento di Chimica Industriale e Ingegneria Chimica "G. Natta," Politecnico di Milano, Milano, Italy

Hans Peter Calis Department of Chemical Engineering, Delft University of Technology, Delft, The Netherlands

Peter Collins Department of Sales & Marketing, Catalyst Technology, Sulzer Chemtech Ltd., Winterthur, Switzerland

Andrzej Cybulski Department of Process Development, CHEMIPAN Research & Development Laboratories, Institute of Physical Chemistry, Polish Academy of Sciences, Warsaw, Poland

Pio Foratti Dipartimento di Chimica Industriale e Ingegneria Chimica "G. Natta," Politecnico di Milano, Milano, Italy

Albert W. Gerritsen Department of Chemical Engineering, Delft University of Technology, Delft, The Netherlands

Gianpiero Groppi Dipartimento di Chimica Industriale e Ingegneria Chimica "G. Natta," Politecnico di Milano, Milano, Italy

Vladimir M. Gryaznov Department of Physical and Colloid Chemistry, Russian University of People's Friendship, Moscow, Russia

Suresh T. Gulati Department of Characterization Science and Services, Corning Incorporated, Corning, New York

J. H. B. J. Hoebink Laboratory of Chemical Technology, Eindhoven University of Technology, Eindhoven, The Netherlands

Said Irandoust Department of Chemical Reaction Engineering, Chalmers University of Technology, Göteborg, Sweden

Sven G. Järås Department of Chemical Engineering and Technology, KTH—Royal Institute of Technology, Stockholm, Sweden

Freek Kapteijn Department of Chemical Engineering, Delft University of Technology, Delft, The Netherlands

Paul J. M. Lebens Department of Chemical Engineering, Delft University of Technology, Delft, The Netherlands

Reginald Mann Department of Chemical Engineering, University of Manchester Institute of Science and Technology (UMIST), Manchester, England

G. B. Marin Laboratory of Chemical Technology, Eindhoven University of Technology, Eindhoven, The Netherlands

P. Govind Menon Department of Chemical Engineering and Technology, KTH—Royal Institute of Technology, Stockholm, Sweden

Jacob A. Moulijn Department of Chemical Engineering, Delft University of Technology, Delft, The Netherlands

Natalia V. Orekhova Topchiev Institute of Petrochemical Synthesis, Russian Academy of Sciences, Moscow, Russia

Guido Saracco Dipartimento di Scienza dei Materiali ed Ingegneria Chimica, Politecnico di Torino, Torino, Italy

Nils-Herman Schöön Department of Chemical Reaction Engineering, Chalmers University of Technology, Göteborg, Sweden

Swan Tiong Sie Department of Chemical Technology Engineering, Delft University of Technology, Delft, The Netherlands

Vito Specchia Dipartimento di Scienza dei Materiali ed Ingegneria Chimica, Politecnico di Torino, Torino, Italy

Jean-Paul Stringaro Catalyst Technology, Sulzer Chemtech Ltd., Winterthur, Switzerland

Kálmán Takács Department of Chemical Technology, Technical University of Budapest, Budapest, Hungary.

Enrico Tronconi Dipartimento di Chimica Industriale e Ingegneria Chimica "G. Natta," Politecnico di Milano, Milano, Italy

Martyn V. Twigg Catalytic Systems Division, Johnson Matthey, Royston, Hertfordshire, England

Jolinde M. van de Graaf Department of Chemical Engineering, Delft University of Technology, Delft, The Netherlands

C. M. van den Bleek Department of Chemical Engineering, Delft University of Technology, Delft, The Netherlands

Gerald E. Voecks Jet Propulsion Laboratory, California Institute of Technology, Pasadena, California

Dennis E. Webster Catalytic Systems Division, Johnson Matthey, Royston, Hertfordshire, England

Anthony J. Wilkins Catalytic Systems Division, Johnson Matthey, Royston, Hertfordshire, England

Xiaoding Xu Department of Chemical Engineering, Delft University of Technology, Delft, The Netherlands

Marcus F. M. Zwinkels Department of Chemical Engineering and Technology, KTH—Royal Institute of Technology, Stockholm, Sweden

1

The Present and the Future of Structured Catalysts—An Overview

Andrzej Cybulski
Polish Academy of Sciences, Warsaw, Poland

Jacob A. Moulijn
Delft University of Technology, Delft, The Netherlands

I. INTRODUCTION

Conventional fixed-bed catalytic reactors have some obvious disadvantages, such as maldistributions of various kinds (resulting in nonuniform access of reactants to the catalytic surface and nonoptimal local process conditions), high pressure drop in the bed, and sensitivity to fouling by dust. The search for means allowing for elimination of these setbacks has lead researchers to *structured catalysts.* Three basic kinds of structured catalysts can be distinguished:

1. *Monolithic catalysts.* These are continuous unitary structures containing many narrow, parallel straight or zigzag passages. Catalytically active ingredients are dispersed uniformly over the whole porous ceramic monolithic structure (so-called *incorporated monolithic catalysts*) or are in a layer of porous material that is deposited on the walls of channels in the monolith's structure (*washcoated monolithic catalysts*). The name *monolith* stems from Greek and means "composed of a single rock." However, the material of construction is not limited to ceramics but commonly includes metals, as well. Although not fully correct, we speak of ceramic and metallic monoliths. Initially, the cross section of channels in monoliths was like a *honeycomb structure*, and this name is still in common use.

2. *Membrane catalysts.* An interaction between the passages in the monolith can occur if the walls are permeable. Such catalysts are called *wall-through monolithic catalysts* or *membrane catalysts.* The catalytically active material is present on or inside the walls of these passages. Radial mass transport occurs by diffusion through the pores of the permeable walls. Consequently, mass fluxes through the walls are often rather small.

3. *Arranged catalysts.* Structured catalysts that allow for a relatively fast mass transport over the reaction zone in the direction perpendicular to flow are classified here as arranged catalysts. Particulate catalysts arranged in arrays belong to this class. Any other nonparticulate catalysts, such as packings covered with catalytically active material, similar in design to those used in distillation and absorption columns and/or static mixers, are also classified into this group.

Table 1 Classification of Structured Catalysts

1. Design
   1.1 Monolithic catalysts; single-passage flow monoliths
   1.2 Membrane catalysts; wall-through catalysts
   1.3 Arrange catalysts
2. Support material
   2.1 Ceramics
   2.2 Metal
3. Mixing conditions
   3.1 A very limited radial mixing inside the channel and no mass exchange between individual channels with resulting zero mixing over the reactor (monolithic catalysts)
   3.2 An intense radial mixing over the cross section of the reactor (arranged catalysts)
   3.2 A very limited radial mixing inside the channel with a limited mass transfer between adjacent passages; a very limited radial mixing over the reactor (membrane catalysts)
4. Mode of operation
   4.1 Steady state (e.g., treatment of industrial off-gases)
   4.2 Nonstationary processes
      • periodic changes (e.g., catalytic mufflers, reverse-flow converters, rotating monoliths)
      • oscillations (e.g., Taylor flow of gas/liquid mixtures through channels)

It is clear from this classification scheme that by *structured catalysts* we mean regular structures free of randomness at a reactor's level, which is characteristic for a randomly packed bed of particles of various shapes. These structures are spatially arranged in a reactor. Structures at or below the level of particles (micropores in zeolites, macropores in MCM-41, or shaped pellets such as lobes, miniliths, and the like) are not dealt with in this book, since randomness of packing will always result in the lack of a uniform structure at the level of a reactor.

The most characteristic features of structured catalysts are given in Table 1. The main difference between the three types of structured catalysts we have just distinguished consists in the rates of radial mixing in the reactor containing the structured catalyst: from zero radial mass transfer in monolithic reactors to a very intense radial mass transfer in reactors with structural catalysts. For the sake of simplicity, reactors containing monolithic or membrane catalysts will be referred to as monolithic or membrane reactors, respectively.

II. MONOLITHIC CATALYSTS

The essence of monolithic catalysts is the very thin layers, in which internal diffusion resistance is small. As such, monolithic catalysts create a possibility to control the selectivity of many complex reactions. Pressure drop in straight, narrow channels through which reactants move in the laminar regime is smaller by two or three orders of magnitude than in conventional fixed-bed reactors. Provided that feed distribution is optimal, flow conditions are practically the same across a monolith due to the very high reproducibility of size and surface characteristics of individual monolith passages. This reduces the probability of occurrence of hot spots resulting from maldistributions characteristic of randomly packed catalyst beds.

Comprehensive reviews on catalytic combustion, including the use of monoliths for automotive converters, have been published in the last decades [1–5]. Reviews on monoliths including also nonenvironmental and noncombustion applications of monoliths were also published [6–8]. The steadily increasing interest in monolithic catalysts is reflected in the literature. This is illustrated by the results of the computer literature search in *Chemical Abstracts* (see Table 2). The first world conference on monolithic catalysts was organized in 1995 in St. Petersburg, Russia. As is usually the case in new developments, monolithic catalysts are covered by patents. This, along with the rather costly way of manufacturing them, has contributed to a relatively high price for monolithic catalysts (two to three times more expensive than particulate catalysts). However, the first patents are now close to expiration. Moreover, many ceramic and metallic supports can be manufactured in bulk, and this reduces their price, thereby increasing their accessibility for many applications. Therefore we expect that much cheaper monolithic catalysts will soon become available, not only in combustion processes that are stimulated by legislation but also for a number of typical chemical processes, just because of the technical and economic advantages of monolithic processes.

Monolithic catalysts have found many applications in combustion and environmental uses, e.g., as afterburners of engine exhausts and for removal of harmful compounds from industrial off-gases. The first important industrial applications of monolithic catalysts were for decolorization of nitric acid tail gas and for car exhaust emission control. Beginning in the late 1960s, investigations on monoliths were expanded by car manufacturers and the industries responsible for emissions of considerable amounts of gaseous air pollutants. The main reason for focusing research on monoliths was their low pressure drop. Through the Clean Air Act, legislators of California stimulated a search for effective afterburner catalysts that would not produce high pressure drop. Conventional particulate catalysts were sufficiently active in the removal of carbon monoxide, unburned hydrocarbons (UHC), and nitrogen oxides (NO_x). However, high pressure drop in catalytic mufflers filled with particulate catalysts resulted in a several percent increase in fuel consumption. The low pressure drop in monoliths was also important for off-gas cleaners. Usually there is an insufficient surplus of pressure before the stack to allow the installation of devices producing high pressure drop. Developments in the production of both ceramic and metallic monolithic supports resulted in an industrial production of monolithic catalysts of long lifetime that met the high requirements of units for efficient oxidizing of CO and UHC. By 1975 the first cars equipped with catalytic converters became available. In 1985 approximately 100 million catalytic mufflers were in use in the United States. Now several hundred million converters are in everyday operation, mostly in the United States and Europe.

Table 2 Results of Computer Search for Publications on Monolithic/Honeycomb Structures

	Number of publications	
Period	Patents and patent applications	Nonpatent papers
1967–1970	5	14
1971–1975	57	79
1976–1980	149	150
1981–1985	208	166
1986–1990	731	234
1991–1995	1035	507

A catalyst in the catalytic converter for engine exhaust treatment is subject to significant and frequent changes in flow rate, gas composition, and temperature. Ceramic refractory materials known in the late '60s were characterized by a rather high thermal expansion coefficient. Those materials could easily crack and rupture during frequent and large temperature changes. The lifetime of monolithic catalysts based on such materials would be impractically short. A breakthrough in the technology of monolithic catalysts for catalytic mufflers was that by Corning, Inc., who had developed monoliths made from cordierite. Cordierite ($2MgO{\cdot}2Al_2O_3{\cdot}5SiO_2$) appeared to have an almost zero thermal expansion coefficient, and this made it essentially insensitive to temperature changes. Together with an advanced extrusion technology for the monoliths' manufacture, this guaranteed Corning commercial success in the field of ceramic supports for catalytic mufflers. Monoliths made from other ceramic materials needed for other applications have been developed and became commercially available, e.g., mullite, titania, zirconia, silicon nitride, silicon carbide, and the like, all of them doped with other compounds if necessary. Manufacturers of ceramic monoliths can provide blocks of large size. Moreover the blocks can be stacked side by side and/or on top of one another to form a structure with the dimensions demanded. Monoliths with channels of ca. 1 × 1 mm are in predominant use with car converters. Monoliths with larger channels, up to 6 × 6 mm, are used if dusty gases are to be processed. Cordierite monoliths are discussed in more detail in Chapter 2 of this book.

The catalyst in automotive converters is inactive until it is warmed up to a certain temperature. During this period carbonaceous matter is deposited on active sites of the catalyst, decreasing its activity. Deactivation of this sort is reversible: The deposit is quickly burned off after the temperature reaches the level at which the catalyst becomes active in combustion processes. The warm-up period is on the order of a few minutes and depends on the thermal capacity of the catalyst. Up to 75% of air pollutants are emitted during the warm-up period. The lower the thermal capacity, the shorter the warm-up period becomes and, as a consequence, the lower is the emission of pollutants. This stimulated the tendency to decrease the mass of the monolith by lowering the thickness of the walls between adjacent channels. Modern cordierte monoliths have walls down to 150 μm thick. A further considerable decrease of wall thickness in ceramic monoliths is doubtful. Metallic monoliths are more promising in this respect. These monoliths are produced from sheets down to 40 μm thick, 50 μm being now the standard. The warm-up period for metallic catalysts has been shortened to several tens of seconds. The heat transfer characteristics of metallic monoliths is claimed to be superior to that of ceramic monoliths. It is comparable with that in packed-bed reactors. This creates a possibility to use metallic monoliths when a thorough control over temperature in tubular reactors is needed. Metallic monoliths are described in Chapter 3 of this book.

In general, both cordierite and metallic monoliths are unsuitable as catalytic supports. To process a monolith into an active monolithic catalyst, a layer of porous catalytic support must be deposited on the walls between channels. γ-Alumina appeared to be the most effective support for automotive catalysts. The alumina layer is deposited by sol-gel technique (so called *washcoating*). Adherence of γ-alumina to cordierite is relatively strong. However, to form the stable γ-alumina layer on a metallic surface, we need to use an appropriate alloy that is appropriately processed before the layer is deposited. Stainless steel containing chromium, aluminum, and yttrium subjected to thermal treatment under oxidizing conditions meets requirements of automotive converters. Aluminum in the steel is oxidized to form γ-alumina needles (*whiskers*) protruding above the metal surface. Whiskers make adhesion of the γ-alumina deposited on such a surface sufficiently strong. The increasingly stringent regulation of emissions to the air made only noble metals suitable for the preparation of automotive catalysts. Platinum with an admixture of rhodium is now a standard catalytic species. Noble metals can be incorporated into the layer of γ-alumina by conventional methods known to those skilled in the art.

Low pressure drop, one of the greatest advantages of monoliths, is also one of the major drawbacks. Contrary to packed beds, monoliths do not damp nonuniformity of flow, which usually appears at the inlet due to the large changes in the diameters of inlet pipe and of the reactor. This nonuniformity is propagated throughout the reactor zone. Therefore, properly designed deflectors must be installed to equalize flow over the cross section before exhausts enter the monolith. To shorten the warm-up period, electrically heated monoliths are being implemented. Presently installed catalytic converters are active for more than 100,000 km. The manufacture and operational characteristics of automotive catalysts for gasoline engines are discussed in Chapter 4 of this book. More general aspects of the manufacture of monolithic catalysts for applications of all types are presented in Chapter 21. The computer-aided design of catalysts, including monolithic catalysts, is the subject of Chapter 22.

Another important environmental problem is the air pollution from stationary sources, such as the industrial plants of power plants. Flue gases from chemical plants contain organic pollutants that must be removed or destroyed. The polluting components of the gases are mainly hydrocarbons, carbon monoxide, sulfur dioxide, and nitrogen oxides. Catalytic oxidation is then a method expected to be efficient, and indeed is a well-established technique in this field. Again, monoliths, with their low pressure drop and a high resistance to plugging, have been found to be a very effective tool in the cleaning of such gases. Monolithic catalysts have been used to incinerate organic components and carbon monoxide in industrial off-gases from various plants, such as phenol plants, paper mills, phthalic and maleic anhydride plants, ethylene oxide plants, synthetic fiber plants, vegetable oil processing plants, and catalytic cracking reactors. Catalysts used for gas incinerations are essentially the same as those for afterburners. When bigger catalyst blocks are needed, individual monoliths are packed in a frame and a number of frames can be stacked together in a block.

Off-gases from power stations, steam generators, etc., where fuels are burned noncatalytically at very high temperatures, contain a lot of nitrogen oxides. NO_x removal is becoming more and more acute because of stricter regulations. Dutch regulations allow emissions of 75 and 60 ppm NO_x for furnaces operated on liquid and gaseous fuel, respectively. The southern Californian limit for gas turbines was cut to 9 ppm NO_x in 1993. Fuels, especially coal, also contain significant amounts of sulfur compounds which are converted to sulfur dioxide in the furnace.

It is important that subsequent oxidation to SO_3 not take place because of the formation of aerosols, which increases particulate emissions. Therefore the preferred mode of operation of selective catalytic reduction (SCR) of NO_x is such that SO_2 remains unoxidized. The gases cleaned in such a way need not be subject to troublesome treatment before they are emitted to atmosphere.

Monoliths of low cell densities (with openings ranging from 3 to 6 mm) are applied for deNOxification of gases at coal-fired power plants. This is due to the high content of dust in the gases. If dust particles were retained in the monolith, the pressure drop would increase greatly. Because of the abrasive action of the dust particles, incorporated-type catalysts are preferred. Often, WO_3 and V_2O_5 are incorporated into TiO_2 in the anatase form. In some cases, Pt/Al_2O_3 and catalysts containing Cr_2O_3, Fe_2O_3, CoO, and/or MoO_3 are used. Zeolitic monoliths are also in use for NO_x removal: More than twenty-five deNOxification plants based on zeolitic monoliths have been put into use in recent years.

Monoliths are formed in blocks of large diameter. Commercial reactors are usually overdesigned to compensate for catalyst deactivation, dust blocking, and maldistribution at the inlet. A good mixing of ammonia with the flue gases must be attained before the gas enters the monolith to minimize the detrimental effects of this maldistribution.

The selective reduction of NO_x is carried out on a very large scale; by now about 56,000 m^3 of catalyst has been installed in Japan and Germany combined, equivalent to approximately $800 million in sales of catalyst. About 1–1.5 m^3 of catalyst is needed per 1 MW of power capacity. A typical coal-fired power station has a power of 800 MW. Accordingly, very large reactors have to be installed, about 1000 m^3, typically containing extruded monolithic catalysts. Unconventional reactors, such as reverse-flow reactors, and rotating monoliths have been designed for cleaning off-gases. Huge rotating monoliths (up to ca. 20 m in diameter) for processing millions of cubic meters of gas per hour are in operation. Chapter 5 deals with problems of monolithic SCR processes, which have become a routine means for removal of NO_x from industrial plants.

Combustion of fuels using conventional noncatalytic methods is carried out at very high temperatures, up to 2300 K, to get complete conversion to carbon dioxide and water. If the temperature of burning is far lower than this limit, unburned hydrocarbons and carbon monoxide remain in the combustion gases. On the other hand, temperatures above 1900 K favor the formation of nitrogen oxides. This means that noncatalytic combustion of fuels will always be associated with environmental problems. Catalytic combustion of fuels can be performed at much lower temperatures with a process rate sufficiently high to complete oxidation and low enough to avoid NO_x formation. Catalytic combustion allows also for the utilization of fuels with low heat value (LHV fuels). The temperature for noncatalytic combustion of such fuels is too low to complete burning within a reasonable time, i.e., in chambers of acceptable size. The huge amounts of gases to be processed in power stations, steam generators, etc., using conventional particulate catalysts would result in a very high pressure drop, with a considerable loss of energy.

Monolithic catalysts provide an excellent opportunity to make catalytic combustion environmentally friendly and energy saving compared to conventional catalytic systems. The temperature at the normal operation of a combustion unit of whatever design is relatively low but can significantly increase in the case of process fluctuations or perturbations. Hence, the thermal stability of monolithic catalysts is of a great importance. At high temperatures, a washcoat can react with the support or undergo phase transformations. This may result in enclosing catalytic components into closed pores or even in destruction of the catalyst. Therefore, the use of catalytically active ceramics of higher refractoriness has been suggested, even at the cost of their lower surface area. This, however, need not be as high as for typical chemical applications. Zirconia is a promising material in this respect. It can be used at temperatures up to ca. 2500 K, is extremely inert to most metals, and exhibits a great structural integrity in operation. If platinum is used as a catalytic species, the operating temperature must not exceed 1450–1500 K to prevent the escape of platinum from the catalyst surface.

Because of the outstanding prospects for catalytic combustions, a lot of R&D work has been carried out in recent decades on this subject. For the above-mentioned reasons, a considerable proportion of the research is dedicated to the development of novel materials for monoliths. Pilot and demonstration plants for monolithic combustion are in operation, and commercialization is within reach. Catalytic combustion is discussed in Chapter 6 of this book, with emphasis given to the application of monoliths for this purpose.

The low heat conductivity of ceramic materials implies poor heat exchange between ceramic monolithic catalysts and the surroundings. Indeed, monolithic reactors with ceramic catalysts are operated at nearly adiabatic mode. This is no limitation in the case of combustions and environmental applications of monolithic catalysts. There are no thermodynamic constraints in these processes: The final products in these reactions, such as CO_2, H_2O, and N_2, are the products desired. The only limitation is the thermal resistance of a catalysts and of the materials of construction for a reactor. On the other hand, selectivity is a problem for a significant proportion of typical chemical catalytic processes. Final products from the viewpoint of thermodynamics are not necessarily desired products. An intermediate or one of many compounds formed in parallel reactions is often the product demanded. Selectivity usually depends strongly on temperature. Ceramic monoliths, being less controllable in this respect, are not the optimal choice for most applications in the chemical industry. Metallic monoliths have comparable heat transfer characteristics with conventional fixed beds of particulate catalysts. In the case of steam reforming of alkanes, suggestion was even made to replace the conventional granulate catalyst with a monolithic metallic catalyst in the zone where the highest heat flux is desired. Together with a great potential for manipulation with selectivity in thin layers of monolithic catalysts, this also makes monolithic catalysts an attractive alternative for noncombustion and nonenvironmental processes. Methanation, steam reforming of alkanes, hydrogenation/dehydrogenation processes, hydrogen manufacture (also in rotating monoliths), naphtha cracking, gasoline synthesis, etc. were successfully studied using monolithic catalysts. These monolithic processes are now at the developmental stage. More details of these processes are given in Chapter 7 of this book.

Mathematical modeling is a tool widely used in process development and optimization. The performance of monolithic converters is a complex function of design parameters, operating conditions, and the properties of both the catalyst and the reaction mixture. An empirical approach to optimization would thus be costly and time consuming. Therefore a lot of research was done on the modeling of monolithic reactors, and this makes up a considerable part of the review by Cybulski and Moulijn [8]. Mathematical modeling appeared to be particularly useful in the search for improvements in catalytic mufflers. Recently, Oh [9] published a review on the modeling of automotive catalytic converters. A particular emphasis in that review was on experimental validation and the practical applications of models. A one-dimensional heterogeneous model was proven to describe the behavior of the monolithic reactor accurately enough for the purpose of simulation and optimization. Automotive catalytic converters are subject to a multitude of highly transient inlet conditions. Therefore, the modeling of the dynamic behavior of such converters is particularly important. Chapter 8 of this book deals with problems of the modeling of monolithic, two-phase reactors, with an emphasis on nonstationary operation of such converters.

In recent years, the use of monoliths for performing multiphase reactions drew the attention of researchers. There are some aspects of monoliths that make them of interest for three-phase reactions. The main advantages are the same as in the case of two-phase processes: the low pressure drop and the short diffusion distance inside the thin layer of the catalyst, resulting in higher catalyst utilization and possibly improved selectivity. When operating in the Taylor flow regime, it is possible to obtain low axial dispersion and high mass transfer rates. In this flow regime, the gas and the liquid form a sequence of distinct plugs, flowing alternately. The gas plugs are separated from the wall by a thin layer of liquid. This has great advantages: (1) the gas bubbles disturb the laminar flow in the liquid plugs and force the liquid to recirculate within a plug, thus improving the radial mass transfer; (2) since all liquid exchange between plugs must take place via the thin liquid film surrounding the bubble, the axial dispersion is reduced; and (3) the thin liquid film provides a short diffusion barrier between the gas and the catalyst, and in addition it enlarges the gas/liquid contact area.

The main features of monolith reactors (MR) combine the advantages of conventional slurry reactors (SR) and of trickle-bed reactors (TBR), avoiding their disadvantages, such as high pressure drop, mass transfer limitations, filtration of the catalyst, and mechanical stirring. Again, care must be taken to produce a uniform distribution of the flow at the reactor inlet. Scale-up can be expected to be straightforward in most other respects since the conditions within the individual channels are scale invariant.

Due to the higher cost of monolithic catalysts, only processes in which the catalyst is reasonably stable and/or easy to regenerate are feasible. There are three fields in which monolithic catalysts are extensively studied: (1) liquid-phase hydrogenations, (2) oxidation of organic species in aqueous solutions like wastewater, and (3) biotechnology (immobilization of living organisms). Several processes are in the developmental stage, and one, the hydrogenation step in the production of H_2O_2 using the alkylanthraquinone process, has reached full scale, with several plants in operation (EKA AKZO/Nobel). Current and potential applications of monolithic reactors for three-phase processes and modeling of three-phase monolithic reactors are presented in Chapters 9 and 10 of this book.

An interesting monolithic configuration has recently been disclosed that can be suitable for three-phase processes carried out in countercurrent mode [10]. This can be particularly important for processes in which both thermodynamic and kinetic factors favor countercurrent operation, such as catalytic hydrodesulfurization. The flooding of a reactor is a considerable limitation for the countercurrent process run in conventional fixed-bed reactors. Flooding will not occur to that extent in the new monolith. A configuration of channels of the new monolith is such that subchannels open to the centerline are formed at the walls. The liquid flows downward, being confined in these subchannels and kept there by surface tension forces. The gas flows upward in the center of the channel. The results of studies on the new monolith concept are presented in Chapter 11 of this book.

III. ARRANGED CATALYSTS

Monolithic catalysts for two-phase processes are characterized by: (1) poor heat and mass transfer between the gas and the outer surface of the catalyst, and (2) no mass exchange between adjacent channels and consequently zero mass transport in the direction perpendicular to flow. The latter, being the predominant contribution to the overall mechanism of radial heat transfer inside the catalyst bed, results in rather poor heat transfer between the monolith and the surroundings. If more intensive heat and mass transfer within the catalyst bed is needed, arranged catalysts are one of the most effective solutions.

Particulate catalyst can be arranged in arrays of any geometric configuration. In such arrays, three levels of porosity (TLP) can be distinguished. The fraction of the reaction zone that is free to the gas flow is the first level of porosity. The void fraction within arrays is the second level of porosity. The fraction of pores within the catalyst pellets is referred as the third level of porosity. *Parallel-passage and lateral-flow reactors* are examples of TLP reactors. In these reactors, particulate catalyst is located in cages, with openings that allow reactants free access inside the cage. The gas flows between cages via straight or slightly twisted paths that produce a very low resistance to flow. Hence, pressure drop in these reactors is much lower than in conventional fixed-bed reactors. The gas entering the cage moves rather slowly, and to a certain degree this limits mass and heat transfer between the gas phase and the outer surface of particles. Therefore the use of reactors of this type is restricted to slow reactions that proceed in the kinetic regime. Slow processes such as hydrodesulfurization and hydrodenitrification of heavy oil fractions are examples of processes for which parallel-flow reactors have found commercially successful applications. Parallel-passage and lateral-flow reactors are discussed in Chapter 12 of this book.

Another configuration of the TLP reactor is the bead-string reactor, presented in Chapter 13 of this book. According to the idea discussed there, particles (*beads*) of the catalyst are arranged in strings. These strings are put in an array in parallel. The gas flows freely through zones formed between the strings. The pressure drop in a bead-string reactor is also low, but mass transfer between the gas and the outer surface of the catalyst particles is much more intensive than in parallel-passage and lateral-flow reactors.

Heat and mass transfer over the whole reaction zone is the most intensive in the case of structural catalysts derived from structural packings and static mixers. These are structures consisting of superimposed sheets, possibly corrugated before stacking. Sheets are covered by an appropriate catalyst support in which active ingredients are incorporated. The structure formed is an *open cross-flow structure*, with intensive radial mixing even for flow in a laminar regime. A very narrow distribution of residence time makes flow through structural packings close to the plug flow pattern. Radial heat transfer is high because of intensive radial convection, which is an important contribution to the overall heat flow in packed beds. A very high voidage of these structures (ca. 90%) guarantees a low pressure drop. Due to the relatively twisty way of reactants in these catalysts, pressure drop is obviously higher than in monoliths of the same voidage. Chapter 14 deals with structural catalysts, by illustrating their capabilities with practical examples of etherification and esterification reactions.

IV. MEMBRANE REACTORS

In the last two decades, membrane technology has found many applications, starting with desalination and including various separation processes in the fields of biotechnology, environmental techniques, and natural gas and oil exploitation and processing. The scope of these applications will depend on the availability of membranes with acceptable permeability, permselectivity, and stability. The essence of reactors containing membranes is that they combine two functions in one apparatus: separation and reaction. Thus, membrane reactors are multifunctional units. Many reversible reactions cannot reach high conversions because of the limits imposed by thermodynamics. The continuous removal, through the wall, of at least one of the products from the reaction mixture can shift the reaction toward the product side, increasing the yield significantly beyond equilibrium conversion. The selectivity of numerous processes is determined by the conditions for the transfer of the reactants and products to and from the catalytic surface. Hence, an easily controlled supply of at least one of the reactants to the reaction mixture through the membrane can affect the selectivity of the process. The careful control of the supply of a reactant (e.g., oxygen) to the reaction zone minimizes the chance of temperature runaway, thereby improving the safety of reactor operation. (Thus, the combination of reaction and membrane separation can result in an increase in the reaction yield beyond what the reaction equilibrium allows and/or a modification in process selectivity.)

The steadily increasing interest in membrane catalysts is seen in the literature. Many extensive reviews on membrane catalysts/reactors have recently been published [11–15]. Membranes in chemical reactors are used mainly in the field of biotechnology, i.e., for low-temperature processes.

Based on material considerations, membrane reactors can be classified into: (1) organic-membrane reactors, and (2) inorganic-membrane reactors, with the latter class subdivided into dense (metals) membrane reactors and porous-membrane reactors. Based on membrane type and mode of operation, Tsotsis et al. [15] classified membrane reactors as shown in Table 3. A CMR is a reactor whose permselective membrane is the catalytic type or has a catalyst deposited in or on it. A CNMR contains a catalytic membrane that reactants penetrate from both sides. PBMR and FBMR contain a permselective membrane that is not catalytic; the catalyst is present in the form of a packed or a fluidized bed. PBCMR and FBCMR differ from the foregoing reactors in that membranes are catalytic.

Many organic membranes have been developed that are now in commercial use, including reverse osmosis (RO), ultrafiltration (UF), microfiltration (MF), and gas phase separation (GS). They are made of polymeric materials such as cellulose acetate (the first generation of organic membranes), polyamide, polysulfone, polyvinylidene fluoride and polytetrafluoroethylene. The major drawback of these membranes, from the viewpoint of reactor technology, is their thermal instability. Generally, the maximum operable temperature for these membranes is approximately 180°C. Accordingly, high-temperature catalytic processes cannot be carried out using membranes of this kind. Moreover, the corrosiveness of the reaction mixtures, especially at severe process conditions, makes organic membranes less attractive for reaction engineering, at least at this stage in the development of membrane manufacturing technology.

The phenomenon of hydrogen permeation through palladium was discovered by Thomas Graham more than 100 years ago. Since then, more nonporous metals and alloys permeable to hydrogen and oxygen have been discovered. Good examples of this kind of membrane are palladium alloys with ruthenium, nickel, or other metals from Groups VI to VII of the periodic table. Palladium alloys are preferred to pure palladium because of palladium brittleness. A very high permselectivity of palladium membranes to hydrogen favors these membranes for use in coupling hydrogenation/dehydrogenation processes. Dehydrogenations are endothermic reactions, and the heat needed to run such reactions can be supplied form the other side of the membrane by combusting permeated hydrogen.

Table 3 Types of Membrane Reactors

Acronym	Description
CMR	Catalytic membrane reactor
CNMR	Catalytic nonpermselective membrane reactor
PBMR	Packed-bed membrane reactor
PBCMR	Packed-bed catalytic membrane reactor
FBMR	Fluidized-bed membrane reactor
FBCMR	Fluidized-bed catalytic membrane reactor

Silver membranes are permeable to oxygen. Metal membranes have been extensively studied in the countries of the former Soviet Union (Gryaznov and co-workers are world pioneers in the field of dense-membrane reactors), the United States, and Japan, but, except in the former Soviet countries, they have not been widely used in industry (although fine chemistry processes were reported). This is due to their low permeability, as compared to microporous metal or ceramic membranes, and their easy clogging. Bend Research, Inc. reported the use of Pd-composite membranes for the water–gas shift reaction. Those membranes are resistant to H_2S poisoning. The properties and performance characteristics of metal membranes are presented in Chapter 16 of this book.

There are now various inorganic microporous membranes commercially available that can be used for separation on a full scale. These membranes can be made from various inorganic materials of required resistance to mechanical and chemical effects over a wide range of pH and temperature. Vycor glass, alumina, and zirconia, appropriately doped, have been extensively studied. Due to large pore diameters, ranging from 4 nm to 5 μm, these membranes are characterized by much higher mass fluxes than are dense membranes. On the other hand, the structure of the available membranes and those under development pose some limits in high-temperature gas separations and for membrane reactors because of the rather low permselectivity of the membranes. Like dense membranes, applications of inorganic membranes as selective catalysts in hydrogenation/dehydrogenation reactions and for the carefully controlled addition of oxygen (in oxidative coupling of methane) were investigated. Inorganic-membrane catalysts are discussed in detail in Chapter 17 of this book. This chapter also deals with dense membranes deposited on inorganic porous membranes to combine the advantages of both types of membrane: high permselectivity and high mass flux. The brittleness of palladium is then less important because the inorganic support increases the mechanical strength of the structure. Another lamellar structure of this type, namely, combinations of a zeolitic catalyst and an inorganic membrane, are presented in Chapter 19 of this book. Due to a very high selectivity in chemical processes, the potential of these membranes is enormous. Zeolitic membranes were tested at high temperatures and appeared to be stable for long runs.

Catalytic filters are devices capable of removing particulate solids from the fluids containing them and simultaneously stimulating the catalytic process in the fluid. The catalyst is in the form of a thin layer applied on the material of the filter. Filters can be either rigid or flexible, and most filters are of tubular or candel form. As filtration progresses, the filter cake grows. After the pressure drop exceeds a limiting value, the cake is removed by a short injection of the fluid in the direction opposite to the flow at filtering. This injection causes the cake to detach from the filtering medium. This technique found numerous applications. Catalytic filters for flue gas cleaning are discussed in detail in Chapter 15 of this book.

One sort of catalytic filter is used as catalytic muffler for cleaning exhausts from diesel engines. Particulate solids of carbonaceous nature are stopped in monoliths on the walls of adjacent channels, 50% of which are closed at one side of the monolith and another 50% of which are closed at the other side. This forces dusty exhausts to pass through the walls in which catalytically active ingredients are incorporated. When the pressure drop exceeds a limiting value, the flow of the gas is stopped and air is passed through the catalytic filter to burn off the carbonaceous deposit (so-called *soot*). The temperature of the *wall-through monolith* increases significantly during the burn-off period. This imposes strict requirements on the material constituting the monolith. It must be both refractory and resistant to frequent steep temperature changes. The present status and prospects of catalytic mufflers for diesel engines are discussed in Chapter 18 of this book.

Permeable walls are also used in cross-flow reactors. The concept behind the cross-flow reactors is, however, somewhat wider than the principles of devices presented heretofore. Cross-flow involves two different fluids flowing perpendicularly to or from each other in a process apparatus. The streams can be separated by a permeable wall or can be combined without such a wall. Cross-flow reactors of various kinds are presented in Chapter 20 of this book.

V. THE FUTURE OF STRUCTURED CATALYSTS

Monolithic catalysts have proven to be superior to conventional catalysts in the field of automotive afterburners. Monolithic catalysts have also good future prospects in this field. However, there is still room for improvement in these applications. Efficient and cheap methods to reduce emissions during the warm-up period of gasoline-fed engines must be developed. Electrically heated monoliths are the promising remedy for this reduction, but experimental and modeling studies on this subject must and will be continued. Catalytic mufflers are a rather expensive part of a car's equipment. Therefore, prolonging of the catalyst's life is also an important problem to be solved. The removal of poisons from engine fuels is one of the ways to achieve this objective. Another method is to optimize the catalyst's composition, including the monolithic support and the washcoat layer. Studies on deactivation profiles along the monolith will help in optimizing the activity profile along the fresh monolithic catalyst. Present mathematical models describe the behavior of monolithic catalysts fairly well. To make optimization more reliable, models have to be modified. The washcoat distribution over the channel's periphery is highly nonuniform. Therefore, the effectiveness factor can be determined only approximately. The zone where internal diffusion is the limiting step is very narrow for the fresh catalyst. However, it is extended in the course of aging. Hence, improvements in modeling that would take this into account are also welcome. Monolithic wall-through catalysts for diesel engines still require considerable improvement. Due to filter/burn-off cycling and the associated significant temperature changes, cracks and ruptures appear in the monoliths. This reduces catalyst life to 20,000–30,000 km. Certainly, monolithic catalyst will be used for cleaning diesel exhausts in the future, but extending its life is crucial. A search for more active catalysts, which would cause the burning of soot at the rate it is deposited on the wall, is needed. New monolithic structures that would not lead to the formation of relatively thick soot deposits might help in solving the problems of soot removal.

The cleaning of flue gases from stationary sources is another field in which the application of monolithic catalysts will certainly rise. There will be no versatile catalyst for cleaning all off-gases. Therefore tailor-made catalysts with zeoliths of various types for specific applications will be developed. Incorporated-type monolithic catalysts are likely to prevail in this field. Since cleaning usually requires a set of equipment items in series (e.g., converter, heat exchangers), multifunctional reactors (reverse-flow reactors, rotating monoliths) will become more common.

Catalytic combustion is another area where monolithic catalysts will find their place. After all material problems (refractoriness and life of the catalyst) and engineering problems are solved, clean, environmentally friendly processes (at present still in the developmental stage) will be implemented at full scale. LHV fuels will be more widely used after all combustion problems are solved. The success in the commercialization of monolithic combustors will result in a simplification of methods for cleaning flue gases.

The use of monolithic reactors for noncombustion and nonenvironmental two-phase processes will be rather limited. Contrary to "environmental" processes, the catalyst must be adjusted to specific process requirements. The cost of developing new catalyst will pay off rather seldom. Therefore the use of monolithic catalysts for two-phase processes will be rather limited. There are better prospects for monolithic catalysts in three-phase processes. Selective hydrogenations, oxidative waste water treatment, and biochemical processes seem to be the first areas in which monolithic processes will find more applications. Arranged catalysts seem to be very promising in this field. This is due to a good heat exchange between the reaction zone of arranged catalysts and the surroundings. Technologies developed to transform metallic and ceramic structures into active catalysts will certainly be applied to arranged catalysts.

The potential of membrane reactors is enormous. However, the wider use of high-temperature inorganic-membrane catalysts is still a challenge, and this task is far from completion. It is limited by material and engineering factors. A breakthrough in this area will require the close cooperation of catalyst scientists, material scientists, and chemical engineers. Highly selective palladium membranes are prohibitively expensive, mass fluxes (consequently, reactor throughput) are very low, and these membranes are sensitive to sulfur and coking. Therefore, palladium membranes have not found many commercial applications. To make palladium membranes applicable to more processes, their resistance to poisoning must be improved and composites with porous inorganic membranes worked out to increase mass flux. Thinner membranes, smaller pore sizes, and sharper pore size distributions are needed. Tubular membranes have been studied the most, and these are likely to prevail in future high-temperature applications. Bundles of such tubes or multichannel monoliths with permeable walls with the higher filtration areas can improve process economics.

Progress is needed in the manufacture of membranes—currently, inorganic membranes are at least five times as expensive as organic membranes. The cost of manufacture could be considerably reduced if membranes found bulk applications. The stability of catalytic membranes of all sorts is a problem. All membranes are sensitive to fouling due to the small size of their pores and to catalytic activity in the formation of deleterious coke, especially at elevated temperatures. Decoking by controlled oxidation is well known in conventional catalytic processes. However, the particular sensitivity of membranes to the thermal stresses occurring during decoking poses a great problem. Membranes are also subject to high temperature gradients at normal operation because of the considerable thermal effects of the reaction to be carried out using membrane processes. Cracks and ruptures can be formed in the membrane, at the connection of membrane moduli with the other parts of the reactor, or at the junction between layers of a composite membrane. Even minor cracks and ruptures decrease the efficiency of the membrane nearly to zero. Sealing the ends of a membrane element and packing the membrane element to a module housing is one of the challenges in the field of high-temperature ceramic membranes. The different thermal expansion coefficients of the material for the membrane and for the housing can cause stresses at joints and their rupture. All these sorts of material and engineering problems must be solved.

REFERENCES

1. J.P. DeLuca and L.E. Campbell, Monolithic Catalyst Supports. Chapter 10 in *Advanced Materials in Catalysis* (J.J. Burton and K.L. Garten, eds.), Academic Press, London, 1977, pp. 293–324.
2. L.D. Pfefferle and W.C. Pfefferle, Catalysis in Combustion, *Catal. Rev. Sci. Eng. 29*: 219 (1987).
3. R. Prasad, L.A. Kennedy, and E. Ruckenstein, Catalytic Combustion, *Catal. Rev. Sci. Eng. 26*: 1 (1984).
4. D.L. Trimm, Catalytic Combustion, *Appl. Catal. 7*: 249 (1983).
5. K.C. Taylor, Automobile catalytic converters. In *Catalysis: Science and Technology*, Vol. 5 (J.R. Anderson and M. Boudart, eds.), Springer-Verlag, Berlin, 1984.
6. S. Irandoust and B. Anderson, Monolithic Catalysts for Nonautomobile Applications, *Catal. Rev. Sci. Eng. 30*: 341 (1988).
7. R. Brand, B.H. Engler, and E. Koberstein, Potential Applications of Monoliths in Heterogeneous Catalysis. Paper presented at Roermond International Conference on Catalysis, Roermond (The Netherlands), June 1990.
8. A. Cybulski and J.A. Moulijn, Monoliths in Heterogeneous Catalysis, *Catal. Rev. Sci. Eng. 36*: 179 (1994).
9. S.H. Oh, Converter Modeling for Automotive Emission Control. In *Computer-Aided Design of Catalysts* (E.R. Becker and C.J. Pereira, eds.), Dekker, New York, 1995, pp. 259–296.
10. T. Sie, J.A. Moulijn, and A. Cybulski, Internally Finned Channel Reactor, Ned. Pat. Appl., No. 92.01923, October 19, 1992.
11. H.P. Hsieh, Inorganic Membrane Reactors—Review, *AIChE Sympos. Ser. 268, 85*: 53 (1989).
12. H.P. Hsieh, Inorganic Membrane Reactors, *Catal. Rev. Sci. Eng. 33*: 1 (1991).
13. J. Shu, B.P.A. Grandjean, A. van Neste, and S. Kaliaguine, Catalytic Palladium-Based Membrane Reactors: A Review, *Can. J. Chem. Eng. 69*: 1036 (1991).
14. G. Saracco and V. Specchia, Catalytic Inorganic-Membrane Reactors: Present Experience and Opportunities, *Catal. Rev. Sci. Eng. 36*: 305 (1994).
15. T.T. Tsotsis, R.G. Minet, A.M. Champagnie, and P.K.T. Liu, Catalytic Membrane Reactors. In *Computer-Aided Design of Catalysts* (E.R. Becker and C.J. Pereira, eds.), Dekker, New York, 1995, pp. 471–552.

zation. The sintered structure was then cooled to 100°C at a controlled rate and removed from the furnace. Such a sintered honeycomb structure was comprised of 400 cells/in.^{2} (400 Cpi^2) and offered some attractive properties, namely, an open frontal area (OFA) of 80%, a geometric surface area (GSA) of 125 in.^{2}/in.^{3}, a hydraulic diameter of 0.0256 in., and a bulk density of 7.9 g/in.^{3}. Furthermore, its coefficient of thermal expansion (CTE) from room temperature to 1000°C was low, 20×10^{-7} in./in./°C, due to a special glass–ceramic composition (lithium alumino-silicate, Corning Code 9455). Mr. Cole was most impressed with the high surface area (per unit volume) of honeycomb structure and suggested that this would make an ideal support for oxidation catalyst. Of course, the other properties were equally beneficial for this application, considering high exhaust temperature and appreciable temperature gradients due to catalytic exotherms. Corning was given the challenge to produce such substrates in huge volume and to do so in a cost-effective manner. Unfortunately, the Cercor® process was too slow to be cost-effective. Furthermore, the ceramic composition had to be more refractory than that of Cercor® to withstand temperatures approaching 1400°C due to misfiring or other engine malfunctions.

With massive R&D effort, Corning invented both a new ceramic composition [9] and a unique forming process [10], which together permitted the manufacture of monolithic cellular ceramic substrates in a cost-effective manner. Under the trademark of Celcor®, these substrates are manufactured by the extrusion process from cordierite ceramic ($2MgO \cdot 2Al_2O_3 \cdot 5SiO_2$) with a low CTE and a high melting temperature (≈1450°C). One of the unique features of these substrates is their design flexibility, discussed in the next section.

II. REQUIREMENTS FOR CATALYST SUPPORTS

The substrate is an integral part of the catalytic converter system. Its primary function is to bring the active catalyst into maximum effective exposure with the exhaust gases. In addition, it must withstand a variety of severe operating conditions, namely, rapid changes in temperature, gas pulsations from the engine, chassis vibrations, and road shocks. As noted earlier, pellets of cylindrical and spherical geometry and honeycomb monoliths both became available for catalyst supports.

A. Pellets vs. Honeycomb Supports

Each type of substrate had its strengths and weaknesses. The pellets were made of porous γ-Al_2O_3 with a density of 0.68 g/cm^3 and a B.E.T. area of 100–200 m^2/g. They measured about $\frac{1}{8}$ in. in diameter and were available in spherical or cylindrical shape of different aspect ratios. They were selected for their crush and abrasion resistance; they also promoted turbulent flow, which improved the contact of reactant gases with noble-metal catalyst deposited predominately on the outer surface. The latter also improved the rate of pore diffusion mass transfer to the catalyst [12]. Pellets were also replaceable by refilling the container after, say, 50,000 miles. However, the pellet converter was very heavy and slow to warm up; see Fig. 2. It also generated high back pressure and had a severe problem of pellet attrition due to their rubbing against one another during vehicle use. The use of low-density pellets to improve light-off performance would aggravate the attrition problem further.

2

Ceramic Catalyst Supports for Gasoline Fuel

Suresh T. Gulati
Corning Incorporated, Corning, New York

I. HISTORICAL BACKGROUND

The initial efforts at understanding the hazards of automotive exhaust began in California in the late 1940s with paying political and scientific attention to photochemical reactions in the atmosphere between hydrocarbons and nitrogen oxides emitted in automobile exhaust. Professor Haagen-Smit of the California Institute of Technology* showed that some hydrocarbons and nitrogen oxides endemic to automobile exhaust reacted in sunlight to produce oxidants, including ozone, which caused cracking of rubber and irritation of the eyes [1]. A concurrent investigation by the Los Angeles Air Pollution Control District verified that aerosols and mists could be produced photochemically by the polymerization of photooxidation products of exhaust hydrocarbons and laid the scientific basis for a serious examination of the composition and health impact of automotive exhaust [2]. These two studies provided the necessary impetus for the eventual development of the automotive catalytic converter.

Both technical and economic difficulties combined with the unavailability of lead-free gasoline,† and the absence of compelling federal legislation for stringent emission standards delayed the development of the catalytic converter until the establishment of the 1966 California Standards. This was soon followed by the 1967 Federal Clean Air Act requiring all 1968 model year (MY) vehicles to meet emission standards legislated by California a year earlier with the use of leaded gasoline. Although this law, known as the Muskie Bill, led to a resurgence of interest in automotive exhaust catalysts, for seven years (up to MY 1974) the emission standards for leaded gasoline were being met primarily by engine modifications consisting of the use of improved carburetors, air pumps, spark

* Since Cal-Tech is located in Pasadena in the northwestern corner of the Los Angeles basin, it felt the full impact of auto-exhaust related smog.

† Both lead compounds and the halide-containing lead scavengers added to gasoline to prevent engine knock at high engine compression ratios, favored for improved thermodynamic efficiency of the engine, caused rapid deactivation of base metal catalysts, which were considered more cost-effective and realistic for the automotive market than were noble metal catalysts.

retardation, thermal afterburners, and exhaust gas recirculation. However, these approaches had a significant negative impact on vehicle performance, drivability, and fuel economy, which could become prohibitive as emission standards became more stringent.

Both Ford and General Motors (GM) engineers initiated studies in 1967 to measure the relative rates of catalyst degradation due to "thermal" and "poison" deactivation, even with unleaded gasoline, as a result of high-temperature exposure resulting from misfueling and/or engine malfunction, e.g., spark plug misfire [3–5]. These studies demonstrated that, when operated on unleaded gasoline, catalytic systems with noble-metal catalyst could be made durable, dependable, and resistant to engine malfunction with little impact on engine performance, including fuel economy. Seven months after the GM studies were made public, Ed Cole, president of General Motors, in a speech at the annual meeting of the Society of Automotive Engineers (SAE) in January 1970, called for a "comprehensive systems approach to automotive pollution control," including the removal of lead additives from gasoline, to make "advanced emissions control systems feasible." A month later, GM announced that all of its cars beginning with the 1971 MY would be designed to operate on fuel of 91 Research Octane Number, leaded or unleaded. This was achieved by reducing the compression ratio to about 8.5. The petroleum industry, which had dragged its feet for a whole decade, responded by marketing unleaded fuel for use with these vehicles. In November 1970, Mr. Cole told the American Petroleum Institute (API) of GM's plans to install control systems including a catalytic converter on all new vehicles by 1975, which would require unleaded gasoline [6].

A. U.S. Clean Air Act

Shortly after Mr. Cole's address to the API, the U.S. Congress enacted the Clean Air Amendment of 1970, which called for 90% reduction in hydrocarbon (HC) and carbon monoxide (CO) emissions by January 1975 and a similar reduction in nitrogen oxides (NO_x) by MY 1976. In absolute terms, these standards were formalized at 0.41 g HC/mile, 3.4 g CO/mile, and 0.4 g NO_x/mile. Similar laws were also passed in California and Japan. Passage of the Clean Air Act gave the final impetus to catalytic converter development, calling for decisions on the configuration of the catalyst support (pellets or monolith), the choice of catalyst (base metal or noble metal), and the optimum composition of the active ingredients to preserve the B.E.T. area* and to promote catalytic activity over 50,000 miles. The earlier concern over the impracticality of the noble metals, due to high-volume usage, high cost, and limited supply, had been overcome by the reassessment of platinum supply vs. catalyst loading [7].

In view of the high exhaust temperature and large temperature gradients due to exothermic catalytic reactions and engine malfunction, automakers sought ceramic catalyst supports with large surface area, good thermal shock resistance, and low cost. Alumina beads, which had been used as catalyst supports in nonautomotive industry, met these requirements. Ceramic monoliths with honeycomb structure offered another alternative, provided they met the objective of low cost and had a low coefficient of thermal expansion to withstand thermal shock. Three companies—W.R. Grace, American Lava Corp.,† and Corning Glass Works‡—developed new compositions and processes for manufacturing these monoliths, but only Corning succeeded in meeting the cost and technical requirements [8]. Corning scientists invented the cordierite ceramic, with a low coefficient of thermal expansion [9], and the extrusion process that provided the flexibility of honeycomb geometry, substrate contour, and substrate size [10]. In addition, these monoliths offered a high geometric surface area, approaching 90 $in.^2/in.^3$ and a use temperature approaching 1200°C. Consequently, the cordierite ceramic substrate has become the world standard and is used in 95% of today's catalytic converters. Although alternate materials such as FeCrAlloy have become available over the past five years, automakers around the world continue to use ceramic substrates to meet the ever-demanding emissions and durability requirements, due to their cost-effectiveness and decades of successful field performance.

* Surface area calculated from gas (N_2) adsorption using the theory of Brunauer, Emmett and Teller.

† Subsidiary of 3M Company.

‡ Now Corning Incorporated.

Although only pelleted catalysts had been certified in California, ceramic honeycomb supports with very attractive properties had been developed, and they provided an intriguing alternative to pellets [8]. Both types of supports had to be durable and resistant to attrition and catalyst poisoning. They had to meet performance requirements that included light-off, high temperature resistance, efficient heat and mass transport, and low back pressure. In addition, they had to meet the space requirements by modifying their shape and size. While each type of support had its advantages and disadvantages, GM and certain foreign automakers chose pellets, whereas Ford, Chrysler, and others decided to use monolithic supports for MY 1975 vehicles.

B. Cercor® Technology

Figure 1 shows the earliest thin-wall ceramic honeycomb structure, invented and manufactured by Corning Glass Works, for use in rotary regenerator cores for gas turbine engines [11]. This product, trademarked Cercor®, was shown to Ed Cole of General Motors in early 1970 by Corning's CEO, Amory Houghton, and president, Tom MacAvoy, for possible automotive applications. This interesting structure was formed by wrapping alternate layers of flat and corrugated porous cellulose paper, coated with a suitable glass slurry, until a cylinder of desired diameter and length was obtained. The unfired matrix cylinder was then processed through a firing cycle up to a temperature approaching 1250°C to effect sintering and subsequent crystalli-

Figure 1 Cercor® ceramic heat regenerator.

Figure 2 Pellet converter. (Courtesy of Delphi.)

A honeycomb support, on the other hand, with γ-Al_2O_3 washcoat is considerably lighter* and, due to its compact size, can be brought to light-off temperature rapidly by locating it closer to the engine. The gas flow is laminar, with relatively large passageways that result in substantially lower back pressure. Furthermore, the honeycomb support can be mounted more robustly, thereby avoiding the attrition problem associated with pellets [13,14]. As pointed out earlier, GM, American Motors, and certain foreign automakers chose the pellet-type catalyst support, whereas others went with the honeycomb catalyst support.

B. Substrate Requirements

Figure 3 shows the key parameters that affect the performance of a catalytic converter. Many of these parameters are influenced by the substrate design, which will be discussed in the next section. As a preface to that, we will list here the requirements an ideal substrate must meet.

1. It must be coatable with a high-B.E.T.-area washcoat.
2. It must have a low thermal mass, a low heat capacity, and efficient heat transfer to permit gaseous heat to heat up the catalyst-carrying washcoat quickly, notably during light-off.
3. It must provide high surface area per unit volume to occupy minimum space while meeting emissions requirements.
4. It must withstand high use temperature.

---

* A 4.2-L pellet converter weighed about 10 kg in 1975 compared with 5 kg for a honeycomb converter.

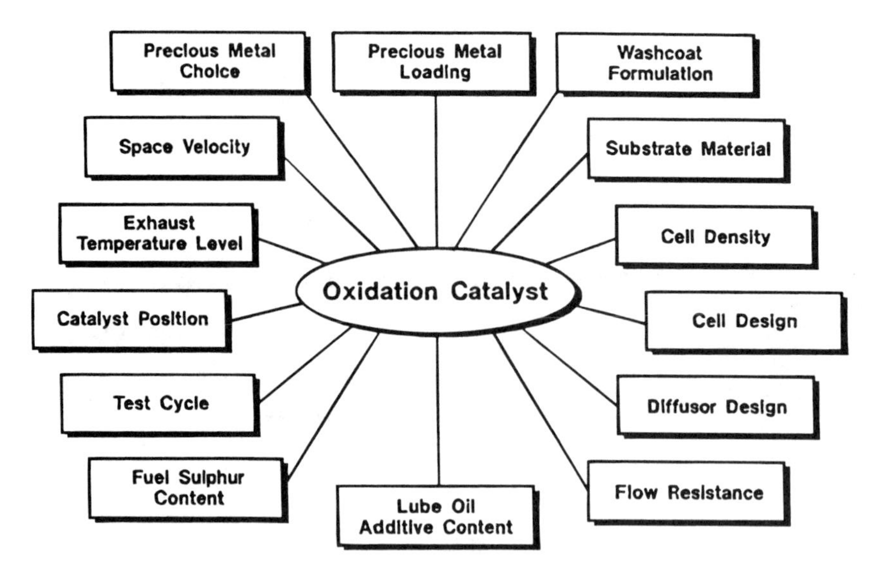

Figure 3 Parameters affecting the performance of the catalytic converter.

5. It must have good thermal shock resistance due to severe temperature gradients arising from fuel mismanagement and/or engine malfunction.
6. It must minimize back pressure to conserve engine power for rapid response to transient loads.
7. It must have high strength over the operating temperature range to withstand vibrational loads and road shocks.

These requirements can be met by optimizing both geometric and physical properties of the substrate, which, in the case of extruded honeycomb substrates, can be independently controlled—a significant design advantage over pellet-type substrates. In view of their design flexibility and other inherent advantages, we will focus on honeycomb substrates in the remainder of this chapter.

III. DESIGN/SIZING OF CATALYST SUPPORTS

The cell shape and size, which can be designed into the extrusion die, affect the geometric properties and hence the size of honeycomb substrate. Two cell shapes, which proved to be cost-effective in terms of extrusion die cost, were the square cell and the equilateral triangle cell, shown in Fig. 4. The cell size has a strong bearing on the cell density (n), the geometric surface area (GSA), the open frontal area (OFA), the hydraulic diameter (D_h), the bulk density (ρ), the thermal integrity (TIF), the mechanical integrity (MIF), the resistance to flow (R_f), the bulk heat transfer (H_s), and the light-off (LOF), which, in turn, affect both the performance and the durability of the catalytic converter. We will present simple expressions for these geometric properties of square and triangular cell substrates [15].

Figure 4 Honeycomb substrates with square and triangular cell structure. (From Ref. 22, courtesy of SAE.)

A. Geometric Properties of Square Cell Substrates

In reference to Fig. 5, the square cell is defined by cell spacing L, wall thickness t, and fillet radius R.* The foregoing geometric properties can readily be expressed in terms of L, t, and R [15,16]:

$$n = \frac{1}{L^2} \qquad \text{cells/in.}^2 \tag{1}$$

$$\text{GSA} = 4n\left[(L - t) - (4 - \pi)\frac{R}{2}\right] \qquad \text{in.}^2/\text{in.}^3 \tag{2}$$

$$\text{OFA} = n[(L - t)^2 - (4 - \pi)R^2] \tag{3}$$

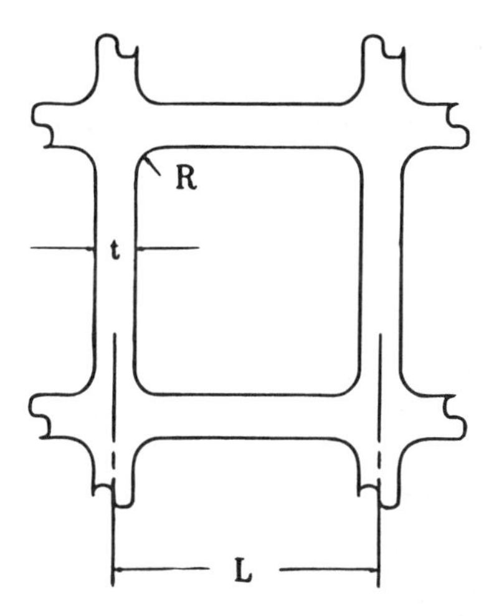

Figure 5 Geometric parameters for the square cell. (From Ref. 15, courtesy of SAE.)

* Note that R normally is not specified, since it varies with die wear; however, we include its effect on geometric properties for the sake of completeness.

$$D_h = 4\left(\frac{\text{OFA}}{\text{GSA}}\right) \quad \text{in.} \tag{4}$$

$$\rho = \rho_c(1 - P)(1 - \text{OFA}) \quad \text{g/in.}^3 \tag{5}$$

$$\text{TIF} = \frac{L}{t}\left(\frac{L - t - 2R}{L - t}\right) \tag{6}$$

$$\text{MIF} = \frac{t^2}{L(L - t - 2R)} \tag{7}$$

$$R_f = 1.775\,\frac{(\text{GSA})^2}{(\text{OFA})^3} \quad 1/\text{in.}^2 \tag{8}$$

$$H_s = 0.9\,\frac{(\text{GSA})^2}{\text{OFA}} \quad 1/\text{in.}^2 \tag{9}$$

$$\text{LOF} = \frac{(\text{GSA})^2}{4\rho_c c_p(1 - P)[\text{OFA}(1 - \text{OFA})]} \tag{10}$$

$$= \frac{(\text{GSA})^2}{4\rho c_p(\text{OFA})} \tag{11}$$

In these expressions, ρ_c denotes the density of cordierite ceramic (41.15 g/in.^{3}), P is the fractional porosity of the cell wall, and C_p is the specific heat of the cell wall (0.25 cal/g°C). TIF is a measure of the temperature gradient the substrate can withstand prior to fracture; MIF is a measure of the crush strength of the substrate in the diagonal direction; R_f is a measure of back pressure; H_s is a measure of steady-state heat transfer, and LOF is a measure of light-off performance. An ideal substrate must offer high GSA, OFA, TIF, MIF, H_s, and LOF values and low D_h, ρ, and R_f values. A close examination of the expressions indicates that certain compromises are necessary in arriving at the optimum substrate, as discussed in the next section.

B. Geometric Properties of Triangular Cell Substrates

Figure 6 defines the parameters L, t, and R for the triangular cell. Its geometric properties are given by the following:

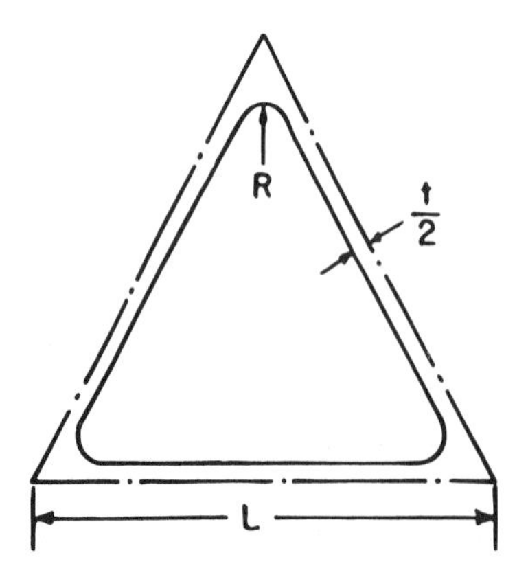

Figure 6 Geometric parameters for the triangular cell. (From Ref. 15, courtesy of SAE.)

$$n = \frac{4/\sqrt{3}}{L^2} \tag{12}$$

$$\mathrm{GSA} = 4\frac{\sqrt{3}}{L^2}\left[(L - \sqrt{3}t) + \left(\frac{2\pi}{3} - 2\sqrt{3}\right)R\right] \tag{13}$$

$$\mathrm{OFA} = \frac{1}{L^2}\left[(L - \sqrt{3}t)^2 - 4\left(3 - \frac{\pi}{\sqrt{3}}\right)R^2\right] \tag{14}$$

$$D_h = 4\left(\frac{\mathrm{OFA}}{\mathrm{GSA}}\right) \tag{15}$$

$$\rho = \rho_c(1 - P)(1 - \mathrm{OFA}) \tag{16}$$

$$\mathrm{TIF} = 0.82\frac{L}{t}\left(\frac{L - \sqrt{3}t - 2\sqrt{3}R}{L - \sqrt{3}t}\right) \tag{17}$$

$$\mathrm{MIF} = \frac{2t^2}{L(L - \sqrt{3}t - 2\sqrt{3}R)} \tag{18}$$

$$R_f = 1.66\frac{(\mathrm{GSA})^2}{(\mathrm{OFA})^3} \tag{19}$$

$$H_s = 0.75\frac{(\mathrm{GSA})^2}{\mathrm{OFA}} \tag{20}$$

$$\mathrm{LOF} = \frac{(\mathrm{GSA})^2}{4c_p\rho(\mathrm{OFA})} \tag{21}$$

The initial substrates in 1975, with square cell configuration, were designed to have 200 cells/in.^{2} with a wall thickness of 0.012 in. They were extruded from cordierite composition (EX-20), which had a fired wall porosity of 35%. Similarly, the triangular cell substrates had a cell density of 236 cells/in.^{2} and a wall thickness of 0.0115 in. They were extruded from a lower-CTE cordierite composition (EX-32), with a fired wall porosity of 40%. The geometric properties of these two substrates, as well as current versions of square cell substrates with cell density of 300 and 400 cells/in.^{2}, are summarized in Table 1.

To facilitate comparison of these four substrates we must first express the substrate requirements in terms of their geometric properties. For good light-off performance, the substrate must have high LOF value. For high conversion efficiency under steady-state driving conditions, the substrate must have high n, high GSA, and high H_s values. For low back pressure, the substrate must have high OFA, large D_h, and low R_f value. And finally, for high mechanical and thermal durability, the substrate must have high MIF and TIF values. A close examination of Table 1 shows the evolution of substrate optimization from 1975 to 1983:

1. The initial substrate, namely, 200/12 □, had low values of n, GSA, and LOF, and yet it met the conversion efficiency requirements, which were less stringent; it had large D_h, low R_f, and modest OFA values, which helped minimize the back pressure and conserve engine power; and it had high MIF and modest TIF values, which met the durability requirements.
2. The subsequent development of 236/11.5 △ and 300/12 □ substrates resulted in significantly higher LOF, GSA, H_s, and MIF values to meet more stringent conversion and durability requirements, but the lower D_h and higher R_f values had an adverse effect on back pressure.

Table 1 Geometric Properties of Honeycomb Substrates (fillet radius $R = 0$)

	Designation*			
	200/12	300/12	400/6.5	236/11.5
Cell Shape	square	square	square	triangle
Wall Porosity (%)	35	35	35	40
L (in.)	0.071	0.058	0.050	0.099
t (in.)	0.012	0.012	0.0065	0.0115
D_h (in.)	0.059	0.046	0.044	0.046
GSA (in.^{2}/in.^{3})	47.0	54.7	69.6	55.9
OFA (%)	68.9	62.9	75.7	63.8
ρ (g/in.^{3})	8.6	9.9	6.7	8.9
$R_f \times 10^{-2}$	120	216	198	200
H_s	2885	4320	5760	3675
MIF × 100	3.5	5.4	1.9	3.4
TIF	5.9	4.8	7.7	7.1
LOF (in °C cal^{-1})	373	471	955	550

*Honeycomb geometry is designated by cell density and wall thickness combination. Thus, 200/12 designates a substrate with 200 cells/in.^{2} and 0.012 in. wall thickness.

3. The current substrate, namely, 400/6.5 □, was developed in the early '80s to meet even more stringent emissions regulations, notably with respect to CO and NO_x, as shown in Table 2; this substrate offered the highest values of LOF, GSA, H_s, OFA, and TIF without any compromise in D_h and R_f values; its MIF value was relatively low, but adequate to meet the durability requirement, as discussed later.

It is clear from this discussion that honeycomb substrates offer the unique advantage of design flexibility to meet the ever-changing performance and durability requirements. Since these requirements can often be conflicting, certain trade-offs in geometric properties may be necessary, as illustrated in Table 1. New advances in honeycomb substrates, in terms of both ceramic composition and cell geometry, necessitated by more stringent performance and durability requirements for 1995+ vehicles [equivalent to low-emission vehicle (LEV) and ultralow-emission vehicle (ULEV) standards] are summarized in Table

Table 2 U.S. Federal Emissions Regulations for Gasoline-Fuel Vehicles (g/mile)

Model year	CO	HC	NO_x
1970	34.0	4.1	4.0
1975	15.0	1.5	3.1
1980	7.0	0.41	2.1
1981	7.0	0.41	1.0
1983	3.4	0.41	1.0

3. A comparison with Table 1 shows that these advanced substrates are designed for close-coupled application, where fast light-off, low back pressure, and compact size are most critical [17,18].

C. Sizing of Catalyst Supports

The size of a catalyst support depends on many factors. Predominant among these are flow rate, light-off performance, conversion efficiency, space velocity, back pressure, space availability, and thermal durability. Other factors, such as washcoat formulation, catalyst loading, inlet gas temperature, and fuel management, can also have an impact on the size of a catalyst support.

Considering the substrate alone, both the conversion efficiency and back pressure depend on its size. The former is related to the total surface area (TSA) of substrate, defined by

$$\text{TSA} = \text{GSA} \times V \tag{22}$$

in which V denotes the substrate volume given by

$$V = A \cdot \ell \tag{23}$$

In Eq. (23), A and ℓ are the cross-sectional area and length of the substrate, respectively. Similarly, the back pressure is related to flow velocity v through the substrate and its length ℓ both of which are affected by substrate size:

$$v = \frac{V_e}{A(\text{OFA})} \tag{24}$$

$$\ell = \frac{V}{A} \tag{25}$$

In Eq. (24), V_e denotes volume flow rate and other terms are as defined previously.

Table 3 Geometric Properties of Advanced Honeycomb Subtrates (fillet radius R = 0)

	Designation		
	300/ 6.7	350/ 5.5	470/ 5
Cell shape	triangle	square	square
Wall porosity (%)	35	24	24
L (in.)	0.088	0.053	0.046
t (in.)	0.0067	0.0055	0.005
D_h (in.)	0.044	0.048	0.041
GSA (in.^{2}/in.^{3})	68.5	67.1	77.5
OFA (%)	75.4	80.5	79.4
ρ (g/in.^{3})	6.8	6.1	6.4
$R_f \times 10^{-2}$	183	153	210
H_s	4680	5035	6765
MIF × 100	1.3	1.2	1.3
TIF	10.8	9.7	9.2
LOF (in °C cal^{-1})	918	888	1140

Experimental data show that the conversion efficiency η depends exponentially on TSA [19]:

$$\eta = 1 - \frac{E_o + E_i \exp(-\text{TSA}/\text{TSA}_o)}{E_o + E_i} \tag{26}$$

In Eq. (26), E_o denotes unconvertible emissions,* E_i denotes convertible emissions, ($E_o + E_i$) denotes engine emissions, and TSA_o denotes the value of TSA that helps reduce the convertible emissions by 63%. Thus, increasing the TSA to 3TSA_o would reduce the convertible emissions by 95%. However, it would also increase the substrate volume by 300%. Obviously, further increases in substrate volume would have very little impact on emissions reduction.

The pressure drop Δp across the substrate depends linearly on flow velocity and its length but inversely on the square of hydraulic diameter [20,21]:

$$\Delta p = C\frac{v\ell}{D_h^2} = \frac{CV_e\ell}{AD_h^2(\text{OFA})} \tag{27}$$

Since the substrate volume controls TSA, which, in turn, affects conversion efficiency, its cross-sectional area A should be maximized and its length ℓ minimized to reduce Δp. Of course, such an optimization of substrate shape will depend on space availability under the chassis.

In practice, both laboratory data and field experience have shown that if the substrate volume is approximately equal to engine displacement, it will meet the conversion requirements. Denoting engine displacement by V_{ed} and engine speed by N (revolutions per minute), the space velocity v_s may be written as

$$v_s = \frac{30NV_{ed}}{V} \quad \text{hr}^{-1} \tag{28}$$

which, for $V = V_{ed}$, becomes

$$v_s = 30N \quad \text{hr}^{-1} \tag{29}$$

The residence time τ for catalytic activity is simply the inverse of space velocity; i.e.,

$$\tau = \frac{120}{N} \quad \text{sec} \tag{30}$$

For typical engine speeds ranging from 1500 to 4000 RPM, the space velocity would range from 45,000 to 120,000 per hour. The corresponding residence time would range from 0.08 to 0.03 sec, which appears to be adequate for catalytic reaction. At lower space velocities the gas temperature is low and requires a longer time for reaction, whereas at higher space velocities the gas temperature is high and requires less time for reaction.

IV. PHYSICAL PROPERTIES OF CATALYST SUPPORTS

The physical properties of a ceramic honeycomb substrate, which can be controlled independent of geometric properties, also have a major impact on its performance and

---

* E_o is estimated to range from 5–10% of total engine emissions.

durability. These include microstructure (porosity, pore size distribution, and microcracking), coefficient of thermal expansion (CTE), strength (crush strength, isostatic strength, and modulus of rupture), structural modulus (also called E-modulus), and fatigue behavior (represented by a dynamic fatigue constant). These properties depend on both the ceramic composition and the manufacturing process, which can be controlled to yield optimum values for a given application, e.g., automotive [18,22,23], diesel [24,25], and motorcycle [26,27].

The microstructure of ceramic honeycombs not only affects physical properties like CTE, strength, and structural modulus, but has a strong bearing on substrate/washcoat interaction, which, in turn, affects the performance and durability of the catalytic converter [28–30]. The coefficient of thermal expansion, strength, fatigue, and structural modulus of the honeycomb substrate (which also depend on cell orientation and temperature) have a direct impact on its mechanical and thermal durability [22]. Finally, since all of the physical properties are affected by washcoat formulation, washcoat loading, and washcoat processing, they must be evaluated before and after the application of washcoat to assess converter durability [28–30].

A. Thermal Properties

The key thermal properties include CTE, specific heat c_p, and thermal conductivity K. The CTE values are strongly dictated by the anisotropy and orientation of cordierite crystal [9,31] as well as by the degree of microcracking [32,33]. The latter is controlled by the composition and firing cycle and can reduce the CTE significantly. As noted earlier, cordierite substrates have an extremely low CTE by virtue of the preferred orientation of anisotropic cordierite crystallites afforded by the extrusion process. Figure 7 shows the average CTE values along the three axes of cordierite crystallite; the complete CTE curves along A, B, and C axes are shown in Fig. 8, along with the calculated mean CTE curve, which represents random orientation. The extrusion process produces the preferred orientation in the raw materials, which upon firing causes alignment in the cordierite so that the lowest CTE axis of the orthorhombic crystal lies in the extrusion direction (also called the axial direction along which the gases flow). Nonaligned crystals generate localized stresses due to expansion anisotropy and lead to microcracking. The combination

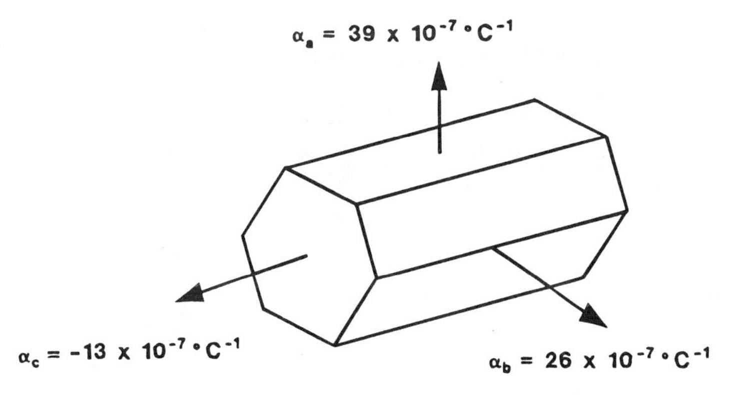

Figure 7 Average CTE values of cordierite crystallite along its three axes (25–800°C).

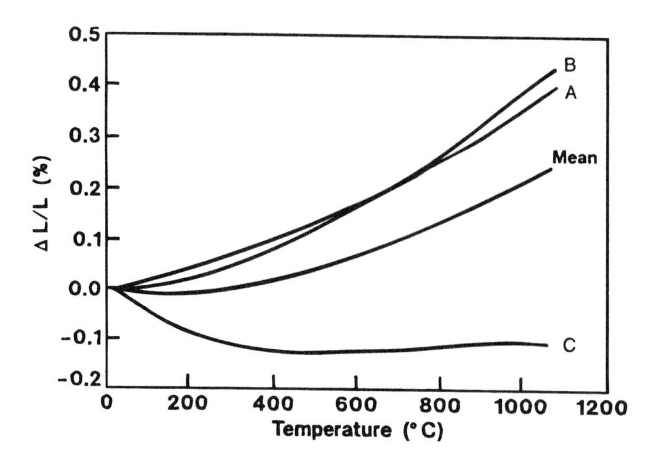

Figure 8 Thermal expansion curves for orthorhombic cordierite crystal.

of preferred orientation and microcracking results in axial CTE values ranging from 1×10^{-7} to 10×10^{-7}/°C (over 25–800°C temperature range), compared with 17×10^{-7}/°C for random orientation. This drastic reduction in CTE value makes the cordierite substrate an ideal catalyst support because the order-of-magnitude-higher CTE of γ-alumina washcoat* can now be managed easily. Tables 4A and 4B list the CTE values in axial and

Table 4A CTE vs. Temperature Data for EX-20, 400/6.5 □ Substrate (10^{-7}/°C)

	Axial CTE α_z		Tangential CTE α_θ	
Temperature, °C	Uncoated	Coated	Uncoated	Coated
400	−1.4	0.5	−0.2	4.2
500	0.8	4.1	2.0	7.5
600	2.6	6.9	3.9	10.3
800	6.1	11.1	7.4	14.3
900	7.3	12.5	8.7	15.6

Table 4B CTE vs. Temperature Data for EX-32, 236/11.5 △ Substrate (10^{-7}/°C)

	Axial CTE α_z		Tangential CTE α_θ	
Temperature, °C	Uncoated	Coated	Uncoated	Coated
400	−4.7	−6.6	1.5	−1.4
500	−2.4	−3.6	3.9	1.6
600	−0.6	−0.6	5.8	4.5
800	3.5	4.7	9.7	9.8
900	5.3	7.0	11.5	12.0
1000	7.0	9.1	13.1	13.9

* Typical CTE of γ-Al_2O_3 ≈ 80×10^{-7}/°C [29].

tangential† directions for 400/6.5 □ and 236/11.5 △ substrates‡ with and without the washcoat. It should be noted that the axial CTE of washcoated substrates seldom exceeds 12×10^{-7}/°C (which is still below that of an uncoated substrate with random orientation), making it resistant to thermal shock. The slightly higher CTE in the tangential direction is attributed to the alignment of high-expansion axes of cordierite crystallite with the webs in the cell junction region, which makes the tangential direction less resistant to thermal shock than the axial direction [35].

The specific heat and thermal conductivity of extruded cordierite substrates are relatively insensitive to wall porosity and substrate temperature. Their average values are [22]:

$$c_p = 0.25 \text{ cal/g°C}$$

$$K = 0.0005 \text{ cal/cm sec°C} \quad \text{in the tangential direction}$$

$$= 0.0010 \text{ cal/cm sec°C} \quad \text{in the axial direction}$$

B. Mechanical Properties

The key mechanical properties include the strength, the E-modulus, and the fatigue constant. The strength is important for withstanding packaging loads, engine vibrations, road shocks, and temperature gradients. Hence, high-strength substrates are more desirable. The E-modulus represents the stiffness or rigidity of the honeycomb structure and controls the magnitude of thermal stresses due to temperature gradients imposed by nonuniform gas velocity and catalytic exotherms. Hence, a low E-modulus, which reduces thermal stresses and increases substrate life, is more desirable. The fatigue constant n represents the substrate's resistance to growth of surface or internal cracks when subjected to mechanical or thermal stresses in service. A high n value implies greater resistance to crack growth and hence is more desirable. Much like thermal properties, mechanical properties are also influenced by the washcoat. Moreover, they vary with temperature and cell orientation.

Strength measurements are generally carried out on multiple specimens, prepared carefully from the substrate, to obtain a representative distribution that accounts for statistical variations associated with specimen preparation and the porous nature of ceramic substrates as well as with their cellular construction. Furthermore, the specimen size should be so chosen as to represent a sufficient number of unit cells, typically 200 to 400 cells across the specimen cross section. Figure 9 shows the specimens and load orientation for measuring the tensile and compressive strengths [22] of extruded cordierite ceramic substrates. The tensile strength in axial and tangential directions is measured in the 4-point flexure test, Fig. 9a, by breaking ten specimens in each direction, per ASTM specifications [36]. The mean values of the modulus of rupture (MOR) at various temperatures are summarized in Table 5A for EX-20, 400/6.5 □ substrates and in Table 5B for EX-32, 236/11.5 △ substrates, respectively, before and after the application of washcoat. These data show, as expected, that the MOR values increase with temperature and washcoat loading. The latter effect is substantial with strength increases of up to 50%. Also, the 15–20% higher MOR values of EX-32, 236/11.5 △ are attributed to higher wall thickness

† Circumferential direction in the cross-sectional plane of the substrate, perpendicular to the axial direction.

‡ The cordierite composition EX-32 for 236/11.5 △ is designed to yield a lower CTE than EX-20, to compensate for the higher E-modulus of the triangular cell substrate [34].

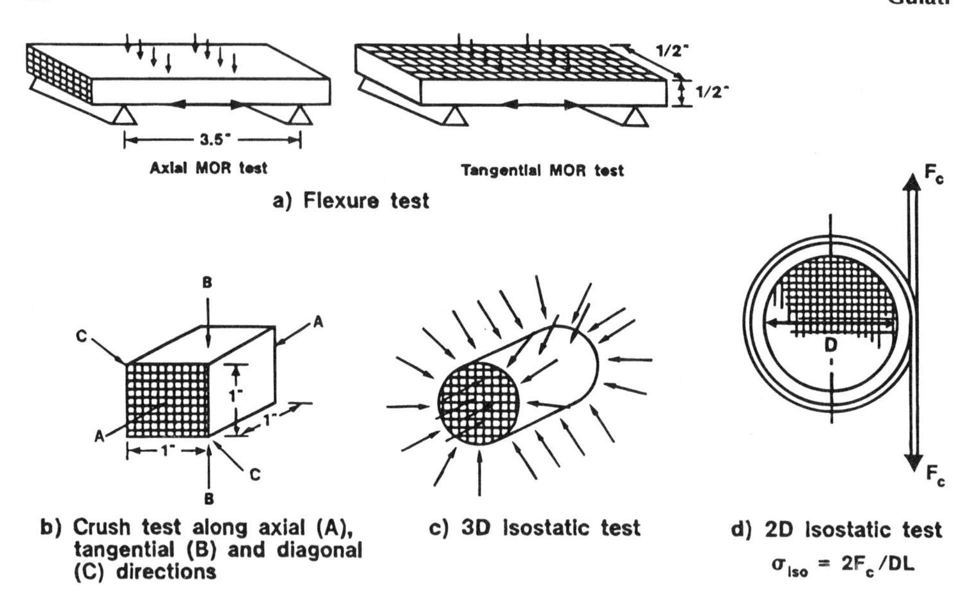

Figure 9 Cell and load orientation of honeycomb specimens for measuring mechanical strength. (From Ref. 22, courtesy of SAE.)

Table 5A MOR Data for Ex-20, 400/6.5 □ Catalyst Support (psi)

	Axial MOR		Tangential MOR	
Temperature, °C	Uncoated	Coated	Uncoated	Coated
25	395	525	200	280
200	370	575	180	260
400	390	585	190	280
600	425	600	200	290
800	450	590	220	300

Table 5B MOR Data for Ex-32, 236/11.5 △ Catalyst Support (psi)

	Axial MOR		Tangential MOR	
Temperature, °C	Uncoated	Coated	Uncoated	Coated
25	475	690	230	295
200	430	675	210	320
400	440	675	220	325
600	500	725	240	330
800	550	765	260	340

Table 6A Room Temperature Compressive Strength of EX-20, 400/6.5 □ Catalyst Support (psi)

	Uncoated	Coated
Crush *A*	3675	4400
Crush *B*	650	780
Crush *C*	55	75
Isostatic	1150	1600

than that of EX-20, 400/6.5 □. It will be shown later that the high MOR values ensure good thermal shock resistance.

The compressive strength, which has a direct relevance to packaging design, is measured in three different ways, namely, (1) uniaxial crushing of 1 in. × 1 in. × 1 in. cube specimens in axial (*A*), tangential (*B*), and diagonal (*C*) directions, (see Fig. 9b); (2) isostatic pressurization of whole substrate (see Fig. 9c); and (3) biaxial compressive strength of whole substrate (see Fig. 9d). Although the uniaxial crush test does not provide the absolute compressive strength of the whole substrate, it is a simple test for assessing the relative compressive strengths along *A*, *B*, and *C* axes. Furthermore, it helps in designing the packaging system for noncircular substrates where the mounting pressure is not uniform. The isostatic test and biaxial compressive test, on the other hand, are more representative of absolute compressive strength and are particularly suited for circular substrates with uniform mounting pressure. Tables 6A and 6B summarize the mean compressive strength of the two substrates, before and after the application of washcoat, at room temperature. The orientation of triangular cell structure defining *A*, *B*, and *C* axes for measuring the crush strength is shown in Fig. 10. It should be noted in Tables 6A and 6B that the triangular cell substrate has significantly higher compressive strength, notably along the *C* axis and under isostatic loading, than the square cell substrate, due to its rigid cell geometry.* Although the superior strength of triangular cell substrate renders it more robust in terms of mechanical durability, its light-off and steady-state conversion efficiency are inferior to those of square cell substrate, thus calling for certain trade-offs. Finally, the typical value of standard deviation for MOR data in Table 5 is ±10% of the mean value, and that for compressive strength data in Table 6 is ±25% of the mean value. MOR measurements above 800°C show that the substrate continues to get stronger up to 1200°C,† after which it behaves like a viscoelastic material, i.e., exhibits permanent deformation without fracture; its

Table 6B Room Temperature Compressive Strength of EX-32, 236/11.5 △ Catalyst Support (psi)

	Uncoated	Coated
Crush *A*	4360	5050
Crush *B*	915	1150
Crush *C*	345	475
Isostatic	1730	2740

* As a consequence of higher cell rigidity, the E-modulus of triangular cell substrate is correspondingly higher, resulting in a similar strain tolerance as that for square cell substrate.

† The strength increase with temperature up to 1200°C is due to the healing of microcracks.

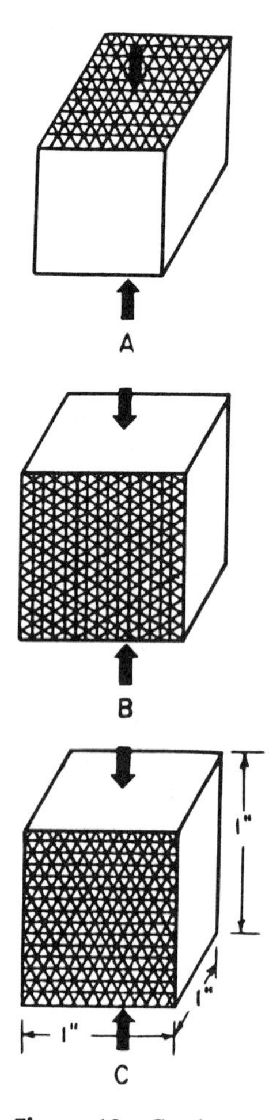

Figure 10 Crush test specimens along *A*, *B*, and *C* axes of triangular cell substrate. (From Ref. 30, courtesy of SAE.)

strength decreases gradually, approaching 40% of its room temperature value at 1400°C, implying that it can still support a load without failing catastrophically [37].

The fatigue behavior of ceramic substrates is relevant to either predicting a safe allowable stress for ensuring the specified life or estimating the life under a specified stress level. The methodology for obtaining the fatigue constant n has been discussed previously [38] and involves MOR measurement at operating temperature and relative humidity at five different stress rates, each spanning one decade a part. The slope of the axial MOR vs. stress rate plot provides the n value; see Fig. 11. The n values at 200°C—when the water vapor in exhaust gas is most active in promoting crack growth due to thermal and mechanical stresses—are shown in Table 7.

According to the power law fatigue model, the safe allowable stress σ_s for a converter life of τ_ℓ is given [39] by

$$\sigma_s = \text{MOR}\left[\frac{\tau_o}{\tau_\ell(n+1)}\right]^{1/n} \tag{31}$$

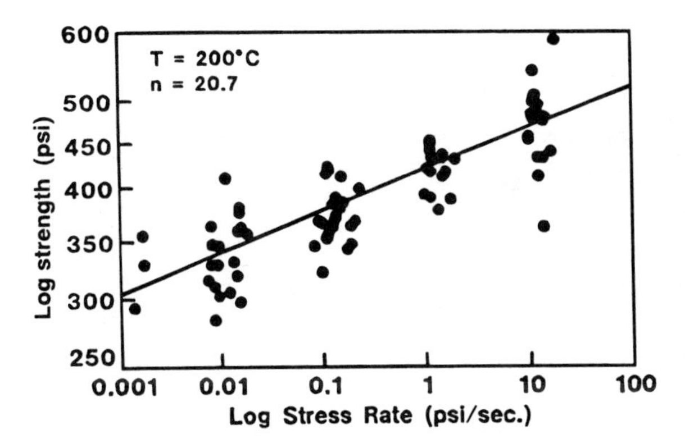

Figure 11 Axial strength as a function of stress rate at 200°C (EX-20, 400/6.5 substrate). (From Ref. 38, courtesy of SAE.)

where τ_o denotes the test duration for measuring MOR, which is typically 40 sec. Assuming a converter life of 100,000 miles at an average vehicle speed of 40 miles/hr, we can estimate the safe allowable stress value as fraction of the substrate's MOR for different values of *n*; see Table 8. Indeed, the higher *n* value of EX-32 substrate permits a higher allowable stress without a concern for crack propagation. Thus, the safe stress for EX-32, 236/11.5 △ substrate is 17% higher than that for EX-20, 400/6.5 □ substrate. In summary, theoretical considerations based on the power law fatigue model require that the net tensile stress in the substrate be kept below 50% of its MOR value to ensure a durability of 100,000 miles.

The E-modulus of extruded ceramic honeycombs is readily obtained by measuring the resonance frequency of MOR bars over a wide temperature range. Such a measurement is carried out in a high-temperature furnace according to ASTM specifications [40]. The E-modulus data (E_z and E_θ) for the two substrates are summarized as function of tempera-

Table 7 Dynamic Fatigue Constant *n* at 200°C

Substrate (uncoated)	*n*	95% Confidence interval
EX-20, 400/6.5 □	20.7	18–24
EX-32, 236/11.5 △	36.1	28–50

Table 8 Safe Allowable Stress σ_s for a 100,000-Mile Life of Ceramic Substrate

n	σ_s/MOR
15	0.38
20	0.48
25	0.55
30	0.61
35	0.65
40	0.68

Table 9A E-Modulus Data for EX-20, 400/6.5 □ Catalyst Support (10^6 psi)

	E_z		E_θ	
Temperature, °C	Uncoated	Coated	Uncoated	Coated
25	1.05	1.40	0.53	0.70
200	1.05	1.60	0.53	0.80
400	1.05	1.70	0.53	0.85
600	1.20	1.75	0.60	0.88
800	1.40	1.82	0.70	0.91

ture in Tables 9A and 9B. In general, the E-moduli increase with temperature and washcoat loading, implying higher thermal stresses at higher temperature. However, washcoat formulation and processing, and substrate microstructure could modify this trend due to interaction at the interface. Also, the triangular cell substrate has higher E-modulus in the axial direction (E_z) due to a higher cell rigidity than the square cell substrate. And finally, the tangential E-modulus (E_θ) is 50% of axial E-modulus (E_z) for the square cell substrate and only 33% of axial E-modulus for the triangular cell substrate, as might be expected from their respective cell geometries.

The foregoing physical properties are key to ensuring the physical durability of catalytic converter over its specified life. The next section demonstrates how these properties interact and affect both the mechanical and the thermal durability of the catalytic converter.

V. PHYSICAL DURABILITY

In addition to stringent emissions limits, automakers are also required to extend the useful life of catalytic converters from 50,000 miles to 100,000 miles. Both the catalytic and physical durabilities must be guaranteed over 100,000 miles. To meet these requirements simultaneously, it is imperative to use the systems approach, in which each component of the converter assembly is carefully designed, tested, and optimized. The converter

Table 9B E-Modulus Data for EX-32, 236/11.5 △ Catalyst Support (10^6 psi)

	E_z		E_θ	
Temperature, °C	Uncoated	Coated	Uncoated	Coated
25	1.53	1.54	0.46	0.54
200	1.53	1.63	0.46	0.56
400	1.71	1.82	0.51	0.64
600	1.80	1.89	0.55	0.66
800	1.88	1.92	0.60	0.67
1000*	2.08	1.95	0.68	0.67

*For temperatures ≥1000°C, the washcoat experiences significant mudcracking due to continued sintering and does not increase the E-moduli any further.

designer is therefore challenged to select appropriate materials, substrate contour, catalyst volume, washcoat loading, and catalyst formulation and to assemble them into a durable package. In view of the variety of materials, configurations, and microstructures employed in various converter components, the task of designing the total system becomes even more complex and requires the cooperative effort of all of the component suppliers to meet the automakers' space, performance, and cost specifications. Such an approach makes best use of each supplier's expertise while keeping other suppliers' constraints in mind, and leads to an optimum converter system that meets the total durability requirements.

A. Packaging Design

The design and size of the substrate are dictated primarily by performance requirements, which were discussed earlier. Next comes washcoat formulation and loading, which must not only provide adequate B.E.T. area for 100,000-mile catalytic durability but also be compatible with the cordierite substrate in terms of enhancing its physical properties, as discussed in the previous section. The choice of precious-metal catalyst and its specific formulation to provide catalytic activity over the desired life depend on the expertise of catalyst companies who stay abreast of worldwide PGM supply-and-demand status as well as the synergy between base metals (which are very effective as stabilizers and promoters) and the catalyst. Indeed, the impact of PGM catalyst on the physical properties of the substrate is negligible compared with that of γ-Al_2O_3 washcoat, so that it does not play as critical a role as other components in optimizing physical durability. Following washcoating and catalyzing, the substrate must be packaged in a robust housing that ensures its physical durability under severe operating conditions for 100,000 miles. Consequently, the packaging design can become the Achilles heel if not dealt with properly.

The typical converter package, shown in Fig. 12, consists of a resilient mat to hold the substrate, end seals to prevent gas leakage (depending on the type of mat), a stainless steel can to house the mat, end seals, and substrate, and a heatshield to protect adjacent components, floor pan, and ground vegetation from excessive thermal radiation [41].

A robust converter package provides positive holding pressure on the ceramic substrate, promotes symmetric entry of inlet gas, provides thermal insulation to the substrate (thereby heating it rapidly and retaining its exothermic heat for catalytic activity while minimizing the radial temperature gradient), and provides adequate frictional force

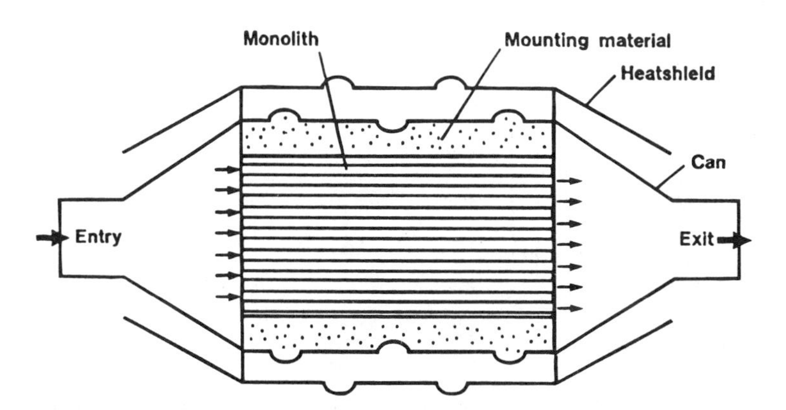

Figure 12 Schematic of converter package. (From Ref. 41, courtesy of SAE.)

at the substrate/mat interface to resist vibrational and back pressure loads that would otherwise result in slippage of the substrate inside the can.

These are complex requirements that call for careful packaging design via selection of durable components like the mat. In earlier designs, the mat was made from stainless steel wiremesh, which did not function as a gas seal, did not provide sufficient holding pressure (notably at high temperature), did not insulate the substrate against heat loss, and did not provide sufficient frictional force at the substrate/mat interface. Consequently, as emissions regulations became more stringent and converters moved closer to the engine, the need for a thermally insulating and resilient mat with good gas sealing and holding pressure capability became apparent. In the early 1980s, 3M Company introduced an intumescent ceramic mat, under the trademark Interam™ [42], containing unexpanded vermiculite, which expands on heating and provides the desired properties for a robust package.

The holding pressure, p_m, during room temperature assembly depends exponentially on the mount density of mat, ρ_m, as follows [41]:

$$p_m = 40{,}000 \exp \frac{-6.7}{\rho_m} \tag{32}$$

This equation estimates p_m in psi if ρ_m is expressed in g/cm^3. The mount density is defined by

$$\rho_m = \frac{W}{g} \tag{33}$$

where W denotes the basis weight of the mat, in g/cm^2, and g denotes the radial gap between substrate and can, in cm. As the mat expands on heating, the holding pressure increases (since it is constrained between the substrate and can) by 200–300% of room temperature value at temperatures approaching 800°C, above which the intumescent property of the mat is lost [21,43].

Figure 13 shows the forces acting on a circular ceramic catalyst support under operating conditions. The holding pressure must be high enough that the frictional force,

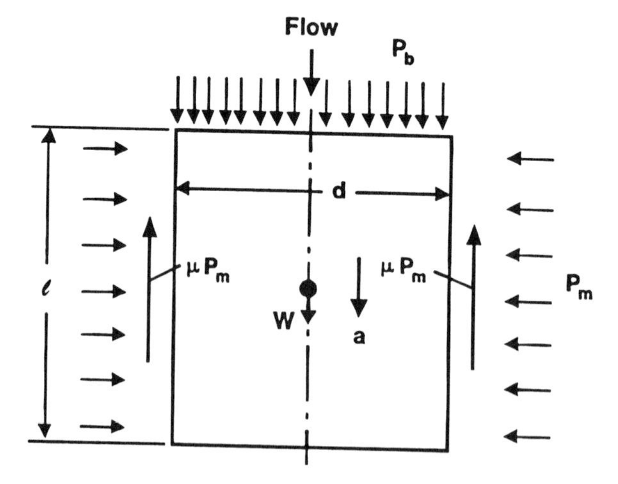

Figure 13 Schematic of ceramic catalyst support subjected to inertia, back pressure, and frictional forces. (From Ref. 44, courtesy of SAE.)

F_f, at the mat/substrate interface exceeds the sum of vibrational (F_v) and back pressure (F_b) forces, to prevent relative motion between substrate and can:

$$F_f \geq F_v + F_b \tag{34}$$

Denoting the friction coefficient between mat and substrate by μ, vibrational acceleration by a, catalyst density by ρ_c, catalyst diameter and length by d and ℓ, respectively, and back pressure by p_b, Eq. (34) yields

$$p_m \geq \left(\frac{d}{4\mu}\right)\rho_c a + \left(\frac{d}{4\mu\ell}\right)p_b \tag{35}$$

Equation (34) helps select the mat and mount density for a given application. For a robust packaging system, the nominal mount density should be 0.95, with a range* of 0.85–1.1 g/cm^3. This corresponds to a nominal mounting pressure of 35 psi, with a range of 15–90 psi, which is adequate to resist both vibrational and back pressure loads [44]. The friction coefficient μ has been measured experimentally and has a value of 0.25. If the mat is not compressed to high enough mount density, it will not act as a good seal. Furthermore, if the temperature and flow velocity of inlet gas are high, the direct impingement of gas could erode the mat, resulting in a loss of holding pressure and the premature failure of the catalyst support. The loose debris from eroded mat can lead to plugging of the catalyst supports, resulting in high back pressure and poor drivability [41]. These potential failure modes may be avoided by (1) selecting a mat with high basis weight, (2) compressing it to high mount density, (3) maintaining the average mat temperature at less than 800°C, (4) promoting convective cooling of the can via sound heatshield design and optimum converter location under the chassis.

Of course, the inlet gas temperature, engine malfunction, and emissions content should also be controlled via proper engine and fuel management to minimize catalytic exotherms and combustion of unburnt fuel within the converter. The high mount density ensures not only the resistance to thermal erosion but also the high holding pressure, which adds to the strength of the catalyst support and enhances its thermal and mechanical durability. The catalyst support, as noted earlier, can withstand an isostatic pressure well in excess of holding pressure, with a safety factor greater than 5. Finally, a mat with a high basis weight enjoys a lower average temperature, thereby preserving its intumescent property, critical for total durability, and is also more accommodating of dimensional tolerances that would otherwise widen the range of mount density and holding pressure.

The can or clamshells are generally made of ferritic stainless steel, AISI 409, with low CTE to minimize changes in mount density due to expansion of the can at operating temperature. Furthermore, since the can is also subjected to holding pressure, its deformation at operating temperature must be minimized for the same reason. To this end, either the can temperature must be kept below 500°C by efficient cooling or a better grade of ferritic stainless steel (e.g., AISI 439) should be used; see Fig. 14. The latter steel also has excellent resistance to high-temperature corrosion, as measured by its weight gain due to oxidation [45]. In addition to limiting the can temperature to less than 500°C, its flexural rigidity must be high, which can be achieved by designing an adequate number of stiffener ribs protruding inward or outward. The noncircular cans, with oval or racetrack contour, have the lowest rigidity along their minor axis and tend to deform excessively,

* The variation of mount density is caused by the tolerance stack-up in the converter components.

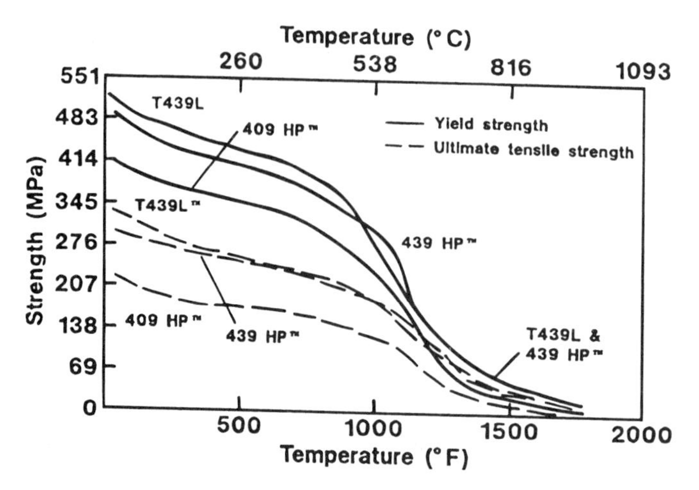

Figure 14 Yield and ultimate tensile strengths of AISI 409 and 439 stainless steels as a function of temperature. (From Ref. 45, courtesy of SAE.)

thereby allowing "blow by" of inlet gas, which not only promotes mat erosion but increases tailpipe emissions. Thus, the stiffener ribs are critical for noncircular cans. In designing inward ribs, care must be taken in controlling the mount density to limit the localized line pressure under the rib to well below the crush strength of the catalyst support. In some applications the can temperature may exceed 600°C, calling for austenitic stainless steel with improved high-temperature corrosion resistance, e.g., AISI 316 steel. However, as shown in Table 10, its CTE is 50% higher than that of ferritic steel, which can reduce the mat mount density significantly. Care must be taken in selecting the initial mount density to compensate for higher thermal expansion of an austenitic stainless steel can.

As mentioned previously, the ideal contour of the catalyst support and its housing is circular, since it results in a uniform holding pressure all around the support. However, space limitations under the chassis or in the engine compartment may require noncircular contours. Figure 15 ranks the four basic contours in order of uniformity of holding pressure, isotropy of clamshell stiffness, minimum can deformation, and radial temperature gradient along the minor axis. Every effort should be made during the early stages of converter design to select the best contour per Fig. 15 for optimum durability.

B. Mechanical Durability

The catalyst support is subjected to mechanical loads during manufacture, during canning, and in service. It must have sufficient strength to sustain these stresses without the onset

Table 10 Properties of Ferritic and Austenitic Stainless Steels for Converter Housing

Property	AISI type 409 (ferritic)	AISI type 316 (austenitic)
Density (lb/in.^{3})	0.28	0.29
E-modulus (10^6 psi)	29	28
CTE (10^{-7}/°C)	120	180
Thermal conductivity (BTU/ft/hr/°C)	84	55

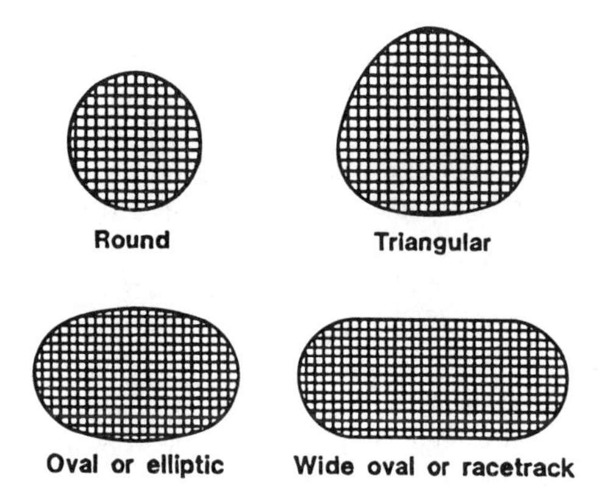

Figure 15 Ranking of contours of catalyst support for optimum durability (From Ref. 41, courtesy of SAE.)

of fatigue degradation. In Section IV.B we reviewed the mechanical strength data of ceramic catalyst supports before and after the application of γ-Al_2O_3 washcoat. The MOR, crush strength, and isostatic strength were all improved by 30–50% following the application of washcoat. Such an enhancement of strength is critical to mechanical durability. If either the washcoat formulation or the calcining process or substrate microstructure are not compatible, or if the substrate/washcoat adhesion is too strong, the expansion mismatch can introduce high stresses in the substrate wall and propagate some of the open pores with sharp tips, thereby degrading the mechanical strength. Such a phenomenon manifests itself in the form of a large scatter in strength data, with a standard deviation approaching 50–70% of mean strength. Catalyst supports with high variability in mechanical strength exhibit premature cracking, during either canning or internal QC tests, and can lead to early failures in the field.

The canning process applies a biaxial compressive stress on the lateral surface of catalyst support via compression of the mat. In the case of a circular substrate, the canning pressure is uniform and well below its biaxial compressive strength p_{2D} [46]. The latter is related to isostatic strength via

$$p_{2D} = p_{3D}\left[1 - \left(\frac{E_\theta}{E_z}\right)\left(\frac{\nu_{z\theta}}{1 - \nu_{r\theta}}\right)\right] \tag{36}$$

where p_{3D} denotes the isostatic strength given in Table 6 and $\nu_{z\theta}$ and $\nu_{r\theta}$ are Poisson's ratios for honeycomb structure, with values of 0.25 and 0.10, respectively. Substituting E_θ/E_z values from Table 9, we find that the biaxial compressive strength ranges from 86–91% of isostatic strength for coated substrates. Using the isostatic strength values in Table 6 we estimate the biaxial compressive strength of 400/6.5 □ catalyst support to be 1400 psi and that of 236/11.5 △ to be 2500 psi. We will compare these with the maximum holding pressure exerted by the mat during canning, which depends on the mount density per Eq. (32). Assuming the maximum mount density of 1.1 g/cm^3, the room temperature holding pressure is estimated to be 90 psi, which at the maximum intumescent temperature of mat may approach 270 psi. This is only 20% and 11% of biaxial compressive strength of 400/6.5 □ and 236/11.5 △ catalyst supports, respectively. Thus, there is a sufficient safety margin in the compressive strength of catalyst supports to sustain canning loads. In the case of a noncircular substrate, the holding pressure is nonuniform, as shown in Fig. 16. The higher pressure is carried by the semicircular portion (whose biaxial strength is very high, as discussed earlier), and the flatter portion experiences much lower pressure due to lower stiffness of the clamshell in that region. When the substrate is not seated or aligned properly in the clamshells, the canning load may lead to localized bending and result in shear crack at the junction of semicircular and flat portions of the contour; as little as 2–5° misalignment can produce shear cracks during canning of 400/6.5 □ catalyst supports.

Chassis vibrations and road shocks are another source of mechanical stresses that the catalyst support must sustain over its useful life. However, these stresses are damped out to a large extent by the converter package—notably its intumescent mat. Thus, the mechanical integrity of the catalyst support depends heavily on the integrity of the mat and the can. It is therefore imperative that the packaging design be made as robust as possible, taking the high-temperature limitations of mat and can materials into account. Vibration tests at 800°C, 80–120-g acceleration, and 100–2000-Hz frequency sweep have shown no evidence of relative motion or mechanical damage in 400/6.5 □ catalyst supports of racetrack contour when mounted with a holding pressure of 100–200 psi [47]. Similarly, no service failures due to impact load from stones and other hard objects have been reported when the catalyst support is properly packaged.

C. Thermal Durability

One of the key durability requirements of ceramic catalysts is to have adequate thermal shock resistance to survive temperature gradients due to nonuniform flow and heat loss

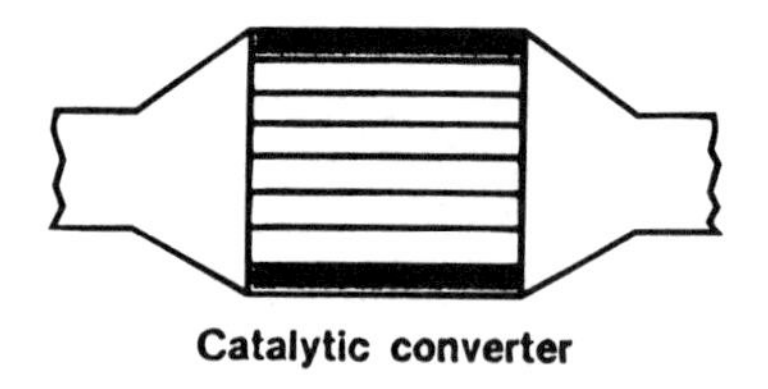

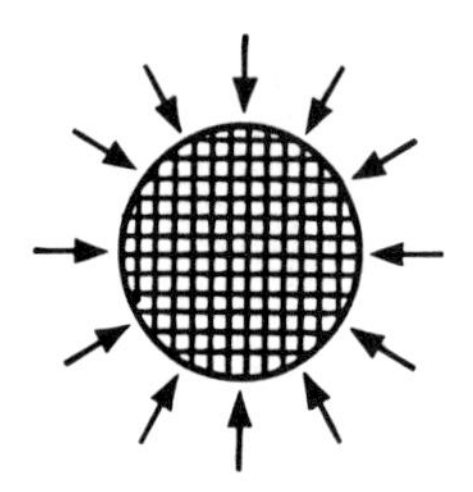

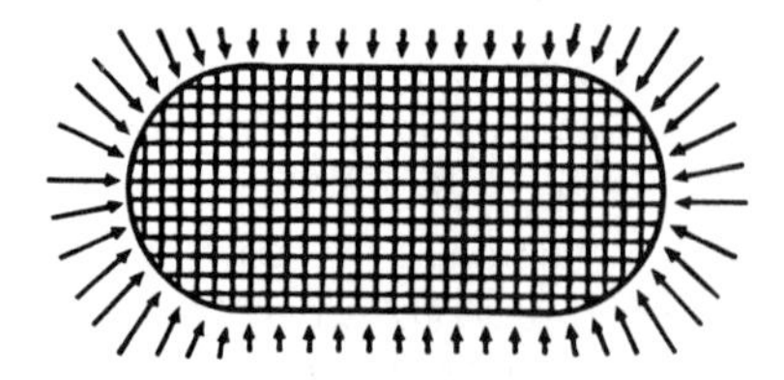

Figure 16 Schematic of pressure distribution during the canning of circular and oval catalyst supports. (Courtesy of SAE, Paper No. 910375.)

to the ambient environment. As can be seen in Fig. 17, the center region of the catalyst support experiences higher temperatures than its periphery, which induces tensile stresses in the outer region.* The magnitude of these stresses depends linearly on the CTE, the E-modulus, and the radial temperature gradient ΔT. These stresses should be kept well below the modulus of rupture of the catalyst support to minimize premature fracture in the radial and axial directions; see Fig. 18. It is, therefore, desirable that the catalyst support exhibit high strength and have a low E-modulus so that it has high thermal integrity to withstand thermal shock stresses.

Figure 19 shows a schematic of a finite element mesh for computing thermal stresses in a circular converter with a prescribed temperature field. As a hypothetical example, we assume a uniform temperature of 1000°C in the central hot region of 3.6-in. diameter and a linear radial gradient of 800°C/in. in the external 1-in.-thick region such that the skin temperature of the 5.6-in.-diameter converter is 200°C. Furthermore, we assume that such a radial temperature profile is constant over the entire 12-in. length and that the axial temperature gradient is negligible. The thermal stresses due to this assumed temperature field are readily computed by using the physical properties similar to those in Section IV [48]. Taking advantage of axial symmetry about the midlength, the axial and tangential stress profiles (σ_z and σ_θ) are shown in Fig. 20 for this hypothetical example. It should be noted that the axial stress (σ_z) reaches its maximum value on the outer surface at the midlength, whereas the tangential (σ_θ) stress remains relatively constant throughout the length of the converter. Since σ_z is caused by the differential elongation of the hotter interior relative to the colder exterior, its maximum value depends largely on the aspect ratio (length/diameter) of the converter, as indicated in Fig. 21. If σ_z approaches the MOR_z value at skin temperature, a ring-off crack can initiate at the surface, as shown in Figure 18b. Similarly, if σ_θ approaches the MOR_θ value, a radial or facial crack can initiate at the inlet or exit faces. To minimize ring-off or facial cracking, the radial temperature

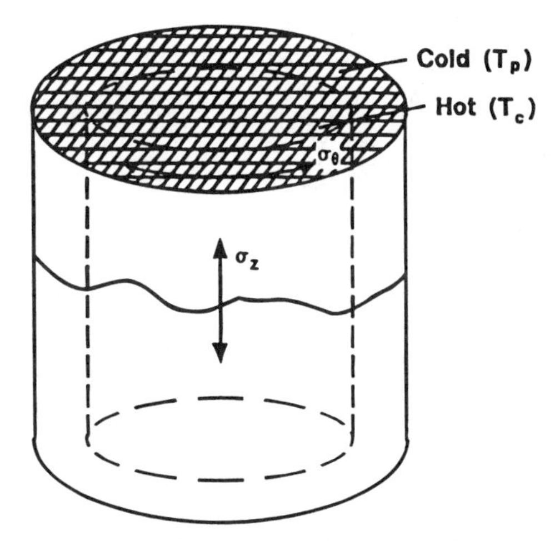

Figure 17 Development of thermal stresses due to the radial temperature gradient. (From Ref. 29, courtesy of SAE.)

* During cool-down, the center region experiences tensile stresses, but these are less detrimental according to failure modes observed in the field.

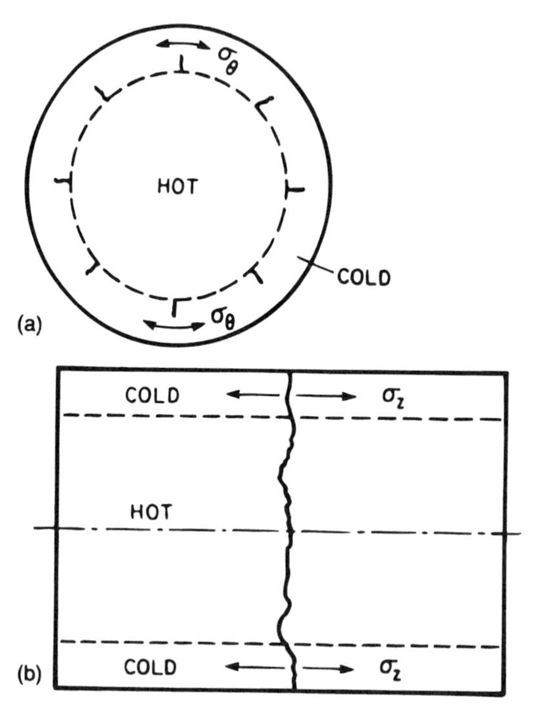

Figure 18 Schematic of radial cracks due to (a) tangential thermal stress and (b) axial thermal stress. (From Ref. 29, courtesy of SAE.)

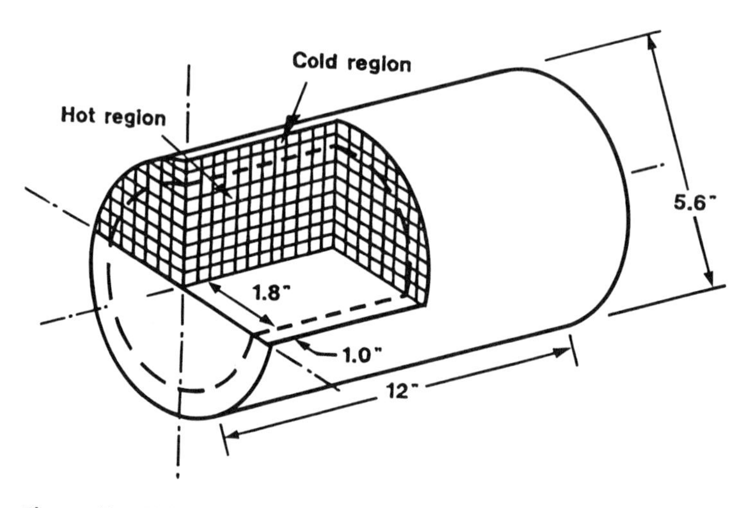

Figure 19 Finite element model for analyzing thermal stresses. (From Ref. 48, courtesy of SAE.)

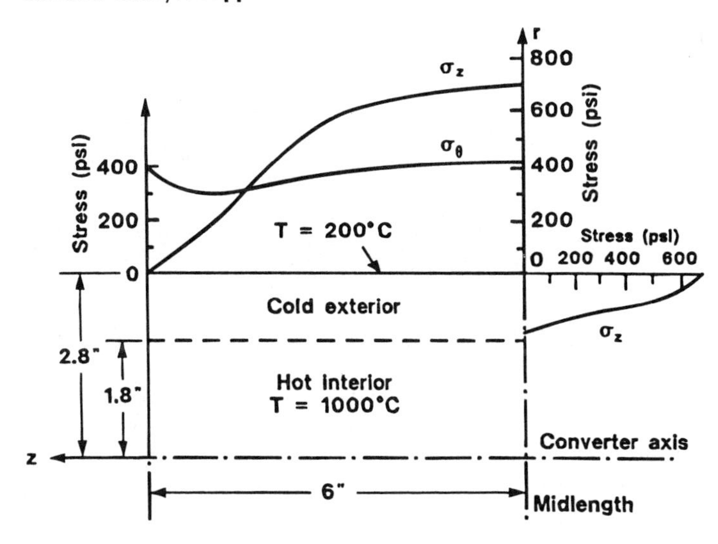

Figure 20 Thermal stress variation along the converter's length. (From Ref. 48, courtesy of SAE.)

gradient should be minimized by good fuel management, proper gas flow, and efficient mat insulation. In particular, ring-off cracking can be eliminated by using a converter with a low aspect ratio. The low aspect ratio is also desirable for minimizing the additional axial stress due to differential expansion of the catalyst support and the clamshell, namely,

$$\sigma_z^* = \mu p_m \left(\frac{L}{D}\right) \tag{37}$$

where μ denotes the friction coefficient between the ceramic mat and the monolith and p_m denotes the mounting pressure at skin temperature. With $\mu = 0.25$, σ_z^* can approach a value of $0.5p_m$ (for $L/D = 2$), which can be 20–30% of the MOR_z value; this should be minimized by limiting the aspect ratio to less than 1.4.

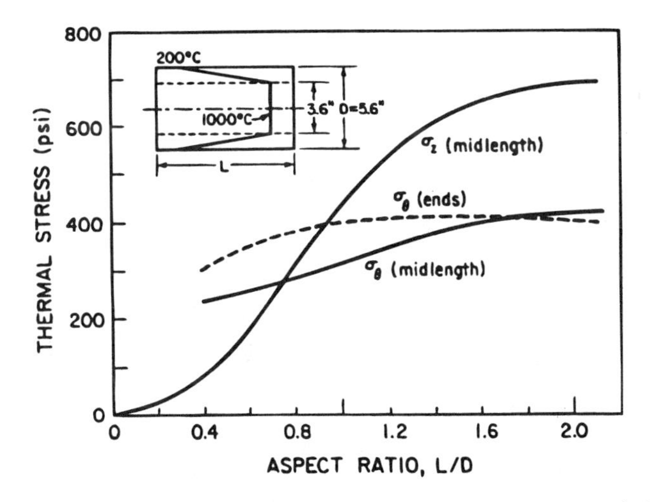

Figure 21 Effect of aspect ratio on thermal stresses. (From Ref. 48, courtesy of SAE.)

An alternate, simple technique for assessing thermal durability is to compute the thermal shock parameter from physical properties. This parameter, defined by Eqs. (38) and (39) is the ratio of mechanical strain tolerance to thermal strain imposed by the radial temperature gradient. The higher this parameter is, the better the thermal shock capability will be.

$$\mathrm{TSP}_z = \frac{(\mathrm{MOR}_z/E_z)}{\alpha_{cz}(T_c - 25) - \alpha_{pz}(T_p - 25)} \tag{38}$$

$$\mathrm{TSP}_\theta = \frac{(\mathrm{MOR}_\theta/E_\theta)}{\alpha_{c\theta}(T_C - 25) - \alpha_{p\theta}(T_p - 25)} \tag{39}$$

In these equations, T_c and T_p denote temperatures at the center and peripheral regions of catalyst support, α_{cz} and α_{pz} denote axial CTE values, and $\alpha_{c\theta}$ and $\alpha_{p\theta}$ denote tangential CTE values at T_c and T_p, respectively. We will compute the TSP values for EX-20, 400/6.5 □ and EX-32, 236/11.5 △ substrates at steady-state operating conditions defined by $T_c = 825°C$ and $T_p = 450°C$. Substituting the physical properties at these temperatures from Tables 4, 5, and 9 into Eqs. (38) and (39), we obtain TSP values summarized in Table 11. Let us make the following observations:

1. Washcoat reduces the TSP of the uncoated substrate, as might be expected from its high CTE.
2. Tangential TSP is generally lower than axial TSP, due to the higher CTE in that direction.
3. TSP values are very similar for the two substrates.

Figure 22 plots the failure temperature, measured in a cyclic thermal shock test [29], for ceramic catalyst supports as a function of their axial TSP values, which were controlled by modifying either the substrate, the washcoat, or the substrate/washcoat interaction. There is an excellent correlation between the failure temperature and the TSP value. Most automakers call for a failure temperature in excess of 750°C, although this may depend on the size of the catalyst and inlet pipe. Thus, a TSP_z value of more than 0.4 is required for the coated substrate. Finally, Fig. 22 shows that the washcoat may reduce the failure temperature of the catalyst support by 100–200°C, a trade-off the automakers are well aware of.

VI. ADVANCES IN CATALYST SUPPORTS

With stricter emission standards and a low back pressure requirement, ceramic substrates have undergone significant developments over the past few years. The thrust in the

Table 11 TSP Values for Catalyst Supports for $T_c = 825°C$ and $T_p = 450°C$

	TSP_z		TSP_θ	
Substrate	Uncoated	Coated	Uncoated	Coated
EX-20, 400/6.5 □	0.70	0.51	0.68	0.40
EX-32, 236/11.5 △	0.65	0.58	0.68	0.66

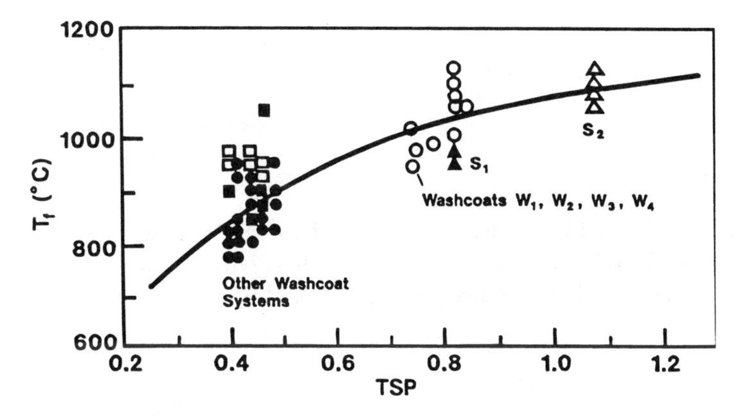

Figure 22 Correlation between axial TSP and failure temperature in a cyclic thermal shock test. (From Ref. 29, courtesy of SAE.)

ceramics area has centered on thin-wall structures, to minimize pressure drop, increase cell density, and increase surface area [20]. The thin-wall structures are extruded from a dense cordierite composition (EX-22) to provide strength and thermal shock resistance equivalent to those of standard cordierite substrates; see Fig. 23. Known as Celcor XT thin-wall substrates, they are available in two different cell sizes*: (1) 350/5.5 □ for low back pressure, and (2) 470/5 □ for higher conversion efficiency and lower back pressure than the standard 400/6.5 □ substrate. The pertinent geometric and physical properties of standard and thin-wall cordierite substrates are summarized in Table 12 [23].

These properties help compare light-off and steady-state conversion activity through heat capacity and GSA values, engine performance through Δp, and physical durability through substrate strength, TIF, and CTE [15]. It is clear in Table 12 that thin-wall substrates enjoy similar or higher GSA than standard substrates, thereby providing equivalent or

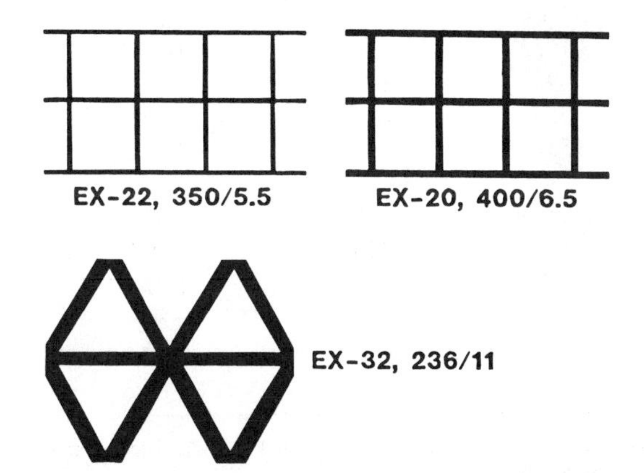

Figure 23 Comparison of standard and thin-wall cordierite substrates. (From Ref. 23, courtesy of Elsevier Science Publishers B.V., Amsterdam.)

* Other cell sizes can also be manufactured for special applications.

improved catalyst activity. In particular, the 470/5 □ substrate offers 11% higher geometric surface area than the 400/6.5 □ standard substrate.

A. Low Back Pressure

The pressure drop values in Table 12 are based on constant flow rate (i.e., given engine conditions) and constant total GSA (i.e., equivalent conversion activity). They are given by Eqs. (40) and (41) for square cell and triangular cell, respectively*:

$$\Delta P_s = \frac{14.2C_s}{(L - t)^3} \tag{40}$$

$$\Delta p_t = \frac{13.3C_t}{(L - t\sqrt{3}/\sqrt{3})^3} \tag{41}$$

Table 12 Nominal Properties of Standard and Thin-Wall Cordierite Substrates

	Composition			
	EX-20 (Std)	EX-32 (Std)	EX-22 (thin-wall)	EX-22 (thin-wall)
Cell structure	400/6.5	236/11.5	350/5.5	470/5
Cell shape	□	△	□	□
Wall thickness, in.	0.0065	0.0115	0.0055	0.005
Open porosity	35%	40%	24%	24%
Mean pore size, μm	3.0	7.0	2.0	2.0
Wall strength, psi	2950	2300	5120	5120
Wall modulus, 10^6 psi	3.78	3.03	6.22	6.22
MIF	0.019	0.034	0.012	0.013
Substrate strength*	$73k$	$64k$	$75k$	$81k$
TIF	7.7	7.1	9.7	9.2
TSR*	$110K$	$105K$	$130K$	$140K$
OFA, %	76	64	81	79
GSA, $in.^2/in.^3$	69.6	55.9	67.1	77.5
Avg. CTE @ 800°C, 10^{-7}/°C	6.0	5.0	4.0	4.0
D_h, in.	0.044	0.046	0.048	0.041
Δp*	$172C$	$191C$	$129C$	$154C$
Wall density, g/cm^3	1.63	1.51	1.91	1.91
Substrate density, g/cm^3	0.41	0.54	0.37	0.39
Heat capacity of substrate, cal/cm^3 °C	0.11	0.12	0.10	0.10

†The constants k, K, and C are normalization constants to help compare strength, thermal shock resistance, and pressure drop characteristics of cordierite substrates with different porosity and cell structures.

* In Eqs. (40)–(42) and in the expressions for C_s and C_t, we have assumed the fillet radius R at the cell corners to be zero.

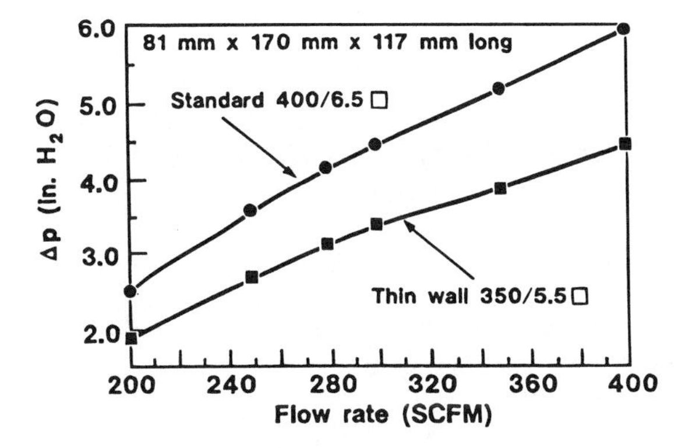

Figure 24 Pressure drop vs. flow rate through uncoated standard and thin-wall ceramic substrate. (From Ref. 23, courtesy of Elsevier Science Publishers B.V., Amsterdam.)

where

$$C_s = \frac{4Q\ell^2}{4A\ell(L - t)/L^2}$$

$$C_t = \frac{4Q\ell^2}{4\sqrt{3}(A\ell/L^2)(L - t\sqrt{3})}$$

in which Q denotes volume flow rate, A and ℓ are cross-sectional area and length of substrate, respectively, and L and t are side length and wall thickness, respectively, of the unit cell. The constancy of total GSA requires that

$$A_s\ell_s \frac{L_s - t_s}{L_t^2} = \frac{A_t\ell_t\sqrt{3}}{L_t^2}(L_t - t_t\sqrt{3}) = \text{constant} \tag{42}$$

which relates the cell dimensions and volume of the four substrates being compared in Table 12.

The linear dependence of back pressure on flow rate as modeled by Eq. (40) was verified experimentally for standard 400/6.5 □ and thin-wall 350/5.5 □ substrates with and without the washcoat. The results of these tests, shown in Figs. 24 and 25, demonstrate

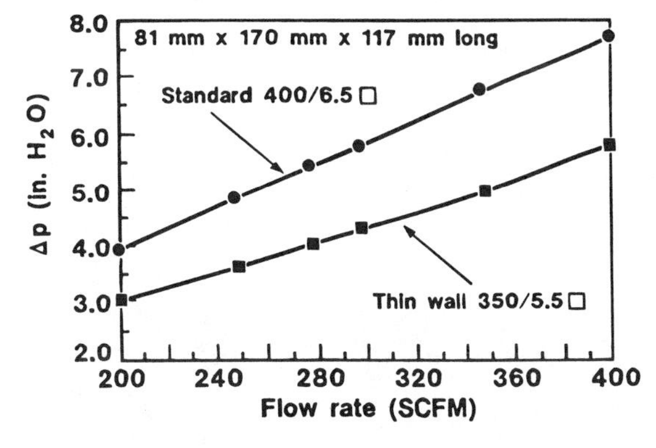

Figure 25 Pressure drop vs. flow rate through coated standard and thin-wall substrates. (From Ref. 23, courtesy of Elsevier Science Publishers B.V., Amsterdam.)

that the thin-wall substrates experience 25% lower back pressure, as predicted. Furthermore, a comparison of Δp for coated vs. uncoated substrates shows that the washcoat increases the back pressure by about 26%. Of course, this will also depend on washcoat loading and microstructure (porosity, surface roughness, etc.), since they both affect the hydraulic diameter and frictional drag.

B. Improved Durability

The dense cordierite composition of EX-22 substrate affords much higher wall strength due to its lower porosity compared with that of standard EX-20 substrate. The wall strength σ_w is an exponential function of porosity P [49]:

$$\sigma_w = \sigma_o \exp - 5P \tag{43}$$

where σ_o denotes the strength of the nonporous cordierite ceramic, with an approximate value of 17,000 psi. Recalling that $P = 0.35$ for EX-20 and $P = 0.24$ for EX-22, we conclude from Eq. (43) that the wall strength of thin-wall composition is nearly twice that of standard cordierite composition. The higher wall strength permits a honeycomb structure with thinner walls compared with those for standard substrate. Since the honeycomb strength depends on the product of the wall strength and the mechanical integrity factor, with the latter being a function of cell geometry, we can optimize the wall thickness of EX-22 substrate by requiring that its strength be equivalent to that of EX-20 substrate. To that end, we recall that

$$\text{honeycomb strength} = k\sigma_w x\text{MIF}, \quad k = \text{constant}^* \tag{44}$$

For equivalent strength,

$$\sigma_{ws}\text{MIF}_s = \sigma_{wt}\text{MIF}_t \tag{45}$$

where σ_{ws} and σ_{wt} denote the wall strength of standard and thin-wall substrates, respectively. Since $\sigma_{wt} = 1.73\sigma_{ws}$, the mechanical integrity factor of a thin-wall substrate can be 42% lower than that of standard substrate. Accordingly, the EX-22, 350/5.5 □ dense cordierite substrate with a wall thickness of 0.0055 in. provides a similar or slightly higher strength than the standard EX-20, 400/6.5 □ substrates; see Table 13. Given their strength equiva-

Table 13 Comparison of Strength and Thermal Shock Parameter of Standard and Thin-Wall Substrates

	EX-20, 400/6.5 □ (std)	EX-22, 350/5.5 △ (thin-wall)
A axis crush strength, psi	3675	6815
B axis crush strength, psi	650	650
C axis crush strength, psi	55	45
Axial MOR, psi	360	460
Axial E-modulus, 10^6 psi	1.05	1.28
MOR/E	3.4×10^{-4}	3.6×10^{-4}
TSP (450–825°C)	0.64	0.83
TSP (600–900°C)	0.70	0.82

* The constant k assumes an identical cell shape of the two honeycombs being compared.

lency, the mechanical durability of EX-22 thin-wall substrate is comparable to that of EX-20, 400/6.5 □ substrate.

The thermal durability is dictated by the MOR, E-modulus, and CTE values. We need only compare the thermal shock parameter (TSP) of EX-22, 350/5.5 □ with that of standard EX-20, 400/6.5 □ substrate. The mechanical strain tolerance (MOR/E) for the two substrates is nearly identical by design; see Table 13. However, the CTE values for EX-22 are significantly lower than those for EX-20 substrate due to composition and process differences; see Fig. 26. Accordingly, the thermal strain due to the hotter center region and the colder peripheral region is significantly lower for EX-22, resulting in higher TSP values. Table 13 compares the TSP values of the two substrates over two different operating temperature regimes, namely, (1) T_c = 825°, T_p = 450°C, and (2) T_c = 900°C, T_p = 600°C. In both cases, the EX-22, 350/5.5 □ substrate enjoys 15–30% higher thermal durability.

Finally, the lower density, higher surface area, lower back pressure, higher thermal shock resistance, and equivalent strength make EX-22 thin-wall substrates ideal for close-coupled application, e.g., preconverters and motorcycle catalysts. Much progress has been reported in their performance and durability recently [17,26,27,50,51].

VII. APPLICATIONS

Ceramic catalyst supports have performed successfully since their introduction in 1976 in passenger cars. Over 600 million units have been installed to date, and these continue to meet emissions, back pressure, and durability requirements. With new developments in substrate composition, washcoat technology, catalyst formulation, and packaging designs, ceramic catalyst supports are finding new and more strenuous applications, including gasoline trucks, motorcycles, and close-coupled converters. The following examples help illustrate their design, performance, and durability.

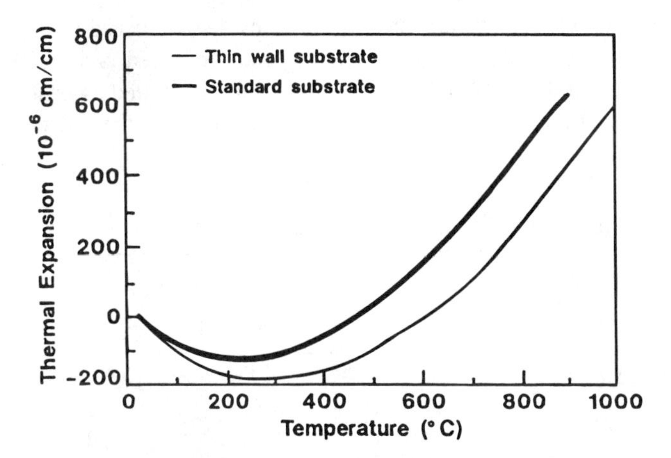

Figure 26 Thermal expansion curves for standard and thin-wall ceramic substrates. (From Ref. 23, courtesy of Elsevier Science Publishers B.V., Amsterdam.)

Table 14 Axial and Tangential MOR Values at 450°C (psi)

	MOR_z (450°C)		MOR_θ (450°C)	
Washcoat code	S_1	S_2	S_1	S_2
W_1	560	545	300	285
W_2	645	530	310	285
W_3	680	580	315	280
W_4	630	550	325	280
Uncoated substrate	505	425	255	220

A. Underbody Converter

In this first example, we will illustrate the effect of washcoat microstructure on thermal integrity of two different 4.66-in.-diameter × 6-in.-long passenger car catalyst supports. The washcoat microstructure was adjusted through formulation and processing conditions of the coating slurry. Similarly, the microstructure of a cordierite substrate with 400/8 square cell structure was adjusted through compositional and process control. We shall denote the two different substrates microstructures by S_1 and S_2 and the four different washcoat microstructures by $W_1, \ldots, W_4$. The compatibility of various substrate/washcoat microstructures was investigated by measuring key physical properties that reflect the catalyst's thermal shock resistance. For this particular passenger car application, the catalyst is required to pass a thermal cycling test, which imposes a center temperature of 825°C and a skin temperature of 450°C. The properties of interest that help assess the substrate/washcoat compatibility are MOR and E at 450°C and the differential expansion strain (DES) over the 450–825°C range, both in axial and tangential directions. These are summarized in Tables 14–16.

Table 16 demonstrates good compatibility between either of the two substrates and the four washcoats. The increase in DES is rather marginal relative to that of the substrate. The impact of these properties on thermal shock parameter is summarized in Table 17. The TSP values are reduced by only 5–10% for S_1 and 10–20% for S_2.

Several catalysts with the above washcoats were thermally cycled at successively higher temperatures until failure occurred. An acoustic technique was employed to detect invisible fractures. The failure temperature (T_f) obtained in this manner is plotted against the TSP value in Fig. 22 for each of these coated monoliths. Also included in these data

Table 15 Axial and Tangential E-Moduli at 450°C (10^6 psi)

	E_z		E_θ	
Washcoat code	S_1	S_2	S_1	S_2
W_1	1.50	1.40	0.73	0.71
W_2	1.48	1.50	0.73	0.73
W_3	1.40	1.45	0.71	0.73
W_4	1.45	1.40	0.72	0.71
Uncoated substrate	1.13	0.92	0.54	0.44

Table 16 Axial and Tangential DES Values over 450–825°C Range (10^{-6} in./in.)

	DES_z		DES_θ	
Washcoat code	S_1	S_2	S_1	S_2
W_1	520	435	610	515
W_2	565	440	610	520
W_3	555	435	625	525
W_4	565	430	610	500
Uncoated substrate	500	390	585	420

are uncoated substrates, S_1 and S_2, indicating superior thermal shock resistance of S_2 relative to S_1. Furthermore, the excellent compatibility between the substrates (S_1 and S_2) and the various washcoats ($W_1, \ldots, W_4$) results in as good a thermal shock resistance of coated catalysts as that of bare substrates.

B. Heavy-Duty Gasoline Truck Converter

For our second example, we will review the impact of a high-temperature washcoat formulation on the durability of EX-32 cordierite substrate with 236/11.5 △ cell structure. The overall dimensions of this oval catalyst, designed for a heavy-duty gasoline truck, are 3.38 in. × 5.00 in. × 3.15 in. long. This particular cordierite composition differs in porosity and mean pore size from EX-20 cordierite, and these are properties that influence the density, the E-modulus, and the tensile strength of the wall material. The microstructural differences in the EX-32, 236/11.5 △ cell structure provide another opportunity to study the substrate/washcoat interaction. This is best done by comparing MOR, E, and α values in axial and tangential directions before and after coating; see Tables 5B, 9B, and 4B.

The MOR data for the EX-32 △ cell monolith are summarized in Table 5B as a function of temperature. Note the beneficial effect of coating on both the axial and tangential MOR values. Compared with the substrate, they are 30–40% higher. This improvement in strength is most likely caused by the reduction in wall porosity and stress concentration at the pores, the filleting of cell corners by the coating, and the reduced microstresses due to the large mean pore size. It is clear that such a beneficial effect on the substrate's strength translates into improved durability of the catalyzed monolith. The data for elastic moduli are shown in Table 9B. Note that the coating has a minimal effect

Table 17 Axial and Tangential TSP Values over 450–825°C Range

	TSP_z (825–450°C)		TSP_θ (825–450°C)	
Washcoat code	S_1	S_2	S_1	S_2
W_1	0.74	0.89	0.67	0.79
W_2	0.79	0.82	0.70	0.75
W_3	0.86	0.92	0.71	0.74
W_4	0.77	0.90	0.74	0.79
Uncoated substrate	0.88	1.16	0.76	1.00

Table 18 Strain Tolerance and Thermal Shock Parameter Data for EX-32, 236/11.5△ Substrate and Catalyst During High-Load Operation of a Heavy-Duty Truck Engine

	Axial direction		Tangential direction	
	ST (10^{-6} in./in.)	TSP	ST (10^{-6} in./in.)	TSP
Uncoated	300	1.73	410	1.93
Coated	400	1.81	490	1.85

NOTE: T_c = 1025°C, T_p = 950°C.

on the rigidity of the monolith. Again, this is highly desirable from a durability point of view. The combination of high strength and low modulus increases the strain tolerance of the cellular structure and makes it more resistant to thermal shock. A plausible explanation for the minimal effect of coating on the substrate's rigidity is the pore size distribution in the wall, which affects the particle size and spatial distribution of the alumina washcoat (inter- vs. intrasubstrate distribution).

The average thermal expansion data in Table 4B show that the coating increases the axial thermal expansion by 30% but leaves the tangential thermal expansion unaltered. This behavior is most likely related to the distribution of washcoat at the corners of the triangular cell. The minimal effect of coating on tangential thermal expansion, however, is good news from a thermal durability point of view, particularly since the substrate expansion is higher in this direction.

The net effect of the above properties on mechanical and thermal durabilities is reflected in the strain tolerance (ST) and TSP values in the temperature range of vehicle operation. The thermocouple data during dynamometer testing at high engine load yielded a center temperature of 1025°C and a skin temperature of 950°C. The ST and TSP values over this temperature range, both in axial and tangential directions, are summarized in Table 18. It is clear from this table that the substrate and washcoat are quite compatible in that the substrate durability is either preserved or enhanced. Indeed, this particular catalyst exhibited no failures during 2000 hours of accelerated durability testing.

C. Close-Coupled Converter

In our last example, we present the design and performance data for a ceramic preconverter system that helps meet tighter emission standards—notably those corresponding to Transition low-emission vehicle (TLEV) and LEV standards. The key requirements of compactness, high surface area, low thermal mass, high use temperature, efficient heat transfer, high-temperature strength, prolonged catalyst activity, and robust packaging design are best met by thin-wall substrates of EX-22 composition.

The pertinent properties of EX-22 substrates with two different cell structures, optimized for light-off performance, are summarized in Table 19. The superior strength of 340/6.3 △ over that of 350/5.5 □ not only permits a robust packaging design, but contributes to long-term durability. Following washcoat and catalyst application, circular substrates, 3.36 in. diameter × 3.15 in. long, were wrapped with 3100-g/m^2 Interam™ 100 mat (with high-temperature integral seals) and assembled in 409 stainless steel cans, using a tourniquet technique, to a mount density of 1.2 g/cm^3. The end cones were then welded on to obtain a robust converter package for high-temperature testing; see Fig. 27.

Table 19 Key Properties of Light-Off Substrates

Property	EX-22, 350/5.5 □	EX-22, 340/6.3 △
Light-off parameter	5075	4790
Back pressure parameter	1.19	0.85
Biaxial compressive strength, psi	825	1465
High-temperature axial MOR @ 900°C, psi	510	650

The mechanical durability of an EX-22, 350/5.5 □ ceramic preconverter was assessed in the high-temperature vibration test, using an exhaust gas generator and an electromagnetic vibration table, under the following conditions:

Exhaust gas temperature	1030°C
Acceleration	45 g's
Frequency	100 Hz

The test was run for a total of 100 hours. The preconverter was cycled back to room temperature and inspected at 5-hr intervals 19 times, with no failure detected in the mounting system. The ceramic substrate maintained its original position, the seals rigidized and remained attached to the mat, and the mat was not eroded. Given the low CTE of EX-22 substrate and the low temperature gradients in the close-coupled location, thermal durability did not pose a problem.

The back pressure was measured for two different preconverter assemblies (all catalyzed) in a chassis dynamometer test [4-liter, 6-cylinder engine with port fuel injection (PFI)]. Both preconverters had identical outside dimensions and catalyst loading but different cell structures. The back pressure, measured with the aid of H_2O monometers, is summarized in Table 20. The differences in back pressure across the two preconverters are attributed to both the back pressure parameter and the frictional drag of catalyzed

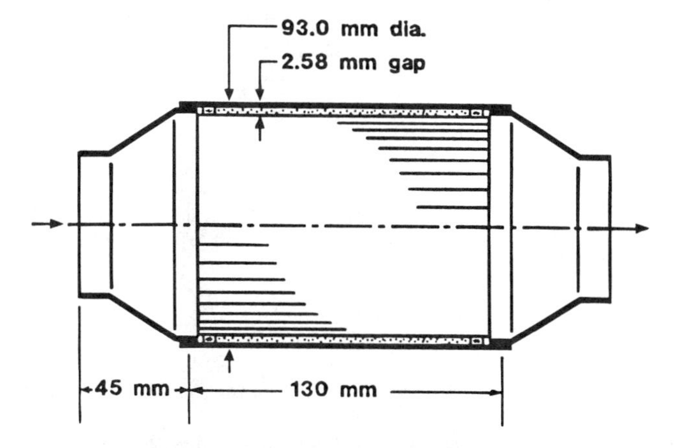

Figure 27 Schematic of packaging design for preconverter. (From Ref. 17, courtesy of SAE.)

Table 20 Back Pressure Data for Two Different Preconverters During Chassis Dynamometer Test @ 50 miles/hr

Cell structure	Back pressure across preconverter (in H_2O)	Total back pressure (in H_2O)
350/5.5 □ Ceramic	12.8	26.8
340/6.3 △ Ceramic	14.2	28.4

Note: 3.66 in.-diameter × 3.54 in.-long preconverter with inlet gas temperature = 700°C.

walls to gas flow. It is clear from Table 20 that the back pressure parameter is inversely proportional to the measured values of back pressure. It should also be noted that the back pressure across the preconverters is about 50% of the total back pressure from exhaust manifold to tailpipe.

The catalytic efficiency was measured on a 1994 vehicle powered by a 4-liter, 6-cylinder engine with port fuel injection. This vehicle's engine out emissions were 2.3 g/mile NMHC (nonmethane hydrocarbon), 16.9 g/mile CO, and 5.9 g/mile NO_x. Three different exhaust configurations were employed during emissions testing using the FTP (federal test procedure) cycle; see Fig. 28. The main converter consisted of two ceramic substrates with a total volume of 3.2 liters. One of these was catalyzed with a Pt/Rh catalyst and the other with a Pd-only catalyst. Figure 29 compares the tailpipe HC emissions with and without the ceramic preconverter (350 □), i.e., for configurations 1 and 3. It shows that the preconverter contributes to emissions reduction during the 40–120-sec interval. The FTP emissions are reduced significantly due to the oxidation of HCs and the generation of exothermic heat for faster light-off of the main converter. This benefit

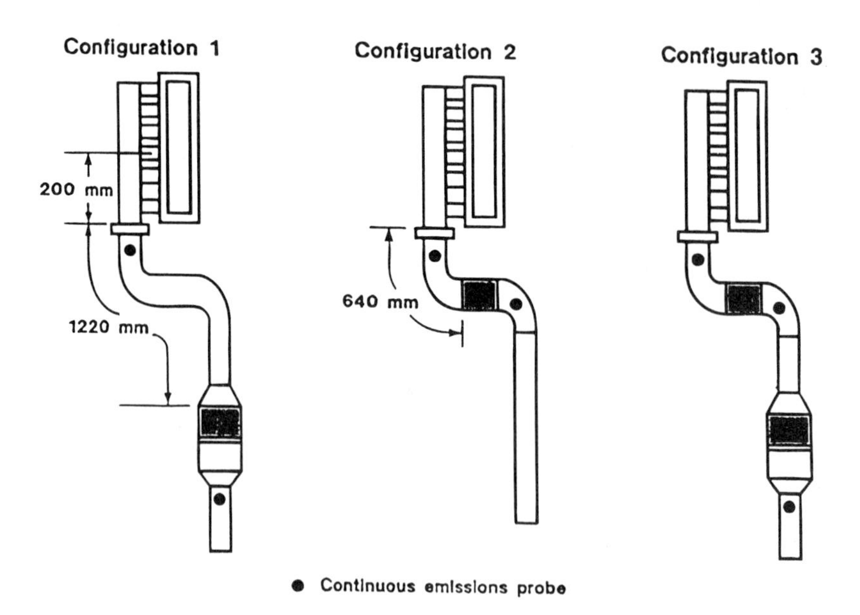

Figure 28 Exhaust configurations during FTP Test: (1) main converter only, (2) preconverter only, (3) main converter plus preconverter. (From Ref. 17, courtesy of SAE.)

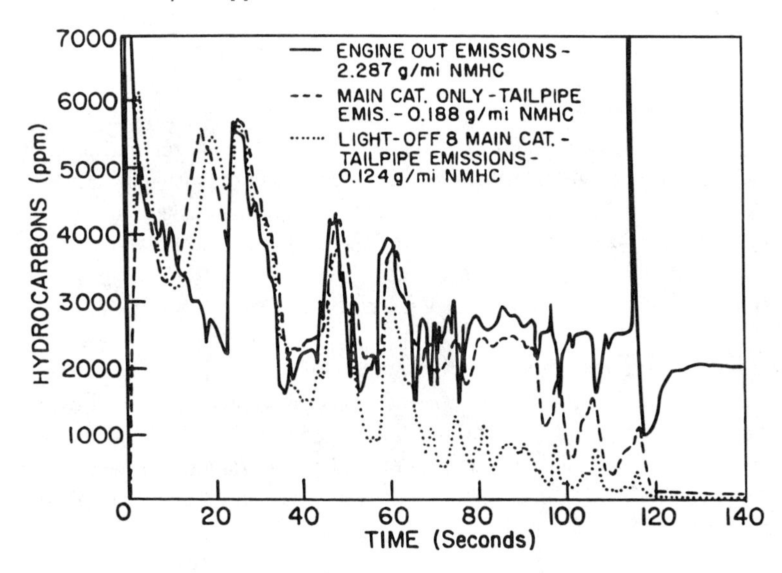

Figure 29 Continuous HC emissions during cold start from engine and tailpipe with main catalyst only vs. main catalyst plus preconverter. (From Ref. 17, courtesy of SAE.)

is attributed to the higher surface area, the lower thermal mass, and the close proximity of the ceramic preconverter to the exhaust manifold, which reduces the time for light-off temperature to 40 seconds after cold start. It is also clear in Fig. 29 that after 120 seconds, the HC emissions for both configurations are identical, implying that the main converter is fully effective.

VIII. SUMMARY

Ceramic catalyst supports offer the advantages of high surface-to-volume ratio, large open frontal area, low thermal mass, low heat capacity, low thermal expansion, high oxidation resistance, high strength, and high use temperature—properties that facilitate quick light-off, high conversion efficiency, low back pressure, good thermal shock resistance, and excellent mechanical durability. Furthermore, the geometric and physical properties of extruded ceramic honeycomb supports can be tailored independently to optimize both their performance and durability. Their microstructure can also be tailored via ceramic composition and the manufacturing process so as to be compatible with high-surface-area γ-alumina washcoat for optimum catalytic and physical durability.

These attributes have made the ceramic catalyst supports ideal for automotive application. Since 1975 over 600 million of these supports have been manufactured and successfully implemented in passenger cars, minivans, jeeps, and gasoline trucks worldwide. New advances in ceramic compositions and high-temperature catalysts have led to improved performance and 100,000-mile durability for close-coupled and manifold-mounted applications. The ceramic converter technology is growing rapidly and finding new applications that must now comply with emissions legislation, e.g., motorcycles, marine engines, and lawn and garden equipment.

This chapter emphasizes the importance of the systems approach during the design phase of the catalytic converter to ensure maximum utilization of chassis or engine space, optimum interaction between substrate and washcoat, and long-term robustness of total converter package. Thus, the systems approach calls for continuous dialogue and prompt feedback among the automaker, substrate manufacturer, catalyst company, mat and seal manufacturers, and canners. In this manner, the automaker's requirements can be best met by taking component suppliers' limitations (e.g., tolerances) into account and arriving at design trade-offs acceptable to all parties. Furthermore, the systems approach provides a rational basis for assessing the probability and warranty cost of converter failure in the field.

Three different converter applications are discussed to illustrate not only the variable operating conditions, but also the effectiveness of the systems approach in optimizing converter design for the passenger car (for both close-coupled and underbody locations) and the heavy-duty gasoline truck. These examples should prove valuable in designing other converter systems for which automakers' requirements and component suppliers' limitations are even more challenging!

ABBREVIATIONS

AISI	American Institute of Steel and Iron
API	American Petroleum Institute
ASTM	American Society for Testing and Materials
B.E.T.	Brunauer, Emmett and Teller (who developed the theory for calculating B.E.T. surface area)
BTU	British thermal unit
CO	carbon monoxide
Cpi^2	cells per square inch
CTE	coefficient of thermal expansion
DES	differential expansion strain
FTP	federal test procedure
GM	General Motors
GSA	geometric surface area
HC	hydrocarbon
LEV	low-emission vehicle
LOF	light-off factor
MIF	mechanical integrity factor
MOR	modulus of rupture
MY	model year
NMHC	nonmethane hydrocarbon
NO	nitrogen oxides
OFA	open frontal area
PFI	port fuel injection
PGM	platinum group metals
QC	quality control
RPM	revolutions per minute
SAE	Society of Automotive Engineers
SCFM	standard cubic feet per minute
ST	strain tolerance

TIF	thermal integrity factor
TLEV	transition low-emission vehicle
TSA	total surface area
TSP	thermal shock parameter
TSR	thermal shock resistance
ULEV	Ultralow-emission vehicle
3M	Minnesota Mining and Manufacturing

REFERENCES

1. Haagen-Smit, A.J. *Ind. Eng. Chem. 44* (1952).
2. Mader, P.P., MacPhee, P.D., Lofbert, R.T., and Larson, G.P. *Ind. Eng. Chem. 44* (1952).
3. Weaver, E.E. SAE Paper No. 690016 (1969).
4. Su, E.C., and Weaver, E.E. SAE Paper No. 730594 (1973).
5. Schwochert, H.W. SAE Paper No. 690503 (1969).
6. Cole, E.N. Address to Annual Meeting of the American Petroleum Institute, New York, New York (1970).
7. Brooks, H.R. Address to the American Metal Market Seminar on Pt and Pd, New York, New York (1968).
8. Bagley, R.D., Doman, R.D., Duke, D.A., and McNally, R.N. SAE Paper No. 730274 (1973).
9. Lachman, I.M., and Lewis, R.M. U.S. Patent 3,885,977 (1975).
10. Bagley, R.D. U.S. Patent 3,790,654 (1974).
11. Hollenbach, R.Z. U.S. Patent 3,112,184 (1963).
12. Summers, J.C., and Hegedus, L.L. *J. Catalysts 51*:185 (1978).
13. Kummer, J.T. *Prog. Energy Combustion Science*, Vol. 6. New York: Pergamon Press (1980).
14. Harned, J.L., and Montgomery, D.L. SAE Paper No. 730561 (1973).
15. Gulati, S.T. SAE Paper No. 881685 (1988).
16. Gulati, S.T. AIAM Seminar on "Catalytic Converters: Fresh Steps." Bangalore, India (1995).
17. Gulati, S.T., Socha, L.S., Then, P.M., and Stroom, P.D. SAE Paper No. 940744 (1994).
18. Gulati, S.T., and Then, P.M. *CAPoC-3 Proc.* Amsterdam: Elsevier (1994).
19. Day, J.P. 8th International Pacific Conference on Automotive Engineering, Yokohama, Japan (1995).
20. Day, J.P., and Socha, L.S. SAE Paper No. 910371 (1991).
21. Gulati, S.T., Ten Eyck, J.D., and Lebold, A.R. SAE Paper No. 930161 (1993).
22. Gulati, S.T. SAE Paper No. 850130 (1985).
23. Gulati, S.T. *CAPoC-2 Proc.* Amsterdam: Elsevier (1991).
24. Gulati, S.T., Lambert, D.W., Hoffman, M.B., and Tuteja, A.D. SAE Paper No. 920143 (1992).
25. Gulati, S.T., and Lambert, D.W. *ENVICERAM '91 Proc.* Cologne, Germany: Deut. Keram. Gessell. (1991).
26. Gulati, S.T., and Scott, P.L. *Two-Wheeler Conf. Proc.* Graz (1993).
27. Gulati, S.T., and Scott, P.L. *Small Eng. Tech. Conf. Proc.*, Pisa, Italy (1993).
28. Gulati, S.T., Summers, J.C., Linden, D.G., and White, J.J. SAE Paper No. 890796 (1989).
29. Gulati, S.T., Cooper, B.J., Hawker, P.N., Douglas, J.M., and Winterborn, D. SAE Paper No. 910372 (1991).
30. Gulati, S.T., Summers, J.C., Linden, D.G., and Mitchell, K.I. SAE Paper No. 912370 (1991).
31. Lachman, I.M., Bagley, R.D., and Lewis, R.M. *Ceram. Bull. 60*:2 (1981).

32. Ikawa, H., Ushimaru, Y., Urabe, K., and Udagawa, S. *Science of Ceramics 14* (1987).
33. Buessem, W.R., Thielke, N., and Sarakaukas, R.V. *Ceram. Age 60*:5 (1952).
34. Gulati, S.T. SAE Paper No. 750171 (1975).
35. Gulati, S.T. SAE Paper No. 830079 (1983).
36. *Annual Book of ASTM Stds.*, Part 17, Designation C158. Philadelphia. ASTM (1975).
37. Gulati, S.T., and Sweet, R.D. SAE Paper No. 900268 (1990).
38. Helfinstine, J.D., and Gulati, S.T. SAE Paper No. 852100 (1985).
39. Evans, A.G. *Int. J. Fract. 10*:2 (1974).
40. *Annual Book of ASTM Stds.*, Part 17, Designation C623. Philadelphia. ASTM (1975).
41. Stroom, P.D., Merry, R.P., and Gulati, S.T. SAE Paper No. 900500 (1990).
42. Langer, R.L., and Marlor, A.J. U.S. Patent 4,305,992 (1981).
43. Gulati, S.T., Ten Eyck, J.D., and Lebold, A.R. SAE Paper No. 922252 (1992).
44. Gulati, S.T. SAE Paper No. 920145 (1992).
45. Gulati, S.T., Sherwood, D.L., and Corn, S.H. SAE Paper No. 960471 (1996).
46. Gulati, S.T., and Reddy, K.P. SAE Paper No. 930165 (1993).
47. Maret, D., Gulati, S.T., Lambert, D.W., and Zink, U. SAE Paper No. 912371 (1991).
48. Gulati, S.T. SAE Paper No. 830079 (1983).
49. Coble, R.L., and Kingery, W.D. *J. Am. Ceram. Soc. 19*:11 (1956).
50. Socha, L.S., Gulati, S.T., Locker, R.J., and Then, P.M. SAE Paper No. 950407 (1995).
51. Locker, R.C., Schad, M.J., and Sawyer, C.B. SAE Paper No. 952414 (1995).

3

Metal and Coated-Metal Catalysts

Martyn V. Twigg and Dennis E. Webster
Johnson Matthey, Royston, Hertfordshire, England

I. INTRODUCTION

Metal crystallites are the active phase in many catalysts, particularly those effective for reactions involving hydrogen or oxygen; Table 1 contains illustrative examples. Bulk metals can be fabricated into shapes suitable to go into a reactor, and, although a low surface area of metal is provided, if reaction rates are high (as with operation at high temperature), than attractive conversions can be achieved. Diffusion effects are minimal, and with short contact times conditions are ideal for selective oxidations; but products are usually limited to species that are thermally stable. Typical catalysts are metal granules and wire gauzes, with products species such as NO, HCN, HCHO, and CH_3CHO. Less thermally stable molecules require reaction at lower temperatures, with compensating higher active-phase surface areas to achieve economic reaction rates.

Table 1 Selected Industrial Catalysts Containing Metallic Active Phases

Active phase	Catalyzed reaction	Catalyst form
Nickel	Methanation (CO to CH_4)	Extrudate/pellets
Nickel	Hydrocarbon steam reforming	Rings
Nickel	Nitroarenes to amines	Powder/pellets
Copper	Methanol synthesis	Pellets
Copper	Water gas shift	Pellets
Copper	Nitroarenes to amines	Pellets
Palladium	Acetylene to ethylene	Pellets
Palladium	Autocatalysts	Monoliths
Iron	Ammonia synthesis	Granules
Silver	Ethylene to ethylene oxide	Rings
Silver	Methanol to formaldehyde	Granules/gauzes
Silver	Ethanol to acetaldehyde	Granules/gauzes
Platinum	Oxidations/VOC	Monoliths
Platinum/rhodium	Autocatalysts	Monoliths
Platinum/rhodium	Ammonia oxidation to NO	Gauzes
Platinum/rhodium	Hydrogen cyanide synthesis	Gauzes

High metal surface area is obtained by having many small metal crystallites dispersed on a high-area refractory oxide support. Precipitated catalysts fabricated in the form of pellets or extruded shapes for use in packed beds suffer the disadvantage of needing catalytic and physical properties provided by the same material [1]. This results in a compromise between strength, activity, and pressure drop characteristics, which can be overcome if these conflicting requirements can be decoupled. One way this has been done is to impregnate active-phase precursors onto a preformed support formulated to optimize desired physical requirements. Examples of catalysts of this type include impregnated low pressure drop ring catalysts used in tubular reactors for steam-reforming hydrocarbons in the production of synthesis gas, and the selective oxidation of ethylene to its epoxide.

However, this degree of decoupling catalytic and physical properties is insufficient in some instances. The most notable is with autocatalysts and other environmental applications where attrition resistance (continual vibration of the converter) and very low pressure-drop are required. As described in the previous chapter, extruded high-strength ceramic monolithic honeycomb substrates are used extensively in this application. Cordierite, an inert substrate having a desirable low coefficient of thermal expansion, is coated with a thin layer of a catalytic formulation comprising highly dispersed active metal(s) on a high-area support together with appropriate promoters. From about 1974, coated monolithic substrates became accepted in the automotive industry as an integral part of emission control systems. Later it was demonstrated that metal monoliths could be coated, and they too gained acceptance in the industry. An initial advantage was that metal structures fabricated from thin foil had cell walls thinner than those made from ceramic material, and this resulted in a lower pressure drop. Thinner-walled ceramic products subsequently became available, but a substantial quantity of metallic substrates are used in automotive applications, as well as other environmental applications. The use of ceramic monoliths is considered in the previous chapter; this chapter considers metallic substrates and other structured-metal catalysts are considered.

II. BULK-METAL CATALYSTS

There are relatively few bulk-metal catalysts, but some are commercially significant. The most important are those fabricated as gauzes. First, however, we discuss some less structured systems. Catalysts in these categories have low metal-surface areas [2] and are particularly sensitive toward poisoning and toward fouling by debris.

A. Metal Powders, Granules, and Shaped Structures

It should be possible to use compressed metal powders as, for example, a catalytic filter, but the present authors are not aware of any industrial examples of this. Raney metals powders, however, have been employed in some liquid-based processes in shallow beds through which reactants pass. Because Raney metals are fine grained, pressure drop can be a problem; so it is more common to use them in an unstructured way in slurry reactors, as, for example, formerly in the oils and fats industry [3]. Raney metals can have high surface areas when freshly prepared, but this decreases quickly in use, particularly when exposed to elevated temperatures. Pressure drop considerations are less significant for beds of metal granules, but there is less effective use of metal than with fine powders. For granules, surface areas in the region of 30–35 $cm^2\ g^{-1}$ are typical for silver used in the conversion of methanol to formaldehyde by oxidative dehydrogenation [4], as shown in Eqs. (1) and (2). During use, surface area increases due to a restructuring of the surface, and surface areas of operating catalysts are typically in the range 40–45 $cm^2\ g^{-1}$. Acetaldehyde is also produced in a similar process from ethanol, and silver-gauze catalysts compete with processes based on the use of granules; see Section II.B.

$$CH_3OH + \tfrac{1}{2}O_2 \rightarrow HCHO + H_2O \tag{1}$$

$$CH_3OH \rightarrow HCHO + H_2 \tag{2}$$

B. Metal-Gauze Catalysts

The classic use of a metal-gauze catalyst is in the oxidation of ammonia with air to nitric oxide for production of nitric acid, which is one of the most efficient selective oxidation processes operated on an industrial scale. The desired selective formation of nitric oxide, reaction (3), is strongly exothermic ($-903\ kJ\ mol^{-1}$); but nonselective burning to water and dinitrogen, reaction (4), competes significantly at pressures above atmospheric pressure. Ammonia slip through the gauze pad (the industrial catalyst comprises several gauzes on top of each other) is doubly detrimental because conversion is decreased by reaction of ammonia with nitric oxide, as in Eq. (5). It is therefore important that mechanical integrity of the gauze pad be maintained to prevent ammonia bypass [5].

$$4NH_3 + 5O_2 \rightarrow 4NO + 6H_2O \tag{3}$$

$$4NH_3 + 3O_2 \rightarrow 2N_2 + 6H_2O \tag{4}$$

$$4NH_3 + 6NO \rightarrow 5N_2 + 6H_2O \tag{5}$$

Industrially, a platinum/rhodium gauze (typically 10% rhodium) is used, although initial work by Ostwald, who filed patients in 1902, made use of platinum alone [6]. His first experiments with platinized asbestos gave small yields, but a coiled strip of platinum afforded up to 85% conversion. This was the first example of a monolithic structured catalyst. Ostwald also described a roll of corrugated platinum strip (some 2 cm wide, weighing 50 g) much like a modern monolithic metallic catalyst substrate (see Section IV.A). However, the disadvantage of having to employ so much platinum was circumvented by Karl Kaiser, who filed patents in 1909 covering preheated air (300–400°C) and a layer of several platinum gauzes [7]. He was the first to use platinum gauze to catalyze ammonia oxidation, and the form he settled on, 0.06-mm diameter woven to 1050 mesh cm^{-2}, is similar to that used today. By 1918, converters were 50 cm in diameter, with several gauzes in a pad operating at 700°C. Today plants run at temperatures up to 940°C, and gauze pads are up to 5 m in diameter, with as many as 15 gauzes, weighing tens of kilograms. Figure 1 shows the installation of a gauze pad in a plant, and Fig. 2 illustrates how the fresh catalyst changes during use as dendritic excrescences of alloy grow from the wire surface. Deepening of the etches so formed results in the fracture of the gauze wire, and physical loss of material is the major mode of deactivation. Another cause of catalyst deactivation in plants running above atmospheric pressure is the physical blanketing of the gauze by rhodium oxide (Rh_2O_3); under appropriate conditions this can be reduced and diffused back into the alloy by annealing at high temperature, but success is usually limited. Typically, platinum/rhodium gauze catalysts used in ammonia oxidation have lives of 50 to 300 days, depending on operating pressure.

Figure 1 Final phase of installing a platinum/rhodium gauze pad in a nitric acid plant.

Another process using platinum/rhodium gauze catalyst is the formation of hydrogen cyanide from methane (13%), ammonia (10–12%), and air (75%), known after its inventor as the Andrussow process [8]. The overall chemical reaction is given in Eq. (6), and this has been very successfully modeled from first principles using 13 simultaneous unimolecular and bimolecular surface reactions [9]. The process is operated at higher temperatures than used in ammonia oxidation, typically 1100–1200°C, so the mechanical strength of platinum/rhodium alloy at high temperatures and its resistance to oxidation play important roles in addition to its catalytic activity. Typical conversions based on ammonia are in the range 60–65%, and the model referred to previously predicts the reactor should be operated at the highest temperature possible, and that selectivity increases only slightly with pressure.

$$NH_3 + CH_4 + 1.5O_2 \rightarrow HCN + 3H_2O \quad (6)$$

There have been claims that platinum/iridium gauzes are better than platinum/

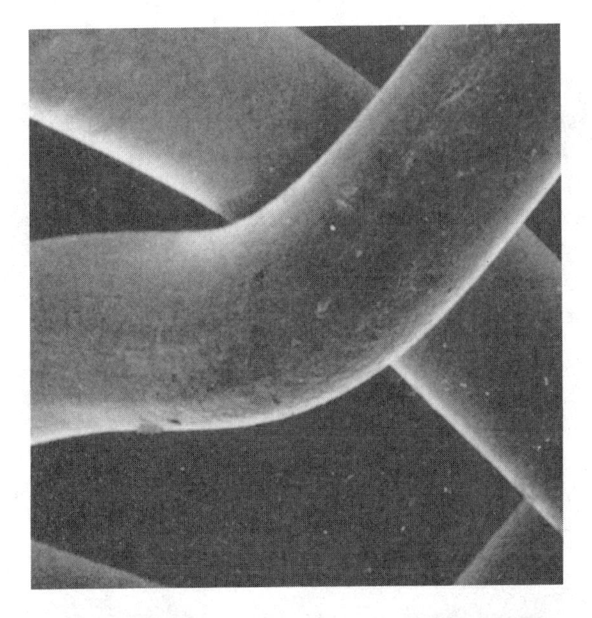

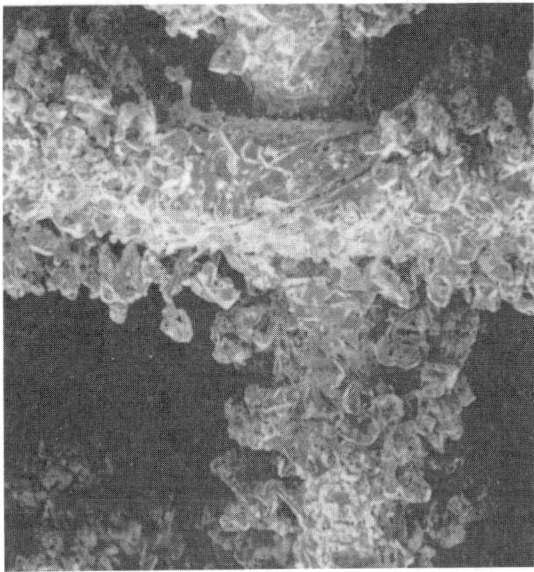

Figure 2 Changes of the surface morphology of a traditional woven platinum/rhodium gauze during use. The surface of fresh catalyst is smooth, but during use dendritic excrescences of alloy grow from the wire surface.

rhodium ones, and that both are better than platinum alone. The gauzes used in the Andrussow process undergo related modes of deactivation to those in ammonia oxidation, but there are some differences that were reviewed by Knapton and others [10]. For instance, gauzes used in ammonia oxidation retain some ductility and can be separated after a period of use. Those used in hydrogen cyanide synthesis are brittle, due to the transformation of the wire into a mass of crystallites; this behavior is shown in Fig. 3. Similar recrystallization

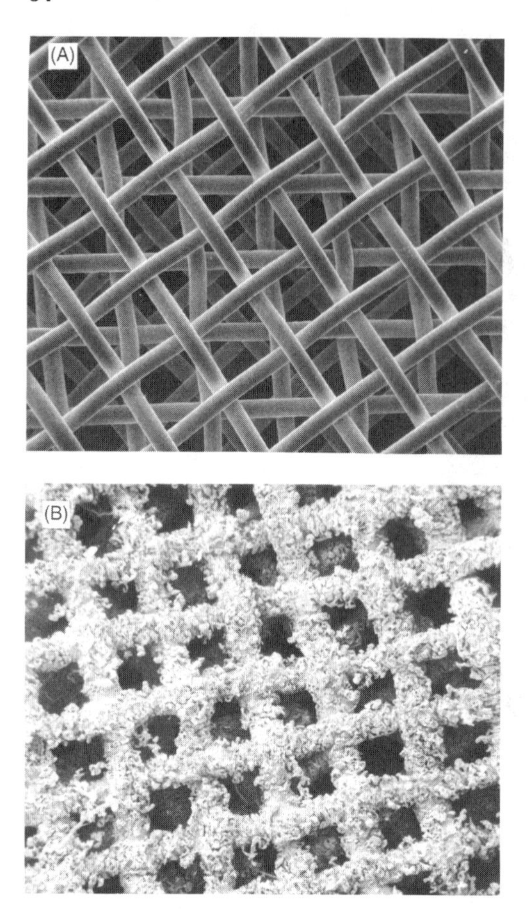

Figure 3 Micrograph of woven platinum/rhodium gauze after prolonged use in the Andrussow process. The gauze has a matte appearance, and the apertures are considerably smaller than in fresh gauze.

processes take place in palladium/nickel gauzes used as getters for the loss of metal from platinum/rhodium gauze in ammonia oxidation [11].

The use of supported platinum-based catalysts in hydrogen cyanide synthesis has been described, and some are in use in certain plants. For instance, Merrill and Perry recommended natural beryl coated with platinum or a platinum alloy [12], and a range of alternative supports were considered by Schmidt and his co-workers [13], including coated foamed ceramic and monolithic substrates. They are claimed to have some advantages in ammonia oxidation [14,15].

Silver gauzes as well as the silver granules discussed in Section II.A are used for the conversion of methanol to formaldehyde and of ethanol to acetaldehyde. Silver gauze is usually thicker than platinum/rhodium gauze used for ammonia oxidation, and it operates at about 630–700°C. Plant designs differ and have different catalyst requirements, but typically the gauze is made from 0.35-mm (350-micron)-diameter wire woven to 20 mesh per linear inch. Up to 150 gauzes may be in an installed pad, with 100–175 kg or more of silver, depending on plant size. The nature of contaminants in the process stream strongly influences gauze life. For instance, low levels of volatile iron contamination [e.g., $Fe(CO)_5$] in methanol produced in early high-pressure methanol plants had a disastrous effect on the subsequent selectivity of its oxidation to formaldehyde. Fortunately, this is no longer a problem with methanol derived from the newer, low-pressure processes using ultraclean synthesis gas. Normally, silver gauzes have lives up to rather less than a year, similar to that of granules.

C. Gauze Improvements

Use of platinum/rhodium gauze is well established for catalytic applications, although the rhodium content has varied somewhat from the original 10% rhodium. Since their introduction into the nitric acid industry they have been manufactured by weaving techniques [16]. However, recently it has been shown that gauze made by computer-controlled knitting techniques has advantages over the traditional material [17]. Figure 4 illustrates

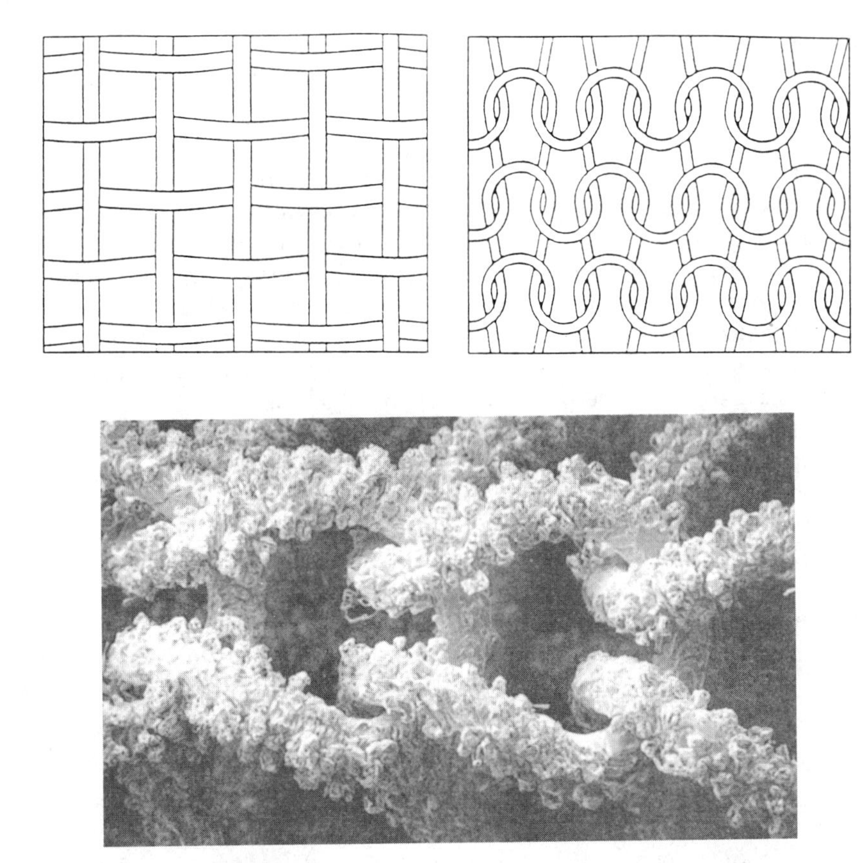

Figure 4 Schematic differences between woven gauzes and the new, knitted gauzes. The lower photograph is a micrograph of the surface developed on a knitted platinum/rhodium gauze after use in an ammonia oxidation plant.

the differences between woven and knitted gauzes, and shows a typical surface structure developed on a used knitted gauze. They are less fragile after use than their woven counterparts, and they have the advantage of stretching more before fracture. Their bulky three-dimensional structure gives them additional surface area, and geometric calculations show the unmasked area of a knitted gauze is 93%, whereas the figure for a woven gauze is some 10% less. Wire life is longer than for traditional gauzes, and mechanical damage leading to metal loss is reduced. Moreover, it appears that gas flow through knitted gauzes is better, and this results in less solid particles' being trapped on the surface, so less rhodium oxide is formed due to retained iron contamination on the surface.

Very recently there has been a further development in the manufacture of platinum/rhodium gauze catalysts. It has been shown [18] that crimping the gauze increases surface exposed to the process stream by a factor of about 1.4. In high-pressure plants, mainly in North America, this enables cost savings by either use of fewer gauzes (and achieves a reduction in pressure drop across the burner) or increased through-put for the same number of installed gauzes.

D. Other Shaped Structures

Like Raney metals, the traditional iron-based catalyst for ammonia synthesis, Eq. (7), contains only low levels of promoters, and the operating catalyst is effectively metal.

$$N_2 + 3H_2 \rightarrow 2NH_3 \tag{7}$$

In keeping with ammonia synthesis being a high-temperature process, the surface areas of reduced catalyst are lower than for Raney nickel, typically about 15 $m^2\ g^{-1}$, and it is used in the form of granules (a few millimeters across) in reactors containing up to 100 metric tons [19]. Patents [20] describe the advantages of special shaped forms, including small monolithic structures. But there appears to be no report of this kind of material being used in commercial plants. There have also been numerous patents disclosing the use of other small shaped-metal catalyst structures, as well as stainless steel shapes carrying a surface layer of active bulk metal. A wide variety of geometric shapes have been considered: spheres, rings/cylinders, saddles, coils, corkscrews, triangles, curlicues, etc. Examples of patented applications include silver (or stainless steel with a surface layer of silver/alkaline earth alloy) for converting ethylene to ethylene oxide [21]. Another example is perforated stainless steel Lessing rings with a surface layer of platinum and palladium, claimed to be advantageous for deep oxidation of ethylene-containing waste streams [22]. An advantage of structured catalysts of these types is their low pressure drop characteristics, but this appears to be no better than gauzes or thin beds of granules discussed previously. Like their coated counterparts discussed in Section IV.B.2, they seem not to be used commercially.

III. COATED-METAL SUBSTRATES

The application of a thin layer of a catalytic formulation to a metal substrate is an approach that provides an elegant means of separating mechanical and catalytic functions; but to be effective, highly active catalysts are required that provide sufficient conversion when present as a thin layer. This section is in two parts: The first deals with the fabrication of metallic monoliths, and the second considers some basic principles associated with applying thin layers of catalytic material to metals, a process known as coating. Applications of coated-metal catalysts are discussed later, in Section IV.

A. Monolith Design and Fabrication

Monoliths have a "multiplicity of parallel channels," and a straightforward way of making metal monolithic substrates is to crimp a strip of metal foil on a pair of rollers having teeth of a predefined profile (usually sinusoidal or triangular) and to combine a crimped sheet with flat foil of similar width to avoid intermeshing. Both are rolled around a spindle or mandril until the right diameter is reached, and fabrication is completed by welding the outermost strips to the one below. Tension of the windings needs to be carefully controlled because, if too loose, the center of the monolith is easily pushed out, and, if too tight, channels may become deformed. Variation in the number of cells per unit area is achieved by varying the pitch and width of the profile on the crimping rolls. Similarly, at least for cylindrical shapes, the dimensions of the resulting monolith can easily be changed. In autocatalyst applications, exhaust gas impinging on the front of the catalyst constantly pulsates, and a major problem is to stop the center of the monolith from being pushed out during use. Ways of overcoming this include forcing pins through the layers perpendicular to the channels, and various forms of welding and brazing between layers or across one or both end faces.

The conventional form of the metallic monolith consists of alternate layers of crimped and flat strip, but other forms have been considered. Many of these reduce the quantity of steel used, which has the double benefit of also reducing weight. Other designs improve the gas/catalyst interface by extending the surface area or introducing local turbulence within the channels. For example, the United Kingdom Atomic Energy Authority (UKAEA) patented a monolith that eliminated the need for a flat strip [23]. Intermeshing of crimped layers was avoided by laying single corrugated strips so the peaks of one layer coincided with those of the layer immediately above or below it or by providing the corrugations at an angle across the strip and rolling in the conventional manner so the troughs of one layer lay across several peaks of the layer underneath, with a relatively small area of contact at each point. Later, a more sophisticated version was developed by A. C. Rochester in what was known as the "herringbone" design [24]. Here corrugations are angled in a pattern reminiscent of a herringbone, and gases can pass from one channel to another so one might expect better heat distribution through the unit. This construction is claimed to have 30% less metal foil than conventional crimped and flat strip designs, and also a higher Nusselt number that increases heat and mass transfer rates resulting from a more tortuous flow path and a greater degree of gas mixing.

There are several reasons why it is desirable to use the thinnest strip possible for automotive substrates. The most immediately obvious is weight, since for a given density of channels per unit of cross-sectional area the weight of a unit is directly proportional to the thickness of the strip. In addition, the thicker strip will reduce the open area and give a higher pressure drop. Finally, the performance of the catalyst is determined by its ability to "light off"; that is the temperature at which conversion efficiency is governed by the temperature at the catalyst surface and not by the temperature of the inlet gas. If there is a substantial heat drain into the substrate from the catalyst surface, then light-off may be inhibited.

Variants of the flat-and-crimped-strip concept have been developed, often to address noncatalytic problems such as extrusion of the monolith center (sometimes referred to as telescoping) resulting from stresses referred to above. In 1986 Emitec introduced the "S-shaped design," a substrate with improved mechanical integrity and better durability. In this concept, metal foil strips are wound around two mandrels in opposite directions before the assembly is inserted into a tubular mantle. The ends of the strips are then joined to the mantle by brazing. The concept is named from the "S" appearance of the end faces, and securing the foil ends to the mantle prevents even large substrate telescoping. The layers are at an angle to the mantle and curve toward the center, carrying expansion forces in that direction and causing the structure to undergo torsional deformation [25]. A further development was the "SM design" introduced in 1991, which could be used for irregular cross sections. Here several stacks of crimped and flat foil layers are wound around a number of centers. The characteristic feature is a radially symmetric radiating structure of the foil layers, the ends of which are attached to the outer mantle at an angle, as in the S shape. Subsequent developments by Emitec, aimed at improved catalyst efficiency to meet future emission legislative requirements, modify the cell shape to change gas flow characteristics. Emitec's approach is to structure the channels to increase the transverse flow in them and to induce a degree of turbulence [26,27]. The ability to modify channel surface and shape is a potential advantage of metallic substrates over their extruded ceramic counterparts, provided it is possible to deliver the level of performance improvement it promises in a cost-effective manner in terms of substrate cost and the subsequent coating. In conventional monoliths, gas flow along individual cells, which have smooth walls, is essentially laminar, and the principal means by which reactants reach the coated wall is by diffusion. To achieve high conversion, a high geometric surface area of the catalyst is needed. If the flow has, by design, a radial component, mass transport to the walls can be increased. Pressure drop compared to the conventional system at the same cell density will increase, and to some degree will be balanced by increased heat and mass transfer coefficients. The compromise might be to reduce the cell density of the monolith or to reduce its length, with a resulting reduction in weight and material usage to offset the increased complexity in preparation of the strip.

Illustrated in Fig. 5 are three potential metal monolith designs that have been described [26,27], called SM, LS, and TS structures. In the TS structure, flow is restricted within a particular channel, but microcorrugations at 90° to the direction of flow cause turbulence and so reduce the thickness of the stagnant layer reactants have to diffuse through. In SM and LS designs there is the opportunity of flow between channels, which provides the possibility of the whole of the catalytic surface being used, even if flow distribution at the catalyst front face is not uniform. However, this may be counterproductive in terms of light-off due to heat dissipation, although benefits would be expected once the catalyst is at normal operating temperature. The extent of these effects would depend on the conflicting requirements of low pressure drop and high geometric surface area. Preliminary data [27] indicated that, at the same cell density, a 14% lower volume of the TS catalyst gives similar hydrocarbon and NO_x control as a conventional monolithic one. With equal catalyst volumes, the TS system gives an average 10% better performance for hydrocarbons and NO_x, whereas at an equal volume, a 300 cells inch^{-2} TS catalyst gives performance equivalent to a 400 cells inch^{-2} conventional monolith.

A radial-flow-converter substrate design has recently been introduced by Bosal. This is based on improving the mass transport of reactants to the walls by increasing cell density to as high as 1,600 cells inch^{-2}, as has been shown by others [28]. Bosal seeks

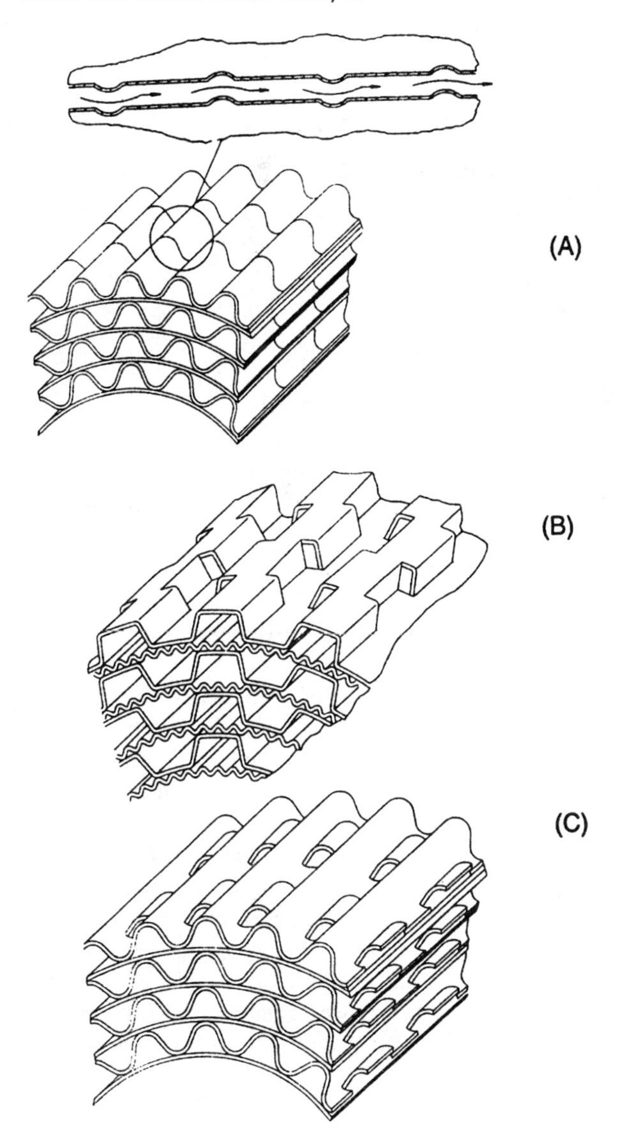

Figure 5 Structural features of some metal monoliths designed to enhance turbulence. A: Transversal structure (TS), in which corrugated layers have microcorrugations at 90° to the direction of flow. B: In the SM structure, flow in channels is split into multiple flow paths. C: In the LS structure, interconnecting flow paths are achieved by means of partial countercorrugation of the corrugated layer. Drawings courtesy of Emitec.

to overcome the high-back-pressure disadvantage of high-cell-density concepts, and weight limitations, by a layout illustrated in Fig. 6, in which the gas flow is presented to a much larger facial area than in conventional monolith designs. This enables the channels to be shorter, and the linear flow rate and pressure drop across each channel are correspondingly lower. Together with a short path between the center of each channel and the active coating of the cell wall, this gives the potential for improved performance and of a smaller

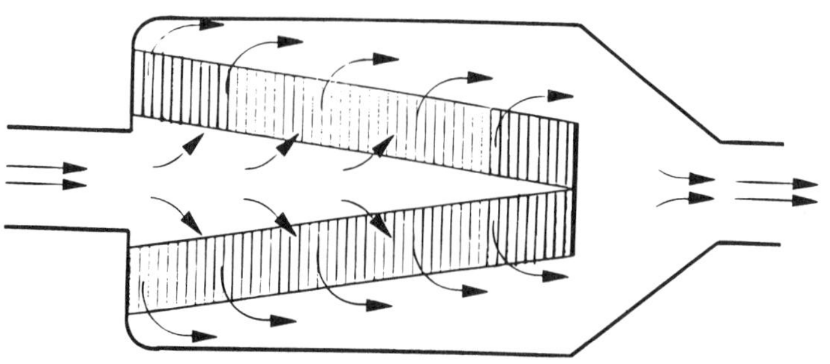

Figure 6 Schematic illustration of the flow pattern in a radial-flow autocatalyst converter of the type designed by Bosal.

volume for equal performance. The possibility also of achieving equal performance with substantially lower precious-metal loadings compared to conventional designs has been pointed out [29]. At the time of preparing this chapter, very few data on aged performance of this system have been reported, and so its longer-term durability remains a question. Nevertheless, data on fresh catalysts seem to give support to the claims of the concept.

A totally different concept was developed by Enga and co-workers [30] at Johnson Matthey to treat exhaust from diesel engines, in which particulate as well as gaseous pollutants need to be catalytically removed. The majority of particulates generated by diesel engines are micron sized and may be a complex mixture of carbon with absorbed hydrocarbons and water. With conventional substrates used in gasoline applications, many of the small particles would pass straight through without interacting with the catalyst surface. In order to overcome this, 0.25-mm Fecralloy wire was used in knitted form wrapped around itself to form a cylindrical shape and then compressed. The level of compression depends on the balance of good particulate-trapping ability and pressure drop. The metallic substrate thus formed was coated with a catalyst formulation to enable the pollutants in the exhaust gas to be oxidized. In early versions of this system, the catalytic blocks were mounted in a modified exhaust manifold so that the gas entered through the circumferential face of the first block and flowed axially through the whole length of the catalyst blocks, as illustrated in Fig. 7A. An improvement on this arrangement

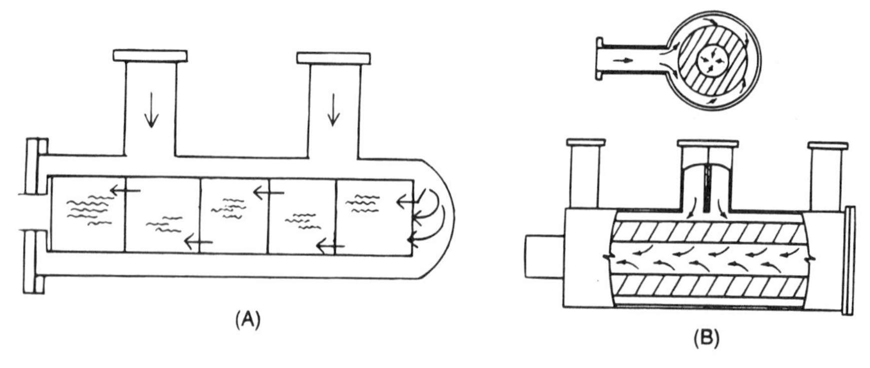

Figure 7 Metal wire-based diesel particulate catalyzed traps. A shows an early design, with exhaust gas flowing through the whole length of the catalyst structure. B illustrates an improved design, with gas flowing radially through a thinner catalyst bed and exiting via a hollow core.

was the formation of blocks having an open central core. These were then butted together under pressure when loaded into the reactor, which had an internal diameter slightly larger than the outside diameter of the catalyst blocks. With this arrangement, gas could be encouraged to flow circumferentially around the outside of the catalyst, then radially through the catalyst, finally exiting through the hollow central core, as shown in Fig. 7B. This design enabled the catalyst facial area to be substantially increased while minimizing the thickness of the catalyst bed and thereby also minimizing the pressure drop across the catalyst, as in Fig. 7B. A further improvement to this system was obtained by reversing the inlet flow to the central core, thereby conserving heat in the exhaust gas and aiding light-off of the catalyst, followed by radial flow through the catalyst and exiting through the outer annulus. Also, the effect of the size and shape of the initial wire on the fluid dynamics and the particulate-collecting abilities of the final unit were investigated, with the conclusion that flattened wire should give superior performance.

B. Principles of Coating Metal Substrates

Ferritic steel foils are most commonly used for fabricating metallic substrates. The advantage of ferritic steels lies not only in their resistance to corrosion, but, when appropriately treated, they have a strongly adhering oxide film on their surface. The composition of steels successfully used is typically 70–90% iron, 10–25% chromium, and up to 10% aluminium, together with other, minor components. When heated to 300–400°C the surface oxide film is chromium rich, but at temperatures above 800°C an alumina-rich surface is developed that endows the steel with excellent high-temperature-oxidation resistance [31], and these steels have been used as furnace elements at temperatures as high as 1200–1300°C. In distinction, austentitic steels, which also have good high-temperature resistance (see Fig. 8), develop an iron-rich surface layer that at high temperature tends to flake off, or "spall," as shown in Fig. 9.

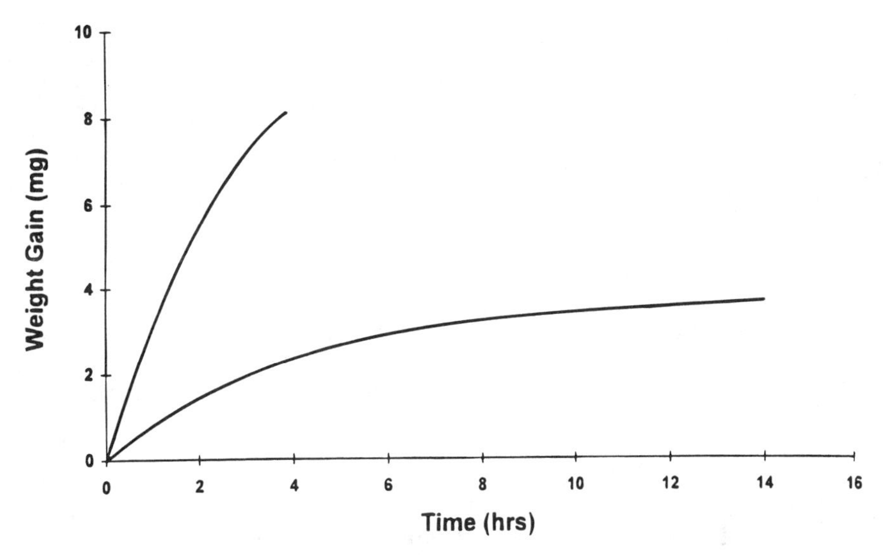

Figure 8 Comparison of thermogravimetric oxidation of austentenitic (upper curve) and aluminum-containing ferritic steels (lower curve) showing weight gain at 1200°C due to surface oxidation as a function of time.

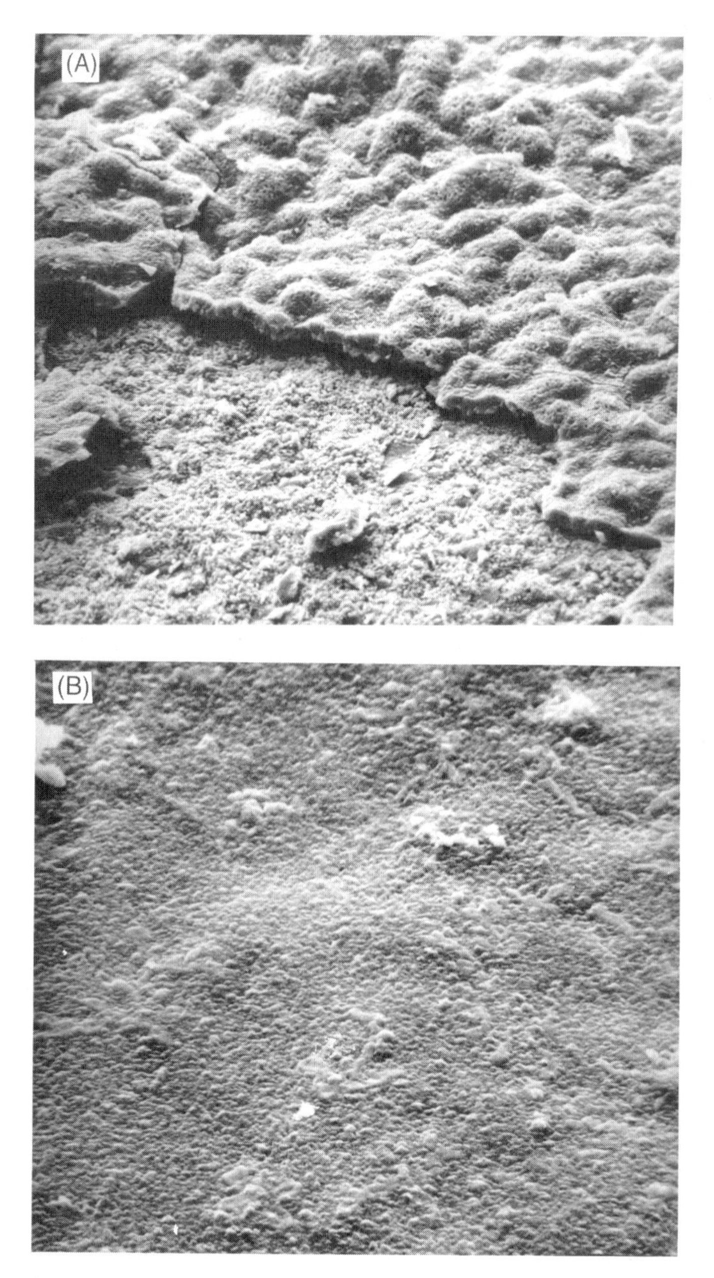

Figure 9 SEM micrographs showing: A the surface of austentenitic and B aluminum-containing ferritic steels after heat treatment in air.

Work on one particular ferritic steel, "Fecralloy," for fabrication of catalyst substrates was pioneered by the United Kingdom Atomic Energy Authority at Harwell and Johnson Matthey, in collaboration with Resistalloy, which developed technology for producing thin strip [32]. This and related alloys, in addition to iron, chromium, and aluminum, contain low levels of elements such as yttrium (0.1–3.0%), thought to enhance the protective properties of the surface alumina layer. Alumina forms by the oxidation of bulk aluminum that migrates to the surface; if the surface layer is broken, it self-heals next time the material is exposed to high temperature. Another advantage associated with the alumina film is its compatibility with alumina-containing washcoat slurries, which substantially eases application of the washcoat and ensures its adhesion to the substrate during the life of the catalyst.

Typical metallic monoliths in use today, illustrated in Fig. 10, have cell densities in the range 15–78 cells cm^{-2} (100–500 cells $inch^{-2}$), corresponding to individual cell dimensions of 1.1–2.5 mm. To minimize additional back pressure due to the catalyst formulation, it is essential that the coating be applied in a controlled manner. It is difficult to define a typical thickness of the catalyst layer (often referred to as "washcoat layer"), since this inevitably differs from one catalyst manufacturer to another, depending upon the precise processing techniques used and the application for the finished catalyst. However, usually it is about as thick as the metal foil, typically 0.04–0.05 mm.

Two methodologies for coating metal substrates have been developed. The first is related to techniques developed for coating ceramic monoliths and involves slurring the components needed to achieve the desired catalytic performance and (for example) pouring it through or sucking it into the channels of the substrate. Excess slurry is removed either by gravitational draining or by applying some form of pressure or vacuum to clear the channels of all but the requisite thickness of material that adheres to the channel walls. Slurry rheology and its flow properties play an important role in this process, even in the

Figure 10 Selection of commercially available metallic monolith substrates. Photograph courtesy of Emitec.

case of ceramic monoliths. However, with ceramics the channel walls have some porosity, and this is important because it causes the removal of some liquid from the wet coating and causes a change in its rheological properties. However, metallic substrates are essentially nonporous and present a different kind of surface to the slurry. Accordingly, both the rheology and the composition of the slurry may need to be modified to achieve adequate coating control. In some cases it might be appropriate to repeat the process to obtain the right loading and composition of the coating. After coating it is normal to calcine the unit, usually in the region 350–700°C, to develop adhesion of the coating within itself and to the channel. The metallic substrate might seem to be at a disadvantage compared to ceramic ones; in the latter case some of the larger pores on the channels' surface fill with washcoat and help adhesion. The thermal expansion characteristics of the coating is also more comparable to ceramic monoliths than to metal. However, these problems are overcome by a combination of suitable chemistry in the coating formulation and the use of metal alloys having the ability to form an adherent and stable alumina surface layer, which acts as a key for the washcoat layer. In practice it is possible to achieve at least as good levels of adhesion on metallic monoliths as can be obtained with their ceramic counterparts.

The second method that can be used for coating metallic substrates cannot be used for ceramics, because it involves coating metal foil before the monolith is constructed. This methodology has the advantage that a number of techniques can be used to apply the thin coating, such as painting, spraying, or dipping, and although slurry rheology is important in this technique, it is not so critical as in conventional washcoating. The required thickness can be applied in either single layers or multiple layers. After completion of the coating process, the monolith can be formed in the conventional way. It might be argued that in this technique the coating at the points where two layers of metal coincide becomes unavailable for use in the ultimate catalytic application. Proponents of the technique have suggested that in the more conventional technique, washcoat accumulation in the corners where the two foils meet also results in some effective loss of active material. However, one major problem arising from coating before constructing the monolith is that the presence of the coating makes welding or brazing difficult. In order to preserve the integrity of the unit under conditions of pulsing exhaust gas flow, retaining pins at right angles to the direction of gas flow have been used [33]. Both types of coating are in current use, although the more conventional coating technique predominates. However, the second approach might be favored at very high cell densities (more than 1,000 inch^{-2}), although an earlier coating technique [34] employing sols of the components could be used with preformed monoliths.

IV. COATED-METAL CATALYST APPLICATIONS

The main feature of monolithic catalysts is a high ratio of geometric surface area to volume and a low pressure drop with distribution of gas flow through a large number of parallel channels. The most significant application for structured catalyst units of this type is in the control of exhaust emissions from cars; this area is discussed here under five subheadings. Nonautomotive applications are discussed later, in Section IV.B.

A. Automotive Applications

Metallic monoliths offer features such as very low back pressure, and they are used when these characteristics are particularly important. They are often fitted to performance cars, and they probably represent 10–15% of the total number of units produced. Although ceramic monoliths predominate (see Chapter 2), there is the potential for an increase in the use of metallic monoliths as, for example, starter catalysts, which are discussed in Section IV.A.3. Reactions involved in automotive emission control are illustrated by Eqs. (8), (9), and (10). Unburnt and partially oxidized hydrocarbons (aldehydes etc.) are converted to water and carbon dioxide according to Eq. (8). Similarly, carbon monoxide is oxidized to carbon dioxide, as in Eq. (9). In contrast, nitrogen oxides react with reducing species (carbon monoxide, hydrocarbons, etc.) to give dinitrogen.

$$H_nC_m + (m + n)O_2 \rightarrow mCO_2 + \frac{n}{2} H_2O \quad (8)$$

$$2CO + O_2 \rightarrow 2CO_2 \quad (9)$$

$$2NO + 2CO \rightarrow N_2 + 2CO_2$$

Since three reaction types are involved, conventional autocatalysts are known as three-way catalysts, often referred to as TWCs; but in order to achieve the contradicting requirements of simultaneous oxidation and reduction, it is necessary to maintain operation close to the stoichiometric point. The use of electronic fuel injection and computer-based engine management systems enables this to be done with the necessary degree of control.

In some respects it is perhaps surprising that metal monoliths did not gain prominence in the early days of automotive catalysis. The engineers in car companies undoubtedly would have been more comfortable at that time with metals rather than ceramics—which was rapidly given the soubriquet of "brick." In addition, when considering some of the more important properties of the monolith, properly designed metallic systems might have advantages over the ceramic concepts with respect to low back pressure, mechanical strength, resistance to thermal degradation, high geometric surface area per unit volume, weight, and ease of handling without damage. For example, the early ceramic monoliths had a cell wall thickness of 0.33 mm, which is quite thick compared to the capabilities for rolled metal foils; the influence of wall thickness on pressure drop and geometric area is illustrated in Fig. 11. So what were the key items that held up the development of metallic forms of the monolithic substrate?

Exhaust gas is quite corrosive, and in addition to the carbonaceous residues arising from burning fossil fuels in air (hydrocarbon, carbon monoxide, and carbon dioxide principally), some oxides of nitrogen are produced, as well as substantial amounts of water vapor. Also, most petroleum fuels contain significant levels of sulfur compounds, which after combustion appear in the exhaust gas as sulfur dioxide. As the exhaust gas leaves the combustion chamber of the engine it is very hot, progressively cooling as the gas approaches the end of the tailpipe. Thus the catalyst, which will normally be located from a few inches to a few feet from the engine manifold, will be exposed to a very acidic and corrosive environment as well as to high temperatures. The majority of steels are unable to withstand such an environment, especially in thin strip form and at high temperature [32], for a period that would need to be in excess of that required to enable the accumulation of 50,000 road miles on a car. This led to the evaluation of special steels, notably ferritic compositions, which for reasons discussed in Section III.B are particularly suitable for being coated with a well-adhering layer of a catalyst formulation.

1. Historical Background

The debate about the contribution of pollutants emitted from motor vehicles and certain stationary sources to photochemical smog began in earnest in the 1950s. Research into

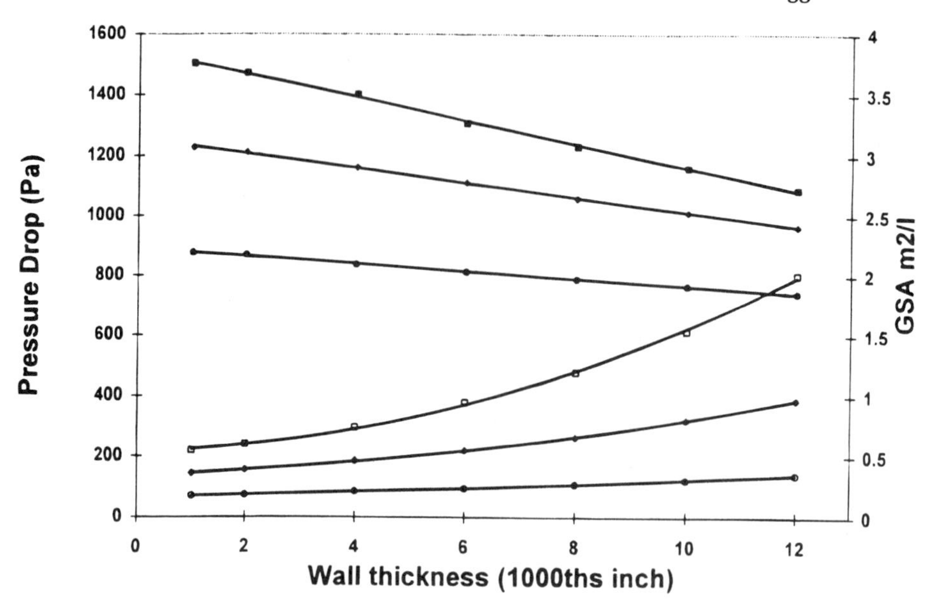

Figure 11 Linear variation of geometric surface area and nonlinear pressure drop across a metal monolith as a function of metal foil thickness for three cell densities: 200, 400, and 600 cells inch^{-2}.

the causes of photochemical smog revealed it was formed by the reaction of oxides of nitrogen with reactive hydrocarbons and oxygen in the presence of sunlight, and surveys suggested that a high proportion of manmade emissions were derived from motor vehicles. This was especially true in the Los Angeles basin in California, where the high number of cars, long hours of sunlight, and geographical features trapping the pollutants were all favorable to the production of photochemical smog.

Extensive research went into finding methods of reducing the levels of pollutants from vehicles, initially by engine modifications. Political pressures from an increasingly powerful and vocal environmental lobby led, in 1970, to the U.S. Clean Air Act. This targeted a reduction of approximately 90% in emissions of hydrocarbons (HC), carbon monoxide, and nitrogen oxides (NO_x) from new vehicles relative to an average late-1960s model-year vehicle. In order to make it feasible, the targeted reduction was spread over several years. Other features included in this legislation were the introduction of unleaded gasoline by 1974; a requirement for the emission control system to be effective for a minimum of 50,000 miles; and a defined test cycle and procedure to standardize the measurement of emissions. It was found that the targeted requirements could not be met by engine modifications alone without penalties in fuel economy and driveability, and by 1975 catalyst-equipped cars began to be produced. Since that time, legislation requiring the furnishing of catalysts has spread to many urbanized countries. Today, worldwide, a majority of new cars are equipped with catalytic converters as standard.

In the earliest catalysts, two basic support configurations were used. The first was thermally stable alumina in the form of cylindrical pellets or spheres, typically 3 mm in diameter, that had for several decades been used in the chemical processes industry, such as petroleum refining. The second type of catalyst support was the so-called monolith, made from metal, as described in this chapter, or a ceramic material such as cordierite ($2MgO \cdot 2Al_2O_3 \cdot 5SiO_2$) the subject of the previous chapter. The monolith has strong thin walls, usually in a grid configuration, that are frequently square or triangular, although other shapes are possible, such as hexagonal. These run along the length of the piece and give a large number of parallel channels (usually 30–65) per square centimeter of the face of the monolith. This provides high geometric surface area combined with low pressure drop. The monolithic type of technology quickly became the dominant form, and today has almost completely displaced the use of catalyst pellets.

2. *Underbody Catalysts*

When catalysts were first used to meet the early American regulations, their location often reflected the existing space available in the design of the underside of the vehicle. Accordingly, most catalysts were fitted beneath the car, and therefore potentially some distance away from the engine. As a consequence of the use of catalysts being new to the automotive industry there was some concern over their ability to last for 50,000 miles. Location in the underfloor position, where the exhaust gas temperature was lower, could be considered a benefit in this respect, and this position soon became a standard. Of course, performance requirements at that time were considerably less than today.

It was at this time that the first realistic attempts to evaluate metallic substrates made from ferritic steels were made. Early ceramic monoliths had 200 or 300 cells per square inch of face, but with relatively thick cell walls in the region of 0.011–0.012 in. (approximately 0.3 mm) and an open area of 70% (200 cpsi) and 60% (300 cpsi). In 1977 Dulieu and co-workers [35] showed that metal monoliths were equally durable as oxidation catalysts for hydrocarbon and carbon monoxide conversion as was the same catalyst formulation on ceramic substrate. Because thin foil (0.05 mm) and higher cell densities (400 cpsi) could be used, a substantially lower volume of the metallic system was possible (45% compared with 100%) without detraction from performance, pressure drop, or weight. However, to overcome the problem of extrusion of the central region of the monolith, the metallic pieces used in these trials were electron-beam welded, one layer to another and to the outer casing, which was an expensive operation. In addition, the ceramic monolith manufacturers responded by increasing the cell density to 400 cpsi and reducing the cell wall thickness to 0.0065 in. (0.015 mm) to match the perceived advantage of the metallic monolith. Nevertheless, a subsequent paper from Volkswagen [36] demonstrated that a 500-cell metallic substrate could still yield advantages in volume requirements and pressure drop compared to a 400-cell ceramic, due to the larger open area of the metallic monolith. Thus, metallic-monolith-based catalysts were fitted as the main or only converter to some Volkswagen production cars, and this continues to the present time.

In certain applications in which the power loss through exhaust back pressure is critically important, the thinner wall of the metal may give results that overcome the increased cost of the metal substrate. Thus, Pelters, Kaiser, and Maus [37] showed that at equal volume of catalyst at the same cell density and catalyst loading, the Porsche 911 Carrera 4 gave about 4 kW more engine power with the metallic support than with ceramic. They also found the coated metallic monolith to be equally as robust as its ceramic counterpart. A few other car manufacturers, notably in Germany, have also used metallic substrate underfloor; Table 2 gives the geometric surface area and thermal capacity of typical uncoated metallic monoliths as a function of cell density and foil thickness. However, the use of metallic substrates in underbody locations is not general, and overall ceramic-based catalysts dominate in this position.

Table 2 Geometric Surface Area (GSA) as a Function of Cell Density for Uncoated Metallic Monoliths

Cell density (cells in^{-2})	Foil thickness (mm)	GSA (m^2/ dm^2)	Thermal capacity (JK^{-1} $liter^{-1}$)
100	0.05	1.79	172
200	0.05	2.67	266
300	0.05	3.07	—
400	0.05	3.68	375
500	0.05	4.01	406
600	0.05	4.32	437
400	0.04	3.69	301
500	0.04	4.04	329
600	0.04	4.33	350

Data from: F-W. Kaiser and S. Pelters, *SAE Paper* 910837 (1991); R. Brück, R. Diewald, P. Hirth, and F-W. Kaiser, *SAE Paper* 950789 (1995).

3. *Starter Catalysts*

One of the first production-scale applications for metal-substrate-based autocatalysts was for starter catalysts in about 1980. In general, starter catalysts are small units fitted close to the engine manifold to achieve light-off as soon as possible after starting the engine. The generation of an exotherm over the starter catalyst increases the temperature of the exhaust gas entering the main catalyst bed, enabling it, in turn, to achieve early light-off. There are a number of requirements associated with coupled catalysts: geometric surface area, cell density, diameter, length, flow distribution, thermal capacity, catalyst coating durability, ability to withstand thermal and mechanical stress, and ability to locate them close to the engine. Some of these factors are interrelated. For example, for fast catalyst heating and optimum utilization of the exhaust gas energy, the minimum possible thermal capacity is required, combined with the maximum active catalyst surface and the highest rate of heat transfer between the catalyst structure and the gas flowing over it [38]. For a given catalyst formulation, the active catalyst area will be related to the geometric surface area of the substrate, which will itself, for a given size and design, be related to cell density. Thus, high cell density would be expected to be beneficial for light-off. The material type, and the amount of it, will contribute to the heat capacity, which needs to be minimized. Thus, an increase in cell density and foil thickness will adversely affect heat capacity in this context. A third factor concerns the influence of the exhaust gas parameters (temperature, flow rate, and gas distribution). At low exhaust gas temperatures, light-off might be depressed as cell densities are increased, since the heat capacity considerations dominate, whereas at higher exhaust gas temperatures the positive benefits of higher cell density become apparent. Similarly, a small-diameter substrate, concentrating the exhaust gas energy onto a small area of face, would seem to be beneficial for light-off by heating up the unit more quickly. However, once the catalyst has reached operational temperature it will subsequently be subjected to the full rigors of high exhaust temperatures when the engine itself has fully warmed up, concentrated on the same small area of face. In severe cases this can lead to partial melting of the monolith and substantial loss of performance of the catalyst coating. In addition to the desirable properties of the monolith indicated above, the properties of the catalyst coating have a particularly major role in the performance of the starter catalyst. Thus, the coating needs to have inherently high activity at low temperature if the unit is to light off quickly, and this must form a major feature in the design of the formulation. The stability over time and accumulated mileage of the low light-off temperature feature is also very important if the overall catalyst system is to maintain its performance. Conversely, as indicated above, the catalyst formulation must also have the property of resistance to very high temperatures, in the region of 1000°C, when the engine is running hot! It is not the purpose of this chapter to discuss catalyst materials, but a further factor must be considered in the catalyst design, namely, that the role of the starter catalyst is to light off quickly and generate an increase in exhaust gas temperature at the inlet to the main catalyst. If the starter catalyst is too efficient or too selective in removing pollutants, the main catalyst may not function correctly. For example, an incorrect balance of reductants could impact detrimentally NO_x removal, or under some circumstances there could be an insufficient level of reactants to generate sufficient exotherm to maintain the catalyst temperature on the second catalyst above the light-off point.

4. Electrically Heated Catalysts

Once lit off, modern autocatalysts are extremely efficient at converting engine-out pollutants. Nevertheless, with increasingly stringent legislative limits being put in place between now and early into the next century, much emphasis is being placed on the performance of the emissions system under cold-start conditions, that is, during the first minute or two of running, because this is where the bulk of the remaining tailpipe emissions arise. It is therefore important that the catalyst reach operating temperature as soon as possible after the engine is started. In a standard system the initial warmup of the catalyst arises from heat transferred from the exhaust gas. If the catalyst could be heated independently, light-off—and hence maximum performance—could be achieved at an earlier time.

Several methods for decreasing the time to light-off have been evaluated. One of these has been the use of electric energy to preheat a metallic substrate. Interestingly, this was anticipated in some of the early UKAEA patents [23], although at the time regulations were not severe enough to make development worthwhile. However, as it became clear that extremely fast control of emissions would be essential for meeting California legislative standards toward the end of the century, interest in methods ensuring rapid increase in catalyst temperature was revived.

Two pioneering companies put major effort into developing the concept of electrically heated catalysts (EHCs), Emitech in Germany and Corning in America. These companies adopted a different approach to the form of the EHC, which was based on the expertise each had gained through development of their more conventional autocatalyst substrates. Thus the Corning EHC is based on their well-established extrusion technology, but in this case applied to powdered metal instead of ceramic oxides, yielding the familiar grid-shaped pattern of square cells. Emitech systems are based on metal strip with flat and crimped layers. Other companies, notably Camet, which also have designs based on metal foil strip, played a part in the early developments of EHCs. More recently, NGK developed a concept based, as are their conventional ceramic substrates, on extrusion technology, but in this case the metal powder is extruded in the form of hexagonal cells.

Many questions have been raised concerning the use of EHC systems in automotive emissions control, but the major concerns are not generally whether with their use the

California and European Stage 3 regulations can be reached, but rather about their power requirements, long-term durability, and cost. Initially quoted power requirements were large, typically involving 200–400 A. Clearly, the drain from a conventional battery would be considerable (equivalent to almost 5 kW), and at first it was thought a second battery would be needed, with the necessity of heavy cabling because of the large currents involved. However, it was not intended that electric heating would continue once the conventional catalyst had achieved operating temperature. It was recognized that an EHC would take time to reach operating temperature, and so it was initially proposed that the EHC be switched on 20–30 sec before cranking the engine, so that the total time current was being drawn would be 50–60 sec. The electric energy needed to heat the EHC (E_{elec}) is given by Eq. (11), and this can be equated to the amount supplied, as in Eq. (12):

$$E_{elec} = MC_p\Delta T \tag{11}$$

$$E_{elec} = \frac{V^2 t}{R} \tag{12}$$

where M = mass (g), C_p = specific heat (Wh g^{-1} °C^{-1}), δT = temperature change (°C), R = resistance (Ω), V = applied voltage (V), and t = time (hr). The key parameters are the EHC mass, its heat capacity, and electrical resistance. Optimizing these properties enabled the power requirement to be reduced to 1–2 kW. For example, in the case of an extruded metal monolith, mass can be reduced by reducing the dimensions (facial area and length), cell density, and wall thickness of the unit. However, limitations are placed on the extent to which this can be achieved in practice. For example, the objective is to heat the exhaust gas flowing through the EHC by the transfer of heat from the hot surface. But if the contact time is too short (short channel length, low cell density), than heat transfer will be restricted, and the exhaust gas temperature might then be too low to enable the main catalyst to reach operating temperature quickly. Figure 12 shows the range of EHC designs from a number of manufacturers using metal-foil and extruded-metal manufacturing techniques. At the time of writing, the authors know of only one commercial application of an EHC, and this involves a small number of vehicles. It will be interesting to see how this approach develops in the future as legislated emission requirements become increasingly stringent.

5. *Motorcycle Catalysts*

Emission regulations covering control of pollution from motorcycles, especially two-stroke engines, have been promulgated in various areas of the world, and the regulated limits can be met with the aid of catalysts [39]. In this case the catalyst needs to be inserted into the exhaust pipe and become part of the exhaust system. Since the design of the exhaust system plays a vital role in the performance of the engine, and because of space constraints, the integration of the catalyst is a significant design consideration.

In this application, metal monolith catalyst systems offer some important advantages over their ceramic counterparts. For example, because they have an integral metal mantle, they can be welded directly into the exhaust system. In addition, the use of thin metal foil provides low back pressure, and the possibility of different cell densities (generally in the range 100–400 cells $inch^{-2}$), monolith volume, and length provides useful design parameters. On small two-stroke engines, the high concentrations of carbon monoxide and hydrocarbons in the exhaust gas cause large exotherms over the catalyst, especially at wide-open throttle, and this must be tolerated by the monolith itself and by the catalyst

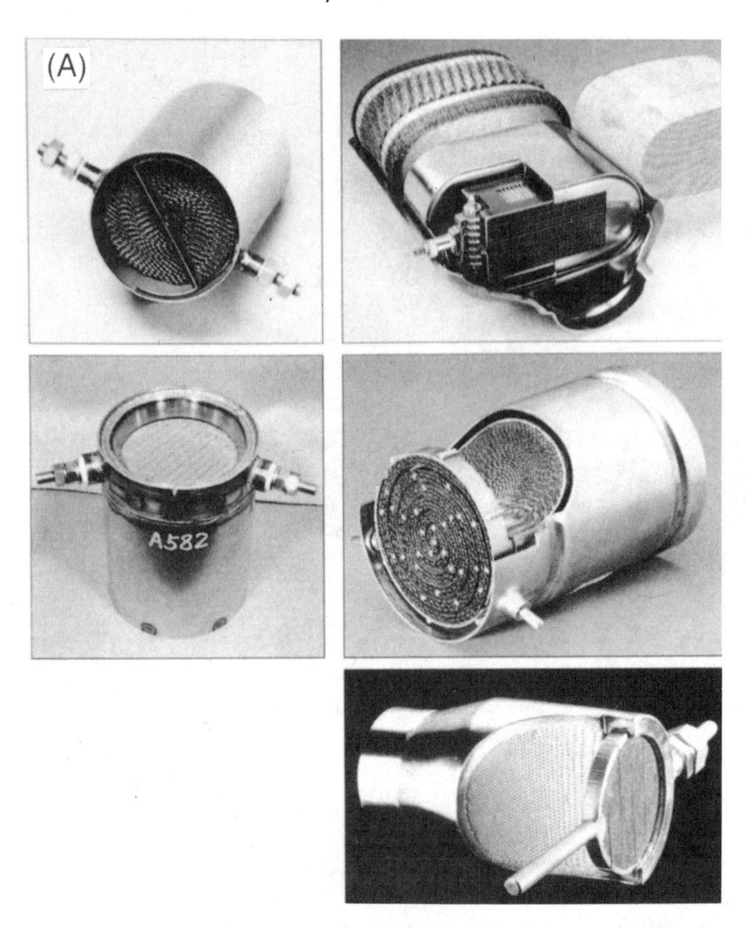

Figure 12 Electrically heated autocatalysts from a variety of manufacturers employing metal-foil and extruded-metal techniques. (Reproduced by permission of Society of Automotive Engineers.)

formulation applied to it. Conventional poisoning of the catalyst is not generally a problem, although oil fouling from two-stroke engines under certain operating conditions is a possibility.

At present the main areas of the world in which catalysts are required on motorcycles by legislation are Europe (Switzerland and Austria) and Taiwan. In Germany, many motorcycles are fitted with catalyst on a voluntary basis.

Discussions of tighter legislation in these countries, and the initiation of legislation in other areas in the future (for example, India), are under way. Consequently this may prove to be a growing application for metallic monoliths. Although engine sizes are much smaller than for cars, the ratio of catalyst volume to engine swept volume is generally in the region of 1.5 (much higher than for most cars), but catalyst volumes are generally very small compared to those used on cars.

B. Nonautomotive Applications

Since its introduction some two decades ago, catalytic control of exhaust emissions from automobiles has had remarkable success, and now over 200 million vehicles worldwide

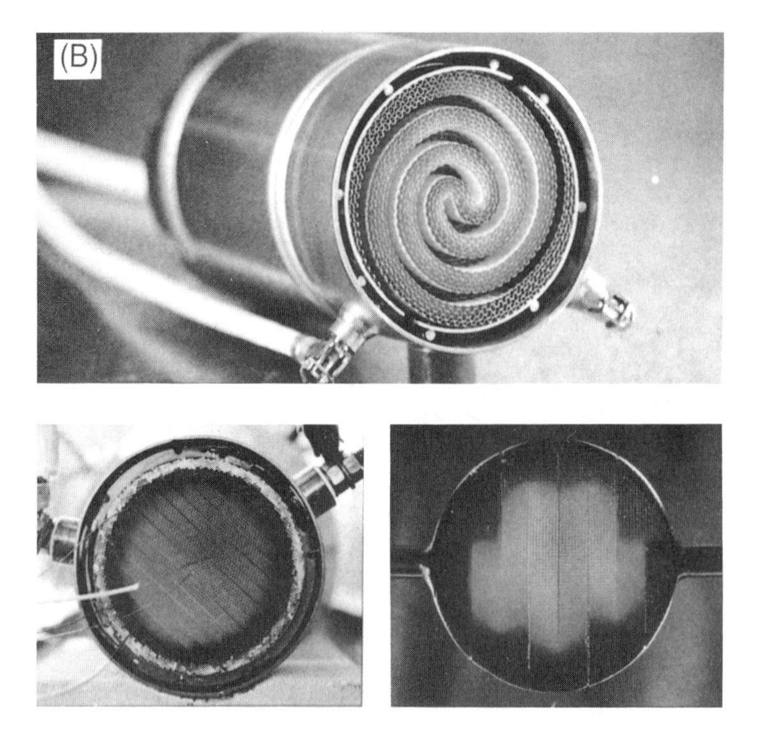

Figure 12 Continued

have been equipped with catalytic converters. Catalytic control has also been used on a variety of stationary industrial sources for many years. Catalysts were first used on stationary sources in the 1940s for energy recovery and odor control. Subsequently, in America the 1970 Clean Air Act gave impetus for catalyst development to embrace a wider range of applications. Catalysts used for stationary internal combustion engines are not considered separately here because in general they do not make particular use of metallic substrates. Catalysts for removal of volatile organic compounds (VOCs) and selective catalytic reduction (SCR) of NO_x emissions make use of coated-metal monoliths; these are considered in Section IV.B.1. The subsequent subsection deals with potential applications for small coated-metal pieces packed either randomly or in an ordered way in large reactors for chemical process industry applications.

1. Environmental Applications

The American 1970 Clean Air Act defined ambient air quality standards (NAAQS) in the United States for: atmospheric ozone, NO_x, lead, carbon monoxide, sulfur oxides, and PM-10 (particulate matter less than 10 μm). The strategy to reduce levels of lead, NO_x, PM-10, and to some extent carbon monoxide was to control emissions from automobiles that included the phasing-out of leaded fuel. As previously noted, ozone is a product of the photochemical reaction of volatile organic compounds with NO_x (photochemical smog), so the balance between organic compounds and NO_x pollutants is important in meeting target ozone levels (e.g., 0.12 ppm). Emissions from stationary sources is an important factor, and limits have been set for them. Because of low pressure drop requirements, coated monolithic catalysts

Table 3 Industries Using Catalytic Systems for Control of VCOs

• Paper printing and coating	• Coil coating
• Metal decorating and printing	• Wood and board printing
• Food processing	• Carpet manufacture
• Food frying	• Tobacco drying
• Animal rendering	• Organic chemical manufacture

are used in the oxidative removal of VOCs and SCR of NO_x. In addition to ceramic monoliths, metallic substrates are being employed increasingly. Due to the scale of operation, these are larger than those used for autocatalysts, and, particularly for metal substrates, construction methods and materials used are different because physical demands are different.

a. Oxidation of VOCs Catalysts of various kinds have been used successfully to destroy VOC emissions from a variety of sources for over 25 years, and modern catalysts provide high destruction efficiencies and long life. Washcoat formulations are not the same as in autocatalysts, and because a wide range of VOCs are encountered in different processes (some typical examples are listed in Table 3), catalyst formulations used in different applications can have very different compositions and include different promoters. Catalysts should be formulated and the system designed for specific applications to achieve optimum effectiveness. For instance, as shown in Table 4, the oxidation of formaldehyde can light off below 150°C, whereas the corresponding temperature for some halogenated compounds might be above 500°C, with benzene and acetone being at considerably lower temperatures. The most commonly used active metals are platinum and palladium for oxidative destruction of VOCs. In applications that involve the removal of both carbon monoxide and NO_x, such as exhaust from internal combustion engines or gas turbines, the catalyst formulation will also include rhodium.

Conventional monoliths that are similar to but larger than those used for autocatalysts can be employed for oxidation of VOCs in small-scale applications, although it is more usual to use larger, square metal monoliths in large-scale applications. Individual monolith units are built into a metal frame to form a catalyst panel that can be several feet on each side and that can easily be slotted into a reactor with the help of a crane. Figure 13 illustrates how square metal monoliths about 50 cm × 50 cm, and 9 cm thick, are built

Table 4 Operating Temperatures for Catalytic Oxidation of VOCs

Contaminant	Operating temperature (°C)
Formaldehyde	100–150
Carbon monoxide	250
Styrene	350
Paint solvents	350
Phenol/formaldehyde	400
Phenol/cresol	400
Ethyl acetate	350–400

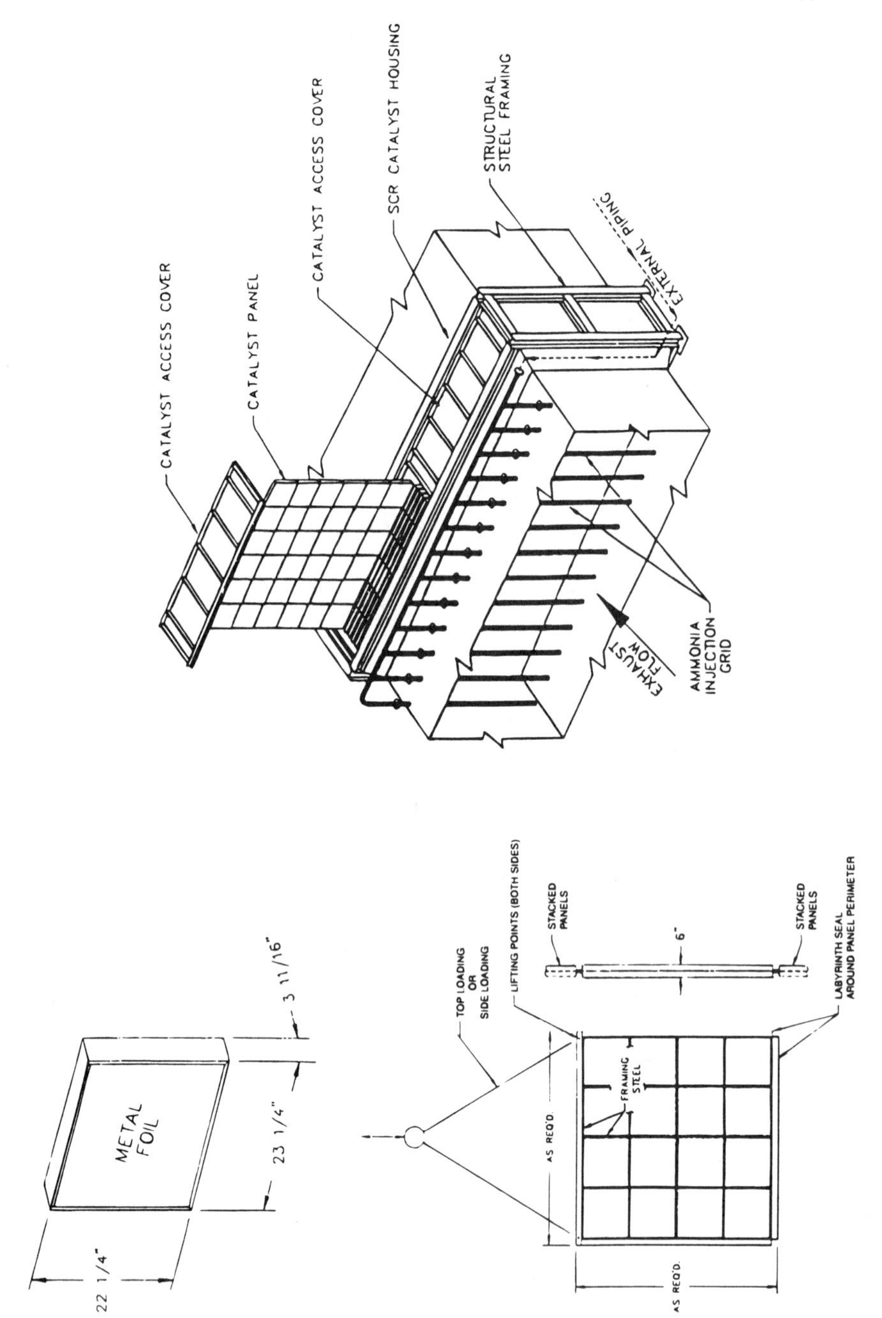

Figure 13 Fabrication of individual square metal monoliths into catalyst panels that are inserted into an industrial SCR unit for control of NO_x emissions. Similar reactors are used for oxidation of VOCs.

into such a panel, in this instance for a SCR unit (reactors of this type containing several panels are used for both VOC oxidation and SCR NO_x reduction (see Section IV.B.1.b). The monoliths are fabricated from stainless steel (see Section III.B) and usually take the form of a corrugated strip of metal foil in combination with a flat strip. Often-quoted features include high surface area per unit volume, good resistance to mechanical and thermal shock, high thermal conductivity that enhances low light-off temperature, and all-metal construction, reducing warping of the monolith housing. The thin steel strip used in the manufacture of these monoliths is subject to corrosive attack in the presence of halide ion under certain circumstances, which could limit their use in the destruction of halogenated hydrocarbons. However, the majority of applications involving the destruction of VOCs are at much lower temperatures than commonly found in automotive applications, and so lower grades of metal can be considered for the production of the monolith blocks.

b. Reduction of NO_x As well as motor vehicles, substantial levels of NO_x are produced from stationary sources. These may include stationary engines, chemical processes, and the generation of electricity. In an overall pollution control strategy, the reduction of the NO_x levels from these sources is extremely important.

There are two distinct types of catalytic NO_x control. In the first, the so-called nonselective catalytic reduction process (NSCR), the reactor is operated in an analogous way to an automotive catalyst, and the catalyst frequently is based on platinum-group metals, typically platinum with rhodium, or palladium, operating at a stoichiometric or slightly rich air–fuel ratio. A number of fuels have been used as the reducing agent for NO_x, for example, hydrogen, natural gas, naphtha, and other hydrocarbons. Applications of the technology include nitric acid tail-gas treatment and NO_x control of exhaust from electric generators, pumps, and compressors. A feature of these applications is that the inlet gas composition is relatively stable in flow and composition, and is "clean" in the sense of not containing significant particulate contamination or catalyst poisons. Both ceramic and metallic substrates can be used in these applications. In general, large beds of catalysts are required; and to minimize pressure drop, larger cell dimensions are normally used than are commonly found for autocatalysts.

The second process is known as selective catalytic reduction (SCR). SCR may also be used in the treatment of nitric acid process tail gas and similar processes, but has achieved prominence through its application to NO_x removal from electricity-generating power stations, especially those that are coal fired. In SCR, a range of reductants for the NO_x can be used; the most common is ammonia. The primary reactions involved are shown in Eqs. (13) and (14). Oxygen is required for this form of NO_x control, and levels of 2–3% are typically needed for optimum catalyst performance.

$$4NO + 4NH_3 + O_2 \rightarrow 4N_2 + 6H_2O \tag{13}$$

$$2NO_2 + 4NH_3 + O_2 \rightarrow 3N_2 + 6H_2O \tag{14}$$

NO_x removal at the 80% level or higher can be achieved in favorable cases, but this can fall to 70%, where there is a possibility of significant fouling or poisoning of the catalyst. Deactivation of these and other environmental catalyst systems has recently been reviewed [40]. Often the SCR system is placed downstream of an oxidation catalyst to permit simultaneous removal of NO_x, hydrocarbons, and carbon monoxide.

Square metal monoliths, as shown in Figure 13, are used in SCR reactors. Catalyst lives of up to more than 10 years are possible, and with proper catalyst management techniques there is no need to replace all of the catalyst at once. For example, vacant slots for additional catalyst are normally built into the reactor. When NO_x conversion decreases, or ammonia slip increases to the permit level, fresh catalyst may be placed into the vacant space, leaving the remainder of the catalyst in place. Subsequently, periodic replacement of a fraction of the total catalyst inventory should be sufficient to maintain the overall reactor performance [41].

Commercial SCR catalyst used in connection with coal-based power stations are generally composed of base metals, since platinum-group metal catalysts are too readily poisoned and have too narrow an operating temperature window for this application. Favored compositions are titania-based together with active components, normally oxides of vanadium, tungsten, or molybdenum. For these systems the optimum reaction temperature is usually in the range 300–400°C.

In the original process developed in Japan for the control of NO_x from power stations, the SCR catalyst was a ceramic monolith. In general, the Japanese power generators use high-quality coal (less than 1% sulfur and 10% ash), and many plants use electrostatic precipitators to take out dust. However, in Europe, especially in Germany, and in America, higher-sulfur-content coal is normal, and the conversion of SO_2 to SO_3 over the SCR catalyst can be a problem. In addition, some SCR reactors have to deal with high dust levels in the gas stream, which can be highly erosive; it was concluded that metallic monoliths are preferred for high-dust applications or where poor flue gas distribution, changes in load, etc. are probable. These factors led to the development of a new form of substrate for power station SCR systems, the so-called "plate type," based on metal foil, and Siemens developed a particular system for high-dust, high-sulfur applications [42] to which they ascribe a number of advantages: less prone to blockage due to their structure, which permits slight vibration of the plates so dust is continually dislodged; higher resistance to erosion than ceramic counterparts, since the metal itself is more resistant to erosion; individual plates are thinner than the walls of a ceramic monolith and so have a lower pressure drop; SO_2 to SO_3 conversion appears to be lower on metallic plate converters, and coatings can formulated to minimize this. The method of separating the plates is shown in Fig. 14. Normally the catalyst unit is made up of modules, which are fitted together to form the larger bed. Distances between the plates vary according to the gas dust burden, being typically 4 mm in low-dust applications (5–15 g m^{-3}), rising to 6–7 mm for high dust applications (40–60 g m^{-3}). Though ceramic (square cell) structures can be made with cell dimensions comparable to the distance between plates, the metallic version still shows better resistance to channel blocking in high-dust atmospheres. However, in low-dust situations the ceramic-based system can perform better due to a higher surface area per unit volume.

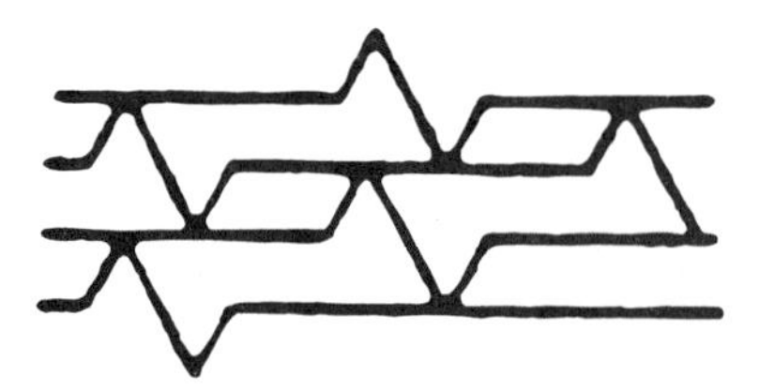

Figure 14 Method of sheet separation in a SCR plate reactor intended for use in dust-containing flue gas. The distance between the plates depends on the amount and nature of the dust particles present.

2. *Small Coated-Metal Structures*

Metallic monoliths for use as autocatalysts are contained within a stainless steel shell or mantle, with the reactant stream flowing through the multiplicity of parallel channels. The combination is a complete catalytic reactor. An alternative arrangement is to use smaller monolithic structures oriented or randomly packed in a large reactor. This approach utilizes the high geometric surface area provided by structured monoliths while allowing appropriate heat transfer with the process stream and the reactor walls for reactions that are strongly endothermic or exothermic, as well as having a desirably low pressure drop compared with, for example, conventional ceramic ring-based catalysts. Moreover, there is less porous material in which secondary reactions can take place (about 10% of a conventional solid porous pellet), so in kinetically controlled reactions coated catalysts should have the advantage of improved selectivity over conventional ones. The most notable process of this kind is selective oxidations, and these aspects of related foam catalysts have been reviewed [43].

Although there is little published in the academic literature in this area, there are a considerable number of relevant patents involving a wide variety of geometrical-shaped substrates. These are usually fabricated from aluminium-containing alloys, such as Kanthal and Fecralloy, which may be coated with a catalytic layer, as described in Section III.B. The following examples illustrate the range of work done in this area.

Hunter [44] described washcoated thin-walled metal half-cylinders, slightly tapered at one end and perforated to give a large number of projections over the outer surface, which were used in carbon dioxide methane reforming, Eq. (15), with ruthenium as the catalytically active phase.

$$CO_2 + CH_4 \rightarrow 2CO + 2H_2 \tag{15}$$

In this application, solar energy drove the endothermic reforming, and the reverse reaction, forming methane, liberates energy at a sufficiently high temperature for it to be useful in, for example, a power station for generating electricity. Work by one of the present authors showed [45] that metal cylinders with openings in their walls performed well in conventional natural-gas steam-reforming, according to the equilibria given in Eqs. (16) and (17).

$$CH_4 + H_2O \rightarrow CO + 3H_2 \tag{16}$$

$$CO + H_2O \rightarrow CO_2 + H_2 \tag{17}$$

Small monoliths with higher geometric area have also been used randomly packed in a heated tube reactor for methane steam-reforming, with good results [46]. They have been shown to be effective in methanation reactions and to have low-pressure-drop characteristics even when compared with perforated metal cylinders described above [47]. Even lower pressure drop can be obtained by ordering small monoliths, rather than randomly packing them, so their channels are aligned with the direction of the process stream [48]. This arrangement does, however, restrict heat transfer to and from the walls of the reactor.

3. *Miscellaneous Applications*

The use of metallic monolithic substrates has been proposed in a variety of situations not referred to elsewhere in this chapter, and a few of these are noted in this section. However, none appear to have reached significant large-scale commercialization.

During the operation of sealed infrared carbon dioxide lasers, the working gas can undergo dissociation according to Eq. (18), and even low concentrations of oxygen formed in this way can cause the steady laser discharge to degenerate into local arcs.

$$2CO_2 - 2CO + O_2 \tag{18}$$

It is therefore very important that any free oxygen and carbon monoxide be recombined as soon as possible. Originally, a thin platinum wire heated electrically to about 1000°C was used to catalyze the recombination reaction. However, the fragile nature of the wire and the power consumed caused problems. Stark and Harris [49] recommended the use of a heated platinum-based catalyst on a Fecralloy substrate for this purpose, which is considerably more efficient than a platinum wire. Oxidation reactions are high-temperature applications for metallic monolithic catalysts, and a particular use that received considerable attention is in flameless gas turbines [50]. Here a metallic catalyst is fitted into a specially designed combustion chamber, where hydrocarbon oxidation takes place at a lower temperature than it would in a flame, and as a result the levels of NO_x formed can be about two orders of magnitude less than from a standard flame combustor. At the same time, hydrocarbon and carbon monoxide emissions are markedly reduced. Other advantages include compatibility with a wide range of fuels, including low-grade chemicals such as waste solvents. There has been much research into the use of systems of this type in aircraft engines to reduce NO_x emissions, and metal monolithic catalysts were also investigated for use in aircraft for ozone decomposition. At the end of the 1970s, jet-engined passenger aircraft had the capability of flying at high altitudes, which are associated with significant fuel economies. However, at about 40,000 feet (particularly over polar routes) ozone concentration is significant, and this is introduced into the cabin by the air conditioning unit, causing headaches and irritation of the eyes, nose, and throat. As a result, a number of catalysts were tested extensively for ozone decomposition in air [51]. In this application, metallic substrates were seen to have advantages of low pressure drop and high geometric surface area per unit weight, weight being an important consideration in aircraft applications. But ceramic-based catalysts are now used in this application [52]. American regulations [53] now require time-weighted aircraft cabin ozone concentrations not to exceed 0.1 ppm.

V. CONCLUSIONS

Metal is the active phase in many major commercial industrial catalysts, and several processes make use of structured bulk metal; wire gauzes are the best examples. Although they were devised more than 90 years ago, they remain central in the manufacture of key chemicals, such as nitric acid and hydrogen cyanide, and improvements in their structures are still being made. Recent developments in this area include knitting rather than weaving, and corrugated gauzes are being introduced in North America. It may be expected that other improvements will be introduced as newer manufacturing techniques are applied to these older structured catalyst systems.

Metal alloys can be fabricated into desirable catalyst structures having high geometric area, low pressure drop, and low weight properties, but bulk metal generally does not have sufficient surface area to provide good catalytic activity. However, technology has been developed to coat such substrates with catalytically active layers containing well-stabilized small crystallites of active metal that give good catalytic performance. The resulting structured catalysts have been used in autocatalyst and other environmental applications, and very large structures are made by welding several units into panels. Although not discussed in this chapter, catalytic heat exchangers and other intensive catalytic reactors have been made using these approaches [54]. It may be expected that in the future more complex structures of this kind will be developed to realize the full potential of structured catalysts.

REFERENCES

1. M.V. Twigg. In: R. Pearce and W.R. Patterson, eds., *Catalysis and Chemical Processes*. Lenonard Hill, London, 1981, pp. 11–34.
2. D.R. Anderson. *J. Catalysis 113*:475 (1988).
3. H.B.W. Patterson. *Hydrogenation of Fats and Oils*. Applied Science Publishers, London, 1983.
4. R.T. Donald. In: M.V. Twigg (ed.), *The Catalyst Handbook*. Manson, London, 1996, pp. 490–503.
5. N.H. Harbord. In: M.V. Twigg (ed.), *The Catalyst Handbook*. Manson, London, 1996, pp. 469–489.
6. F.W. Ostwald. *Chem. Z. 27*:457 (1903); British patents 698 (1902), 8300 (1902).
7. K. Kaiser. German patent 271,517 (1909).
8. L. Andrussow. U.S. patent 1,934,838 (1933).
9. N. Waletzko and L.D. Schmidt. *AIChE J. 34*:1146 (1988).
10. A.G. Knapton. *Platinum Metals Revs. 22*:131 (1978); L.D. Schmidt and D. Luss. *J. Catalysis 22*:269 (1971).
11. Y. Ningz. Yang and H. Zhao. *Platinum Metals Rev. 39*:19 (1995).
12. D.R. Merrill and W.A. Perry. U.S. patent 2,478,875 (1948).
13. D.A. Hickman, M. Huff, and L.D. Schmidt. *Ind. Eng. Chem. Res. 32*:809 (1993).
14. L.E. Campbell. U.S. patents 5,256,387 (1993), 5,217,939 (1993).
15. C.D. Keith. U.S. patent 3,428,424 (1969).
16. B.T. Horner. *Platinum Metals Revs. 35*:58 (1991). For recent surface area measurements of Pt/Rh gauzes see: E. Bergene, O. Tronstad, and A. Holmen, *J. Cat., 160*, 141 (1996).
17. B.T. Horner. *Platinum Metals Revs. 37*:76 (1993).
18. W.A. Hochella and S.A. Heffernen. W.O. patent 22,499 (1992).
19. J.R. Jennings (ed.). *Catalytic Ammonia Synthesis Fundamentals and Practice*. Plenum Press, New York, 1991.
20. A. Nielson, S.S. Bergh, and H. Topsoe. U.S. patent 3,243,386 (1966); P.J. Davidson, J.F. Davidson, and F. Kirk. European patent 86308349 (1986).
21. W.J. McClements, Hopewell, and R.C. Datin. U.S. patent 2,974,150 (1961).
22. E.C. Betz. U.S. patent 3,994,831 (1976).
23. J.A. Cairns and M.L. Noakes. British patent 1,546,097 (1976).
24. G.L. Vaneman. *Studies in Surface Science and Catalysis, Catalysis and Automotive Pollution Control II*, A. Crucq, ed. Elsevier, Amsterdam, *71* 537 (1991).
25. H. Bode, W. Maus, and H. Swars. IMechE Proceedings: *Worldwide Emission Standards and How to Meet Them*, 77 (1991).
26. W. Held, M. Rohlfs, W. Maus, H. Swars, R. Brück, and F.W. Kaiser. *SAE Paper* 940932 (1994).
27. R. Brück, J. Diringer, U. Martin, and W. Maus. *SAE Paper* 950788 (1995).
28. M. Luoma, P. Lappi, and R. Lylykangas. *SAE Paper* 930940 (1993).

29. F. Bonnefoy, F. Petitjean, and P. Steenackers. *Studies in Surface Science and Catalysis, Catalysis and Automotive Pollution Control III*, A. Frennet and J.M. Bastin, eds. Elsevier, Amsterdam, *96* 335 (1995).
30. B.E. Enga, M.F. Buchman, and I.E. Lichtenstein. *SAE Paper* 820184 (1982); B.E. Enga and J.F. Plakosh. *SAE Paper* 850018 (1985).
31. U.S. Atomic Energy Commission. British patent 1,045,993 (1965).
32. A.S. Pratt and J.A. Cairns. *Platinum Metals Revs. 21*:74 (1977).
33. M. Määthänen and R. Lylykangas. *SAE Paper* 900505 (1990).
34. J.A. Cairns. British patent 1,522,191 (1976).
35. C.A. Dulieu, W.D.J. Evans, R.J. Larbey, A.M. Verrall, A.J.J. Wilkins, and J.H. Povey. *SAE Paper* 770299 (1977).
36. P. Oser. *SAE Paper* 880319 (1988).
37. S. Pelters, F.W. Kaiser, and W. Maus. *SAE Paper* 890488 (1989).
38. R. Brück, R. Diewald, P. Hirth, and F-W. Kaiser. *SAE Paper* 95079 (1995).
39. B.H. Engler, E. Koberstein, and U. Plotzke. *Proc. 5th Int. Pacific Conf. on Automotive Eng.*, 271.1; *SAE Paper* 89127 (1989).
40. J.R. Kittrell, J.W. Eldridge, and W.C. Conner. In: J.J. Spivey, ed., *Catalysis, Royal Society of Chemistry. 9*, 126 (1992).
41. S.M. Cho and S.Z. Dubow. Paper presented at the Annual Meeting of the American Power Conference, Chicago, 13–15 April 1992.
42. G.W. Spitznagel, K. Huttenhofer, and K.J. Beer. In: J.N. Armor, ed. *Environmental Catalysis, Amer. Chem. Soc. Symposium Series 552*:172 (1994). P.A. Lowe and W. Ellison. In: J.N. Armor, ed. *Environmental Catalysis, Amer. Chem. Soc. Symposium Series 552*:190.
43. M.V. Twigg and J.T. Richardson. In: G. Poncelet, J. Martens, B. Delmon, P.A. Jacobs, and P. Grange, eds. *Preparation of Catalysts VI*. Elsevier, Amsterdam, 1995, p. 345.
44. J.B. Hunter. U.S. patent 4,349,450 (1982).
45. M.V. Twigg. European patent 0,082,614 (1986).
46. J.D. Rankin and M.V. Twigg. European patent 0,021,736 (1985).
47. C.J. Wright. U.S. patent 4,388,277 (1983).
48. C.J. Wright. G.B. patent 2,103,953 (1983).
49. D.S. Stark and M.R. Harris. *J. Phys. E. 21*:715 (1988).
50. H.J. Jung and E.R. Becker. *Platinum Metals Rev. 31*:162; (1987); B.E. Enga and D.T. Thompson. *Platinum Metals Rev. 23*:134 (1979).
51. A.E.R. Budd. *Platinum Metals Rev. 24*:90 (1980).
52. R. Heck, R. Farrauto, and H. Lee. *Catalysis Today 13*:43 (1992).
53. Airplane Cabin Ozone Contamination. Code of Federal Register, 14 CFR Parts 25 and 121. Washington, D.C., U.S. Goverment Printing Office, 1980.
54. A. Pinto and M.V. Twigg. European patent 0,124,226 (1984).

4

Autocatalysts—Past, Present, and Future

Martyn V. Twigg and Anthony J. J. Wilkins
Johnson Matthey, Royston, Hertfordshire, England

I. INTRODUCTION

The development of the internal combustion engine, in the form of the gasoline spark-ignition engine used in automobiles, has provided society with tremendous mobility over recent decades. Indeed, the benefits to the individual are so strong that in many countries it can be said the car is a key component of modern society. The desired internal combustion reaction is the oxidation of hydrocarbon fuel to carbon dioxide and water, according to Eq. (1):

$$4H_mC_n + (m + 4n)O_2 \rightarrow 2mH_2O + 4nCO_2 \qquad (1)$$

In practice, however, combustion is not completely efficient, and for this reason, as well as other physical effects, unburned hydrocarbons and partially combusted hydrocarbon oxygenates, such as aldehydes, may be present at varying levels in raw engine exhaust gas. In the pollution control arena these species are referred to differently in different parts of the world. The "hydrocarbons" are usually designated by the abbreviation HC. But in Europe, total hydrocarbons (THC), including the most difficult to oxidize, methane, are measured and reported; partially oxidized oxygenates are not measured. In America, methane is excluded, but oxygenates are included in nonmethane organic gases (NMOG) analysis. Carbon monoxide is also present as a partial oxidation product, which is formed according to Eq. (2). During the power stroke, moreover, under the conditions of high pressure and temperature, nitrogen and oxygen react in the engine cylinder and establish an equilibrium with nitric oxide, as in Eq. (3):

$$4H_mC_n + (m + 2n)O_2 \rightarrow 2mH_2O + 4nCO \qquad (2)$$

$$N_2 + O_2 \rightarrow 2NO \qquad (3)$$

At some stage, as the product gases expand and cool rapidly en route to the exhaust system, this equilibrium is frozen, and although thermodynamically unstable at low temperatures, appreciable amounts (e.g., up to 3,500 ppm) of nitric oxide can be present in the exhaust gas from an engine.

The three major pollutants from the internal gasoline engine are HC, CO, and NO_x, and they are the cause of significant environmental concern. However, catalysts, particularly in the form of monolithic honeycomb structures, have played a major role in rending those pollutants harmless by converting them to water, carbon dioxide, and nitrogen. This chapter puts into historical context how this was achieved, and describes some of the future trends in this important area of environmental protection.

II. HISTORICAL DEVELOPMENT

A. Background

As early as the 1940s, significant environmental problems were occurring, and with increasing frequency in some parts of the world, most notably America, especially the Los Angeles basin, which experiences frequent ambient temperature inversions. By the 1950s this had been related [1] to the photochemical interaction of hydrocarbons, oxygen, and nitrogen oxides in the atmosphere, forming ozone-containing photochemical smog, Eq. (4):

$$HC + NO_x + h\nu \rightarrow O_3 + \text{other products} \quad (4)$$

The finger was already pointing toward the motor car as a major source of manmade emissions. In time, political pressures exerted by the environmental lobby resulted in the Clean Air Act of 1970, which laid down a program with the target of achieving a 90% reduction in emissions relative to an uncontrolled average 1960-model-year vehicle. Initially, some improvements were made by engine modifications. However, the 1975 U.S. federal and California limits could not be met by engine modifications alone, and the catalytic converter was shown to be the best way forward, provided that unleaded fuel could be made available countrywide. Considerable research into catalytic systems was undertaken by both industry and academics. Catalysts using base metals such as nickel, copper, cobalt, and iron seemed initially to be attractive on the basis of cost. However, these catalysts were adversely affected by sulfur and residual traces of lead in the fuel, and the catalysts eventually chosen were based on the platinum-group metals, which were well known and used in the chemical process industry worldwide [2].

For a few years, the emission limits could be met by oxidation of the carbon monoxide and hydrocarbons emitted by the engine, as in Eqs. (1) and (5):

$$2CO + O_2 \rightarrow 2CO_2 \quad (5)$$

The most common catalyst, the so-called COC (conventional oxidation catalyst), was based on platinum and palladium on an alumina support. However, as legislation tightened further, it became necessary to control the NO_x emissions also. This brought two further requirements: closer control around the stoichiometric air–fuel ratio, and the addition of a further catalytic metallic component, rhodium, to the catalyst formulation to enable the NO_x to be selectively reduced to nitrogen, as in Eqs. (6) and (7):

$$2m\text{NO} + 4\text{H}_m\text{C}_n \rightarrow (4n + m)\text{N}_2 + 2m\text{H}_2\text{O} + 4 \quad (6)$$

$$2NO + 2CO \rightarrow N_2 + 2CO_2 \quad (7)$$

Early concepts to achieve this centered around two-catalyst systems. The engine was run slightly rich to enable the reduction of NO_x over the first, rhodium-containing, catalyst, then air was introduced between the two catalysts to enable the second catalyst to behave as a COC and to oxidize CO and hydrocarbons, However, it was important for the first catalyst to reduce NO_x to N_2 with high selectivity. For example, if any ammonia was formed, it would be reoxidized on the second catalyst back to NO_x according to Eq. (8). This illustrates the importance of selectivity in autocatalyst design, and it is considered further in Section VII.A.

$$4NH_3 + 5O_2 \rightarrow 4NO + 6H_2O \tag{8}$$

Some European car manufacturers, notably VW and Volvo, used Pt/Rh catalysts as oxidation catalysts. Provided the engine management system gave them reasonably close control around stoichiometric, it was shown that these catalysts could give a degree of NO_x control. By 1979, oxygen sensors had been developed and were placed in the gas flow close to the exhaust manifold to provide feedback control of the fueling, so conditions could be maintained at or around the stoichiometric point. This enabled relatively good, consistent catalytic performance, and the use of platinum–rhodium catalysts to control HC, CO, and NO_x simultaneously became the preferred system.

Because it controlled all three major pollutants on one catalyst, the concept was christened the three-way catalyst—now normally called TWC. Early TWCs had a narrow operating range over which all three pollutants were removed to an acceptable degree, and they were almost universally fitted to U.S. cars from about 1980. Since then considerable technical effort has gone into improving the performance and widening the operating air–fuel ratio window of the catalyst. Initially this was necessary to provide control with carbureted cars, where the fueling management was relatively crude and slow in response. The technical effort in catalyst improvement has subsequently gone hand in hand with increasing sophistication in the development of fuel management systems, particularly fuel-injection methods. Thus, key technologies enabling the introduction of effective TWCs were oxygen sensors and electronic fuel injection, now closely controlled by a microprocessor system.

B. Legislative Requirements

Tables 1 and 2 summarize the development of the legislation in the United States from the time when serious attention was paid to reducing emissions from automobiles. From 1977 in California, where there were some particularly pressing air-quality problems [3], they were allowed to legislate lower levels of emissions than in the rest of the United States. The emission numbers are generated by driving the vehicle on a chassis dynamometer (rolling road) to a well-defined test drive cycle (see Fig. 1). This cycle represents conditions typical on a U.S. freeway and has a maximum speed of 55 mph. The emissions are collected from the tailpipe as soon as the ignition is switched on, analyzed, and the pollutant concentrations calculated. It can be seen from the table that emission limits are decreasing steadily, especially from 1993 onward, and that over the period there has been a major reduction in the emission levels. With the introduction of each successive emissions band, a small percentage of vehicles are required to meet the next band. This culminates in a requirement for a small number of vehicles to emit zero emissions in the year 2000, which can be achieved currently only by electric vehicles.

Emission legislation in Europe was such that the furnishing of catalysts was not required until 1993. The legislative developments since that date are summarized in Table 3 for gasoline and in Table 4 for diesel vehicles. In some respects the European test protocol is similar to that of the United States, but the test drive cycle itself is different

Table 1 U.S. Federal and California Emission Standards

Year	Area	Emissions (g/mile, FTP Test)		
		HC	CO	NO_x
Typical precontrol values		15	90	6.00
1970	All	4.10	34	4.00
1975	All	1.50	15	3.10
1977	Fed.	1.50	15	2.00
1977	Calif.	0.41	9.00	1.50
1980	Fed.	0.41	7.00	2.00
1981	All	0.41*	3.40	1.00
1993	Calif.	0.25**	3.40	0.40
1994	Fed.	0.25**	3.40	0.40
2003	Fed.	0.125***	1.70	0.20

*Equivalent to 0.39 g/mile NMHC.
**NMHC = Nonmethane hydrocarbons, i.e., all hydrocarbons excluding methane.
***NMOG = Nonmethane organic gases, i.e., all hydrocarbons and reactive oxygenated hydrocarbon species such as aldehydes, but excluding methane.

Table 2 California Emission Standards Post-1993

Year	Emissions (g/mile, FTP Test)		
	HC	CO	NO_x
1993	0.25*	3.40	0.40
1994 (TLEV)	0.125**	3.40	0.40
1997 (LEV)	0.075**	3.40	0.20
1997 (ULEV)	0.04**	1.70	0.20
1998 (ZEV)	0.00	0.00	0.00

*NMHC = Nonmethane hydrocarbons, i.e., all hydrocarbons excluding methane.
**NMOG = nonmethane organic gases, i.e., all hydrocarbons and reactive oxygenated hydrocarbon species such as aldehydes, but excluding methane.

(as shown in Fig. 2) and represents a more European style of driving. For 1993 and 1996 legislation, the vehicle is allowed to idle for 40 sec before sampling of the tailpipe gases begins. For the year 2000 and beyond, the sampling begins as soon as the ignition is switched on, as in the U.S. drive test cycle.

It is not easy to compare the emissions values obtained during the two cycles. In addition to the obvious differences between the driving protocols, the total amount of all the hydrocarbons are measured in Europe, whereas methane is not included in the United States. Also, all new production vehicles have to meet every new European legislative mandate as soon as it is introduced, compared to a phasing in strategy in California. However, it is generally considered that future legislation will require similar technical solutions for both markets.

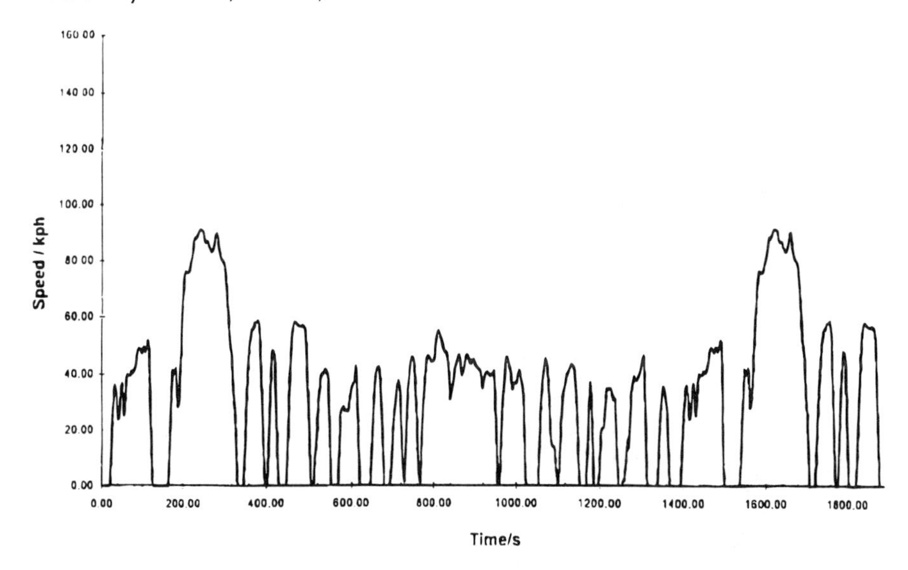

Figure 1 Speed/time trace for the U.S. federal test procedure (FTP). The American test cycle for passenger cars and light-duty vehicles from 1975 model year.

Table 3 European Emission Limits, Gasoline Vehicles

	Emissions g/km			
Year	Total HC	HC + NO_x	CO	NO_x
1993		0.97	2.72	
1996 (Stage II)	0.50		2.20	
2000 (Stage III)*	0.20		2.30	0.15
2005 (Stage IV)**	0.10		1.00	0.08

*Proposed; the test has no 40-sec idle before sampling begins.
**Indicative.

III. CATALYST TYPES

In general there are two ways of producing highly dispersed, catalytically active materials: precipitation of active precursors and support components from solution, and impregnation of preformed support, such as high-area pellets or rings [4]. Almost since the beginnings of autocatalyst technology, impregnation techniques were used, because this enabled costly materials such as platinum-group metals to be located only a short distance from the gas/solid interface, where they are most effective, thereby minimizing the amounts needed. However, many other considerations have to be taken into account in the design of autocatalysts; these are discussed in this section.

The catalyst used on a vehicle has to present a low-pressure resistance to the engine, which if too high could cause problems in running the engine to maximum efficiency

Table 4 European Emission Limits, Diesel Vehicles

	Emissions g/mile		
Year	HC + NO_x	CO	NO_x particulate
1993	0.97	2.72	0.14
1996 (Stage II) IDI	0.70	1.00	0.08
DI	0.90	1.00	0.10
2000 (Stage III)*	0.56	0.64	0.50
0.05			
2005 (Stage IV)**	0.30	0.50	0.25
0.03			

* Proposed; no 40-sec idle before sampling begins.

** Indicative.

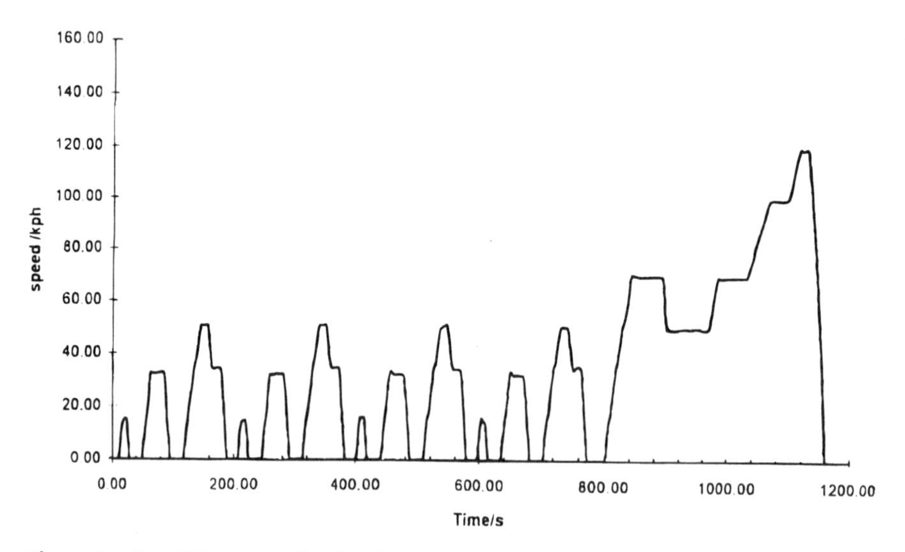

Figure 2 Speed/time trace for the European test procedure (ECE/EUDC). The high-speed extra urban driving cycle (EUDC) was added in 1993.

and power. The catalyst must also not suffer attrition as a result of the pulsating gas flow and natural vibration associated with a moving vehicle. Initially, catalyst pellets were used that had the advantages of being relatively cheap and readily available [5]. They were used in the U.S. market, particularly by General Motors. Pellet catalyst beds were known to be effective and to be widely used in the chemical process industries. In the case of automotive pollution control they were encased in a relatively wide, flat container. Although pellets met a number of the criteria for a successful substrate material, they were prone to attrition through rubbing together in the high space velocity and pulsing flow of the exhaust gas, leading to some loss of surface material. In this event there could be a loss of active material, and the attrited powder could result in an increase in back pressure, or if the powder was able to find its way out of the container it could be considered as an unregulated emission. As a result, pellets gradually ceased to be used as a catalyst support in the automotive emission control scene, and they have been completely replaced by monolithic structures.

A. Monolithic Honeycombs

The most widely used form of autocatalyst substrate has been, and probably will continue to be for the immediate future, the so-called monolith. Although the concept of the monolith has been around a lot longer than autocatalysts, it is very well suited to the application. There can be various forms of the monolithic concept, including those based on ceramic or metallic foams, and also a version built up from intermeshed wire. In practice, however, the one that has found most favor in meeting all of the requirements, at least to a substantial degree, is the form incorporating a series of channels passing along the length of the piece. These have generally been made either by an extrusion technique, which readily allows different sizes, shapes, and lengths to be manufactured, or by a layering technique, which alternates flat sheets with a corrugated or ribbed sheet.

The design of the substrate must provide a maximum superfacial surface area that can be presented to the exhaust gas, since it is upon this surface that the catalytic coating is applied and on which the pollutant and reactant gases must impinge in order to react. Another important consideration is that the mechanical strength and resistance to damage through vibration be sufficiently high, since the catalytic unit is designed to last the life of the car under normal conditions. To some extent, the requirement for high surface area and low back pressure goes against the requirement for high strength, and so a compromise has to be reached. Similarly, the catalyst is subjected to a wide range of temperatures throughout its life, from below zero to possibly in excess of 1000°C. More important, the rate of change of temperature is often very rapid, both increasing and decreasing, during operation. Obvious examples are accidental malfunctions, such as misfire, and design features, like deceleration fuel-cutoff for fuel economy. For these reasons the substrate has to be resistant to thermal deterioration and has to have the right thermal expansion properties. Ideally, also, the substrate, which represents the major part of the weight of the catalyst, should be light, to minimize its effect on total vehicle weight and fuel economy. It should also be as inexpensive as possible. Finally, since the application of the catalyst to the preformed monolith is by a chemical coating process, ease of handling and reproducibility of processing are also important features of the substrate.

B. Substrate Materials

There is the possibility to make substrates in various materials: Alumina is an obvious possibility, but monoliths formed from alumina are particularly susceptible to thermal shock problems, and they readily crack during rapid temperature excursions. Silicon carbide and boron nitride are other possible materials having good properties, but they are expensive.

By far the most successful material has been compositions that when extruded and fired at high temperature form cordierite, $2MgO \cdot 2Al_2O_3 \cdot 5SiO_2$; in excess of 90% of substrates on cars today are made in this way. These cordierite-containing monoliths are discussed in detail in Chapter 2. For some applications, monolithic substrates made of metal are preferred. One advantage of this is that very thin (0.05-mm) foil can be used, giving very low back pressure and high surface area, combined with good mechanical strength characteristics when appropriately fabricated. However, to withstand the very corrosive environment in the car exhaust, the thin foil has to be stable to corrosion at high temperatures; normally, an iron–chromium–aluminum ferritic steel is required. Use of these special alloys inevitably increases the cost of this form of substrate. Nevertheless, in some circumstances the benefits outweigh the cost; see Chapter 3. Properties of both types of substrate are summarized in Table 5.

IV. THREE-WAY CATALYST COMPOSITIONS

Today's three-way catalysts are based on a broad composition comprising platinum-group metals (PGMs), usually platinum, palladium, and rhodium in some combination, alumina and ceria, together with support stabilizers, catalyst promoters, and components to modify the chemistry taking place in the catalyst to improve selectivity. Commonly used elements in these contexts are nickel, barium, lanthanum, and zirconium. While most commercial catalysts contain one or more of these minor elements, all contain PGMs, alumina, and ceria, although the proportions vary in different catalysts. For almost all cars fitted with autocatalyst since the mid-1970s, their principal catalyst element has been based on the platinum-group metals, in preference to cheaper and more abundant base metals, because they provide the best performance over the life of a vehicle. Base metals are generally not as active or as stable as the platinum-group metals used. In addition, base metals are more susceptible to loss of performance through poisoning by sulfur, trace lead, and residues from oil additives that are present in the exhaust gas. Since legislation is set to get tighter, and the duty the catalyst is required to do grows, it seems likely the PGMs will continue to be preferred. Moreover, there are more than adequate reserves of these metals to enable this to happen, and freshly mined material is now being supplemented by recycled material from scrapped cars as they become available [6].

Although the primary driving force in autocatalyst development is performance, cost is, of course, also a major consideration. Early oxidation catalysts used platinum and palladium, since both are good oxidation catalysts under lean operating conditions. When

Table 5 Physical Properties of Typical Ceramic and Metal Monolithic Substrates Used for Autocatalysts

	Ceramic support	Metal support
Wall thickness (mm)	0.15	0.04
Cell density ($inch^{-2}$)	400	400
Open facial area (%)	76	92
Specific surface (m^2 $liter^{-1}$)	2.8	3.2
Specific weight (g $liter^{-1}$)	410	620
Weight without shell (g $liter^{-1}$)	550	620
Thermal conductivity (cal sec^{-1} cm^{-1} K^{-1})	3×10^{-3}	4×10^{-2}
Thermal capacity (kJ kg^{-1} K)	0.5	1.05
Density (kg $liter^{-1}$)	2.2–2.7	7.4
Thermal expansion (K^{-1})	0.7×10^{-6}	0–15
Maximum working temperature (°C)	1,200–1,300	1,500

the need for NO_x conversion arose, the preferred solution was to use platinum and rhodium, which gave better conversion of all three pollutants than either metal alone. As the technology has improved, enhanced-performance, lower-cost palladium/rhodium three-way catalysts have been developed for some applications, especially smaller cars. Subsequently, palladium-based systems have been enhanced more to achieve better hydrocarbon removal [7]. While palladium-containing catalysts do generally have better hydrocarbon performance, this may be offset by lower performance for NO_x and CO, so the choice of platinum/rhodium or palladium/rhodium may be affected by the balance of HC/CO/NO_x from a particular engine. Palladium is also more sensitive to poisoning, for example, by sulfur or lead in the exhaust gas. In some areas this may be one of the more important criteria. But lead levels are now close to zero in markets where catalysts are used, and fuel sulfur levels are generally tending to fall. In order to get the best overall performance from the PGMs it is becoming common to combine the advantages of platinum, palladium, and rhodium. In some cases this is in a single catalyst incorporating all three metals. But many cars already incorporate more than one catalyst containing different PGMs.

Alumina has always had a role in autocatalysts, and it has several functions. First, it forms a much higher surface area support for the catalytically active components than does the bare substrate. This enables the catalytic metal to be highly dispersed in the form of very small (initially less than 10 nm) crystallites, and this high dispersion results in high catalytic performance. Similarly, because, for example, platinum has a reasonably strong affinity for alumina, and the concentration of metal crystallites on the surface is low, the sintering of the active metal into larger, less active crystallites at high temperatures is inhibited. In addition, alumina tends to absorb materials that would normally poison the catalytically active sites (lead is an example) and enables retention of performance.

As with alumina, ceria has several roles to play within the catalyst formulation. It has some effect on stabilizing alumina surface area at high temperatures, and it is also capable of stabilizing the dispersion of platinum in these systems, important because the effect is particularly marked in the 600–800°C region, where many present-day catalysts operate. In addition, ceria allows two other more directly performance-related phenomena to take place: oxygen storage and the water gas shift reaction shown in Eq. (9):

$$CO + H_2O \rightarrow H_2 + CO_2 \tag{9}$$

In the case of the former, the ability of ceria to store some oxygen when the exhaust gas mixture is lean of stoichiometric and to release it when the mixture goes rich enables CO and hydrocarbon adsorbed on the catalyst during rich excursions to be oxidized by stored oxygen when there is insufficient oxygen in the exhaust gas. This improves oxidation performance of the catalyst under rich operation. Ceria also encourages the "water gas shift" reaction, which affords hydrogen, which also improves catalyst performance under rich operation. Exact catalyst formulations are proprietary, but it is clear from published information [8] that the forms and the way in which these elements are incorporated into the catalyst are very important, as, for example, is the manipulation of the chemistry during the impregnation stages.

Although the Pt/Rh catalyst generally retains good efficiency for hydrocarbon control, it is desirable to enhance the efficiency of catalysts for hydrocarbon removal in the light of the severe future U.S. hydrocarbon standards. Developments in palladium catalyst technology have resulted in a significant improvement in hydrocarbon control. These catalysts will see application in close-coupled converter systems, either as a single palladium catalyst or as a palladium catalyst in combination with rhodium [9] and in some cases with platinum also. In some cases, catalysts of different compositions can be used in sequence, with advantage in applications where the volume of catalyst required is sufficient to allow more than one piece to be used. In Europe there is still some concern about the adverse effects of low levels of lead in fuel on Pd-only systems and of sulfur on their NO_x performance. Accordingly, the trend is toward Pd/Rh and Pd/Pt/Rh trimetal catalysts to mitigate these effects.

V. CATALYST COATING PROCESSES

There are two general approaches to washcoating a substrate to produce an active catalyst. *Washcoating* is the term used for the application of the thin layer of active catalyst to the substrate. First, a layer of oxides can be applied and fixed to the substrate by a high-temperature treatment, during which ceramic bonding between the washcoat and the substrate takes place. The active components are then added, in an impregnation step, or alternatively the active components can be incorporated into the washcoat before it is applied to the substrate. The latter approach involves less processing with the substrate but requires additional preparative work before that stage. Both techniques appear to have been employed commercially, and two or more washcoats may be applied on top of each other. This might be done to ensure good physical separation of components that interact in a negative way. It is important to ensure that the catalytic coating adheres firmly to the substrate, especially during thermal cycling. So a reasonable match between the thermal expansion of the substrate and the coating is important, and this is true of alumina. In the case of metal substrates (see Chapter 3), where there would be a major mismatch of thermal expansion, aluminum-containing ferritic steels are used, and these give a self-healing surface film of alumina when heated to high temperature. The washcoat has an affinity for this oxide surface.

The type of alumina used in the washcoat has a significant impact on the stability of its surface area. As aluminas are heated to higher temperatures they go through a series of phase changes, accompanied by loss of surface area, until at around 1100°C a low-surface-area alpha-alumina is formed [10]. Other components in the catalyst formulation can have an impact on the rate of these changes, in both a positive and a negative sense. Thus at high temperatures, which would be achieved, for example, with close-coupled catalysts (see Section X), it is important to use a stabilized alumina. A number of chemical elements are known to do this; examples commonly used are barium oxide, zirconia, and ceria.

VI. CATALYST CANNING

When substrates have been coated, the active catalyst unit needs to be incorporated into a canister, which physically protects the catalyst and also enables it to be fit into the exhaust train. Since the catalyst is designed to last the life of the car, the quality of the steel used is important, for holes in the exhaust system could lead to emissions escaping before being converted over the catalyst. In addition, the bulk of catalysts are based on ceramic substrates and have a significantly different thermal expansion characteristic to the exterior steel shell. It is usual to have a special interlayer between the shell and the catalyst that has the flexibility to allow for differential expansion as the temperature changes. The most common material used for this purpose is an expandable ceramic mat, but in some applications a knitted stainless steel mesh material is used [11]. These serve a dual purpose, for they also prevent movement of the catalyst due to pulsing of the exhaust gas from the engine. A further consideration is the correct design of the inlet and outlet cones, to ensure the best presentation of the exhaust gas flow to the face of the catalyst. Finally, the catalyst has to be fitted in the space available on the vehicle concerned.

VII. AUTOCATALYSTS IN OPERATION

In the chemical industry most catalytic reactors operate at constant pressure, mass flow rate, and temperature. Such steady-state converters are often maintained without major changes in their operating parameters over long periods. They are well understood and are relatively straightforward to model mathematically. Autocatalysts also must have longevity; but in marked contrast they operate in a very transient mode, which makes their design and development more complex. Modeling their behavior is significantly more difficult. These aspects are only touched upon in this section, which highlights the main features associated with the in-service operation of practical autocatalysts.

A. Kinetics of Operation

Formulation strongly affects important catalyst properties, such as light-off characteristics, and the air–fuel ratio window in which it is able to operate satisfactorily. Operating conditions, such as exhaust gas temperature and mass flow rate, and the ratio of air to fuel in the exhaust at a given instant are important in determining the actual efficiency of a particular catalyst in use. When the vehicle engine is first started, the exhaust gas and catalyst are normally at ambient temperature. The catalyst will typically begin operating when it reaches between 250 and 300°C, and the point at which it begins to operate is called the *light-off temperature*. The speed at which it reaches this temperature is dependent on how far from the engine the catalyst is sited in the exhaust system, the thermal mass of the exhaust manifold and pipes leading to the catalyst, how fast the exhaust gas warms up, and the concentration of reducing species (especially CO) and oxygen in the exhaust gas. In practice, the light-off temperature is time, temperature, and reactant concentration dependent, and is under kinetic control. Before light-off, negligible conversion occurs; but once the light-off temperature is reached, conversion rises rapidly to close to 100%, as shown in Fig. 3. This rise in conversion is accompanied by an increase in temperature over the catalyst, the extent of which depends on the concentration of HC, CO, and NO_x being converted. Light-off times are reduced by ensuring that the engine, and hence the exhaust gas, warms up quickly and that the air–fuel ratio changes quickly from its rich value at start-up to stoichiometric when there is more O_2 available for CO and HC combustion. Once the catalyst is at operating temperature, the reactions are mass-transfer controlled, and conversions between 95 and 100% are normal. A plot of conversion as a function of air–fuel ratio for a typical TWC is shown in Fig. 4. Around the stoichiometric point, maximum conversion of HC, CO, and NO_x is achieved within a narrow operating "window." If the air–fuel ratio is slightly rich (excess fuel), HC and CO conversion is depressed, but NO_x reduction is enhanced. Conversely, if the air–fuel ratio is slightly lean (excess air), NO_x conversion is significantly less, and oxidation of CO and HC is strongly favored. Under normal vehicle operation, the air–fuel ratio oscillates between slightly

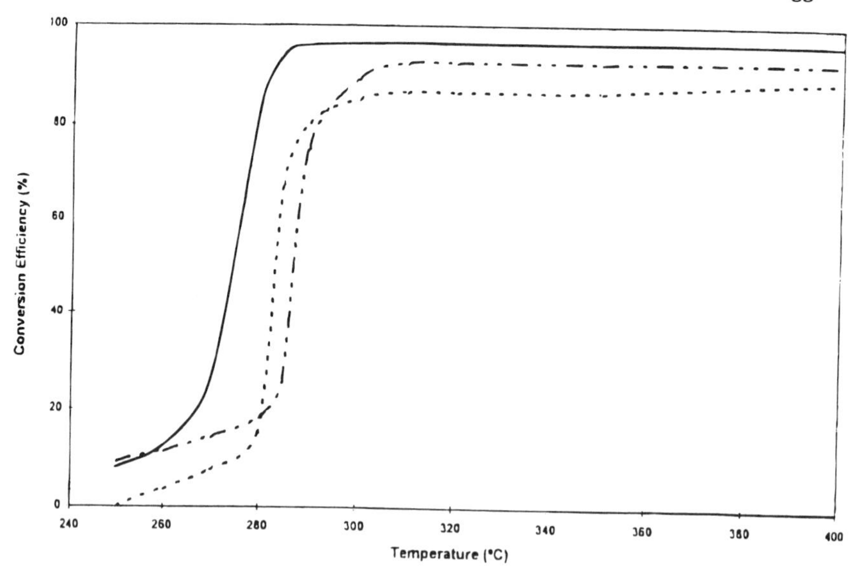

Figure 3 Schematic light-off curve for a gasoline autocatalyst, showing conversion efficiency vs. temperature for HC, CO, and NO_x. *Key*: CO ------, HC - - - -, NO_x .._._.._.._.

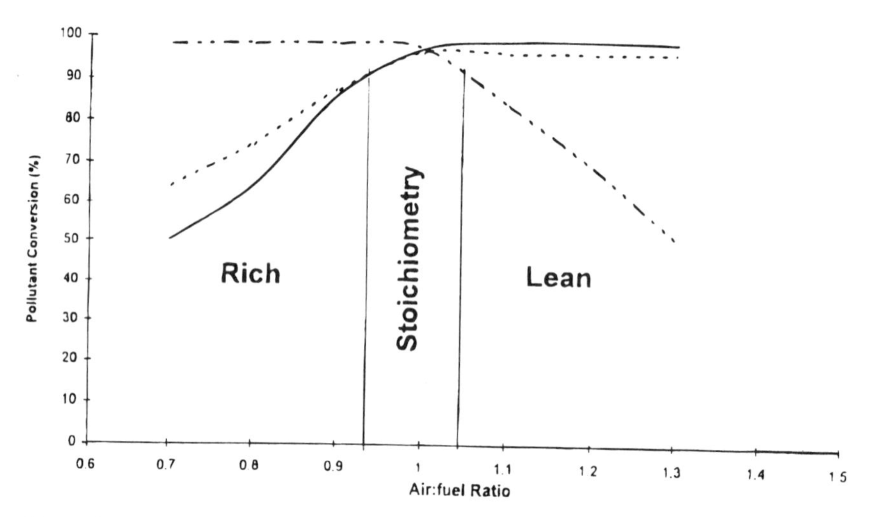

Figure 4 Schematic conversion of HC, CO, and NO_x over a three-way autocatalyst as a function of lambda (measured air-to-fuel ratio divided by the stoichiometric ratio; lambda = 1 is the stoichiometric point). The engine management system controls the air/fuel ratio around lambda = 1. *Key*: CO ------, HC - - - -, NO_x -.-.-..

rich and slightly lean. However, when the vehicle is accelerated or decelerated, these perturbations can be larger and can significantly change the exhaust gas composition. For instance, during accelerations the air–fuel ratio may be enriched to improve performance, and as an economy measure fuel is often cut off during decelerations. Under demanding transient conditions such as these, conversion over the catalyst has to change rapidly to prevent breakthrough of particular pollutants. To improve air–fuel ratio control, a lambda sensor is fitted in the exhaust stream prior to the catalyst, which monitors O_2 partial pressure and provides an electrical signal to the electronic control module that changes the quantity of fuel fed to the injectors. This maintains the air–fuel ratio close to the stoichiometric point, for maximum catalyst operating efficiency. Continuous improvements in the electronic control strategies ensure that once light-off is achieved, virtually no increase in emissions is measured over the remainder of the test cycle, as illustrated in Fig. 5. The graphs represent the cumulative emissions of pollutants during the test cycle. With an extremely efficient control strategy, there is an initial increase in emissions while the catalyst is cold. However, once light-off has taken place, there is only a small increase in emissions over the remainder of the cycle, represented by a virtually zero slope of the line. When the control strategy is less exact, the cumulative emissions continue to increase after light-off, represented by a positive slope of the line and higher tailpipe emissions.

B. Thermal Stability

A major challenge for future catalysts is the development of even more thermally stable materials, due to the fact that under typical close-coupled operating conditions (see Section X.B.2), the sintering of catalyst materials is significantly enhanced. In future emission systems, it is probable that the catalyst maximum operating temperature will be increased to 950°C or above. Then significant surface area loss and crystallite growth can be expected to occur. This has resulted in extensive research and development into methods of dispersing and stabilizing the key catalytic materials for higher-temperature operation. Incorporation of these materials into new catalyst systems has resulted in a significant improvement in high-temperature durability when aged at temperatures up to 1050°C [12].

C. Effects of Poisons

In the search to make catalyst systems ever more effective, the effects of the composition of the fuel can play an important role. In the case of two-stroke engines, the components

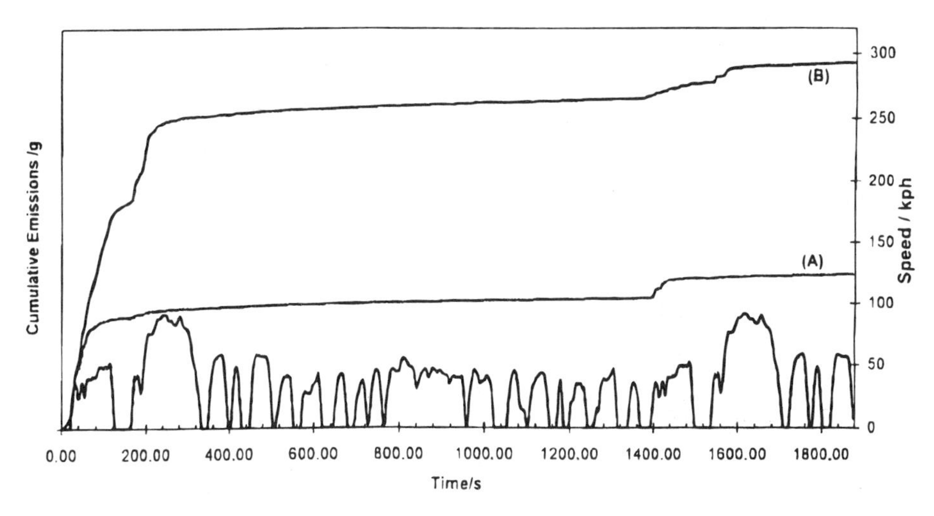

Figure 5 Effect of engine control on tailpipe emissions. A well-calibrated gasoline engine results in significantly lower emissions after the catalyst (A) than does the vehicle with a poor engine calibration (B).

in the lubrication oil can also be very important. Two of the key "poisons" in gasoline fuels are sulfur and lead, and the effect of these elements can be substantial. In U.S. unleaded fuel, the levels of lead have fallen over the long period this fuel has been available to levels that are on the borderline of detection. However, sulfur levels vary widely across the country and are typically in the range 75–750 ppm. In Japan, both lead and sulfur levels are very low. However, in Europe there is variation in both lead and sulfur levels across the member states, but the lead levels should have reached a low level by the time Eurostage III legislation is in force. The effects of sulfur, present in the exhaust gas as SO_2, on catalyst performance can vary, depending on both catalyst formulation and pollutant molecule being treated. In general, the greatest sensitivity is found for palladium as the catalytic element, and in NO_x removal [13,14].

Lubricating-oil consumption in modern engines is generally very low (0.1 liters per 1,000 km), and their contribution to catalyst deactivation is small. However, with the requirement for extended catalyst durability and extended drain periods for oils, there is considerable interest about the effect of the oil additives on catalyst life. The chief component of the oil that affects catalyst durability is phosphorus, which is usually present in the form of zinc dialkyl-dithiophosphate (ZDDP). Both combusted and uncombusted forms of ZDDP can reach the catalyst, resulting in different effects on activity depending on the temperature of operation. The level of phosphorus in the oil and the amount of alkaline earth metals present (such as calcium) can dictate the extent to which phosphorus can be deposited on the catalyst. However, studies have shown quite clearly that well-formulated lubricants and well-designed catalysts ensure that the antiwear properties of the oils are maintained and that catalyst-equipped vehicles meet the emission standards required [15,16].

D. Unregulated Emissions

The decision to eliminate nickel in catalyst formulations for the European market resulted in a difficult challenge for the catalyst designer, who must maximize reduction of nitrogen oxide while minimizing the reduction of sulfur dioxide to hydrogen sulfide. Tackling this selectivity issue has sometimes resulted in some loss of TWC performance as H_2S formation is suppressed. The specific advantage of nickel oxide has been its ability to absorb hydrogen sulfide and then to release the stored sulfur as SO_2 under oxidizing conditions. This allows the catalyst designer to formulate the catalyst composition for maximum NO_x reduction activity. It has therefore been necessary to identify effective substitutes for nickel to suppress hydrogen sulfide emissions. Progressive improvements in catalyst technology for H_2S suppression have been achieved over the past several years. The current development formulations are both effective in suppressing hydrogen sulfide emissions and maintaining high converter efficiency [17,18].

VIII. DIESEL CATALYSTS

Diesel exhaust usually demands special aftertreatment technologies. In the following two subsections, removal of hydrocarbons and carbon monoxide and a novel approach to the elimination of soot particles are discussed. The effective removal of NO_x under the lean conditions prevailing in diesel exhaust has not yet been commercialized, and some approaches being investigated are discussed in Section X.C. Some aspects of diesel exhaust treatment are also discussed in Chapter 18.

A. Diesel Oxidation Catalysts

There is concern about the levels of carbon dioxide emissions from cars, as well as the three main pollutants from combustion: HC, CO, and NO_x. Lower CO_2 emissions result from improved fuel economy that can be obtained in several ways. An important, more fuel efficient approach is through the use of so-called lean-burn engines that operate with excess air rather than with stoichiometric air–fuel mixtures.

The classic lean-burn engine is the diesel engine, and there have been many developments in advanced combustion systems for diesel engines. These include more precise fuel injector design, better combustion chamber geometry, and optimized swirl, together with improved injection timing and turbocharging, which have all contributed to better-performing, cleaner engines. Nevertheless, as the emission standards progressively tighten there is an increasing need to reduce engine out-emissions further.

Oxidizing hydrocarbons and CO is favored in the lean (oxygen-rich) exhaust gas, but the problem of treating NO_x emissions under such conditions is particularly difficult. Moreover, as a direct result of fuel efficiency, exhaust gas temperatures are generally quite low, so highly active catalysts with low light-off temperatures are required to effect the desired conversion of hydrocarbons and CO [19]. Diesel fuel contains sulfur compounds, which are converted to SO_2 during combustion and may be converted over an oxidation catalyst to SO_3, which can react with washcoat components, forming sulfates. These sulfates provide a mechanism for storing sulfur in the catalyst. Sulfuric acid formed from SO_3 can adsorb onto soot and contribute to the weight of particulates emitted. While it is possible to devise catalyst formulations that inhibit sulfate formation, these often also exhibit inhibition of the desired HC/CO/VOF oxidation reactions and can undesirably increase light-off temperature.

While diesel exhaust gas temperatures under start-up and low-load conditions can be very low compared with a gasoline engine, they can be substantially higher under high-speed/full-load conditions. Accordingly, the catalyst must retain low-temperature performance after having been exposed to high temperatures, and so thermal durability is an important catalyst characteristic. Thus the key parameters required for diesel oxidation catalysts are: high oxidation activity at low temperature, low sulfate production and storage, and good thermal stability over extended mileage.

By optimizing catalysts, these targets have been achieved. For instance, Fig. 6 shows results for aged catalysts tested on a 6-liter diesel engine in the ECE R49 test (a static engine test). The fuel used contained 0.05% sulfur, and the maximum catalyst inlet temperature during the cycle was 540°C. The catalysts contained platinum at two different loadings, 10 g ft^{-3} and 2.5 g ft^{-3}. Figures 6a and 6b show that the lower-loaded catalyst is less efficient for CO removal, but, compared with the higher-loaded catalyst, has almost equivalent performance for HC oxidation. This is mirrored in Fig. 6c, which shows a similar efficiency for the removal of particulate soluble organic fraction (SOF). However, the higher-loaded platinum catalyst generates SO_3 and hence forms sulfate. The lower-loaded catalyst does not form sulfate (Fig. 6d), so the net effect on particulate emissions (Fig. 6e) is that the higher-loaded catalyst is worst due to the sulfate contribution.

As sulfur levels in diesel fuel are lowered, sulfate formation will become less important and oxidation of HC and CO will be facilitated, but the classic problem of NO_x

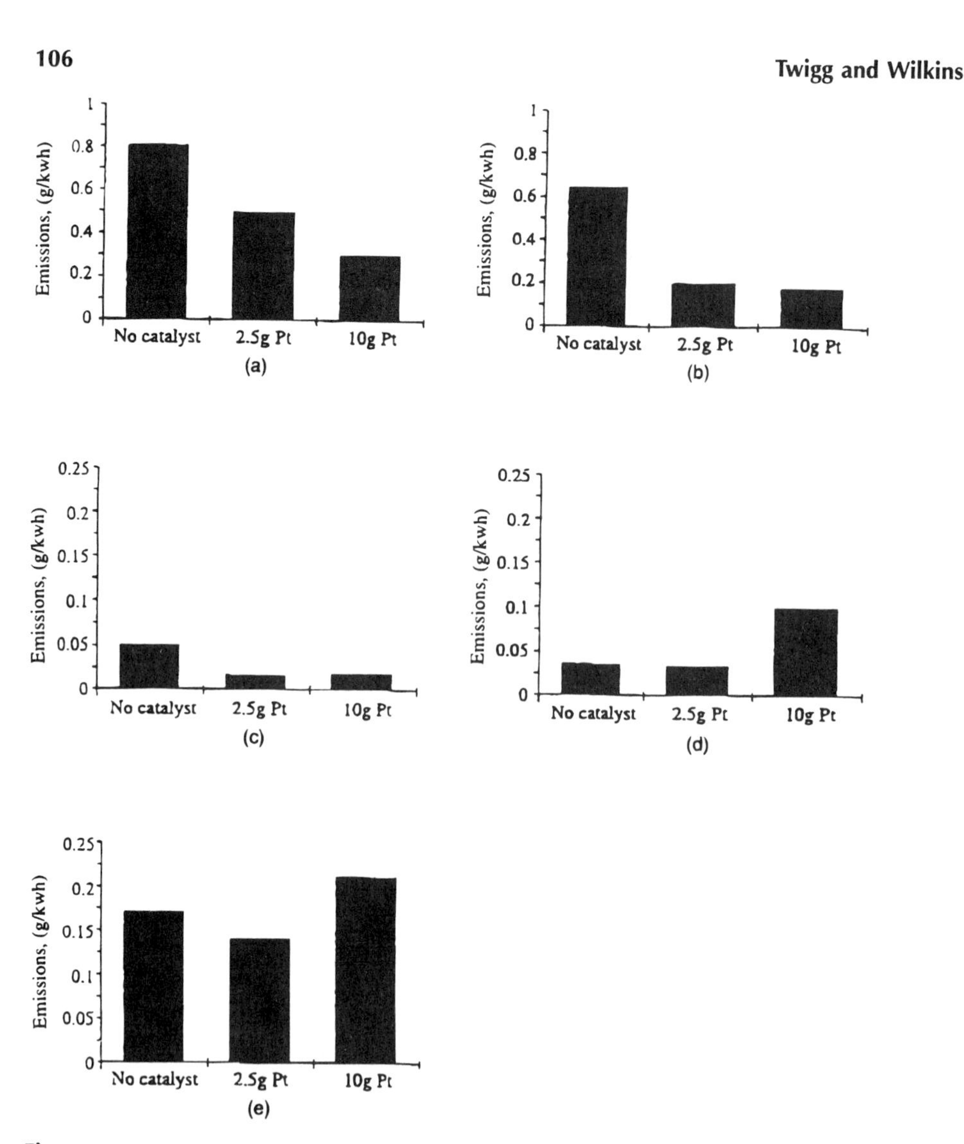

Figure 6 Effect of platinum content (g ft^{-3}) in diesel oxidation catalysts on tailpipe emissions in the ECE R49 test: carbon monoxide (a), hydrocarbon (b), soluble organic fraction (c), sulfur trioxide (d), and particulates (e). The fuel sulfur content was 500 ppm.

removal under lean conditions remains. However, significant advances are being made in the area of lean-NO_x removal, and these are discussed in Section X.C.

B. The Continuously Regenerating Trap

The inherent differences in the fuels and combustion processes in a spark-ignition gasoline-fueled engine and a diesel engine result in significant differences in the nature of their exhaust gases. Diesel exhaust always contains excess oxygen (very lean operation), and it contains a far higher amount of particulate material, commonly referred to as soot. Over recent years there have been growing concerns about the soot itself and its soluble organic fraction, SOF. These are condensible hydrocarbons originating from the fuel and lubricating oil that dissolve in light organic solvents. SOF is known to contain undesirable polyaromatic hydrocarbons, and the effects of oxidation catalysts on the toxicity of diesel exhaust relating to these components is well known [20]. In principle, SOFs can be adsorbed on the surface of an oxidation catalyst, where they are destroyed. The oxidation of polyaromatic hydrocarbons is clearly beneficial, but an oxidation catalyst has little effect on the particulate material.

Filtration techniques have been widely explored to remove particulates from diesel exhaust, but, until recently, removing the collected material from, for example, a ceramic wall filter has caused problems associated with the high ignition temperature of soot (about 600°C). In some situations the additional temperature rise due to soot oxidation can be sufficient to actually melt the ceramic!

A new approach is to combust the accumulating soot at a lower temperature by reaction with nitrogen dioxide to give carbon dioxide. The nitrogen dioxide is generated from nitric oxide by passing the exhaust gas over a special oxidation catalyst, which is followed by a particulate trap. The process of soot trapping and soot destruction are continuous at temperatures above 250°C. The patented system based on this technology is referred to as a continuously regenerating trap, or CRT [21], and is shown schematically in Fig. 7. If high-sulfur fuel is used, the soot removal process is inhibited and the efficiency of the CRT is reduced. Therefore, operation with low-sulfur fuel is usually necessary. Already a large number of CRTs have been fitted to heavy-duty diesel vehicles in several countries, and successful demonstration of the CRT has been achieved over several million kilometers of operation.

IX. CATALYSTS FOR COMPRESSED NATURAL GAS ENGINES

There is interest in compressed natural gas (CNG), which is largely methane, as a "clean fuel" for passenger and heavy-duty vehicles. However, relative cost and ease of distribution and availability are additional factors to be considered in order for its use to become widespread. In contrast to U.S. legislation, in Europe methane is not excluded from the hydrocarbon emission measurement, and so it would appear that the European requirements

Figure 7 Simplified schematic diagram of a continuously regenerating trap (CRT). The platinum catalyst oxidizes hydrocarbons and carbon monoxide, and also nitric oxide to nitrogen dioxide, which is used to oxidize soot retained in the filter. In the illustration this is a ceramic-wall flow filter, which has alternate channels blocked at the front inlet and rear outlet faces.

are more demanding for CNG-fueled vehicles than in America, because methane is likely to be the main hydrocarbon pollutant. Methane, with no carbon—carbon bonds, is one of the most unreactive hydrocarbons, so it might be more difficult for CNG-fueled vehicles to meet the required limits, especially when HC and NO_x are not combined in the legislative emission limits. However, CNG vehicles generally give lower emissions, especially CO, than their gasoline-fueled counterparts [22]. Palladium-containing catalysts are usually good for methane removal, but platinum and rhodium also have good activity. So in common with three-way catalysts, a combination of some or all of these elements is normally incorporated into catalysts for CNG applications.

Figure 8 shows results from a CNG-fueled light-duty van, with an aged Pd/Rh catalyst in an underfloor position. This vehicle runs with retarded ignition, to increase exhaust gas temperature, over the ECE portion of the European cycle. The results show that it meets the presently proposed European 2000 standards when run over both the Euro II and Euro III test drive cycles, and for CO and NO_x gives results better than are currently outlined for Euro 2005 standards.

X. FUTURE TRENDS

Legislative emission requirements are continuing to tighten, as noted in Section II.B, and, in response, catalyst technology and engine management systems have been improved to meet them. These demands continue as even lower emissions are required, and since almost all of the HC, CO, and NO_x are removed by the catalyst when it is hot and operating stoichiometrically, the vast majority of emissions result from when the catalyst has not yet reached its operating temperature. The means of addressing what is called the "cold-

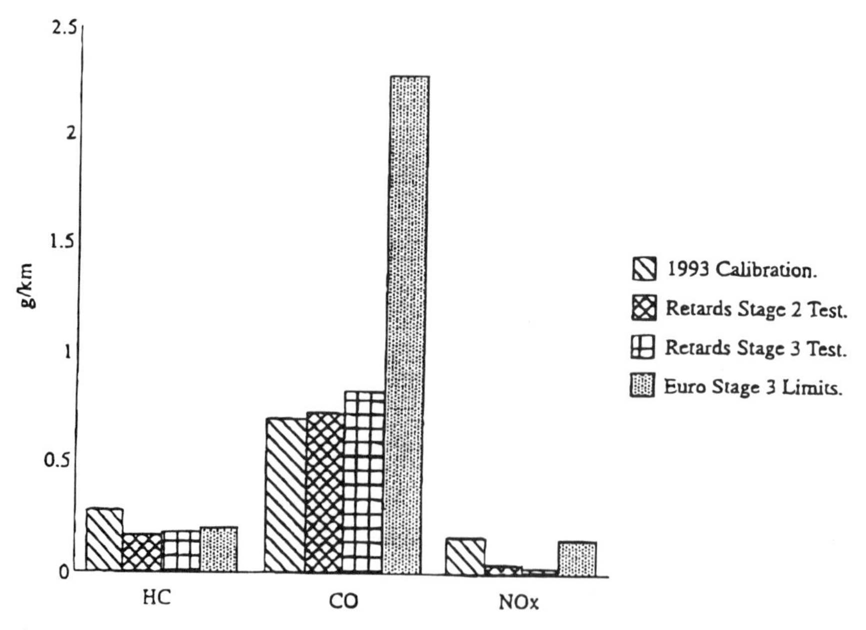

Figure 8 Performance of an optimized palladium/rhodium catalyst fitted on a natural-gas-fueled vehicle.

start problem" are discussed in the following subsections. Additional new requirements will also have to be met. For example, the performance of the catalyst in use will have to be monitored, and this will be done with two oxygen sensors in the exhaust gas, one before and one after the catalyst, as described in Section X.A.

There is also a growing requirement for reduced carbon dioxide emissions, corresponding to improved fuel economy. This could be achieved by smaller cars with smaller engines, and some further engine improvement, and in this context the use of lean-burn combustion strategies could provide some immediate fuel economies. However, lean-burn engine exhaust gas (such as from a diesel engine) is oxygen rich, making NO_x reduction chemically very difficult to achieve. The important topic of lean-NO_x control is covered in Section X.C.

A. On-Board Diagnostics

The objective of emission systems and emission standards is to reduce the incidence of pollutants from motor vehicles in the atmosphere. It is therefore important to know whether, in use, the system is operating to an acceptable level and whether all or part of the emission system needs replacement. In order for the monitoring systems to be effective, it is critically important that the measuring system itself be extremely stable and be affected only by the parameter it is set to measure. Thus, for U.S. legislation, the ideal system would be a hydrocarbon sensor; but presently available hydrocarbon sensors are not sufficiently stable and may be affected by variations in other components in the exhaust gas during normal operation. The most frequently used diagnostic system is the dual oxygen sensor, which measures the voltage generated from sensors in front of and behind the catalyst [23]. The difference between the two sensor outputs is intended to be a measure of stored oxygen available for oxidation of pollutants, e.g., HC. In order to improve the performance of three-way catalysts, materials that behave as oxygen storage components have, since the earliest days, been included in TWCs. A typical example is ceria: As described earlier, this works by storing oxygen from the gas phase during periods of lean operation and than releasing it when oxygen is deficient. This encourages oxidation reactions of HC and CO during these periods [24], and absorption of oxygen from slightly lean exhaust gas will enhance NO_x reduction. This oxygen storage/release affects the oxygen content measured by the rear sensor, and when correctly correlated with catalyst performance can provide a measure of catalyst deterioration. However, sensor voltages are not linear with respect to oxygen content, and these and other factors make on-board diagnostics of catalyst performance a complex process. The catalyst components are now required to be adjusted to meet the requirements, not only of catalyst performance, but of the system for measuring it.

B. Lower-Emission Requirements

From Table 2 it is seen that ultra-low-emission vehicles (ULEV) will have to meet 0.04 g nonmethane hydrocarbons per mile traveled compared to 0.25 g per mile in 1993. In contrast, the CO and NO_x emissions will only halve between 1993 and ULEV standards. While this does not mean that the CO and NO_x standards are "easily" achievable, it highlights the greater importance attributed to the control of HC by the regulatory bodies in the United States. Similarly, the European standards for hydrocarbons are promulgated to reduce progressively at each stage of the legislation. In addition, for the first time the

HC and NO_x limits will be separated rather than be combined as at present. It has long been recognized that the key to HC control at this level is the cold start before the catalyst reaches operating temperature. In particular, most of the hydrocarbon emissions (key to meeting the California standards in later years) arise from the first 40–60 sec of the FTP test cycle. Therefore, current research is aimed toward substantially reducing the emissions during the cold-start period. A number of approaches are being investigated, both in terms of the development of new catalyst technology and in the design of the system incorporating the catalyst.

1. *Electrically Heated Catalysts*

As an alternative to generating heat on the catalyst surface or from the engine, the heat can be supplied directly to the catalyst by an electric current. Systems incorporating this concept have been shown to achieve very low tailpipe emission standards [25–27]. In general these systems incorporate an electrically heated metal monolith with a catalyst coating, followed immediately by the main volume of catalyst. However, because the gas flow through the EHC is high, the power requirement is correspondingly high. Through development, the power required has been reduced from about 5 kW (at 200 A) to currently around 1.5–2 kW, with the target being less than 1 kW. The power is only required at start-up, but could cause a substantial drain on the battery.

2. *Catalyst Systems*

An alternative approach is to use only catalysts. A number of advanced TWCs have been developed with improved high-temperature durability, which enables catalysts to be located close to the engine, reducing catalyst warm-up time [28]. A significant advance was the development of high-loaded palladium-containing catalysts with superior performance compared to platinum–rhodium catalysts. These improved catalysts, are based on palladium alone or on palladium combined with rhodium or on a combination of all three metals. The total precious-metal loading and their ratio can be varied to give the required performance at optimum cost [29,30]. The catalyst position can also be varied. The most commonly considered locations are: underfloor only, starter with underfloor, and close-coupled only.

a. Underfloor Catalyst Figure 9 shows the benefits of high-loaded palladium catalysts compared with Pt/Rh catalyst in the underfloor position on a 1.2-liter European car in the Euro III drive cycle. The catalysts (1.66-liter volume) were aged on an engine equivalent to 80,000-km road durability. Hydrocarbon emissions were lower for the palladium-containing catalysts than for the Pt/Rh catalyst. Nevertheless, they all fail the hydrocarbon emission requirement, because the catalyst did not light off quickly enough (Fig. 10).

b. Starter with Underfloor If a small "starter catalyst" is situated closer to the engine than the underfloor catalyst, it will light off first, and the exotherm produced will enable the main catalyst to reach operating temperature sooner than would otherwise be the case. This was illustrated for the 1.2-liter-engine vehicle referred to in the previous section. A combination of a trimetal starter catalyst (0.6-liter 30 cm from the manifold) and the previous underfloor catalyst was tested. The results obtained are shown in Fig. 11 and Table 6. The starter alone had better performance than just the underfloor catalyst, and it met the HC and CO Euro 2000 requirements. While NO_x emission was reduced, it still exceeded the standard. Combining starter and underfloor catalyst gave substantially further improvement in all three pollutants, especially HC and NO_x, so that emissions met the requirements and approached the proposed 2005 limits. This excellent result was

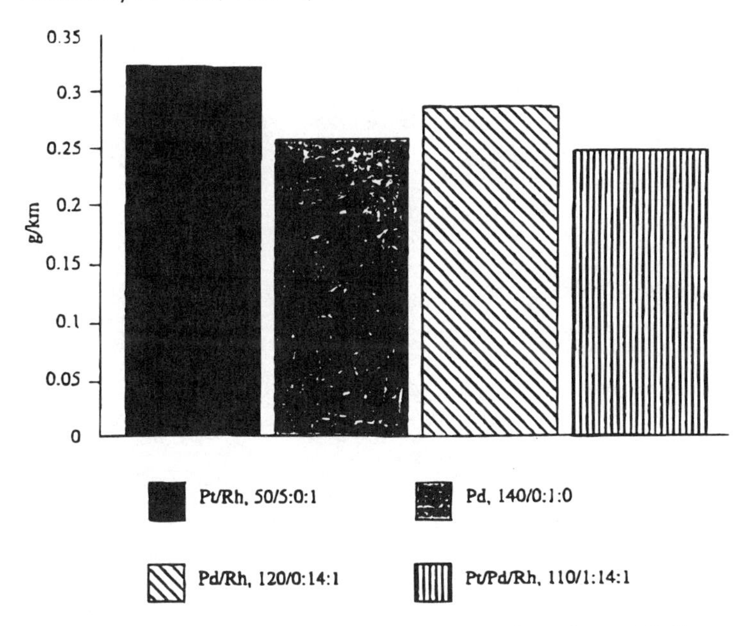

Figure 9 Hydrocarbon emissions in the ECE/EUDC test from a 1.2-liter-engine vehicle fitted with catalysts containing different platinum-group metals located under the floor.

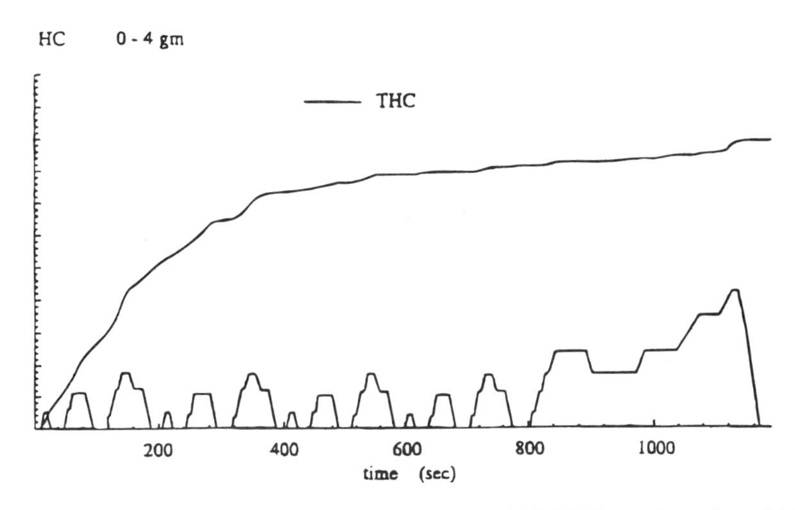

Figure 10 Cumulative THC emissions in the ECE/EUDC test from the vehicle used in Fig. 9 with an underfloor catalyst.

obtained by increasing the total catalyst volume. A test on the same vehicle with half the underfloor catalyst volume and the same starter catalyst gave similar results for HC and CO (Table 6). The NO_x performance, although still better than either the starter or full-sized underfloor alone, was not so good.

c. *Close-Coupled Catalysts* The effect of having all the catalyst close to the engine on the vehicle used previously is shown in Fig. 12. With 1.2 liters of catalyst in

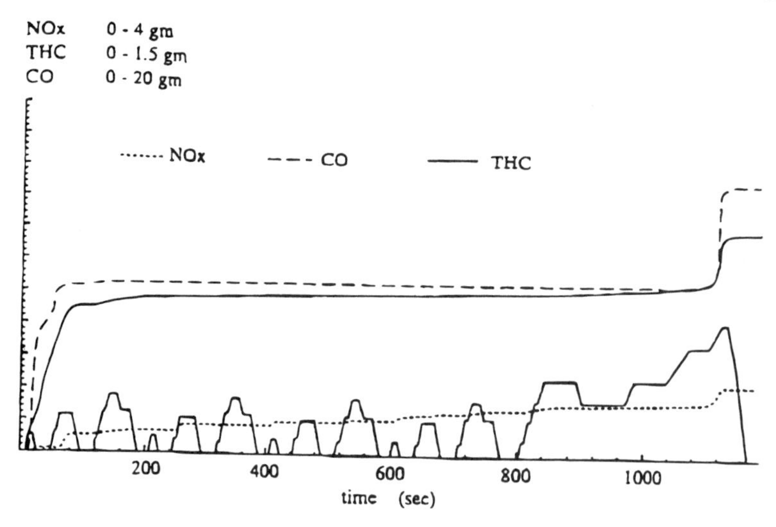

Figure 11 Cumulative THC, CO, and NO_x emissions in the ECE/EUDC test from the vehicle used in Fig. 9 fitted with starter and underfloor catalysts.

Table 6 Effect of Catalyst Volume and Exhaust System Location on Residual Emissions

	Emissions (g/km)		
	HC	CO	NO_x
Underfloor (1.66 liter)	0.246	1.996	0.470
Starter (0.60 liter)	0.117	1.600	0.237
Starter (0.60 liter) with underfloor (1.66 liter)	0.076	1.179	0.089
Starter (0.6 liter) with reduced underfloor (0.80 liter)	0.082	1.242	0.171

NOTE: A 1.2-liter European vehicle was tested over the modified ECE/EUDC drive cycle.

this position, the trends are the same as for catalysts underfloor: The palladium-containing catalysts have better hydrocarbon performance than does the Pt/Rh catalyst and, in spite of the low total catalyst volume, gave the best HC figures. They are well inside the suggested Euro year 2005 targets. The CO and NO_x figures generally meet Euro year 2000 targets, and in the case of Pd/Rh and trimetal catalysts almost meet Euro year 2005 standards on this car. Figure 13 shows the close-coupled catalyst lights-off earlier than in the underfloor case, and makes substantial inroads into the "cold-start" hydrocarbon emissions.

3. *Low-Light-Off Catalysts*

With catalysts that are very effective at low temperatures there is the possibility of locating them at position remote from the engine and avoiding excessively high temperature. The potential use of HC traps upstream of the catalyst requires the development of such technology. To achieve this objective, substantial enhancements in catalyst light-off must be achieved to produce a significant effect during an emission test. New catalysts under

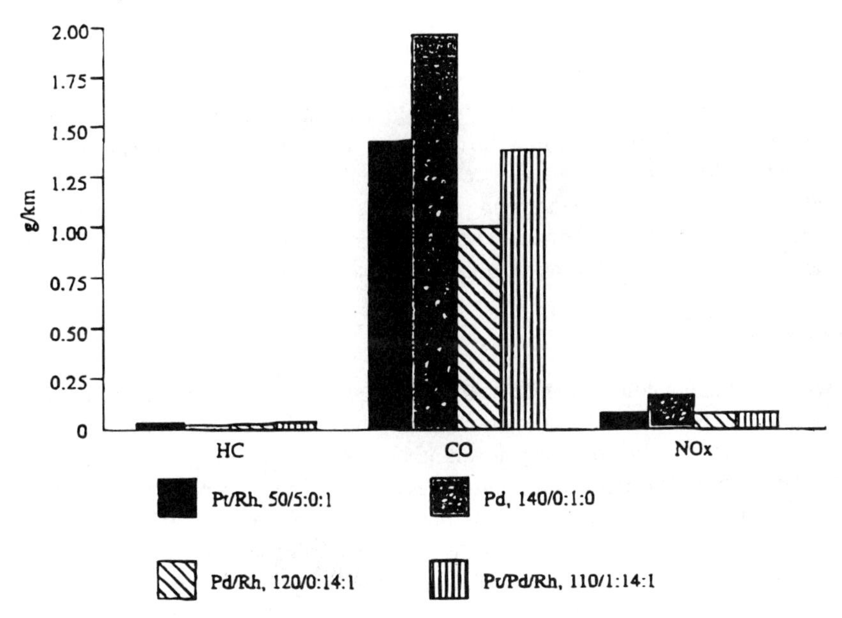

Figure 12 THC, CO, and NO_x emissions in the ECE/EUDC test from the vehicle used in Fig. 9 fitted with a close-coupled catalyst.

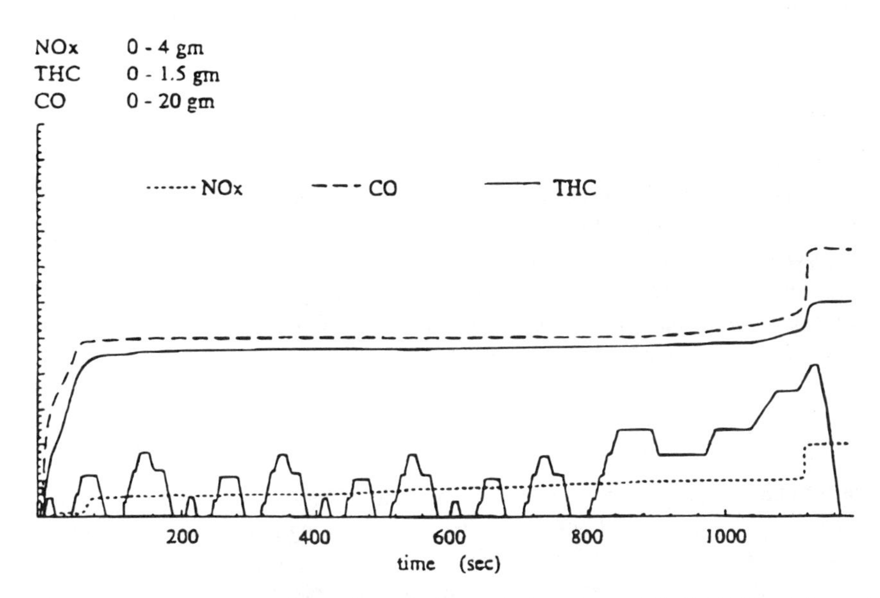

Figure 13 Cumulative THC, CO, and NO_x emissions in the ECE/EUDC test from the vehicle used in Fig. 9 fitted with a close-coupled catalyst.

development have demonstrated the potential of significantly lowering the light-off temperature in comparison to a conventional platinum/rhodium three-way catalyst. Installation of this catalyst on a vehicle has revealed promising advantages in cold-start control of HC and CO emissions. As a result of new formulations such as these, additional options now emerge whereby low-light-off catalysts could be used to improve effectiveness under the low-speed portion of the European driving cycle or could be utilized to help meet the TLEV and LEV standards in California. They may also be used in conjunction with electrically heated catalysts or HC traps to meet ULEV standards at the end of the decade.

4. *Hydrocarbon Traps*

A possible way of reducing cold-start hydrocarbon emissions is the use of a material that traps hydrocarbon species during the cold start, before the catalyst is lit off, then desorbs them when the catalyst has lit off. Early forms [31] of this system had two monoliths (bricks). The first is a hydrocarbon trap, and the second a TWC. To protect the trap, a bypass valve was incorporated so that it was not exposed to very hot exhaust gas. A potential problem is that the thermal mass of the trap delays light-off of the rear catalyst so the trap desorbs hydrocarbon before the TWC has reached working temperature. Also, in many smaller cars space constraints limit the use of extra bricks. A better option is to combine the adsorbing material and catalyst formulation on a single monolith to maximize the overlap of the desorption of stored hydrocarbon and catalyst light-off temperatures. A key factor is the stability of the trapping material; also important is the potential effect of the trapping material on the performance of the catalyst. Such a system is under development [32], and suitably stable trap materials have been identified that have good overlap of the desorption temperatures and TWC light-off.

5. *Burner-Assisted Warm-Up*

This system is conceptually similar to the EHC in that extra heat is provided from a separate source. During start-up, some fuel is injected and combusted in a burner in the exhaust system. The heat produced ensures that the exhaust gas has a sufficiently high temperature to enable the catalyst to light off, but the provision of combusted fuel is required only during the cold-start phase. Continuous burning would have a negative effect on fuel economy. Again the question of valving and control systems is key to this strategy as well as to overall safety considerations [33,34].

C. Operation Under Lean Conditions

Although most of this chapter is concerned with conventional four-stroke gasoline engines operating under stoichiometric conditions, there is an accelerating interest in the development of engines that run part or all the time at air–fuel ratios substantially lean of stoichiometry. Vehicles equipped with such engines (for example, diesel, lean-burn, two- and four-stroke gasoline, and CNG-fueled engines) can have improved fuel economy and hence lower CO_2 emissions, but they must meet regulations for HC, CO, NO_x, and particulate emissions. Emission control catalysts developed for stoichiometric control are not appropriate for lean-burn engines; in particular, control of NO_x presents a different set of problems and requires different types of catalyst technologies from those used for stoichiometric TWC operation.

1. NO_x Control Under Lean Conditions

Control of HC and CO under lean operating conditions should be straightforward, but the reduction of NO_x under these strongly oxidizing conditions is not. Nevertheless, especially in Europe, NO_x emissions are a major concern, and legislative proposals now being discussed will require the removal of some NO_x from lean-burn engines. So far, three approaches have been tried; two of these have met with some level of success.

a. Direct Decomposition of NO_x Since Iwamoto [35] showed that copper-exchanged ZSM-5 zeolite had stable steady-state activity for NO decomposition selectively to dinitrogen, much work has been done on these systems to make them work at high space velocities and the low NO_x levels in the presence of steam encountered in automotive applications. The influence of the degree of copper exchange in the zeolite, excess oxygen levels, and the presence of sulfur oxides and water contribute to the poor conversion achieved in real applications. As a result, direct decomposition of NO_x is not yet a viable approach.

b. NO_x Reduction Under Net Oxidizing Conditions. For many years NO_x emissions from power stations and chemical operations have been controlled by the so-called SCR (selective catalytic reduction) process, using ammonia as the reducing agent over a catalyst, as illustrated in Eqs. (10) and (11):

$$4NH_3 + 4NO + O_2 \rightarrow 4N_2 + 6H_2O \tag{10}$$

$$4NH_3 + 2NO + O_2 \rightarrow 3N_2 + 6H_2O \tag{11}$$

Held and co-workers at Volkswagen [36] showed that significant levels of NO_x could be reduced under lean automotive conditions by reductants (for example, urea, hydrocarbons) over a Cu-ZSM-5 catalyst. Many such copper catalysts have been screened, but durability is a problem. Two systems have been thoroughly researched and refined: platinum on modified alumina or zeolite for low-temperature operation (150–250°C), and Cu-ZSM-5 for higher temperatures (300–450°C). These temperature ranges are quite narrow compared with the operating window of a three-way catalyst. The lower limit is determined by catalyst activity, and the upper one by competing direct oxidation of the reductant. Nevertheless, they fall into a suitable range for use in diesel applications, which form a major part of the current lean-burn vehicle fleet. Ideally, since engine emissions contain species (HC, CO) capable of reducing NO_x, these could be used to achieve the removal of NO_x. However, in many cases the level of hydrocarbon in the exhaust gas is not sufficient to achieve the necessary level of NO_x control. Alternatively, small amounts of added hydrocarbon, most conveniently the fuel used in the engine, can be injected into the exhaust gas prior to the catalyst to effect the necessary reaction with the NO_x.

Figure 14 shows the NO_x performance of three generations of platinum-based lean-NO_x catalysts on a 1.9-liter TDI diesel bench engine at a space velocity of 45,000 hr^{-1}. Before testing, catalysts were conditioned (1 hr at 300°, 400°, and finally 500°C), or aged in four cycles of a 13-hr procedure with a maximum temperature (750°C) for 5 hr. Figure 14 shows that NO_x conversion is 20–30%; and if low levels of fuel are injected into the exhaust gas, some 20% additional NO_x conversion results (Fig. 15). When the catalyst was evaluated on a similar 1.9-liter TDI diesel and run over the European test cycle, the catalyst inlet temperature was in the range 130–200°C for most of the ECE cycle but rose to 250–350°C during the EUDC section. Thus, during parts of the test cycle the catalyst temperature was outside the optimum operating temperature range for both the platinum and the copper catalysts. Therefore, the challenge is to develop catalysts with wider

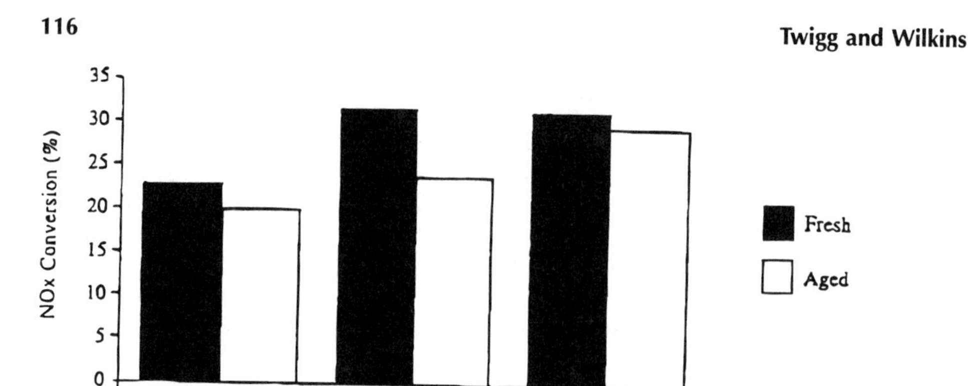

Figure 14 Optimal passive NO_x performance (without secondary fuel injection) for fresh and aged catalysts on a 1.9-liter TDI diesel bench engine.

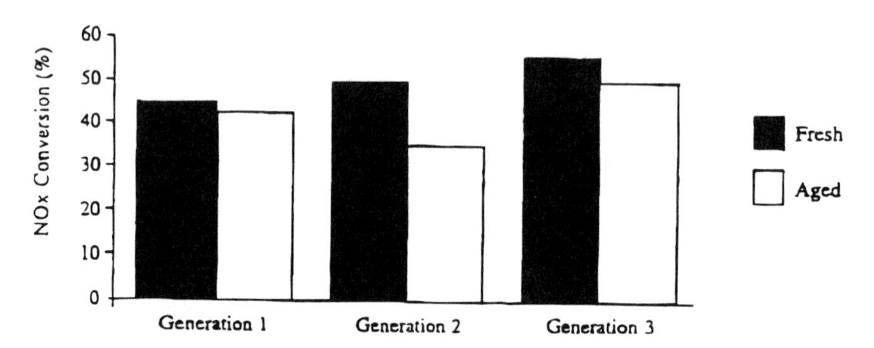

Figure 15 Optimal active NO_x performance (with secondary fuel injection) for fresh and aged catalysts on a 1.9-liter TDI diesel bench engine.

effective operating temperature ranges while maintaining or improving selectivity toward NO_x reduction.

c. NO_x Storage and Release. A further means of reducing NO_x in high-oxygen exhaust gas is to incorporate NO_x-adsorbent materials into catalyst formulations. During use, NO_x is absorbed and then, when appropriate, reduced under controlled conditions. An approach has been to oxidize nitric oxide catalytically to NO_2, which is then taken up as a nitrate, for example, by reaction with a suitable basic oxide. If the exhaust gas is then switched to reducing for a short period, the stored nitrate becomes unstable at low temperatures and releases NO and oxygen [37–39]. If the catalyst also contains a component for reducing the released NO, such as rhodium, the tailpipe NO_x will be dramatically lowered. The efficiency of this system can be high but is very dependent on several factors. For example, the capacity of the NO_2-adsorbing component, the rate at which the NO_x can be adsorbed, and the rate at which NO is desorbed when the system is switched to reducing conditions are key parameters. However, the first step is oxidation of NO to NO_2. Platinum is an efficient catalyst for this, and temperature-programed desorption measurements have confirmed that adsorbent capacity is substantially higher for NO_2 than for NO. Under the richer condition, the released NO can react with, for example, CO over rhodium catalyst, as shown in Eq. (13):

$$MNO_3 \rightarrow NO + \tfrac{1}{2}O_2 + MO \quad (12)$$

$$NO + CO \rightarrow CO_2 + \tfrac{1}{2}N_2 \quad (13)$$

A major advantage of the lean-burn engine over a stoichiometric engine is the potential of increased fuel economy and lower CO_2 emissions. Clearly, making the exhaust gas richer, to regenerate stored NO_x, will offset some of the fuel economy gain, and so it is important to minimize the time of regeneration and to maximize the period spent lean. Taking the stoichiometry during the regeneration phase to the stoichiometric position, as shown in Fig. 16, results in NO_x breakthrough, implying that the rate of release of NO from the adsorbent is faster than the subsequent reduction. However, if the mixture is richened still further, NO_x breakthrough is lessened. Furthermore, as shown in Fig. 17, the regeneration time required decreases markedly as the regeneration becomes richer, and so these parameters can be optimized. Two factors still need resolution: The first is that most of the good NO_x adsorption components also store SO_x, yielding sulfates that are often substantially more stable than the corresponding nitrates. Consequently, the amount of adsorbent available for NO_x adsorption progressively reduces if it is sulfated. Second, there is the possibility of thermal deactivation of the adsorption component through long-term use. Both of these factors are critical for successful implementation, and both are being actively developed.

XI. CONCLUSIONS

Autocatalysts have made dramatic contributions to reducing tailpipe HC, CO, and NO_x emissions from vehicles powered by internal combustion engines. Catalyst systems are now available that convert more than 95% of each pollutant from the exhaust gas of gasoline engines operating under stoichiometric conditions. These achievements resulted from catalyst developments and the integration of advanced catalysts with the electronic control of the engine such that the total system minimizes engine emissions and optimizes aftertreatment effectiveness. Diagnostic measurements of catalyst performance can now be made during the life of the vehicle and indicate malfunction.

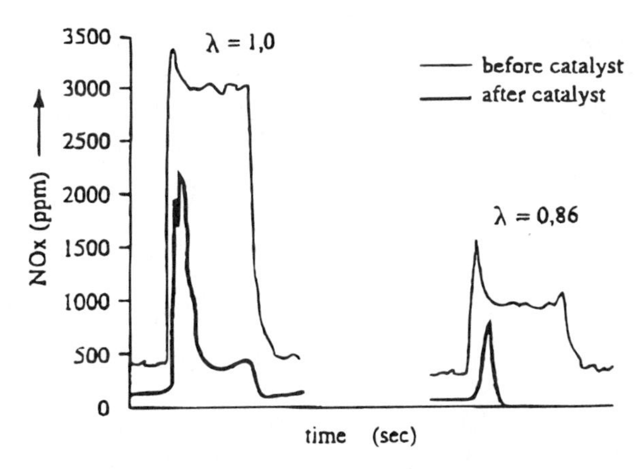

Figure 16 Effect of regeneration conditions (reduction potential) on the performance of an experimental NO_x trap.

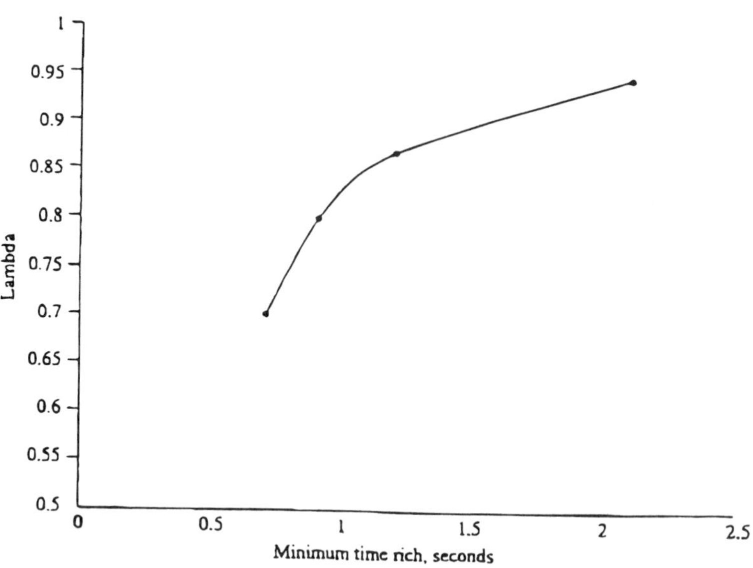

Figure 17 Effect of air/fuel ratio on the regeneration time required for an experimental NO_x trap operating at a space velocity of 30,000 hr^{-1}.

The situation does not remain static, however, and the future has many challenges for autocatalysts. Tightening legislation has dictated that even the emissions during start-up, that are not converted by the catalyst because it is too cold, have to be addressed by techniques such as low-light-off catalysts in combination with heat management, preheated catalysts, hydrocarbon traps, and the like. Progress is being made in the development of emission control catalysts to meet these challenges from tightening emission legislation, and will lead to new developments in catalysts throughout the rest of this decade. It is likely that palladium and rhodium will be the primary ingredients in future autocatalysts and that palladium will continue to emerge as an important contributor for the control of hydrocarbon emissions.

New horizons are being opened for NO_x control under lean operation for lean-burn diesel engines and gasoline engines. Both copper and platinum-group metal catalysts are under development, and significant issues with respect to the durability of such catalyst systems remain to be resolved. Other approaches include the storage of NO_x as nitrate followed by its periodic reduction under optimal conditions. These developments represent some of the most challenging and intriguing possibilities for the catalyst designer in the future.

REFERENCES

1. A.J. Haagen-Smit and M.M. Fox. *Ind. Eng. Chem. 48*:1484 (1956).
2. K.C. Taylor. *Automobile Catalytic Converters*. Springer-Verlag, Berlin, 1984, pp. 24–25.
3. A.C. Stern. *J. Air Pollution Control Assoc. 32*:44 (1982); T.C. Austin, R.H. Cross, and P. Helnen. *SAE Paper* 811231 (1981).
4. M.S. Spencer. In: M.V. Twigg, ed., *Catalyst Handbook*. Manson, London, 1996, pp. 38–48.

5. M.L. Church, B.J. Cooper, and P.J. Wilson. *SAE Paper* 890815 (1989).
6. *Platinum—Interim Review.* Johnson Matthey, London, November 1996.
7. J.S. Hepburn, K.S. Patel, M.G. Meneghel, and H.S. Gandhi. *SAE Paper* 941058 (1994).
8. B. Harrision, A.F. Diwell, and C. Hallett. *Platinum Metals Rev. 32*:73 (1988).
9. R.J. Brisley, G.R. Chandler, H.R. Jones, P.J. Anderson, and P.J. Shady. *SAE Paper* 950259 (1995).
10. J.T. Richardson. *Principles of Catalyst Development.* Plenum Press, New York, 1989, pp. 104–106.
11. D. Kattge. *SAE Paper* 880284 (1988).
12. G.J.J. Bartley, P.J. Shady, M.J. D'Aniello, G.R. Chandler, R.J. Brisley, and D.E. Webster. *SAE Paper* 930076, (1993); Z. Hu and R.M. Heck. *SAE Paper* 950254 (1995).
13. P. Beckwith, P.J. Bennett, C.L. Goodfellow, R.J. Brisley, and A.J.J. Wilkins. *SAE Paper* 940310 (1994).
14. P.J. Bennett, P. Beckwith, S.D. Bjordal, C.L. Goodfellow, R.J. Brisley, and A.J.J. Wilkins. *ISATA Paper* 96EN043 (1996).
15. Y. Niura and K. Ohkubo. *SAE Paper* 85220 (1985).
16. P.S. Brett, A.L. Neville, W.H. Preston, and J. Williamson. *SAE Paper* 890490 (1989).
17. T.J. Truex, H. Windawi, and P.C. Ellgen. *SAE Paper* 872162 (1987); R.S. Petrow, G.T. Quinlan, and T.J. Truex. *SAE Paper* 890797 (1989).
18. A.F. Diwell, S.E. Golunski, and T.J. Truex. A. Crucq, Ed., *Catalysis and Automotive Pollution Control II.* Elsevier, Amsterdam, 1991, p. 417.
19. J. Jochheim, D. Hesse, T. Duesterdiek, W. Engeler, D. Neyer, J.P. Warren, A.J.J. Wilkins, and M.V. Twigg. *SAE Paper* 962042 (1996).
20. K.F. Hansen, F. Bak, M. Andersen, H. Bejder, and H. Autrup. *SAE Paper* 940241 (1994); B.T. McClure, S.T. Bagley, and L.D. Gratz. *SAE Paper* 920834 (1992).
21. P.N. Hawker. *Platinum Metals Rev. 39*:2 (1995); B.J. Cooper and S.A. Roth. *Platinum Metals Rev. 35*:178 (1995); B.J. Cooper and J.E. Thoss. *SAE Paper* 890404 (1989); B.J. Cooper, H.J. Jung, and J.E. Thoss. U.S. patent 4,902,478; European patent 341,832.
22. N. Fricker, H.E. Janikowski, and G.P. Stover. *The Institution of Gas Engineers 57th Autumn Meeting*, Harrogate, Communication 1474, 1991.
23. W. Clemmens, M. Sabourin, and T. Rao. *SAE Paper* 900062 (1990).
24. J.S. Hepburn and H.S. Gandhi. *SAE Paper* 920831 (1992).
25. J.E. Kubsch. *SAE Paper* 941996 (1994).
26. H. Mizuno, F. Abe, S. Hashimoto, and T. Kondo. *SAE Paper* 940466 (1994).
27. K.P. Reddy, S.T. Gulati, D.W. Lambert, P.S. Schmidt, and D.S. Weiss. *SAE Paper* 940782 (1994).
28. G.J.J. Bartley, P.J. Shady, M.J. D'Aniello, G.R. Chandler, R.J. Brisley, and D.E. Webster. *SAE Paper* 930076 (1993).
29. R.J. Brisley, G.R. Chandler, H.R. Jones, P.J. Anderson, and P.J. Shady. *SAE Paper* 950259 (1995).
30. A. Punke, U. Dahle, J.J. Tauster, H.N. Rabinowitz, and T. Yamada. *SAE Paper* 950255 (1995).
31. J.K. Hochmuth, P.L. Burk, C. Tolentino, and M.J. Mignano. *SAE Paper* 930739 (1993).
32. R.J. Brisley, N.R. Collins, and D. Law. European patent application 95308884 (1995).
33. P. Oeser, E. Mueller, G.R. Haertel, and A.O. Schuerfeld. *SAE Paper* 940474 (1994).
34. J.S. Hepburn, A.A. Adamczyk, and R.A. Pawlowicz. *SAE Paper* 942072 (1994).
35. M. Iwamoto, M. Yokoo, K. Sasaki, and S. Kagawa. *J. Chemical. Soc. Farad. Trans. 77*:1629 (1981).
36. W. Held, A. Koenig, T. Richter, and L. Puppe. *SAE Paper* 900496 (1990).

37. N. Mijoshi, S. Matsumoto, K. Katoh, T. Tanaka, J. Harada, and N. Takahara. *SAE Paper* 950809 (1995).
38. W. Bögner, M. Krämer, B. Krutsch, S. Pischinger, D. Voigtländer, G. Wenninger, F. Wirbeit, M.S. Brogan, R.J. Brisley, and D.E. Webster. *J. Appl. Catal. B*:153 (1995).
39. M.S. Brogan, R.J. Brisley, A.P. Walker, D.E. Webster, W. Bögner, N.P. Fekete, M. Krämer, B. Krutzsch, and D. Voigtländer. *SAE Paper* 952490 (1995).

5

Monolithic Catalysts for the Selective Reduction of NO_x with NH_3 from Stationary Sources

Alessandra Beretta, Enrico Tronconi, Gianpiero Groppi, and Pio Forzatti
Politecnico di Milano, Milano, Italy

I. INTRODUCTION

A. Background

Emission of nitrogen oxides from stationary sources, primarily from thermal power plants, is regulated differently in the various industrialized countries. National NO_x emission limits in Japan correspond to 60, 130, and 200 ppm for new large gas-fired, oil-fired, and coal-fired power plants, respectively. In Europe, representative NO_x emission limits from Austria, Belgium, and Germany are 50, 75, and 100 ppm for the cases of gas, oil, and coal-power plants producing more than 300 MWth; less stringent limits are set for existing plants, ranging from 100 to 300 MWth. In the United States the federal emission limits are less severe. However, local NO_x emission standards have been introduced in California and in states on the East Coast; they include very high levels of NO_x abatement, with the tolerated limits of emission for gas turbines and combined cycle plants burning natural gas being 5–9 ppm.

The control of NO_x emissions includes techniques for modifying the combustion stage (primary measures) and for treating the effluent gases (secondary measures). The primary measures, which are extensively applied, guarantee NO_x reduction levels on the order of 50–60%, which may not fit the most stringent limits. Among the secondary measures, one well-established technology is the selective catalytic reduction (SCR) process, which is applied worldwide due to its efficiency and selectivity [1–3]. The SCR reaction consists of the reduction of NO_x to innocuous water and nitrogen, by means of NH_3 (either liquified or in aqueous solution) as reducing agent; a precursor of ammonia such as urea can also be used. The catalyst is used in form of monoliths, with geometry and design specific for the different applications. Developed and first applied in Japan during the late 1970s, the SCR process has undergone a wide diffusion in Europe, beginning in 1985. Selective catalytic reduction systems are currently operating in several European countries, including Austria, Denmark, France, Germany, Italy, Luxemburg, The Netherlands, Sweden, and Switzerland, for a total capacity of about 60,000 MWe [4–7]. This technology presently accounts for 90–95% of the $DeNO_x$ flue gas treatment in Europe and Japan. The United States, where the present SCR applications are confined to gas turbines (for a total of 5000 MWe), has a wide potential for the diffusion of SCR systems, due to the future introduction of more stringent local regulations about NO_x emissions.

In this chapter, an overview of the key factors that affect the operation of the SCR catalysts is presented. Data are given concerning the characteristics of commercial monolithic catalysts. The chemical and mass transfer phenomena that control the SCR reaction are addressed, together with steady-state modeling of the monolithic reactor. It is also shown that the study of the effect of the morphological and geometrical properties of the catalyst, the effect of feed composition, and the effect of the interaction between the $DeNO_x$ reaction and SO_2 oxidation offers space for improving both the catalyst and the reactor design. Finally, the main aspects connected to dynamic operation of the SCR catalysts are presented.

B. SCR Reaction and Catalysts

The SCR process consists of the reduction of NO_x (typically 95% NO and 5% NO_2 v/v) with NH_3. The reaction stoichiometry is usually represented as: $4NO + 4NH_3 + O_2 \rightarrow 4N_2 + 6H_2O$. This reaction is selectively effected by the catalyst, since the direct oxidation of ammonia by oxygen is prevented. In the case of the treatment of sulfur-containing gas streams, the $DeNO_x$ reaction is accompanied by the catalytic oxidation of SO_2 to SO_3. Oxidation of SO_2 is highly undesirable because SO_3 is known to react with water and residual ammonia to form ammonium sulfates, which can damage the process equipment.

Metal-oxide-based systems are the most widely used SCR catalysts. They have largely replaced the previous noble-metal catalysts employed in the early SCR installations, which are more active but less selective and less tolerant to poisoning. The commercial metal oxide catalysts consist of homogeneous mixtures of titanium dioxide, tungsten trioxide (or molybdenum trioxide), and divanadium pentoxide. TiO_2 in the anatase form is used as high-surface-area carrier ($\approx$80% w/w) to support the active components. Vanadium oxide is responsible for the activity of the catalyst in the reduction of NO_x and for the undesired oxidation of SO_2 to SO_3 as well. Accordingly, the V_2O_5 content is usually low (less than 1% w/w in high-sulfur applications). WO_3 is employed in larger amounts ($\approx$10% w/w) in order to increase the acidity of the catalyst; porous tungsten oxide also imparts superior thermal stability to the catalyst. Silico-aluminates and glass fibers are used as ceramic additives to improve the catalyst strength [8–11].

The recent introduction of zeolite catalysts in SCR applications (gas-fired cogeneration plants) has to be mentioned. Iron- or copper-containing zeolites guarantee high DeNOxing performances up to temperatures of 600°C, where metal oxides become thermally unstable [12,13]. The use of zeolite-based NO_x reduction catalysts with distinct structures has been covered in the patent literature, namely, mordenite, clinoptilotite, faujasite (both types X and Y), and pentasil [14,15].

Special catalysts have also been studied and developed for low-temperature and clean-gas applications. They comprehend both metal-oxide-based and noble-metal-based catalysts and apply novel design concepts, as presented in a subsequent paragraph.

Several proposals have been advanced in the literature concerning the mechanism of the SCR reaction over vanadia-based catalysts. They converge in suggesting that the

$DeNO_x$ reaction involves a strongly adsorbed NH_3 species and a gaseous or weakly adsorbed NO species, but differ in their identification of the nature of the adsorbed reactive ammonia (protonated ammonia vs. molecularly coordinated ammonia), of the active sites (Brønsted vs. Lewis sites) and of the associated reaction intermediates [16,17]. Concerning the mechanism of SO_2 oxidation over DeNOxing catalysts, few systematic studies have been reported up to now. Svachula et al. [18] have proposed a redox reaction mechanism based on the assumption of surface vanadyl sulfates as the active sites, in line with the consolidated picture of active sites in commercial sulfuric acid catalysts [19]. Such a mechanism can explain the observed effects of operating conditions, feed composition, and catalyst design parameters on the $SO_2 \rightarrow SO_3$ reaction over metal-oxide-based SCR catalysts.

C. Process Configurations

Figure 1 shows the three possible locations of the SCR unit: immediately after the boiler (high-dust arrangement, HD), after the particulate removal (low-dust arrangement, LD), and after the sulfur dioxide removal (tail-end arrangement, TE) [20].

The HD configuration is used mostly in coal-fired applications because the temperature of the flue gas between the economizer and the air preheater is optimal for the catalyst activity ($\approx$300–400°C) and because dust removal is usually accomplished by means of cold electrostatic precipitators. Ammonia slip must be kept at low levels (<5 ppm or even below 2–3 ppm), and SO_2 oxidation must be as low as 0.5–1.0% in order to minimize the formation and deposition of ammonium sulfates in the heat exchanger and in the fly ash.

The LD configuration typically is applied in the plants where the residual dust content and the resistivity of the dust are compatible with the use of hot electrostatic precipitators. One of the advantages of this system is the reduced deterioration of the monolithic catalyst.

The TE configuration is used in the case of "clean" applications, where most of the dust and SO_2 has been removed from the flue gas. Due to the reduced concern for oxidation of SO_2, higher vanadia contents can be used in TE arrangements than in HD or LD arrangements. This results in a smaller catalyst volume and in a slower deterioration rate of the catalyst. Besides, this avoids any plugging of the air preheater and NH_3 contamination of both fly ash and desulfurization wastewater. However, the costs associated with the reheating of the flue gas and with the use of a gas–gas heat exchanger might outweigh the savings from the simplified reactor design and from reduced catalyst volume and maintenance.

There is a greater operating experience in HD plants [3,21–27] than in LD and TE plants, which are dedicated to specific applications. Typically, design specifications for HD systems include 80–85% NO_x reduction efficiency (with $\alpha = NH_3/NO$ feed ratios of 0.8–0.85) and ammonia slip lower than 3–5 ppm.

The activity of the catalysts decreases with time, depending on the flue gas conditions to which they are exposed. The main causes of catalyst deterioration are:

Sintering and rutilation of titania after very long-term operation in gas firing
Poisoning of the catalyst active sites by alkali metals in oil firing
Plugging of catalyst pores by calcium compounds in coal firing
Poisoning of the catalyst by arsenic in the case of wet-bottom boilers
Accumulation of phosphorous components in lubricating oil in the case of diesel engines

High Dust (HD) configuration

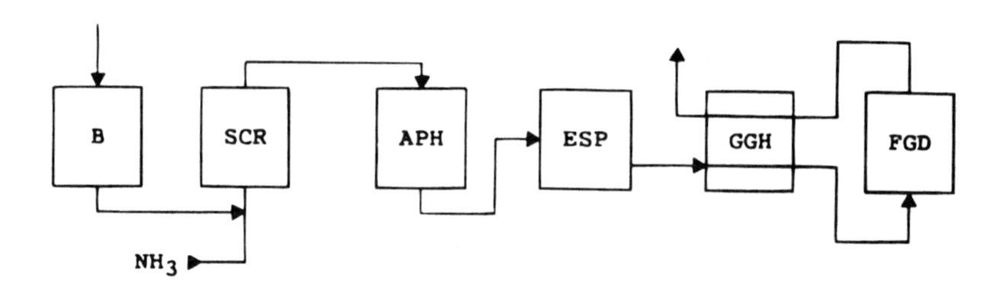

Low Dust (LD) configuration

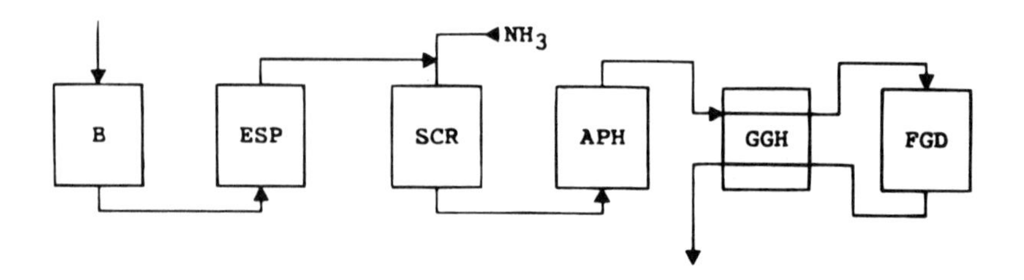

Tail End (TE) configuration

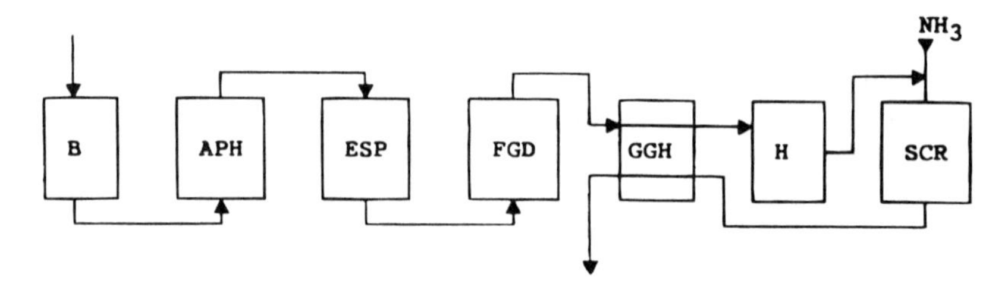

Figure 1 SCR process configurations. FGD: flue gas desulfurization; B: boiler; APH: air preheater; ESP: electrostatic precipitator; GGH: gas–gas heat exchanger; H: heater. (From Ref. 20.)

Other possible causes of deactivation include:

Incrusting and plugging of the surface (e.g., deposition of fly ash)
Deposition of ammonium sulfates within the catalyst pores
Erosion

In spite of all these possible deactivation mechanisms, longtime stable activity is always reported for SCR catalysts: 16,000 and 24,000 hours of operation are typically guaranteed by catalyst suppliers for HD and TE systems, respectively, but even longer catalyst lives are observed in practice. Besides, economy is significantly influenced by optimized strategies for the addition of extra catalytic material and the replacement of spent catalyst.

II. COMMERCIAL MONOLITH-SHAPED SCR CATALYSTS

The original SCR catalysts were offered in the form of pellets or spheres and used in clean or low-dust applications. They were soon replaced by parallel-flow catalysts, namely, honeycomb monoliths, plates, and coated metal monoliths, as shown in Fig. 2. In the case of the SCR process, the main advantages of monolith-shaped catalysts with respect to conventional packed beds are as follows:

The flow through the straight channels of the monolith matrix is subject to very low pressure drops; in a packed bed with the same external geometric surface area the pressure drop may be greater by two or three orders of magnitude.

The monolithic supports typically have higher geometric surface areas than do packed beds; this is specifically advantageous for the SCR process. Due to its high rates, the $DeNO_x$ reaction suffers from strong intraporous diffusional limitations and is confined only to a thin outer layer of catalyst, so that the NO_x reduction efficiency is controlled by the catalyst geometric area rather than by its volume.

Monolithic structures are attrition resistant and show a low tendency to fly-ash plugging, which is most important in the treatment of flue gases.

A. Honeycomb Catalysts

Selective catalytic reduction honeycomb monoliths (typically characterized by a square channel section) are obtained by extruding a mass of pastelike catalytic material. The ele-

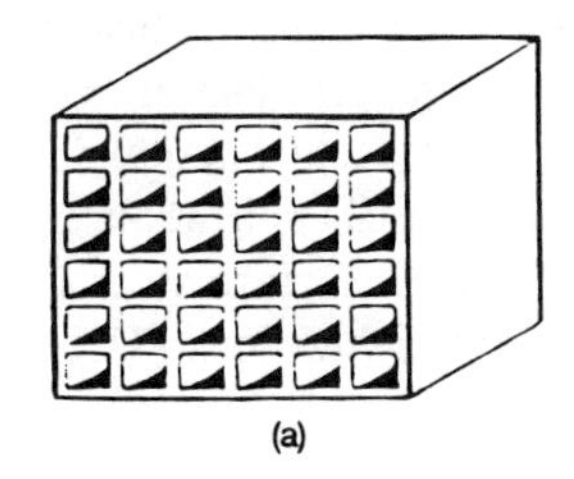

(a)

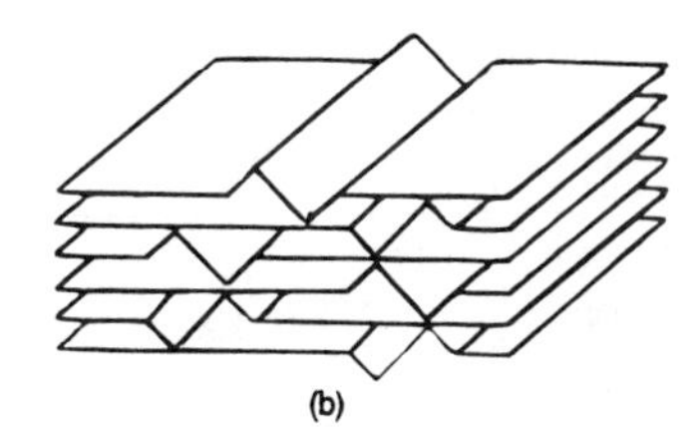

(b)

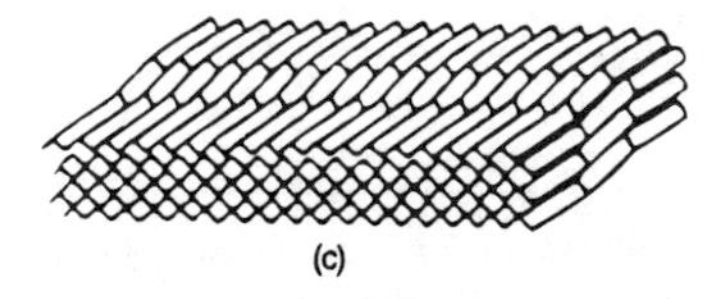

(c)

Figure 2 Types of monolithic SCR catalysts: (a) honeycomb monolith; (b) plate-type catalyst; (c) coated metal monolith.

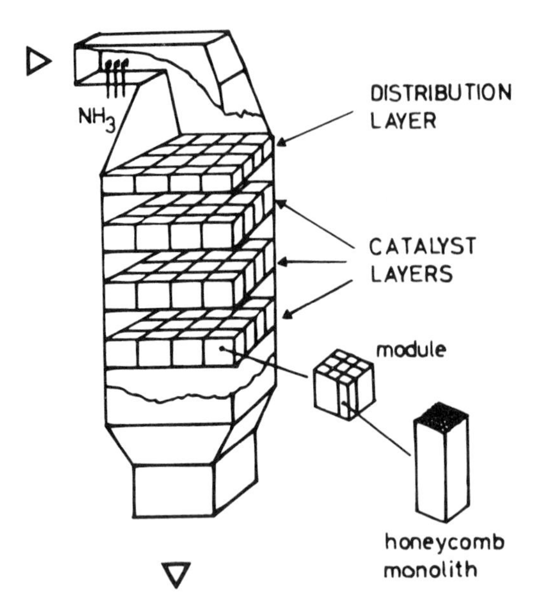

Figure 3 Structure of SCR monolith reactor.

ments usually have dimensions 150 mm × 150 mm × (350–1000) mm, and they are assembled into standard steel-cased modules, which are then placed inside the reactor in form of layers, as shown in Fig. 3. Honeycomb catalysts are used in HD, LD, and TE applications; the geometrical characteristics of the monolith are specific to the process configuration. Typical data of commercial honeycomb catalysts are listed in Table 1. Note that in order to minimize erosion phenomena and catalyst plugging, the monoliths for HD applications have larger channel openings and wall thicknesses, and thus lower geometric surface areas, than do monoliths used in low-dust environments. In the case of clean applications (TE), the

Table 1 Typical Data for Honeycomb Catalysts

	Catalyst Type		
	HD	LD	TE
Element size, mm	150 × 150	150 × 150	150 × 150
Length, mm	500–1000	500–1000	500–1000
Number of cells	20 × 20	25 × 25	40 × 40
Cell density, cells/cm^2	1.8	2.8	7.1
Wall thickness (inner), mm	1.4	1.2	0.7
Pitch, mm	7.4	5.9	3.7
Geometric surface area, m^2/m^3	430	520	860
Void fraction	0.66	0.63	0.66
Specific pressure drop, hPa/m	2.3	3.7	8.2

Note: HD = high-dust configuration, LD = low-dust configuration, TE = tail-end configuration. Pressure drops are evaluated at gas linear velocity of 5 m/s. Based on product information by BASF, Frauenthal, Hüls, Mitsubishi, and Siemens.

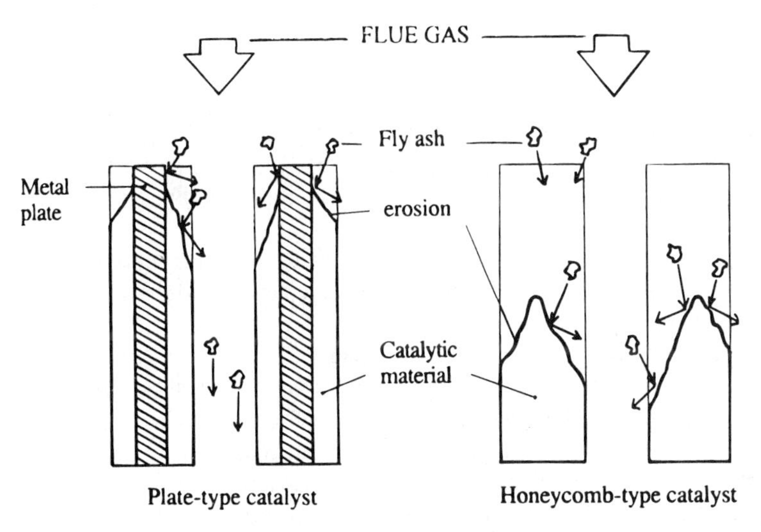

Figure 4 Mechanism of erosion in plate-type and honeycomb catalysts.

absence of dust makes possible the use of monoliths with very small openings and thin walls, which results in extremely large geometric surface areas.

B. Plate-Type Catalysts

Plate-type catalysts are obtained by depositing the catalytic material onto a stainless steel net or perforated metal plate. As in the case of honeycomb matrices, the plates are assembled in modules and inserted into the reactor in layers. They are preferentially used for high-dust and high-sulfur applications, as in coal-fired power plants. In fact: (1) with respect to honeycombs, plate-type monoliths are less prone to blockage owing to their structure, which permits slight vibration of the individual plates; (2) the metal support makes the plates more resistant to erosion than the all-ceramic materials; as shown in Figure 4, as soon as the inlet section of the channel exposes the metal sheet, erosion proceeds no further; (3) the plates are very thin, so only a small portion of the cross section is obstructed and pressure drops are very low. Table 2 reports typical geometric data for commercial plate-type catalysts.

Table 2 Typical Data for Plate-Type Catalysts

Element size, mm	500 × 500
Length, mm	500–600
Wall thickness (inner), mm	1.2–0.8
Pitch, mm	6.9–3.8
Cell density, cells/cm^2	2–6.5
Geometric surface area, m^2/m^3	285–500
Void fraction	0.82–0.8
Specific pressure drop, hPa/m	1.0–2.7

Note: Pressure drops are evaluated at gas linear velocity of 5 m/s. Based on product information by Hitachi, Zosen and Siemens.

C. Other Catalysts

Composite ceramic monolith catalysts are also included among the commercial SCR systems. They are manufactured by depositing a layer of catalytic ingredients onto strong, thin-walled ceramic honeycomb supports (usually cordierite). They may suffer from erosion problems in the presence of dust, and their use may be preferably limited to a clean environment.

Coated metal monoliths have a similar structure, but in this case the support is represented by thin metal foils; they are characterized by large cell densities and are used mostly in dust-free applications.

It has been mentioned that special low-temperature SCR catalysts, also used in structured configurations, have been developed by several companies. Hitachi Zosen has proposed a vanadium–titanium thick sheet-type catalyst reinforced with ceramic fibers ("ceramic papers") for low temperatures and clean-gas applications [28]. Shell has presented a new proprietary DeNOxing technology that comprises a novel low-temperature catalyst consisting of silica granules impregnated with titanium and vandadium oxides as well as novel structured reactor concepts (parallel-flow reactor and lateral-flow reactor) for housing the catalysts [29]; these are discussed further in Chapter 11. Grace has developed a catalytic system in which the active components (vanadia or platinum) are supported over a proprietary metal monolith (CAMET); the CAMET catalyst system has been studied specifically for natural gas and oil-fired turbine applications [30].

III. STEADY-STATE MODELING OF SCR MONOLITHIC REACTORS

Only recently have efforts been devoted to a detailed chemical engineering analysis of the operation and design of monolith SCR catalysts.

A. Kinetics of the DeNO$_x$ Reaction

The rate of the DeNO$_x$ reaction is first order in respect to NO concentration and essentially independent of NH_3 concentration when ammonia is in excess. However, in SCR industrial applications a substoichiometric feed ratio ($\alpha = NH_3/NO < 1$) is employed in order to minimize the slip of unconverted ammonia and the formation of ammonium sulfates. A kinetic dependence on ammonia is apparent when NH_3 becomes the limiting reactant. Several authors have proposed kinetic expressions for the SCR reaction that account for the observed dependences [31–37]. In particular, the simplest expressions are based on Eley–Rideal kinetics; in line with the mechanistic studies, they assume that the reaction occurs between strongly adsorbed ammonia and gas-phase NO. Beekman and Hegedus [36] have proposed and fitted to experimental data obtained over commercial SCR catalysts the following kinetic expression:

$$r_{NO_x} = k_{NO_x} C_{NO_x} \frac{K_{NH_3} C_{NH_3}}{1 + K_{NH_3} C_{NH_3}} \tag{1}$$

which reduces to first order in respect either to NO or to NH_3 in the limiting cases of $K_{NH_3} C_{NH_3} >> 1$ and $K_{NH_3} C_{NH_3} << 1$, respectively. Equation (1) neglects the influence of oxygen and water concentrations; this simplification is correct in the concentration range of industrial interest (O_2 > 2%, H_2O > 5% v/v) [31]. Similar kinetics have been adopted by Lefers et al. [38] and Tronconi and Forzatti [39].

B. Inter- and Intraphase Mass Transfer Limitations in $DeNO_x$ Reaction

Both in laboratory and power plant conditions the SCR reactor works under combined intraparticle and external diffusion control, due to the high reaction rate and the laminar flow regime in the monolith channels.

The analysis of gas–solid interphase mass transfer has been rigorously addressed by Tronconi and Forzatti [39] through the development of a three-dimensional model of the SCR reactor accounting for cross-sectional concentration profiles of the reactants inside the single channel. It has been shown that the rate of external mass transfer is affected both by the geometry of the monolith duct (being optimal for a circular section) and by the reaction rate at the wall, as illustrated in Fig. 5. In the specific case of monoliths with square channels, which is most frequent in SCR applications, it has been shown, however, that the latter dependence can be neglected and a proper estimation of the local Sherwood number, Sh, can be found in the Nusselt number obtained from solution of the Greatz–Nusselt problem with constant wall temperature [39]. Thus, based on the analogy of mass transfer with heat transfer in laminar flow within square ducts, a simple one-dimensional model can be formulated for the SCR reactor that is in close agreement with the rigorous multidimensional model. Application of the same analogy to the development of simplified SCR reactor models for cases of channel geometry other than the square introduces errors in the estimation of the interphase mass transfer rate. The errors range within ±20%, depending on the kinetics and the channel geometry; on the other hand, it has to be noted that they affect to a much less extent the estimation of NO conversion.

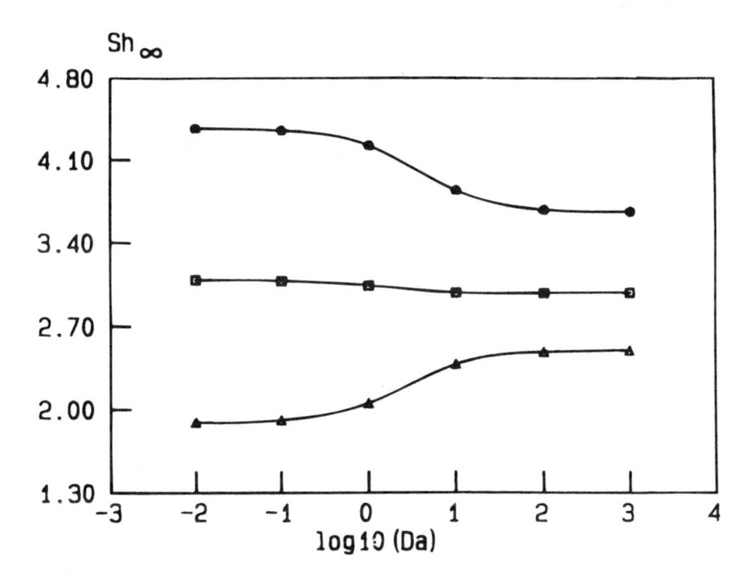

Figure 5 Influence of the Damkohler number, Da, and the monolith channel geometry on the asymptotic Sherwood number, Sh_∞. ● = Circular channel; □ = Square channel; △ = Equilateral triangular channel. (From Ref. 39.)

Actually, it has been shown that the use of one-dimensional models is in general legitimate and sufficiently accurate for design calculations.

Recent studies of intraphase mass transfer in SCR reactors proposed in the literature [36,40] have been based on the adoption of the Wakao–Smith random pore model [41,42] to describe NO and NH_3 diffusion inside the catalyst walls. Solution of differential mass balances for diffusion and reaction in the intraporous region confirmed that the presence of steep internal concentration gradients prevail under industrial SCR conditions, resulting in extremely low effectiveness factors (0.02–0.05 according to Ref. 40). The concentration of the limiting reactant (ammonia in industrial SCR operation) drops to zero within a very thin layer of catalyst confined to the surface of the monolith wall. The scarce utilization of the catalysts motivates the development of novel morphological configurations that can improve the effective diffusivities of the reactants. In a later section the analysis of the effect of pore size distribution on the process efficiency will be addressed.

C. Steady-State Model Equations

Analyses of monolith reactors specific for SCR applications are limited in the scientific literature. Buzanowski and Yang [43] have presented a simple one-dimensional analytical solution that yields NO conversion as an explicit function of the space velocity; unfortunately, this applies only to first-order kinetics in NO and zero-order in NH_3, which is not appropriate for industrial SCR operation. Beeckman and Hegedus [36] have published a comprehensive reactor model that includes Eley–Rideal kinetics and fully accounts for both intra- and interphase mass transfer phenomena. Model predictions reported compare successfully with experimental data. A single-channel, semianalytical, one-dimensional treatment has also been proposed by Tronconi et al. [40]. The related equations are summarized here as an example of steady-state modeling of SCR monolith reactors.

Dimensionless Material Balances for NO and NH_3 in the Gas Phase

$$\frac{dC^*_{NO}}{dz^*} = -4Sh_{NO}(C^*_{NO} - C^*_{NO,wall}); \quad \text{at } z^* = 0 \quad C^*_{NO} = 1 \tag{2}$$

$$\alpha - C^*_{NH_3} = 1 - C^*_{NO}; \quad \text{at } z^* = 0 \quad C^*_{NH_3} = \alpha \tag{3}$$

with

$$\mathrm{Sh}_{NO} = 2.977 + 8.827(1000z^*)^{-0.545} \exp(-48.2z^*) \tag{4}$$

Dimensionless Material Balances for NO and NH_3 in the Solid Phase

$$\mathrm{Sh}_{NO}(C^*_{NO} - C^*_{NO,wall}) = \mathrm{Sh}_{NH_3}(C^*_{NH_3} - C^*_{NH_3,wall})\frac{D_{e,NH_3}}{D_{e,NO}} \tag{5}$$

$$\mathrm{Sh}_{NO}(C^*_{NO} - C^*_{NO,wall}) = \mathrm{Da}\left[C^{*2}_{NO} - Y_0^2 + 2(S_1 - S_2)\left(C^*_{NO} - Y_0 - S_2 \ln\frac{C^*_{NO} + S_2}{Y_0 + S_2}\right)\right]^{1/2} \tag{6}$$

with

$$S_1 = \frac{D_{e,NH_3}}{D_{e,NO}} C^*_{NH_3} - C^*_{NO} \quad \text{and} \quad S_2 = S_1 + \frac{D_{e,NH_3}}{D_{e,NO}} \frac{1}{K^*_{NH_3}} \tag{7}$$

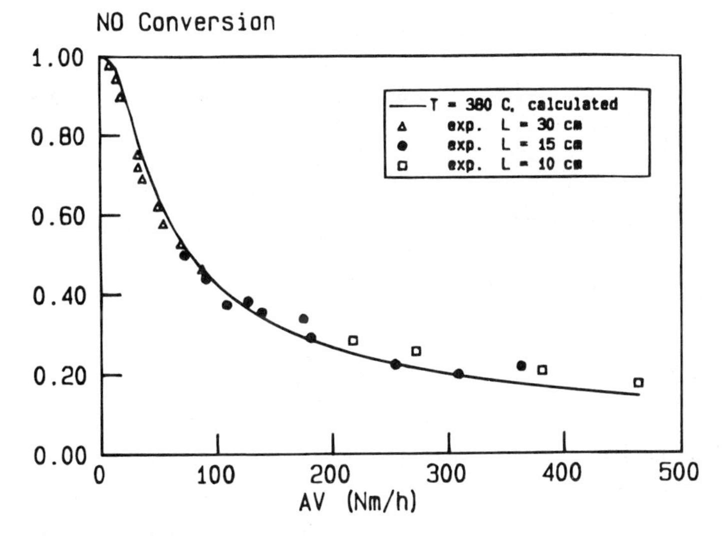

Figure 6 Comparison of experimental and calculated effects of monolith length and area velocity AV = volumetric flow rate/geometric surface area on NO conversion. Channel hydraulic diameter = 6 mm, $\alpha = 1.2$, T = 380°C, feed = 500 ppm NO, 500 ppm SO_2, 2% v/v O_2, 10% $H_2O + N_2$. (From Ref. 40.)

and:

$$\text{if } S_1 \geq 0, \quad \text{then} \quad Y_0 = 0; \quad \text{otherwise}, \quad Y_0 = -S_1 \tag{8}$$

In Eqs. (2)–(8), concentrations of NO and NH_3 are normalized with respect to the inlet NO concentration C^{O}_{NO}, the dimensionless axial coordinate is:

$$z^* = \frac{z/d_h}{\text{ReSc}}$$

(d_h being the hydraulic channel diameter);

$$\text{Da} = \frac{(k_{NO}D_{e,NO})^{1/2}d_h}{D_{NO}}$$

is a modified Damköhler number; $K^*_{NH_3} = K_{NH_3}C^{O}_{NO}$ is the dimensionless NH_3 adsorption constant; D_i is the molecular diffusivity of species i; $D_{e,i}$ is the effective intraporous diffusivity of species i evaluated according to the Wakao–Smith random pore model [41]. Equation (4) is taken from Ref. 39. Equations (6)–(8) provide an approximate analytical solution of the intraporous diffusion-reaction equations under the assumption of large Thiele moduli (i.e., the concentration of the limiting reactant is zero at the centerline of the catalytic wall); the same equations are solved numerically in Ref. 36.

This model was successfully compared with laboratory data of NO conversion over commercial honeycomb SCR catalysts, as shown, for example, in Fig. 6.

D. Interaction Between $DeNO_x$ Reaction and SO_2 Oxidation

As mentioned earlier, the $DeNO_x$ reaction over vanadia-based catalysts is accompanied, in the case of sulfur-containing gas, by the oxidation of SO_2 to SO_3. This aspect has seldom been considered in the modeling of SCR systems. Beeckman and Hegedus [36] have adopted first-order kinetics for SO_2 oxidation, as usually done in the industrial literature. However, Svachula et al. [18] have demonstrated, through an extensive experimental study, that the $SO_2 \rightarrow SO_3$ reaction involves: (1) a variable kinetic order with respect to SO_2 concentration, decreasing from 1 at low SO_2 concentrations to lower values at greater C_{SO2}, (2) asymptotic zero-order kinetics with respect to oxygen, (3) inhibiting effects of water and ammonia, (4) slight promoting effect of NO_x. Based on a redox reaction mechanism, Svachula et al. [18] and Tronconi et al. [44] have developed the following kinetic expression to account for all of these effects:

$$r_{SO_2} = \frac{k_1 C_{SO_2} C_{SO_3}(1 + bC_{NO})}{1 + k_2 C_{SO_3} + k_3 C_{SO_2} C_{SO_3} + k_4 C_{SO_3} C_{H_2O} + k_5 C_{SO_3} C_{NH_3} + k_6 C_{SO_2} C_{SO_3}/\sqrt{C_{O_2}}} \tag{9}$$

Tronconi et al. [44] have included Eq. (9) in a complete model for $DeNO_x$ reaction and SO_2 oxidation, showing that, opposite to the former reaction, the latter one occurs in a chemical regime due to its very low rate. Dedicated experiments have indicated in fact that SO_2 oxidation involves the whole catalyst volume; also, application of literature criteria [45] have confirmed the absence of external and intraporous gradients of SO_2 concentration [18].

The dependence of SO_2 oxidation on the local concentrations of NO and NH_3 establishes a kinetic interaction with NO reduction. This is evident in Fig. 7, which shows laboratory data obtained under simultaneous occurrence of $DeNO_x$ reaction and SO_2 oxidation, as in industrial conditions. The data are compared with the predictions of the complete model given in [44]. Two regimes characterize SO_2 oxidation: the regime of high SO_2 conversions for $\alpha < 1$, and the regime of low SO_2 conversions for $\alpha > 1$. The transition between the regimes occurs close to $\alpha = 1$, in correspondence with the onset of the inhibiting effect exerted by the unreacted ammonia. This trend is well reproduced

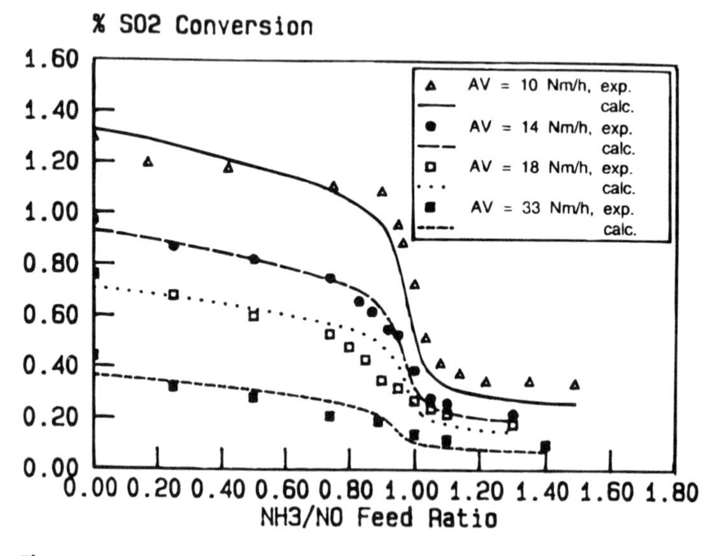

Figure 7 Experimental and calculated SO_2 conversions over a commercial SCR monolith catalyst as a function of α = NH_3/NO feed ratio and of the area velocity AV. (From Ref. 47.)

by the model predictions (the kinetic parameters were estimated on the basis of independent studies over the same catalyst), which supports the adequacy of rate expression (9). It is worth stressing that a simple first-order dependence on SO_2 concentration cannot reproduce the experimental data of Fig. 7, but would result in a constant trend of SO_2 conversion.

IV. SCR CATALYST AND REACTOR DESIGN

As previously discussed, the physico-chemical phenomena occurring in a monolith SCR catalyst are relatively well understood by now; the mathematical models describing such phenomena can then be applied with confidence to the rational design of SCR catalysts and processes.

A. Effect of Catalyst Morphology

Given the strong influence of intraporous diffusional resistances, NO conversion is a sensitive function of the morphological properties of the catalyst; Beeckman and Hegedus [36] first investigated the potential for catalyst improvement offered by pore structure optimization. These authors have developed a new type of catalyst following the indications of a mathematical model of NO_x reduction over a monolith-shaped vanadium–titanium–oxide catalyst. Specifically, they identified the optimal pore structure as a bimodal one consisting of 0.25-cm^3/cm^3 micropore porosity, 0.45-cm^3/cm^3 macropore porosity, 8-nm micropore diameter, and 1000-nm macropore diameter; the optimization was constrained by assuming a total porosity of 0.7 cm^3/cm^3 in order to satisfy the requirements on the mechanical resistance of the catalyst. The optimal morphological configuration represents the best compromise between large specific surface areas (associated with high fractions of micropores with small diameter) and high intraporous diffusivities for NO and NH_3 (associated with high fractions of macropores with large diameter). A 50% improvement in NO reduction activity was predicted by the model for the optimized morphology and verified in laboratory experiments over a vanaolia-titania-silica catalyst. The authors also predicted that the optimal pore structure would not affect the extent of SO_2 oxidation, based on simple first-order kinetics in SO_2 concentration.

Beretta et al. [46] have addressed the optimization of the SCR catalyst morphology, taking into account the above-discussed kinetic interaction between NO reduction and SO_2 oxidation. The authors have confirmed that catalysts with relatively low surface areas are in general desired in order to minimize SO_2 oxidation and that high fractions of macropores are necessary for optimal $DeNO_x$ performances, due to enhanced effectiveness factors of the catalyst. In correspondence with the specific case of equimolar flows of NO and NH_3, however, SO_2 conversion is affected not only by the catalyst surface area, but also by the pore size distribution. As shown in Fig. 8, in fact, for a fixed value of surface area, the inhibiting effect of ammonia on SO_2 oxidation can be modulated by varying the morphological parameters that control the relative rates of NO and NH_3 intraporous diffusion; namely, an increment of microporosity favors the preferential diffusion of ammonia inside the monolith wall and is accompanied by a decrease of SO_2 conversion.

After these results, Tronconi et al. [44] addressed a systematic study of the morphological properties that favor the onset of an excess of ammonia inside the catalyst and of the lower limit of α where this can exist. The authors found that at α as low as 0.81, the

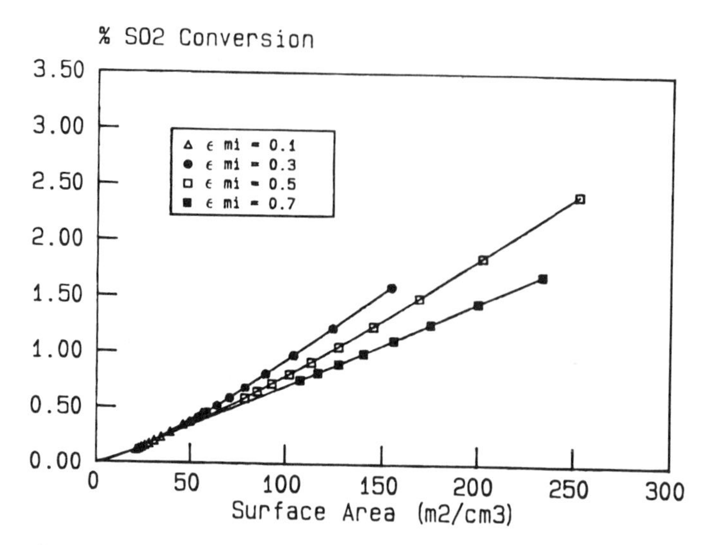

Figure 8 Calculated dependence of SO_2 conversion on catalyst surface area and pore volume distribution. Case of stoichiometric feeds: $\alpha = 1$; AV = 11 Nm/h; ϵ_{mi} = volume fraction of micropores. (From Ref. 40.)

diffusion of ammonia inside the whole catalyst thickness (and the consequent inhibition of SO_2 oxidation) can be affected in principle, at least in a small portion of the monolith length, by adopting microporous morphologies. Unfortunately, such structures would be highly inefficient from the point of view of NO_x reduction; also, the large surface area associated with the presence of micropores would cause an intrinsic undesired promotion of SO_2 conversion and prevail on the opposite desired effect of inhibition induced by ammonia. It was then concluded that, in the range of substoichiometric feed ratios typically used in industrial conditions, the design of pore structure offers no room for exploiting the interaction between $DeNO_x$ reaction and SO_2 oxidation in order to minimize SO_3 formation.

B. Effect of Monolith Geometry

Technical constraints are often imposed on the design of the monolith geometry by the extrusion process, as well as by the mechanical properties of the extrudate; the specific SCR application (e.g., high-dust vs. low-dust) is also crucial for the definition of the catalyst geometrical features. Here, attention is paid to the influence that the monolith parameters (wall thickness, channel size, channel shape) have on both $DeNO_x$ reaction and SO_2 oxidation in order to advance guidelines for optimization of the catalyst geometry.

It is well known that though NO conversion is unaffected by the thickness of the monolith wall beyond a small critical value, SO_2 conversion increases linearly with increasing wall thickness. This is indicated in Fig. 9; such trends reflect the different influence of internal diffusional resistances on $DeNO_x$ reaction and SO_2 oxidation, which, as discussed previously, are respectively confined to a superficial layer of the catalyst and active inside the whole wall. Consequently, the design of SCR monoliths should pursue the realization of very thin catalytic walls; Fig. 9, for example, shows that reducing the catalyst half-thickness from 0.7 mm to 0.2 mm does not alter the DeNOxing performance but causes a decrease of SO_2 oxidation as significant as 78%.

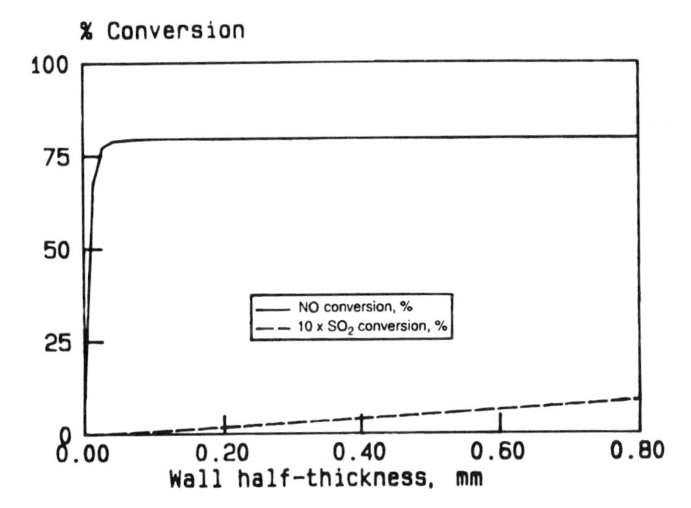

Figure 9 Calculated effect of the half-thickness of the monolith catalyst walls on NO and SO_2 conversions. $\alpha = 0.8$; AV = 11 Nm/h. (From Ref. 44.)

Tronconi and Forzatti [39] have shown that increasing the channel hydraulic diameter d_h adversely affects the NO conversion by enhancing the resistances to transport of the reactants from the bulk gas phase to the catalyst surface. Conversion of SO_2, instead, is unaffected by d_h due to the absence of diffusional limitations. However, the realization of small channel openings guarantees the achievement of a specific NO_x abatement level with a reduced amount of catalyst and is consequently associated with a lower SO_2 conversion [44]. It must be noted that selection of the channel size has to account for plugging and erosion problems, too.

Interphase resistances are also involved in the effect of the channel shape on NO_x conversion. Due to peripheral nonuniformities of the reactant concentrations at the catalyst wall, cornered ducts are characterized by slower overall process rates than the circular ones [39].

C. Effect of Ammonia Inlet Maldistribution

The process sensitivity to inhomogeneities in the ammonia distribution has been investigated in the literature [26,40]. The representative situation was examined in which the target feed ratio $\alpha = 0.8$ prevails in one-third of the total monolith channels, whereas $\alpha - \delta\alpha$ is established in one-third and $\alpha + \delta\alpha$ in the remaining third. The calculated percentage increments of catalyst volume required to secure NH_3 slip below 5 ppm for growing $\delta\alpha$ are plotted in Fig. 10, assuming the reactor performance under a homogeneous distribution of α as a reference. Such a trend is confirmed when more detailed discretizations are adopted to represent the distribution of α; large deviations of α from the ideal value cause dramatic drops in DeNOxing activity and have to be avoided in order to minimize the catalyst load. Concerning SO_2 oxidation, the reaction is partially moderated inside those channels where α values close to 1 (high concentrations of NH_3) prevail; nevertheless, the increase of the reactor volume necessary to grant the desired NO_x

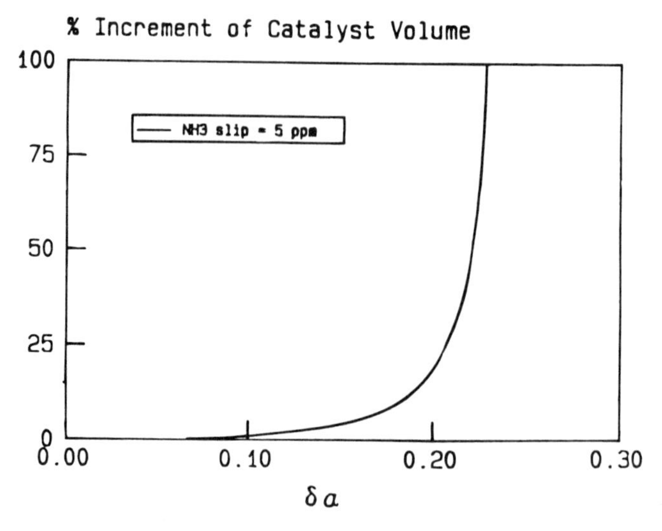

Figure 10 Influence of $\delta\alpha$, the maldistribution parameter, on the extra catalyst volume required to keep the NH_3 slip below 5 ppm. (From Ref. 40.)

conversion predominates over the local inhibiting effects due to the excess ammonia, and results in almost proportional enhancements of SO_3 formation.

D. An Alternative Reactor Configuration

Orsenigo et al. [47] have proposed an alternative reactor design suitable in principle to exploit NH_3 inhibition for minimizing SO_3 formation in the SCR process. This is based on the idea of splitting the NO_x-containing feed stream in substreams fed separately to the SCR reactor; in this way, a portion of the catalyst volume can operate with an excess of ammonia, while the overall NH_3/NO feed ratio is still substoichiometric.

Figure 11 shows schematically the simplest design accomplishing this purpose; the configuration consists of two reactors in series, with catalyst volumes V_1 and V_2. The total gas feed is partitioned in two streams, with flow rates Q_1 and Q_2, respectively. Q_1 is fed to the first reactor after being mixed with ammonia (the whole amount), whereas Q_2 is simply combined with the stream entering the second reactor. Depending on the ratio $Q_1/(Q_1 + Q_2)$, an excess of ammonia can be attained in the first reactor, with an overall $\alpha < 1$. This is expected to reduce the extent of SO_3 formation while still securing low values of NH_3 slip.

Simulations have been carried out assuming: overall area velocity AV = 10 Nm/hr, overall $\alpha = 0.8$, α in the first reactor = 1, feed concentrations = 500 ppm NO, 1000 ppm SO_2, kinetic parameters and catalyst morphological properties [44]. For a given split ratio $R = V_1/(V_1 + V_2)$ of the catalyst volume, the constraint on the overall α determines uniquely the partitioning of the flow rates. The results are shown in Fig. 12, where the calculated ammonia slip and the SO_3 outlet concentration are plotted as functions of R. The limiting case $R = 0$ corresponds to a reference conventional configuration where the total feed stream is fed to the second reactor. On increasing R, the outlet SO_3 concentration decreases progressively due to the increase of the catalyst volume (in the first reactor), which operates with a stoichiometric feed ratio, thus favoring the inhibiting action of

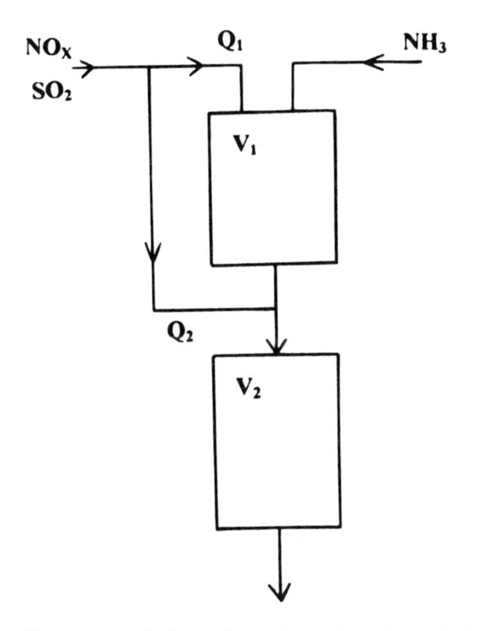

Figure 11 Split-feed configuration of the SCR reactor.

ammonia. The DeNOxing activity is enhanced by low values of the split ratio; the presence of a small portion of catalyst with higher concentrations of NH_3 promotes the kinetics of the process with respect to the reference conditions. However, with $R > 0.35$, the overall performance of the system is adversely affected with respect to a conventional configuration by the incomplete NO conversion associated with the operation of the first reactor. Consequently, the NH_3 slip exhibits a minimum with increasing split ratio. It must be noted that the undesired precipitation of ammonium sulfates is controlled by the product of the outlet concentrations of SO_3 and NH_3 due to thermodynamic reasons [48]. Figure

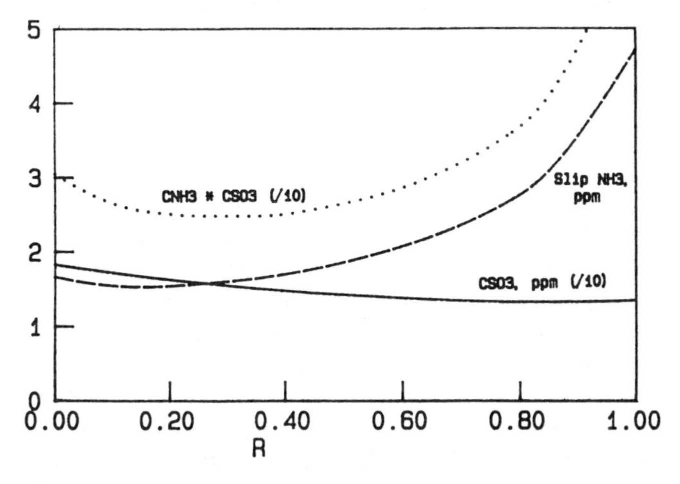

Figure 12 Calculated effect of catalyst-volume split ratio R on the outlet concentrations of SO_3 and NH_3 and on their product. (From Ref. 47.)

12 shows that such a product also goes through a minimum, located at $R \cong 0.3$. This optimum reactor configuration has higher DeNOxing activity than the conventional design and secures a 20% reduction of the potential for ammonium sulfates formation. Even though these are just theoretical results, multifeed reactor configurations have in effect been proposed as innovative lines of SCR catalysis [49].

V. UNSTEADY OPERATION OF SCR CATALYSTS

In the last few years a growing interest has been focused on unsteady operation of monolith SCR catalysts. This is due in essence to the following reasons.

1. Industrial SCR monolith reactors for the reduction of NO_x in the flue gases of thermal power stations often operate under transient conditions associated with, e.g., startup and shutdown of the plant, as well as with load variations; predictive control systems are expected to help reduce the levels of polluting emissions during such transients.
2. Prediction of the dynamic behavior of the catalyst is essential for controlling the ammonia injection in SCR systems applied to stationary diesel engines.
3. Russian scientists have demonstrated theoretically and experimentally the feasibility of reverse-flow SCR processes in Matros-type adiabatic reactors; this periodic operation may be advantageous for abatement of NO_x in low-temperature exhaust gas streams.
4. Experimental evidence has been published showing that unsteady SCR operation can also bring about enhanced NO_x reduction efficiencies without incrementing the NH_3 slip.

For all of these reasons, a thorough understanding of the NH_3 adsorption–desorption phenomena on the catalyst surface is a prerequisite. In fact, typical SCR catalysts can store large amounts of ammonia, whose surface evolution becomes the rate-controlling factor of the reactor dynamics. Also, mathematical modeling appears to be even more useful for the analysis and development of unsteady SCR processes than in the case of steady-state operation.

A. Unsteady Kinetics of NH_3 Adsorption–Desorption and Reaction over SCR Catalysts

Rate expressions for adsorption–desorption of NH_3 over a V_2O_5-based catalyst were provided by Noskov et al. [50]. They would assume an ideal Langmuir-type catalyst surface, which is probably adequate only for limited ranges of operating conditions.

Andersson et al. [51] determined the kinetics of adsorption–desorption of ammonia on a $V_2O_5/\gamma\text{-}Al_2O_3$ catalyst by TPD experiments. The composite shape of their NH_3 TPD curves clearly pointed out that either NH_3 adsorption does not occur only on a single site or the adsorption energy varies with the surface coverage θ. The authors analyzed their data on the basis of the latter assumption, and estimated accordingly the parameters in rate expressions for adsorption and desorption of NH_3. However, a strong correlation was apparent between the estimates of adsorption and desorption parameters, suggesting that equilibrium was approached during the TPD runs due to the low flow rates; consequently, extrapolation of their kinetics to different experimental conditions appears to be questionable.

Tronconi et al. [52] studied the kinetics of NH_3 adsorption–desorption over binary (V_2O_5/TiO_2) and ternary (V_2O_5–WO_3/TiO_2) model catalysts in powder form by transient response techniques. Perturbations both in the NH_3 concentration and in the catalyst temperature were imposed. In a typical run, a rectangular step feed of ammonia in helium as the carrier gas was admitted to the flow microreactor at constant temperature (in the range 220–350°C), followed by its shutoff; eventually, the catalyst temperature was increased according to a linear heating schedule in order to complete the desorption of ammonia. The NH_3 concentration in the reactor effluents was monitored by a quadrupole mass spectrometer. Typical results of a run over a binary catalyst at 220°C are shown in Fig. 13.

The transient response experiments were analyzed by a dynamic isothermal PFR model, and estimates of the relevant kinetic parameters were obtained by global nonlinear regression over all runs. It was found that a simple Langmuir approach could not represent the data accurately, and surface heterogeneity had to be invoked. The best fit was obtained using a Temkin-type adsorption isotherm with coverage-dependent desorption energy:

$$E_d = E_d^{\circ}(1 - \alpha\theta) \tag{10}$$

combined with the rate expressions:

$$r_{AD} = k_{AD}C_{NH_3}(1 - \theta) \tag{11}$$

$$r_{DES} = k_{DES}^{\circ} \exp\left(\frac{-E_d}{RT}\right)\theta \tag{12}$$

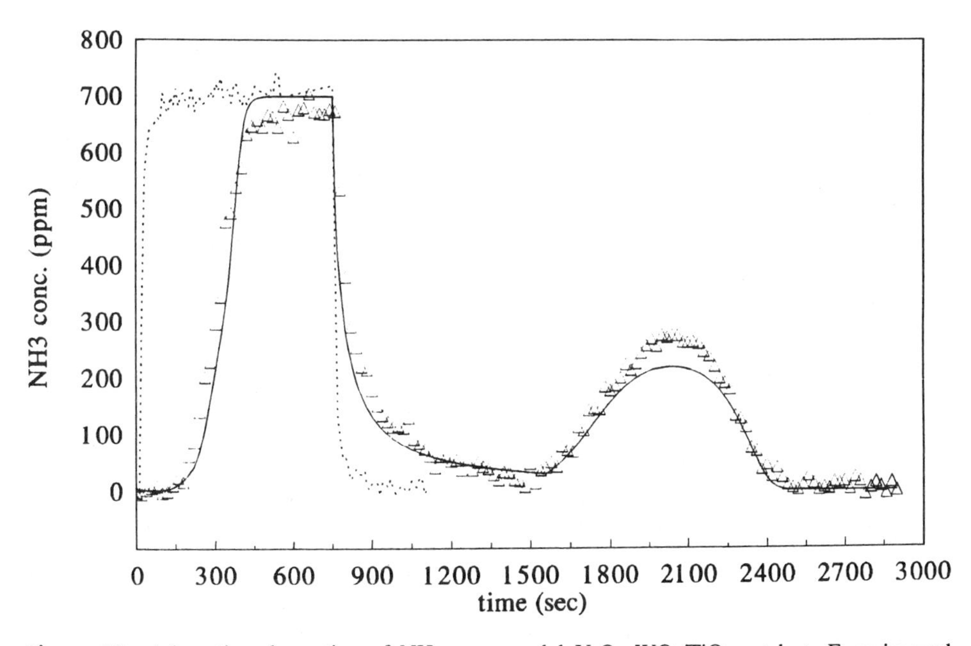

Figure 13 Adsorption–desorption of NH_3 on a model V_2O_5–WO_3/TiO_2 catalyst: Experimental data and model fit (solid line). T = 220°C. The dots represent the inlet NH_3 concentration; the triangles show the outlet NH_3 concentration. The TPD run started at t = 1500 sec. (From Ref. 52.)

The solid line in Fig. 13 represents a typical fit of the transient data. A nice agreement with experiment is apparent both in the case of the NH_3 rectangular step feed and in the case of the subsequent TPD run. The optimal parameter estimates yield an activation energy for desorption at zero coverage (E_d°) close to 100 kJ/mol, in agreement with literature values, and indicate a nonactivated NH_3 adsorption process. This is in line with the spontaneity of adsorption of a basic molecule, like ammonia, over the acid catalyst surface [17]. It is worth emphasizing also that the estimated kinetics account quite satisfactorily for large variations in the NH_3 surface coverage (θ = 0–0.8) and in the catalyst temperature (T = 220–500°C).

The same transient response techniques have been successfully applied in our laboratory to investigate also the kinetics of the surface reaction of NO with preadsorbed NH_3.

B. Mathematical Modeling of SCR Monolith Catalysts Under Unsteady Conditions

Andersson et al. [51] have proposed a one-dimensional plug-flow dynamic model of a single channel in SCR monolith catalysts used to remove NO_x from mobile diesel exhausts. In this model, axial conduction was neglected along with the accumulation terms for heat and mass in the gas phase, whereas gas–solid heat and mass transfer coefficients were estimated according to the correlations in Ref. 39. The model was developed for coated monoliths, so no account was given of intraporous diffusional resistances. Thus, the model consisted of two energy balances for the gas and the solid phases, two mass balances for NO and NH_3 in the gas bulk, two interphase continuity relationships for NO and NH_3, and one mass balance for adsorbed ammonia. Accumulation terms were included only in the equations for the solid temperature and for the NH_3 surface coverage. The rate expressions for NH_3 adsorption–desorption determined by TPD runs were included in the model equations, and the rate expression for the surface $DeNO_x$ reaction was evaluated from pilot reactor data. The model simulations were compared to results of a dynamic test with a 12-dm^3 engine equipped with a honeycomb SCR catalyst operating with a stoichiometric injection of ammonia. A good match between calculated and measured NO conversions was achieved, but some deviations were observed for temperature and ammonia slip, requiring some adjustment of the kinetic parameters for NH_3 desorption. Notably, the model could be run in parallel with the experiments; accordingly, this type of model can be used to predict the catalyst NO reduction efficiency in real time, and appears suitable to control the NH_3 injection strategy in SCR applied to mobile diesel engines.

Tronconi et al. [52] have validated against experiment a more complex model, accounting also for the reaction and diffusion of NO and NH_3 inside the porous walls of homogeneous isothermal honeycomb SCR catalysts. The model equations are as follows:

Material Balance of Adsorbed NH_3

$$\Omega \frac{\partial \theta}{\partial t} = r_{AD} - r_{NO} \qquad (13)$$

Material Balance of Gaseous NO and NH_3 in the Porous Catalyst Matrix

$$\frac{D_{e,\mathrm{NH_3}}}{S^2}\frac{\partial^2 C_{\mathrm{NH_3}}}{\partial x^{*2}} - r_{\mathrm{AD}} = \frac{\partial C_{\mathrm{NH_3}}}{\partial t} \tag{14}$$

$$\frac{D_{e,\mathrm{NO}}}{S^2}\frac{\partial^2 C_{\mathrm{NO}}}{\partial x^{*2}} - r_{\mathrm{NO}} = \frac{\partial C_{\mathrm{NO}}}{\partial t} \tag{15}$$

Continuity at the Gas–Solid Interface

$$k_{\mathrm{mat,NH_3}}(C^b_{\mathrm{NH_3}} - C^S_{\mathrm{NH_3}}) = \frac{D_{e,\mathrm{NH_3}}}{S}\frac{\partial C_{\mathrm{NH_3}}}{\partial x^*}\bigg|_{x^*=1} \tag{16}$$

$$k_{\mathrm{mat,NO}}(C^b_{\mathrm{NO}} - C^S_{\mathrm{NO}}) = \frac{D_{e,\mathrm{NO}}}{S}\frac{\partial C_{\mathrm{NO}}}{\partial x^*}\bigg|_{x^*=1} \tag{17}$$

Material Balances of Gaseous NO and NH_3 in the Monolith Channel

$$\frac{\partial C^b_{\mathrm{NH_3}}}{\partial t} = -\frac{v}{L}\frac{\partial C^b_{\mathrm{NH_3}}}{\partial z^*} - \frac{4}{d_h}k_{\mathrm{mat,NH_3}}(C^b_{\mathrm{NH_3}} - C^S_{\mathrm{NH_3}}) \tag{18}$$

$$\frac{\partial C^b_{\mathrm{NO}}}{\partial t} = -\frac{v}{L}\frac{\partial C^b_{\mathrm{NO}}}{\partial z^*} - \frac{4}{d_h}k_{\mathrm{mat,NO}}(C^b_{\mathrm{NO}} - C^S_{\mathrm{NO}}) \tag{19}$$

Rate Expressions

$$r_{\mathrm{NO}} = k_{\mathrm{NO}}C_{\mathrm{NO}}\theta \tag{20}$$

$$r_{\mathrm{AD}} = k_{\mathrm{AD}}C_{\mathrm{NH_3}}(1-\theta) - k_{\mathrm{DES}}\theta \tag{21}$$

In Eqs. (13)–(27), Ω is the adsorption capacity of the catalyst; θ is the NH_3 surface coverage; D_e is the effective intraporous diffusivity; S is the monolith wall half-thickness; L is the monolith length; v is the gas velocity in the monolith channels; k_{mat} are gas–solid mass transfer coefficients; and d_h is the hydraulic diameter of the monolith channels. The resulting system of coupled PDEs, with suitable initial and boundary conditions, was solved numerically by discretizing the unknown solutions [i.e., $\theta(x, z, t)$, $C_{NO}(x, z, t)$, and $C_{NH3}(x, z, t)$] along the axial and radial coordinates using orthogonal collocation techniques, and integrating the resulting set of ODEs in time with a library routine based on Gear's method. All the parameters appearing in the model equations were either estimated from the above reported study of transient NH_3 adsorption–desorption and reaction or estimated independently. The model was used to simulate laboratory data obtained over a commercial SCR high-dust monolith catalyst during start-up (NH_3 injection) and shutdown at different temperatures and NH_3/NO molar feed ratios. Figure 14 compares some of such data with model predictions. To achieve the agreement shown in the figure, the kinetic parameters had to be slightly modified with respect to those determined over the model catalyst in powder form; the estimates are reported in the figure caption.

The validated model was applied to simulation of typical transients occurring during the operation of industrial SCR monolith reactors. Simulation of reactor start-up and shutdown showed in all cases that the change in NO outlet concentration is considerably

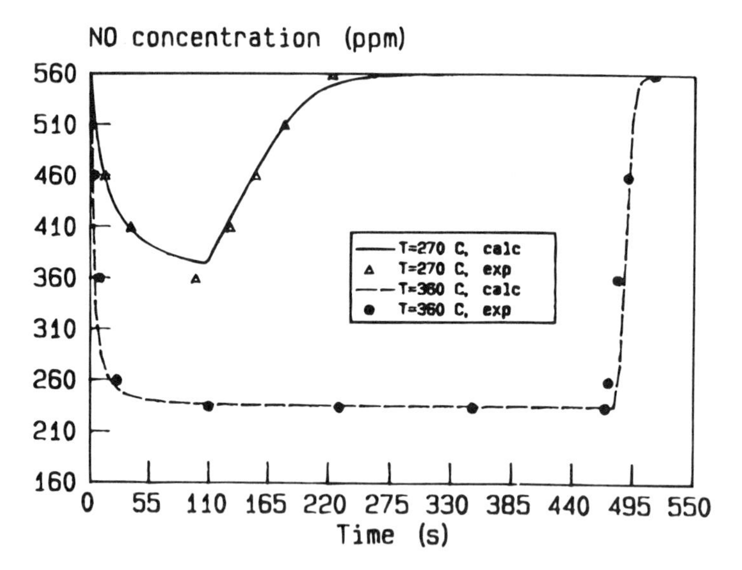

Figure 14 Start-up and shutdown of an SCR honeycomb reactor: Comparison of laboratory data with model predictions of outlet NO concentration C°_{NO} = 560 ppm, α = 0.8, AV = 33 Nm/h, T = 270 and 360°C. Kinetic parameters used in the simulations: $k^{\circ}_{ad} = 3.4\ 10^5\ m^3\ mol^{-1}\ s^{-1}$; E_a = 9.0 kcal/mol; $k^{\circ}_{des} = 4.1 \times 10^7\ sec^{-1}$; E°_d = 29.5 kcal/mol; α = 0.315; $k_{NO} = 9.8 \times 10^9 \exp(-9360/T)\ sec^{-1}$. (From Ref. 52.)

delayed with respect to the variation of the inlet NH_3 concentration. This is unfavorable for a feedback control system using the ammonia feed as the control variable, and makes the adoption of a predictive dynamic model quite attractive.

C. Reverse-Flow SCR

Matros [53] has proposed and demonstrated the use of adiabatic fixed-bed reactors with periodic flow reversal to achieve autothermal operation with a low overall exothermicity, thus allowing for greater energy efficiency than traditional steady-state processes. Such reactors are now being applied industrially for SO_2 oxidation and for catalytic combustion of volatile organic compounds. Matros and co-workers [50,54] also investigated theoretically and experimentally the application of reverse-flow operation to SCR treatment of low-temperature gases, which usually requires costly gas–gas heat exchangers in traditional flow sheets.

The basic mode of reverse-flow operation consists of the initial preheating of the catalyst bed to a temperature high enough to guarantee ignition. Then the cold reacting gas is fed to the reactor, where in a narrow heat exchange zone it is heated up to the light-off temperature, and the reaction ignites. Since the inlet gas temperature is too low to provide stable steady-state ignition, the heat front slowly creeps in the gas flow direction. If the flow direction is reversed before extinction by means of switching valves, a new reaction front ignites and moves into the fixed bed from the opposite side, and the old front is pushed back. After a large number of flow reversals, a steady state of periodic operation is established, with reaction fronts moving back and forth in a completely symmetric way.

Figure 15 shows two different reverse-flow configurations that have been proposed and investigated for SCR applications. A classic flowsheet for reverse-flow operation is sketched in Fig. 15a, consisting of one layer of catalyst that is placed between two layers of inert material serving as regenerative heat exchangers. Ammonia is added directly to the feed gas before the switching valves. The use of this configuration appears to be limited to the treatment of dry or low-concentrated gases due to the possible deposition of ammonium salts in the cold inert packing and in the cold zones of the catalyst bed. The alternative configuration sketched in Fig. 15b has been proposed to overcome this problem. It consists of two separated catalyst beds placed between two layers of inert material. NH_3 is injected in the interspace between the two catalyst beds, i.e., the hot, ignited zone of the reactor. Deposition phenomena can be avoided if NH_3 is completely consumed before reaching the cold zones.

Mathematical modeling was used to investigate the potential of the latter configuration in meeting process requirements [50]. A two-phase, one-dimensional model for packed-bed reactors was used that takes into account interphase heat and mass transfer, gas-phase convection, heat capacity and conductivity of the solid phase, chemical reactions at the catalyst surface, and catalyst effectiveness factor. SO_2 free gases were examined, and two competitive processes were considered, i.e., NH_3 oxidation by NO_x and by O_2. To account for the dynamics of the reaction kinetics, an unsteady-state mass balance of adsorbed ammonia was included, based on Langmuir-type rate expressions for NH_3 adsorption–desorption. Simulations confirmed the efficiency of central NH_3 injection, which provides high NH_3 coverages only in the hot zone of the catalyst layers, thus avoiding formation of ammonium salts, and high NO_x conversions. An industrial application of this technology has been operating since 1989 in Russia, where a process stream from an oleum plant is treated containing (% v/v): NO_x 0.1–0.8; O_2 5.2–5.6; H_2O 2–2.2; and balance N_2. Outlet NO_x concentrations not exceeding 15–35 ppm have been claimed.

The general advantages reported for monoliths in SCR applications still hold for reverse-flow operation. Furthermore, it has been claimed that the use of honeycomb structures in reverse-NO_x processes allows for autothermal treatment of gas stream with even lower NO_x concentrations with respect to packed beds. On the basis of a general two-phase model, it has been evaluated that an adiabatic temperature rise of less than 10°C (NO_x < 500 ppm) permits a stable periodic regime with monoliths, whereas at least 15°C is required with packed beds. Reverse-flow SCR on honeycomb structures has been

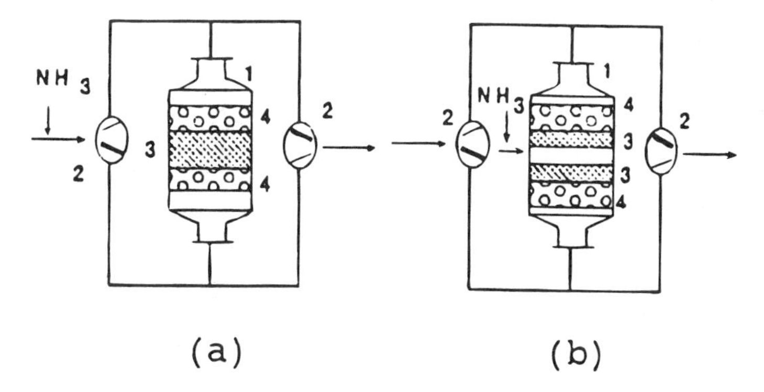

Figure 15 Flowsheet diagrams for a reverse-flow SCR. 1 = reactor, 2 = switching valves, 3 = catalyst, 4 = inert layer. (From Ref. 50.)

experimentally tested at the pilot scale. In these experiments the processing of gases with NO_x concentration up to 1300 ppm was shown to be possible at inlet temperatures of 30–40°C.

In summary, periodic reverse-flow SCR seems advantageous for NO_x abatement in the specific cases of low-temperature streams with relatively high NO_x concentrations. Moreover, the problems associated with the treatment of SO_2- and dust-containing gases have not been addressed so far. Accordingly, the present features of this technology do not fit power plants applications.

An alternative periodic reverse-flow process has been proposed that takes advantage of the NH_3 storage capacity of the catalyst under unsteady conditions [55]. In this process the pellet catalyst is periodically saturated using $\alpha > 1$. The feed is then switched to $\alpha < 1$, but NH_3 waves creep in the flow direction in a similar way to the propagation of heat waves mentioned above. The NH_3 front is maintained within the catalyst bed by periodic flow reversal, thus preventing NH_3 slippage. According to the experimental data reported in Ref. 55, this process solution grants a 30% saving of catalyst load with respect to stationary operation, yielding the same NO_x reduction efficiency and the same NH_3 slip due to the excess of adsorbed ammonia prevailing throughout the catalyst mass.

D. SCR by Ljungstroem Air Heater

The use of the Ljungstroem air heater as SCR reactor is another strategy for NO_x abatement in power plants that is based on unsteady-state periodic operation [56]. Indeed, catalytic reactors in SCR technology require considerable volumes that are not always available when existing boilers have to be retrofitted. Moreover the high costs of traditional SCR installations favor alternative solutions requiring limited modifications of existing plants.

The Ljungstroem air heater consists of a rotating cylinder into which packets of metal sheets are inserted. Profiled metal sheets have large geometric areas to provide effective heat exchange; thus they can be coated with a layer of SCR-type catalyst. Accordingly, a Ljungstroem air heater can be operated simultaneously as a regenerative heat exchanger and as an SCR catalytic reactor. No additional equipment is needed, and, in principle, NO_x reduction can be performed without major modifications of existing plants. Notably, during rotation of the cylinder, the coated metal sheets are alternately in contact with countercurrent flows of cold inlet fresh air and hot exhaust gases. Thus a periodic cycling of catalyst temperature profiles occurs, forcing transient operation of the reactor. It should be noted that strict process constraints on operating temperature and space velocity, the latter being higher than in conventional SCR processes, have to be faced in the actual application of this technology.

Two alternative configurations have been proposed and tested at full scale in power plants. In the first one, NH_3 is injected into the flue gases, and reaction proceeds similarly to traditional steady-state configurations except for forced nonisothermal temperature profiles of the catalyst. A full-scale retrofitted air heater has been installed on a 215-MWe power unit equipped with a gas/oil boiler. About 70% of NO_x removal efficiency was obtained with flue gases containing low NO_x concentrations (150 ppm). In the second configuration, NH_3 is injected on the cold air side. NH_3 is adsorbed on the catalyst and transported by Ljungstroem rotation to the hot flue gases, where reaction takes place. In this design, NH_3 slip is avoided since any excess NH_3 is carried to the boiler and oxidized. Experiments carried out in a plant equipped with a boiler fired with brown coal have shown that over 30% of NO_x efficiency abatement can be reliably achieved.

NOTATION

AV	area velocity
C_i	gas-phase concentration of species i
Da	modified Damköhler number $[= (k_{NO}D_{e,NO})^{1/2}d_h/D_{NO}]$
$D_{e,i}$	effective intraporous diffusivity of species i
d_h	hydraulic diameter of monolith channels
D_i	molecular diffusivity of species i
E_a	activation energy of adsorption
E_d	activation energy of desorption
k_i°	preexponential factor for rate constant of species i
k_i	rate constant of species i
$k_{mat,i}$	gas–solid mass transfer coefficient of species i
L	monolith length
r_{AD}	rate of adsorption
r_{DES}	rate of desorption
Re	Reynolds number $(= \rho v d_h/\mu)$
r_{NO}	rate of $DeNO_x$ reaction
r_{SO_2}	rate of SO_2 oxidation
S	half-thickness of monolith wall
Sc	Schmidt number $(= \mu/\rho D)$
Sh_i	Sherwood number of species i $(= k_{mat,i}d_h/D_i)$
T	temperature
t	time
v	gas linear velocity
x	monolith transverse coordinate
z	axial distance along the monolith catalyst

Greek Symbols

α	NH_3/NO feed molar ratio
ϑ	catalyst surface coverage
μ	gas viscosity
ρ	gas density
Ω	NH_3 adsorption capacity of the monolith catalyst

Subscripts and Superscripts

*	dimensionless quantity
°	inlet conditions

ACKNOWLEDGEMENT

Financial support from CNR–Attivita' di Comitato, Rome, is gratefully acknowledged.

REFERENCES

1. H. Bosch and F. Janssen, Catalytic reduction of nitrogen oxides. A review of the fundamentals and technology, *Catal. Today 2*:369 (1988).
2. P. Forzatti, F. Bregani, Riduzione catalitica selettiva degli NO_x con NH_3 nei fumi emessi dalle centrali termiche, *La Termotecnica 11*:57 (1991).
3. H. Gutberlet, B. Schallert, Selective catalytic reduction of NO_x from coal-fired power plants, *Catal. Today 16*:207 (1993).
4. J. Ando. NO_x abatement for stationary sources in Japan, EPA-600/7-83-027, May 1983, p. 177.
5. F. Nakajima, Air pollution control with catalysis—Past, present, and future, *Catal. Today 10(1)*:1 (1991).
6. B. Schonbucher, *Reduction of nitrogen oxides from coal-fired power plants by using the SCR process. Experiences in the Federal Republic of Germany with pilot and commercial scale DeNO$_x$ plants*, Joint Symposium on Stationary NO_x Control, San Francisco, Calif., March 6–9, 1989.
7. F. Janssen, R. Meijer, Quality control of DeNO$_x$ catalysts, *Catal. Today 16(2)*:157 (1993).
8. K. Kartte, H. Nonnenmaker, U.S. Patent No. 3,279,884, 1996.
9. M. Kunichi, H. Sakurada, K. Onuma, S. Fujii, Ger. Offen. 2,443,262, 1975.
10. F. Nakajima, M. Tacheuci, S. Matsuda, S. Undo, T. Mori, Y. Watanabe, M. Inamuri, U.S. Patent No. 4,085,193, 1978.
11. I. Wachs, E. Hardcastle, *Proceedings of 9th International Congress on Catalysis*, Chemical Institute of Canada, Ottawa, 1988, p. 1449.
12. Kiovski, P.B. Koradia, C.T. Lim, Evaluation of a zeolitic catalyst for NO_x reduction with NH_3, *Ind. Eng. Chem. Prod. Res. Develop. 19*:218 (1980).
13. J.P. Chen, M.C. Hausladen, R.T. Yang, Delaminated Fe_2O_3-pillared clay: Its preparation, characterization, and activities for selective catalytic reduction of NO by NH_3, *J. Catal. 151*:135 (1995).
14. H. Inaba, Y. Kamino, S. Onitsuka, T. Watanabe, Jpn. Kokai Tokkyo Koho 79,132,472, 1978.
15. P.B. Koradia, J.R. Kiovsky, Ger. Offen. 3,000,383, 1979.
16. M. Inomata, A. Miyamoto, Y. Murakami, Mechanism of the reaction of NO and NH_3 on vanadium oxide catalyst in the presence of oxygen under dilute gas condition, *J. Catal. 62*:140 (1980).
17. L. Lietti, P. Forzatti, G. Ramis, G. Busca, F. Bregani, Potassium doping of vanadia/titania de-NOxing catalysts: Surface characterization and reactivity study, *Appl. Catal. B—Environm. 3*:13 (1993).
18. J. Svachula, L.J. Alemany, N. Ferlazzo, P. Forzatti, E. Tronconi, F. Bregani, Oxidation of SO_2 to SO_3 over honeycomb DeNoxing catalysts, *Ind. Eng. Chem. Res. 32*:826 (1993).
19. B.S. Balzhinimaev, V.E. Ponomarev, N.P. Belyaeva, A.A. Ivanonv, G.K. Boreskov, Studies of fast relaxations in SO_2 oxidation on active components of vanadium catalysts, *React. Kinet. Catal. Lett. 30*:23 (1991).
20. P. Forzatti, L. Lietti, Recent advances in De-NOxing catalysis for stationary applications, *Heter. Catal. Rev. 3*:33 (1996).
21. K. Goldschmidt, VKR full-scale SCR experience on hard coal fired boilers, *Proceedings Joint Symposium on Stationary NO$_x$ Control*, New Orleans, Louisiana, March 23–26, 1987.
22. P. Necker, Experience gained by Neckarwerke from operation of SCR DeNOx units, *Proceedings Joint Symposium on Stationary NO$_x$ Control*, San Francisco, Calif., March 6–9, 1989.
23. H. Kuroda, I. Morika, T. Murotaka, F. Nakashjima, Y. Kato, A. Kato, Recent developments in the SCR system and its operational experiences, *Proceedings Joint Symposium on Stationary NO$_x$ Control*, San Francisco, Calif., March 6–9, 1989.

24. T. Mori, N. Shimizu, Operating experience of SCR systems at EPDC's coal-fired power stations, *Proceedings Joint Symposium on Stationary NO_x Control*, San Francisco, Calif., March 6–9, 1989.
25. M. Novak, H.G. Rych, Design, operation, and testing of the SCR NO_x reduction plants at the Duernrohr power station, *Proceedings Joint Symposium on Stationary NO_x Control*, San Francisco, Calif., March 6–9, 1989.
26. L. Balling, D. Hein, De-NO_x catalytic converters for various types of furnaces and fuels. Development, testing, and operation. *Proceedings Joint Symposium on Stationary NO_x Control*, San Francisco, Calif., March 6–9, 1989.
27. W.L. Prins, Z.L. Nuninga, Design and experience with catalytic reactors for SCR-DeNO_x, *Catal. Today 16*:187 (1993).
28. Hitachi Zosen product information (1990).
29. R. Samson, F. Goudria, O. Maaskant, T. Gilmore, The design and installation of a low-temperature catalytic NO_x reduction system for fired heaters and boilers, *Proceedings Fall International Symposium of the American Flame Research Committee*, San Francisco, Calif., October 8–10, 1990.
30. C.J. Pereira, K.W. Plumlee, M. Evans, CAMET metal monolith catalyst system for cogen applications, *Proceedings 2nd International Symposium on Turbomachinery, Combined Cycle Technologies and Cogeneration—IGTI*, Vol. 3, p. 131, 1991.
31. J. Svachula, N. Ferlazzo, P. Forzatti, E. Tronconi, F. Bregani, Selective reduction of NO_x by NH_3 over honeycomb DeNOxing catalysts, *Ind. Eng. Chem. Res. 32*:1053 (1993).
32. H.G. Lintz, T. Turek, Intrinsic kinetics of nitric oxide reduction by ammonia on a vanadia–titania catalyst, *Appl. Catal., A: General 85*:13 (1992).
33. C.U.I. Odenbrand, A. Bahamonde, P. Avila, J. Blanco, Kinetic study of the selective reduction of nitric oxide over vanadia–tungsta–titania/sepiolite catalyst, *Appl. Catal., B: Environ. 5*:117 (1994).
34. V. Tufano, M. Turco, Kinetic modelling of nitric oxide reduction over a high-surface area V_2O_5–TiO_2 catalyst, *Appl. Catal., B: Environ. 2*:9 (1993).
35. S.C. Wu, K. Nobe, Reduction of nitric oxide with ammonia on vanadium pentoxide, *Ind. Eng. Chem. Prod. Res. Dev. 16(2)*:136 (1977).
36. J.W. Beeckman, L.L. Hegedus, Design of monolith catalysts for power plant NO_x emission control, *Ind. Eng. Chem. Res. 30*:969 (1991).
37. J. Maragozis, Comparison and analysis of intrinsic kinetics and effectiveness factors for the catalytic reduction of NO with ammonia in the presence of oxygen, *Ind. Eng. Chem. Res. 31*:987 (1992).
38. J.B. Lefers, P. Lodder, G.D. Enoch, Modelling of selective catalytic denox reactors—strategy for replacing deactivated catalyst elements, *Chem. Eng. Technol. 14*:192 (1991).
39. E. Tronconi, P. Forzatti, Adequacy of lumped parameter models for SCR reactors with monolith structure, *AIChE J. 38*:201 (1992).
40. E. Tronconi, P. Forzatti, J.P. Gomez Martin, S. Malloggi, Selective catalytic removal of NO_x: A mathematical model for design of catalyst and reactor, *Chem. Eng. Sci. 47*:2401 (1992).
41. N. Wakao, J.M. Smith, Diffusion in catalyst pellets, *Chem. Eng. Sci.* 17:825 (1962).
42. R.S. Cunningham, C.J. Geankoplis, Effects of Different Structures of Porous Solids on Diffusion of Gases in the Transition Region, *Ind. Eng. Chem. Fund. 7*:535 (1968).
43. M.A. Buzanowski and R.T. Yang, Simple design of monolith reactor for selective catalytic reduction of NO for power plant emission control, *Ind. Eng. Chem. Res. 29*:2074 (1990).
44. E. Tronconi, A. Beretta, A.S. Elmi, P. Forzatti, S. Malloggi, A. Baldacci, A complete model of SCR monolith reactors for the analysis of interacting NO_x reduction and SO_2 oxidation reactions, *Chem. Eng. Sci. 49*:4277 (1994).
45. D.E. Mears, Tests for transport limitations in experimental catalytic reactors, *Ind. Eng. Chem. Proc. Res. Develop. 10*:541 (1971).

46. A. Beretta, E. Tronconi, L.J. Alemany, J. Svachula, P. Forzatti, Effect of morphology on the reduction of NO_x and the oxidation of SO_2 over honeycomb SCR catalysts. *New Developments in Selective Oxidation II* (V. Cotés Corberan and S. Vic Bellon, Eds), Amsterdam: Elsevier, 1994, p. 869.
47. C. Orsenigo, A. Beretta, P. Forzatti, G. Svachula, E. Tronconi, F. Bregani, A. Baldacci, Theoretical and experimental study of the interaction between NO_x reduction and SO_2 oxidation over $DeNO_x$–SCR catalysts, *Catal. Today, 27*:15 (1996).
48. S. Matsuda, T. Kamo, A. Kato, F. Nakajima, Deposition of ammonium bisulfate in the selective catalytic reduction of nitrogen oxides with ammonia, *Ind. Eng. Chem. Prod. Res. Develop. 21*:48 (1982).
49. E. Hums, M. Joisten, R. Muller, R. Sigling, H. Spielmann, Innovative lines of SCR catalysis: NO_x reduction for stationary diesel engine exhaust gas and dioxin abatement for waste incineration facilities, *Catal. Today, 27*:29 (1996).
50. A.S. Noskov, L.N. Bobrova, Y.S. Matros, Reverse-process for NO_x-off gases decontamination, *Catal. Today 17*:293 (1993).
51. S.L. Andersson, P.L.T. Gabrielsson, C.U.I. Odenbrand, Reducing NO_x in diesel exhausts by SCR technique: Experiments and simulations, *AIChE J. 40*:1911 (1994).
52. E. Tronconi, L. Lietti, P. Forzatti, S. Malloggi, Experimental and theoretical investigation of the dynamics of the SCR-$DeNO_x$ reaction, *Chem. Eng. Sci., 51*:2965 (1996).
53. Y.S. Matros, Catalytic processes under unsteady-state conditions, *Studies in Surface Science and Catalysis*, Vol. 43 (B. Delmon and J.T. Yates, Eds.), Amsterdam: Elsevier, 1989.
54. Y.S. Matros, G.A. Bunimovich, A.S. Noskov, The decontamination of gases by unsteady-state catalytic method. Theory and practice, *Catal. Today 17*:261 (1993).
55. K. Hedden, B. Ramanda Rao, N. Schon, Selektive katalytische Reduktion von Stickstoffmonoxid mit Ammoniak unter periodisch wechselnden Reaktionsbedingungen, *Chem. Ing. Tech. 65*:1506 (1993).
56. M. Kotter, H.G. Lintz, T. Turek, Selective catalytic reduction of nitrogen oxide by use of the Lijungstroem air-heater as reactor: A case study, *Chem. Eng. Sci. 47*:2763 (1992).

6

Catalytic Fuel Combustion in Honeycomb Monolith Reactors

Marcus F. M. Zwinkels, Sven G. Järås, and P. Govind Menon
KTH—Royal Institute of Technology, Stockholm, Sweden

I. INTRODUCTION

One of the most promising applications for monolith catalysts is catalytic fuel combustion. Outstanding features of monolith catalysts, such as low pressure drop at high mass throughputs, and high mechanical strength, are particularly well utilized in catalytic combustion. The most important feature of catalytic fuel combustion is the catalytic activation of fuel and/or oxygen molecules, allowing complete combustion of gaseous or gasifiable fuels at temperatures below 1500°C, which can be compared to common flame temperatures of more than 2000°C. The relatively low temperatures in a catalytic combustor lead to a practically insignificant thermal formation of nitrogen oxides, which is the main advantage of catalytic fuel combustion. The opportunities offered by catalytic fuel combustion have attracted strong attention, most of which has been directed toward the application in gas turbine combustors.

This chapter deals with the specific research and development issues related to the use of a catalyst at temperatures up to 1500°C in a catalytic combustor. A short survey of emission problems related to combustion and strategies for emission reduction is followed by a discussion of catalysis in combustion. Specific aspects of monolith combustion catalysts, such as material problems and combustor design, are then treated briefly.

A. Emissions from Combustion Processes

Since the second half of the nineteenth century, combustion of fuels in engines has been used to generate mechanical power. This mechanical power is used for transportation or conversion into electric energy. Combustion engines, such as spark ignition engines, diesel engines, and gas turbines, have undergone continuous development since their introduction. However, it was not before the 1950s that an awareness of the environmental effects related to combustion of fuels evolved. Phenomena such as acid rain, smog, greenhouse effect, and ozone depletion are now recognized as serious problems.

The just-mentioned problems are related to the typical emissions from most combustion processes: carbon monoxide (CO) and unburned hydrocarbons (UHC), both resulting from incomplete combustion, and nitrogen oxides ($NO_x = NO + NO_2$). NO_x may be produced from molecular nitrogen from the combustion air or from nitrogen-containing compounds in the fuel.

The desire to reduce emissions from various combustion processes has continually gained importance since the 1950s. This has inspired the development of technologies for improving the combustion process or cleaning the combustion products. Catalysis has played and will play a dominating role in this development [1].

The most prominent example of exhaust cleanup technology that has been developed is the three-way catalyst for automobiles, which simultaneously eliminates UHC, CO, and NO_x [2], which has been dealt with in detail in Chapter 4. Abatement of emissions from sources other than automobiles has been developed and commercialized during the last decades as well. A striking example is the catalytic abatement of volatile organic compounds (VOC) from industrial processes [3]. Furthermore, selective catalytic reduction (SCR) of nitrogen oxides by ammonia over, typically, a V_2O_5–TiO_2–SiO_2 catalyst, is now widely applied to remove NO_x in effluent gas from heavy-oil- and coal-burning boilers and electric power plants [4].

Nojiri et al. [5] report in a recent review that 250 plants in Japan alone and over 140 plants elsewhere in the world are applying SCR technology today. More detailed aspects of SCR have been dealt with in Chapter 5. Currently, removal of NO_x in oxidizing atmospheres, e.g., diesel exhaust gases, circumventing the use ammonia is one of the major challenges in catalysis R&D [6,7].

The methods just discussed for reduction of emission of pollutants have one thing in common: They are all tailpipe solutions.

B. Preventive Strategies for Emission Reduction

Much effort has been directed at reducing emissions by improving or modifying the combustion process itself. Two approaches have been dominating the research and development efforts in this field: upgrading the fuel and improving the combustion.

Typical examples of the first approach, upgrading or improving the fuel, are the catalytic removal of sulfur and aromatic compounds in automotive fuels [8,9]. A shift from the use of coal to the use of natural gas as a fuel in many industrial applications has led to reduced emissions, due to the favorable composition of natural gas as compared to coal. However, future developments of combustion processes will most likely include the use of more low-grade fuels, such as heavy fuel oils [10]. The use of coal will increase again, which can be related to its relative abundance. Finally, low-Btu fuels, such as gasified biomass or gasified coal will play an important role [11].

The second approach for reduction of emissions from combustion processes is the improvement of the combustor itself. Improved fuel–air mixing, steam injection, advanced flow patterns, sophisticated injection systems, and flue gas recirculation have resulted in significant reduction of emissions of NO_x, CO, and UHC. However, continual development is needed to meet toughening future regulations.

Besides control of emissions from combustion engines, high engine efficiency becomes increasingly important. Gas turbines are continually gaining ground, representing both a power generation technology with low emission levels as compared to other combustion engines and potentially high efficiencies. The most serious pollutants from gas turbines are nitrogen oxides. This is especially the case when clean fuels, such as natural gas, are used. NO_x is inevitably formed at the high temperatures occurring in gas turbine combustors. In recent years several approaches have been tested for reducing the emissions from gas turbines. Catalytic combustion, being the most promising option for emission reduction, has received considerable attention during the last 10 years and has been reviewed by various authors [12–17]. For a better understanding of the low-NO_x potential of catalytic combustion and the problems of the technology, the next section discusses NO_x production during combustion and approaches for decreasing NO_x production.

C. Production of NO_x in Combustion Processes

The formation of NO_x in combustion processes has been extensively studied, and the possible pathways are well documented [10,18–20]. The predominant compound at high emission levels is nitric oxide (NO). Nitrogen dioxide (NO_2) is more dominant at lower temperatures, e.g., at low-load conditions.

The three different mechanisms according to which NO can be produced are as follows.

Thermal NO: formed by oxidation of atmospheric nitrogen, which is generally accepted to proceed through the Zeldovich chain mechanism [18,21]. The overall reaction and the elementary reactions are:

$$O_2 + N_2 \Leftrightarrow 2NO \tag{1}$$

$$O_2 \Leftrightarrow 2O \tag{2}$$

$$O + N_2 \Leftrightarrow N + NO \tag{3}$$

$$N + O_2 \Leftrightarrow NO + O \tag{4}$$

The level of thermal NO_x emissions is nearly linearly dependent on residence time and rises exponentially with flame temperature; i.e., $NO_x \propto e^{0.009T}$ [10,22]. This is illustrated in Fig. 1, showing the NO equilibrium concentration as a function of temperature [20].

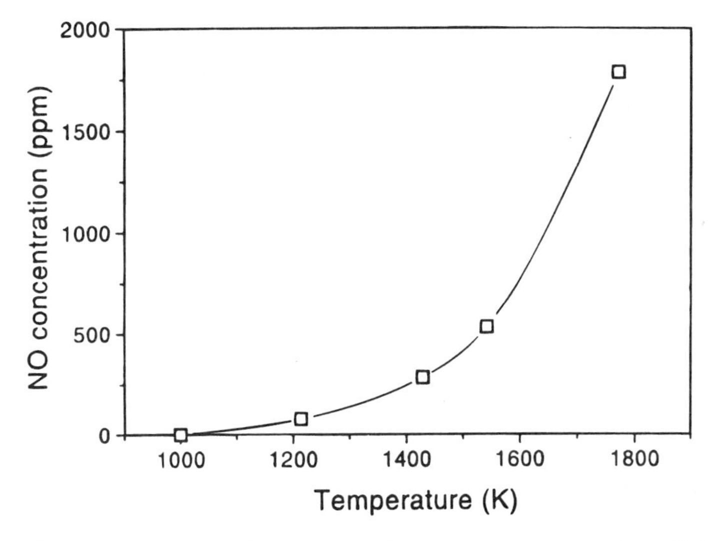

Figure 1 NO equilibrium concentration versus temperature for a 40:1 mixture of N_2/O_2. (From Ref. 20.)

Excess oxygen increases the production rate, but excess air lowers the flame temperatures, which leads to a decrease in the NO_x production rate. Hence, maximal NO_x production is observed at near-stoichiometric conditions.

Prompt NO_x: formation through fast radical reactions at the flame front. Hydrocarbon fragments are believed to attack molecular nitrogen [10,23] under fuel-rich conditions. The formation of hydrogen cyanide is believed to be an important first step:

$$\begin{aligned} CH + N_2 &\rightarrow HCN + N \\ HCN + \text{oxidant} &\rightarrow \cdots \qquad (5) \\ &\rightarrow NO + \cdots \qquad (6) \end{aligned}$$

It is difficult to quantify the amount of prompt NO_x produced in flames, but the values are likely to be between 0 and 30 ppmv in gas turbine combustors, i.e., at around 15 vol% O_2 [18]. Low temperatures and fuel-rich conditions favor high, prompt NO_x levels.

Fuel NO_x: oxidation of nitrogen-containing compounds in the fuel. Fuel NO_x levels are only slightly dependent on temperature, in contrast to thermal NO_x. The conversion of fuel nitrogen to NO_x is high under lean conditions and decreases at higher nitrogen content of the fuel. The nature of the nitrogen compound also effects the conversion to NO_x, with high-boiling-point nitrogen compounds giving high yields of NO_x [10].

Thermal NO_x is the dominating type of NO_x produced in gas turbine combustors, which often use fuels with low nitrogen content. All efforts regarding emission reduction should thus aim for a reduction of thermal NO_x. This is typically achieved by reducing the flame temperature and residence time. However, this results in increased CO and UHC levels, which causes many solutions to represent some sort of compromise.

It should be noted that the dominance of thermal NO_x is not observed for fuels with high nitrogen contents, such as gasified biomass or coal. Most of the NO_x produced from such fuels originates from fuel nitrogen [24].

D. Low-NO_x Gas Turbine Combustors

A gas turbine is an external combustion engine, which means that mechanical power is generated in two steps, with heat as an intermediate. Heat is produced from the chemical energy in the fuel in the combustion chamber. The conversion of heat into mechanical energy is then achieved in the turbine. A schematic view of a simple open-cycle gas turbine unit is given in Fig. 2 [e.g., 25,26].

The main components of any gas turbine unit are the air compressor, the combustion chamber, and the turbine. Compressed air is fed to the combustion chamber, where it is mixed with the fuel. The complete combustion of the fuel generates heat, which is taken up by the combustion gas. The hot gases from the combustion chamber are cooled by mixing them with bypass air in order to reach a suitable turbine inlet temperature, after which they expand through the turbine and to the atmosphere. Some of the bypass air may be used to cool the turbine blades. The work produced in the turbine is used to power the compressor and to generate mechanical power. The mechanical power is used to produce electricity in a generator.

The performance of a gas turbine is dependent mainly on the turbine inlet temperature (TIT), the pressure ratio (p_2/p_1), and the efficiencies of the compressor, the turbine, and,

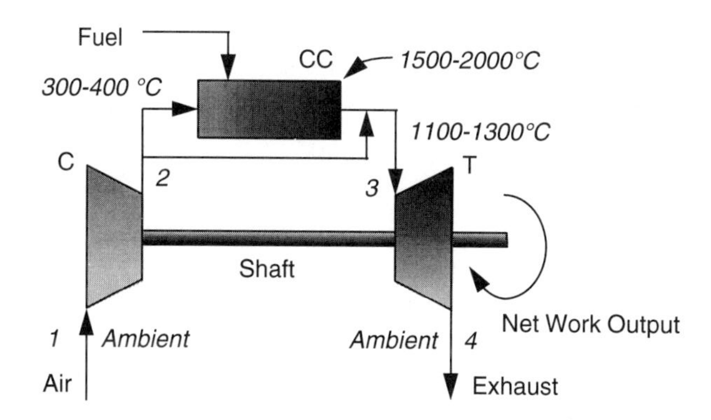

Figure 2 Open-cycle gas turbine unit. C: compressor; CC: combustion chamber; T: turbine.

in regenerative gas turbines, the heat exchanger. An increased TIT leads to improved efficiency. In practice, the TIT is limited by the maximum operating temperature of the construction materials. The efficiency is also a function of the pressure ratio; i.e., an optimum in the cycle efficiency is obtained at a specific pressure ratio for each maximum cycle temperature. The optimum cycle efficiency is increased and shifted to higher pressure ratios for higher TIT values.

The TIT of gas turbines installed today are typically around 1100°C. However, you can observe a continuous tendency of the TIT to increase. Next-generation large gas turbines are expected to have a TIT of around 1300°C. This may be achieved by cooling the high-pressure stage of the turbine or by using ceramic materials in the turbine wheel.

Often the maximum temperatures in the combustion chamber exceed the TIT. Dilution air is mixed with the hot combustion gases downstream of the combustion chamber to obtain a suitable temperature. Therefore, the gas turbine efficiency is not necessarily related to the maximum temperature in the combustor. This fact is important for catalytic combustors, operating at relatively low temperatures, as will be discussed in the next section.

As mentioned previously, the main pollutants from gas turbines are CO, UHC, and NO_x. The first two, of which CO usually occurs in higher concentrations, are related primarily to incomplete combustion, which is a measure of the combustion efficiency in the combustion chamber. Hence, all measures that improve the combustion, will have a positive influence on the CO and UHC emissions.

Gas turbines have continually developed toward higher efficiency and lower emissions. Various approaches have been followed in the development of low-emission gas turbine combustors [18,27]. The following approaches are regarded as potential options for future low-emission combustors.

1. *RQL Combustion: Rich Burn—Quick Quench—Lean Burn*

A partial combustion is achieved in a first zone with excess fuel, after which air is added under intense turbulence. The final combustion occurs under lean conditions in the second zone. The rich conditions in the first zone ensure low NO_x levels and combustion stability, whereas the large air excess in the second zone, and hence the low temperature, avoids the formation of thermal NO_x. This approach yields 50% reduction in NO_x emissions today [27] as compared to conventional diffusion flame combustors. The problem with this approach is the transition from rich to lean conditions through the stoichiometric point. This transition should be fast enough to prevent the occurrence of high-temperature zones, which would lead to thermal NO_x formation.

2. Lean-Premix Combustion

Air and fuel are intimately mixed prior to entering the combustion zone, and hence the local air–fuel ratios can be controlled and variations minimized. The homogeneous air–fuel mixture is combusted at a very high air–fuel ratio, and hence involves only a small adiabatic temperature rise. Therefore, the maximum temperature in the combustor is kept at levels at which the thermal NO_x concentration is low. Lean-premix systems suffer from stability problems at ultralean conditions, i.e., low temperatures, and may therefore require stabilization by a diffusion flame [28]. It was already mentioned previously that optimizing both CO/UHC and NO_x is difficult. This can be deduced from Fig. 3, showing the NO and CO emissions versus adiabatic flame temperature for a hypothetical lean-premix combustor. At low temperatures, the CO emissions increase rapidly, due to flame instability (the comparable increase in NO_x levels at high temperatures has already been discussed in the previous section). It is clear that a temperature zone exists where low levels of both CO and NO_x may be obtained.

3. Catalytic Combustion

A catalytic combustor is basically a lean-premix combustor, in which the combustion is stabilized by a catalytic surface. Hence, the expression *catalytically ignited thermal combustion* or *catathermal combustion* is also used [15]. The catalyst stabilizes the combustion at low temperatures, which broadens the window in which both CO and NO_x are sufficiently low; cf. Fig. 3. The next section briefly discusses the prominent features of catalytic combustion.

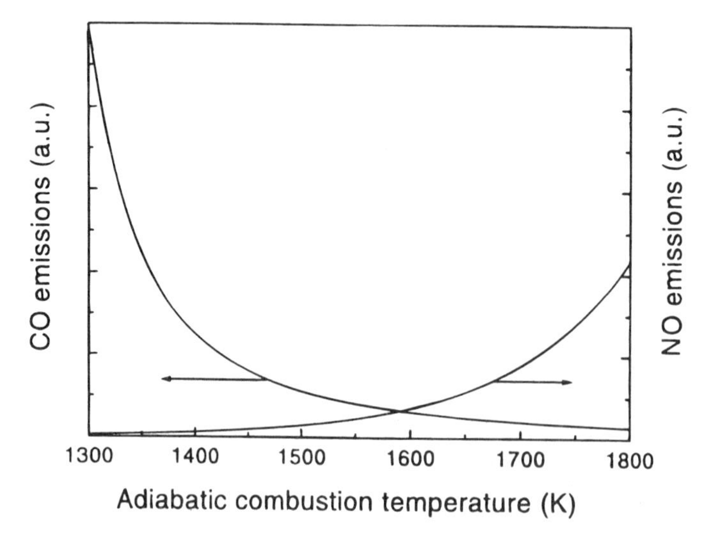

Figure 3 Influence of adiabatic flame temperature on CO and NO emissions for a hypothetical lean premix combustor. (From Ref. 26.)

E. Catalytic Combustion for NO_x Control

The previous section clearly demonstrated the major challenge presented by the desire to reduce emissions from gas turbine combustors. This challenge is to minimize the NO_x emissions while maintaining low levels of CO and UHC—in other words, to achieve high combustion efficiency. Catalytic fuel combustion is one of the most promising methods for achieving this goal. The catalyst lowers the activation energy for the combustion reactions. Hence, the combustion may start at much lower temperatures than would be possible for homogeneous gas-phase combustion. This means that complete combustion of a certain air–fuel mixture can be achieved at much lower flame temperatures when combusted over catalyst as compared to homogeneous combustion. These low flame temperatures result in the main advantage of catalytic combustion: *complete combustion of a fuel with low CO and UHC emissions at temperatures so low that the formation of thermal NO_x is practically avoided.* Other advantages of catalytic combustion over homogeneous gas-phase combustion are better combustion stability, especially for lean mixtures, and hence a wider operating window of fuel–air ratios. Moreover, a more even radial temperature profile can be obtained in a catalytic combustor. Finally, a catalytic combustor will produce less noise than a flame combustor.

An advantage offered by catalytic combustion that may be important for heat generation is the potentially improved heat transfer, since heat is generated on a solid surface. Heat generation systems operating at catalyst temperatures up to 800–1000°C may thus be controlled more efficiently through external cooling of the catalyst.

However, most of the following discussion will deal with catalytic combustion for gas turbine applications. Figure 4 shows a schematic view of an open-cycle gas turbine unit with a catalytic combustor (cf. a conventional unit in Fig. 2.).

It is essential to mention here that the reduction of the adiabatic combustion temperature in a catalytic combustor as compared to a conventional gas-phase combustor does not lead to a decrease in the efficiency of the gas turbine. The turbine inlet temperature, which largely determines the efficiency of a gas turbine, is the same for turbines with catalytic or homogeneous combustors. On the other hand, the necessary cooling of the hot gases after the combustion zone in a conventional combustor is omitted in a catalytic combustor. However, the turbine blades still need to be cooled with air, as for a gas

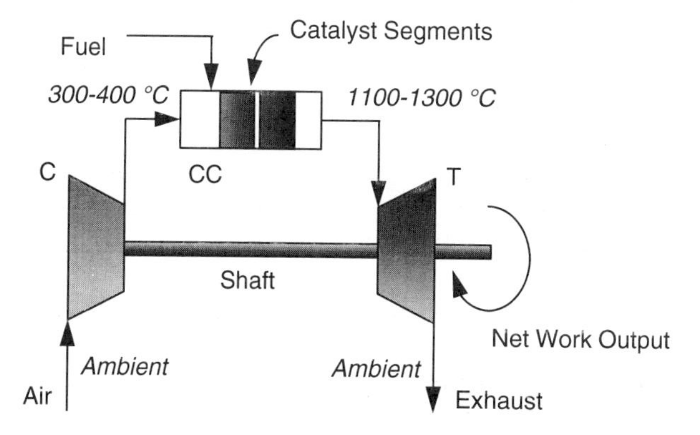

Figure 4 Open-cycle gas turbine unit with catalytic combustor. C: compressor; CC: combustion chamber; T: turbine. (From Ref. 26.)

turbine with a homogeneous combustor. This will be necessary until turbine materials are developed that have sufficiently high maximum operating temperatures.

II. CATALYSIS IN COMBUSTION

Complete catalytic oxidation has received increasing attention since the late 1950s as a method for cleaning the exhaust gases from combustion processes and other industrial activities, such as painting, printing, and the synthesis of organic compounds. Catalysts have been developed for the removal of small amounts of a number of volatile organic compounds, VOC. Moreover, much effort has been directed toward the elucidation of oxidation mechanisms. An excellent survey of this field was recently published by Spivey [3].

Catalytic combustion of fuels was introduced in the beginning of the 1970s [29,30]. During the first decade of the development of catalytic combustion, the United States had a leading position, with large projects at, amongst others, Westinghouse and Engelhard. However, interest in the United States faded, and since the beginning of the 1980s Japan has caught up and taken the lead. This may be concluded from reading the patent literature [17]. Recent years have seen a more global interest in catalytic combustion as a way to reduce NO_x emissions from combustion processes. Large research programs are going on in the United States, Japan, and Europe, most often focusing on the development of catalytic gas turbines. In 1994, during the Second International Workshop on Catalytic Combustion, held in Tokyo, the latest developments were presented and discussed. The first commercial catalytic gas turbine is expected within 5 years.

This section briefly discusses some aspects of catalytic combustion mechanisms, i.e., surface reaction kinetics and heterogeneous–homogeneous reactions. Based on this discussion and the previous section, the extreme demands on combustion catalysts are presented. Finally, the role of mathematical modeling of this complex catalytic system is examined.

A. Catalytic Steering of Free-Radical Reactions

Homogeneous gas-phase oxidations and combustion of hydrocarbons proceed normally via free-radical reactions. When a solid surface, for instance, of a catalyst, is additionally present, conditions are fulfilled for the occurrence of heterogeneous–homogeneous reactions. The term *heterogeneous–homogeneous reactions* is often used "to denote different variants of a process in which a solid surface participates collectively with a surrounding volume phase, regardless of the consequences of its action" (Kiperman [31, p. 37]). Such reactions can begin on a solid catalyst surface and then propagate or go to completion, with some or all of the active species escaping to the gas phase. The escape of intermediate species to the volume must be aided by the formation of weak bonds between these species and the catalyst surface—a condition often met at high temperatures, as in oxidation or combustion. Indeed, heterogeneous–homogeneous reactions occur readily in oxidation catalysis, as indicated in Table 1.

Azatyan [32,33] has shown that in chain processes, adsorbed atoms and radicals not only recombine but also react with the gas-phase species, thereby producing new radicals to be supplied to the gas phase. Hence, heterogeneous reactions of radicals may

Table 1 Typical Temperatures for the Transfer of Reactive Intermediates from the Catalyst Surface to the Gas Phase in Catalytic Oxidations

Reactant	Catalyst	Observed lower limit of transfer to volume (°C)
H_2	Pt	>500
CO	Pt	180
CH_4	Pd	200
	SiO_2, Al_2O_3	650
$CH_4 \rightarrow C_2H_6$	MgO, $Li_2O \cdot MgO$	>500
CH_4, C_2H_4, C_2H_6 C_3H_6	Bi_2O_3, Bi–Mo, PbO, MgO	500
Hydrocarbons C–C_8	Al_2O_3, Cr_2O_3	400
C_3H_6	BiO_3-MoO_3	450
Benzene*	Pt	80–130
	Pd	220
Toluene*	Pt	180
Styrene*	Pt	240
CH_2O	Al_2O_3	650
CH_3OH	Ag, Ag-pumice	500
	Pt	650
CH_3OH	Gr IV metal oxides	500

*Complete oxidation to CO_2 and H_2O.
Adapted from Ref. 31.

cause chain termination as well as chain propagation and chain branching. These reactions of chain propagation differ fundamentally from the normal heterogeneous chain initiation, but both these types of reactions can still be classified as heterogeneous–homogeneous. Their joint occurrence gives an added incentive and importance to the use of catalyst surfaces as potential steering or controlling tools in chain processes, including combustion.

Reviewing the available data in the literature, Kiperman [31] has discussed (1) the simplest variants of mechanisms for heterogeneous–homogeneous reactions, (2) the main regularities of these reactions under steady-state and non-steady-state conditions, and (3) examples illustrating specific features of these reactions.

The escape of intermediate species from the catalyst surface to the volume is a necessary condition for the occurrence of heterogeneous–homogeneous reactions. This can sometimes depend critically on the initial ratio of the reactants. For instance, Ismagilov et al. [34] observed that the complete oxidation of alcohols and amines was possible via the gas-phase formation of radical-like species if the ratio of initial components had a near stoichiometric value, but not when there is an excess of oxygen. Another essential factor, which determines reaction propagation in the volume, is the form and pore structure of the catalyst. The radical-like species could be deactivated or converted into final products if their mean free path is less than the pore size in the catalyst. If a catalyst granule is 0.01 cm in size and has through-pores, a given radical species can undergo 1000 collisions in passing through the pore at a pressure of 1 bar. Similarly, a rough estimate can show that 10^{13} species can leave a catalyst surface of total area 1 m^2 and outer surface area of 0.01 m^2, assuming that the fractions of heterogeneous and heterogeneous–homogeneous reactions are equal to 1% each and that the rate of the latter is 100 times faster (the turnover number is taken to be 0.01 sec^{-1}). This is sufficient for reaction transfers from the catalyst surface to the volume and for the chain mechanism of their further propagation in the volume (cf. Kiperman [31]).

One important conclusion from Kiperman's analysis concerns the impact on the kinetics of the interrelations between the two components of heterogeneous–homogeneous reactions. This problem is discussed by Golodets [35] and Garibyan and Margolis [36], who followed the relationship between the beginning of ignition of hydrogen oxidation in the volume and the activity of oxide catalysts initiating the transfer of the reactions from the catalyst surface to the volume. The conclusion was that low-activity catalysts are more useful for the transfer from the surface to the volume. At the same time, acceleration of the process in the volume must compensate for low rates of surface reactions. These conclusions are of direct interest and relevance to catalytic combustion at high temperatures.

Kiperman [31] also warns that detection of free radicals in the postcatalyst volume in itself cannot serve as concrete proof of their direct participation in the process. The relation also has to be revealed between the nature of these radicals formed in the volume and the intermediates of the true heterogeneous component of the reaction. Obviously, sophisticated analytical and characterization procedures are needed to elucidate the nature of the species reacting on and desorbing from a catalytic surface. A powerful tool to study adsorption and desorption of radicals from surface is laser-induced fluorescence, applied to hydroxyl and oxygen radicals by a number of researchers; cf. Ref 37. Such techniques will continue to aid in the elucidation of heterogeneous–homogeneous mechanisms.

It was proposed by Pfefferle and Pfefferle [15] that a catalytically active surface might not only generate species that can desorb and react in the gas phase, but also act as a sink for radicals. This would imply that the fast homogeneous radical reactions would be retarded in the presence of catalyst. Moreover, they found that the consumption of fuel and oxygen molecules by heterogeneous reaction at a catalytic surface could retard the homogeneous ignition. For example, ignition of an ethane–air mixture over a heated quartz surface could occur at temperatures lower than for a heated platinum surface for ethane–air ratios over 0.35 [38]. The reader is referred to Pfefferle and Pfefferle [15,39] for reviews of the earlier work in this area.

The catalytic formation or destruction of gas-phase radicals is an important phenomenon, since it may have an effect on the formation of NO_x. This was recently demonstrated by Griffin et al. [40], who studied the formation of NO_x in a catalytic combustor with downstream homogeneous combustion. They found that when the adiabatic flame temperature and total residence time were kept constant, the concentration of NO_x decreased if the contribution of the catalytic conversion increased as compared to the homogeneous combustion. Measurements were performed at flame temperatures between 1300 and 1500°C. Their results confirmed the idea posed earlier by Markatou et al. [41] that higher hydroxyl and oxygen radical concentrations in the gas phase would cause higher NO_x levels.

An interesting coupling between heterogeneous and homogeneous reactions is provided by thermocatalytic oxidation or combustion, which consists of combined thermal and catalytic processes in sequence. This is in a way the converse of catathermal oxidation or combustion, involving the catalytic steering of free-radical reactions. The first step of the thermocatalytic process is thermal oxidation, as in a flame, followed by a catalytic oxidation. Zieba et al. [42] have reported the higher efficiency (by 10–30% in conversion) of the thermocatalytic process for the elimination of volatile organic compounds (VOCs), when H, O, and OH radicals are generated by a hydrogen flame just under a supported-Pt or mixed-oxide combustion catalyst. At constant temperature of the catalyst, the conversion of methane increased as the flame was brought nearer to the catalyst bed. Radicals formed in the postflame zone have an activating effect on the reactions occurring in the catalyst bed, whereas water vapor released in the hydrogen/air flame exhibits an inhibiting effect on the subsequent catalytic conversion.

B. Mechanism and Kinetics

Complete catalytic oxidation of hydrocarbons to carbon dioxide and water may proceed according to different mechanisms, depending on the type of catalyst and on process conditions, such as hydrocarbon concentration and temperature. Usually, total oxidation is thought to take place through either the Langmuir–Hinshelwood mechanism (LH), with both oxygen and the hydrocarbon chemisorbed on the catalyst surface, or the Eley-Rideal mechanism (ER), in which chemisorbed oxygen reacts with gaseous species. Participation of lattice oxygen is also possible for metal oxides at elevated temperatures according to the Mars–Van Krevelen mechanism, which was elegantly shown by Arai and coworkers [43].

The kinetics and mechanism of methane combustion have been the subject of many investigations, e.g., Refs. 43–47, because of the importance of natural gas as a potential fuel for catalytic combustors. Under conditions expected in catalytic combustors, i.e., excess oxygen, a first order in methane is generally observed [48], whereas a variety of orders has been observed for other hydrocarbons [13]. The actual mechanism appears to be quite complex and depends on the fuel used. For instance, inhibiting effects are observed for the products carbon dioxide and water in methane combustion over supported palladium catalysts [49,50]. The inhibition of methane adsorption and the formation of a surface palladium hydroxide were proposed to explain the observation.

Another relevant issue in methane combustion over palladium is the significant increase in activity that is observed with time on stream [46,49]. This may be explained by the removal of chloride ions from palladium precursor [46] but was also observed for chlorine-free precursors [49]. A change in the morphology of the palladium oxide on the surface during operation may be the cause for the activity enhancement.

An interesting aspect of supported palladium is the fact that palladium oxide is the thermodynamically stable phase below 1052 K, whereas the metallic palladium is stable above this temperature. It was found that the oxidic form was much more active in methane combustion than the metallic form [51,52]. This phenomenon is actually used in controlling the catalyst temperature in one of the proposed combustor concepts (See Section IV.B) and is treated in more depth in Section III.A.

In short, catalytic oxidation of hydrocarbons is a structure-sensitive reaction, and its mechanism is strongly dependent on the type of catalyst and on process conditions. This means that the morphology of the active phase will affect the catalyst activity, and hence the preparation procedure will have a strong influence on catalyst performance.

It is worth pointing out some differences between complete catalytic oxidation for abatement of hydrocarbon emissions and catalytic fuel combustion for the generation of heat and electric power. The most important difference is the operating temperature, which is between 100 and 500°C for catalytic oxidation and between 300 and 1000–1400°C for catalytic fuel combustion. The other main difference is the concentration of the combustible compound. In catalytic oxidation, trace amounts of hydrocarbons may be removed, which results in near-isothermal operation. Fuel concentration in catalytic combustion may be such that the adiabatic temperature rise over the catalyst may be as high as 500–1000°C.

A typical performance plot for a combustion catalyst is given in Fig. 5. We can distinguish four different regions in the reaction rate versus temperature plot for a combustion catalyst. In region A, the kinetics of the chemical reaction control the rate of combustion. As the rate increases with rising temperature, the buildup of heat results in catalyst light-off (B). In region C, the heterogeneous reaction rate is high and the transport of fuel and air to the catalyst surface will limit the overall rate. At still higher temperatures, homogeneous combustion may start playing an increasingly important role (D).

The coupling of heterogeneous reactions on the catalyst surface and homogeneous gas-phase reactions, as discussed in the previous section, is important for the design and operation of a catalytic combustor with maximum temperatures over 900°C, which is the case for gas turbine combustors. It is worth pointing out that the ignition of the fuel–air mixture over the catalyst at much lower temperatures than possible for homogeneous gas-phase combustion is the reason why catalytic combustors can operate at flame temperatures as low as 1100°C [53]. Hence, the formation of thermal NO_x, which is the most important type of NO_x for gas turbine combustors, is practically avoided.

C. Extreme Demands on Combustion Catalysts

The use of a catalyst in a combustion environment places severe demands on the material properties. The desired properties of a combustion catalyst can be deduced from Figs. 4 and 5. The catalyst must have high combustion activity, enabling catalyst ignition at the lowest possible temperature, i.e., at the compressor outlet temperature. Arai and coworkers showed the importance of a high catalyst surface area to the low-temperature activity and, hence, the ignition temperature [43]. Theoretical calculations performed in our laboratory confirmed the importance of catalyst surface area and porosity to the ignition of combustion catalysts [54]. The high surface area must be maintained during

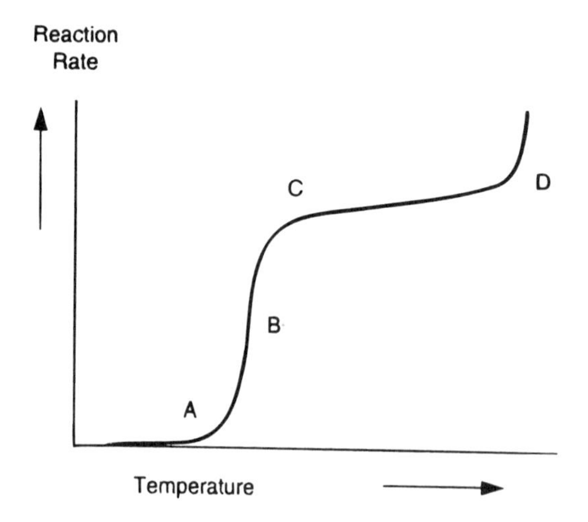

Figure 5 Activity versus temperature plot for a combustion catalyst. A: kinetic regime, B: light-off, C: mass transfer control, D: homogeneous reactions. (From Ref. 15.)

extended operation. This implies that the catalyst must stand temperatures up to 1000–1400°C in an atmosphere containing oxygen and steam. The catalyst must be designed for rapid ignition; hence, the thermal mass of the catalyst must be low. On the other hand, high thermal mass facilitates reignition after catalyst extinction. The reaction rate under mass-transport-controlled conditions is determined by the catalyst support, which must be optimized for maximum internal and external mass transport rates. Moreover, the catalyst must be able to withstand the strong and rapid temperature fluctuations that may occur in a gas turbine combustor. Hence, the thermal shock resistance must be high, which is favored by a low thermal expansion coefficient, high mechanical strength, and high thermal conductivity. The catalyst must favor complete burnout of the fuel for low UHC and CO emissions at temperatures low enough to minimize NO_x formation. Finally, the catalyst must allow large mass throughputs at low pressure drop.

A combustion catalyst must thus simultaneously fulfill requirements of high activity at combustor inlet conditions and high stability at the maximum temperatures occurring in the catalytic combustor. Unfortunately, these are contradictory demands. This was demonstrated by McCarty and Wise [55]. Figure 6, taken from their study, shows the relationship between methane oxidation activity and the stability of various perovskite-type materials ($LaMO_3$). The trade-off between activity and stability is clear.

The contradictory demands of high activity and high stability impose the need to apply reaction engineering to the design of catalytic combustors. Promising approaches are presented in the next section.

Low pressure drop was already mentioned as one of the desired properties of a combustion catalyst. Hence, the monolithic or honeycomb type of catalyst is considered

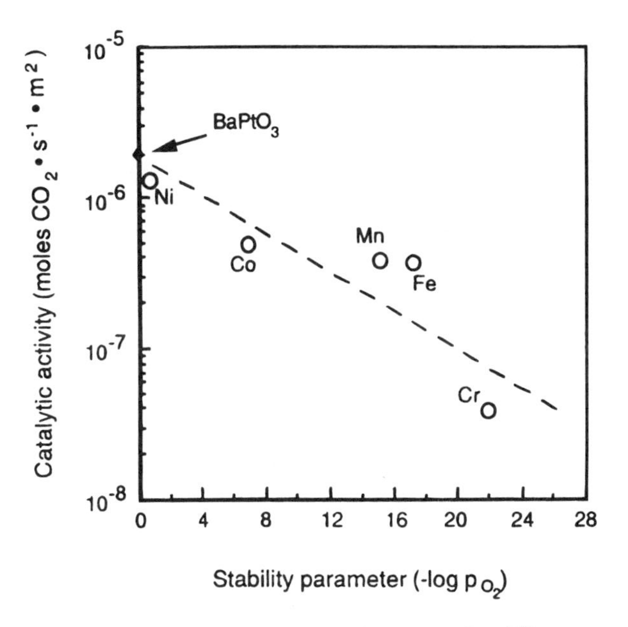

Figure 6 Specific methane oxidation rate and stability parameter of various perovskites $LaMO_3$ (M = Ni, Co, Mn, Fe, Cr). The stability parameter is expressed as the equilibrium oxygen pressure of the solid at 1273 K. (From Ref. 55.)

to be the best configuration for the catalytic gas turbine combustor. The main advantages of monolithic reactors, low pressure drop and absence of internal mass transfer limitations, were recently argued for by Villermaux and Schweich [56].

D. Modeling of Monolith Combustion Catalysts

The preceding sections show that catalytic fuel combustion is a process in which complex kinetics for heterogeneous and homogeneous reactions are combined with mass and heat transfer effects. This leads to difficulties in predicting the behavior of combustion catalysts under real conditions. Therefore, mathematical modeling is a powerful tool to assist experimental work, to interpret results, and to aid in the design of catalytic combustors.

Modeling of catalytic combustors has been the subject of a number of studies. The models used varied in degree of complexity and could therefore answer various types of questions. General issues of modeling monolith catalytic reactors are discussed in Chapter 8 of this book and in the reviews of Irandoust and Andersson [57] and Cybulski and Moulijn [58]. Hence, only topics that are specific to the modeling of catalytic combustion in monolith catalysts are considered here. A description of some important aspects of different types of models are as follows.

1. *Type of Kinetic Expressions Used*

Many models use an apparent reaction rate [59], with an Arrhenius-type temperature dependence and, as a result, consider the diffusion and reaction on the interior surface of the catalyst only under certain restrictive assumptions. It is important to incorporate mass transfer effects within the washcoat in the model to give a realistic description of the processes in the monolith [54,60].

Both studies show that at relatively low temperatures, i.e., during ignition of the catalyst, the rate-limiting step shifts from chemical kinetics to diffusion in the washcoat. This is clear from Fig. 7, computed using a one-dimensional model by Nakhjavan et al. [54]. Figure 7A shows the Thiele modulus and Fig. 7B an external diffusion limiting factor F versus dimensionless axial position in the reactor at various times on-stream for the catalytic combustion of propene in monolith reactors. The time is defined as the time after injection of the fuel in a preheated air flow.

$$F = \frac{c_{g,\text{bulk}} - c_{g,\text{surface}}}{c_{g,\text{bulk}}} \cdot 100\%$$

It is clear that both the Thiele modulus and F increase continuously during the ignition of the catalyst, representing the shift from control by chemical kinetics to control by washcoat diffusion and external mass transfer. At near–steady state ($t = 25$ sec) the process is almost completely controlled by external mass transfer, indicated by the F value of 90%.

2. *Description of Homogeneous–Heterogeneous Interactions*

Many models use only simple kinetics to describe the gas-phase reactions without taking radical transfer between the two phases into account. This may be an oversimplification, since radical transfer may strongly influence the results of the calculations. Moreover, homogeneous combustion kinetics, even without the presence of a catalyst, are extremely complex, with already over 300 known primary reactions for methane.

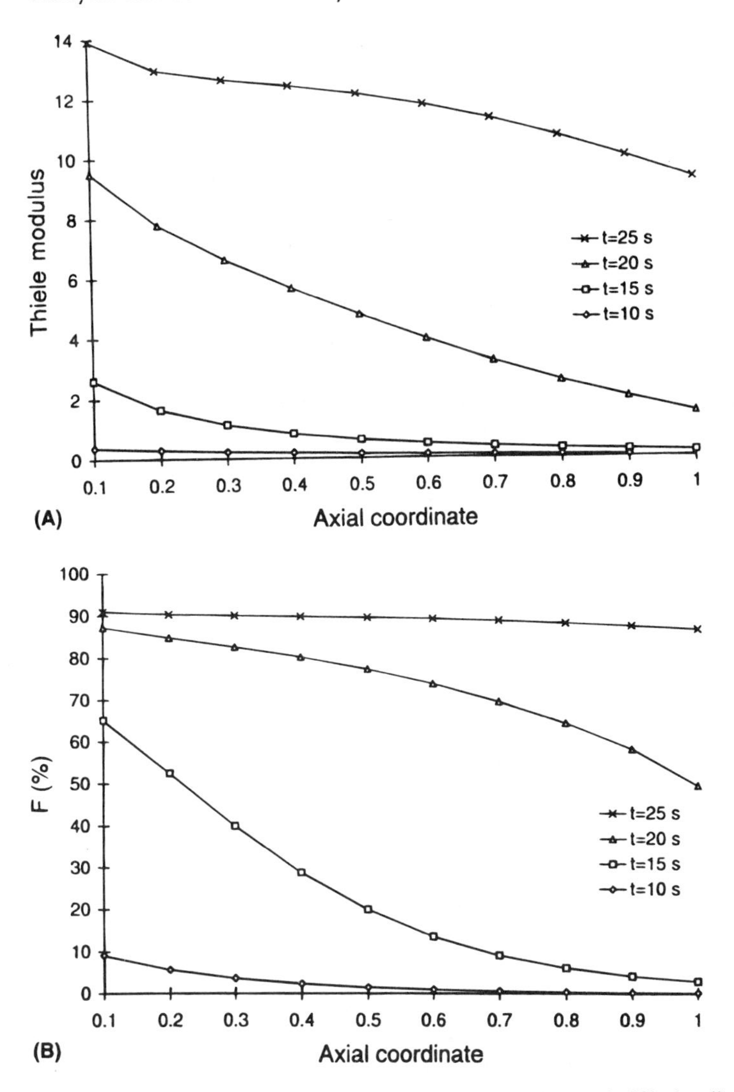

Figure 7 (A) Thiele modulus along the reactor. (B) External diffusion limiting factor F along the reactor. Fuel: propene; equivalence ratio 0.47; gas inlet: 660 K; monolith preheat temperature 400 K. t represents the time after the start of fuel injection into the air flow. Reprinted from Ref. 54, copyright (1995), with kind permission from Elsevier Science Ltd.

Markatou et al. [41] modeled methane combustion over heated catalytic and inert surfaces, using coupled heterogeneous and homogeneous mass, energy, and momentum balances and 46 elementary gas-phase reactions. They showed the importance of the description of hydroxyl radical desorption to the modeling results. Hydroxyl radicals are important in homogeneous combustion reaction schemes, and hence their adsorption and desorption from surfaces is important for understanding catathermal combustors; thus they have received considerable attention. For example, desorption of hydroxyl radicals generated on a platinum foil during hydrogen oxidation were studied by Fridell et al. [61]. Such experimental and theoretical model studies may be a support in the development of monolith catalytic combustors. Pfefferle pointed out recently that simple models without coupling may work well in cases where most of the fuel conversion takes places on the surface of the catalyst. However, if chemical coupling with the gas phase is important or if a large part of the heat is released in the gas boundary layer, more complex models are needed that describe the transfer of heat and mass between the surface and the gas phase more accurately [62].

3. *Channel Interaction*

Only few studies have dealt with the heat transfer between monolith channels in catalytic fuel combustion and the effect of this heat transfer on temperature profiles and thermal stresses [63,64]. Furuya et al. [64] used simulation of the thermal stress caused by radial gas temperature inlet variations and developed monolith designs with improved resistance to such stress. Worth et al. [63] showed the importance of the conductivity of the monolith material to the radial temperature profiles in monolith combustion catalysts. Metal monoliths, with higher thermal conductivity, showed much flatter temperature profiles than ceramic monoliths, indicating that thermal stress and risk for hot spots will be less of a problem for the former than for the latter.

More extensive work has been carried out to reveal channel interaction effects in monolithic automotive exhaust catalysts. Zygourakis and Aris [65,66] modeled radial monolith temperature profiles under transient conditions and showed the impact of inlet flow maldistribution on the reactor performance. Such effects are of great importance for monolithic combustion catalysts as well, since they encounter much larger temperature gradients.

4. *Flow Description*

Models may be one-, two-, or three-dimensional. Of course, more advanced models describe the flow in the channels more accurately, but this is not required in all types of investigations. Section 2 in Chapter 8 provides an in-depth discussion of relevant flow phenomena that have to be taken into account when modeling monoliths. One-dimensional models can provide interesting information on mass and heat transfer effects in the washcoat [54]. However, if the development of a laminar flow in the monolith channels is to be simulated, a two-dimensional model is needed. This is the case if multimonolith combustors are investigated [67]. This specific design, discussed in detail in Section IV.A, has a number of monoliths in series, which enables variation of monolith materials and cell density. Moreover, it leads to improved mass transfer by induced turbulence at the entrance of each monolith segment.

Development of models that describe the flow in monolith channels in catalytic converters has received great attention [68,69]. See also Ref. 58 and references therein. As with channel interaction, the models applicable to monolith combustion catalysts are similar to those for catalytic converters.

It is obvious that including all aspects in one single model would require large computational capacity. Therefore, it is necessary to optimize the complexity of the model for the purpose of the investigation, as discussed in Section 3 of Chapter 8.

III. DESIGN OF HONEYCOMB MONOLITH COMBUSTION CATALYSTS

A. Pros and Cons of Catalysis at High Temperatures

As already discussed in Section II.C, the demands on combustion catalysts for gas turbines are unusually severe because of the extreme process conditions (operation for at least 1 year at temperatures above 1000°C in atmospheres containing oxygen and steam, thermal shocks, very high space velocities, erosion due to abrasive particles or dust, etc.). The material properties of the catalyst may not be able to cope with these conditions. Catalyst deactivation is possible due to sintering, phase changes, and structural transformations. Surface enrichment or depletion of the different components of the catalyst may occur due to several causes, as is well known in industrial catalysis [70].

As against these inherent limitations, high-temperature operation offers a few advantages. For instance, catalyst poisoning by traces of sulfur, halogens, phosphorus, arsenic, lead, alkali metals, etc., are so common in catalytic processes at low or medium (<500°C) temperatures; at higher temperatures, such poisons are readily decomposed on and/or desorbed from the catalyst surface; hence, they do not pose a serious problem. Furthermore, the specific activity of the catalyst is also not very critical, since the catalyst often only has to trigger off the reaction, which can then proceed to completion in flames by free-radical mechanisms. Kiperman's kinetic analysis of heterogeneous–homogeneous reactions, discussed in Section II.A, even suggests that low-activity catalysts are more useful for the transfer of the reactions from the surface to the volume or gas phase.

A typical example of the unusual and very subtle problems encountered in developing catalysts for high-temperature catalytic combustion is given in a very recent study [71] by the Kyushu University (Japan) group of Arai, who pioneered the work on substituted hexa-aluminates as catalytic materials [72–75]. The catalytic properties of Pd supported on hexa-aluminates ($Sr_{0.8}La_{0.2}XAl_{11}O_{19}$, where X = Al and Mn) were studied [71] for use in high-temperature catalytic combustion. The activity of the supported Pd catalyst increased initially with a rise in temperature, but decreased at higher temperatures (about 700°C). The drop in catalytic activity became pronounced when the supported Pd particles were sintered into large agglomerates after calcination above 1000°C. From in situ XRD, temperature-programed desorption (TPD) and oxidation (TPO), it was revealed that the activity drop accompanies a dissociation of PdO onto metallic Pd species. Since the dissociation of PdO is thermodynamically dependent on oxygen partial pressure, the temperature at which the conversion drop appeared is influenced by the oxygen concentration in the fuel–air mixture. The dissociation of PdO seems to remove adsorbed oxygen species necessary for the catalytic combustion.

Very similar results were also obtained by Farrauto et al. [51] from a study of the high-temperature catalytic chemistry of supported Pd for the combustion of methane. Palladium oxide supported on alumina decomposes in two distinct steps in air at atmospheric pressure. The first step occurs between 750 and 800°C and is believed to be a decomposition of Pd–O species dispersed on bulk Pd metal, designated (PdO_x/Pd). The second decomposition occurs between 800 and 850°C, and it behaves like crystalline palladium oxide (PdO). To form the oxide once again, metallic Pd has to be cooled down to 650°C, thus causing a hysteresis gap of 150°C. Above 500°C, catalytic methane oxidation can occur only as long as the palladium oxide phase is still present. Above 650°C, metallic Pd cannot chemisorb oxygen, and hence it is catalytically inactive toward methane oxidation.

A significant drop in catalytic activity for catalytic combustion of methane due to the above-mentioned PdO decomposition or the inability of metallic Pd to chemisorb oxygen above 650°C, however, can be effectively avoided by using a catalytically more active Mn-substituted hexa-aluminate (X = Mn) as a catalyst support [71]. The catalytic activity of this Mn-hexa-aluminate compensates for the drop in activity of Pd so that a stable combustion reaction can be attained in a whole temperature range. Thus, the use of catalytically active support materials is one possible solution to overcome the unstable activity of the supported Pd catalyst in high-temperature combustion. This is an area of research where the terms *active catalyst, catalyst support*, and *promoter* no longer have their conventional meaning. This will be clear from the discussion in the following sections on the different components in monolithic combustion catalysts. The following paragraphs will highlight mainly materials aspects of monolithic combustion catalysts. A more detailed discussion on this subject can be found in the review by Zwinkels et al. [17].

B. Monolith Substrate

Monolith substrate materials suitable for use in high-temperature combustion are either various ceramics or certain alloys. In Section II.C, the demands on a combustion catalyst were already summarized. The substrate should have a high thermal shock resistance, but at the same time the melting temperature should be higher than the maximum anticipated operating temperature. A number of interesting materials are summarized in Table 2, accompanied by thermal expansion coefficients and thermal conductivities.

Materials such as aluminum titanate and silicon carbide appear to be promising for high-temperature catalytic combustion. However, problems such as extrudability, the application of washcoats, and reaction with deposited washcoats are not solved yet. For instance, when hexa-aluminate, presented in the introduction to this section, was applied to silicon carbide monoliths, solid-state reactions occurred at 1200–1400°C [76], causing exfoliation of the coating and the formation of new phases. The application of an intermediate mullite layer was suggested as an approach to hinder these solid-state reactions.

Metal monoliths have interesting properties, such as high tolerance for mechanical stress and vibrations and high thermal conductivity. Moreover, the cell walls may be thinner as compared to their ceramic counterparts. However, the maximum operating temperature of metal monoliths is not as high as for various ceramics. This is not a problem for certain combustor designs that limit the temperature in (part of) the catalyst

Table 2 Possible Monolith Substrate Materials

Material	Maximum temperature (°C)	Thermal expansion coefficient $\times 10^6$ ($°C^{-1}$)	Thermal conductivity ($W \cdot m^{-1} \cdot °C^{-1}$)	Source product data
Cordierite	1400	1	1	[12]
Mullite	1450	5 (@ 1000°C)	4.4–5 (20–1000°C)	C&C*
Dense alumina	1500	8		[12]
Aluminum titanate	1500	2–2.4 (20–1400°C)	880 (@ 100°C)	C&C*
Silicon carbide	1450	4.9 (20–1400°C)	170–35 (20–1000°C)	C&C*
Silicon nitride	1200	3 (20–1400°C)	20 (20–1000°C)	C&C*
Fecralloy®	1100	15 (20–1000°C)	12–23 (20–800°C)	Sandvik†

*C&C: Céramiques et Composites, France
†Sandvik: AB Sandvik Steel, Sweden; Fecralloy® is a trademark of Johnson-Matthey.

(cf. Section IV). Fecralloy®, one the most interesting materials for use as metal monoliths, consists of roughly 15% Cr, 5% Al, traces of rare earth metals, and the balance being Fe. This alloy, if treated in the proper way, attains an alumina whisker-covered surface, as seen in Fig. 8 [77]. This surface renders the alloy highly oxidation resistant and also acts as an anchor for the washcoat, which thus adheres strongly to the substrate.

C. Washcoat

The most important property for a washcoat, apart from the just-mentioned adherence and inertness with respect to the substrate, is a stable, large surface area. The surface area must be maintained during extended operation in which the catalyst is exposed to maximum temperatures of 1000–1400°C in a steam-containing atmosphere. Steam is well known to accelerate sintering of porous materials [78].

The importance of a large surface area in the high-temperature application that catalytic fuel combustion represents has not always been accepted. However, it has been shown experimentally [43] and theoretically [54] that a large surface area is beneficial for the ignition and for the activity at relatively low temperatures, at which the chemical reaction is determining the rate. High activity at such low temperatures, $\leq$700°C, is vital for stable combustion. For high-temperature operation, when mass transfer and heat transfer are controlling the overall rate, the importance of the surface is much less pronounced. For instance, Matsuura et al. [79] showed that the influence of the surface area on the temperature needed for 10% conversion of methane over Pt/MgO was stronger than on the temperature for 90% conversion.

γ-Al_2O_3 is the most common large-surface-area washcoat material used in applications such as three-way exhaust catalysts. However, the γ-phase is converted into the

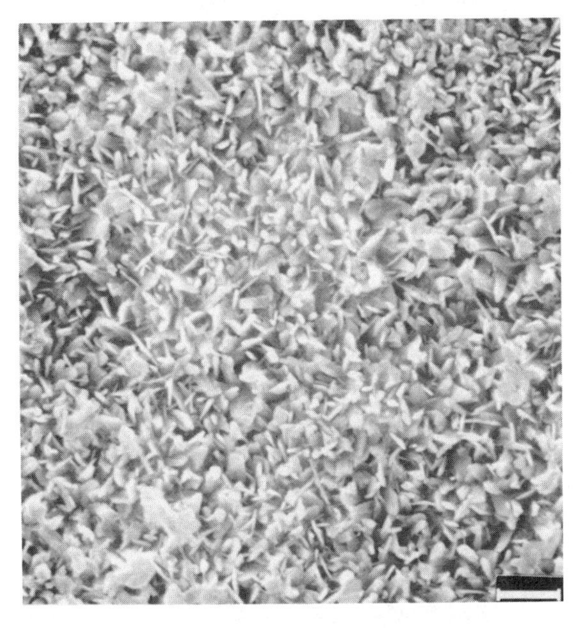

Figure 8 Alumina whisker-covered surface of a Fecralloy®-type metal monolith. The bar represents 1 μm. (From Ref. 77.)

stable α-phase at temperatures over 1000°C, involving sintering of primary particles, accompanied by loss of surface area. Most of the development work with respect to washcoat materials has focused on the modification or stabilization of alumina [17]. Modification with alkaline earth metals, such as barium [80], rare earth metals (mainly lanthanum [81–83], or silicon [84] proved to be more or less successful. In all cases it was attempted to retard diffusion of supposedly either oxygen or aluminum ions in the lattice or on the surface, which would decrease the rate of surface loss. More examples are given in Ref. 17.

The incorporation of stabilizing elements into the alumina lattice was proposed by Arai and co-workers in a series of papers [72–75]. They prepared alumina, doped with Ba, La, or Sr, by sol-gel synthesis from alkoxides, and achieved high thermal stability up to 1600°C, at which temperature, for instance, $BaO \cdot Al_2O_3$ still had a surface area of over 10 m^2/g [73]. The possibility to incorporate transition metals into the lattice, giving active combustion catalysts, was already touched on in the introduction to this section. Groppi et al. [85] recently reported on the synthesis of similar materials by easier coprecipation of nitrates. The results obtained with the two different preparation routes were comparable. Recently, Lowe et al. [86] reported on the synthesis of oxide gel powders with various compositions from alkoxide precursors. They found $La_2O_3 \cdot 11Al_2O_3$ to be the most promising candidate, with good resistance to sintering in moist air. The benefit of sol-gel synthesis, apart from the homogeneity of the product, is the high porosity and large surface area of the gel obtained. Lowe et al. reason that if sintering is to occur during the operation of a catalytic combustor, we should start with the largest initial surface area possible, allowing the longest catalyst life.

Most likely, the higher the degree of homogeneity in the washcoat material, the higher the resistance to sintering. Sol-gel synthesis provides a method to produce mixed oxide materials with homogeneity on the atomic scale [87] and should thus be explored further, especially with respect to novel compositions.

D. Active Material

The active material in a combustion catalyst typically is either a Pt-group element or a transition metal oxide. These compounds have long been known from the complete oxidation of methane [44], other hydrocarbons, and CO [3]. However, simple metal oxides and precious metals have several drawbacks. Precious metals are prone to severe sintering during operation in a catalytic combustor over 900–1000°C. Moreover, precious metals may become volatile in a certain form at these temperatures. Simple metal oxides, supported on a refractory oxide, also may undergo strong deactivation at the above temperatures. The most striking problem is the solid-phase reaction between active materials, such as the oxides of cobalt, copper, and nickel, and alumina washcoat [88], giving spinels with much less activity than the corresponding supported metal oxides [89].

The reaction between the active compound and the washcoat can be avoided following two approaches. The first approach is to stabilize the active phase by applying it as a complex oxide. The spinel-type oxides, mentioned above, are relatively inactive. On the other hand, perovksites, often AMO_3 (A = rare earth metal, e.g., La, Sr; M = transition metal, e.g., Co, Cr, Fe), exhibit promising behavior, which has attracted much attention [55,90,91]. Perovskites may also be used as unsupported oxides, but their surface area is small and not particularly thermostable. The activity of perovskites is dependent mostly on the M cation, but the A cation also has a significant effect [91]. Partial substitution of the A cation sometimes improves activity and stability [92]. The reader is referred to the extensive review by Fierro [93].

E. The Systems Approach

The homogeneity of the washcoat is beneficial for high thermal stability. Obviously, the same can be claimed for the washcoat-active phase system. Heterogeneous washcoats with specific sites on which active compounds are deposited will eventually undergo solid-state reactions during operation at temperatures over 1000°C. Hence, a washcoat in which the active sites are uniformly distributed in the lattice should be a superior solution. As mentioned, perovskites may be used as neat oxides, but their thermal stability is not sufficient. On the other hand, the hexa-aluminates, developed by Arai and co-workers, exhibit excellent thermal stability and have great potential for use as active washcoats. This was shown for $Sr_{0.8}La_{0.2}MnAl_{11}O_{19-\alpha}$ [94], which was heated in air at 1300°C for over 6000 hours, after which the crystal structure remained unchanged. During the aging, the surface area had dropped from 18 to 4 m^2/g, and there was still significant, yet low, activity left.

These types of materials will probably play an important role in the further development of combustion catalysts, especially if their activity can be improved. Moreover, they can be extruded to give active monoliths directly. This, however, still generates problems with respect to mechanical strength and thermal shock properties due to the relatively high thermal expansion coefficient [64].

IV. PRACTICAL APPLICATIONS OF HONEYCOMB MONOLITHS IN CATALYTIC COMBUSTORS

Catalytic combustors can be applied in any kind of gas turbine, (1) with large multican combustors [95–97], (2) for cogeneration systems [53], or (3) small regenerative gas turbines for automotive applications [98].

Most of the work on catalytic combustion applications has dealt with the use of natural gas, since this is a clean gaseous fuel with low nitrogen content. However, a number of other hydrocarbon fuels have been investigated as well [99–101], such as propane, gasoline, diesel fuel, and kerosenes.

An interesting type of fuel for which catalytic combustion is particularly suited is low-heat-value (LHV) gas, for example, generated by gasification of coal or biomass [102]. An LHV gas consists of a mixture of a number of components, including, among others, carbon monoxide, hydrogen, light hydrocarbons (methane), carbon dioxide, water, and nitrogen. The combustion behavior of LHV gases was recently reported by Groppi et al. [103] using manganese-containing hexa-aluminate catalysts and Zwinkels et al. [104], who studied supported platinum catalysts. Their results show that the catalyst type, the reaction conditions, and the fuel composition have a strong effect on the combustion behavior of the LHV gas.

An LHV gas may contain up to a few thousand ppm of ammonia, produced from fuel-bound nitrogen during the gasification of a solid fuel. One of the major challenges in the catalytic combustion of LHV gases is to circumvent the formation of NO_x from this ammonia. The selectivity for this reaction is strongly dependent on the air–fuel ratio in the catalytic combustor and on the catalyst type [102,105]. Clark et al. [102] and Tucci

[100] reported on similar multistage catalytic combustor approaches that minimize the conversion of ammonia to NO_x. These include a rich catalytic zone followed by a secondary fuel injector and a lean catalytic zone, i.e., similar to the RQL (rich burn–quick quench–lean burn) idea discussed in Section I.C. Only about 10% of the ammonia was converted to NO_x under optimal conditions [102].

It can be concluded from the previous sections that the perfect combustion catalyst has not been found yet, and it is most likely very hard to develop. Hence, reaction engineering must help to circumvent the inherent compromise between activity and stability and the limitations of material science as of today. In this section, the principles of the most promising approaches are outlined. Figure 9 shows schematically the three currently most promising approaches in catalytic combustor design.

A. Multimonolith Combustor

The multimonolith combustor, demonstrated by Osaka Gas, has a segmented catalyst zone. The stability of the catalyst increases on the way to exiting the combustor, whereas the activity is high for the first segment and decreases thereafter. In this manner, the first segment is not exposed to the maximum temperatures occurring at the combustor outlet, and hence very active catalysts, such as palladium, may be used in the first segment. The other segments, of which the number is limited by space only, are made of thermostable hexa-aluminate materials in the case of Osaka Gas. Their excellent thermal stability can thus be combined with the high activity of a palladium catalyst.

This system was recently demonstrated in a real 150-kW engine during more than 200 hours at 1100°C combustor outlet temperature [53] at NO_x levels below 15 ppmv at 15% oxygen. A schematic view of a multimonolith combustor is given in Fig. 10 [106].

The advantage of a multimonolith combustor is the simplicity of the system, with only one fuel inlet. The problem, of course, is the availability of thermostable materials for the high-temperature segments, especially for gas turbines that require a maximum combustor temperature higher than the 1100°C in the Osaka Gas design. Moreover, the conversion in the first segment must be limited in some way, since too high a conversion leads to temperature runaway and catalyst deactivation.

B. Partial Catalytic Combustor

The partial catalytic combustor is an approach that is followed by, amongst others, Catalytica, Engelhard, and Johnson-Matthey [99,107,108]. The idea is to add all fuel to the catalyst but to combust only part of it in the catalyst zone. The rest is heated in the

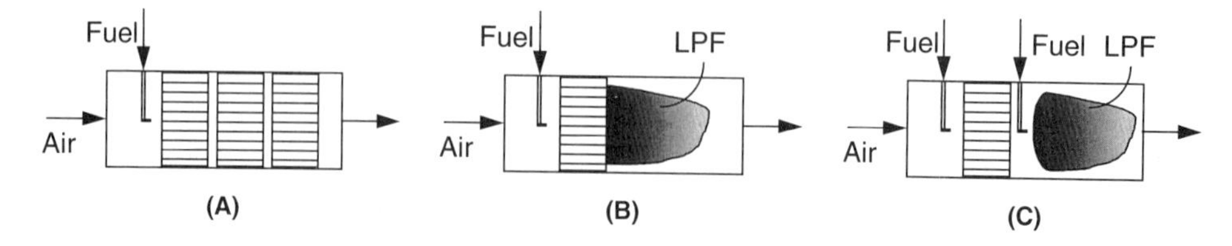

Figure 9 Simplified representation of different catalytic combustor concepts. A: Multimonolith combustor; B: partial catalytic combustor; C: hybrid combustor. LPF = lean premixed flame.

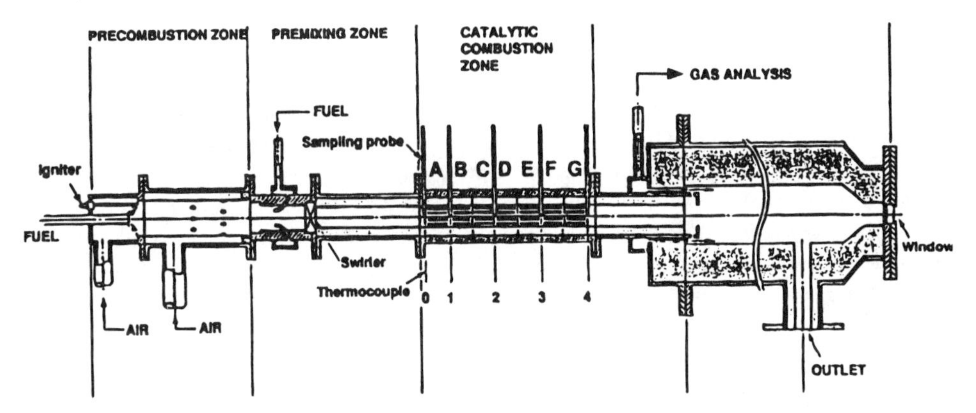

Figure 10 Osaka Gas catalytic combustor. A, B: Pd powder (0.02 μm) on cordierite; C–G: extruded cation-substituted hexa-aluminate. (From Ref. 106.)

catalyst only to a temperature high enough to support homogeneous gas-phase combustion downstream of the catalyst. The major advantage of this approach is that the maximum temperature of the catalyst can be kept at 800–900°C, which means that catalyst materials are already available today. Furthermore, it is easier to increase the combustor outlet temperature to 1300–1500°C, which will cause material problems for the multimonolith concept.

Just as for the multimonolith combustor, the problem inherent in the partial catalytic combustor is the control of the conversion in the catalyst. Too high a conversion in the catalyst leads to temperature runaway and catalyst deactivation. This is claimed to be solved by a combination of catalyst chemistry and monolith design. The PdO–Pd transition is used to fine-tune the activity by using the fact that metallic palladium, which is formed at temperatures over approximately 700°C, depending on pressure and oxygen concentration, is less active for combustion than is PdO. Moreover, the use of metal monoliths in this approach allows the coating of only half of the channels, i.e., every other channel [109,110]. The uncoated channels thus act as a bypass for part of the fuel–air mixture and at the same time cool the catalyst. Another solution was claimed to be the application of a diffusion barrier on top of the washcoat, thus hindering high reaction rates near the catalyst outlet. The idea was recently tested at full scale showing NO_x levels of 10–15 ppmv at 15% oxygen [111].

C. Hybrid Combustor

The third approach is the hybrid combustor, which was developed at Toshiba and presented in a series of publications [95–97]. The difference with the partial catalytic combustor is that only part of the fuel is added upstream of the catalyst. This fuel is nearly completely combusted over the catalyst, bringing the temperature up to approximately 800–900°C. At this point the rest of the fuel is added and then combusted homogeneously. The advantages of this approach are the same as for the partial catalytic combustor. However, the problem with catalyst overheating is less pronounced here, since complete conversion of all the fuel added to the catalyst is allowed. On the other hand, the additional fuel injection downstream of the catalyst renders the system much more complex and harder to control. The idea is shown schematically in Fig. 11 [97]. Operation was demonstrated at NO_x levels below 10 ppm at actual exhaust conditions.

It needs to be mentioned here that there is no clear dividing line between any two of the three alternatives. The partial combustor and the hybrid combustor may both be equipped with a multimonolith catalyst zone. Furthermore, the temperature in the hot segments of a multimonolith combustor will be so high that homogeneous combustion takes place in the monolith channels. It is not clear what the importance of the catalytic activity and catalyst surface area is under such conditions. There is still much ambiguity about this aspect of high-temperature catalytic combustors.

V. CONCLUDING REMARKS

Catalytic combustion for gas turbines is an area that has developed rapidly during the last two decades. Several novel approaches for catalytic combustors have shown high potential, often as a result of combined materials science and reaction engineering. However, a substantial effort is still needed before commercial gas turbines with catalytic combustors become available on the market.

There are three important outstanding issues that need concerted R&D. The first issue is the further development of materials with thermal stability under combustion condition at the high temperatures prevailing in a gas turbine combustion chamber. Promising materials have been developed, but none fulfills the demand of a lifespan of at least 1 year. Besides, the most promising materials, such as the family of the hexa-aluminates, must be available in a honeycomb monolith shape, either as washcoat or directly extruded. Much work still needs to be done to optimize the preparation of monolithic thermostable catalysts.

The second issue is the improvement of the low-temperature performance of combustion catalysts, i.e., the activity at combustor inlet conditions. All the proposed catalytic combustor designs available today need a pilot flame, or a heat exchanger in the case of recuperative gas turbines, to heat the compressed combustion air to a temperature sufficient for ignition of the catalyst. The possibility of avoiding this pilot flame is considered very important, since it would further reduce NO_x emissions. The catalyst surface area and washcoat loading are very important for the low-temperature activity.

The third outstanding issue in the development of catalytic gas turbine combustors is the development of a complete system, i.e., monolith and washcoat design, fuel inlets, and

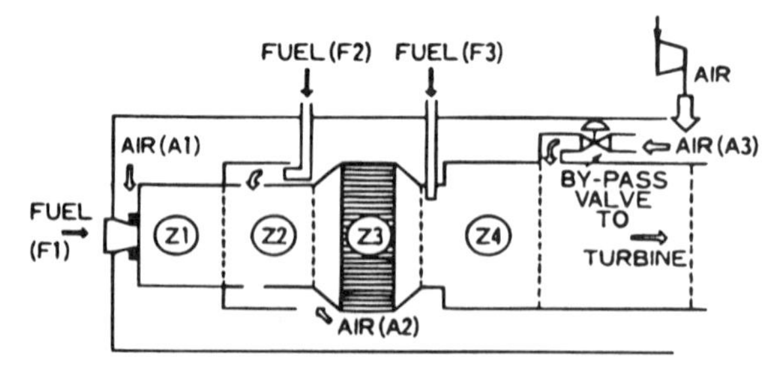

Figure 11 Toshiba hybrid catalytic combustor. A1–A3: air inlets; F1–F3: fuel inlets; Z1: precombustion zone; Z2: premixing zone; Z3: catalyst zone; Z4: gas-phase combustion zone. (From Ref. 97.)

control approaches. The three combustor designs discussed in this chapter are promising but leave room for improvement.

ACKNOWLEDGMENT

The catalytic combustion work at the Department of Chemical Engineering and Technology–Chemical Technology at the Royal Institute of Technology is financially supported by NUTEK, the Swedish National Board for Industrial and Technical Development.

REFERENCES

1. J.N. Armor, Materials needs for catalysts to improve our environment, *Chem. Mater. 6*:730 (1994).
2. K.C. Taylor, Automobile catalytic converters, *Catalysis—Science and Technology* (J.R. Anderson and M. Boudart, Eds.), Springer-Verlag, Berlin, 1984, p. 119.
3. J.J. Spivey, Complete catalytic oxidation of volatile organics, *A Specialist Periodic Report—Catalysis* (G.C. Bond and G. Webb, Eds.), The Royal Society of Chemistry, Cambridge, U.K., 1989, p. 157.
4. H. Bosch and F. Janssen, Catalytic reduction of nitrogen oxides. A review on the fundamentals and technology, *Catal. Today 2*:369 (1987).
5. N. Nojiri, Y. Sakai, and Y. Watanabe, Two catalytic technologies of much influence on progress in chemical process development in Japan, *Catal. Rev. Sci. Eng. 37*:145 (1995).
6. S. Sato, Y. Yu-u, H. Yahiro, N. Mizuno, and M. Iwamoto, Cu-ZSM-5 zeolite as highly active catalyst for removal of nitrogen monoxide from emission of diesel engines, *Appl. Catal. 70*:L1 (1991).
7. M. Konno, T. Chikahisa, T. Murayama, and M. Iwamoto, Catalytic reduction of NO_x in actual diesel engine exhaust, *SAE Paper*:920091 (1992).
8. N.L. Occelli and R.G. Anthony, *Hydrotreating Catalysts: Preparation, Characterization and Performance*, Elsevier, Amsterdam. 1989.
9. B.C. Gates, J.R. Katzer, and G.C.A. Schuit, *Chemistry of Catalytic Processes*, McGraw-Hill, New York, 1979.
10. A.K. Gupta and D.G. Lilley, Review: The environmental challenge of gas turbines, *Journal of the Institute of Energy 65*:106 (1992).
11. L. Lamarre, Activity in IGCC worldwide, *EPRI Journal July/August*:6 (1994).
12. D.L. Trimm, Catalytic combustion (review), *Appl. Catal. 7*:249 (1983).
13. R. Prasad, L.A. Kennedy, and E. Ruckenstein, Catalytic Combustion, *Catal. Rev. Sci. Eng. 26*:1 (1984).
14. J.P. Kesselring, Catalytic combustion, *Advanced combustion methods* (F.J. Weinberg, Ed.), Academic Press, London, 1986, p. 327.
15. L.D. Pfefferle and W.C. Pfefferle, Catalysis in combustion, *Catal. Rev. Sci. Eng. 29*:219 (1987).
16. H. Arai and M. Machida, Recent progress in high-temperature catalytic combustion, *Catal. Today 10*:81 (1991).
17. M.F.M. Zwinkels, S.G. Järås, P.G. Menon, and T.A. Griffin, Catalytic materials for high-temperature catalytic combustion, *Catal. Rev. Sci. Eng. 35*:319 (1993).
18. A.W. Lefebvre, *Gas Turbine Combustion*, McGraw-Hill, New York, 1983.
19. G.A. Lavoie, J.B. Heywood, and J.C. Keck, Experimental and theoretical study of nitric oxide formation in internal combustion engines, *Comb. Sci. Technol. 1*:313 (1970).
20. Z.R. Ismagilov and M.A. Kerzhentsev, Catalytic fuel combustion—A way of reducing emission of nitrogen oxides, *Catal. Rev. Sci. Eng. 32*:51 (1990).

21. J. Zeldovich, The oxidation of nitrogen in combustion and explosions, *Acta Physicochimica URSS 21*:577 (1946).
22. R.E. Jones, Gas turbine engine emission—Problems, progress and future, *Prog. Energy Comb. Sci. 4*:73 (1978).
23. C.P. Fenimore, Formation of nitric oxide from fuel nitrogen in ethylene flames, *Comb. and Flame 19*:289 (1972).
24. J. Leppälahti, P. Simell, and E. Kurkela, Catalytic conversion of nitrogen compounds in gasification gas, *Fuel Processing Technology 29*:43 (1991).
25. T.D. Eastop and A. McConkey, *Applied Thermodynamics for Engineering Technologists,* Longman, Harlow, Essex, UK, 1993.
26. M.F.M. Zwinkels, Ph.D. Thesis, KTH—Royal Institute of Technology, Stockholm, 1996.
27. R. Egnell and R. Gabrielsson, *Alternativa motorer*, NUTEK, Stockholm, 1991.
28. B. Becker, F. Bonsen, and G. Simon, Simple and reliable combustion control system. *ASME Paper 90-TG-173*:7 (1991).
29. W.C. Pfefferle, Belgian Patent 814, 752, Engelhard Corp., Iselin, NY, 1974.
30. W.C. Pfefferle, U.S. Patent 3,928,961, Engelhard Corp., Iselin, NY, 1975.
31. S.L. Kiperman, Kinetic peculiarities of the gas-phase heterogeneous–homogeneous reactions, *Kinet. Catal. (Engl.) 35*:37 (1994).
32. V.V. Azatyan, Reversible change of heterogeneous factors in branched chain processes, *Kinet. Katal. (Rus.) 23*:1301 (1982).
33. V.V. Azatyan, Chain processes and non-steady state nature of surfaces, *Usp. Khim. 54*:33 (1985).
34. Z.R. Ismagilov, S.N. Pak, L.G. Krishtopa, and V.K. Yermolaev, Role of free radicals in heterogeneous complete oxidation of organic compounds over IV period transition metal oxides, Proc. 10th Int. Congr. Catal., Budapest, 1992, p. 231.
35. G.I. Golodets, Possible causes of nonstationary phenomena in gas-phase radical chain oxidation in the presence of transition-metal oxides, *Kinet. Catal. (Engl.) 28*:1074 (1987).
36. T.A. Garibyan and L.Y. Margolis, Heterogeneous–homogeneous mechanism of catalytic oxidation, *Catal. Rev. Sci. Eng. 31*:355 (1989–90).
37. F. Gudmundsson, E. Fridell, A. Rosén, and B. Kasemo, Evaluation of OH desorption rates from Pt using spatially resolved imaging of laser-induced fluorescence, *J. Phys. Chem. 97*:12828 (1993).
38. L.D. Pfefferle, T.A. Griffin, M. Winter, D.R. Crosley, and M.J. Dyer, The influence of catalytic activity on the ignition of boundary layer flows—Part 1: Hydroxyl radical measurements, *Comb. Flame 76*:325 (1989).
39. W.C. Pfefferle and L.D. Pfefferle, Catalytically stabilized combustion, *Prog. Energy Comb. Sci. 12*:25 (1986).
40. T. Griffin, W. Weisenstein, A. Schlegel, S. Buser, P. Benz, H. Bockhorn, and F. Mauss, Investigation of the NO_x advantage of catalytic combustion, Proc. 2nd Int. Workshop Catalytic Combustion, April 18–20, Tokyo (H. Arai. ed.), Catalysis Society of Japan, Tokyo, 1994, p. 138.
41. P. Markatou, L.D. Pfefferle, and M.D. Smooke, A computational study of methane–air combustion over heated catalytic and non-catalytic surfaces, *Comb. Flame 93*:185 (1993).
42. A. Zieba, T. Banaszak, and R. Miller, Thermal-catalytic oxidation of waste gases, *Appl. Catal. A. 124*:47 (1995).
43. H. Arai, T. Yamada, K. Eguchi, and T. Seiyama, Catalytic combustion of methane over various perovskite-type oxides, *Appl. Catal. 26*:265 (1986).
44. R.B. Anderson, K.C. Stein, J.J. Feenan, and L.J.E. Hofer, Catalytic oxidation of methane, *Ind. Eng. Chem. 53*:809 (1961).
45. C.F. Cullis and B.M. Williat, Oxidation of methane over supported precious metal catalysts, *J. Catal. 83*:267 (1983).

46. N. Mouaddib, C. Feumi-Jantou, E. Garbowski, and M. Primet, Catalytic oxidation of methane over palladium supported on alumina—Influence of the oxygen-to-methane ratio, *Appl. Catal. A. 87*:129 (1992).
47. R. Burch and F.J. Urbano, Investigation of the active state of supported palladium catalysts in the combustion of methane, *Appl. Catal. A. 124*:121 (1995).
48. K. Otto, Methane oxidation over Pt on γ-alumina: Kinetics and structure sensitivity, *Langmuir 5*:1364 (1989).
49. F.H. Ribeiro, M. Chow, and R.A. Dalla Betta, Kinetics of the complete oxidation of methane over supported palladium catalysts, *J. Catal. 146*:537 (1994).
50. R. Burch, F.J. Urbano, and P.K. Loader, Methane combustion over palladium catalysts: The effect of carbon dioxide and water on activity, *Appl. Catal. A. 123*:173 (1995).
51. R.J. Farrauto, M.C. Hobson, T. Kennelly, and E.M. Waterman, Catalytic chemistry of supported palladium for combustion of methane, *Appl. Catal. A. 81*:227 (1992).
52. R.J. Farrauto, J.K. Lampert, M.C. Hobson, and E.M. Waterman, Thermal decomposition and reformation of PdO catalysts; support effects, *Appl. Catal. B. 6*:263 (1995).
53. H. Sadamori, T. Tanioka, and T. Matsuhisa, Development of a high temperature combustion catalyst system and prototype catalytic combustor turbine test results, Proc. 2nd Int. Workshop Catalytic Combustion, 18–20 April, Tokyo (H. Arai. ed.), Catalysis Society of Japan, Tokyo, 1994, p. 158.
54. A. Nakhjavan, P. Björnbom, M.F.M. Zwinkels, and S.G. Järås, Numerical analysis of the transient performance of high-temperature monolith catalytic combustors; Effect of catalyst porosity, *Chem. Eng. Sci. 50*:2255 (1995).
55. J.G. McCarty and H. Wise, Perovskite catalysts for methane combustion, *Catal. Today 8*:231 (1990).
56. J. Villermaux, and D. Schweich, Is the catalytic monolith reactor well suited to environmentally benign processing? *Ind. Eng. Chem. Res. 33*:3025 (1994).
57. S. Irandoust and B. Andersson, Monolithic catalysts for nonautomobile applications, *Catal. Rev. Sci. Eng. 30*:341 (1988).
58. A. Cybulski and J.A. Moulijn, Monoliths in heterogeneous catalysis. *Catal. Rev. Sci. Eng. 36*:179 (1994).
59. H.H. Lee, *Heterogeneous Reactor Design*, Butterworth, Boston, 1985.
60. R.E. Hayes and S.T. Kolaczkowski, Mass and heat transfer effects in catalytic monolithic reactors, *Chem. Eng. Sci. 49*:3587 (1994).
61. E. Fridell, U. Westblom, M. Aldén, and A. Rosén, Spatially resolved laser-induced fluorescence imaging of OH produced in the oxidation of hydrogen on platinum, *J. Catal. 128*:92 (1991).
62. L.D. Pfefferle, Modeling heterogeneous–homogeneous reactions and transport coupling for catalytic combustion systems, Proc. 2nd Int. Workshop Catalytic Combustion. Tokyo, 18–20 April, (H. Arai. ed.), Catalysis Society of Japan, Tokyo, 1994, p. 78.
63. D.J. Worth, S.T. Kolaczkowski, and A. Spence, Modelling channel interaction in a catalytic monolith reactor, *Trans. IChemE. 71*:331 (1993).
64. A. Furuya, T. Nishida, and T. Matsuhisa, Thermal stress analysis for high temperature combustion catalyst honeycomb, Proc. 2nd Int. Workshop Catalytic Combustion, Tokyo, 18–20 April, (H. Arai. ed.), Catalysis Society of Japan, Tokyo, 1994, p. 70.
65. K. Zygourakis and R. Aris, Heat transfer in the array of passages of a monolith reactor, AIChE 75th Annual Meeting, Los Angeles, 1982.
66. K. Zygourakis, Transient operation of monolith catalytic converters: A two-dimensional reactor model and the effects of radially nonuniform flow distributions, *Chem. Eng. Sci. 44*:2075 (1989).
67. Y. Tsujikawa, S. Fuji, H. Sadamori, S. Ito, and S. Katsura, Numerical simulation of 2D flow of catalytic combustor, Proc. 2nd Int. Workshop Catalytic Combustion, Tokyo, 18–20 April, (H. Arai. ed.), Catalysis Society of Japan, Tokyo, 1994, p. 96.

68. L.C. Young and B.A. Finlayson, Mathematical models of the monolith catalytic converter: Part I. Development of model and application of orthogonal collocation, *AIChE J. 22*:331 (1976).
69. L.C. Young and B.A. Finlayson, Mathematical models of the monolith catalytic converter: Part II: Application to automobile exhaust, *AIChE J. 22*:343 (1976).
70. P.G. Menon, Diagnosis of industrial catalyst deactivation by surface characterization techniques, *Chem. Rev. 94*:1021 (1994).
71. K. Sekizawa, M. Machida, K. Eguchi, and H. Arai, Catalytic properties of Pd-supported hexaaluminate catalysts for high-temperature catalytic combustion, *J. Catal. 142*:655 (1993).
72. M. Machida, K. Eguchi, and H. Arai, Effect of additives on the surface area of oxide supports for catalytic combustion, *J. Catal. 103*:385 (1987).
73. M. Machida, K. Eguchi, and H. Arai, Preparation and characterization of large surface area $BaO{\cdot}6Al_2O_3$, *Bull. Chem. Soc. Jpn. 61*:3659 (1988).
74. M. Machida, K. Eguchi, and H. Arai, Catalytic properties of $BaMnAl_{11}O_{19-\alpha}$ (M = Cr, Mn, Fe, Co, and Ni) for high-temperature catalytic combustion, *J. Catal. 120*:377 (1989).
75. M. Machida, K. Eguchi, and H. Arai, Effect of structural modification on the catalytic property of the Mn-substituted hexaaluminates. *J. Catal. 123*:477 (1990).
76. K. Eguchi, H. Inoue, K. Sekizawa, and H. Arai, Thick film coating and fiber spinning of hexa-aluminate compounds for catalytic combustion, Proc. 2nd Int. Workshop on Catalytic Combustion, Tokyo, 18–20 April (H. Arai. ed.), Catalysis Society of Japan, Tokyo, 1994, p. 60.
77. M.F.M. Zwinkels, S.G. Järås, and P.G. Menon, Preparation of combustion catalysts by washcoating alumina whiskers-covered metal monoliths using a sol-gel method, *Catalyst Preparation VI* (G. Poncelet, J. Martens, B. Delmon, P.A. Jacobs, and P. Grange., Eds.), Elsevier, Amsterdam, 1995, p. 85.
78. D.L. Trimm, Thermal stability of catalyst supports, *Catalyst Deactivation V* (C.H. Bartholomew, and J.B. Butt., Eds.), Elsevier, Amsterdam, 1991, p. 29.
79. K. Matsuura, Y. Hashimoto, O. Takayasu, K. Nitta, and Y. Yoshida, Heat-stable ultrafine single-crystal magnesium oxide and its character as a support material for high-temperature combustion catalysts, *Appl. Catal. 74*:273 (1991).
80. I. Amato, D. Martorana, and B. Silengo, *Sintering of Pelleted Catalysts for Automotive Emission Control*, Plenum Press, New York, 1975, p. 187.
81. H. Schaper, E.B.M. Doesburg, and L.L. v. Reijen, The influence of lanthanum oxide on the thermal stability of gamma alumina catalyst supports, *Appl. Catal. 7*:211 (1983).
82. I.I.M. Tijburg, J.W. Geus, and H.W. Zandbergen, Application of lanthanum to pseudo-boehmite and γ-Al_2O_3, *J. Mater. Sci. 26*:6479 (1991).
83. J.S. Church, N.W. Cant, and D.L. Trimm, Stabilization of aluminas by rare and alkaline earth ions, *Appl. Catal. A. 101*:105 (1993).
84. B. Beguin, E. Garbowski, and M. Primet, Stabilization of alumina by addition of lanthanum, *Appl. Catal. 75*:119 (1991).
85. G. Groppi, M. Bellotto, C. Cristiani, P. Forzatti, and P.L. Villa, Preparation and characterization of hexaaluminate-based materials for catalytic combustion. *Appl. Catal. A. 104*:101 (1993).
86. D.M. Lowe, M.I. Gusman, and J.G. McCarty, Synthesis and characterization of sintering resistant aerogel complex oxide powders, *Catalyst Preparation VI* (G. Poncelet, J. Martens, B. Delmon, P.A. Jacobs, and P. Grange, Eds.), Elsevier, Amsterdam, 1994, p. 445.
87. L.L. Hench and J.K. West, The sol-gel process, *Chem. Rev. 90*:33 (1990).
88. M.C. Marion, E. Garbowski, and M. Primet, Catalytic properties of copper oxide supported on zinc aluminate in methane combustion, *J. Chem. Soc., Faraday Trans. 87*:1795 (1991).
89. M.A. Quinlan, H. Wise, and J.G. McCarty, *Basic Research on Natural Gas Combustion Phenomena—Catalytic Combustion*, SRI International, Menlo Park, CA, 1989.
90. B. de Collongue, E. Garbowski, and M. Primet, Catalytic combustion of methane over bulk and supported $LaCrO_3$ perovskites, *J. Chem. Soc., Faraday Trans. 87*:2493 (1991).

91. A. Baiker, P.E. Marti, P. Keusch, E. Fritsch, and A. Reller, Influence of the A-site cation in $ACoO_3$ (A = La, Pr, Nd, and Gd) perovskite-type oxides on catalytic activity for methane combustion, *J. Catal. 146*:268 (1994).
92. P. Salomonsson, T. Griffin, and B. Kasemo, Oxygen desorption and oxidation-reduction kinetics with methane and carbon monoxide over perovskite type metal oxide catalysts, *Appl. Catal. A. 104*:175 (1993).
93. J.L.G. Fierro, Structure and composition of perovskite surface in relation to adsorption and catalytic properties, *Catal. Today 8*:153 (1990).
94. H. Arai, K. Eguchi, M. Machida, and T. Shiomitsu, Heat resistance of hexaaluminate catalyst for high-temperature catalytic combustion, *Catalytic Science and Technology 1*:195 (1991).
95. T. Furuya, S. Yamanaka, T. Hayata, J. Koezuka, T. Yoshine, and A. Ohkoshi, Hybrid catalytic combustion for stationary gas turbine—Concept and small scale test results, *ASME Paper 87-GT-99* (1987).
96. T. Kawakami, T. Furuya, Y. Sasaki, T. Yoshine, Y. Furuse, and A. Ohkoshi, Feasibility study on honeycomb ceramics for catalytic combustor, *ASME Paper 89-GT-41* (1989).
97. T. Yoshine, S. Yamanaka, T. Furuya, and Y. Hara, Combustion characteristics of a gas turbine catalytic combustor, 19th CIMAC Int. Cong. on Comb. Engines, Florence, Italy, 1991.
98. R. Lundberg, AGATA—A European ceramic gas turbine for hybrid vehicles, *ASME Paper 94-GT-8*: (1994).
99. B.E. Enge and D.T. Thompson, Catalytic combustion applied to gas turbine technology, *Plat. Met. Rev. 23*:134 (1979).
100. E.R. Tucci, Use catalytic combustion for LHV gases, *Hydrocarbon Proc. 3*:159 (1982).
101. S. Maruko, T. Naoi, and K. Onoe, Multistage catalytic combustion systems, Proc. 2nd Int. Workshop Catalytic Combustion, 18–20 April, (H. Arai. ed.), Catalysis Society of Japan, Tokyo, 1994, p. 142.
102. W.D. Clark, B.A. Folsom, W.R. Seeker, and C.W. Courtney, Bench scale testing of low-NO_x LBG combustors, *Trans. ASME 104*:120 (1982).
103. G. Groppi, A. Belloli, E. Tronconi, and P. Forzatti, Catalytic combustion of CO–H2 on Mn-substituted hexaaluminates, JECAT '95, Lyon-Villeurbanne, 26–28 April 1995, p. 257.
104. M.F.M. Zwinkels, G.M.E. Heginuz, B.H. Gregertsen, K. Sjöström, and S.G. Järås, Catalytic combustion of gasified biomass over Pt/Al_2O_3, Accepted for publication in *Appl. Catal. A.* (1996).
105. C. Jung-Min Sung, L.A. Kennedy, and E. Ruckenstein, The effect of nitrogen content on the oxidation of fuel bound nitrogen in a transition metal oxide catalytic combustor, *Comb. Sci. Tech. 41*:315 (1984).
106. H. Sadamori and A. Matsuhisa, Research and development on a high-temperature combustion catalyst for gas turbines, 10th Symp. on Catal. Comb., November 1, Japan, 1990.
107. R.A. Dalla Betta, J.C. Schlatter, M. Chow, D.K. Yee, and T. Shoji, Catalytic combustion technology to achieve ultra low NO_x emissions: Catalyst design and performance characteristics, Proc. 2nd Int. Workshop Catalytic Combustion, 18–20 April, Tokyo (H. Arai, ed.), Catalysis Society of Japan, Tokyo, 1994, p. 154.
108. L.M. Quick and S. Kamitomai, Catalytic combustion reactor design and test results, Proc. 2nd Int. Workshop Catalytic Combustion 18–20 April, Tokyo (H. Arai, ed.), Catalysis Society of Japan, Tokyo, 1994, p. 132.
109. R.A. Dalla Betta, N. Ezawa, K. Tsurumi, J.C. Schlatter, and S.G. Nickolas, U.S. Patent 5, 183, 401, Catalytica, Tanaka, 1993.
110. R.A. Dalla Betta, K. Tsurumi, and N. Ezawa, U.S. Patent 5, 232, 357, Catalytica, Tanaka, 1993.
111. K.W. Beebe, M.B. Cutrone, R.A. Dalla Betta, J.C. Schlatter, S.G. Nickolas, Y. Furuse, and T. Tsuchiya, Development of a catalytic combustor for a heavy-duty utility gas turbine, IGTC, 22–27 October, Yokohama, 1995, p. 251.

7

Unconventional Utilization of Monolithic Catalysts for Gas-Phase Reactions

Gerald E. Voecks
Jet Propulsion Laboratory, Pasadena, California

I. INTRODUCTION

Since the adoption of honeycomb monolith catalysts for the treatment of automobile Otto-cycle exhaust emission, there has been a steadily increasing number of investigations into the potential for alternative uses of this type of support. During the ensuing research into improving the performance of the auto emission catalyst, the most prominent parallel effort was application of this type of catalyst to control the direct oxidation of fuels, rather than exhaust emission treatment, for Brayton- and Rankine-cycle engines. Catalytic control of these two different oxidation reactions was achieved in large part because of the unique, low pressure drop through the catalyst bed and because of the excellent uniformity in flow, heat, and mass transfer properties relative to conventional packed-pellet beds. However, actual operating conditions were quite different, i.e., combustion of a low-heating-value gas mixture in hot automobile exhaust over a broad range of very high flow rates, compared to combustion of high-heating-value fuels over a narrower range of flows at the inlet to turbine combustors.

These applications encouraged researchers to investigate how to improve the operating conditions of other catalytic reactions by taking advantage of supporting catalysts on the honeycomb monolithic supports. Many of these early efforts in the 1970s were directed toward other high-temperature, exothermic reactions that led to exploring two-stage combustion in which the first stage is fuel-rich, allowing the second stage to be ultralean and very low in NO_x emissions. Subsequently, other applications that involved reactants other than the typical air and hydrocarbon fuel were investigated, resulting in demonstrations of operations with very promising potential for commercial use. Many of these novel applications are described in a recent review article [1]. The application of different types of monolithic catalysts to these and various other chemical processes are discussed in this chapter.

II. UNCONVENTIONAL APPLICATIONS

The majority of the investigations of monolithic catalysts deal with gas/gas reactions, although there are a limited number of efforts in which liquid or liquid/gas conditions have been explored. Gas/gas processes, discussed below, are organized into three categories: hydrogen production, synthesis, and dissociation. Applications of monoliths to liquid/gas reactions, in which liquid is sustained throughout the monolithic catalyst operation, are discussed in a later chapter.

A. Hydrogen Production Reactions

The production of hydrogen from hydrocarbon feedstock has been practiced industrially for many years. However, the recent demand for hydrogen for novel applications or small-scale operations necessitates reactor designs that lead to more efficient operation and more rapid response to meet transient demands. One example of this is hydrogen production from a variety of hydrocarbons for use by fuel cells, such as phosphoric acid or polymeric membrane types, to generate electricity. Operating in a load-following capacity for backup electrical power or for transportation requires hydrogen to be produced on rapid demand. These applications have unique needs, including optimizing efficiency with respect to (1) integration within the system operations to utilize waste heat and (2) atypical duty schedules, i.e., not steady-state operation. Fuel-rich partial oxidation and steam reforming are two such processes, and both have been demonstrated to benefit from the use of monolithic catalyst beds.

1. *Partial Oxidation*

Fuel-rich partial oxidation is a very attractive hydrogen production system because of its simplicity of operation. There is a distinct difference in operation between the fuel-rich combustion process used specifically for hydrogen production and that used for the first stage of two-stage combustors. In the latter, which is typically associated with heat engine or boiler operation, the combined hydrogen, unconverted hydrocarbons, and carbon monoxide, as well as the heat produced in the first stage, are used in the ensuing combustion process, thereby reducing the need to maximize the hydrogen yield. For processes in which hydrogen yield is paramount and waste heat is either secondary or unusable in the operation, catalytic control is required to optimize hydrogen and carbon monoxide production.

Based on the thermodynamic equilibrium prediction of product gases from fuel-rich combustion of a gasolinelike hydrocarbon in air (Fig. 1), the narrow operating range for maximizing hydrogen yield, without forming carbon, can be seen. The optimum hydrogen yield is at an air/fuel (mass) ratio of approximately 5.3:1 under ambient inlet conditions. However, the fuel-rich flammability limit is exceeded under ambient conditions, and either no reaction or gas-phase "cracking" would take place without the presence of a suitable catalyst. Stable noncatalytic operations at an air/fuel ratio limit on the order of 8.5:1 would reduce the hydrogen in the product stream by approximately 50% (from about 24 to 12% of the product) and would also raise the exhaust stream temperature by approximately 140 K, from 1420 to 1560 K. To achieve this type of control over carbon formation and to optimize the hydrogen yield over variable operating conditions for such applications as lean-burn internal combustion engine operation [2] or fuel cells [3], controlled catalyst performance was required. Figures 2 and 3 compare the temperature profiles and the hydrogen product yield, respectively, for a conventional packed-pellet-bed catalyst and a ceramic honeycomb monolith catalyst. In this case the fuel was a hydrocarbon liquid, e.g., aviation fuel, which was vaporized and mixed with air prior to

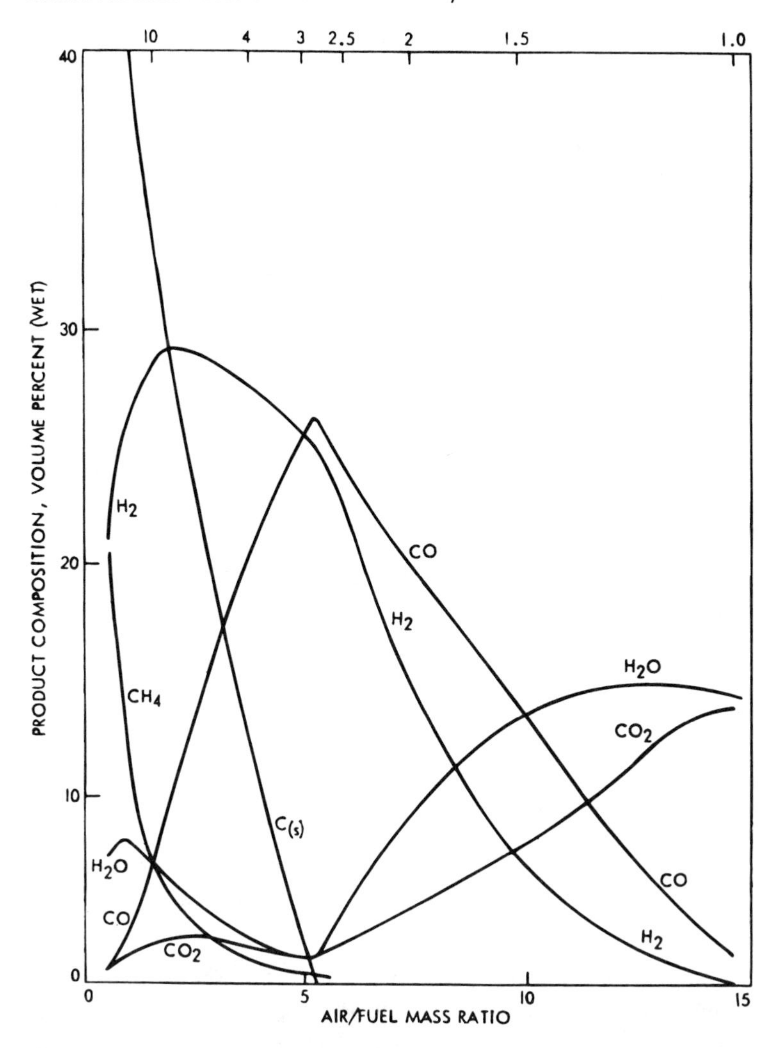

Figure 1 Theoretical equilibrium product composition for Indolene ($CH_{1.92}$)/air adiabatic reaction at 300 K and 303 kPa. (From Ref. 2.)

contact with the catalyst bed. The monolith support used was the Corning Celcor product (46.5 cells/cm^2) with alumina washcoat and a nickel catalyst loading of approximately 12% as compared to 25% on the commercial pellets. A uniform-cell-density, multipiece, honeycomb monolith bed, 14.2 cm in diameter and 20.3 cm long, was used in this case; a tight metal mesh packing around the nickel-catalyzed monolith pieces ensured no bypass. The honeycomb monolith catalyst performance was superior to the packed bed with respect to axial heat transfer based on the very small increase in maximum bed temperature that resulted from increases in total air–fuel flows (i.e., increase in total heat release), over a range of approximately 6:1 in this operation.

In another case of controlled oxidation, the application of a variable-cell-density monolith bed was demonstrated, utilizing a segmented bed of three 2.54-cm-long sections. The result combined monoliths with cell sizes ranging from 6.35 mm (inlet) to 4.76 mm

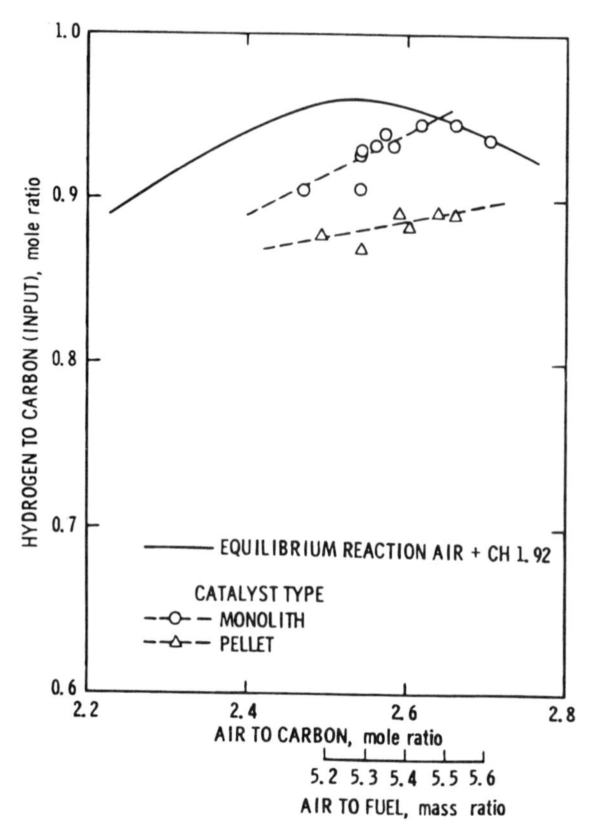

Figure 2 Hydrogen (product)/carbon (fuel) mole ratio as a function of air/carbon mole ratio for monolith and pellet catalyst beds. (From Ref. 2.)

(center) to 3.18 mm (exit) (equivalent to approximately 1.9, 4.3, and 8.7 cells/cm^2, respectively). This design was responsible for holding a virtually flat temperature, as shown in Table 1, over the entire 7.62-cm length for methane and propane fuels under both fuel-rich (28–40% of theoretical stoichiometric mass of air, or percent TA) and fuel-lean conditions (250–320% TA) [4]. The platinum catalyst exhibited stable activity over a series of tests that accumulated over 100 hours and included many start-up/shutdown transients.

Many desirable characteristics of honeycomb monolithic catalyst beds were illustrated by these two examples: (1) catalyst loading lower than on pellets was adequate to maintain activity; (2) the axial heat transfer rate resulted in rapid heat dissipation from the reaction zone; (3) a variation of void-to-catalyst-surface ratio with new turbulent interfaces at each interval introduced a convenient control over the extent of reaction. Two other variables that were investigated are loading of catalyst over bed length and type of catalyst used; however, the comparative analyses have not been as systematically demonstrated for the particular operation described in these tests.

A recent investigation [5–7] of the oxygen–methane reaction for synthesis gas production addressed the effect of various monolith designs on catalyst control of the reaction rate and temperatures. Catalyst bed geometry and gas flow rates strongly affect mass transfer at the catalyst boundary layer and, therefore, the selectivities of fast reactions

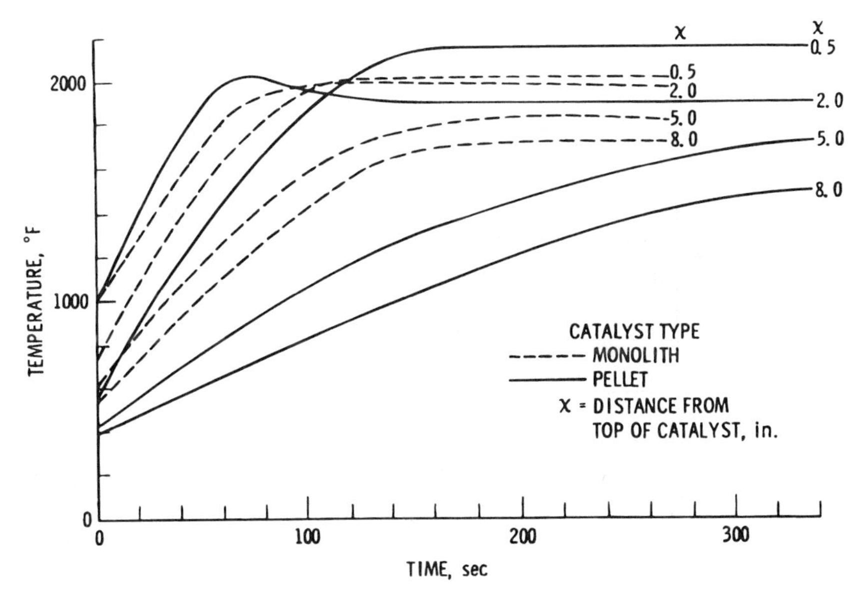

Figure 3 Temperature profiles of the monolith and pellet catalyst beds during light-off and steady-state operation. (From Ref. 2.)

that occur in series. The partial oxidation reaction of methane requires high rates of mass transfer to achieve maximum selectivity to carbon monoxide and hydrogen rather than to carbon dioxide and water. A detailed understanding of the effects of the monolith geometry on this reaction would provide the basis for industrial catalyst configuration and reactor design considerations. In this work three "monolith" configurations were included: (1) 40-mesh or 80-mesh Pt–10% Rh woven-wire-gauze assembly (1–10 layers thick); (2) alpha-phase alumina foam (spongelike) structure of 30–50 pores per inch (ppi) with 1.7–14% (by weight) Pt or Pt–Rh catalyst directly on the support; (3) cordierite honeycomb monoliths, containing 62 cells/cm^2 with 12–14% (by weight) Pt catalyst directly on the support. The dimensions of each monolith bed were from 0.15–2.0 mm thick and 18 mm in diameter for the gauze assemblies and from 2–20 mm thick and 17 mm in diameter for the ceramic supports. Residence times of 10^{-4} to 10^{-2} sec (approximately 2–5 standard liters per minute, 13 to 33 cm/sec velocity) were employed to investigate the effect of short contact times at 140 kPa on the oxidation reaction selectivity, i.e., maximum hydrogen production. Temperatures in these tests were controlled by preheating the gas mixture and by the extent and selectivity of reaction. Thermodynamic equilibrium predicted carbon formation at temperatures below 1273 K and methane/oxygen molar ratios of 2:1 (29.6% in air).

A representative comparison of the effect of the catalyst bed geometry on methane conversion and product selectivity over a range of methane/air ratios is shown in Fig. 4. Unlike typical supported catalysts, where the catalyst is well-dispersed and submicrometer-sized, the noble-metal catalysts in these methane oxidation reactions were basically films with micrometer-sized surface features. (Other tests on both extruded cordierite and alumina foam monoliths with lower catalyst loading resulted in similar carbon monoxide production but lower hydrogen yields than those illustrated in the figure, which provided evidence that the reaction is catalyst-dependent and not initiated by the monoliths or gas

Table 1 Axial Temperature Profiles Through Variable-Geometry Monolith Catalyst (Pt) Bed with Natural Gas and Propane Fuels

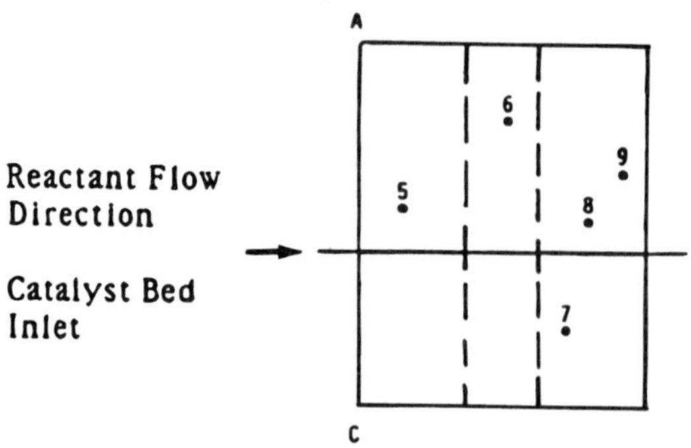

	Fuel-rich						Fuel-lean					
Natural	Run 1122-2			Run 1124-2			Run 1201-1			Run 1201-4		
Gas	*TC*	*K*	*(°F)*	*TC*	*K*	*(°F)*	*TC*	*K*	*(°F)*	*TC*	*K*	*(°F)*
	5	1273	(1831)	5	1518	(2273)	5	1144	(1600)	5	1457	(2162)
	6	1303	(1886)	6	1501	(2242)	6	1432	(2118)	6	1591	(2404)
	7	1342	(1956)	7	1647	(2504)	7	1486	(2214)	7	1620	(2456)
	8	1359	(1986)	8	—	—	8	1434	(2121)	8	—	—
	9	1333	(1940)	9	1597	(2414)	9	1441	(2134)	9	1561	(2349)

	Run 1123-2		
Propane	*TC*	*K*	*(°F)*
	5	1359	(1986)
	6	1367	(2001)
	7	1309	(1896)
	8	1284	(1852)
	9	1293	(1867)

Run no.	Preheat (K)	Space velocity (hr^{-1})	Theoretical air (%)
1122-2	622	21.100	28
1224-2	607	37.800	40
1223-2	489	39.100	37
1201-1	656	105.000	227
1201-4	660	92.500	196

Note: From Ref. 4.

phase.) In addition to differences in hydrogen and carbon monoxide selectivity at comparable methane conversion, there was a major difference in temperature profile characteristics between the monolith supports. Surface temperature and reaction product variations with respect to methane/air inlet composition are shown in Fig. 5 for the extruded cordierite supported catalyst. Large axial temperature gradients (inlet to exit), expected from ceramic monoliths, were observed. These gradients are probably the result of the combined effects

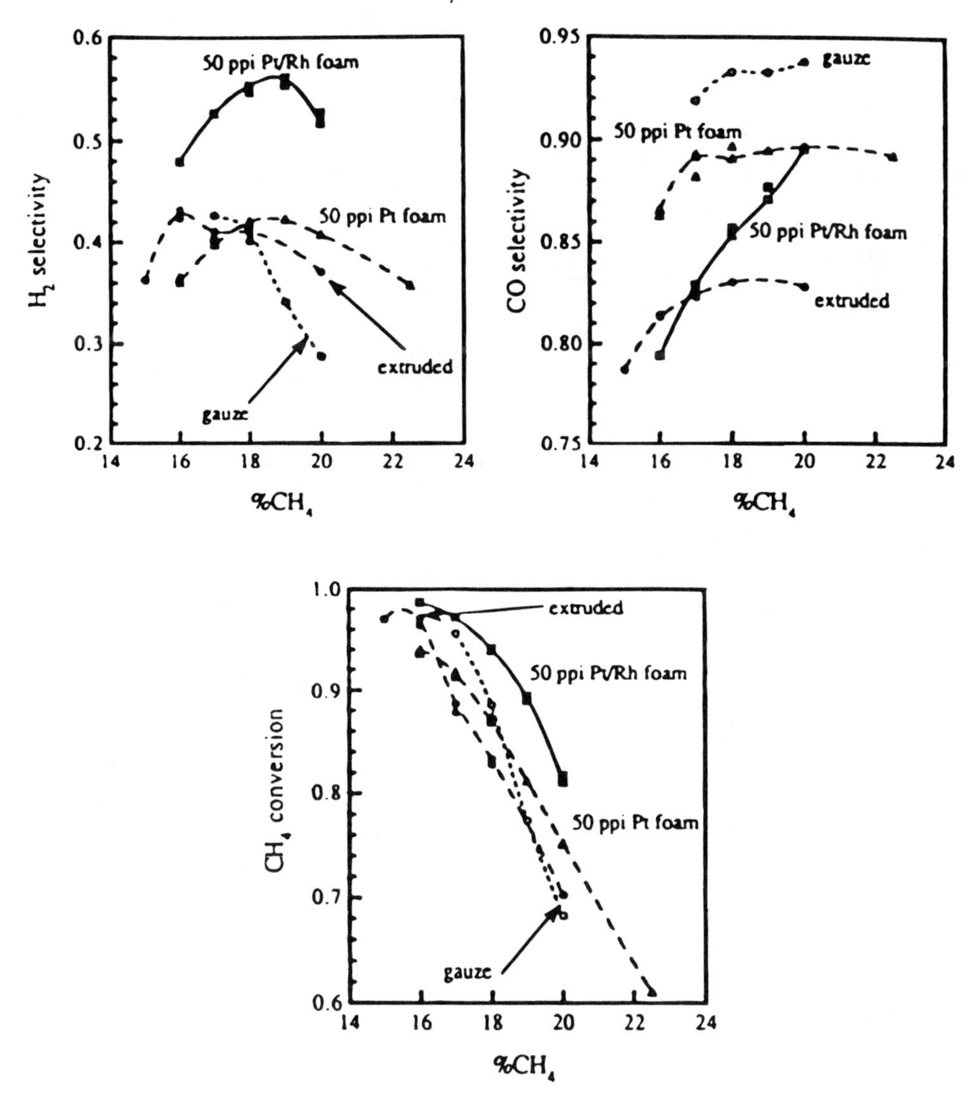

Figure 4 Product selectivities and CH_4 conversion for the following samples: five layers of 40-mesh Pt–10% Rh gauze; a 7-mm-long, 14-wt%-Pt extruded monolith with 400 cells/in.^{2}; a 7-mm-long, 12-wt%-Pt, 50-pores-per-inch (ppi) foam monolith; and a 7-mm-long, 9.9-wt%-Pt and 9.9-wt%-Rh, 50-ppi foam monolith. All experiments were performed at 4–5 slpm total flow. (From Ref. 7.)

of (1) the occurrence of most of the oxidation reaction on or near the front face of the catalyst bed, resulting in a higher temperature at the upstream surface than at the downstream face, (2) the subsequent heat loss from the front face, and (3) the lower thermal conductivity of ceramic monoliths as compared to metal gauze.

In this highly exothermic reaction, kinetic control is a function of the catalyst characteristics and surface temperature, and the mass transport control and boundary layer thickness is based on the catalyst bed geometry and flow velocity. Both the gauze and ceramic foam have tortuous flow paths, but the extruded cordierite monolith has microchan-

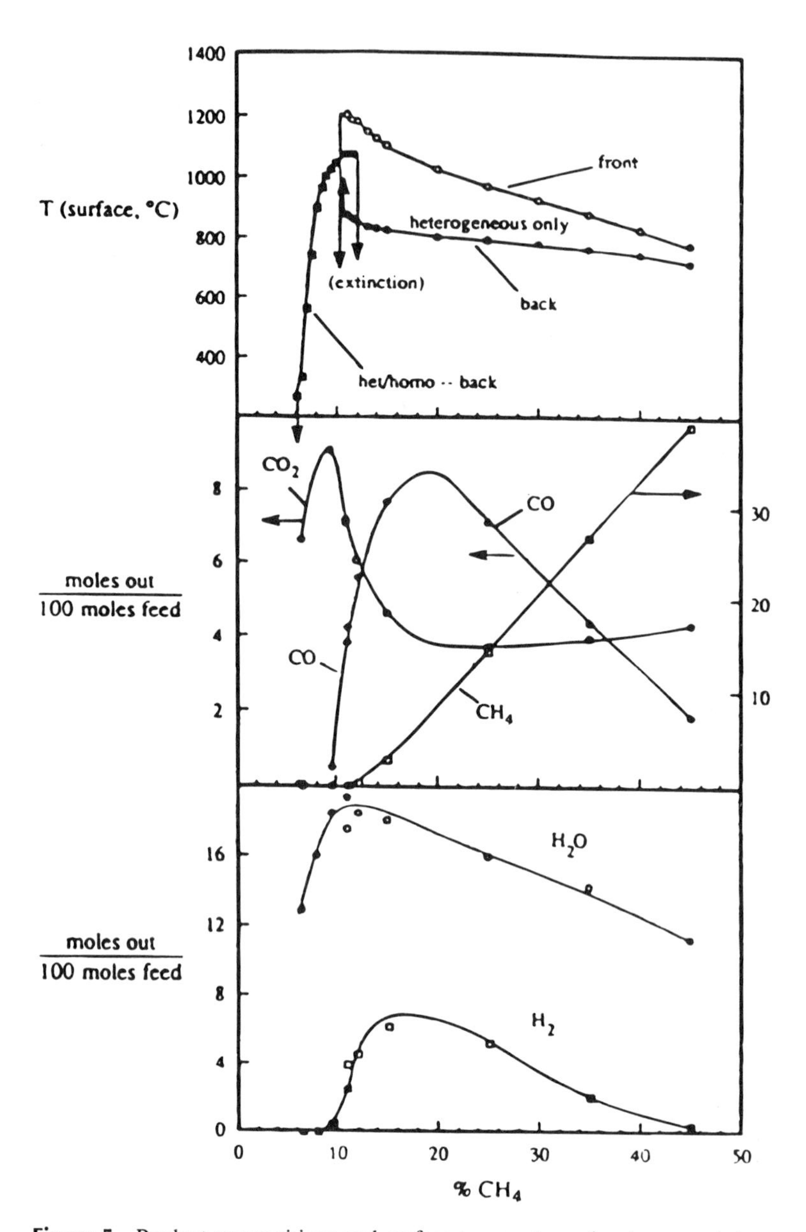

Figure 5 Product compositions and surface temperatures (upstream and downstream ends) for a 7-mm-long, 12-wt%-Pt extruded monolith (400 cells/in.^{2}) at 2.5-slpm total flow. The H_2 and CO selectivities are lower than those reported in previous experiments because the reactor was not insulated. A heterogeneous steady-state reaction is maintained for rich (>12%) compositions, while at leaner compositions, a combination of heterogeneous and homogeneous reactions (a blue flame) is observed, with multiple ignited steady-states for compositions between 10 and 12% CH_4. (From Ref. 7.)

nels in which laminar flow will develop. However, by sustaining catalytic activity at extremely high flow rates, it was possible to maintain a high rate of formation of the selective partial oxidation products while controlling complete conversion (which is mass transfer limited) to carbon dioxide and water, thereby increasing the methane conversion. Control of this highly exothermic reaction appeared to be very dependent on the support, and in the cases illustrated here the characteristics of the foam ceramic support seemed to be very appropriate for selective oxidation of methane for synthesis gas production. The combination of high space velocities and small reactor size, in addition to proper catalyst choice for product selectivity, is significant for adapting this approach to commercial practice. Clearly, only monolithic supports with void fractions that allow for high flow rates and low pressure drop through the bed, as compared to packed-pellet or ring beds, can offer a solution to control this type of reaction.

2. *Steam Reforming*

Steam reforming of hydrocarbons, a reaction in which the activation of steam and a hydrocarbon fuel is necessary to promote the endothermic conversion to hydrogen and carbon oxides, has historically used packed catalyst beds. Two reaction regimes exist in this process: (1) the strongly endothermic reaction region at the inlet, where the bulk of the hydrocarbon is converted to a carbon oxide and lighter molecular weight hydrocarbons, and (2) the reaction toward the exit, where high temperatures (approximately 1220 K) are necessary to convert the residual methane to hydrogen and carbon oxides. Heat energy for the reaction is supplied by (1) preheat of the steam/hydrocarbon reaction mixture at the inlet and (2) external combustion processes with subsequent heat transfer across the reactor tube wall. At the inlet section, the rate-limiting process is heat transfer from the hot combustion gases through the heated reactor wall and catalyst bed. The temperature profile through the catalyst bed from reactor wall to centerline can drop to the point where hydrocarbon cracking and carbon lay-down can occur if heating and reactant feed rates are not carefully controlled. In typical industrial practice, the tube diameter and reactant flow rates are very limited because of the damage that could occur to the heated reactor wall, and to the catalyst inside the reactor at the wall, as a result of overheating.

Control of the temperature throughout the reforming catalyst bed can be established by use of a monolithic catalyst. The heat transfer control can be accomplished by combining three effects that monolithic catalyst beds can impact significantly: (1) direct, uniform contact of the catalyst bed with the reactor wall will enhance conductive heat transfer; (2) uniformity of catalyst availability to the reactants over the length of the flow will provide continuity of reaction; and (3) coordination of void-to-catalyst ratio with respect to the rate of reaction will moderate gas-phase cracking relative to catalytically enhanced hydrocarbon–steam reactions. This combination provides conditions for a more uniform reaction over the catalyst bed length.

Temperature control was demonstrated in a series of experimental tests [8–10]. In one series of tests, both ceramic and metal honeycomb monolith supports were used [8,9]. The ceramic supports used in these tests were 46.5-cells/cm^2 (square cells) Celcor monoliths 6.2 cm in diameter and 10.2 cm long, whereas the metal honeycomb monolith supports (made of Kanthal, a high-temperature alloy) had 38.8 cells/cm^2 (hexagonal cells) and were 6.3 cm in diameter and 5.1 cm long. The ceramic and metal monoliths were washcoated with gamma alumina, and the nickel catalyst loadings were 13 and 5.6% (average), respectively. In comparison, the conventional catalyst used for reference was a packed bed of 0.64-cm right cylinders that had 19% nickel, by weight. A combination of five Celcor monoliths achieved a catalyst bed length of 51 cm, the same as was used for the pellet bed. The void fraction of the packed bed was 30%, whereas that of the monoliths was 70%, a result of the large cell size of the monoliths. The hydrocarbon used in these tests was *n*-hexane. The impact of various combinations of total reactant flow rates (0.79 L/sec to 2.37 L/sec), wall temperatures (1089 and 1200 K), and steam-to-carbon ratios (3.0, 2.5, 2.0, 1.5, and 1.0) was examined with respect to hexane conversion in these tests. Initial tests led to degradation of the cordierite Celcor structure, and carbon formation and cell blockage occurred. The high temperature and concentration of steam at the inlet section were suspected to have caused a reaction with the silica in the cordierite that resulted in a weakening of the support. By replacing the inlet section (first 20%) of the cordierite monolith bed with Kanthal monoliths, stable, enhanced steam reforming of hexane and a much improved thermal profile were reported on the "hybrid monolith," as illustrated in Fig. 6.

Conversion efficiency is definitely affected by the large void fraction, which is apparent in the results from changes in the total throughput, or space velocity (0.56 versus 1.11 sec^{-1}), shown in Fig. 7. In this comparison, the concentration of unconverted hexane increased tenfold when the flow rate was doubled. The impact of improvements in conductive heat transfer, combined with the mass transfer limitations associated with the cell size and honeycomb design, and a catalyst loading that was nearly one-half that of commercial pellet catalysts (average, 11.5% versus 19.2%) suggested that both carbon formation and steam/hydrocarbon reactions were better controlled with monolithic supports under the conditions employed. This comparison was made where the extent of the endothermic reaction is equal between the pellet bed and the hybrid cordierite/metal monolith bed.

PRODUCT COMPOSITION COMPARISON

	FUEL FLOW lb/hr	INLET TEMP °F	WALL TEMP °F	S/C	SV hr^{-1}	H_2	CO	CO_2	CH_4	C_6H_{14}
PELLET BED	1.5	1000	1700	2.5	2000	66.8	19.6	10.8	0.12	0.03
HYBRID MONOLITH BED	1.6	1000	1700	2.5	2000	66.1	20.0	10.1	0.19	0.03

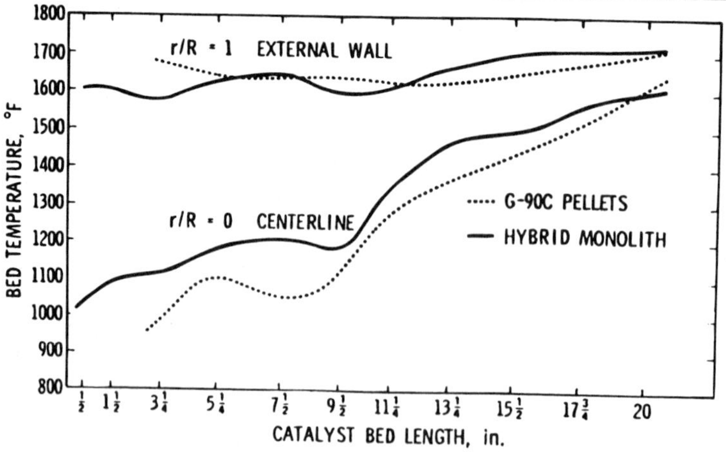

Figure 6 Axial bed-temperature profiles and product gas composition comparison for steam reforming of *n*-hexane on the hybrid monolith and G-90C pellet bed. (From Ref. 8.)

PRODUCT COMPOSITION COMPARISON

	FUEL FLOW lb/hr	INLET TEMP °F	WALL TEMP °F	S/C	SV hr^{-1}	H_2	CO	CO_2	CH_4	C_6H_{14}
HYBRID MONOLITH BED	3.01	1000	1500	2.5	4000	64.9	12.5	16.0	1.97	0.69
HYBRID MONOLITH BED	1.51	1000	1500	2.5	2000	67.7	16.4	13.7	1.42	0.062

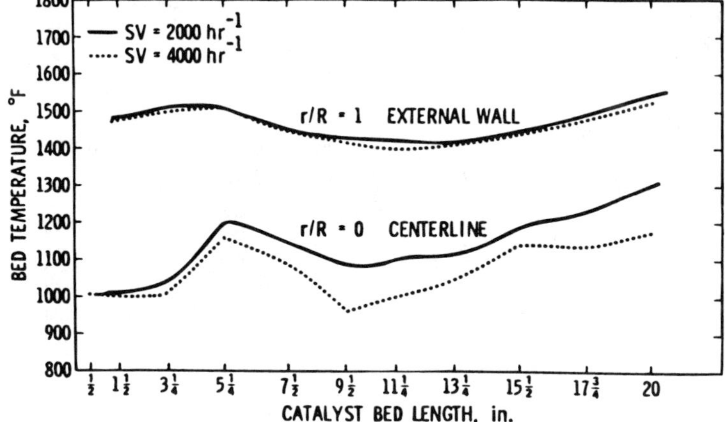

Figure 7 Effects of space velocity on axial bed-temperature profiles and product composition for steam reforming of *n*-hexane on the hybrid monolith. (From Ref. 8.)

However, carbon formation and destruction of the cordierite support were both found to have taken place over the course of various test conditions. These findings indicated that while the heat transfer into the monolithic catalyst bed improved, (1) the gas-phase hexane cracking reaction that produced carbon precursor species (due to the high void fraction and mass transfer limitation) still existed, and (2) the combination of sustained high temperature and high steam density on cordierite warranted use of only metal monoliths for this application. However, the relatively low loading of nickel in the monolith catalyst and the mass transfer limitation still resulted in equivalent conversion under conditions similar to those found in industrial practice.

A subsequent set of steam reforming experiments with metal monolithic supported catalysts was performed to determine the effect of void fraction alone. Catalyst loadings were 6.4 (average) and 19% by weight, and void fractions were 70 and 30% for the monolith and pellets, respectively. No carbon was observed in the first series of tests, which are shown in Fig. 8. In spite of the higher void fraction, the metal monolithic catalyst was superior in hexane conversion and in the suppression of methane production, a common undesirable by-product when catalyst temperatures are below the level of the steam-reforming activation energy. As in the previous tests, higher flow rates resulted in significantly poorer performance for the monolith catalyst than for the pellet bed unless the steam-to-carbon ratio was increased. An increase in steam was demonstrated to have helped to control the gas-phase reactions of hydrocarbon cracking and carbon formation.

These results became the basis for investigating another modification to improve the performance of a monolith catalyst bed for steam reforming, i.e., use of a metal monolith support that, although made up of parallel cells, was not honeycomb in design and had a multipath flow [10] (see Chapter 14). This modification would be expected to

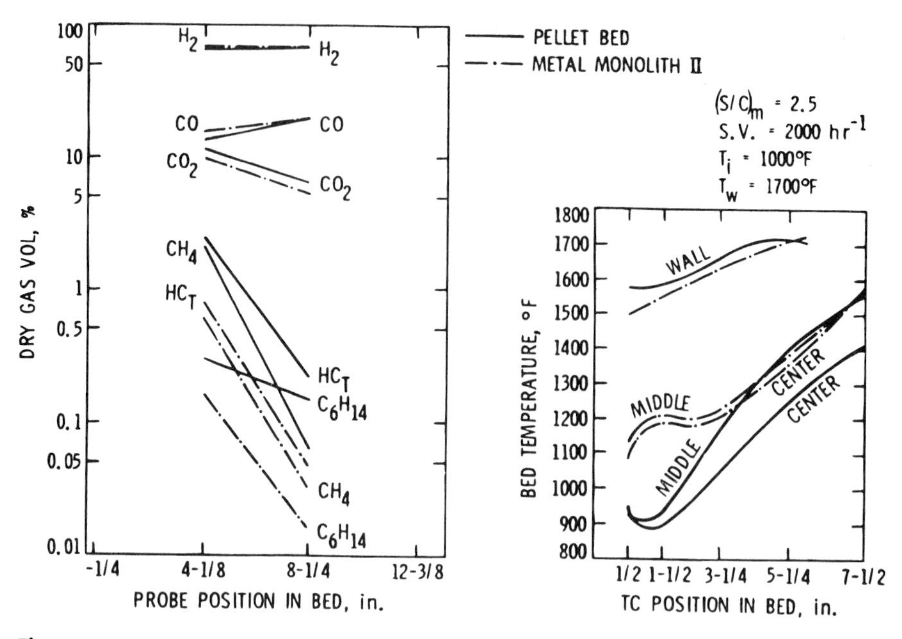

Figure 8 Steam reforming of hexane at flow rates of 2.0 and 0.64 lb/hr of water and hexane, respectively. Axial bed-temperature and composition profiles for a metal monolith (250 cells/in.^{2}), consisting of Kanthal support/γ-Al_2O_3 washcoat/NiO catalyst, and a packed bed of Girdler G-90C pellets ($\frac{1}{4}$ in. $\times$ $\frac{1}{4}$ in.) of alumina impregnated with nickel. (From Ref. 9.)

result in a combination of enhanced convective heat transfer and reduced mass transfer limitation. An example of the configuration chosen to verify improvements in reforming is shown in Fig. 9. The catalyst bed was custom-made, consisting of three layers, one immediately on top of the next, of 62 very tightly packed 2.65-cm-wide metal strips. Each strip contained two parallel, offset rows of "ridges" that were open on either end. In order to inhibit "sandwiching" of the strips and collapsing of the flow channels in each section of the monolith bed, the individual strips were aligned so that the ridges were offset by one-half the spacing between them. The high-temperature alloy strips were alumina-washcoated and catalyzed with nickel (10% by weight of washcoat). The layers of strips were in tight contact with the metal wall of the reactor, to enhance conductive heat transfer, and were oriented orthogonally to each other to retain alignment in each of the three segments. This bed design provided for gases to flow through three sequentially connecting rows of two-dimensional, tortuous-path segments, each containing two off-setting rows of open-ended ridges, over a total 8.6-cm-long bed. Unlike the situation in a honeycomb design, the gases flowing through the bed could mix radially in two dimensions and contact the reactor wall, thereby providing for convective and conductive heat transfer through the axial flow path. This pattern also provided a tortuous flow path for the reactants that would enhance wall contact. Results of steam reforming methane in the metal monolithic catalyst bed were compared to a packed bed of commercial catalyst in the shape of 0.64-cm right cylinders. All the physical and operating parameters are shown in Table 2.

A series of tests at four different flow rates (0.20 L/sec to 0.50 L/sec) were reported during which both axial and radial temperature profiles were recorded along with the

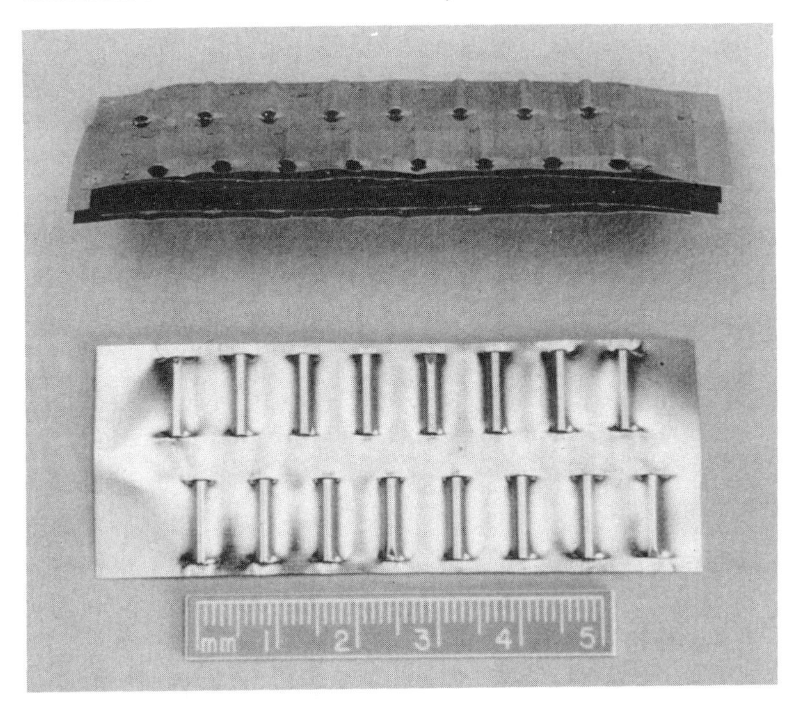

Figure 9 Configuration of the metal strips and the packing arrangement used in each layer of the metal monolith bed. (From Ref. 10.)

extent of methane conversion. Figures 10 and 11 show the axial and radial thermal profiles for the pellet bed and the monolith bed, respectively, at the highest flow conditions (0.50 L/sec, gaseous inlet hourly space velocity of 8400) for a steam-to-carbon ratio of 3:1. These illustrate the variations in catalyst bed temperature ($X/R = 1$ at the inside wall, $X/R = 0$ at the centerline) that correspond to this flow rate with external reactor wall temperatures of about 1140 K.

The relative methane conversions for the entire series of tests are shown in Fig. 12. From these data the heat transfer coefficients that were calculated (Table 3) show a significant improvement of the monolith bed over the packed bed. It is apparent that a combination of (1) the endothermic reaction rate and (2) heat transfer via conduction and convection are balanced over these flow rates in this monolith bed. However, based on the unfavorable shift in the extent of methane conversion from the monolith to the packed bed as flows increase, shown in Fig. 12, it appears that the mass transfer of reactants to the catalyst surface in this monolith bed may be rate-limiting at flow rates at and above 0.50 L/sec, probably a function of the high void fraction of this bed design.

It is important to note that, although this inlet flow rate is lower than the inlet rates (0.78 to 1.6 L/sec) found to be carbon-free when a honeycomb monolith bed was compared to a pellet bed in the steam reforming of hexane, this bed length is also shorter. In addition to the obvious differences in bed configuration, other factors contributing to this difference in maximum flow rate may be (1) the high rate of reaction associated with the hexane "cracking reaction" (i.e., ease of carbon–carbon bond breakage), (2) lower wall temperatures (about 100°) in the methane work, (3) higher steam-to-carbon ratio (3.0:1 compared

Table 2 Physical Parameters and Operating Conditions for Steam Reforming of Methane

Parameters	Conventional packed-bed catalyst	Metal supported catalyst bed
Reactor diameter, m (in.)	6.2×10^{-2} (2.4)	6.1×10^{-2} (2.4)
Bed height, m (in.)	7.6×10^{-2} (3.0)	8.6×10^{-2} (3.4)
Catalyst	G90B, 1/4-in. × 1/4-in. cylinder, nickel catalyst	Alumina-washcoat steel-alloy support impregnated with nickel
Reactor volume, m^3 (ft^3)	232×10^{-6} (82×10^{-4})	246×10^{-6} (87×10^{-4})
Void volume, m^3 (ft^3)	81×10^{-6} (29×10^{-4})	221×10^{-6} (78×10^{-4})
Void fraction	0.35	0.90
Uncorrected total geometric surface area, m^2 (ft^2)	0.20 (2.2)	0.50 (5.4)
Steam/carbon ratio	3.0	3.0
Operating pressure, N/m^2 (psia)	2×10^5 (29.2)	2×10^5 (29.2)
No. of thermocouples	48	52
Inlet gas preheat temperature, °C (°F)	316 (600)	316 (600)

Note: From Ref. 10.

to 2.5:1), and (4) different catalyst activity. However, since no kinetic data were obtained, a direct comparison between the methane and hexane experiments cannot be made.

The conclusions drawn from the investigations regarding monolithic catalyst supports were: (1) Heat transfer, both axially and radially, was enhanced; (2) convective heat transfer and tortuous flow paths contributed to enhancing the endothermic reaction; (3) the void fraction was too high to enhance surface reaction rate relative to gas-phase reaction rates; and (4) a washcoated metal monolith support material was inert to high-temperature steam. A better understanding of the correlations between catalyst activity, void fraction, and enhanced heat transfer rates is integral to optimizing the design of the monolith support for this application.

3. *Autothermal Reforming*

Autothermal reforming is a term adopted for the process in which a mixture of air and steam serves as the oxidant in the conversion of hydrocarbon fuels to a hydrogen-rich product. This process has also been reported to become more efficient as a result of the use of monolithic catalyst beds [11]. An example of this has been the demonstration of a modified version of the fuel-rich partial oxidation process in which noble metal catalysts were used in place of nickel on ceramic monoliths [2]. In earlier reports where packed catalyst beds were used, the concept to control carbon formation, which was predicted by thermodynamic equilibrium at low air-to-fuel ratios, was demonstrated by introducing steam, in addition to air, as an oxidant.

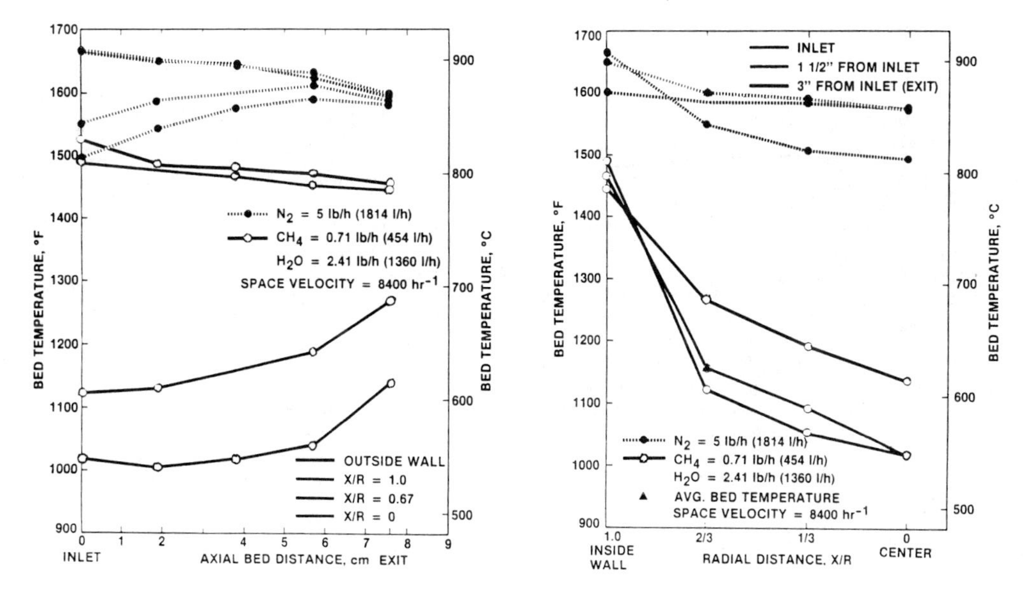

Figure 10 Axial (left) and radial (right) temperature profiles of the packed-commercial-pellet bed during steam reforming of methane and nitrogen flow conditions at gaseous-inlet hourly space velocities of 8400 hr^{-1} [X/R = distance from centerline/inside radius of reactor]. (From Ref. 10.)

The Engelhard two-bed reactor design for autothermal reforming is shown schematically in the patent drawing in Fig. 13. The first bed, used for partial oxidation of the fuel (natural gas feedstock), had a monolith bed length-to-diameter ratio (22.9:1.9 cm) of 12 and operated at a linear velocity of 7.78 cm/sec. The second segment of the reactor was designed to serve as a steam reformer and incorporated a packed catalyst bed that had a length-to-diameter ratio (23.5:7.62 cm) of 3. The reactor design was intended to oxidize enough of the hydrocarbon feed in the inlet portion to generate heat and a gaseous mixture of hydrogen, carbon oxides, and unconverted lighter hydrocarbon species. The hydrocarbon species, in combination with the excess steam, were subsequently reformed in the second catalyst bed to additional hydrogen and carbon oxides. All the energy for the steam reforming process was provided by the inlet preheat and the exothermic partial oxidation process in the first bed. No evidence of catalyst support degradation or carbon formation was reported.

The monolithic support for this reactor was of a honeycomb design, which required that the flow velocity be high enough that the exothermic reaction rate between the fuel and air that was catalyzed with a noble metal would be mediated by axial heat transfer through the bed and by the steam diluent. Under these conditions, heat was radiated

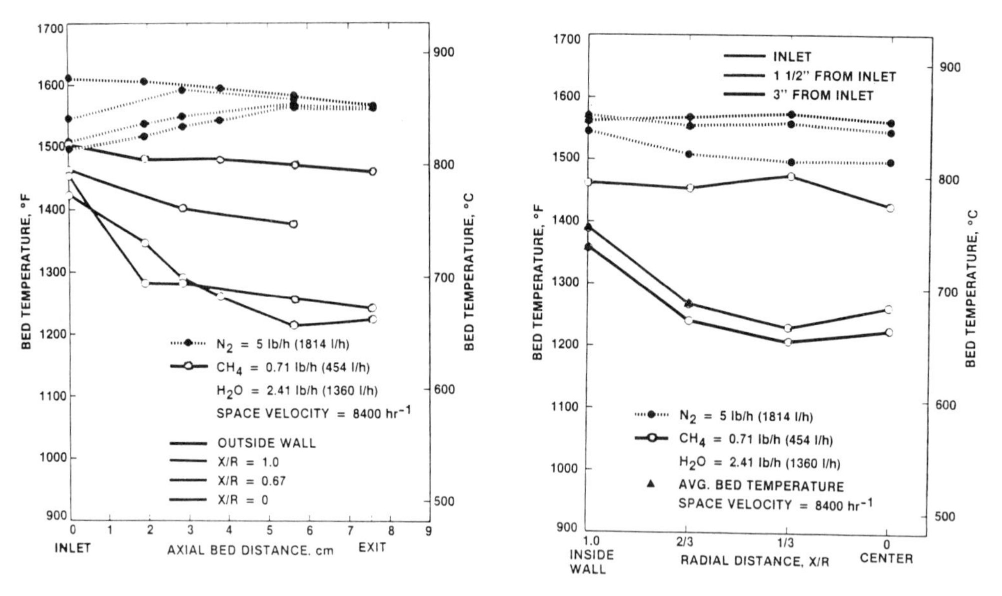

Figure 11 Axial (left) and radial (right) temperature profiles of the metal monolith bed during steam reforming of methane and nitrogen flow conditions at gaseous-inlet hourly space velocities of 8400 hr^{-1}. (From Ref. 10.)

upstream, serving to preheat the incoming gas stream and facilitate the hydrocracking and oxidation reactions, much the same as in the partial oxidation process discussed earlier. This concept for producing a hydrogen-rich synthesis gas was suggested for possible use with fuel cells, for methanol production [12], as well as for liquid fuel production (Fischer–Tropsch process) [13] when natural gas feedstock is used.

B. Synthesis Reactions

1. *Methanation*

The application of monolith catalysts to a variety of commercial synthesis processes has been investigated because of the potentially smaller size and lower pressure drop through the chemical reactors. One of the earliest of these investigations was for methanation, the chemical reaction between carbon monoxide and hydrogen to produce methane selectively. In a detailed study [14] a comparative evaluation involved the use of nickel catalyst on (1) spherical alumina pellets (0.32-cm diameter), (2) alumina washcoated (10–20% by weight) cordierite monoliths with 31- and 46-cells (square)/cm^2 density, (3) an alumina

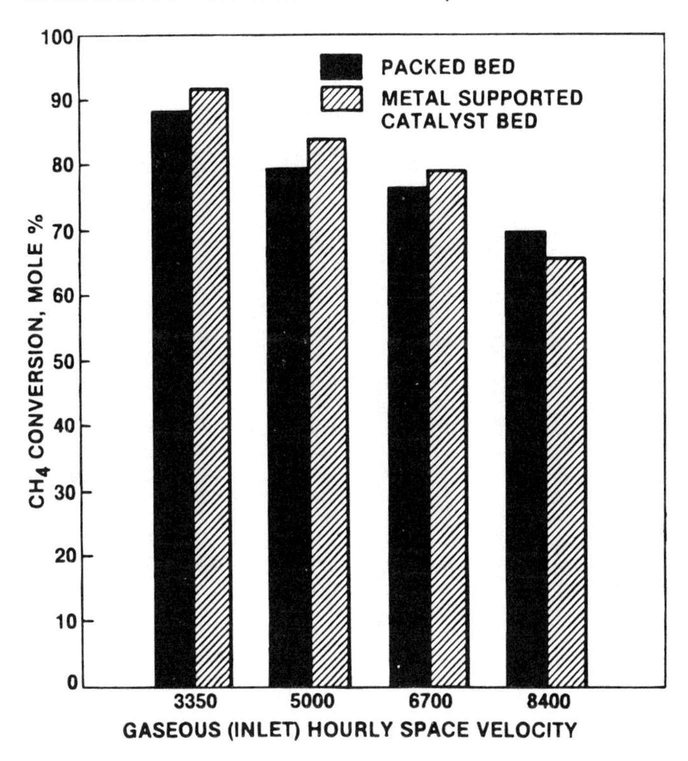

Figure 12 Comparisons of the methane conversion between the packed commercial pellet bed and the metal monolith bed during steam reforming at different gaseous-inlet hourly space velocities. (From Ref. 10.)

Table 3 Calculated Heat Transfer Coefficients (*U*) from Reactive (Methane Steam Reforming) and Nonreactive (Nitrogen) Test Data

	U, Packed bed		*U*, Metal supported catalyst bed	
Flow rates	W/m^2 °K	(BTU/hr) (ft^2) (°F)	W/m^2 °K	(BTU/hr) (ft^2) (°F)
N_2 = 726 L/hr	32.4	5.7	38.6	6.8
CH_4 = 182 L/hr, H_2O = 542 L/hr	129	22.7	244	43
N_2 = 1814 L/hr	54.5	9.6	74	13
CH_4 = 454 L/hr, H_2O = 1360 L/hr	216	38	306.6	54

Note: From Ref. 10.

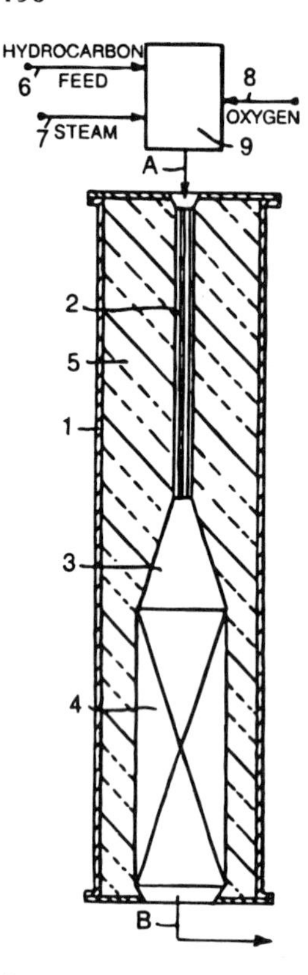

Figure 13 Monolith catalyst bed configuration used in demonstration of autothermal reforming of hydrocarbons. The monolith catalyst bed through which the hydrocarbon, oxygen, and steam mixture (A) passes is No. 2, and the pellet steam reforming bed is No. 4. (From Ref. 11.)

monolith of 37-cells (triangular)/cm^2 density, and (4) one commercial catalyst. Two types of tests were conducted. One type was differential measurements of activity at 500–525 K and a space velocity of 8.33 sec^{-1}. The second type was conversion/temperature relationship measurements over a temperature range of 475–725 K and space velocities of 4.17, 8.33, and 13.9 sec^{-1} (to evaluate the effect of geometry on the mass-transfer-influenced reaction rate). Both included tests at pressures of 140, 1000, and 2500 kPa. The sample size evaluated in all tests was 2.5 cm in diameter and 1.3 cm long, using a gas mixture of 95% nitrogen, 4% hydrogen, and 1% carbon monoxide.

Data on differential measurement tests are shown in Table 4. The turnover numbers for the monolith catalysts at 140 kPa, relative to hydrogen chemisorptive uptake, were 40–100% higher than the pellets and 50–500% higher at the elevated pressures. In addition, the γ-phase alumina monolith catalyst was less active than the alumina-washcoated monoliths despite nearly twice as much nickel loading. Furthermore, the particular geometry of the monolith cell appeared to make no significant contribution to conversion activity. A characteristic of the catalyst in both monolithic and pellet form is the trend of decreasing energy of activation with increasing metal loading; however, the monolithic catalysts had higher activation energies than the pellets, in general. In these short (1.3-cm) catalyst bed tests, the monolith supported catalysts achieved 90% conversion of carbon monoxide at lower temperatures than the pellet-supported catalysts, and the methane yield was also greater.

Table 4 Low-Conversion, Differential Reaction Rate Data at 140 kPa, 500 K, and 525 K, 30,000 hr^{-1} Gaseous Hourly Space Velocity (GHSV), with a Gas Mixture Containing 95% N_2, 4% H_2, 1% CO

Catalyst	H_2 uptake (μmol/g)	% CO conversion at 500 K	CH_4 turnover number[a] at 500 K	CH_4 turnover number[a] at 525 K	Apparent activation energy (kJ/mol)
Tests performed at 140 kPa					
3% Ni/10% Al_2O_3/monolith[b]	32	6.2	5.2	12.8	82
6% Ni/20% Al_2O_3/monolith[c]	66	10.6	3.8	11.5	110
11% Ni/19% Al_2O_3/monolith[c]	105	14.4	2.7	9.2	106
12% Ni/20% Al_2O_3/monolith[c]	76	13.4	3.7	9.8	106
19% Ni/20% Al_2O_3/monolith[c]	75	10.9	4.5	12.5	91
20% Ni/20% Al_2O_3/monolith[c]	65	13.5	4.3	11.7	87
25% Ni/Al_2O_3 monolith[d]	179	15.4	3.0	7.4	79
3% Ni/Al_2O_3 spheres[e]	38	2.4	1.6	5.7	112
6% Ni/Al_2O_3 spheres	74	4.7	2.5	8.9	112
14% Ni/Al_2O_3 spheres	159	14.6	2.8	6.5	73
~40% Ni/Al_2O_3 (Girdler G-87)	167	14.7	1.8	4.4	77
25% Ni/Al_2O_3 monolith-crushed	179	12.7	2.5	7.4	95
Tests performed at 1000 kPa					
3% Ni/10% Al_2O_3/monolith[b]	46	15.6	15	47	106
6% Ni/Al_2O_3 spheres[f]	74	6.4	3.1 ± 0.4	12.2 ± 1.8	124 ± 14
Tests performed at 2500 kPa					
6% Ni/20% Al_2O_3 monolith[c]	66	43	9.3	40	131
3% Ni/Al_2O_3 spheres	38	11	6.5	27	105
14% Ni/Al_2O_3 spheres	187	49	4.7	11	90

Note: From Ref. 14.

[a] The units of turnover number are molecules/site sec × 10^3.

[b] 46.5 squares/cm^2.

[c] 31 squares/cm^2.

[d] Pure γ-Al_2O_3 monolith (Corning Glass Works) having 37 triangular channels per cm^2.

[e] All catalyst spheres shown in this table have an average diameter of 0.32 cm.

[f] Data shown are based on the average of four different samples. Deviations are expressed as standard deviations—about ±10% for the turnover numbers.

Summary data for different conversion/temperature conditions are provided in Tables 5–7. Rate constants were calculated from these data, and it was determined that although the operations at 140 kPa were influenced by mass transfer, this was not the rate-limiting step; however, the reaction was mass transfer limited at 1000 kPa. The higher carbon monoxide conversion values and methane production observed for the monolith-supported nickel compared to pellets were explained to be due to the provision by the monoliths of smaller pore diffusion resistance and higher mass transfer rates at higher temperatures, primarily a result of shorter diffusion paths in thin alumina coatings on the monolith walls.

Although the monolith catalyst demonstrated higher activity and selectivity than the pellet catalyst in these tests, there are two important points to be made regarding the operations: (1) The higher space velocity tests on the monolith catalysts were also at

Table 5 Conversion-Temperature and Activity Data for Methanation of CO over Pellet and Monolith Nickel Catalysts at Low Pressure (140 kPa) with a Reaction Gas Mixture Containing 95% N_2, 4% H_2, 1% CO

Catalyst	GSA (cm^2/cm^3)	Temperature for 90% CO conversion	Yield[c] at 90% CO conversion		Rate of methane production (g mol/cm^3 cat. sec × 10^6)
			CH_4	CO_2	
Tested at GHVS of 15,000 hr^{-1}					at 600 K
19% Ni/19% Al_2O_3 31 □/cm^2 monolith		555 K	0.91	0.07	1.8
3% Ni/Al_2O_3 0.32 cm spheres		600	0.78	0.21	1.3
14% Ni/Al_2O_3 0.32 cm spheres		543	0.84	0.11	1.6
~40% Ni/extrudites 1.0 cm × 0.32 cm diam. G-87 (Girdler)		500	0.83	0.14	1.45
Tested at GHVS of 30,000 hr^{-1}					at 600 K
12% Ni/20% Al_2O_3/monolith, 31 □/cm^2	12.5	584 K	0.87	0.12	2.9
25% Ni/Al_2O_3 monolith	14.9	570	0.80	0.11	2.9
14% Ni/Al_2O_3 0.32 cm spheres	8.7	638	0.82	0.16	2.2
~40% Ni/Al_2O_3 extrudates 1.0 cm × 0.32 cm diam. G-87 (Girdler)	11.3	668	0.77	0.17	2.2
Tested at GHSV of 50,000 hr^{-1}					at 700 K
3% Ni/10% Al_2O_3/monolith, 46.5 □/cm^2	17.8	625	0.70	0.19	4.0
3% Ni/Al_2O_3 0.32 cm spheres[a]	8.7	698 (58.9)[b]	0.64[b]	0.24[b]	2.3
6% Ni/Al_2O_3 0.32 cm spheres	8.7	685 (80.0)[b]	0.52[b]	0.24[b]	3.0

Note: From Ref. 14.

[a] Data shown for pellets are the average of four runs on four different samples of the same catalyst; reproducibility was within ±8%. For monolith samples, rates for 2–3 duplicate samples agreed within 5%.

[b] Sample did not reach 90% conversion at any temperature; the temperature at which maximum conversion was reached is listed along with maximum % conversion in parenthesis. The yields were calculated at maximum CO conversion.

[c] Yield is the fraction of converted CO appearing as a given product.

higher pressures, and (2) the catalyst bed was very short. The first point addresses the issue concerning the advantage of a very low pressure drop through monolithic supports. A low pressure differential allows for an increase in overall operating pressures and flow rates without a system penalty of increased pressure losses such as occur in packed-pellet beds. The second point addresses the issue regarding the flow pattern in a long honeycomb bed, where laminar flow will develop. Such a pattern is not observed in a short bed, where the turbulence developed from entering the channels is a factor and provides additional mixing at the walls. The conclusion was that a honeycomb monolith catalyst used for methanation of synthesis gas (a hydrogen–carbon monoxide gas mixture) could result in a more compact reactor, operate at higher throughput without high back pressure, and selectively produce more methane at high conversion rates.

Table 6 Conversion-Temperature and Activity Data for Methanation of CO over Pellet and Monolith Nickel Catalysts at Intermediate Pressure (1000 kPa) with a Reaction Mixture Containing 95% N_2, 4% H_2, 1% CO and at a GHSV of 30,000 hr^{-1}

Catalyst	GSA (cm^2/cm^3)	Temperature for 90% CO conversion	Yield at 90% CO conversion		Rate of methane production at 650 K (g mol/cm^3 cat. sec × 10^6)
			CH_4	CO_2	
3% Ni/10% Al_2O_3/monolith, 31 □/cm^2	15.5[c]	540 K	0.90	0.07	5.8
3% Ni/10% Al_2O_3/monolith, 46.5 □/cm^2	17.8[d]	590 K	0.88	0.06	6.0
6% Ni/Al_2O_3 0.32-cm spheres[a]	8.7	719 (88)[b]	0.83[b]	0.11[b]	4.4 ± 0.24

Note: From Ref. 14.

[a] Data shown are the average of four runs using four different samples of the same catalyst; for the rate based on volume the standard deviation was $\pm 0.24 \times 10^{-6}$ g mol/cm^3 cat. sec or, in other words, ±5.5%.

[b] The average CO conversion did not reach 90% at any temperature. The average temperature at which maximum CO conversion occurred and the maximum percent conversion (in parenthesis) are listed. The yields were calculated at maximum CO conversion.

[c] Estimated assuming 18% loss of GSA due to coating.

[d] Measured from enlarged photograph.

Table 7 Conversion-Temperature and Activity Data for Methanation of CO over Pellet and Monolith Nickel Catalysts at High Pressure (2500 kPa) with a Reaction Mixture Containing 95% N_2, 4% H_2, 1% CO

Catalyst	Temperature for 95% CO conversion	% yield at 95% CO conversion		Rate of methane production at 600 K (g mol/cm^3 cat. sec × 10^6)
		CH_4	CO	
Tested at 30,000 hr^{-1} GHSV:				
6% Ni/20% Al_2O_3 monolith, 31 □/cm^2	518 K	99%	1%	3.6
3% Ni/Al_2O_3 spheres 0.32-cm diameter	615	96	3	2.9
14% Ni/Al_2O_3 spheres 0.32-cm diameter	548	97	1	3.4
Tested at 50,000 hr^{-1} GHSV:				
11% Ni/20% Al_2O_3 monolith, 31 □/cm^2	588	97	1	5.4

Note: From Ref. 14.

In another brief examination [15] of the impact of monolith supports for methanation catalysts, a comparison between nickel and ruthenium catalysts was made utilizing a metal (Fecralloy) support. The conversion tests were run at 673 K, 5400 kPa, 3.47 sec^{-1}, and with a gas composition of 62% hydrogen, 18% carbon monoxide, and 20% water vapor. A ruthenium pellet catalyst that was run in comparison was approximately twice as active as ruthenium on the monolith. However, the difference in product (methane) selectivity was 97% for the metal monolith catalyst and 83% for the pellet bed. In the comparison between nickel and ruthenium, shown in Fig. 14, the ruthenium was more active and selective. The lack of impact on activity or selectivity as a result of steam addition to the reactant mixture provided useful practical data as well. No further details regarding the catalyst characteristics were provided.

2. *Methanol to Gasoline*

Another synthesis process proposed to receive benefits from operating with monolith catalysts is the conversion of methanol for gasoline production [16,17]. The catalyst used was the ZSM-5 zeolite. However, rather than binding the catalyst onto the wall by use of a washcoat, it was uniformly crystallized on the cordierite honeycomb (62 cells/cm^2) wall surfaces (up to 30% by weight), similar to the method described in the patent assigned to Lachman and Patil [18]. The effects of methanol partial pressure on conversion and temperature on hydrocarbon selectivity were determined. Three regimes of mass transfer resistances are experienced in this reaction: reactant transfer to the reactor walls within the monolith channels through the laminar flow, diffusion resistance at the surface between zeolite crystals on the walls, and diffusion into the zeolite molecular-size pores to the active sites within the crystals, where the reaction rate limit is anticipated.

It was pointed out that diffusion effects are less severe for thin zeolite layers on monolithic substrates than for pellets, such as were used for performance comparison. Methanol partial pressure variations led to results similar to other work in which it has been found that lower pressures favor olefin formation. As the temperature is raised, light

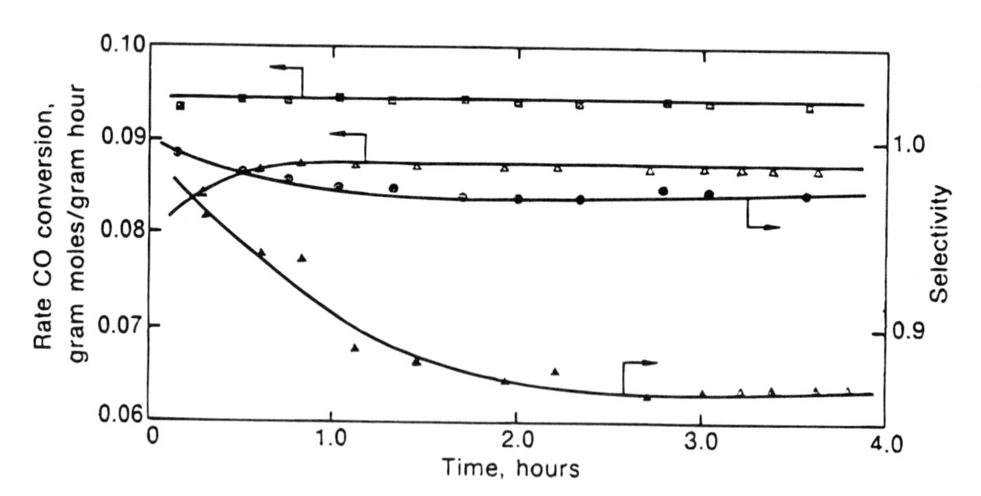

Figure 14 Comparison of methanation rates and selectivity for converting CO to CH_4 for ruthenium- and nickel-catalyzed metal monoliths. (Circles and squares: ruthenium; triangles: nickel.) (From Ref. 15, reprinted with permission of Gulf Publishing Co.)

olefins and methane production increase as a result of hydrocarbon cracking processes. Test results indicate that while methanol is converted to gasoline-range hydrocarbons whose composition mix lies between that resulting from fixed- and fluid-bed catalysts, the percentage of durene in the product is higher than that in either of them. Catalyst deactivation was observed after 4 hours of continuous operation, but most of its activity was restored after the coke was burned off with air. It was reported that because of the size of the zeolite crystals on the support walls, surface reactions may dominate over the shape-selective aspect of the zeolite, which is limited by the diffusion of reactants into, and products out of, the zeolite microcrystals at the wall.

3. *Hydrogen Cyanide Production*

A synthesis procedure for producing hydrogen cyanide from the partial oxidation of ammonia and methane with oxygen was studied using monolithic catalyst supports like those used in the analysis described earlier for synthesis gas production [6]. Layers of noble metal gauzes, as well as foam and extruded ceramic monoliths catalyzed with films of platinum, were analyzed comparatively for hydrogen cyanide synthesis under very high flow rates of 0.1–1.0 m/sec. The catalyst bed was less than 25 mm long in order to minimize the residence time of the reactants in the catalyst bed. High flow rates were used in order to maintain high mass transfer conditions of reactants at the catalyst surface. An important aspect of the comparison of monolithic catalysts included different pore sizes of the pore geometry and different channel sizes of the extruded honeycomb geometry. A list of these monoliths and the comparative selectivity for product are shown in Table 8.

The conversion and product selectivity is optimized when the gas mixing is improved, which reduces the boundary layer thickness at the catalyst surface and increases the mass transfer coefficient for a given channel dimension. The laminar flow through the honeycomb channels in the extrudate monolith results in a relatively thick boundary layer

Table 8 Optimal Selectivity of HCN Synthesis for Gauze, Extruded Monolith, and Foam Monolith Catalysts at ~1400 K

Catalyst	Characteristic channel dimension d (mm)	Selectivity at optimum $\frac{HCN}{NH_{3_o} - NH_3}$	Optimal conversion $\left(\frac{HCN}{NH_{3_o}}\right)_{opt}$
40-mesh Pt-10% Rh gauze	0.6	0.85	0.65
50-ppi[a] foam monolith (Pt coated)	0.5	0.70	0.28
30-ppi foam monolith (Pt coated)	0.8	0.50	0.29
2300-csi[b] extruded monolith (Pt coated)	0.5	0.37	0.15
1000-csi extruded monolith (Pt coated)	0.8	0.30	0.15
400-csi extruded monolith (Pt coated)	1.3	0.04	0.05

Note: From Ref. 6.

[a] ppi, pores per inch.

[b] csi, cells per square inch.

through which reactants must diffuse to reach the surface catalyst. The foam structure enhances the gas stream mixing that results in higher mass transfer rates to the surface and a higher product selectivity. A decrease in channel diameter decreases the boundary layer thickness and increases the product selectivity in both foam and honeycomb monolith structures, as illustrated by the results in Table 8. Conversion efficiency is still limited because of the mass transfer limitation that results from laminar flow. The impact of the axial temperature profile and the length-to-channel diameter ratio for the extrudate monolith apparently are not of sufficient magnitude to warrant additional information, despite the expected variations in turbulence that are present at the entrance of the extrudate monolith relative to the length.

4. *Chlorination/Oxychlorination*

A rather different approach has been described for use of honeycomb monolith catalysts in the chlorination and oxychlorination of hydrocarbons [19]. The reactors described are tube bundles with inside diameters up to 25 mm and lengths up to 2 m. In the reactor, fifteen individual monolithic catalysts are charged, one upon the other, with a 3-mm glass sphere spacer between them. The monoliths (10–100 channels/cm^2) are impregnated with mixtures of copper chloride and potassium chloride that vary in concentration and ratios from inlet to exit for reaction control. An example cited described the inlet reactant mixture as a combination of hydrogen chloride, olefin (such as ethylene), and air introduced at the rates of 0.0278, 0.0161, and 0.0158 standard l/sec, respectively. Two additional quantities of air, 0.0158 and 0.0083 L/sec each, are added at locations one-third and two-third, respectively, along the linear distance through the bed. Multiple reactor tubes are arranged in bundles and coolant is circulated around the tubes to assist in controlling the exothermic reaction. Use of monolith catalysts helped to reduce the quantity of hydrocarbon converted to oxides and to increase the yield of 1,2-dichloroethane.

5. *Hydrogenation/Dehydrogenation*

Hydrogenation and dehydrogenation reactions, with benzene and cyclohexane, respectively, have been investigated with nickel and platinum catalysts on monoliths [20–23]. These two compounds have a peculiar relationship in that the cyclohexane was considered as a potential hydrogen storage compound for transportation use at one time. In this cyclic operation, cyclohexane would be dehydrogenated onboard a vehicle to provide a hydrogen source for clean combustion. The benzene product would be stored and subsequently rehydrogenated to cyclohexane for reuse. Rapidity of reaction rates (i.e., fast rate of reaction at low temperatures, high mass velocity, and thermal control throughout the catalyst bed) without coking was a highly desirable characteristic for the catalysts. Although slower than at higher temperatures, platinum (0.24%) on a washcoated ceramic honeycomb (36 cells/cm^2) was observed to be active for benzene hydrogenation at mass velocities between 0.10 and 0.18 kg/sec at temperatures between 373 and 413 K with hydrogen-to-benzene molar ratios between 5 and 10. An estimation of the heat transfer coefficient in the monolith catalyst of 32.2–37.7 W/m^2 K seemed to be consistent with the mass velocities. The reaction mechanism has been established to be of the Eley–Rideal type.

6. *Sulfur Dioxide Production*

In an another effort to utilize the advantages of monoliths to control selectivity of product, a patent [24] was issued for monolithic supported catalysts for selectively converting hydrogen sulfide to sulfur dioxide. Oxidation of hydrogen sulfide or organic sulfide reactants in packed beds typically have long contact times, due to activity limitations, that produce mixtures of sulfur dioxide and sulfur trioxide. One type of monolithic catalyst reported in this patent is called a mass catalyst, in which the catalytically active materials—i.e., titanium dioxide, iron oxide, and platinum—were extruded in monolithic form. The channel shape could be any of the common geometries with void fractions between 50 and 70%, and the catalytically active element or elements, some as oxides, preferably range from 60–99% of the total weight. A precious metal was included on the order of 0.05–1% by weight. One reported combination was titanium dioxide as the monolithic support with iron oxide and platinum added to comprise 4 and 0.25% by weight of iron and platinum, respectively.

Alternatively, the catalyst could be supported on any of the conventional monoliths of ceramic or metal composition with washcoat and impregnated catalyst. These catalysts were to be operated at temperatures between 523 and 823 K and at space velocities between 0.56 and 2.78 sec^{-1} for residence times ranging from 0.5 to 2.5 sec. Conversion of gas streams containing hydrogen sulfide (or sulfide mixtures of carbon disulfide and carbonyl sulfide with hydrogen sulfide) below 1% (volume), oxygen in air at an oxygen/sulfur molar ratio ranging from 2.3:1 to 4.0:1, water at concentrations between 3 and 30%, and the balance nitrogen were used as the evaluation environment. One monolith configuration used in the evaluation was a square mass catalyst, 20 mm on a side, with 169 square channels 1.4 mm on a side. The relative conversion selectivity was 30/1170 sulfur trioxide/sulfur dioxide from the mass monolith catalyst, compared to 450/750 from normal titania extrudate catalyst, thus illustrating the superior performance of the monolith catalyst approach.

C. Gasification Reactions

1. *Hydrazine*

The rapid production of large volumes of high-temperature gases under pressure is yet another application where monolith supports have been investigated for controlling chemical reactions. One example of this application is in the area of aerospace, for use as gas generators for rocket thrusters or ignitors and for turbine engine restart.

Within these operations are many different parameters that are unique to either monopropellant or bipropellant use as well as the duty of the generator, but each has two similarities: the residence time and the "delay" time, i.e., the time span for the catalyst to reach optimum gas producing conditions after propellant injection. Both times are typically desired to be on the order of milliseconds or less.

The reason for demanding the catalyst to reach activation so rapidly is one of propellent conservation. Such operations can require space velocities approaching three orders of magnitude higher than typical industrial or automotive applications. Reaction conditions may reach temperatures of 1300 K and pressures as high as 2800 kPa, depending on ignitor/thruster design, mono- or bipropellant, propellant type, and engine thrust. The difficulties encountered are much the same as those observed in the control of methane oxidation to form synthesis gas or hydrogen cyanide synthesis from ammonia. Specifically, high conversion and reaction rates are experienced, as are short residence times and high temperature excursions. There are, however, two major difficulties unique to the aerospace applications: (1) no catalyst bed or reactant preheat may be available to establish a catalyst preactivation temperature; (2) typical operations may be of the pulsed-mode, variable-duration type rather than steady state.

One investigation of such an application [25] examined a proof-of-concept comparison in catalytic activity between a conventional packed bed of Shell 405 (25 × 30 mesh alumina particles), a metal honeycomb monolith, and a metal sponge monolith. Iridium was the catalyst in each case, and a hydrogen-rich mixture of hydrogen and oxygen was the propellant. Short residence time and high flow rate were desired to prevent flashback or combustion after ignition occurred. The packed-particle bed of Shell 405 had a low void fraction that resulted in a back pressure of up to 1000 kPa (compared to a void fraction of up to 90% in the honeycomb and sponge monoliths that resulted in negligible back pressure). The catalyst beds used in the tests were 5 cm long and 1.25 cm in diameter; mixtures of hydrogen and oxygen were flowed through them at rates between 1.33 to 2.21 g/sec. The honeycomb bed was designed such that the alumina-washcoated channels were not parallel to the axial flow but rather at a 45° angle, and the 5-cm-long bed was segmented into eight 0.64-cm lengths in order to create new regions of turbulence throughout the flow path. The sponge catalyst bed was a tungsten support with an alumina washcoat and approximately 8 to 12 pores per centimeter.

While activity was demonstrated at ambient pressures for all three beds, only the packed-particle and sponge beds also demonstrated activity against pressures simulating space operations, e.g., approximately 14 Pa. The most probable cause for the differences observed in activity was the reduction in mass transfer limitation in the sponge and packed beds due to the tortuos path of the gas mixture through the beds. Alternative designs of the honeycomb bed that may have provided acceptable activity were not examined in this preliminary investigation.

Of comparative interest was another reported investigation into a packed-bed replacement for rapid gasification, but no honeycomb monolith was included. A catalyst on a metal foam structure was designed and operated for hydrazine thrusters [26] in which hydrazine, a monopropellant, is dissociated into nitrogen and hydrogen. As in the former case, a replacement for the iridium/alumina Shell 405 packed bed was addressed. By use of a tungsten metal foam support for the alumina/iridium catalyst, the activity was demonstrated to be nearly equal to that of the Shell 405, but with approximately one-third the iridium required. Chamber pressures of 1360 kPa were developed in catalyst beds 1.3 cm in diameter and 2 cm long. During certain test conditions, oscillations were observed and determined to be due to a rate of gas expansion that exceeded the void fraction downstream of the maximum temperature. To accommodate the expansion rate of gases through the bed during ignition, a staging of the pore size of the foam support was proposed. In another approach to this application [27], a metal gauze was employed, and the reactant inlet was designed to flow into the center of the bed and radially out through the gauze bed, rather than through a cylindrical catalyst bed, as in the former example.

2. *Steam/Naphtha Cracking*

In quite a different application, a novel approach for producing olefins via a hydrocarbon-steam cracking process, without the use of a catalyst, was demonstrated to benefit from the use of a honeycomb monolithic catalytic reactor [28]. A typical problem associated with cracking processes of this type is maintaining the appropriate combination of heat transfer and residence time, which, if not balanced, will lead to either poor conversion or coking of the reactor. The cracking reactor described in this report was a ceramic heat exchanger/reactor design built of silicon carbide, a material that provides a number of desirable features: (1) high thermal conductivity, (2) excellent thermal shock resistivity, and (3) high melting point. The heat exchanger/reactor module was fabricated by extrusion. Its design, shown in Figure 15, provided for a single-pass flow for the process gas stream through 61.5-cm-long square channels that had a surface/volume ratio of 600. Flue gas from an external combustor flowed into the top of the heat exchange cavities that surround the reactor channels and flowed parallel to the reacting steam/naphtha mixture. Process stream residence times between 5 and 200 msec assisted in the reaction control. The hydrocarbon feed, up to 1.4 g/sec, was diluted (about 1/1 naphtha/steam on a weight basis) and preheated with steam prior to entry into the reactor, where temperatures can exceed 1120 K. Extensive modeling of the reactor was also completed.

Table 9 presents typical results from an operation extending beyond 7200 sec in duration and compares them to the yield from the predictive model simulation. The conditions under which these results were obtained were flow rates of 0.57 g/sec (both for naphtha and for steam) and 17.8 g/sec (flue gas) at inlet temperatures of 723 K on the process side and 1298 K on the flue gas side. This design shows considerable promise for controlling product composition and conversion with short residence time.

III. SUMMARY

Over the past few years, the potential for use of monolithic supports of various geometries and materials for a wide range of gaseous reaction processes has become more widely explored. The advantages of using monolithic supports over the conventional packed-bed reactor designs, in terms of enhanced heat and mass transfer, offer significant opportunities to control chemical reactions much more tightly over a wide range of conditions. The

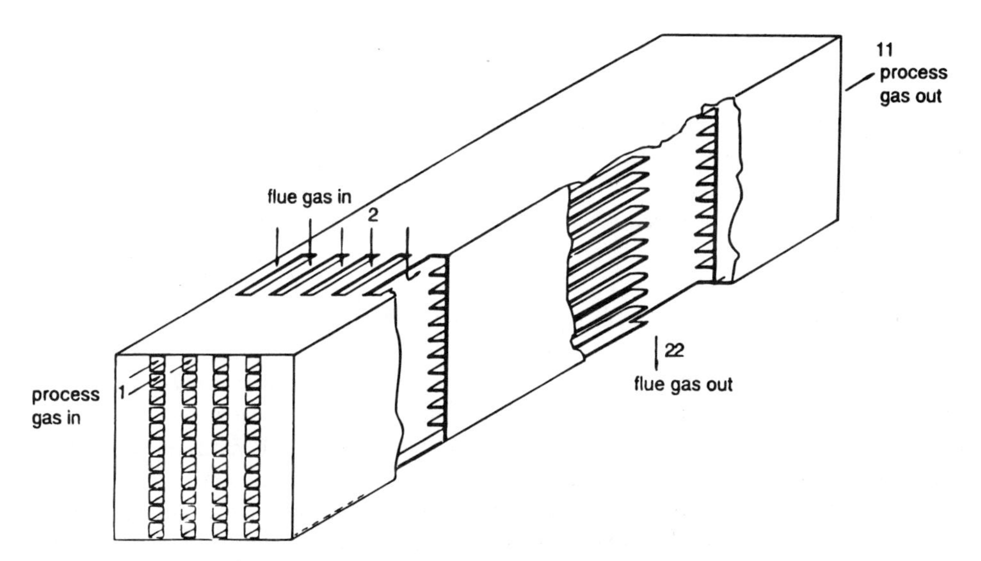

Figure 15 Silicon carbide heat exchange/reactor design for steam/naphtha cracking process. (From Ref. 28, reprinted with permission of the American Institute of Chemical Engineers.)

Table 9 Comparison Between Simulated and Experimental Yields of Naphtha/Steam Cracking in Honeycomb Reactor*

Yields (wt. %)	Simulated	Experimental
H_2	1.00	0.86
CH_4	16.81	16.15
C_2H_2	1.33	1.14
C_2H_4	36.15	36.01
C_2H_6	2.66	2.63
C_3H_4	2.15	1.87
C_3H_6	13.28	13.54
C_3H_8	0.19	0.33
1,3-C_4H_6	5.27	5.39
1-C_4H_8	0.60	
2-C_4H_8	0.37	
i-C_4H_8	1.47	
i-C_4H_{10}	0.03	
n-C_4H_{10}	0.10	
sum C_4H_8 + C_4H_{10}	2.57	2.41
Benzene	7.14	7.38
Fuel gas	17.81	17.01
C_2-fraction	40.14	39.78
C_3-fraction	15.62	15.74
C_4-fraction	7.84	7.80
Gasoline	16.68	18.16
Fuel oil	1.91	2.18
Conversion (mol %)	99.05	

Note: From Ref. 28.

* Total residence time (above 600°C) = 201 msec; residence time in ceramic honeycomb structure = 129 msec; residence time in two outlet zones = 72 msec.

range of geometries and materials available for fabricating monoliths allows tailoring of the reactor design to achieve better yield and selectivity of products than has been possible with conventional approaches.

Trade-offs in performance advantages between the honeycomb multichannel supports and others that offer variable flow patterns are necessary for each application. More compact and efficient reactors will provide many novel opportunities not only to improve chemical processes in wide industrial practice, but to develop novel processes for more efficient operation in the near future.

Ideas on how to make various monolithic supports more efficient while at the same time making the entire industrial processing operation more environmentally and efficiently oriented illustrate how important it is to focus efforts on bringing the benefits of monolithic catalysts into the chemical processing industry. Novel approaches to the cleanup of volatile organic compounds, such as the integration of photocatalysis with a monolithic reactor to oxidize acetone [29] and for the destruction of toluene and xylene [30], may also have broader potential in industrial applications in the future. A better understanding of how the unique properties associated with the various combinations of geometry, materials, and designs that affect heat and mass transfer throughout a reactor will lead to significant improvements in future industrial practice.

REFERENCES

1. A. Cybulski and J.A. Moulijn, Monoliths in heterogeneous catalysis, *Catal. Rev.-Sci. Eng. 36(2)*:179 (1994).
2. G.E. Voecks and D.J. Cerini, Application of rich catalytic combustion to aircraft engines, Proceedings of Third Workshop on Catalytic Combustion, Asheville, NC, pp. 477–490 (1979).
3. K.H. Chen, J. Houseman, and C.N. Jennings, *Experimental Study on Catalytic Partial Oxidation of No. 2 Fuel Oil with Air*, Jet Propulsion Laboratory Report 778-3 (1977).
4. J.P. Kesselring, W.V. Krill, H.L. Atkins, R.M. Kendall, and J.T. Kelly, *Design Criteria for Stationary Source Catalytic Combustion Systems*, Report EPA-600/7-79-181, pp. 7–71 (1979).
5. D.A. Hickman and L.D. Schmidt, Synthesis gas formation by direct oxidation of methane over monoliths, Symposium on Catalytic Selective Oxidation, Washington, DC, pp. 1263–1267 (1992).
6. D.A. Hickman and L.D. Schmidt, The role of boundary layer mass transfer in partial oxidation selectivity, *J. Cat. 136*:300 (1992).
7. D.A. Hickman and L.D. Schmidt, Synthesis gas formation by direct oxidation of methane over Pt monoliths, *J. Cat. 138*:267 (1992).
8. M. Flytzani-Stephanopoulos and G.E. Voecks, *Autothermal Reforming of n-Tetradecane and Benzene Solutions of Naphthalene on Pellet Catalysts, and Steam Reforming of n-Hexane on Pellet and Monolithic Catalyst Beds*, Final Report, DE-AI03-78ET-111326, pp. 74–119 (1980).
9. M. Flytzani-Stephanopoulos and G.E. Voecks, *Conversion of Hydrocarbons for Fuel Cell Applications, Part I: Autothermal Reforming of Sulfur-Free and Sulfur-Containing Hydrocarbon Liquids; Part II: Steam Reforming of n-Hexane on Pellet and Monolithic Catalyst Beds*, Final Report DOE/ET-11326-1, Jet Propulsion Laboratory Publication 82-37, pp. 75–120 (1981).
10. G.E. Voecks, *Comparisons of a Metal Supported Catalyst Bed to a Conventional Packed Bed for Steam Reforming Methane, Topical Reports on Experimental Efforts in Support of DOE Phosphoric Acid Fuel Cell Program*, Final Report, Jet Propulsion Laboratory Publication D-5614, pp. 12–38 (1988).
11. R.M. Yarrington, I.R. Feins, and H.S. Hwang, Evaluation of steam reforming catalysts for use in autothermal reforming of hydrocarbon feed stocks, Proceedings of National Fuel Cell Seminar, San Diego, CA (1980).
12. R.M. Yarrington and W. Buchanan, Preparation of normally liquid hydrocarbons and a synthesis gas to make the same, from a normally gaseous hydrocarbon feed, U.S. Pat. 5,023,276, Washington, DC (1991).
13. W.T. McShea, III and R.M. Yarrington, Method of methanol production, U.S. Pat. 4,927,857, Washington, DC (1990).
14. G.A. Jarvi, K.B. Mayo, and C.H. Bartholomew, Monolithic-supported nickel catalysts: I. Methanation activity relative to pellet catalysts, *Chem. Eng. Commun. 4*:325, (1980).
15. E.R. Tucci and W.J. Thompson, Monolith catalyst favored for methanation, *Hydrocarbon Processing* (Feb.):123 (1979).
16. J.E. Antia and R. Govind, Conversion of methanol to gasoline-range hydrocarbons in a ZSM-5 coated monolithic reactor, *Ind. Eng. Chem. Res. 34*:140 (1995).
17. J.E. Antia and R. Govind, Applications of binderless zeolite-coated monolithic reactors, *Appl. Cat. A: General 131*:107 (1995).

18. I.M. Lachman and M.D. Patil, Method of crystallizing a zeolite on the surface of a monolithic ceramic substrate, U.S. Pat. 4,800,187, Washington, DC (1989).
19. R. Strasser, L. Schmidhammer, K. Deller, and H. Krause, Chlorinating reactions and oxychlorination reactions in the presence of honeycomb monolithic catalyst supports, U.S. Pat. 5,099,085, Washington, DC (1992).
20. A. Parmaliana, C. Crisafulli, R. Maggiore, J.C.J. Bart, and N. Giordano, Catalytic activity of honeycomb catalysts. I. The benzene-cyclohexane (de)hydrogenation reaction, *React. Kinet. Catal. Lett. 18(3–4)*:295 (1981).
21. A. Parmaliana, A. Mezzapica, C. Crisafulli, S. Galvagno, R. Maggiore, and N. Giordano, Benzene hydrogenation on nickel/honeycomb catalysts, *React. Kinet. Catal. Lett. 19(1–2)*:155 (1982).
22. A. Parmaliana, M.E. Sawi, G. Mento, U. Fedele, and N. Giordano, A kinetic study of the hydrogenation of benzene over monolithic-supported platinum catalyst, *Appl. Cat. 7*:221 (1983).
23. A. Parmaliana, M.E. Sawi, U. Fedele, G. Giordano, F. Frusteri, G. Mento, and N. Giordano, A kinetic study of low temperature hydrogenation of benzene over monolithic-supported platinum catalyst, *Appl. Cat. 12*:49 (1984).
24. T. Chopin, J-L. Hebrard, and E. Quemere, Monolithic catalysts for converting sulfur compounds into sulfur dioxide, U.S. Pat. 5,278,123, Washington, DC (1994).
25. G.E. Voecks and R.A. Bjorklund, *Catalytic Ignitors for Hydrogen–Oxygen Thrusters*, NASA Quarterly Progress Report, Jet Propulsion Laboratory internal report (1985).
26. E.W. Schmidt, *Monolithic Catalyst Beds for Hydrazine Reactors*, Final Report 73-R-360 (1973).
27. W.K. Burke, Monolithic high activity catalyst bed for a catalytic gas generator, U.S. Pat 4,938,932, Washington, DC (1990).
28. G.J. Heynderickx, G.F. Froment, P.S. Broutin, C.R. Busson, and E.J. Weill, Modeling and simulation of a honeycomb reactor for high-severity thermal cracking, *AIChE J. 37(9)*:1354 (1991).
29. M.L. Sauer and D.F. Ollis, Acetone oxidation in a photocatalytic monolith reactor, *J. Cat. 149*:81 (1994).
30. B. Sanchez, M. Romero, A. Vidal, B. Gabrellas, J. Blanco, and P. Avila, Destruction of toluene and xylene using concentrated solar photons, Proceedings of the ASME/JSME/JSES International Solar Energy Conference, San Francisco, CA, pp. 123–129, (1994).

8

Modeling of Monolithic Reactors for Automotive Exhaust Gas Treatment

J. H. B. J. Hoebink and G. B. Marin
Eindhoven University of Technology, Eindhoven, The Netherlands

I. INTRODUCTION

The use of monoliths as catalytic reactors focuses mainly on applications where low pressure drop is an important item. When compared to fixed beds, which seem a natural first choice for catalytic reactors, monoliths consist of straight channels in parallel with a rather small diameter, because of the requirement of a comparably large surface area. The resulting laminar flow, which is encountered under normal practical circumstances, does not show the kinetic energy losses that occur in fixed beds due to inertia forces at comparable fluid velocities. Despite the laminar flow, monolith reactors still may be approached as plug-flow reactors because of the considerable radial diffusion in the narrow channels [1].

As such a monolith is different from a fluidized bed, which also shows a limited pressure loss, but in combination with serious axial dispersion in the emulsion phase at least. A major distinction between monolith and fluidized-bed reactors, however, concerns heat transfer rates to the surroundings. The latter are applied when heat effects due to reaction are to be compensated, while monolithic reactors can be considered as adiabatic for most practical purposes. In automotive exhaust gas treatment, adiabatic behavior is even stimulated by insulation of the reactor wall to promote a fast light-off of the reactor after a cold engine start.

In this chapter, modeling of monolith reactors will be considered from a first-principles point of view, preceded by a discussion of the typical phenomena in monoliths that should be taken into account. General model equations will be presented and subsequently simplified, depending on the subprocesses that should be described by a model. A main lead will be the time scales at which these subprocesses occur. If they are all small, the process operates in the steady state, and all time-dependent behavior can be discarded. Unsteady-state behavior is to be considered if the model should include the time scale of reactor startup or if deactivation of the catalyst versus time-on-stream has to be addressed. A description of fully dynamic reactor operation, as met when cycling of the feed is applied, requires that all elementary steps of a kinetic model with their corresponding time scales are incorporated in the reactor model.

Although the present considerations are valid for monolith reactors in general, independent of the actual chemical reactions, details will refer mostly to the application of monoliths in automobile exhaust gas treatment, which has received most attention in the past and still is dominant in the practical use of monoliths. Several reviews treat the extensive literature on monoliths, among which is a very recent one [2].

II. OVERVIEW OF RELEVANT TRANSPORT PHENOMENA

The route from reactant to product molecule in a monolith reactor comprises reactant transport from the bulk gas flow in a channel toward the channel wall, simultaneous diffusion and reaction inside the porous washcoat on the channel wall, and product transport from the wall back to the bulk flow of the gas phase.

Laminar flow is the usual flow regime met in monolith reactors, given that the typical Reynolds number has values below 500. The radial velocity profile in a single channel develops from the entrance of the monolith onward and up to the position where a complete Poiseuille profile has been established. The length of the entrance zone may be evaluated from the following relation [3]:

$$\frac{L_e}{d_c} \leq 0.06\mathrm{Re} \tag{1}$$

It is usually neglected because typically it is less than 10% of the reactor length. Experimental work [4] has shown that the development of the velocity profile barely influences the reactor performance.

The distribution of gas over all parallel channels in the monolith is not necessarily uniform [2,5,6]. It may be caused by a nonuniform inlet velocity over the cross-sectional area of the monolith, due to bows in the inlet tube or due to gradual or sudden changes of the tube diameter. Such effects become important, because the pressure loss over the monolith itself is small. Also, a nonuniformity of channel diameters could be a cause at the operative low Reynolds numbers, as was reported for packed beds [7]. A number of devices were proposed to ensure a uniform inlet velocity [5,8], which indeed increases the total pressure drop.

A convenient simplification is the approximation of laminar flow by plug flow with axial dispersion, which is allowed [1] if

$$\frac{D_{\mathrm{mol}}L}{\bar{v}d^2} >> \frac{1}{28} \tag{2}$$

Equation 2 expresses whether radial diffusion, which in the case of laminar flow is due to molecular diffusion, is fast enough to outlevel radial concentration profiles. This approximation usually holds for monolithic reactors because of the rather small channel diameter. The corresponding axial dispersion coefficient can be calculated [1] from the following:

$$\frac{D_{\mathrm{ax}}}{\bar{v}d} = \frac{1}{192}\frac{\bar{v}d}{D_{\mathrm{mol}}} \tag{3}$$

and can be neglected in many practical situations.

Pressure drop over the monolith may be calculated using the friction factor relation [9]:

$$4f = \frac{A}{\mathrm{Re}}\left(1 + 0.0445\mathrm{Re}\,\frac{d_h}{L}\right)^{0.5} \tag{4}$$

Values of the constants A were reported for different channel geometries. The value 64 for circular channels corresponds to fully developed laminar flow in circular tubes.

Transport of heat or mass to the wall of a single, circular channel under laminar flow conditions is known as the classical Graetz problem [10]. For heat transport only, the energy equation contains axial convection and radial conduction:

$$\rho c_p 2\bar{v}\left(1 - \frac{4r^2}{d_c^2}\right)\frac{\partial T}{\partial x} = \lambda\left(\frac{\partial^2 T}{\partial r^2} + \frac{1}{r}\frac{\partial T}{\partial r}\right) \tag{5}$$

The inlet boundary condition is

$$T = T^{\mathrm{in}}, \qquad \text{at } x = 0 \tag{6}$$

The boundary condition at the channel center is

$$\frac{\partial T}{\partial r} = 0, \qquad \text{at } r = 0 \tag{7}$$

and at the wall, in the case of a constant heat flux through the wall,

$$\lambda\frac{\partial T}{\partial r} = q, \qquad \text{at } r = \frac{d_c}{2} \tag{8}$$

or, in the case of a constant wall temperature,

$$T = T_w, \qquad \text{at } r = \frac{d_c}{2} \tag{9}$$

The solution of Eqs. (5) to (8) or (9) presents the temperature distribution in the channel, from which, as a function of the axial distance, either the wall temperature or the heat flux through the wall may be calculated, depending on whether Eq. (8) or (9) was applied. The results allow calculation of a local heat transfer coefficient, α, or local Nusselt number Nu:

$$\mathrm{Nu} = \frac{\alpha d_c}{\lambda} = \frac{-d_c \left.\dfrac{\partial T}{\partial r}\right|_{d_c/2}}{T_m - T_w} \tag{10}$$

with T_m the cup-mixing temperature of the gas phase. The following correlation was obtained [11]:

$$\mathrm{Nu}(x) = 1.36\left(\frac{d_c}{2x}\,\mathrm{Re}Pr\right)^{1/3}, \qquad \text{if } \frac{2x}{d_c} \le 0.001\mathrm{Re}Pr \tag{11}$$

If the channel length is sufficiently long, temperature profiles become similar, leading to a constant value of the Nusselt number. Equation 11 might be replaced by the limit value of the Nusselt number:

$$\mathrm{Nu} = 3.65, \qquad \text{if } \frac{2x}{d_c} \geq 0.1\mathrm{Re}Pr \tag{12}$$

which holds for the case of constant wall temperature [12]. The heat transfer correlations can be summarized as average Nusselt numbers along the length of the monolith [9]:

$$\mathrm{Nu} = c\left(1 + 0.095\mathrm{Re}Pr\,\frac{d_c}{L}\right)^{0.45} \tag{13}$$

The value of the constant c depends on the channel shape and was calculated for various channels geometries [13,14, see also 2].

Considerations along the above lines lead to analogous correlations for the Sherwood number for the description of mass transfer in a single channel. The application of the rather simple Nusselt and Sherwood number concept for monolith reactor modeling implies that the laminar flow through the channel can be approached as plug flow, but it is always limited to cases in which homogeneous gas-phase reactions are absent and catalytic reactions in the washcoat prevail. If not, a model description via distributed flow is necessary.

Some other phenomena, however, also require the distributed-flow description. Young and Finlayson [13], who modeled a monolith reactor via distributed flow, using the kinetics of Voltz et al. [15] for the oxidation of carbon monoxide, have shown that at some axial position, reaction light-off occurs, accompanied by a sudden increase of both the Nusselt and Sherwood numbers. Upstream, Nu and Sh decrease gradually according to a constant-wall-flux approach, Eq. (8); downstream, a constant wall temperature, Eq. (9), is appropriate. The ignition is caused by a rate increase at low concentrations of CO, which may even lead to multiple steady states [16]. Therefore the former situation corresponds with high concentrations in the washcoat, e.g., controlled by kinetics, while in the latter case the rate is limited by mass transfer, e.g., low concentrations in the washcoat. Similar results were reported [17] for the reaction between NO_x and NH_3. It is obvious that ignition behavior leads to strong local gradients and changes of the local interface concentration and temperature, which cannot be described by the application of a relatively simple heat or mass transfer correlation.

Experimental work under reaction conditions with monoliths was performed by Ullah et al. [18], who found:

$$\mathrm{Sh} = 0.766\left(\mathrm{Re}\ Sc\,\frac{d_c}{L}\right)^{0.483} \tag{14}$$

Almost similar results were obtained experimentally by Votruba et al. [19], who studied evaporation of water and hydrocarbons from porous monoliths. These results predict Nu and Sh values clearly lower than does Eq. (13), and moreover suggest that Nu or Sh values would fall under their theoretically predicted lower limit at a low Reynolds number [16,20]. It is not unlikely that the discrepancy is due to a maldistribution of flow over the different monolith channels, as a result of the low pressure drop, similar to the effect signalized for fixed beds at low Reynolds numbers [7]. Experimental work [4], which was carried out with an inert fixed bed in front of the monolith reactor to assure an even distribution, gave data that come quite near to the results of Hawthorne, Eq. (13) [2].

Concerning the washcoat, the occurrence of concentration or temperature gradients should be considered separately for any specific case, along the lines that have been developed for estimation of their significance in catalytic fixed-bed reactors [21]. For concentration gradients, the well-known concept of the effectiveness factor, based on the Thiele or Weiss modulus, is very useful for steady-state operation of the reactor. Since the washcoat is very thin, typically 25 μm, it may be approached as a flat plate. It should be kept in mind, however, that the distribution of washcoat over the monolith matrix is not necessarily homogeneous, which might require the involvement of more than one washcoat dimension [16]. A criterion for estimating the significance of heat transport limitation was published by Mears [22]. But as with catalyst pellets in fixed beds, temperature gradients inside the washcoat will be negligible in most situations since the thickness of the washcoat layer is rather small and its heat conductivity is relatively high.

The influence of heat losses through the reactor wall have been studied [5,23]. Radial temperature gradients inside the monolith material can often be neglected, because the operation is usually adiabatic. This means that modeling of one single channel is adequate. Any nonuniform flow distribution may be incorporated into a reactor model by integration of the single channel performance over the whole cross section of the reactor.

III. GENERAL MODEL EQUATIONS

The application of monoliths as catalytic reactors has several aspects that ask for a different approach when setting up a reactor model. A useful concept in this respect is to distinguish between the time scales of the various subprocesses involved in the reactor's operation.

As in the normal practice of chemical reactors, we might be interested in the reactor performance under steady-state operation, which requires a rather simple model, consisting of only an energy equation and continuity equations for the components involved.

Model equations have to be extended with accumulation terms in case reactor startup behavior has to be described as well. Several situations can occur in practice, each requiring an appropriate approach. In applications dealing with automobile exhaust gas treatment, monolith transients after a cold start of the engine are studied in order to enhance reactor light-off. Heating of the monolith by the hot engine gases has a relatively large time scale, on the order of 100s, meaning that transients have to be included in the energy equation for the solid phase. A quasi-steady-state approach is allowed for the gas-phase energy equation and for the continuity equations, since here accumulation effects decay on the time scale of the gas residence time in the monolith, which usually is a few seconds at most [24,25]. In catalyst deactivation studies, the accumulation of deposited poisons should be taken into account [26], which is rather slow and allows to assume a steady state for all other processes.

Reactor control models for monoliths require a more detailed study of the time scales of all occurring subprocesses, because of their dynamic character. Under dynamic circumstances, the rates of the individual elementary steps of a catalytic cycle, such as adsorption, surface reaction, and desorption, are not equal to each other anymore, since the time scales of the corresponding processes may differ by many orders of magnitude. Therefore, accumulation effects on the catalyst surface have to be taken into account as well, which demands that continuity equations for surface species be included in the model. Such aspects may even play a role in the steady state if the kinetics depend on rate-determining steps that change according to the concentration level of the reactants and/or products. Continuity equations for surface species are also needed if the model is to describe transient or periodic operation of the reactor to study the potential of improved reactant conversion or product selectivity. This phenomenon is often referred to as *cycling of the feed* [27–32] and is known especially for car exhaust gas converters, where it results spontaneously from the lambda-sensor-based control of the air/fuel ratio [33,34].

Model equations are presented here for a general situation that is not specific to automobile exhaust gas converters and includes transient reactor behavior. The following assumptions are made:

1. Only a single channel is considered. The channel has a hydraulic diameter d_h, and its wall consists of the porous washcoat with a layer thickness δ_w. The monolith material, surrounding the washcoat, has a thickness δ_m, which corresponds to half of the spacing between channels. As a result, cylindrical coordinates can be used.
2. Reactions may occur heterogeneously on the catalytic material in the washcoat and homogeneously in the fluid phase.
3. The reactor is adiabatically operated.
4. Axial heat conductivity and axial diffusion in the fluid phase are neglected because of the usually large convective transport.
5. Conduction in the washcoat is described with an effective heat conductivity, and diffusion is Fickian with an effective diffusivity.

For one circular channel, the energy equation for the fluid phase is

$$\rho C_p \frac{\partial T}{\partial t} = -\rho C_p v \frac{\partial T}{\partial x} + \frac{\lambda}{r}\frac{\partial}{\partial r}\left(r\frac{\partial T}{\partial r}\right) + \sum_i (-\Delta_f H_i) R_i^h \tag{15}$$

The continuity equation for component i in the fluid phase is

$$\frac{\partial C_i}{\partial t} = -v\frac{\partial C_i}{\partial x} + \frac{D_i}{r}\frac{\partial}{\partial r}\left(r\frac{\partial C_i}{\partial r}\right) + R_i^h \tag{16}$$

Because the flow is one-dimensional, only the axial velocity v is involved in the momentum equation, which contains the axial pressure gradient and the viscous friction losses due to the radial velocity profile:

$$\rho\frac{\partial v}{\partial t} = -\rho v\frac{\partial v}{\partial x} + \frac{\mu}{r}\frac{\partial}{\partial r}\left(r\frac{\partial v}{\partial r}\right) - \frac{\partial p}{\partial x} \tag{17}$$

For the washcoat, the energy equation contains axial and radial heat conduction and heat production due to the catalytic reactions:

$$\rho^w C_p^w \frac{\partial T^w}{\partial t} = \lambda^w \frac{\partial^2 T^w}{\partial x^2} + \frac{\lambda^w}{r}\frac{\partial}{\partial r}\left(r\frac{\partial T^w}{\partial r}\right) + (1-\epsilon^w)\sum_i (-\Delta_f H_i) R_i^w + \epsilon^w \sum_i (-\Delta_f H_i) R_i^h \tag{18}$$

The continuity equation in the washcoat concerns axial and radial diffusion, and net production due to homogeneous reactions in the pores and catalytic reactions, that ultimately result from the difference between adsorption and desorption rates:

$$\epsilon^{w}\frac{\partial C_i^w}{\partial t} = D_{ei}^w\frac{\partial^2 C_i^w}{\partial x^2} + \frac{D_{ei}^w}{r}\frac{\partial}{\partial r}\left(r\frac{\partial C_i^w}{\partial r}\right) + \epsilon^w R_i^h + a_{\text{cat}}R_i^w \tag{19}$$

The monolith material requires an energy equation only:

$$\rho^m C_p^m \frac{\partial T^m}{\partial t} = \lambda^m \frac{\partial^2 T^m}{\partial x^2} + \frac{\lambda^m}{r}\frac{\partial}{\partial r}\left(r\frac{\partial T^m}{\partial r}\right) \tag{20}$$

For surface species, the continuity equations involve adsorption, desorption, and net production of component i from surface reactions:

$$L_t \frac{\partial \theta_i}{\partial t} = R_i^w \tag{21}$$

Equation 21 does not account for the diffusion of species on the catalyst surface. The following initial and boundary conditions apply, if it is assumed that a cold and empty reactor is fed on time $t = 0$ with a hot stream of reactants:

$$t = 0 \wedge 0 \le x \le \mathrm{L}$$

$$\begin{aligned} &0 \le r \le R && T = T_0,\quad C_i = 0,\quad v = 0 \\ &R \le r \le R + \delta_w && T^w = T_0,\quad C_i = 0,\quad \theta_i = 0 \\ &R + \delta_w \le r \le R + \delta_m && T^m = T_0 \end{aligned} \tag{22}$$

$$\begin{aligned} &t > 0 \\ &(z = 0) \wedge (0 \le r \le R) && T = T^{\text{in}},\quad C_i = C_i^{\text{in}},\quad v = v^{\text{in}}, \\ &(z = 0) \vee (z = L) \wedge (R \le r \le R + \delta_w) && \frac{\partial T^w}{\partial x} = 0,\quad \frac{\partial C_i^w}{\partial x} = 0 \\ &(z = 0) \vee (z = L) \wedge (R + \delta_w \le r \le R + \delta_m) && \frac{\partial T^m}{\partial x} = 0 \end{aligned} \tag{23}$$

$$\begin{aligned} &t > 0 \\ &(r = 0) \wedge (0 \le z \le L) && \frac{\partial T}{\partial r} = 0,\quad \frac{\partial C_i}{\partial r} = 0,\quad \frac{\partial v_r}{\partial r} = 0 \\ &(r = R) \wedge (0 \le z \le L) && \lambda^w \frac{\partial T^w}{\partial r} = \lambda\frac{\partial T}{\partial r},\quad D_{ei}^w\frac{\partial C_i^w}{\partial r} = D_i\frac{\partial C_i}{\partial r} \\ & && v = 0,\quad T = T^w,\quad C_i = C_i^w \\ &(r = R + \delta_w) \wedge (0 \le z \le L) && \lambda^m\frac{\partial T^m}{\partial x} = \lambda^w\frac{\partial T^w}{\partial r},\quad \frac{\partial C_i}{\partial r} = 0,\quad T^m = T^w \\ &(r = R + \delta_m) \wedge (0 \le z \le L) && \frac{\partial T^m}{\partial r} = 0 \end{aligned} \tag{24}$$

The model description just proposed has never been validated or used in simulation studies. It does incorporate, however, all phenomena that have been reported as potentially important for monolith reactor modeling. It is obvious that many simplifications can be made for specific applications, and this indeed has been done in the literature.

A major simplification is the adoption of fully developed laminar flow in the channels, since the length of the entrance region can usually be neglected [3]. It means that continuity Eq. (17) can be discarded, and that the velocity v in Eqs. (15) and (16) can be replaced by the radial parabolic velocity profile according to Poisseuille. The approximation of laminar flow via plug flow with axial dispersion allows the use of Sherwood and Nusselt numbers to describe the mass and heat transfer from the bulk flow toward the washcoat if ignition effects are absent. The consequence is that in Eqs. (15) and (16) the terms describing radial temperature and concentration profiles are not needed anymore, and that the velocity v can be replaced by the mean velocity over the cross-sectional area. Any nonuniform distribution of flow over the different channels can be accounted for by applying Eqs. (15)–(24) to each of the channels, and combining all parallel outlet flows into one single reactor effluent. In practice nonuniformities in the flow distribution are often neglected. Another approximation that is widely accepted concerns equal properties of the washcoat and the monolith material in the case of a ceramic honeycomb.

The numerical solution of the reactor model, consisting of a set of partial differential equations, is most commonly achieved by application of orthogonal collocation in the space coordinates [13,35,38]. The resulting coupled ordinary differential equations may be integrated in time by using routines from the NAG Fortran library.

IV. RESULTS OF MONOLITH REACTOR MODELING

As mentioned in the introduction, the following discussion on modeling results takes as a lead that distinction should be made between steady-state models, unsteady-state models, and dynamic models. The results mentioned focus mainly on automotive exhaust gas treatment, which application has been widely studied, with major emphasis on the oxidation of carbon monoxide.

A. Steady-State Operation

A lumped-parameter approach seems adequate for practical purposes if ignition effects are absent. Under steady-state conditions, following the approach of Lie et al. [35], the continuity equation for component i in the gas phase is

$$\Phi_m^{\mathrm{sup}} \frac{d}{dx}\frac{C_i}{\rho} = -\rho k_{f,i} a_v \left(\frac{C_i}{\rho} - \frac{C_i^w}{\rho} \right) \tag{25}$$

When neglecting diffusion limitation inside the pores of the washcoat and assuming catalytic reactions only, the continuity equation is

$$\rho k_{f,i} a_v \left(\frac{C_i}{\rho} - \frac{C_i^w}{\rho} \right) = -a_{\mathrm{cat}} R_i^w \tag{26}$$

The corresponding boundary conditions are

$$\frac{C_i}{\rho} = \frac{C_i^{\text{in}}}{\rho}, \qquad \text{at } x = 0 \tag{27}$$

Assuming isothermal operation, some results are shown for the case of exhaust gas treatment. Typical data for the parameters are shown in Table 1.

A kinetic model [15], with adapted kinetic parameters [25], was used, which accounts for oxidation by oxygen of carbon monoxide, propene, methane, and hydrogen, and also includes inhibition effects caused by nitrogen oxide. The following net production rates were applied in Eq. (26):

$$R_{CO} = -\frac{k_1}{ADS}\frac{C^w_{CO}}{\rho}\frac{C^w_{O_2}}{\rho} \qquad \frac{\text{mol}}{\text{m}^2_{\text{Pt}}\text{sec}} \tag{28}$$

$$R_{C_3H_6} = -\frac{k_2}{ADS}\frac{C^w_{C_3H_6}}{\rho}\frac{C^w_{O_2}}{\rho} \qquad \frac{\text{mol}}{\text{m}^2_{\text{Pt}}\text{sec}} \tag{29}$$

$$R_{CH_4} = -\frac{k_3}{ADS}\frac{C^w_{CH_4}}{\rho}\frac{C^w_{O_2}}{\rho} \qquad \frac{\text{mol}}{\text{m}^2_{\text{Pt}}\text{sec}} \tag{30}$$

The rate of hydrogen oxidation was set equal to the rate of CO oxidation:

$$R_{H_2} = -\frac{k_1}{ADS}\frac{C^w_{H_2}}{\rho}\frac{C^w_{O_2}}{\rho} \qquad \frac{\text{mol}}{\text{m}^2_{\text{Pt}}\text{sec}} \tag{31}$$

The adsorption term ADS in the denominator is

$$ADS = \left(1 + K_1\frac{C^w_{CO}}{\rho} + K_2\frac{C^w_{C_3H_6}}{\rho}\right)^2\left(1 + K_3\left(\frac{C^w_{CO}}{\rho}\right)^2\left(\frac{C^w_{C_3H_6}}{\rho}\right)^2\right)\left(1 + K_4\frac{C^w_{NO}}{\rho}\right)^{0.7} \tag{32}$$

Rate and equilibrium coefficients used were as follows, with data as shown in Table 2:

$$k_i = A\exp -\frac{E_a}{T_s} \tag{33}$$

$$K_i = B\exp\frac{D}{T_s} \tag{34}$$

Table 1 Reactor Parameters for Steady-State Isothermal Simulation

Active sites surface area	2.5×10^4	$\text{m}^2_{\text{Pt}}\text{m}_R^{-3}$
Washcoat external surface area	2.4×10^3	$\text{m}^2_i\text{m}_R^{-3}$
Channel diameter	1.0×10^{-3}	m_R
Reactor length	0.15	m_R
Reactor porosity	0.6	—
Nusselt, Sherwood number	3.65	—
Washcoat porosity	0.4	—
Washcoat thickness	2.5×10^{-5}	m

Table 2 Rate Parameters for Eqs. (33) and (34)

i	A $kg^2mol^{-1}m_{Pt}^{-2}sec^{-1}$	E_a J mol^{-1}	B $kg^q mol^{-q}$	D J mol^{-1}
CO	5.252×10^{10}	12,556	1.83	961
C_3H_6	1.091×10^{12}	14,556	5.82	361
CH_4	5.744×10^{7}	19,000	2.45×10^{-6}	11,611
NO			1.34×10^{4}	3,733

Note: For q-values, see Eq. (32).

Figure 1 shows concentration profiles along the reactor axis, as predicted by the model for typical inlet concentrations 0.6 vol% CO, 468 ppm C_3H_6, 50 ppm CH_4, 0.2 vol% H_2, 500 ppm NO, 0.585 vol% O_2. The methane conversion was negligible. The figure reflects the expected behavior that methane and carbon monoxide are the exhaust gas components that are most difficult to oxidize. Figure 2 indicates that at higher temperatures, diffusion limitation inside the washcoat must be taken into account, since concentrations of CO in the bulk gas phase and in the washcoat become significantly different at increased reaction rates. The results reflect the general agreement in the literature that below light-off, catalytic mufflers operate in a regime, limited by kinetics, whereas mass transfer limitation occurs above the light-off temperature [36,37].

When considering nonisothermal conditions, the energy equations have to be added to Eqs. (25)–(27). For the gas phase

$$\phi_m c_p \frac{dT}{dx} = -\alpha a_v (T - T_s) \tag{35}$$

For the energy equation of the solid phase it is assumed that washcoat and monolith have the same material properties and that radial temperature gradients are negligible because of the small washcoat layer thickness, and because of adiabatic operation [37],

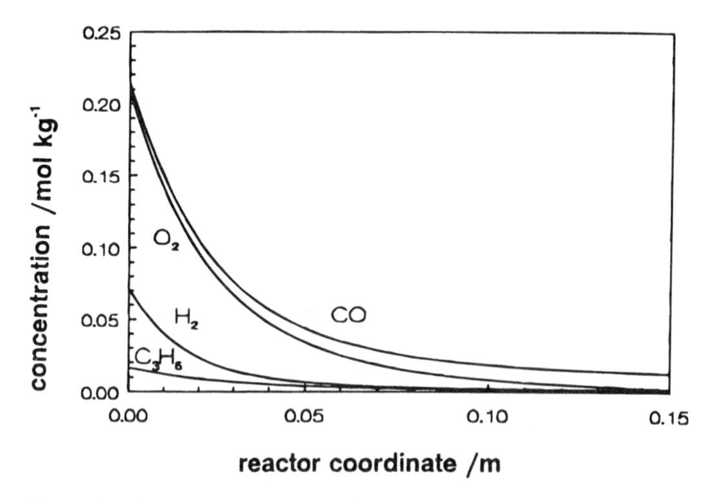

Figure 1 Axial concentration profiles for CO, C_3H_6, H_2, and O_2, isothermal model, $T = 773$ K, $\phi_m^{sup} = 6$ kg m_r^{-2} sec^{-1}.

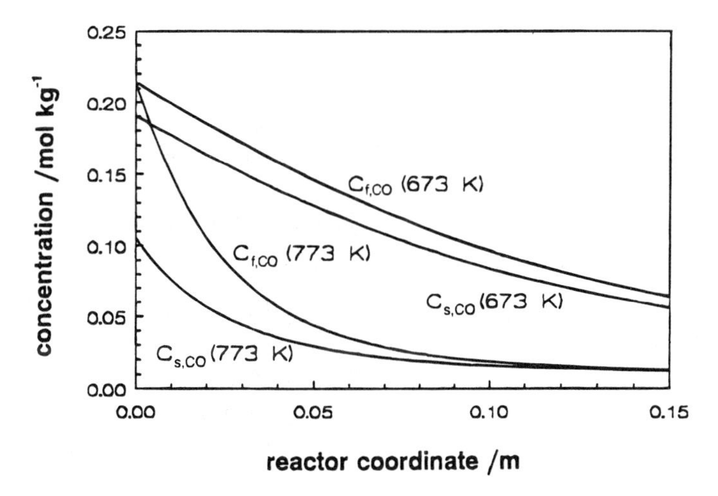

Figure 2 Axial profiles in fluid phase (= f) and in washcoat pores (= s), two temperatures, $\phi_m^{sup} = 6$ kg m_r^{-2} sec^{-1}.

$$\lambda^s(1 - \epsilon)\frac{d^2T^s}{dx^2} = -\alpha a_v(T - T^s) - a_{cat}\sum_i(-\Delta_f H)_i R_i^w \tag{36}$$

The corresponding boundary conditions are:

$$T = T^{in}, \qquad \text{at } x = 0 \tag{37}$$

$$\frac{dT^s}{dx} = 0, \qquad \text{at } x = 0 \quad \text{and} \quad x = L \tag{38}$$

Typical data for the parameters are included in Table 1. Figure 3 shows axial temperature profiles in the solid phase, calculated with and without the incorporation of axial solid conduction in the model for CO oxidation kinetics of Voltz et al. [15] below the light-off temperature, and at relatively low mass flow rate, where the effect of heat

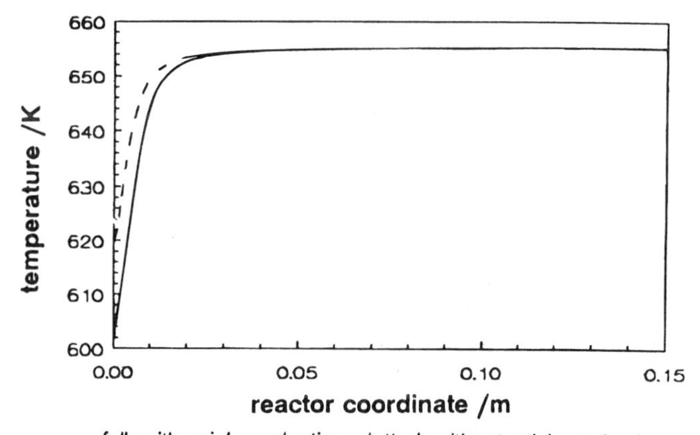

Figure 3 Axial temperature profiles in solid phase with and without axial solid heat conduction, $T^{in} = 598$ K, $\phi_m^{sup} = 0.33$ kg m_r^{-2} sec^{-1}.

conduction is expected to be largest. The influence of heat conduction is rather small in the steady state, and may be neglected [19] if

$$Pe_s = \frac{Lv\rho C_p \epsilon}{\lambda^m(1 - \epsilon)} < 200 \tag{39}$$

Axial heat conduction in the solid phase is important, however, when reactor light-off or hot spots in the reactor have to be considered [38].

Nievergeld et al. [39] used a kinetic model [40] to study axial profiles of surface species in a monolith reactor under steady-state conditions for the combined reactions $CO + O_2$ and $CO + NO$ over Rh/Al_2O_3 catalyst. The reactor model of Lie et al. was used, with reactor parameters as were shown in Table 1. The kinetic parameters are taken from Oh et al., except for the activation energy of CO desorption, which was considered constant at 112 kJ mol^{-1}, rather then depending on the surface coverage. A stoichiometric equivalence ratio was defined relative to the stoichiometry of the global reaction:

$$\phi = \frac{2\overline{C_{O_2}^{in}} + \overline{C_{NO}^{in}}}{\overline{C_{CO}^{in}}} \tag{40}$$

Figure 4 shows the steady-state gas-phase concentrations of the reactants versus the axial reactor coordinate, with CO as limiting reactant. Mass transfer limitation is negligible, since the feed temperature is below light-off, and the concentrations of the reactants in the pores of the washcoat (not shown) reach about 95% of the values of the gas-phase concentrations. The CO conversion near the outlet of the reactor is 100%, while the conversion of NO is only 9%. The surface coverages as a function of the axial reactor coordinate are shown in Fig. 5. The oxygen coverage does not exceed 10^{-5} as long as the CO conversion is not complete. As shown in Fig. 6, the production rates of both CO_2 and N_2 increase toward the outlet of the reactor until the surface concentration of adsorbed CO becomes zero. The relatively high surface coverage of CO at the inlet of the reactor is inhibiting the adsorption of the other reactants and the dissociation of NO, resulting in low production rates for CO_2 and N_2. Toward the outlet, the adsorption rate of CO

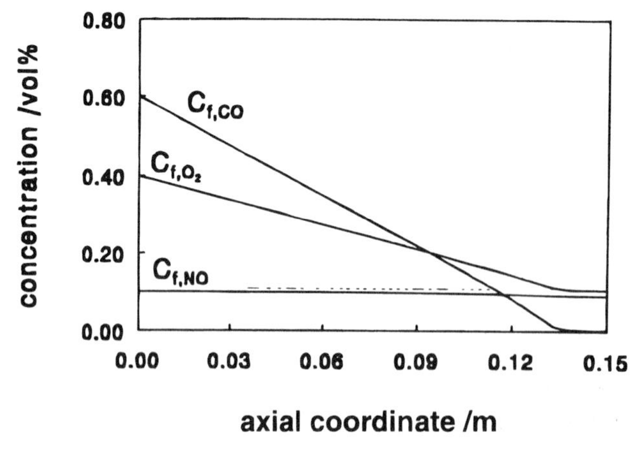

Figure 4 Fluid concentrations C_f as a function of axial coordinate. T = 490 K, ϕ_m^{sup} = 0.83 kg m_r^{-2} sec^{-1}, ϕ = 1.5.

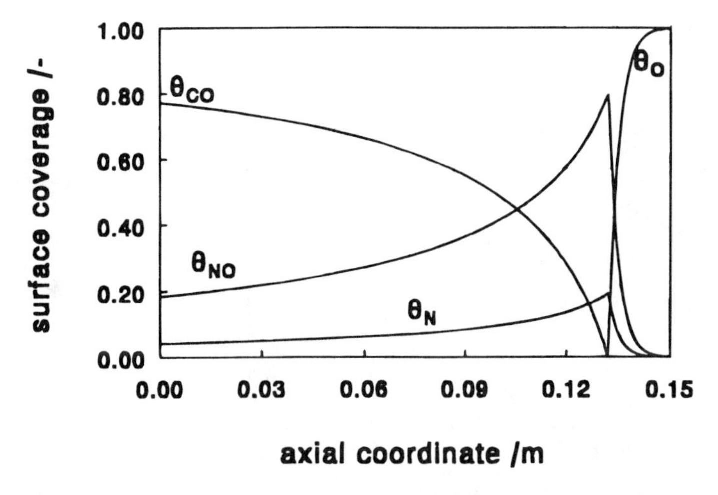

Figure 5 Surface coverages as a function of axial coordinate. Conditions, see Fig. 4.

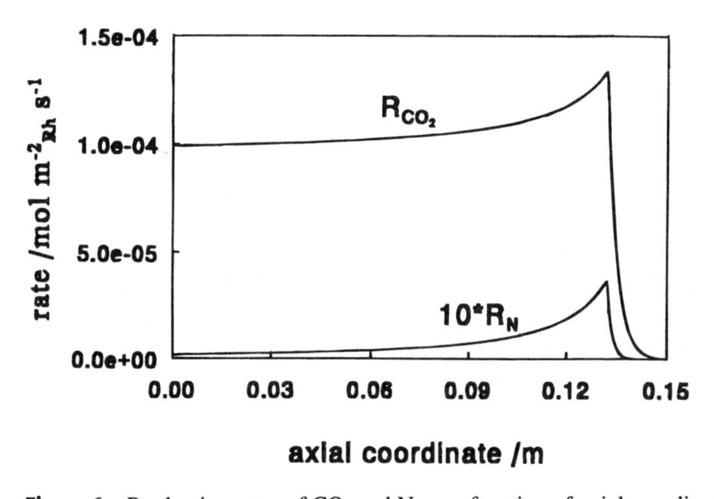

Figure 6 Production rates of CO_2 and N_2 as a function of axial coordinate. Conditions, see Fig. 4.

decreases due to the decrease of the gas-phase concentration, and as a result more free sites become available for the adsorption of O_2 and adsorption and dissociation of NO. The resulting higher oxygen coverage causes the production rate of CO_2 to increase, while the increasing coverage of NO and N leads to a higher production rate of N_2.

Figure 7 shows the steady-state temperature profiles of the bulk gas-phase T and the solid-phase (washcoat plus ceramic) T_s as a function of the axial reactor coordinate. Simulations have been performed at a feed temperature above the light-off temperatures of the global reactions; i.e., CO and NO conversions are higher than 50%. The surface coverages, not shown, decrease monotonously toward the outlet for CO, O, and NO, while the nitrogen coverage has a maximum in the second half of the reactor. The differences with Fig. 5 are explained by the higher temperature. As shown in Fig. 7, the heat production in the washcoat leads to a solid-phase temperature that is higher than the temperature of

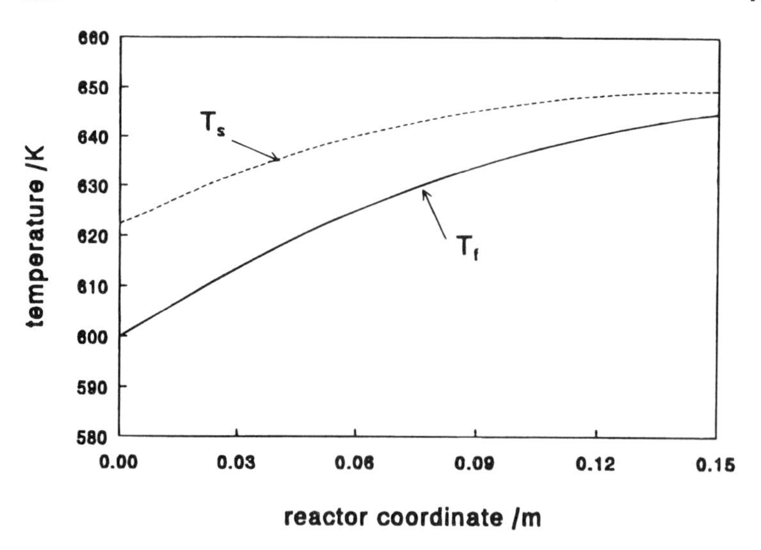

Figure 7 Temperature of the fluid and solid phase along reactor axis. T^{in} = 600 K, ϕ_m^{sup} = 16.7 kg m_r^{-2} sec^{-1}, ϕ = 1.

the bulk gas phase. The temperature difference becomes smaller near the outlet of the reactor, since the CO_2 formation rate decreases toward the outlet, as shown in Fig. 8.

The CO coverage is not inhibiting the adsorption of the other reactants now, since the higher temperature leads to a higher CO desorption rate. As a result, the lower concentrations of the reactants toward the outlet lead to decreasing adsorption rates and a lower production rate of CO_2. This is in contrast to Fig. 6, in which an increasing CO_2 production toward the outlet is shown. The surface reaction between NO* and N* is the most important step in the N_2 formation in the first half of the reactor. In the last part, the recombination of two N adatoms becomes an extra and dominant contribution in the

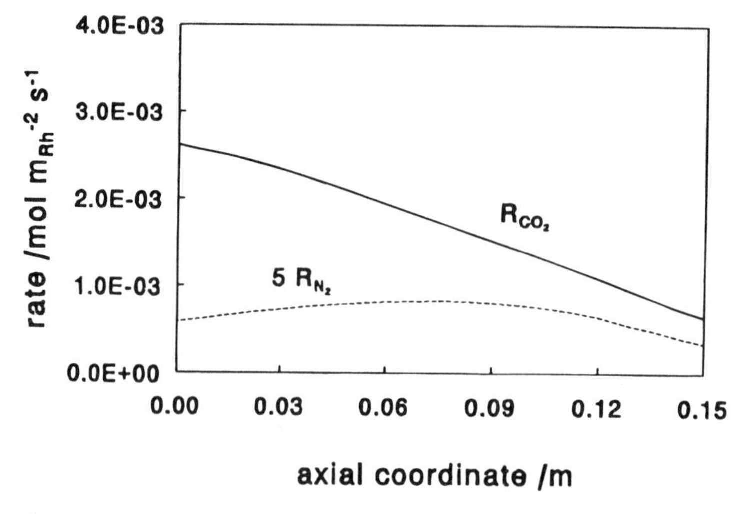

Figure 8 Production rates of CO_2 and N_2 as a function of axial coordinate. Conditions, see Fig. 7.

N_2 formation, leading to a lower N adatom coverage and a maximum in the production rate of N_2, as shown in Fig. 8.

Figure 9 shows the concentration of CO in the bulk gas phase and in the pores of the washcoat. At a feed temperature of 600 K, considerable mass transfer limitation occurs. The concentration in the pores of the washcoat at the inlet of the reactor is about 60% of the bulk gas-phase concentration. The concentrations of NO and O_2 in the pores of the washcoat, not shown, have values of about 55% of the bulk gas-phase concentration for the former and 80% for the latter.

With increasing feed temperature, the production rates and the corresponding heat development become gradually more pronounced in the first part of the reactor. Therefore heat transfer between solid and gas gets larger near the reactor inlet and smaller near the outlet, causing a rather uniform solid temperature at high feed temperatures, as depicted in Fig. 10.

Simulations performed at the same conditions, but without axial heat conductivity, showed identical temperature profiles as the ones given in Fig. 10, while results at a low mass flow, typically 0.83 kg m_r^{-2} sec^{-1}, showed a temperature of the solid phase at the inlet of the reactor that is lower than the temperature calculated from the model with axial heat conductivity. This phenomenon was also observed by Lie et al. [35] and indicates that axial heat conduction in the solid phase can be neglected under steady-state conditions when the fluid flow is large enough.

B. Unsteady-State Operation

Unsteady-state operation of a monolith was modeled to study the light-off behavior of catalytic mufflers as well as the deactivation of the catalyst.

The response of the reactor on a step change in the inlet temperature was studied [25] for the simultaneous oxidation of CO, propene, methane, and hydrogen, using adapted kinetics [15]. Accumulation of heat in the solid phase was considered as transient, while

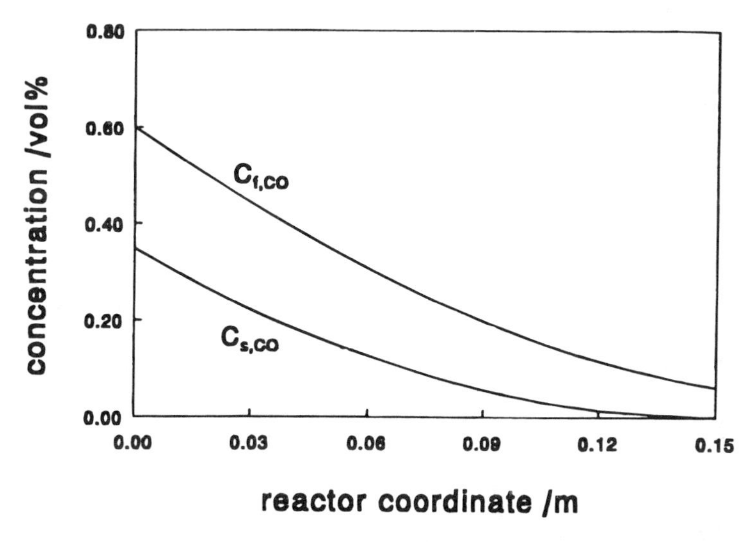

Figure 9 CO concentration in bulk gas and washcoat pores versus the axial coordinate. Conditions, see Fig. 7.

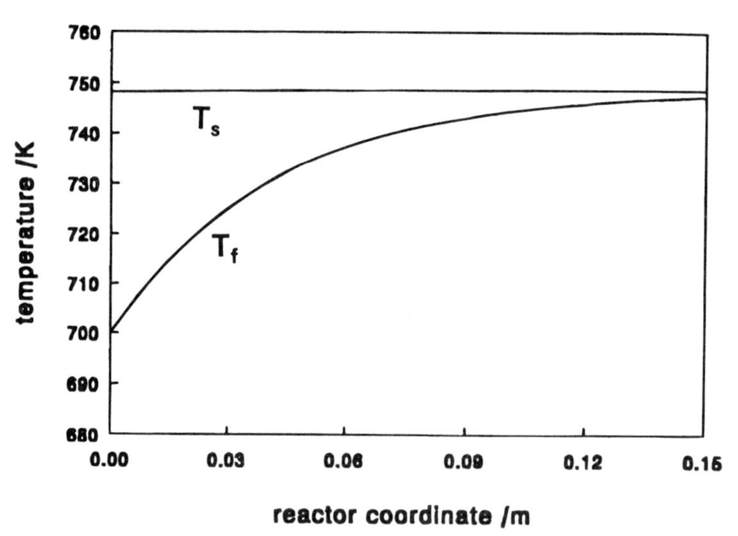

Figure 10 Temperature of the bulk gas phase and the solid phase versus the axial coordinate. T^{in} = 700 K. Other conditions, see Fig. 7.

all other subprocesses were taken as quasi-stationary. The model was a lumped-parameter version, and axial heat conduction in the solid phase was incorporated. Reaction light-off is enhanced if a shorter reactor is used with many narrow channels that contain relatively more catalyst at the entrance. The first and second point were confirmed by T'ien [24], who used a similar approach with gas-phase reactions included as well, and who found also that a high porosity of the monolith is favorable for a fast light-off.

Oh and Cavendish [25] noted that a sudden decrease of the inlet temperature may cause a local temperature rise that exceeds the adiabatic temperature rise based on the feed composition. The same result, obtained with other models [13,37], was ascribed to fast diffusion of hydrogen. But contrary to these authors, Oh and Cavendish also find this so-called wrong-way behavior in the absence of hydrogen. Due to decreasing temperatures in the inlet part of the monolith, much reactant may reach the hotter part of the reactor, leading to a very fast reaction and corresponding heat production. The resulting hot spot moves in time to the outlet of the reactor and disappears.

Chen et al. [23] extended the model of Oh and Cavendish to account for radial temperature gradients on reactor scale as a result of a nonuniform flow distribution over the channels and heat losses at the outer reactor wall due to convection and radiation. At steady state the reactor center is isothermal in axial direction, but radial gradients near the outer wall are considerable. During the transient heating, axial gradients are present in the center as well. A similar study [5] showed that heating up the reactor takes more time if the inlet velocity profile is not flat. Heat losses through the reactor wall decrease the maximal possible steady-state conversion but hardly affect the heating-up period. A nonuniform flow distribution causes a nonuniform catalyst deactivation, which in turn leads to worse transient heating behavior. Leclerc et al. [41] report a higher steady-state conversion when the flow distribution is uniform.

For automotive exhaust gas treatment, Oh [42] presented a transient, one-dimensional single-channel model that predicts the temperature and species concentrations as a function of axial position and time in both solid and gas phase. Uniform flow distribution was shown to be a reasonable assumption, since the model gave similar performance to the more extended model of Chen et al. [23]. Mass and heat accumulation in the gas phase was neglected in the calculations, as well as mass accumulation in the solid phase. Heat and mass transfer were described with the asymptotic Nusselt and Sherwood numbers. Axial conduction in the solid phase was taken into account, but not diffusion limitation in the washcoat because of the typical thinness of the layer. The results of model calculations were compared to federal test procedure results with a vehicle engine, and showed good agreement with respect to the warming-up behavior and the cumulative emission of hydrocarbons.

Catalyst deactivation in a monolith reactor was studied [26], assuming poisoning of active sites by phosphorus. Axial and radial phosphorus deposition profiles were simulated as a function of time with a two-dimensional model. Light-off of the deactivated monolith was studied as well. It was shown that deactivation occurred more slowly when using a larger washcoat layer thickness, a larger BET-surface area, thinner channel walls, or higher catalyst loading. Similar results were obtained by Pereira et al. [43].

C. Dynamic Operation

Many modeling studies and experimental investigations have demonstrated that the intentional unsteady operation of reactors can profoundly influence conversion and/or selectivity. Several reviews were published [28,29,31,44–46]. The effects for automotive exhaust gas converters were recently discussed [30].

During periodic operation, the system is forced to follow changes in the input. So-called *cycling of the feed* is the case where oscillations are applied to the concentrations of the reactor feed. Whether the input is followed perfectly depends on the dynamic behavior of the system. The most important parameter describing the dynamic behavior is the characteristic response time τ_c: A small value of τ_c corresponds to a fast-responding system. Based on the period T_m of the forced oscillation and characteristic response time of the system τ_c, three different periodic operations can be distinguished [27,45]:

1. Quasi-steady-state periodic regime ($T_m >> \tau_c$). The input variable varies rather slowly compared to the dynamics of the system, and the system follows the input variable almost exactly. The time-averaged performance of the reactor is calculated applying the quasi-steady-state approximation to the state of the system and averaging out the resulting performances at any time.

2. Dynamic regime ($T_m \approx \tau_c$). When the period of the oscillation is of the order of the system's characteristic response time, the system is in intermediate or dynamic periodic operation. The transient behavior of the system has to be determined to predict the effects of periodic operation. Dynamic reactor operation may result in considerably higher performance if resonance phenomena are involved, and therefore this range of operation is of particular interest for optimization of the reactor.

3. Relaxed steady-state or sliding regime ($T_m << \tau_c$). When the input varies rapidly relative to the characteristic response time, the state oscillates with a very small amplitude. The quasi-steady-state approximation can be applied to the state using the time-averaged value of the control. The performance of the system can be predicted using the performance in comparable steady-state operation.

The CO oxidation over noble metal catalyst has been studied most intensively under both steady-state operation and periodic operation. Cutlip [47] studied the CO oxidation on a Pt/Al_2O_3 catalyst in a gradientless reactor. The observed time-averaged reaction rates achieved values that were twenty times the steady-state rate at a modulation period of 1 minute, when the time-averaged feed stream composition was stoichiometric. This rate enhancement during periodic operation was explained by a better balance between the surface coverages of adsorbed CO and O with respect to the stoichiometry of the surface reaction. Under steady operation, the major part of the active sites will be covered by CO, which limits the availability of adsorbed oxygen, while under oscillating conditions the coverages of CO and O adatoms have periodically comparable values, which lead to a higher time-averaged CO_2 production rate. It indicates that individual elementary steps of a kinetic model have different rates and that accumulation of surface species should be taken into account when time-variant behavior occurs. Various other studies were performed, using different types of reactors [33,48–53] and reactions [54–58].

Lie et al. [35] modeled the CO oxidation by O_2 on a Pt/Al_2O_3 catalyst with an isothermal monolithic converter in order to assess the effect of cyclic feeding on the performance of the reactor. The kinetic model of Herz and Marin [59] was used, which consists of a closed sequence of elementary steps. The reactor model is essentially as described by Eqs. (25)–(27), but now includes accumulation terms for all three phases: the gas phase, the pores of the washcoat, and the catalyst surface.

$$\epsilon\rho \frac{\partial}{\partial t}\left(\frac{C_i}{\rho}\right) = -\phi_m^{\text{sup}} \frac{d}{dx}\frac{C_i}{\rho} - \rho k_{f,i} a_v \left(\frac{C_i}{\rho} - \frac{C_i^w}{\rho}\right) \tag{41}$$

$$\epsilon^w \rho \frac{4\epsilon\delta_w}{d_c}\frac{\partial}{\partial t}\frac{C_i^w}{\rho} = \rho_f k_{f,i} a_v \left(\frac{C_i}{\rho} - \frac{C_i^w}{\rho}\right) + a_{\text{cat}} R_i^w \tag{42}$$

The boundary conditions at $x = 0$ result from the oscillating feed at the inlet of the reactor, for CO given by:

$$C_{CO}(0, t) = \overline{C}_{CO}^{in}(1 + A_0 \sin 2\pi f t) \tag{43}$$

CO is oscillating in counterphase with O_2, so the time-dependent feed concentration of O_2 is given by

$$C_{O_2}(0, t) = \overline{C}_{O_2}^{in}[1 + A_0 \sin (2\pi f t + \pi)] \tag{44}$$

As initial conditions for concentrations and surface coverages, the values corresponding with the steady state at the same time-averaged feed were taken. Values for the reactor model parameters were shown in Table 1.

Typical oscillations of the gas-phase concentrations are shown in Fig. 11, with the corresponding surface coverages shown in Fig. 12. Rates and conversion are plotted in Fig. 13. The most striking feature in Fig. 11 is that maxima in the dioxygen concentration at the reactor inlet turn into minima at the outlet. The maxima of the surface coverage of oxygen, however, become more pronounced near the reactor outlet. Notice the relatively low surface coverage of oxygen at this temperature, which is below light-off. For CO the amplitudes of bulk gas concentration and surface coverage increase toward the reactor outlet, and no phase shift occurs. Remarkable is the strong periodic reduction of the CO surface coverage near the reactor outlet. The time-averaged conversion is larger than the steady-state conversion, shown at time $t = 0$. This effect is more pronounced at high space times. The periodically enhanced reaction rate shows two maxima around the minimum coverage of adsorbed CO as the changing surface coverages of CO and O oppose each other. In fact, the rate would be maximum when the surface coverage of both CO and O equals 0.5. The results are explained as follows.

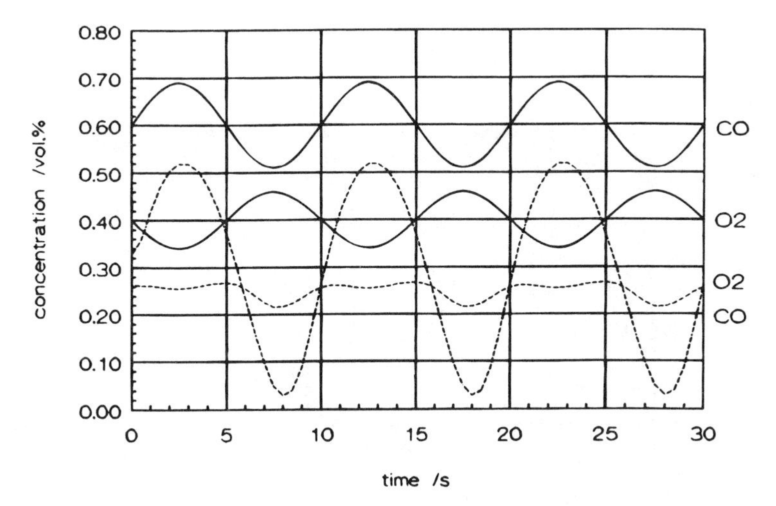

Figure 11 Oscillations of fluid-phase CO and O_2 at reactor inlet (drawn) and outlet (dotted). 573 K, $\phi_m^{sup} = 0.6$ kg m_r^{-2} sec^{-1}, $\phi = 1.5$, $A_0 = 15\%$.

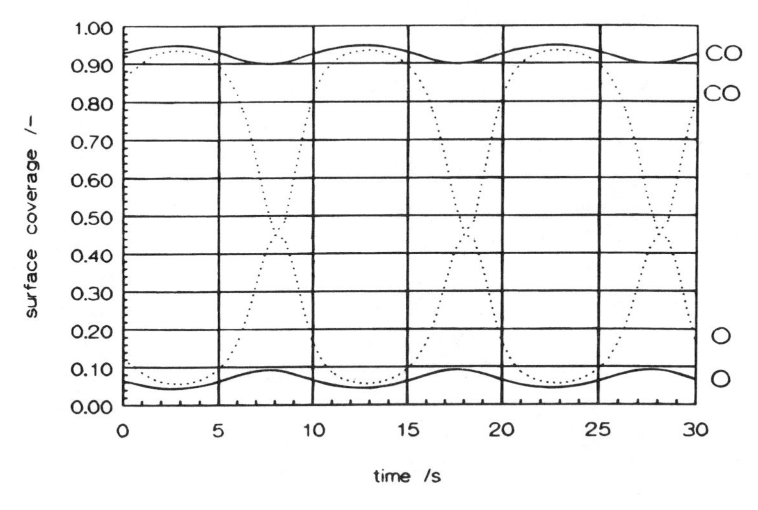

Figure 12 Surface coverages of CO and O at inlet (drawn) and outlet (dotted) of reactor. Conditions as in Fig. 11.

In the steady state, corresponding to time $t = 0$, the catalyst's surface is almost completely covered with CO, resulting in a relatively low reaction rate due to CO inhibition of the adsorption of the second reactant. When, due to oscillation, the CO gas-phase concentration increases, even more CO adsorbs on the surface, which decreases further the reaction rate. When, in contrast, the CO gas-phase concentration decreases, free sites become available for O_2 adsorption and the rate increases correspondingly. The fact that a CO decrease in the bulk gas phase is accompanied by an increase in O_2 gives an extra enhancement of the rate and, hence, of the CO conversion. As a result the amplitude of

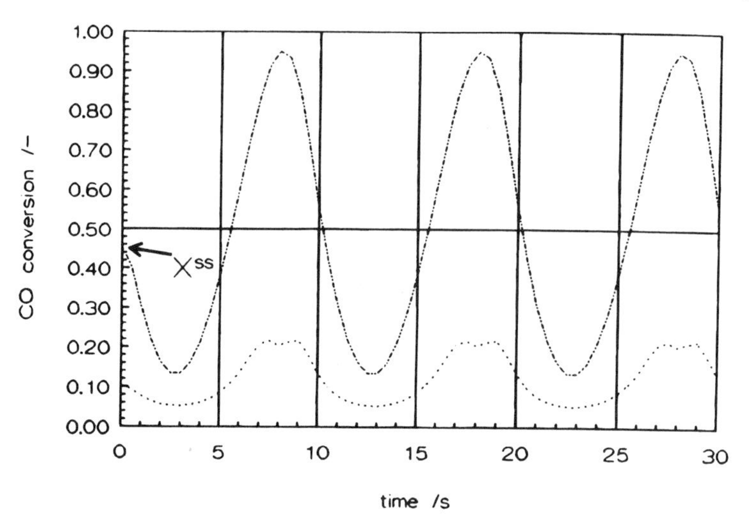

Figure 13 CO_2 production rate (dotted), conversion (broken), and time-averaged conversion (drawn). Time zero represents steady state. Conditions as in Fig. 11.

the CO oscillations is reinforced. The oxygen amplitude is correspondingly attenuated. Moreover, a phase shift of π radians in the oscillation of the oxygen gas-phase concentration is observed at larger axial coordinates. The periodic reaction rate enhancement overcompensates the periodic decreases, leading to a positive effect on the time-averaged conversion as compared to the steady-state conversion. This phenomenon is attributed to the nonlinear character of the kinetics.

Above the light-off temperature, the results are quite different. The CO surface coverage is relatively low at the reactor inlet and decreases further toward the outlet due to lower gas-phase concentrations. As a result, more oxygen can be adsorbed, leading to higher oxygen coverage near the outlet. In the first half of the period the CO coverage increases with increasing gas-phase concentration, and vice versa in the second half. Oxygen shows a similar behavior. Because a growing CO coverage is accompanied by a diminishing oxygen coverage, a rather constant reaction rate results during the first half of the period. However, when the CO coverage decreases further during the second half, θ_O reaches its maximum and the reaction rate follows the decrease of θ_{CO}. The net result is that the time-averaged conversion of CO is less than the steady-state conversion.

It is clear from the foregoing that the temperature has a large influence on the time-averaged conversion and its deviation from the steady-state conversion. Lie et al. [35] found that the enhancement is largest when CO and O_2 oscillate in counterphase, and that it increases with increasing amplitude. An increase of the time-averaged conversion is possible only if the surface is almost completely covered with CO at steady state. Therefore, for CO oxidation a positive effect of cycling is to be expected only under the light-off temperature, since such a situation occurs only at low temperatures. Above light-off, the CO and O coverages are comparable at steady state and cycling of the feed does not improve the performance. Cho and West [60] and Cho [48] observed the same effects experimentally for the CO oxidation in an integral fixed-bed reactor.

Nievergeld et al. [39] considered nonisothermal operation by adding the energy equations to the Lie et al. model:

$$\epsilon\rho C_p \frac{\partial T}{\partial t} = -\phi_m^{\text{sup}} c_p \frac{\partial T}{\partial x} - \alpha a_v (T - T_s) \tag{45}$$

$$(1 - \epsilon)\rho_s C_{ps} \frac{\partial T_s}{\partial x} = \lambda_s (1 - \epsilon) \frac{d^2 T_s}{dx^2} + \alpha a_v (T - T_s) + a_{\text{cat}} \sum_{i=1}^{4} (-\Delta_r H)_i r_i \tag{46}$$

The corresponding boundary conditions are:

$$T = T^{\text{in}}, \quad \text{for } x = 0 \tag{47}$$

$$\frac{dT_s}{dx} = 0, \quad \text{for } x = 0 \text{ and } x = L \tag{48}$$

The reactions considered were simultaneous CO oxidation and NO reduction, using a kinetic model for rhodium catalyst [25]. Initial and boundary conditions for the concentrations of reactants were the same as given by Lie et al., NO oscillating in phase with oxygen. Kinetic and reactor parameters used were mentioned in Section 8.IV.A.

Simulation results with cyclic feeding at a frequency of 0.1 Hz are shown in Figs. 14–16. Figure 14 shows the gas-phase concentrations at the outlet of the reactor as a function of time. The amplitude of the CO oscillation relative to the time-averaged value increases from 15% to 76% toward the outlet of the reactor, while the amplitudes of NO and O_2 decrease from 15% to 10%. The maxima in the oxygen oscillation at the reactor inlet turn into minima at the outlet, as discussed before for CO oxidation over Pt/Al_2O_3 by Lie [35]. The surface coverages of CO, O, and NO at the inlet, not shown, oscillate in phase with the corresponding gas-phase concentrations and with amplitudes relative to their time-averaged values of 7%, 30%, and 24%. The surface coverage of N at the inlet oscillates in phase with NO, with an amplitude of only 4%.

Figure 15 shows the surface coverages at the outlet. The oxygen coverage, which is not shown, oscillates in phase with adsorbed NO around a time-averaged value of 1.6 $\times$ 10^{-5} and an amplitude of 151%. The amplitudes of all coverages increase toward the outlet, while the minima of adsorbed CO and the maxima of the other species become sharper, leading to relaxation type of oscillations.

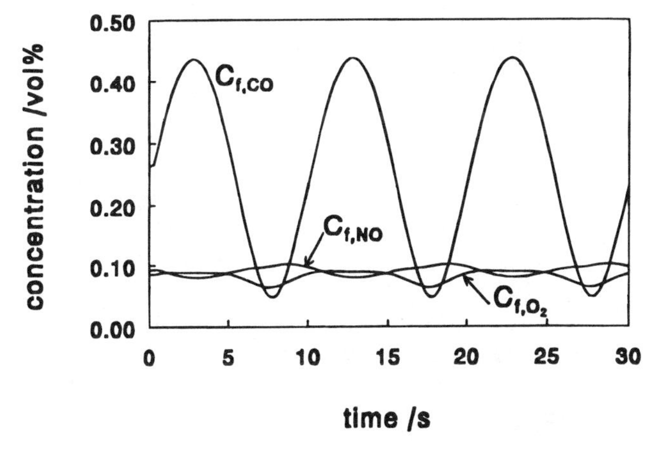

Figure 14 Outlet gas-phase concentrations as a function of time. T = 505 K, G = 10^{-2} kg/sec, f = 0.1 Hz, A_0 = 0.15, ϕ = 1.

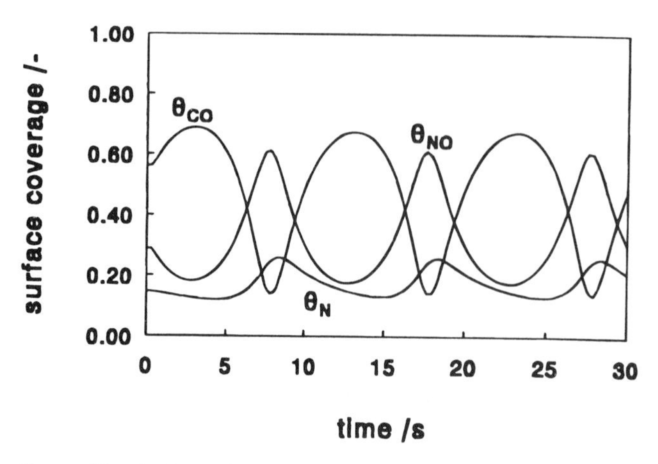

Figure 15 Outlet surface coverages as a function of time. T = 505 K, G = 10^{-2} kg/sec, f = 0.1 Hz, A_0 = 0.15, ϕ = 1.

Figure 16 shows the production rate of both CO_2 and N_2 at the outlet of the reactor. The steady-state conversions of CO and NO are 56% and 7.4%, respectively, while the time-averaged conversions during cycling of the feed are 57.5% and 8.9%. These differences are explained as follows.

At steady state, the high surface coverage of CO causes a relatively low production rate of CO_2 and N_2. When the gas-phase concentration of CO increases owing to the oscillation, the surface becomes even more occupied with adsorbed CO, while the corresponding decrease of the concentrations of O_2 and NO results in a decrease of the other surface species. The result is a further decrease of the production rates of CO_2 and N_2. In contrast, when the CO gas-phase concentration is decreased, free sites become available for O_2 and NO adsorption and for dissociation of NO, with higher production rates as a result. The fact that the CO decrease in the bulk gas is accompanied by an increase of O_2 and NO gives an extra enhancement of the production rates of CO_2 and N_2 and hence

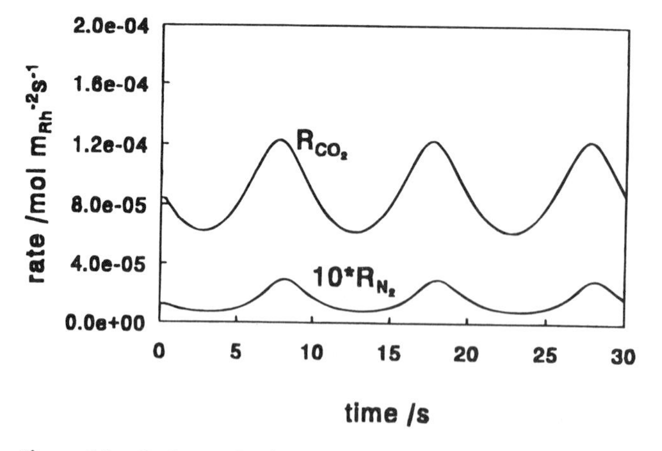

Figure 16 Outlet production rates as a function of time. T = 505 K, G = 10^{-2} kg/sec, f = 0.1 Hz, A_0 = 0.15, ϕ = 1.

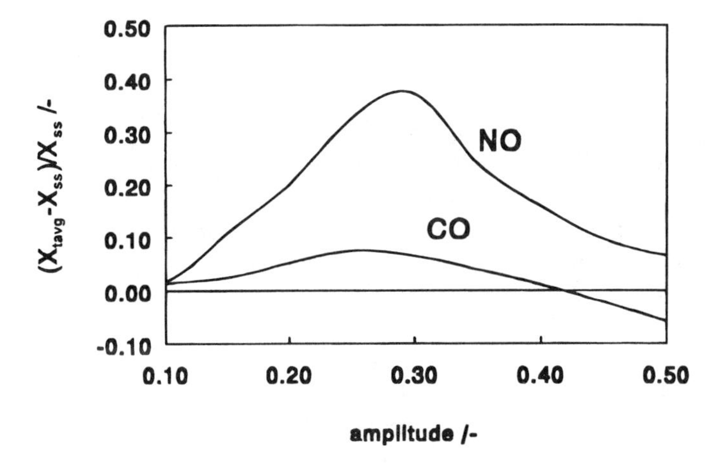

Figure 17 Increase of time-averaged conversion over steady-state conversion versus amplitude. $T = 500$ K, $\phi_m^{sup} = 1.7$ kg m_r^{-2} sec^{-1}, $f = 0.1$ Hz, $\phi = 1$.

of the conversions of CO and NO. The positive feedback from the CO coverage to the CO gas-phase concentration results in a reinforcement of the CO oscillation, whereas the negative feedback from the O and NO coverages to the corresponding gas-phase concentrations causes attenuation of the corresponding oscillations. The periodic enhancement of the reaction rates overcompensates the periodic decrease, leading to a positive effect on the time-averaged conversions of both CO and NO as compared to the steady-state conversions. This phenomenon is again attributed to the nonlinear character of the kinetics.

The relative change of the conversion as a function of the amplitude and the temperature is shown in Figs. 17 and 18. Increasing the amplitudes, see Fig. 17, enhances the advantage of the cyclic feeding at 0.1 Hz up to 30% for NO and 8% for CO.

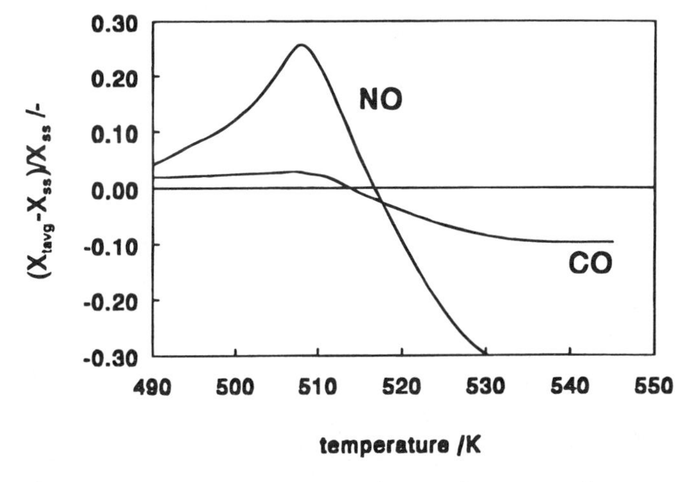

Figure 18 Increase of time-averaged conversion over steady-state conversion versus temperature. $A_0 = 0.15$. Other conditions, see Fig. 17.

At higher amplitudes, periodic stoichiometric limitation by CO causes decreasing time-averaged conversions. Higher temperatures are beneficial when oscillating the feed (Fig. 18) as long as the local CO conversion is incomplete during a cycle. This is in contrast to CO oxidation in the absence of NO [35], where beneficial effects were observed only at temperatures below the light-off. This was attributed to the comparable steady-state degrees of coverage by CO and oxygen at high temperatures. The presence of NO causes a much lower degree of coverage by oxygen over the complete temperature range of interest. Further increase of the temperature again leads to a stoichiometric limitation by CO during a larger fraction of the cycle time, ultimately leading to time-averaged conversions below the steady-state conversions.

The influence of the modulation frequency, see Fig. 19, shows beneficial effects in the frequency range 0–10 Hz, which is largest for NO reduction, with a maximum around the typical frequency of 1 Hz as met in the practice of automotive exhaust gas converters.

V. CONCLUDING REMARKS

Modeling of monolith reactors from first principles presents a valuable tool in the design of such reactors and in the analysis of the underlying phenomena. The results presented show that the reactor behavior can be adequately described and understood by a combination of the reactor's transport characteristics and the intrinsic kinetics obtained with a laboratory reactor of another type. As such we can generalize monolith models to other reaction networks, e.g., extend the given description of the dynamic operation for combined CO oxidation and NO reduction in the automotive exhaust gas converter to include other reactions, like the oxidation of various hydrocarbons and of hydrogen. The availability, however, of a proper kinetic model is a definite prerequisite.

Distinction of the various time scales at which subprocesses proceed is useful to determine the necessary complexity or possible simplification of the reactor model. A reactor study under dynamic conditions asks for a detailed kinetic model on the level of

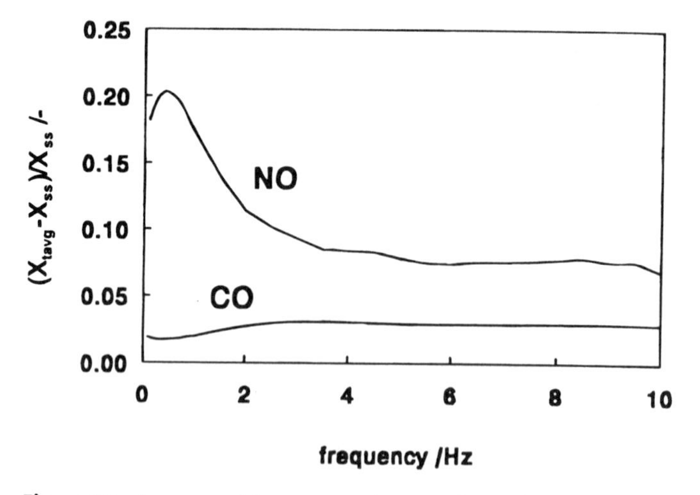

Figure 19 Increase of time-averaged conversions relative to steady-state conversions versus frequency. T = 489 K, A_0 = 0.15. Other conditions, see Fig. 17.

elementary steps, since the time scales of these steps become involved in the overall reactor performance. It is likely that for exhaust gas converters such a complete kinetic model also becomes necessary when modeling the steady-state performance. The ongoing demand for zero-emission vehicles may mean that the kinetics along the reactor axis, can no longer be described with a single rate-determining step, as done with, e.g., Langmuir–Hinshelwood-type kinetics, because the concentrations in the reactor develop from relatively high at the entrance to extremely low at the outlet.

Modeling of the monolith dynamics presents an opportunity to improve the performance of catalytic mufflers by involving the benefits of cyclic operation via an appropriate control strategy.

LIST OF SYMBOLS

Roman Letters

Symbol	Description	Unit
a_{cat}	catalytic surface area per unit reactor volume	$m_{Rh}^2\ m_R^{-3}$
a_i	collocation coefficient	
a_v	geometric surface area per unit reactor volume	$m_l^2\ m_R^{-3}$
A	preexponential factor	sec^{-1}
A_0	amplitude in fraction of C_f^{in}	
c_p	specific heat	$J\ kg^{-1}\ K^{-1}$
C	concentration	$mol\ m_f^{-3}$
d	diameter	m
d_c	internal diameter of channel	m_R
D	diffusion coefficient	$m_f^3\ m_i^{-1}\ sec^{-1}$
E_A	activation energy	$kJ\ mol^{-1}$
f	frequency	Hz
G	gas flow	$kg\ sec^{-1}$
$\Delta_f H$	enthalpy of formation	$kJ\ mol^{-1}$
$k_{f,i}$	mass transfer coefficient for species i	$m_f^{-3}\ m_i^{-2}\ sec^{-1}$
$k_{r,i}$	reaction rate coefficient for species i	sec^{-1}
L	reactor length	m_R
L_t	surface concentration of active sites	$mol\ m_{Rh}^{-2}$
M	molar mass	$kg\ mol^{-1}$
Nu	Nusselt number	
Re	Reynolds number	
S_i^0	sticking coefficient for component i on clean surface	
t	time	sec
T	temperature	K
T_m	modulation period	sec
v	velocity	$m\ sec^{-1}$
x	axial coordinate	m_R
X	conversion	

Greek Letters

Symbol	Description	Unit
α	heat transfer coefficient	$W m_i^{-2}\ K^{-1}$
δ_w	thickness of washcoat	m_R

ϵ	void fraction	$m_f^3\ m_R^{-3}$
ϵ^w	washcoat porosity	$m_f^3\ m_w^{-3}$
θ	surface coverage	$mol\ mol_{Rh}^{-1}$
λ	thermal conductivity	$W\ m^{-1}\ K^{-1}$
ρ	density	$kg\ m_f^{-3}$
ϕ	stoichiometric equivalence ratio	
ϕ_m^{sup}	superficial mass flow	$kg\ m_R^{-2}\ sec^{-1}$

Subscripts

a	adsorption
cat	catalytic surface area
d	desorption
f	fluid phase
h	hydraulic
i	referring to reactant i or interface
mol	molecular
r	reactor
s	pores in the washcoat or solid phase
w	washcoat

Superscripts

in	inlet
ss	steady state
$t$avg	time-averaged

REFERENCES

1. G.I. Taylor, Dispersion of soluble matter in solvent flowing slowly through a tube, *Proc. Roy. Soc. A219*:186 (1953).
2. A. Cybulski and J.A. Moulijn, Monoliths in heterogeneous catalysis, *Catal. Rev.-Sci. Eng. 36*:179 (1994).
3. D.F. Sherony and C.W. Solbrig, Analytical investigation of heat or mass transfer and friction factors in corrugated duct heat or mass exchanger, *Int. J. Heat Mass Transfer 13*:145 (1970).
4. M.A.M. Boersma, J.A.M. Spierts, W.J.G. Van Lith, and H.S. Van der Baan, The oxidation of ethene in an empty and packed tubular wall reactor operating in the reaction- and diffusion-controlled regimes, *Chem. Eng. J. 20*:177 (1980).
5. K. Zygourakis, Transient operation of monolith catalytic converters: A two-dimensional reactor model and the effects of radially nonuniform flow distributions, *Chem. Eng. Sci. 44*:2075 (1989).
6. G. Eigenberger and U. Nieken, Katalytische Abluftreinigung: Verfahrenstechnische Aufgaben und neue Lösungen, *Chem.-Ing.-Techn. 63*:781 (1991).
7. E.U. Schlünder, On the mechanism of mass transfer in heterogeneous systems—in particular in fixed beds, fluidized beds and on bubble trays, *Chem. Eng. Sci. 32*:845 (1977).
8. J.S. Howitt and T.C. Sekella, Flow effects in monolithic honeycomb automotive catalytic converters, *SAE Paper 740244* (1974).
9. R.D. Hawthorn, Afterburner catalysis—Effects of heat and mass transfer between gas and catalyst surface, *AIChE Symp. Ser. 70*:428 (1974).

10. E.R.G. Eckert and J.F. Gross, *Introduction to Heat and Mass Transfer*, McGraw-Hill, New York, 1963.
11. J.R. Sellars, M. Tribus and J.S. Klein, Heat transfer to laminar flow to a round tube or flat conduit—The Graetz problem extended, *Trans. Am. Soc. Mech. Eng. 78*:441 (1956).
12. L.P.B.M. Janssen and M.M.C.G. Warmoeskerken, *Transport Phenomena Data Companion*, Edward Arnold, London, 1987.
13. L.C. Young and B.A. Finlayson, Mathematical models of the monolith catalytic converter. Part I. Development of the model and application of orthogonal collocation, *AIChEJ. 22*:331 (1976).
14. R.K. Shah and A.L. London, *Laminar Flow Forced Convection in Ducts: A Source Book for Compact Heat Exchanger Analytical Data* (Adv. Heat Transfer, suppl. 1), Academic Press, London, 1978.
15. S.E. Voltz, C.R. Morgan, D. Liedermann, and S.M. Jacob, Kinetic study of carbon monoxide and propylene oxidation on platinum catalysts, *IEC Prod. Res. Dev. 12*:294 (1973).
16. R.E. Hayes and S.T. Kolaczkowski, Mass and heat transfer effects in catalytic monolith reactors, *Chem. Eng. Sci. 49*:3587 (1994).
17. E. Tronconi and P. Forzatti, Adequacy of lumped parameter models for SCR reactors with monolithic structure, *AIChEJ. 38*:201 (1992).
18. U. Ullah, S.P. Waldrum, C.J. Bennett and T. Truex, Monolithic reactors: Mass transfer measurements under reacting conditions, *Chem. Eng. Sci. 47*:2413 (1992).
19. J. Votruba, J. Sinkule, V. Hlavacek, and J. Skrivanek, Heat and mass transfer in honeycomb catalysts, *Chem. Eng. Sci. 30*:117 (1975).
20. S. Irandoust and B. Andersson, Monolithic catalysts for nonautomobile applications, *Catal. Rev.-Sci. Eng. 30(3)*:341 (1988).
21. G.F. Froment and K.B. Bischoff, *Chemical Reactor Analysis and Design*, Wiley, New York, 1990.
22. D.E. Mears, Diagnostic criteria for heat transport limitations in fixed bed reactors, *J. Catal. 20*:127 (1971).
23. D.K.S. Chen, E.J. Bissett, S.H. Oh, and D.L. Van Ostrom, *A Three-Dimensional Model for the Analysis of Transient Thermal and Conversion Characteristics of Monolithic Catalytic Converters*, SAE paper 880282 (1988).
24. J.S. T'ien, Transient catalytic combustor model, *Combustion Sci. Techn. 26*:65 (1981).
25. S.H. Oh and J.C. Cavendish, Transients of monolithic catalytic converters: Response to step changes in feedstream temperature as related to controlling automobile emissions, *IEC Prod. Res. Dev. 21*:29 (1982).
26. S.H. Oh and J.C. Cavendish, Design aspects of poison-resistant automobile monolithic catalysts, *IEC Prod. Res. Dev. 22*:509 (1983).
27. Y.S. Matros, *Catalytic Processes Under Unsteady-State Conditions*, Elsevier, Amsterdam, 1989.
28. A. Renken, Unsteady state operation of continuous reactors, *Int. Chem. Eng. 24*:202 (1984).
29. P.L. Silveston, Catalytic oxidation of carbon monoxide under periodic operation, *Can. J. Chem. Eng. 69*:1106 (1991).
30. P.L. Silveston, Automotive exhaust catalysis under periodic operation, *Catalysis Today 25*:175 (1995).
31. P. Silveston, R.R. Hudgins, and A. Renken, Periodic operation of catalytic reactors, introduction and overview, *Catalysis Today 25*:91 (1995).
32. F.M. Dautzenberg, J.N. Helle, R.A. Van Santen, and H. Verbeek, Pulse-technique analysis of the kinetics of the Fischer–Tropsch reaction, *J. Catal. 50*:8 (1977).
33. K.C. Taylor and R.M. Sinkevitch, Behavior of automotive exhaust catalyst with cyclic feed streams, *IEC Prod. Res. Dev. 22*:45 (1983).

34. J.A. Moulijn and F. Kapteijn, *Chemisch Magazine* 499 (1989).
35. A.B.K. Lie, J. Hoebink, and G.B. Marin, The effects of oscillatory feeding of CO and O_2 on the performance of a monolithic catalytic converter of automobile exhaust gas: A modelling study, *Chem. Eng. J. 53*:47 (1993).
36. R.H. Heck, J. Wei, and J.R. Katzer, Mathematical models of monolithic catalysts, *AIChEJ. 22*:477 (1976).
37. J.W. Kress, N.C. Otto, M. Bettman, J.B. Wang, and A. Varma, Diffusion-reaction of CO, NO and O_2 in automotive exhaust catalysts, *AIChE Symp. Ser. 76*:202 (1980).
38. L.C. Young and B.A. Finlayson, Mathematical models of the monolith catalytic converter. Part II. Application to automobile exhaust, *AIChEJ. 22*:343 (1976).
39. A.J.L. Nievergeld, J.H.B.J. Hoebink, and G.B. Marin, The performance of a monolithic catalytic converter of automobile exhaust gas with oscillatory feeding of CO, NO and O_2: A modelling study, *Stud. Surf. Sc. Cat.*, accepted (1995).
40. S.H. Oh, G.B. Fisher, J.E. Carpenter, and D.W. Goodman, Comparative kinetic studies of CO–O_2 and CO–NO reactions over single crystal and supported rhodium catalysts, *J. Catal. 100*:360 (1986).
41. J.P. Leclerc, D. Schweich, and J. Villermaux, A new theoretical approach to catalytic converters, in *Catalysis and Automotive Pollution Control II* (A. Crucq, ed.), Elsevier, Amsterdam, 1991, p. 465.
42. S.H. Oh, Catalytic converter modeling for automotive emission control, in *Computer-Aided Design of Catalysts* (E.R. Becker and C.J. Pereira, eds.), Dekker, New York, 1993, p.
43. C.J. Pereira, J.E. Kubsh, and L.L. Hegedus, Computer-aided design of catalytic monoliths for automobile emission control, *Chem. Eng. Sci. 43*:2087 (1988).
44. J.E. Bailey, Periodic operation of chemical reactors: A review. *Chem. Eng. Commun. 1*:111 (1973).
45. J.E. Bailey, Periodic phenomena, *Chemical Reactor Theory: A Review* (L. Lapidus and N.R. Amundson, eds.), Prentice-Hall, Englewood Cliffs, NJ, 1977, pp. 758–813.
46. R. Yadav and R.G. Rinker, The efficacy of concentration forcing, *Chem. Eng. Sci. 44*:2191 (1989).
47. M.B. Cutlip, Concentration forcing of catalytic surface rate processes, *AIChEJ. 25*:502 (1979).
48. B.K. Cho, Performance of Pt/Al_2O_3 catalyst in automobile engine exhaust with oscillatory air/fuel ratio. *IEC. Res. 27*:30 (1988).
49. X. Zhou, Y. Barshad, and E. Gulari, CO oxidation on Pd/Al_2O_3. Transient response and rate enhancement through forced concentration cycling, *Chem. Eng. Sci. 41*:1277 (1984).
50. X. Zhou and E. Gulari, CO oxidation on Pt/Al_2O_3 and Pd/Al_2O_3 transient response and concentration cycling studies, *Chem. Eng. Sci. 41*:883 (1986).
51. Y. Barshad and E. Gulari, A dynamic study of CO oxidation on supported platinum, *AIChEJ. 31*:649 (1985).
52. G. Vaporciyan, A. Annapragada, and E. Gulari, Rate enhancements and quasi-periodic dynamics during forced concentration cycling of CO and O_2 over supported Pt–SnO_2, *Chem. Eng. Sci. 43*:2957 (1988).
53. W.R.C. Graham and D.T. Lynch, CO oxidation on Pt: Variable phasing of inputs during forced composition cycling, *AIChEJ. 36*:1796 (1990).
54. A.K. Jain, P.L. Silveston, and R.R. Hudgins, Forced composition cycling experiments in a fixed-bed ammonia synthesis reactor, *Chem. Reaction. Eng. 9*:97 (1982).
55. A. Nappi, L. Fabbricino, R.R. Hudgins, and P.L. Silveston, Influence of forced feed composition cycling on catalytic methanol synthesis, *Can. J. Chem. Eng. 63*:963 (1985).
56. H.A. El Masry, The claus reaction: Effects of forced feed composition cycling, *Appl. Catal. 16*:301 (1985).

57. M.R. Prairie and J.E. Bailey, Experimental and modelling investigations of steady state and dynamic characteristics of ethylene hydrogenation on Pt/Al_2O_3, *Chem. Eng. Sci. 42*:2085 (1987).
58. L. Chiao, F.K. Zack, and R.G. Rinker, Concentration forcing in ammonia synthesis: Plug-flow experiments at high temperature and pressure, *Chem. Eng. Comm. 49*:273 (1987).
59. R.K. Herz and S.P. Marin, Surface chemistry models of carbon monoxide on supported platinum catalysts, *J. Catal. 65*:281 (1980).
60. B.K. Cho and L.A. West, Cyclic operation of Pt/Al_2O_3 catalyst for CO oxidation, *IEC Fundam. 25*:158 (1986).

9

The Use of Monolithic Catalysts for Three-Phase Reactions

Said Irandoust
Chalmers University of Technology, Göteborg, Sweden

Andrzej Cybulski
Polish Academy of Sciences, Warsaw, Poland

Jacob A. Moulijn
Delft University of Technology, Delft, The Netherlands

I. INTRODUCTION

The monolith honeycomb structure is widely used as a catalyst support for gas treatment applications such as the cleaning of automotive exhaust gases and industrial off-gases [1,2]. In these applications, in which large volumetric flows must be handled, monoliths offer certain advantages, such as low pressure drop and high mechanical strength.

In the last 15 years, the use of monoliths has been extended to include applications for performing multiphase reactions. Particular interest has been focused on the application of monolith reactors in three-phase catalytic reactions, such as hydrogenations, oxidations, and bioreactions. There is also growing interest in the chemical industries to find new applications for monoliths as catalyst support in three-phase catalytic reactions.

These applications, with the gas and liquid flowing cocurrently, often require high surface-to-volume ratios such as can be provided by monoliths. Due to the presence of three phases, the interfacial transport is of major importance for the reactor design. Here, the hydrodynamics of the system is an important factor in bringing the phases into mutual contact. Gas-phase species must diffuse through the liquid layer to the catalyst surface where the reaction occurs. Generally, the diffusion rate through the liquid layer is small compared to the rate of reaction. Hence, mass transfer limitations in the gas–liquid interphase, in the liquid film surrounding the catalyst, and finally in the porous catalyst may be of great importance.

Liquid-phase hydrogenations are common processes in both large-and small-scale industry. Here, a great emphasis is generally placed upon the reaction selectivity and conversion. Two conventional reactors used in hydrogenation processes are slurry reactors and trickle-bed reactors. The main features of the monolith reactors are a combination of the advantages of these conventional reactors, avoiding their disadvantages, such as high pressure drop, mass transfer limitations, filtration of the catalyst, and mechanical stirring.

Oxidation of organic and inorganic species in aqueous solutions can find applications in fine chemical processes and wastewater treatment. Here, the oxidant, often either air or pure oxygen, must undergo all the mass transfer steps mentioned above in order for the reaction to proceed. During the last decade, increased environmental constraints have resulted in the application of novel processes to the treatment of waste streams. An example of such a process is wet air oxidation. Here, the simplest reactor design is the cocurrent bubble column. However, the presence of suspended organic and inert solids makes the use of monolith reactors favorable.

The specific features of monolith reactors also attracted the attention of specialists in the field of biotechnology. In general, column reactors, packed-bed reactors, fluidized-bed reactors, and slurry reactors are often used as bioreactors. Monolith reactors are a very attractive alternative for fermentation processes, immobilization of living organisms, etc. They can be used with biocatalysts such as enzymes, microorganisms, and animal cells. For fermentation systems with gas evolution, the use of a monolith as a bioreactor is the most appropriate choice. The packed-bed columns and fluidized-bed reactors are not recommended due to channeling and plugging of liquid flow in the former and to lower conversion caused by back-mixing in the latter.

In this chapter, after a general description of possible flow patterns in monolith channels, the main features and properties of monoliths will be discussed. Following this, the monolith reactor will be compared to some other conventional reactors that are widely used. Next, applications of monolith reactors in catalytic gas–liquid processes will be summarized. Finally, some ideas concerning the future needs in this field will be presented.

II. GENERAL DESCRIPTION OF TWO-PHASE FLOW IN MONOLITHS

Many flow patterns have been described for two-phase, gas–liquid flow through narrow tubes. Figure 1 shows different types of gas–liquid flow in vertical capillaries.

In *bubble flow*, the nonwetting gas flows as small bubbles dispersed in the continuous, wetting liquid. The presence of bubbles in the liquid phase causes increased radial dispersion in the liquid phase within separate channels and better mass transfer as compared to single liquid flow. When large bubbles, of diameter close to that of the tube, separate liquid slugs and some small bubbles are entrained within the liquid slugs, we speak of *slug flow*. *Segmented flow*, which is also called *plug flow* or *Taylor flow* (after Sir Geoffrey Taylor, who studied this type of flow in the 1960s [4]), is characterized by an absence or minor amount of small gas bubbles in liquid slugs separated by gas plugs. The length of the gas plugs is usually greater than the tube diameter. A thin liquid film separates the Taylor bubbles from the wall. Segmented flow can exist in a wide range of liquid and gas loadings. For example, Irandoust et al. [5] observed segmented flow in a 1.5-mm-ID capillary for gas holdups ranging from 20 to 75%. Above a certain liquid loading no gas will mix in, while at low liquid loadings the liquid slugs become too short to keep the flow segmented. At cocurrent downflow with frictional pressure drop balanced by the gravity force, the total superficial velocity is approximately constant and independent of the gas and liquid holdup.

In *annular flow*, the liquid phase is flowing along the wall of a tube as a film while the gas flows as a cylinder in the center of the tube. Some small droplets may be dispersed

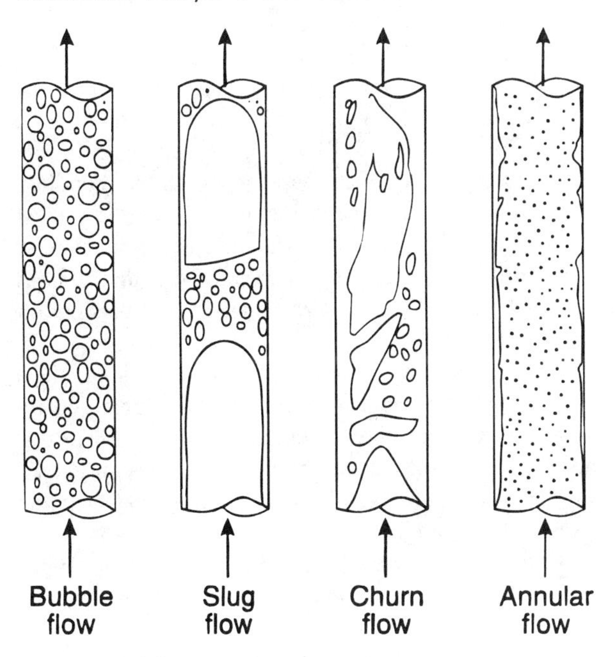

Figure 1 Flow patterns in vertical two-phase flow. (From Ref. 3.)

in the gas phase. The liquid film can be either smooth or wavy. This flow pattern is observed at relatively low liquid loading and high gas holdup. In *droplet flow*, the gas phase is the continuous phase, while the liquid is dispersed as small droplets in the gas phase. *Dispersed flow* is very similar to droplet or annular flow and therefore is referred to as either one of these. Most of the liquid is entrained as liquid drops in the gas phase, but the liquid film covering the wall still exists.

In *churn flow*, the liquid slug is too short to provide a stable liquid bridge between two consecutive gas bubbles. Churn flow is somewhat similar to slug flow. It is, however, chaotic and frothy. The gas bubbles become narrow, and their shape is distorted. The continuity of the liquid in the slug between the consecutive gas bubbles is destroyed by a high local gas concentration in the slug. As a consequence, the liquid slug falls. At a lower level, the liquid reaccumulates and is again lifted by the gas. The alternating motion of the liquid is typical for churn flow.

In *froth flow* the gas and the liquid are intimately mixed as a froth, with the liquid forming very thin films surrounding the gas bubbles. This flow is created by passing the gas through a glass frit. The froth flow is usually unstable but can be stabilized by surfactants.

The difference in flow types caused by the difference in flow direction is small, due to the relatively slight influence of the gravitational force in capillary tubes. Fukano and Kariyasaki [6] have shown that in capillaries, due to the predomination of surface tension force over the gravitational force, intermittent flow types occur in a much wider range of gas and liquid loadings.

The flow pattern in a capillary can change with an increase or decrease of the gas and liquid flow rates. With a given liquid flow rate, increasing the gas velocity will change the flow pattern from bubbly flow, via slug flow and churn flow, to annular flow.

Next, some transitions are discussed in brief.

Annular flow–segmented flow transition. If the flow is initially annular in the intermediate region, the annular ring may grow in thickness while it flows further along the tube. Eventually, the ring can become so thick that it fills the cross section, resulting in the split of a gas bubble. Depending on the length of the capillary or on the monolith, transition from annular flow to segmented flow may occur. Figure 2 shows the typical transition areas between annular, segmented, and bubble flow. The transition from annular flow to segmented flow will occur when the liquid loading increases. Note that besides the transition area, another area exists where the flow pattern obtained is dependent on the method of introducing the liquid into the tube.

Bubble flow–slug flow transition. Transition from dispersed bubbles to slugs requires a process of coalescence. As the gas flow rate is increased, the bubble density increases. This closer bubble spacing results in an increase in agglomeration. However, increased liquid flow rate can cause a breakup of larger bubbles, and this might be sufficient to prevent the transition. The maximum bubble void fraction at which the transition happens is around 0.25 (see Refs. 5 and 3).

Slug flow–segmented flow transition. Taylor bubbles will be formed when the gas flow rate is increased to such an extent that it forces bubbles to become closely packed and to agglomerate into Taylor bubbles.

Slug flow–churn flow transition. As the gas flow is increased even more, a transition to churn flow occurs. The subjective discrimination between slug flow and churn flow makes it difficult to identify the transition exactly. Taitel et al. [3] use the definition that is based on the behavior of the liquid film between the Taylor bubble and the wall. In this case the churn flow is characterized as the condition where oscillatory motion of the liquid is observed.

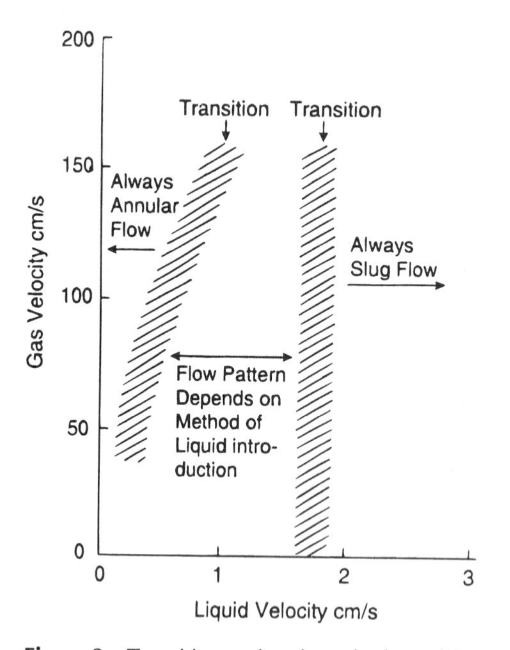

Figure 2 Transition regime in a single capillary. (From Ref. 7.)

Churn flow–segmented flow transition. This transition is an entry region phenomenon associated with the existence of slug flow further along the pipe, and is described by Taitel et al. [3] as follows:

> At the inlet the gas and liquid which are introduced form liquid plugs and Taylor bubbles. A short liquid plug is known to be unstable, and it falls back and merges with the liquid slug coming from below, causing it to approximately double its length. In this process, the Taylor bubble following the liquid slug overtakes the leading Taylor bubble and coalesces with it as the slug between the two bubbles collapses. This process repeats itself, and the length of the liquid slugs as well as the length of the Taylor bubbles increase as they move upward, until the liquid slug is long enough to be stable and form a competent bridge between consecutive Taylor bubbles.

Churn flow or slug flow–annular (dispersed) flow transition. The flow becomes annular when the gas flow rates are enhanced to a certain point. The high gas flow rate causes a wavy interface of the liquid film. As a consequence, parts of the waves will enter the gas core as entrained drops. This results in an upward flow direction of the liquid, due both to interfacial shear and to drag on the waves and drag on the droplets. Annular flow can therefore exist only when the gas velocity is sufficient to lift the droplets in the gas core. The minimum gas velocity required to suspend a drop is determined from the balance between the gravity and drag forces on the drop.

The flow pattern may change with length. Satterfield and Özel [7] observed that as bubbles and slugs flow down, a liquid ring frequently appears within a bubble. The ring grows in thickness and finally fills the complete cross section of a capillary, thus splitting the bubble in two.

III. FEATURES OF MONOLITH REACTORS IN CATALYTIC GAS–LIQUID REACTIONS

A monolithic support consists of a large number of narrow, parallel channels separated by thin walls. Unlike the situation in conventional packed-bed reactors, the monolith channels have a well-defined geometry. The channels may have a variety of cross-sectional shapes, such as square, sinusoidal, circular, triangular, and hexagonal. The walls may contain the catalytically active material, but more frequently a washcoat consisting of a thin layer of a porous oxide is deposited onto the channel wall. Owing to its porosity it has a large surface area on which the catalytically active material is fixed. The open structure of the monolith allows high flow rates with low pressure drop.

Since there is no radial bulk transport of fluid between the monolith channels, each channel acts basically as a separate reactor. This may be a disadvantage for exothermic reactions. The radial heat transfer occurs only by conduction through the solid walls. Ceramic monoliths are operated at nearly adiabatic conditions due to their low thermal conductivities. However, in gas–liquid reactions, due to the high heat capacity of the liquid, an external heat exchanger will be sufficient to control the reactor temperature. Also, metallic monoliths with high heat conduction in the solid material can exhibit higher radial heat transfer.

In catalytic gas–liquid reactions, the gas and liquid are fed into one end of the monolith and flow cocurrently either upwards or downwards in one of several flow regimes. The performance of a monolith reactor is highly dependent upon the prevailing flow pattern in the monolith channels. The desired flow pattern through the monolith channels is segmented flow. This flow pattern consists of liquid slugs well separated from each other by distinct gas slugs. Such flow provides a thin liquid film between the gas slugs and the channel walls and a good recirculation within the liquid slugs. The radial mixing within the liquid slugs and the thin liquid film with a large surface area increase the mass transfer from the gas slugs to the surface of the catalyst. Due to the very thin liquid film in this flow type, the axial dispersion is very low. More details on the hydrodynamics of this flow type can be found in Chapter 10.

Optimum performance of the monolith reactor requires uniform and stable distribution of gas and liquid over the cross section of the monolith. Because the monolith consists of many small channels, it may be difficult to obtain a good distribution of the gas and liquid flows within the monolith. This is very important for the monolith reactor, since an uneven inlet distribution would be propagated throughout the reactor. On the other hand, if the inlet distribution is appropriate, no nonuniformity will occur along the reactor.

Internal and external mass transfer resistances are important factors affecting the catalyst performance. These are determined mainly by the properties of the fluids in the reaction system, the gas–liquid contact area, which is very high for monolith reactors, and the diffusion lengths, which are short in monoliths. The monolith reactor is expected to provide apparent reaction rates near those of intrinsic kinetics due to its simplicity and the absence of diffusional limitations. The high mass transfer rates obtained in the monolith reactors result in higher catalyst utilization and possibly improved selectivity.

One unique feature of the monolith reactor is the possibility of having an internal recirculation of the gas flow without the use of a pump [5,8–10]. This self-recirculation is possible due to the very low net pressure drop across the monolith. In a monolith reactor with downflow operation in slug flow regime, the fluids are not driven through the channels by an external pressure, but pulled through by gravity. This corresponds to a total superficial velocity of about 0.45 m sec^{-1}. When liquid is added to the channel at a lower rate, gas will be entrained to make the total velocity 0.45 m sec^{-1}.

The monolith reactors offer some additional operational advantages, such as higher tolerance against bed plugging and the ease of reactor design. Scaling up of monolith reactors with slug flow is expected to be straightforward. As already mentioned, the only critical part of the scaling up is the inlet flow distribution [5].

In summary, the main advantages of using monolith reactors in catalytic gas–liquid reactions are as follows:

Very low pressure drop
Excellent mass transfer properties
High surface/volume ratio
Short diffusion distances
Low axial dispersion (with segmented flow)
Good contact areas between the phases involved
Ease of reactor scale-up

The main drawbacks of monolith reactors are:

Higher catalyst manufacturing cost
Short residence time
Poor heat transfer
Difficulty of distributing the fluid uniformly over the reactor cross section

The balance of advantages and drawbacks of the monolith reactors is positive, making this reactor type very attractive for applications in multiphase processes. The modeling of monolith reactors and some concepts for reactor design are presented in Chapter 10.

IV. COMPARISON BETWEEN MONOLITH REACTORS AND SOME CONVENTIONAL REACTORS

Of primary interest for the industrial application of monolith reactors is to compare them with other conventional three-phase reactors. Two main categories of three-phase reactors are *slurry reactors*, in which the solid catalyst is suspended, and *packed-bed reactors*, where the solid catalyst is fixed. Generally, the overall rate of reactions is often limited by mass transfer steps. Hence, these steps are usually considered in the choice of reactor type. Furthermore, the heat transfer characteristics of chemical reactors are of essential importance, not only due to energy costs but also due to the control mode of the reactor. In addition, the ease of handling and maintenance of the reactor have a major role in the choice of the reactor type. More extensive treatment of conventional reactors can be found in the works by Gianetto and Silveston [11], Ramachandran and Chaudhari [12], Shah [13,14], Shah and Sharma [15], and Trambouze et al. [16], among others.

A. Monolith Reactors Versus Slurry Reactors

Slurry reactors are widely used in the chemical process industry due to their superior mass transfer characteristics. Catalytic hydrogenation of unsaturated fatty oils and catalytic oxidation of olefines are among practical examples in which slurry reactors are utilized.

The catalyst particles used in slurry reactors are usually quite small, providing very short diffusional distances within the catalyst. The values of the gas–liquid volumetric mass transfer coefficient, $k_L a$, range between 0.01 and 0.6 sec^{-1} in slurry reactors [12]. The corresponding values in a monolith reactor range between 0.05 and 0.30 sec^{-1} [17]. It should be noticed that most of the studies of $k_L a$ in slurry reactors have been performed on gas–liquid systems with no solid particles. The presence of solid particles may have a negative effect on the gas–liquid mass transfer. According to Schöön [18], the values of volumetric liquid–solid mass transfer coefficient, $k_s a_s$, usually range from 1 to 4 sec^{-1} for slurry reactors; the corresponding values for the monolith reactors lie between 0.03 and 0.09 sec^{-1} [19]. However, the difference in $k_s a_s$ between slurry and monolith reactors is leveled out somewhat, since the kinetically slow step in a slurry process is generally the mass transfer across the liquid film around the gas bubbles [18].

When the heat transfer is considered, the slurry reactors are more efficient, due to large liquid holdup and a relatively high flow velocity of the reaction mixture at the heat exchange surface. Also, it is relatively easy to arrange heat-exchanging devices in the slurry reactors as compared to monolith reactors.

The major disadvantages of using slurry reactors are the problems connected with agitation, filtration of solid catalyst, and reactor scale-up. Although slurry reactors provide the possibility of a rapid replacement of the decayed catalyst, the use of monolith reactors will eliminate costly catalyst recovery steps in industrial operation. The recovery steps are often a hazard when the catalyst is pyrophoric.

The scale-up of monolith reactors is expected to be much simpler. This is due to the fact that the only difference between the laboratory and industrial monolith reactors is the number of monolith channels, provided that the inlet flow distribution is satisfactory. In slurry reactors, scale-up problems might appear. These are connected with reactor geometry, low gas superficial velocity, nonuniform catalyst concentration in the liquid, and a significant back-mixing of the gas phase.

Among other drawbacks of slurry reactors are the high axial dispersion and the low solid fraction that can be held in suspension. Concerning the flow capacity, it should be emphasized that the slurry reactors are less sensitive to gas flow rates, whereas in monolith reactors the gas flow rate is limited by the restrictions of slug flow.

The comparison between slurry and monolith reactors is summarized in Table 1. Based on the known features of slurry and monolith reactors, it can be concluded that the slurry reactors are preferable for mass-transfer-limited processes as far as the overall process rates are concerned. However, due to the low concentration of solid catalyst in slurry reactors, the productivity per unit volume in these reactors is not necessarily higher than that of monolith reactors. For processes occurring in kinetic regime, the monolith reactors are preferable due to their easier operation. The productivity of slurry reactors might apparently be increased by increasing the catalyst concentration. However, suspensions with a high concentration of fine catalyst particles behave as non-Newtonian liquids, with all the negative consequences in heat and mass transfer.

B. Monolith Reactors Versus Packed-Bed Reactors

Another type of multiphase catalytic reactor is the packed-bed reactor, where the catalyst particles constitute a stationary bed. The way of introducing the gas and liquid reactants into the reactor categorizes the different types of packed-bed reactors. These reactors are

Table 1 Comparison Between Typical Slurry, Trickle-bed, and Monolith Reactors for Catalytic Gas–Liquid Reactions

Property	Slurry*	Trickle-bed	Monolith
Particle/channel diameter (mm)	0.01–0.1	1.5–6.0	1.1–2.3
Volume fraction of catalyst	0.005–0.01	0.55–0.6	0.07–0.15
External surface area (m^2/m^3)	300–6000	600–2400	1500–2500
Diffusion length μm	5–50	100(shell)–3000	10–100
Superficial liquid velocity (m sec^{-1})			
Test reactor	—	0.0001–0.003	0.1–0.45
Full scale	—	0.001–0.02	0.1–0.45
Superficial gas velocity (m sec^{-1})			
Test reactor	—	0.002–0.045	0.01–0.35
Full scale	0.03–0.5	0.15–3.0	0.01–0.35
Pressure drop (kPa m^{-1})	~6.0	50.0	3.0
Volumetric mass transfer coefficient (sec^{-1})			
Gas–liquid, k_La	0.01–0.6	0.06	0.05–0.30
Liquid–solid, k_sa_s	1–4	0.06	0.03–0.09

* Catalyst load 1%, stirrer rate 500–1000 rpm.

commonly used in catalytic hydroprocessing and industrial hydrogenation and oxidation reactions.

The most common type of packed-bed reactor is the trickle-bed reactor with liquid and gas downflow. Here, the gas constitutes the continuous phase, while the liquid flows in the form of a liquid film over the solid particles. In packed-bed reactors, it is possible to attain high volumetric-catalyst loads. A major disadvantage of the conventional packed-bed reactors, however, is the high pressure drop in the reactor. In monolith reactors, the pressure drop is up two orders of magnitude lower than in packed-bed reactors. A high pressure drop will result in both high pumping-energy costs and nonuniform partial pressure of gaseous reactant, affecting the reactor performance. The pressure drop in the reactor can be reduced by using larger catalyst particles. This is at the expense of intraparticle diffusion limitations. Bed-plugging is also encountered in conventional packed-bed reactors, leading to loss in production capacity.

The flow distributions over the cross section of the reactor can also be a problem in packed-bed reactors. An additional source of trouble is connected with the flow maldistribution inside the packed-bed reactors caused by inhomogeneity of packing. In monoliths, however, the flow within the channels is stable, provided that the flow is properly distributed at the reactor inlet.

The scale-up of packed-bed reactors is a difficult task due to the possibility of obtaining poor catalyst wetting, channeling, and bypassing. Again, scaling-up of monolith reactors has proved to be simpler [5]. It implies just a multiplication of the numbers of channels.

The heat transfer problems in packed-bed reactors are connected with poor radial mixing. These problems are more pronounced in packed-bed reactors with low liquid rates. In such cases, the catalyst will not be completely wetted, resulting in hot spots and in some cases temperature runaway effects.

Other known disadvantages of packed-bed reactors are low external and internal mass transfer rates. For trickle-bed reactors, a representative value for both k_La and k_sa_s is 0.06 sec^{-1} [18]. As was pointed out previously, the corresponding values for monolith reactors are higher due to the enhanced radial mass transfer in liquid slugs and to shorter diffusion length in both the liquid film and the solid catalyst.

The optimal operating conditions of a monolith reactor will be different from those of a packed-bed reactor, due primarily to the differences in flow pattern. Comparing the flow capacities of the monolith and trickle-bed reactors (Table 1), we find that the liquid flow capacity is about the same for the two reactor types. Trickle-bed reactors, however, can be operated at much higher gas velocities than monolith reactors. In the monoliths, the requirement of obtaining slug flow limits the use of the higher gas velocities required by the reaction. The problem can be overcome by sectioning the monolith and allowing for a free access of gaseous reactant to each section.

A comparison between monolith and trickle-bed reactors is shown in Table 1. The superiority of the monolith reactor over the conventional packed-bed reactor is mainly due to much lower pressure drop, the ease of scaling up, and higher mass transfer rates in the former option.

V. APPLICATION OF MONOLITH REACTORS IN THREE-PHASE PROCESSES

There are three main fields of (potential) applications of monolithic catalysts/reactors: hydrogenations, oxidations, and bioprocesses. Pioneers in studying monolithic processes (hydrogenations, mass transfer and flow phenomena) were scientists from the Chalmers University of Technology, Sweden. Extensive studies on monoliths have also been performed at the Delft University of Technology in The Netherlands (hydrogenations), the University of Tulsa in Oklahoma, U.S.A. (oxidations, two-phase flow through narrow channels), and Kyushu University in Fukuoka, Japan (bioprocesses).

A. Hydrogenations

Hydrogenations have been the most studied gas/liquid reactions in monolith reactors. The main difference between these hydrogenations was the degree of influence of the mass transfer steps. Research on this subject was started at the Chalmers University of Technology, using monoliths with channels of sinusoidal shape.

Hatziantoniou and Andersson [20] studied hydrogenation of nitrobenzoic acid in a palladium-impregnated monolith in the segmented flow regime with downflow of reactants. The monolith was made of a mixture of glass, silica, and minor amounts of other oxides reinforced by asbestos fibers. Channels had a cross-sectional area of 2 mm^2. The monolith was impregnated with palladium chloride, dried and reduced to give 2.5 wt.% Pd. The reaction was studied at 309–357 K and 0.116–0.42 MPa and at flow rates of both phases of 0.03–0.051 m sec^{-1}. The effectiveness factor for the monolithic catalyst ranged from 0.081 to 0.115, whereas for 5-mm spheres in a trickle bed it amounted to 0.021–0.024. This can be attributed to the shorter diffusion length in the monolith (<0.15 mm) than in pellets (2.5 mm in a typical trickle bed). For the whole trickle-bed reactor, the effectiveness factor is lower still, due to the more restricted access of reactants to some parts of the catalytic surface.

Hatziantoniou et al. [21] carried out hydrogenation of nitrobenzene and *m*-nitrotoluene in a monolith reactor with segmented flow. They used the same monolithic carrier, but the catalyst was richer in palladium: 5.3 wt.% Pd. Experiments were performed at 346–376 K and 0.59–0.98 MPa and at velocities of both phases of 0.017–0.042 m sec^{-1}. Slurry hydrogenation was also performed, with the aim to provide data for the evaluation of mass transfer coefficients in the monolith. The authors concluded that the dominating mass transfer step for hydrogen is the direct transport from the gas plugs to the channel wall. In one experiment, this mode corresponded to 70% of the total hydrogen transported. The volumetric mass transfer coefficient ranged from 0.00014 to 0.00023 m sec^{-1}.

Irandoust and Andersson [22] studied hydrogenation of 2-ethylhexenal in a monolith. Slurry experiments with powdered monolithic catalyst were also performed. The reaction was investigated at 413–433 K and 0.4–0.98 MPa and at velocities of both phases of 0.023–0.085 m sec^{-1}. The monolithic process proceeded much faster, as is shown in Fig. 3. This large difference might be due to deactivation of the catalyst during grinding.

Competitive hydrogenation of thiophene and cyclohexene was the subject of research by Irandoust and Gahne [23]. A monolith made of γ-Al_2O_3 of cell density 31 cells/cm^2 with channels of cross-sectional area 1.7 m^2 was used. Molybdenum and cobalt were subsequently incorporated into the monolithic structure to give 12.3 wt.% MoO_3 and 4.6 wt.% CoO, i.e., as in a commercial pelleted catalyst. A reactor 25.4 mm in diameter was operated at 509–523 K and 3–4 MPa. The average linear velocity of both gas and liquid ranged from 0.0175 to 0.035 m sec^{-1}. The kinetic parameters were found to be in good agreement with the data reported in the literature. The effect of diffusion limitations was evaluated to be negligible on the basis of the Weisz modulus, which amounted to 0.32. Thus, the monolith reactor is an excellent alternative to hydrotreating in trickle-bed

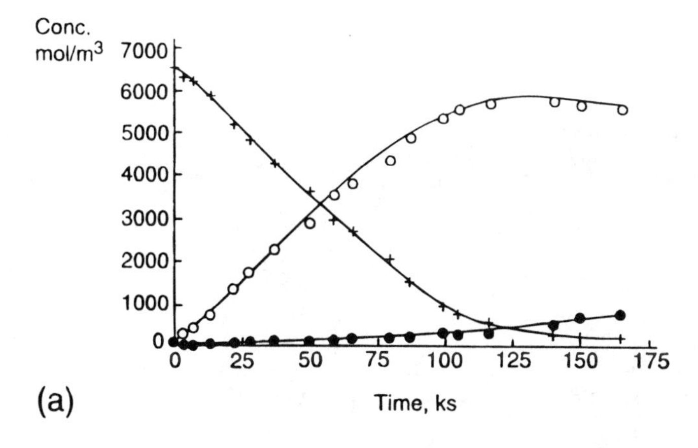

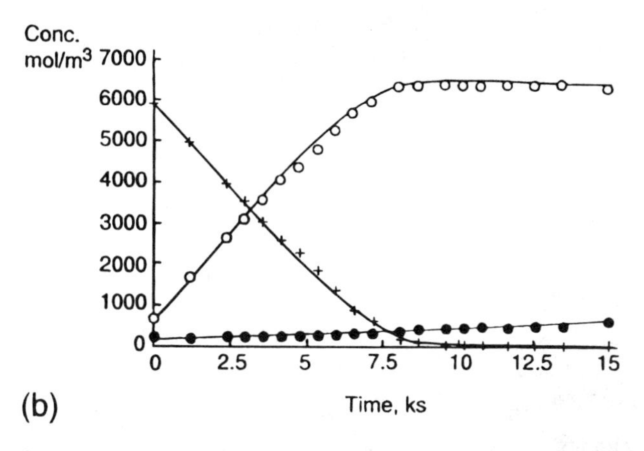

Figure 3 Hydrogenation of 2-ethylhexenal: 433 K, 0.98 MPa. (a) Slurry reactor; (b) monolith reactor. + 2-ethylhexenal; ○ 2-ethylhexanal; ● 2-ethylhexanol. (From Ref. 22.)

reactors. Edvinsson and Irandoust [24] used the same monolithic Co–Mo catalyst to perform hydrodesulfurization (HDS) of dibenzothiophene (DBT). The reactor was operated at 543–573 K and 6–8 MPa. They estimated kinetic parameters in a Langmuir–Hinshelwood model. The rate of HDS of DBT was approximately one-fourth that of thiophene.

Soni and Crynes [25] studied hydrodesulfurization and hydrodenitrification of raw anthracene oil and Synthoil liquid in monolithic and trickle-bed reactors. Co–Mo monolithic catalyst specimens 25.4 mm long and 1 cm in diameter with 3.37 wt.% CoO and 7.25 wt.% MoO_3 were tested. Monoliths were stacked with redistributors in between, to a total length of 35.5 cm. The monolithic through-reactor was operated with downflow at 544 K and 10.6 MPa. The reactors were compared on the basis of unit surface area. Results of some runs are presented in Fig. 4. The monolithic catalyst activity was higher than that for pellets. This was probably due to the differences in average pore radii and the intraparticle diffusion length; for monoliths these quantities were 8.0 nm and 0.114 mm, versus 3.3 nm and 1 mm for pellets. Effectiveness factors for the monolith and pellets were evaluated to be 0.94 and 0.216, respectively.

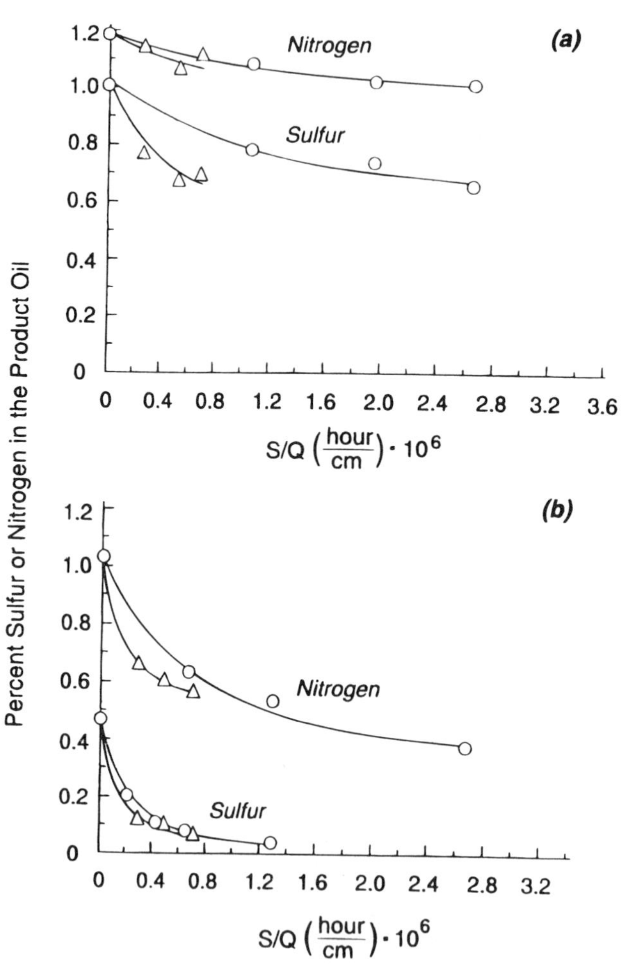

Figure 4 HDS and HDN reaction extent as a function of surface area/volumetric rate of oil: 643 K, 10.2 MPa. (a) Anthracene oil; (b) Synthoil liquid. ○ pellets, △ monolith. (From Ref. 25.)

Another study dealing with hydrotreatment using monolithic catalysts was carried out by Scinta and Weller [26]. These authors studied the hydrodesulfurization and liquification of a high-sulfur coal in batch autoclave using tetralin as a solvent. Three different configurations of Co–Mo–γ-Al_2O_3 monolithic catalysts, 31 and 46 square channels/cm^2 and 36 triangular channels/cm^2, were tested. The monoliths examined had a pore diameter of 22.5–25 nm and a composition of 3 wt.% CoO and 15 wt.% MoO_3. These monoliths were placed in a special holder that was mounted in the stirrer shaft. The monolith channels were parallel to the direction of stirring. It was concluded that the most favorable monolith was the one with 31 square cells/cm^2, increasing the fraction of oil, decreasing the yields of asphaltene and gas, and increasing the consumption of hydrogen. All catalysts tested reduced the sulfur in the oil and asphaltene. The oil showed lower levels of sulfur and nitrogen and higher H/C ratios than the asphaltenes, indicating that the conversion of asphaltene to oil involves C-S and C-N bond breakages.

Mazzarino and Baldi [27] investigated hydrogenation of α-methylstyrene into cumene, using a ceramic monolith coated with Pd (1% of active metal). They studied the process for upflow and downflow of reactants. Temperature ranged from 303 to 323 K, and partial hydrogen pressure was 0.02–0.1 MPa. The liquid flow rate was varied from 0.0005 to 0.0034 m sec^{-1}, and the gas flow rate was up to 0.0012 m sec^{-1}. The authors found a better monolith performance with upflow of reactants. The sensitivity of the process to the liquid flow rate was rather slight, while that to the gas flow rate was significant. Comparison between reaction rates in the monolith and the trickle bed is shown in Fig. 5. It follows from this graph that the monolithic catalyst performs better than pellets in a higher range of gas flow rate.

Otten [28], van der Riet et al. [29] and Xu et al. [30] studied hydrogenation of benzaldehyde to benzyl alcohol over nickel and palladium, alumina- and carbon-wash-coated cordierite monolithic catalysts of cell density 62 cells/cm^2. They used a gradientless reactor of the Berty type [31]. It was operated batchwise with respect to the liquid phase and semibatch as far as hydrogen is concerned: The gas was supplied at the rate it was consumed. The process was investigated at temperatures ranging from 390 to 450 K and under pressure from 11 to 21 bar. Flow rates of both phases were impossible to determine. In any case, the speed of rotation of a stirrer circulating the reaction mixture was high enough to make external mass transfer negligible. As expected, the activity of palladium catalysts on alumina washcoat was higher (up to ten times more) than for nickel catalysts—the ratio of rates varied with the overall conversion of benzaldehyde, decreasing with the increase of the conversion. The activity of catalyst that incorporated palladium into the layer of carbon/alumina deposited on ceramic substrate was more than two times higher than for palladium/alumina-washcoated ceramic monolith.

The activity of nickel monolithic catalyst was compared with that of commercial particulate Engelhard Ni-catalyst Ni-707 of size 3.2 mm. Both reactions were performed at the same process conditions (420 K, 16 bar, 1250 RPM, initial amount of benzaldehyde = 1.37 mol, the same amount of nickel, 35.5 g of NiO/Al_2O_3 in samples investigated).

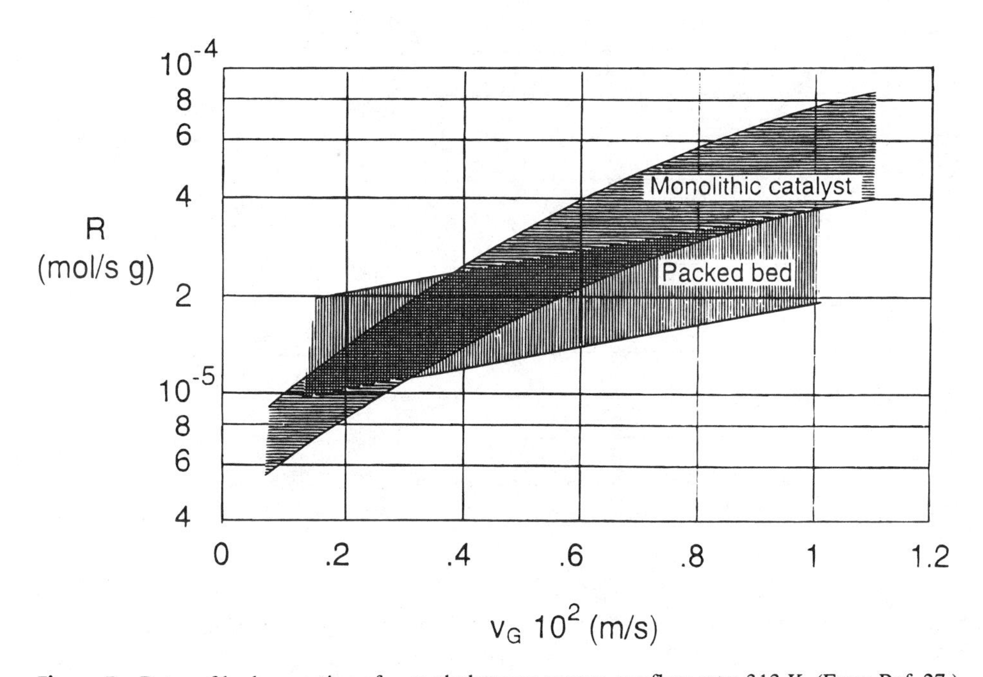

Figure 5 Rates of hydrogenation of α-methylstyrene versus gas flow rate: 313 K. (From Ref. 27.)

Both catalysts were activated at the optimum conditions determined using TPR. The rates at the maximum selectivity to benzyl alcohol were compared. In the presence of particulate catalyst, the rate amounted to 0.00095 mol/($g_{Ni,t}$ · min), while for monolithic catalyst the rate was approximately 0.00175 mol/($g_{Ni,t}$ · min), i.e., about two times more. The diffusion length in the nickel monolith is much shorter than in the 3.2-mm particles. This resulted in a lower effectiveness factor for the nickel pellets, and hence in a lower reaction rate. Selectivity of both catalysts with respect to benzyl alcohol was nearly the same, at least within the precision of analytical methods used: 94.9% for pelleted catalyst and 95.1% for monolithic catalyst. We may therefore conclude that the selectivity is not controlled by internal diffusion but by the surface properties of the catalysts.

Preliminary kinetic studies have been performed. The Langmuir–Hinshelwood rate expression was used to correlate results of experiments as it was indicated by the shape of kinetic curves (see Fig. 6). However, the reaction order with respect to hydrogen appeared to be dependent on temperature, while activation energy depends on pressure (9.6 kJ/mol at 11 bar and 35.5 kJ/mol at 21 bar). Therefore the rate of benzaldehyde consumption was approximated using the following simple power law equation:

$$r_{\text{benzaldehyde}}(\text{mol } g_{\text{Me}}^{-1} \text{ sec}^{-1}) = 0.000167 \cdot e^{(-3000/T)} \cdot P^{1.7}$$

(with P in bars), which allows for only the rough evaluation of rates at various process conditions. Clearly, more kinetic studies are needed to find the proper rate equation.

An extensive investigation was carried out on the potential of monolith reactors in competitive hydrogenation of mixtures containing alkenes, alkadienes, aromatics, and functionalized aromatics such as styrene. At present, hydrogenation of such mixtures is performed in trickle-bed reactors using Pd/Al_2O_3 or Ni/Al_2O_3 catalysts. Because intrinsic hydrogenation rates are very high, intraparticle diffusion plays an essential role in the

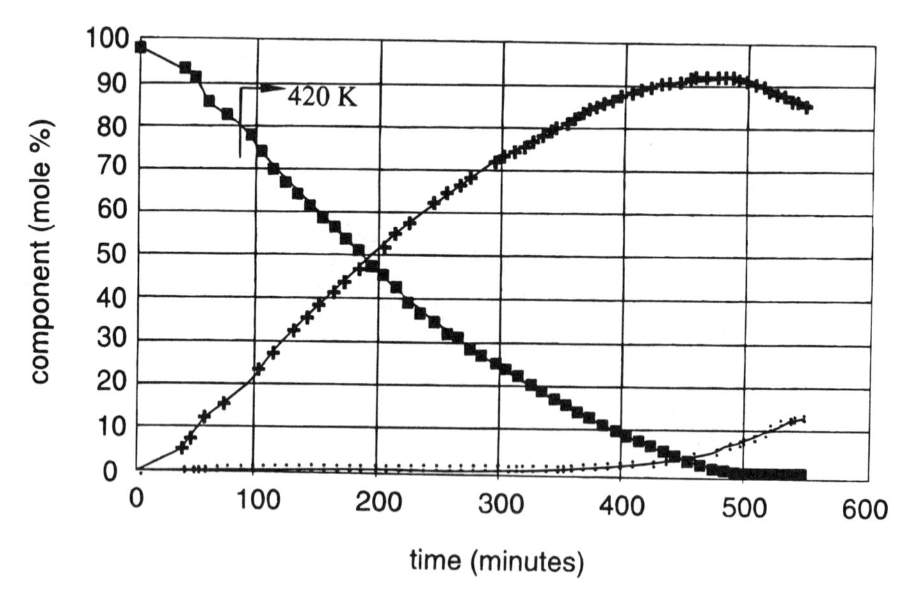

Figure 6 Hydrogenation of benzaldehyde over Ni-monolithic catalyst: 420 K, 16 bar, 1250 RPM. ■ benzaldehyde, + benzyl alcohol, : toluene. (From Ref. 28.)

process. Therefore monolithic catalysts seemed to be promising in application to this process.

To assess the potential of monolith reactors, model experiments have been performed. A typical example is the competitive hydrogenation of styrene and 1-octene in toluene over a monolithic Pd catalyst [32,33]. The palladium catalyst was chosen as being the most active in this hydrogenation [34], although other metals are also of interest because of better stability and resistance to poisons. The experiments were carried out in a bench scale setup, shown in Fig. 7. With respect to the unsaturated hydrocarbons it operates in the batch mode, whereas the hydrogen pressure is kept constant and in that sense it is a semibatch reactor. The operating conditions were in the range of 40–60°C and 6–15 bar; the liquid-to-gas ratio was varied between 1:1 and 1:6, and the superficial velocity between 0.5 and 1.25 m sec^{-1}. The catalyst layer had a thickness of 15 μm. The reactions occurring are given in Fig. 8. Besides hydrogenations, isomerization reactions take place. The desired reactions are indicated by the bold arrows, i.e., the production of ethylbenzene and internal olefins. The formation of octane and the hydrogenation of the aromatic rings are not desired.

Typical results are given in Fig. 9. It is clear that styrene reacts away at the highest rate; simultaneously with the disappearance of styrene, ethylbenzene is formed. The subsequent reaction of ethylbenzene into ethylcyclohexane does not proceed at a measurable rate. The isomerization of 1-octene into internal olefins is slower than the hydrogenation of styrene but faster than the hydrogenation into octane. The serial reaction network for the isomerizations is clear from Fig. 10. They all go through a maximum in agreement with serial kinetics. As expected, the maxima occur in the sequence 2-octene < 3-octene < 4-octene.

The results can be interpreted in terms of Langmuir–Hinshelwood–Hougen–Watson kinetics. Styrene adsorbs more strongly than octenes and, as a consequence, only after styrene has been converted does the formation of octanes proceed at a high rate. The

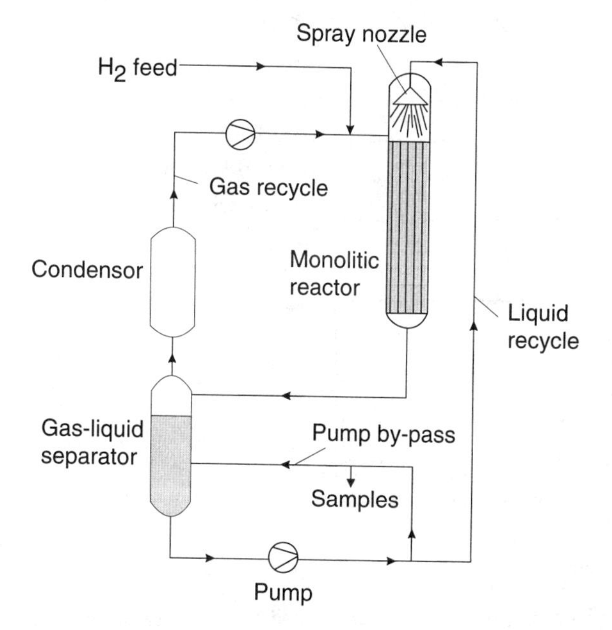

Figure 7 Test equipment for studying monolithic hydrogenations. (From Ref. 33.)

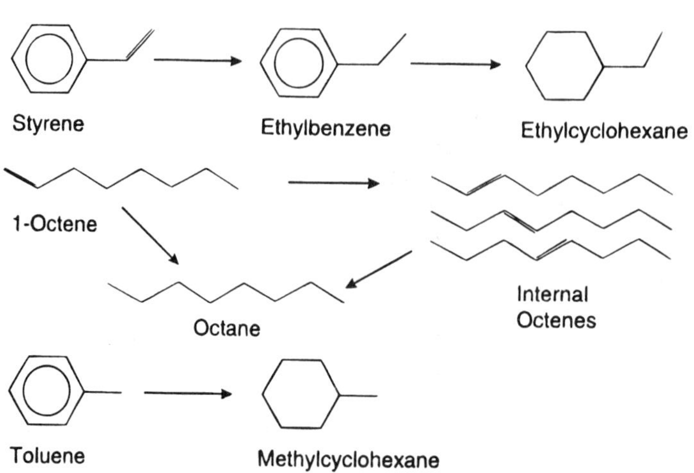

Figure 8 Reaction network of hydrogenation of styrene/1-octene mixtures. (From Ref. 33.)

selectivity for styrene conversion is evaluated by a comparison with the conversion of octenes; see Fig. 11. The straight line in the figure refers to the case of selectivity independent of conversion: Both styrene and octenes react away in the same proportion. In reality, a high selectivity exists, as is shown in the figure. A large part of the conversion of octenes takes place only when the conversion of styrene is over 90%. The selectivity to internal octenes is evaluated as a function of the conversion of styrene in Fig. 12. Rather good selectivities are obtained. Conversion of aromatic ring was always (much) less than 0.2%.

The results showed that the activity is improved by increasing the linear velocities and liquid-to-gas ratios. The absence of internal diffusion limitations was checked by means of the Weisz–Prater criterion. It appeared that internal gradients were negligible.

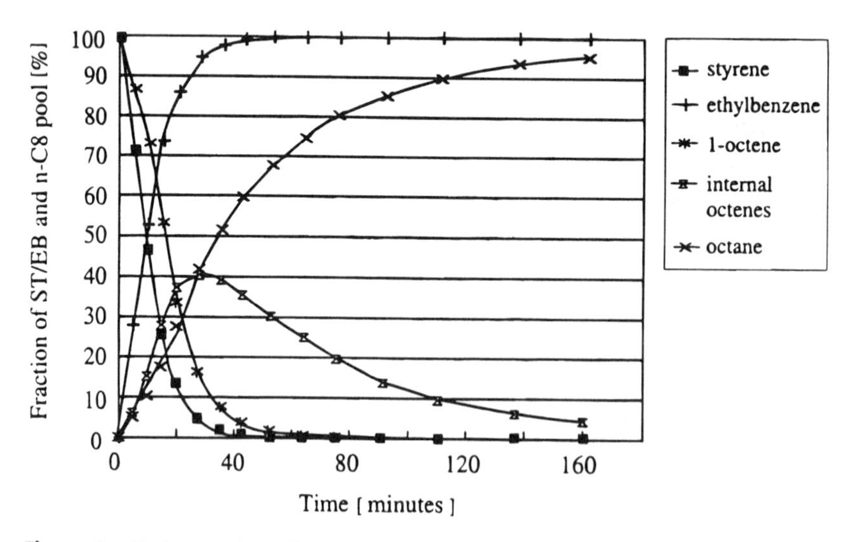

Figure 9 Hydrogenation of styrene/1-octene mixtures: a text case; concentrations of styrene, ethylbenzene, 1-octene, octane, and internal octenes versus time. (From Ref. 33.)

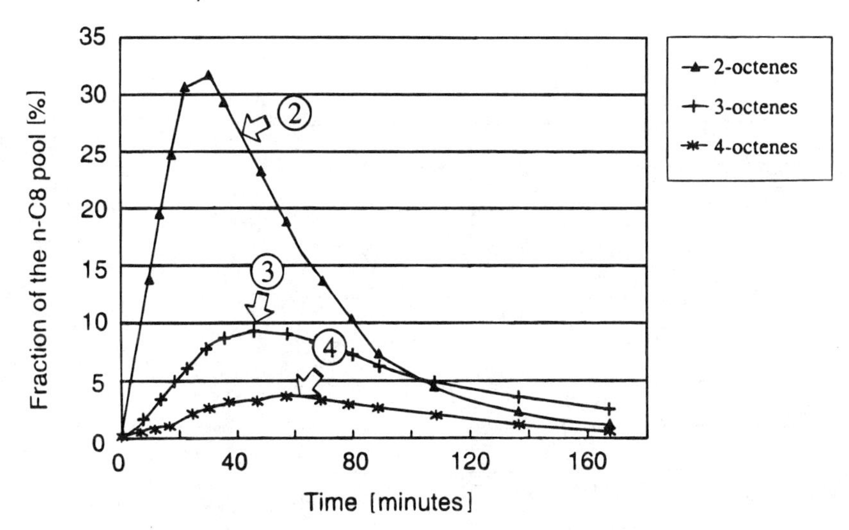

Figure 10 Hydrogenation of styrene/1-octene mixtures: a test case; concentrations of 2-octenes, 3-octenes, and 4-octenes versus time. (From Ref. 33.)

However, external mass transfer of hydrogen might play a role. An important conclusion was that the liquid distribution over the individual channels was not uniform. Of course, in the regular monolithic structure an initial maldistribution will be preserved throughout the reactor. Consequently, the conditions at the inlet of the reactor are critical. In this case nonuniform wetting of the channels was concluded to be the reason for the increase of process rate with increasing flow rate. In spite of this nonideality, the reactor performed satisfactorily: Rates and selectivities were comparable to or higher than those reported

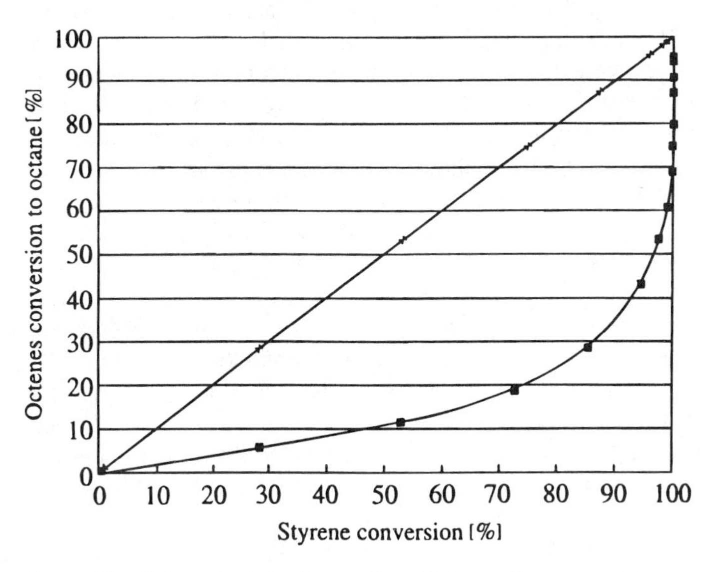

Figure 11 Selectivity in hydrogenation of styrene/1-octene mixtures: a test case; octenes versus styrene conversion (aromatics hydrogenation always much less than 0.2%). (From Ref. 33.)

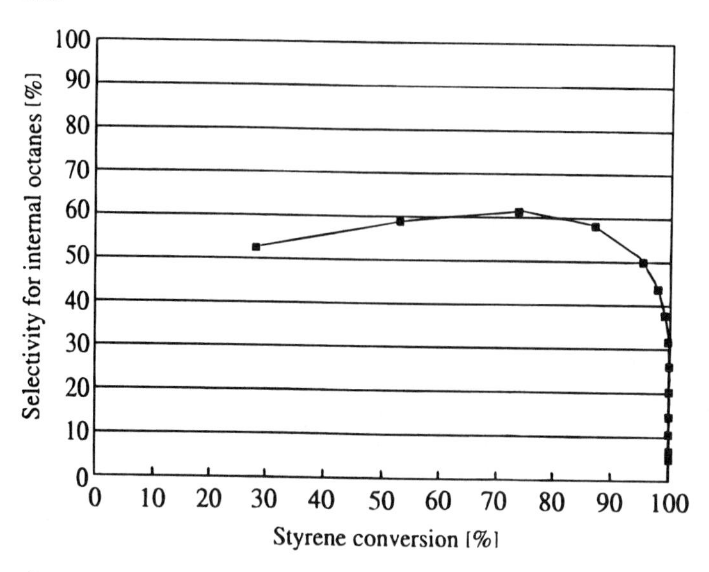

Figure 12 Selectivity in hydrogenation of styrene/1-octene mixtures: a test case; selectivity of internal octenes versus styrene conversion. (From Ref. 33.)

for conventional trickle beds. It was concluded that monolith reactors are very promising, the more so because considerable improvements can be expected when further optimizations—in particular, feed distribution and dedicated catalyst development—have been realized.

Edvinsson et al. [35] studied catalytic hydrogenation of acetylene in the presence of heptane, which was used to remove continuously the green oil formed on the catalytic surface by polymerization. They used monolith supports made of γ-alumina sintered to α-alumina and cordierite washcoated with α- or γ-alumina. Palladium was incorporated with the final content of 0.04 wt.% and metallic dispersion of 14%. At first all monoliths were studied under nonprocess conditions. Based on the results of those investigations, the catalyst on α-alumina support was chosen for studying the process. A monolith of 22 mm in diameter and 40 mm in length was used. The process of hydrogenation of mixtures of 3% acetylene, 28% ethylene, 6 or 11% hydrogen, and nitrogen (to the balance) was studied at 303–313 K and 1.3–2 MPa. The gas flow rate was up to 33.3 $cm^3\ sec^{-1}$ (at NTP, 273 K, and 1 atm), while the liquid flow rate was up to 3.3 $cm^3\ sec^{-1}$. As shown in Fig. 13, both activity and selectivity of the catalyst vary with time on stream: Selectivity drops by approximately 30% with the increase of acetylene conversion from approximately 20 to about 90 mole%. The highest selectivity was for α-alumina supported catalyst with an average pore diameter of 0.08 μm. The higher selectivity found is attributed to the relatively large pores and accordingly better mass transfer. The authors also studied aging of the catalyst. Activity and selectivity decreased during the first 60 hr and then leveled out. This study was extended by Asplund et al. [36], who utilized the same reactor and catalyst. They found the process for selective hydrogenation of acetylene very sensitive to deactivation, in spite of the removal of hydrocarbon deposits by heptane. The reason might be the formation of a strongly bound, highly unsaturated coke that increases the rate of ethylene hydrogenation.

Large-scale production of hydrogen peroxide has been put on stream using monoliths for hydrogenation of alkylanthraquinones [37,38,5]. Irandoust et al. [5] studied the process

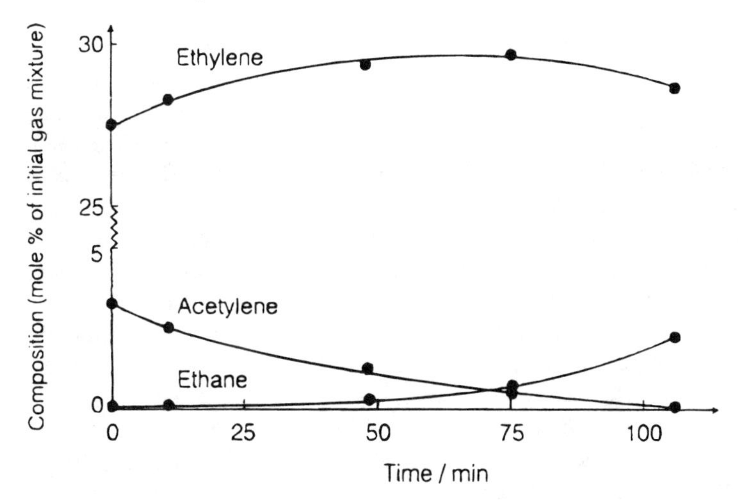

Figure 13 Hydrogenation of acetylene: 303 K, 2 MPa; initial gas composition (mole%): 3% C_2H_2, 28% C_2H_4, 11% H_2, balance N_2.(From Ref. 35.)

in both the pilot plant reactor and the full-scale reactor, where the scale-up factor was 1:20. The process was investigated with downward and upward flow of reactants. Alkylanthraquinones were hydrogenated at 323 K and 0.4 MPa. The total flow rate due to gravity was evaluated to be 0.45 m sec^{-1} in the pilot reactor, while that observed was 0.42 m sec^{-1}. In the industrial reactor, the production capacity, and consequently the reaction rate, was essentially constant over the broad range of the liquid flow rate: from 0.12 to 0.33 m^3 m^{-2} sec^{-1}, which was the maximum possible loading. This observation indicates the nearly constant total flow rate, which is very similar to that in the pilot unit. A uniform distribution of both phases over the reactor cross section was found to be of major importance. An appropriate distribution resulted in a nearly constant capacity of pilot plant and industrial reactors, indicating no scale-up effects. However, the production capacity in the case of downward flow of the reactants was 30–50% higher than that of upward flow at the same superficial velocity. The very high interface surface area between phases, in combination with the very thin liquid film, resulted in the high performance of the monolithic catalyst with segmented flow.

Cybulski et al. [39] have studied the performance of a commercial-scale monolith reactor for liquid-phase methanol synthesis by computer simulations. The authors developed a mathematical model of the monolith reactor and investigated the influence of several design parameters for the actual process. Optimal process conditions were derived for the three-phase methanol synthesis. The optimum catalyst thickness for the monolith was found to be of the same order as the particle size for negligible intraparticle diffusion (50–75 μm). Recirculation of the solvent with decompression was shown to result in higher CO conversion. It was concluded that the performance of a monolith reactor is fully commensurable with slurry columns, autoclaves, and trickle-bed reactors.

B. Oxidations

Kim et al. [40] made an attempt to oxidize phenol in water solutions using a monolith reactor. Alumina-washcoated cordierite monoliths (62 cells/cm^2) impregnated with copper salts to give 10 wt.% (based on alumina weight) were used. A reactor consisted of two sections. In the lower one, the gas was dispersed in the liquid, producing froth. The froth passed upward through monolithic structures 20 cm long and 4 cm in diameter. The process was studied at temperatures up to 393 K under pressure up to about 0.5 MPa. The liquid flow rate was varied from 0.4 to 2 $cm^3\ sec^{-1}$, and the gas flow rate ranged from 33.3 to 150 $cm^3\ sec^{-1}$. Operation of the monolith reactor produced little reaction (conversion $< 15\%$) but substantial vaporization of phenol. Conversion was relatively unaffected by the flow rate of liquid but showed a maximum for a certain gas flow rate. This maximum may be related to the transition from bubble flow to segmented flow.

Crynes et al. [41] continued the study of Kim et al. [40]. The novel monolithic froth reactor, with a monolithic section 0.42 m long and 5 cm in diameter, was used. Cordierite monoliths with a cell density of 62 cells/cm^2 were stacked, one on top of another, to provide a structure 0.33 m long. The monoliths, washcoated with γ-alumina and impregnated with CuO, were tested at 383–423 K and 0.48–1.65 MPa. The liquid flow rate was varied from 0.4 to 3.5 $cm^3\ sec^{-1}$, and the gas flow rate ranged from 15.8 to 50 $cm^3\ sec^{-1}$. Phenol in a concentration of 5000 ppm was typically oxidized with air. The reaction rate versus the liquid flow rate showed a distinct maximum of approximately 2 mol $g_{cat}^{-1}\ sec^{-1}$ at about 1.7 $cm^3\ sec^{-1}$, while the dependence of the reaction rate on the gas flow rate was rather weak, with a tendency to decrease as the flow rate increased.

The effect of temperature on the reaction rate was like the Arrhenius function (Fig. 14), with activation energy of 67 kJ mol^{-1}, indicating that the reaction proceeded in the kinetic regime (and, consequently, with negligible external mass transfer resistance). The activation energy is only slightly below those reported in the literature as the intrinsic values. The reaction rates increased with pressure up to approximately 1.1 MPa and thereafter approached the constant value of about 2 mol $g_{cat}^{-1}\ sec^{-1}$. In the former range,

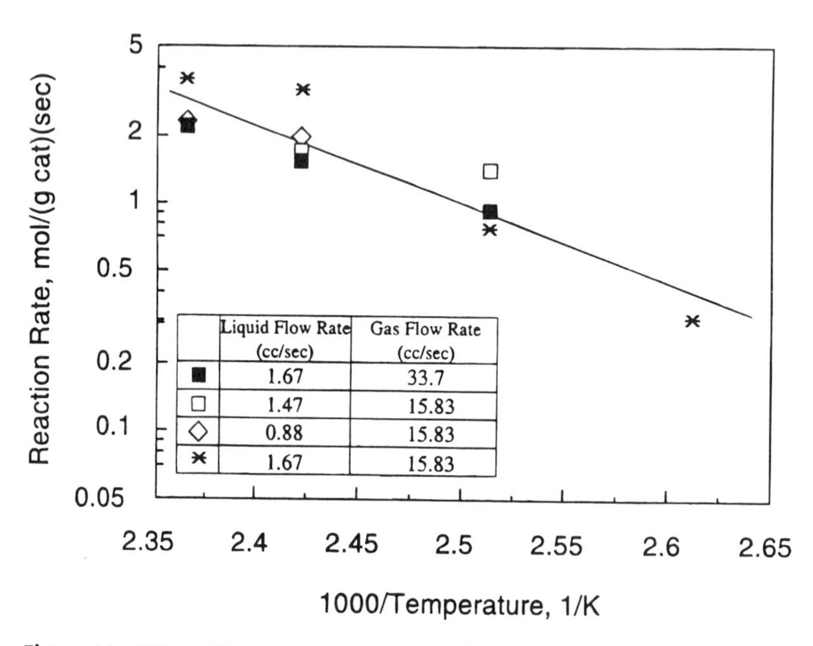

Figure 14 Effect of temperature on the overall reaction rate at different fixed liquid and gas flow rates and pressures. (From Ref. 41.)

the influence of rising oxygen concentration on the reaction rate is observed. In the latter region, the pattern of flow might be changed, and external mass transfer resistance started to be the rate-limiting step. A loss of about 20% of copper during experiments was reported.

Kobe Steel Co. [42] has patented a monolithic process for oxidation of Fe^{2+} to Fe^{3+} in acidic aqueous solutions using monolith made of carbon. The monoliths were prepared by mixing active carbon with a binder, extrusion, and thermal treatment. Slices 150 mm in diameter of cell density 20–60 cells/cm^2 were tested. Monolithic slices 30 mm thick were stacked and separated one from another with turbulizers. The reactor was operated in the countercurrent mode with the gas flowing upward. The liquid was recirculated. The liquid flow rate was varied from 250 to 333 $cm^3\ sec^{-1}$, and the gas flow rate ranged from 83 to 250 $cm^3\ sec^{-1}$. Pressure was up to 0.31 MPa. The oxidation efficiency was 34–80% at circulation time of 1800 sec, and rose to 89–93% at 3600 sec.

Luck et al. [43] reported results of investigations on wet air oxidation of waste streams containing biosolids (mostly carboxylic acids) using a monolith reactor. A 5 wt.% CuO catalyst deposited on titania monolith doped with other metal oxides was studied. The honeycomb structures contained square channels of width 3.56 mm, with a wall thickness of 0.65 mm. The horizontal reactor had a 50-mm ID and was 6 m long. The process was investigated at 508 K, under a pressure of 4.0–4.7 MPa. The typical flow rates were 11.7 $cm^3\ sec^{-1}$ (NTP, 273 K, and 1 atm) for pure oxygen and 2.2 $cm^3\ sec^{-1}$ for the sludge of COD in the range 5–10 kg m^{-3}. COD removal reached up to 74%, while in the blank, noncatalytic tests it was 58%. Traces of copper detected in the supernatant after reaction indicated the transfer of Cu from the catalyst to the reaction mixture.

C. Biotechnology

Benoit and Kohler [44] studied the performance of immobilized catalase incorporated into a silanized cordierite monolith using H_2O_2 as oxidant. An immobilized monolith of 15-mm diameter stacked in two, three, or four sections 95 mm long was investigated. The authors found immobilized monolith to be quite promising for processing very viscous liquids (e.g., in sucrose inversion), in processes where high air flow rates are required (e.g., in the manufacture of gluconic acid), and in processes where plugging of packed beds is expected (e.g., in the hydrolysis of lactose in milk).

Ghommidh et al. [45] investigated the performance of *Acetobacter* cells immobilized in a monolith reactor operated with a pulse flow. A very high productivity up to 2.9×10^{-3} kg $m^{-3}\ sec^{-1}$ of acetic acid was achieved due to the very intensive transfer of oxygen from the gas to the solid. Oxygen transfer in the microbial film was evaluated as controlling the reactor productivity.

Lydersen et al. [46] studied the use of ceramic matrix for the large-scale culture of animal cells. They found that monoliths provide an even distribution and growth of a wide variety of cells in densities equal to or greater than those obtained with conventional methods. The monolithic process was found particularly useful for scale-up from 0.9 m^2 to 18.5 m^2 of surface area, with the same efficiency of surface utilization. Conventional methods showed several limitations in this process. The limitations are due to problems in scaling up, the sensitivity of some cells to the shear forces when stirring, and the difficulties in separating the resulting suspensions. With a monolith reactor, there is no need for either stirring or separating the cells from the spent medium. It was proved that cells grown on the ceramic are readily harvested. The ceramic matrix with cells deposited was subject to patent application [47]. The ceramic was seeded with human foreskin fibroblasts in a suspending medium. After completion of seeding, the serum-free medium in the monolith was replaced with MEM–Hanks growth medium. The maximum cell growth was achieved by recirculation of the medium at a rate of 0.007 mm sec^{-1} during an early lag phase and at 0.0014–0.11 mm sec^{-1} during the rapid-growth phase.

Ariga et al. [48] have investigated the behavior of the monolith reactor in which *Echerichia coli* with β-galactosidase or *Saccharomyces cerevisiae* was immobilized within a thin film of κ-carrageenan gel deposited on the channel wall. The effects of mass transfer resistance and axial dispersion on the conversion were studied. Those authors found that the monolith reactor behaved like the plug-flow reactor. The residence-time distribution in this reactor was comparable to four ideally mixed tanks in series. The influence of gas evolution on liquid film resistance in the monolith reactor was also investigated. It was shown that at low superficial gas velocities, the gas bubble may adhere to the wall, which decreases the effective surface area available for the reaction. The authors concluded that the reactor was very effective in the reaction systems accompanied by gas evolution, such as fermentations.

Shiraishi et al. [49,50] immobilized glucoamylase of *Rhizopus delemar* in monolith structures and used them for saccharification of soluble starch. The process was studied at first in a batch reactor at 50°C and 4.5 bar. The simplified kinetic model was developed. A continuous process was realized in a monolith reactor consisting of 10 pieces stacked on top of each other, where the blocks were rotated by $\pi/4$ on their axes. The reaction rate at a glucose concentration of 460 g dm^{-1} was approximately two times higher than in a conventional industrial process. Conversion of 47% was reached at a space time of 12 hr. The half-life of enzyme was 79 days.

Shiraishi et al. [51] studied a process of oxidation of glucose to maltose in the presence of β-amylase and debranching enzyme (pullulanase or isoamylase). Experiments were performed with glucose concentrations of 5–100 kg m^{-3} at 450°C and pH = 4.8. External mass resistance was found negligible. Again, stacked monolithic blocks were used. Production of maltose by this method was very efficient.

Shiraishi et al. [52,53] have reached the very high productivity of 26.3 kg gluconic acid m^{-3} hr^{-1} by aerobic transformation of glucose using a strain of *Gluconobacter* immobilized on a ceramic honeycomb monolith. The reactor was operated at a glucose concentration of 100 kg m^{-3}, a residence time of 3.5 hr, and an aeration rate of 900 cm^3 min^{-1}. Glucose conversion at these conditions was 94%, and the yield of gluconic acid amounted to 84.6%.

Kawakami et al. [54,55] performed extensive investigations on enzymatic oxidation of glucose and enzymatic hydrolysis of *N*-benzoyl-L-arginine ethyl ester. The glucose oxidase G6500 and the catalase C3515 from *Aspergillus Niger* and the trypsin T0134 from *Govine panceas* were immobilized on silanized cordierite monoliths of cell density 12, 31, and 62 cells/cm^2. A reactor with square cross section 22 × 22 mm and length of 220–330 mm was used. Both upflow and downflow modes of operation were studied. The authors found the conversion of glucose to be independent of the gas velocity above 2 cm sec^{-1}. The conversion for upflow operation was higher regardless of the process conditions (see Fig. 15). The overall effectiveness factor (including both external and internal mass transfer) was estimated to be more than 0.3 for the upflow operation, while it was less than 0.1 for the downflow operation, probably because of the severe external mass transfer limitations. At low liquid velocities, the reactor for upflow was operated in a slug-flow regime, while for downflow the annular regime was observed. In spite of this, the volumetric gas–liquid mass transfer coefficient in the monolith reactor was much

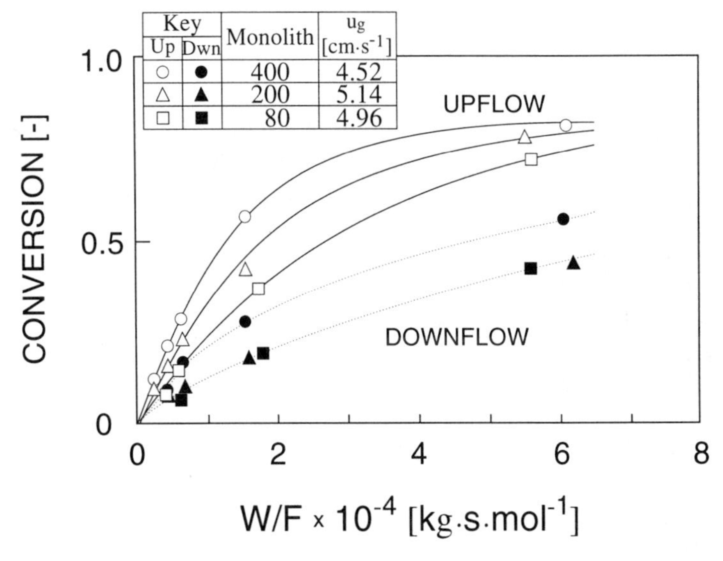

Figure 15 Conversion of glucose versus W/F in immobilized-enzyme monolith reactor with different channel sizes. (From Ref. 55.)

higher than that evaluated for the process carried out in a trickle bed formed of 2-mm particles of a specific surface area equivalent to that of the monolith reactor under consideration.

Papayannakos et al. [56] studied the kinetics of lactose hydrolysis by free and immobilized β-galactosidase. An effective immobilization technique to bind β-galactosidase on ceramic monoliths was presented. The ceramic monoliths used for enzyme immobilization had an average pore size of 10 μm, a specific area of 0.15 $m^2\ g^{-1}$, and a porosity of 0.23 $cm^3\ g^{-1}$. The monolithic support had a height of 7.68 cm and a diameter of 2.27 cm. The shape of monolithic channels was square, with a side length of 3 mm and a wall thickness of 0.3 mm. The cell density of the monolith was 9 cells/cm^2. The immobilized lactase showed considerably enhanced thermal stability compared with the free enzyme and did not suffer from enzyme leaching. The kinetic investigations revealed that the intrinsic kinetics of hydrolysis with immobilized lactase were consistent with those of soluble lactase. It was also concluded that pore diffusion limitations were not important, whereas external mass transfer effects could be present. The performance of a laboratory continuous-flow immobilized-lactase reactor system was also studied. It was shown that at low volumetric flow rates, the external mass transfer limitations reduced the overall efficiency of the catalyst. Simulations showed that by using an apparent effectiveness factor of 0.65, the experimental results could be predicted well.

VI. FUTURE PERSPECTIVES

The monolith reactors have found broad applications in gas-phase processes and are at the development stage for catalytic gas–liquid processes. Currently, there is only one process, hydrogenation of alkylanthraquinones in production of hydrogen peroxide, operating on a commercial scale.

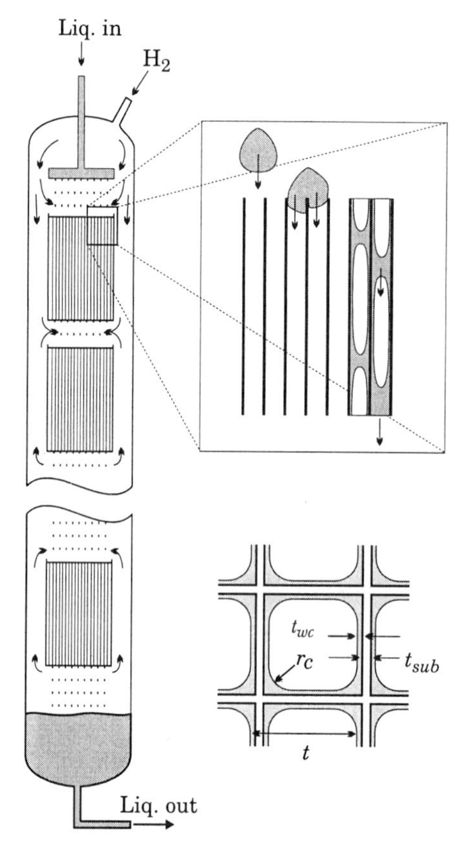

Figure 16 The monolith reactor. (From Ref. 9.)

The wide use of monoliths within the area of emissions control has directed the material development toward high-temperature applications, giving factors like refractoriness, thermal conductivity, and thermal expansion the highest priority. The recent applications of monoliths in chemical processes generally require materials with high surface area in preference to the thermal properties, since these processes are typically operated at steady state at lower temperatures. Another factor affecting the characteristics of monoliths is the cell density. Besides increased contact areas, longer residence times are possible using monoliths of higher cell densities. The latter point is a result of the increased drop in frictional pressure, which reduces the linear velocity of gravity-driven flow. Metallic monoliths offering good thermal properties can be manufactured at higher cell densities than ceramic ones, and are also expected to find applications within catalytic gas–liquid reactions.

The most crucial step in the design of monolith reactors is the proper distribution of fluids over the reactor cross section. However, the available information on the gas–liquid distribution over monoliths is limited, and additional research and development are needed in this field. It should be noted that distributing the fluid at the top of the reactor is not the end of the problem, and redistributions may be needed. This is due to the fact that the monoliths can only be manufactured in short pieces, and the desired length of the catalyst in the reactor is usually obtained by stacking monolith pieces on top of each other (Fig. 16). Hence, the gas and liquid flows need to be collected together and spread out again.

Another important point to be considered is the scale-up of monolith reactors. Although the available information indicates that scale-up is straightforward, more studies need to be performed to develop design methods for monolith reactors. This is of crucial importance, since the successful implementation of monolith reactors in the chemical industry will require detailed design and scale-up procedures.

It is the authors' opinion that, due to the unique features of monolith reactors, the applications of this reactor type for catalytic gas–liquid reactions are likely to increase in the near future.

REFERENCES

1. S. Irandoust and B. Andersson, Monolithic catalysts for nonautomobile applications, *Catal. Rev. Sci. Eng. 30(3)*:341 (1988).
2. A. Cybulski and J.A. Moulijn, Monoliths in heterogeneous catalysis, *Catal. Rev. Sci. Eng. 36(2)*:179 (1994).
3. Y. Taitel, D. Bornea, and A. Dukler, Modeling flow pattern transitions for steady upward gas–liquid flow in vertical tubes, *AIChE J. 26(3)*:345 (1980).
4. G. Taylor, Dispersion of a viscous fluid on the wall of a tube, *J. Fluid Mechanics 10*:161 (1961).
5. S. Irandoust, B. Andersson, E. Bengtsson, and M. Siverström, Scaling up of a monolithic catalyst reactor with two-phase flow, *Ind. Eng. Chem. Res. 28*:1489 (1989).
6. K. Fukano and A. Kariyasaki, Characteristics of gas–liquid flow in a capillary tube, *Nucl. Eng. Des. 141*:59 (1993).
7. C.N. Satterfield and F. Özel, Some characteristics of two-phase flow in monolithic catalyst structures, *Ind. Eng. Chem. Fundam. 16(1)*:61 (1977).
8. R.K. Edvinsson and A. Cybulski, A comparative analysis of the trickle-bed and the monolithic reactor for three-phase hydrogenations, *Chem. Eng. Sci. 49(24B)*:5653 (1994).
9. R.K. Edvinsson and A. Cybulski, A comparison between the monolithic reactor and the trickle-bed reactor for liquid-phase hydrogenations, *Catal. Today 24*:173 (1995).
10. R.K. Edvinsson, *Monolith reactors in three-phase processes.* PhD dissertation, Chalmers University of Technology, Göteborg, 1994.
11. A. Gianetto and P.L. Silveston, *Multiphase Chemical Reactors: Theory, Design, Scale-Up*, Hemisphere, New York, 1986.
12. P.A. Ramachandran and R.V. Chaudhari, *Three-Phase Catalytic Reactors*, Gordon and Breach, New York, 1983.
13. Y.T. Shah, *Gas–Liquid–Solid Reactor Design*, McGraw-Hill, New York, 1979.
14. Y.T. Shah, Design parameters for mechanically agitated reactors, *Adv. Chem. Eng. 17*:1 (1991).
15. Y.T. Shah and M.M. Sharma, Gas–liquid–solid reactors, in *Chemical Reaction and Reactor Engineering* (J.J. Carberry and A. Varma, eds.), 1987, p. 667.
16. P. Trambouze, H. van Landeghem, and J.P. Wauquier, *Chemical Reactors, Design/Engineering/ Operation*, Gulf, Houston, 1988.
17. S. Irandoust, S. Ertlé, and B. Andersson, Gas–liquid mass transfer in taylor flow through a capillary, *Can. J. Chem. Eng. 70*:115 (1992).
18. N.-H. Schöön, Recent progress in liquid-phase hydrogenation: With aspects from microkinetics to reactor design, Proceedings of the 6th Natl. Sympos. Chem. React. Eng., Warsaw, Poland, and the Second Nordic Symposium on Catalysis, Lyngby, Denmark, 1989.
19. S. Irandoust, *The monolithic catalyst reactor*, PhD dissertation, Chalmers University of Technology, Göteborg, Sweden, 1989.

20. V. Hatziantoniou and B. Andersson, The segmented two-phase flow monolithic catalyst reactor. An alternative for liquid-phase hydrogenations, *Ind. Eng. Chem. Fundam. 23*:82 (1984).
21. V. Hatziantoniou, B. Andersson, and N.-H. Schöön, Mass transfer and selectivity in liquid-phase hydrogenation of nitro compounds in a monolithic catalyst reactor with segmented gas–liquid flow, *Ind. Eng. Chem. Process Des. Dev. 25*:964 (1986).
22. S. Irandoust and B. Andersson, Mass transfer and liquid-phase reactions in a segmented two-phase flow monolithic catalyst reactor, *Chem. Eng. Sci. 43(8)*:1983 (1988).
23. S. Irandoust and O. Gahne, Competitive hydrodesulfurization and hydrogenation in a monolithic reactor, *AIChE J. 36(5)*:746 (1990).
24. R. Edvinsson and S. Irandoust, Hydrodesulfurization of dibenzothiophene in a monolithic catalyst reactor, *Ind. Eng. Chem. Res. 32*:391 (1993).
25. D.S. Soni and B.L. Crynes, A comparison of the hydrodesulfurization and hydrodenitrogenation activities of monolith alumina impregnated with cobalt and molybdenum and a commercial catalyst, *ACS Sympos. Ser. 156*:207 (1981).
26. J. Scinta and S.W. Weller, Catalytic hydrodesulfurization and liquefaction of coal. Batch autoclave studies, *Fuel Process. Technol. 1(4)*:279 (1978).
27. I. Mazzarino and G. Baldi, Liquid-phase hydrogenation on a monolithic catalyst, in *Recent Trends in Chemical Reaction Engineering* (B.D. Kulkarni, R.A. Mashelkar, and M.M. Sharma, eds.), Vol. II, Wiley Eastern Ltd., New Delhi, 1987, p. 181.
28. E. Otten, *Hydrogenation of Benzaldehyde over Monolithic Catalysts*, Masters thesis, Delft University of Technology, Delft, The Netherlands, 1994.
29. A.I.J.M. van der Riet, E. Otten, A. Cybulski, J.A. Moulijn, W.Ch. Glasz, and A. Stankiewicz, Hydrogenation of aromatic aldehydes over monolithic catalysts, Proceedings of the EUROPA-CAT-II, S7, P13, p. 537, Maastricht, Belgium, September 3–8, 1995.
30. X. Xu, H. Vonk, A.I.J.M. van de Riet, A. Cybulski, J.A. Moulijn, and A. Stankiewicz, Monolithic catalysts for selective hydrogenation of aromatic aldehyde, Proceedings of the Conference on Catalysis Science and Technology, Tianjin, China, September 12–15, 1995.
31. J.M. Berty, S. Lee, K. Sivegnanam, and F. Szeifert, Diffusional kinetics of catalytic vapor-phase reversible reactions with decreasing total number of moles, *Inst. Chem. Eng., Sympos. Ser. 87*:455 (1984).
32. H.A. Smits, *Selective hydrogenation of model compounds for pyrolysis gasoline over a monolithic palladium catalyst*, SPE thesis, Delft University of Technology, Delft, The Netherlands, 1994.
33. H.A. Smits, A. Cybulski, J.A. Moulijn, W.Ch. Glasz, and A. Stankiewicz, Selective hydrogenation of styrene/1-octene mixtures over a monolithic palladium catalyst, Proceedings of the EUROPACAT-II, S7, O2, p. 515, Maastricht, Belgium, September 3–8, 1995.
34. J.F. Le Page, *Applied Heterogeneous Catalysis: Design, Manufacture and Use of Solid Catalysts*, Editions Technip, Paris, 1987.
35. R. Edvinsson, A.M. Holmgren, and S. Irandoust, Liquid-phase hydrogenation of acetylene in a monolithic catalyst reactor, *Ind. Eng. Chem. Res. 34*:94 (1995).
36. S. Asplund, C. Fornell, A.M. Holmgren, and S. Irandoust, Catalyst deactivation in liquid- and gas-phase hydrogenation of acetylene using a monolithic catalyst reactor, *Catal. Today 24*:181 (1995).
37. T. Berglin and N.-H. Schöön, Selectivity aspects of the hydrogenation stage of the anthraquinone process for hydrogen peroxide production, *Ind. Eng. Chem. Process Des. Dev. 22*:150 (1983).
38. T. Berglin and W. Herrmann, Hydrogen Peroxide, Swedish Pat. No. 431,532 (1984); Eur. Pat. No. EP 102,934 (1986).
39. A. Cybulski, R.K. Edvinsson, S. Irandoust, and B. Andersson, Liquid-phase methanol synthesis: Modelling of a monolithic reactor, *Chem. Eng. Sci. 48(20)*:3463 (1993).

40. S. Kim, Y.T. Shah, R.L. Cerro, and M.A. Abraham, Aqueous phase oxidation of phenol in a monolithic reactor, Proceedings of AIChE Annual Meeting, Miami Beach, Dec. 1992.
41. L.L. Crynes, R.L. Cerro, and M.A. Abraham, Monolith froth reactor: Development of a novel three-phase catalytic system, *AIChE J. 41(2)*:337 (1995).
42. Kobe Steel Co, Jap. Pat. No. 87-97636 (1987).
43. F. Luck, M. Djafer, and M.M. Bourbigot, Catalytic wet air oxidation of biosolids in a monolithic reactor, Proceedings of the European Symposium on Catalysis in Multiphase Reactors, Lyon, France, 7–9 December 1994.
44. M.R. Benoit and J.T. Kohler, An evaluation of a ceramic monolith as an enzyme support material, *Biotechnol. Bioeng. 17*:1617 (1975).
45. G. Ghommidh, J.M. Navarro, and G.A. Durand, A study of acetic acid production by immobilized *Acetobacter* cells: Oxygen transfer, *Biotechnol. Bioeng. 24(3)*:605 (1982).
46. B.K. Lydersen, G.G. Pugh, M.S. Paris, B.P. Sharma, and L.A. Noll, Ceramic matrix for large-scale animal cell culture, *BIO/TECHNOLOGY* (January):63 (1985).
47. I.M. Lachman, L.A. Noll, B.K. Lydersen, W.H. Pitcher, G.G. Pugh, and B.P. Sharma, Cell Culture Apparatus and Process Using an Immobilized Cell Composite, U.S. Pat. Appln. No. 464,011 (1983); Eur. Pat. Appln. EP 121981 (1984).
48. O. Ariga, M. Kimura, M. Taya, and T. Kobayashi, Kinetic evaluation and characterization of ceramic honeycomb-monolith bioreactor, *J. Ferment. Technol. 64(4)*:327 (1986).
49. F. Shiraishi, K. Kawakami, S. Kono, A. Tamura, S. Tsuruta, and K. Kusunoki, Characterization of production of free gluconic acid by gluconobacter suboxydans adsorbed on ceramic honeycomb monolith, *Biotechnol. Bioeng. 33(11)*:1413 (1989).
50. F. Shiraishi, K. Kawakami, K. Kato, and K. Kusunoki, Hydrolysis of soluble starch by glucoamylase immobilized on ceramic monolith, *Kagaku Kogaku Ronbunshu 9(3)*:316 (1983).
51. F. Shiraishi, K. Kawakami, T. Kojima, A. Yuasa, and K. Kusunoki, Maltose Production from soluble starch by β-Amylase and debranching enzyme immobilized on ceramic monolith, *Kagaku Kogaku Ronbunshu 14(3)*:288 (1988).
52. F. Shiraishi, K. Kawakami, and K. Kusunoki, K., Saccharification of starch in an immobilized glucoamylase monolithic reactor, *Kagaku Kogaku Ronbunshu 12(4)*:492 (1986).
53. F. Shiraishi, K. Kawakami, A. Tamura, S. Tsuruta, and K. Kusunoki, Continuous production of free gluconic acid by *Gluconobacter suboxydans* IFO 3290 immobilized by adsorption on ceramic honeycomb monolith: Effect of reactor configuration of further oxidation of gluconic acid to kato-gluconic acid, *Appl. Microbiol. Biotechnol. 31(5–6)*:445 (1989).
54. K. Kawakami, K. Adachi, N. Minemura, and K. Kusunoki, K., Characteristics of a honeycomb monolith three-phase bioreactor: Oxidation of glucose by immobilized glucose oxidase, *Kagaku Kogaku Ronbunshu 13(3)*:318 (1987).
55. K. Kawakami, K. Kawasaki, F. Shiraishi, and K. Kusunoki, Performance of a honeycomb monolith bioreactor in a gas–liquid-solid three-phase system, *Ind. Eng. Chem. Res. 28*:394 (1989).
56. N. Papayannakos, G. Markas, and D. Kekos, Studies on modelling and simulation of lactose hydrolysis by free and immobilized β-galactosidase from *Aspergillus niger*, *Chem. Eng. J. 52*:B1 (1993).

10

Modeling of Monolith Reactors in Three-Phase Processes

Bengt Andersson and Said Irandoust
Chalmers University of Technology, Göteborg, Sweden

Andrzej Cybulski
Polish Academy of Sciences, Warsaw, Poland

I. INTRODUCTION

Monolith reactors have recently found applications in performing catalytic three-phase reactions (see Chapter 9). There is also growing interest in the chemical industries for this novel type of multiphase reactor. A proper modeling of the monolith reactor is a necessary step in order to estimate the overall performance of the reactor.

In this chapter, first, the existing correlations for three-phase monolith reactors will be reviewed. It should be emphasized that most of these correlations were derived from a limited number of experiments, and care must be taken in applying them outside the ranges studied. Furthermore, most of the theoretical work concerns Taylor flow in cylindrical channels (see Chapter 9). However, for other geometries and flow patterns we have to rely on empirical or semiempirical correlations. Next, the modeling of the monolith reactors will be presented. On this basis, comparisons will be made between three basic types of continuous three-phase reactor: monolith reactor (MR), trickle-bed reactor (TBR), and slurry reactor (SR). Finally, for MRs, factors important in the reactor design will be discussed.

II. MODELING OF SINGLE CHANNELS

A. Hydrodynamics of Segmented Flow

1. *Basic Relationships*

Taylor bubbles (gas plugs) in a vertical tube move under the influence of surface tension, inertia, gravitation, and viscous effects. For a Newtonian fluid with constant viscosity and density these phenomena can be described by the Navier–Stokes equations for circular geometry using cylindrical coordinates:

$$\frac{\partial u_l}{\partial z} + \frac{1}{r}\frac{\partial}{\partial r}(r v_l) = 0 \tag{1}$$

$$\rho_l\left(\frac{\partial u_l}{\partial t} + v_l\frac{\partial u_l}{\partial r} + u_l\frac{\partial u_l}{\partial z}\right) = -\frac{\partial p}{\partial z} + \mu_l\left[\frac{1}{r}\frac{\partial}{\partial r}\left(r\frac{\partial u_l}{\partial r}\right) + \frac{\partial^2 u_l}{\partial z^2}\right] + \rho_l g \tag{2}$$

$$\rho_l\left(\frac{\partial v_l}{\partial t} + v_l\frac{\partial v_l}{\partial r} + u_l\frac{\partial v_l}{\partial z}\right) = -\frac{\partial p}{\partial r} + \mu_l\left[\frac{\partial}{\partial r}\left(\frac{1}{r}\frac{\partial}{\partial r}(r v_l)\right) + \frac{\partial^2 v_l}{\partial z^2}\right] \tag{3}$$

The velocity boundary conditions are simple, due to the rigid, impermeable tube walls and to nonslip velocity at the wall. These conditions can be expressed as

$$u_l|_{\text{wall}} = v_l|_{\text{wall}} = 0 \tag{4}$$

$$\left.\frac{\partial v_l}{\partial r}\right|_{\text{wall}} = 0 \tag{5}$$

and

$$\left.\frac{\partial u_l}{\partial r}\right|_{\text{center}} = \left.\frac{\partial v_l}{\partial r}\right|_{\text{center}} = 0 \tag{6}$$

The stress boundary conditions are much more complicated, due to the unknown physical properties of the interface layer. The motion at the interface is affected through the action of surface tension gradients caused by a variation in the surface contaminants along the surface. For a "clean" liquid–gas curved interface, the liquid pressure p_l at the interface is given by the Young and Laplace relation:

$$p_l = p_g - \sigma\left(\frac{1}{R_1} + \frac{1}{R_2}\right) \tag{7}$$

The principal radii of curvature R_1 and R_2 at a given point at the interface are expressed as:

$$\frac{1}{R_1} = \frac{1}{r[1 + (\partial r/\partial z)^2]^{1/2}} \tag{8}$$

$$\frac{1}{R_2} = -\frac{\partial^2 r/\partial z^2}{r[1 + (\partial r/\partial z)^2]^{3/2}} \tag{9}$$

2. *Bubble Shape*

In segmented flow in cylindrical capillaries, Taylor bubbles consist of a cylindrical part and two caps at the front and the rear menisci. The form of the caps may be axisymmetric or nonaxisymmetric. In a balance over a long inviscid bubble surrounded by a moving incompressible and viscous fluid, capillary, viscous, inertial, and gravity forces are taken into account. The latter three, relative to capillary force, are expressed in the following dimensionless numbers:

$$\text{Ca} = \frac{\mu_l U_m}{\sigma} \tag{10}$$

$$\text{Bo} = \frac{\Delta\rho g d_t^2}{\sigma} \tag{11}$$

$$\mathrm{We} = \frac{\rho_l d_t U_m^2}{\sigma} \tag{12}$$

where Ca, Bo, and We are the Capillary, Bond, and Weber numbers, respectively. The Bond number is also known as the Eötvös number (*Eö*). In sufficiently narrow tubes, such as capillaries, the surface tension is predominant. It is therefore usually assumed that We $<<$ 1. Usually, too, it is assumed that at sufficiently small Bond numbers the shape of the bubble is a function of the capillary number alone.

a. Cylindrical Geometry. According to the Young and Laplace relation, small tube radius and high surface tension cause a large difference between the outside and the inside of a gas bubble; i.e., the bubble becomes more rigid. This will make it difficult for the liquid to pass by the long bubble. Under these conditions, where the capillary number approaches zero, the viscous stresses modify the static profile of the bubble only very close to the wall. Thus, we will observe an (almost) axisymmetric flow pattern. If the capillary number is increased further, the liquid film will become thicker, and the caps will no longer be completely hemispherical. In the region of low capillary numbers, the bubble caps are far from being hemispheres. Their shape becomes more complex, as illustrated in Fig. 1.

b. Square Geometry. At low capillary numbers, nonaxisymmetric flow exists. The gas bubble flattens out, leaving liquid regions in the corners separated by thin flat films. The transition from nonaxisymmetric to axisymmetric bubble regime was found by Thulasidas *et al.* [2] to take place at Ca = ca. 0.04. The Taylor bubble then consists of a cylindrical part and two hemispherical caps. As in cylindrical geometry, the shape of the caps changes and the liquid film becomes thicker with increasing capillary number. This is shown in Fig. 2. Coating of a monolith channel will, however, make the cross-sectional area in between square and circular, thus reducing the capillary numbers for the transition area.

In nonpure systems there will also be surface tension gradients, due to the accumulation of surface-active material at the gas–liquid interface. This surface tension gradient will introduce a tangential momentum transfer between the gas and the liquid "skin friction."

3. Streamline Patterns in Liquid Slugs

In segmented flow, the subsequent liquid slugs are separated from each other by the Taylor bubbles. Thus, each liquid slug is enclosed by the two ends of the adjacent Taylor bubbles.

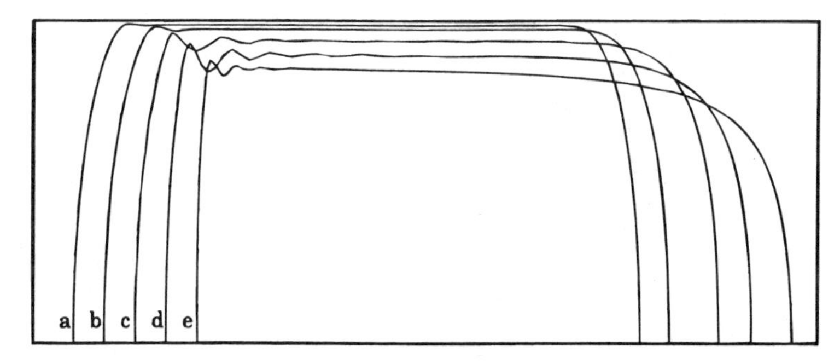

Figure 1 Shape of Taylor bubbles. Ca varies according to (a) 0.001, (b) 0.0032, (c) 0.01, (d) 0.03, and (e) 0.06. Re = 1000. (From Ref. 1.)

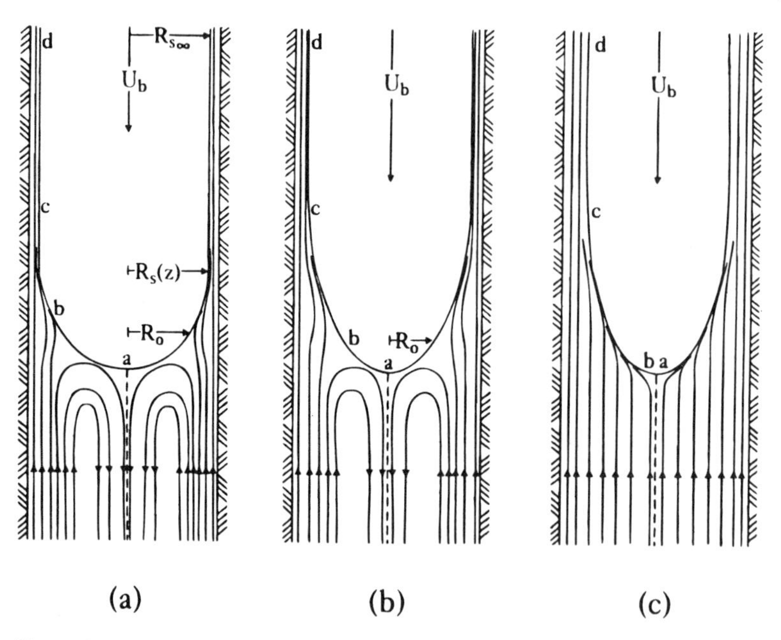

Figure 2 Evolution of the bubble profile and flow field surrounding the bubble. The interface is divided into three regions: a cap region *ab*, a transition region *bc* and a uniform region *cd*. The evolution of the interface and flow field streamlines with increasing capillary number is displayed for (a) Ca = 0.15, (b) Ca = 0.3, and (c) Ca = 0.6. (From Ref. 3.)

Whether the bubble ends are hemispherical or not depends on the capillary number. In all cases, though, the caps are virtually static. In this region the formation of any parabolic streamline profile in the liquid slug is prevented. This results in the recirculation of liquid within the slug. Characteristic streamline patterns within liquid slugs that are separated by Taylor bubbles consisting of two hemispherical bubble ends are shown in Fig. 3. The influence of the capillary number on the velocity profile is shown in Fig. 2. This figure, obtained for the square channels, reveals that at high capillary numbers recirculation within the liquid may be absent. However, very thin liquid films and hemispherical bubble ends are usually observed in segmented flow, and this is the most preferred situation with respect to mass transfer.

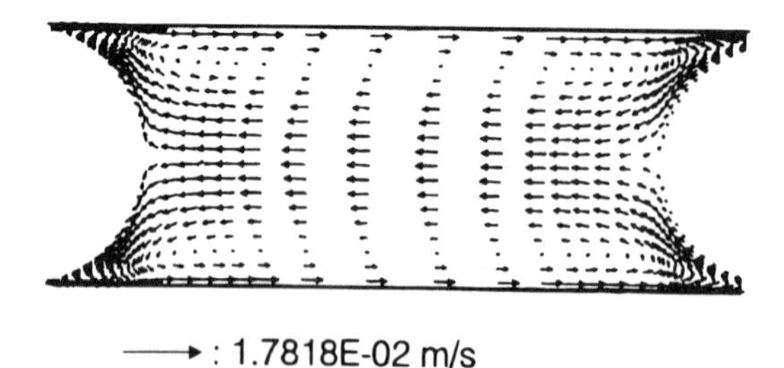

Figure 3 Predicted velocity vectors. (From Ref. 4.)

4. *Velocity of Taylor Bubbles*

The slip velocity of Taylor bubbles flowing through the capillary is higher than the average velocity of the liquid phase. The cylindrical part of the Taylor bubble does not flow through the tube as a closed piston, but is separated from the wall by a thin liquid film. Motion in which the ends of the bubble formed two hemispherical caps filling the complete tube cross section would involve infinite viscous stresses at the wall. The only physical forces tending to maintain the shape of the ends against these stresses are those due to uniform pressure within the bubble and to surface tension force. The influence of gravitational force is assumed negligible, due to the small tube radius. The region where the cylindrical part of the Taylor bubble is separated from the wall by the thin liquid film has uniform pressure without any tangential or axial stress on the free surface. During flow along the tube, the bubble sweeps out liquid volume in front of it. The average velocity of the liquid is thus less than that of the bubble, or, conversely, the bubble velocity is higher than the liquid velocity. The ratio of the excess velocity of the Taylor bubbles and the average velocity is expressed as

$$W = \frac{U_g - U_m}{U_m} \tag{13}$$

and is related to the film-thickness/tube-diameter ratio by

$$1 - W = \left(1 - \frac{2\delta_f}{d_t}\right)^2 \tag{14}$$

or if $\delta_f << d_t$, then

$$W \cong \frac{4\delta_f}{d_t} \tag{15}$$

The excess velocity of the Taylor bubbles can thus be calculated if the thickness of the liquid film is known from measurement or predictions, which are discussed in the next paragraph. If the liquid film is very thin compared to the tube diameter (as is usually true), the excess velocity will be small, too.

5. *Liquid Film Thickness*

The thickness of the film can be correlated with the properties of the liquid and the operating conditions. In all reports, the assumption is made that $\delta_f \ll d_t$. In most references, the correlation is expressed as

$$\frac{\delta_f}{d_t} = a_0 \mathrm{Ca}^{a_1} \tag{16}$$

where a_0 and a_1 are coefficients. The values of these coefficients reported in the literature are listed in Table 1. Taylor [9] found his experimental data to be in reasonable agreement with Eq. (16) using coefficients reported by Fairbrother and Strubbs [8]. Marchessault and Mason [10] expressed the results of their experiments with the following correlation:

Table 1 Values of the Coefficients in Eq. (16)

a_0	a_1	Conditions	Reference
$\frac{2}{3}$	$\frac{2}{3}$	$Ca < 0.002$ horizontal flow film at rest	Bretherton [5]
$\frac{2}{3}$	$\frac{2}{3}$	vertical flow	Concus [6]
$\frac{2}{3}\pi$	$\frac{2}{3}$		Snyder and Adler [7]
$\frac{1}{4}$	$\frac{1}{2}$	$0.001 < Ca < 0.015$	Fairbrother and Stubbs [8]

$$\frac{\delta_f}{d_t} = (0.445 - \frac{0.25}{\sqrt{U_g}}) \cdot \sqrt{\text{Ca}} \tag{17}$$

For low fluid velocities, the film thickness calculated with this equation will not differ significantly from that calculated using the Fairbrother and Stubbs equation.

Irandoust and Andersson [11] developed another empirical expression for the film thickness:

$$\frac{\delta_f}{d_t} = 0.18\ [1 - \exp(-3.1\text{Ca}^{0.54})] \tag{18}$$

which is valid for $10^{-3} < \text{Ca} < 1.9$. This model is generally in accord with the experimental data of Marchessault and Mason [10] and of Taylor [9]. Their results show that, for a given system and flow rate, the liquid film thickness is independent of the length of the Taylor bubbles and liquid slugs. This is true for bubbles one tube diameter long and longer. They also found that the direction of flow had a negligible effect on the measured values of the film thickness. For small Ca, Eq. (18) can be approximated by

$$\frac{\delta_f}{d_t} = 0.56\text{Ca}^{0.54} \tag{19}$$

Thulasidas et al. [2] found for capillary numbers $\text{Ca} < 0.3$ that upward flow gave a thicker liquid film than downward or horizontal flow, for both circular and square geometry.

B. Pressure Drop

In two-phase flow through narrow channels, the pressure drop depends on the type of flow. For example, it is expected that at the same conditions and for a fixed total flow rate, bubble flow will give rise to a higher pressure drop than segmented flow does. This is due to the fact that the liquid phase is the main contributor to the frictional pressure drop. The pressure drop in annular flow may be either higher or lower than in segmented flow. In annular flow, the liquid film substantially reduces the effective diameter of the channel. The pressure drop is therefore caused essentially by the gas flow. The flow pattern that exhibits the lowest pressure drop depends on the gas flow rate. Since transition of the flow pattern may occur during flow, pressure drop is not only of interest in itself but also gives information about the type of flow.

The total pressure drop in segmented flow, which is of the greatest interest for monolith applications for three-phase processes, may be represented as the sum of the frictional pressure drop, hydrostatic head, and capillary pressure:

$$\Delta P_{\text{tot}} = \Delta P_{\text{fr}} + \Delta P_{\text{st}} + \Delta P_c \tag{20}$$

The capillary pressure will have an effect only at high capillary numbers when the curvatures of the front and rear ends of the Taylor bubble are not symmetrical. At low velocity and in narrow channels, the frictional pressure drop is viscosity-dominant and can be calculated using the Hagen–Poisseuille equation

$$\left(\frac{\Delta P}{L}\right)_{\text{fr}} = \frac{32}{d_t^2}(\mu_l U_l + \mu_g U_g) \tag{21}$$

Irandoust *et al.* [12] studied pressure drop in segmented flow. They assumed a negligible contribution of the gas phase to the frictional pressure gradient, which reduces Eq. (21) to

$$\left(\frac{\Delta P}{L}\right)_{\text{fr}} = \frac{32}{d_t^2}\mu_l U_l \tag{22}$$

A good agreement was observed between the measured pressure drop and that calculated from Eq. (22). This confirms the assumption regarding the negligible role of the gas phase in frictional pressure drop at the range of their operating conditions.

However, data of Satterfield and Özel [13] showed that the gas flow rate influences the pressure drop at high gas flow rates. They also studied pressure drop in a series of stacked monoliths. The approximate relationship for the pressure drop due to the orifice effects was added to Eq. (20).

$$\Delta P_{\text{orif}} = \frac{N_c\,(U_{\text{orif}}^2 - U_{\text{tot}}^2)}{2}\,[f_l\rho_l + (1 - f_l)\rho_g] \tag{23}$$

In this equation, it is assumed that both the gas and the liquid move in slugs at the same velocity through the orifice. The contribution of the orifice effects to the total pressure drop is small and is of significance only in monolith blocks with high cell density and at high liquid flow rates. The frictional and static pressure drops prevail. The orifice effects may arise if blocks of monoliths stacked on top of each other are used. The contribution of these effects will depend on the extent of obstruction and the length of individual blocks.

C. Mass Transfer

Mass transfer in Taylor flow is usually characterized as localized into zones of mass transfer resistance [14]. In between these zones, perfect mixing is assumed. Using this concept, the following zones are shown in Fig. 4: (1) mass transfer from the cylindrical part of the Taylor bubbles through the liquid film to the wall, given by

$$N_{GS}A_{GS} = k_{GS}A_{GS}(c^* - c_{\text{wall}}) \tag{24}$$

(2) mass transfer from the hemispherical ends of the Taylor bubble into the liquid, given by

$$N_{GL}A_{GL} = k_{GL}A_{GL}(c^* - c_{\text{bulk}}) \tag{25}$$

and (3) mass transfer of gas from the liquid slugs to the channel wall, given by

$$N_{LS}A_{LS} = k_{LS}A_{LS}(c_{\text{bulk}} - c_{\text{wall}}) \tag{26}$$

The amount of gas transferred to the liquid slug from the hemispherical ends is equal to the mass transferred from the liquid slug to the wall; i.e.,

$$N_{LS}A_{LS} = N_{GL}A_{GL} \tag{27}$$

The total mass transfer of gas from the gas phase to the wall is given by

$$N \cdot A = N_{GS}A_{GS} + N_{GL}A_{GL} \tag{28}$$

or

$$N \cdot A = \left(k_{GS}A_{GS} + \frac{k_{GL}A_{GL}k_{LS}A_{LS}}{k_{GL}A_{GL} + k_{LS}A_{LS}} \right)(c^* - c_{\text{wall}}) \tag{29}$$

1. *Mass Transfer Through the Liquid Film*

Mass transfer according to this mechanism is described by Eq. (24). The mass transfer area for the cylindrical part of the gas bubble is given by

$$A_{GS} = \pi\delta_f(d_t - \delta_f) \tag{30}$$

For a fast reaction occurring on a solid wall, the concentration profile in the liquid film reaches a stationary value for a thin film and long bubbles moving at low velocity. In such a case, the mass transfer coefficient through a plate of infinite length is given by

$$k_{GS} = \frac{D}{\delta_f} \tag{31}$$

where δ_f is the thickness of the film, which can be calculated using correlations given in the previous paragraph. Edvinsson [1] estimated a time constant for diffusion at typical operating conditions in a monolith using the Einstein relationship:

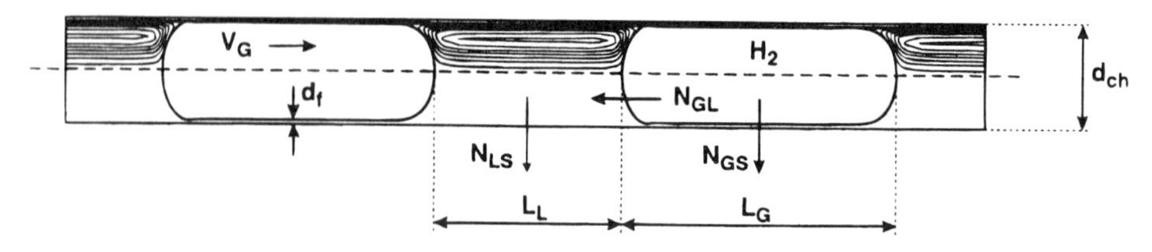

Figure 4 Model for transport of the reactant from the gas phase to the outer surface of the monolith wall. (From Ref. 1.)

$$\tau_D = \frac{\delta_f^2}{2D} \tag{32}$$

and found it to be of order 10^{-1} sec.

2. *Mass Transfer from "Spherical" Ends of the Gas Bubble*

Mass transfer according to this mechanism is described by Eq. (25). The mass transfer area of the hemispherical ends of the gas bubble is given by

$$A_{GL} = \pi d_b^2 \tag{33}$$

Clift et al. [15] derived empirical relationships for the mass transfer coefficients from rigid spheres. Since the ends of Taylor bubbles can be roughly approximated as hemispheres and rigidity is also a quite reasonable assumption, these equations can be applied for the mass transfer from the ends of the Taylor bubbles. For $1 \leq \text{Re} \leq 400$,

$$\frac{\text{Sh} - 1}{\text{Sc}^{1/3}} = \left(1 + \frac{1}{\text{Re} \cdot \text{Sc}}\right)^{1/3} \text{Re}^{0.41} \tag{34}$$

and for $100 < \text{Re} \leq 2000$;

$$\text{Sh} = 1 + 0.724\text{Re}^{0.48}\text{Sc}^{1/3} \tag{35}$$

both with Sherwood number defined as

$$\text{Sh} = \frac{k_{GL} d_b}{D} \tag{36}$$

Irandoust and Andersson [16] developed the following empirical formula for this mass transfer coefficient:

$$\text{Sh} = 0.41\sqrt{\text{Re} \cdot \text{Sc}} \tag{37}$$

Irandoust et al. [17] have modified an expression given by Clift et al. [15]:

$$k_{GL} = 0.69\,(1 + 0.724\text{Re}^{0.48}\text{Sc}^{1/3})\,\frac{D}{d_b} \tag{38}$$

Kawakami et al. [18] studied mass transfer from the gas to the liquid in a biochemical process at upflow and downflow operation. They compared volumetric mass transfer coefficients for monoliths with those for trickle-bed reactors, based on equivalent geometric surface area. The finding was that mass transfer coefficients for monoliths are several times higher than for trickle beds (see Fig. 5).

3. *Mass Transfer Between the Liquid and the Wall*

Hatziantoniou and Andersson [19] developed an empirical model for mass transfer between the liquid phase and the wall:

$$\text{Sh} = 3.51\left(\frac{\text{Re} \cdot \text{Sc}}{L_t/d_t}\right)^{0.44}\left(\frac{L_t}{d_t}\right)^{-0.09} \tag{39}$$

with Sherwood number defined as

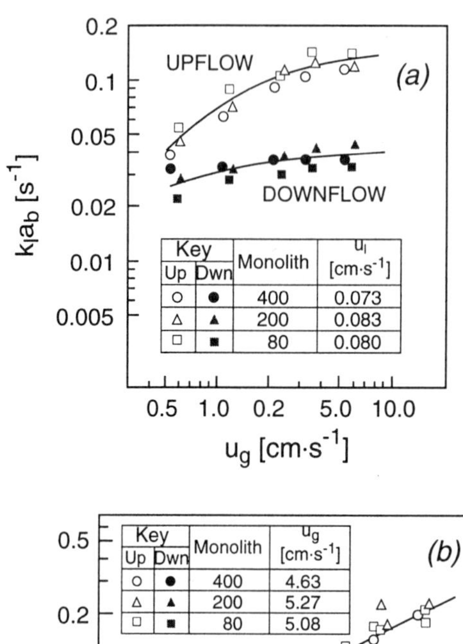

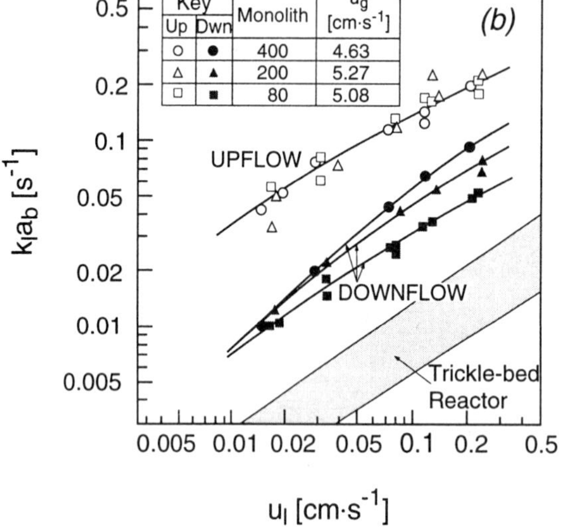

Figure 5 Effects of superficial (a) gas and (b) liquid velocities on volumetric gas–liquid mass transfer coefficients, in monolith reactors with different channel sizes. (From Ref. 18.)

$$\mathrm{Sh} = \frac{k_{LS} d_t}{D} \tag{40}$$

According to Irandoust and Andersson [16], the mass transfer coefficient can be evaluated from

$$\mathrm{Sh} = 1.5 \times 10^{-7} \mathrm{Re}^{1.648} \mathrm{Sc}^{0.177} \left(\frac{\delta_f}{d_t} \right)^{-2.338} \tag{41}$$

4. *Axial Dispersion*

Guedes de Carvalho et al. [20] studied axial mixing in slug flow. They found the following correlation for the dispersion coefficient:

$$\tau D L_t^2 = \frac{0.173}{\tau\phi} + 0.50(\tau\phi)(L_w L_t)^2 \tag{42}$$

where the length equivalent to fully mixed wake, L_w, is related to the channel diameter d_t by

$$L_w = \alpha d_t \tag{43}$$

and α varies from 2.3 to 2.8 for experimental conditions. Axial dispersion was also the subject of a study by Pedersen and Horvath [21]. Irandoust and Andersson [4] simulated the flow pattern, showing significant backflow within bubbles. It is, however, generally accepted that the longitudinal mass dispersion can be neglected and plug flow assumed. This is due to the very small thickness of the liquid film around the gas bubbles, although a little axial mass transport between these bubbles occurs.

In a multichannel monolith with cocurrent downflow, each channel will have the same residence time and a residence-time distribution close to an ideal tubular reactor. But due to nonuniform flow distribution, the gas–liquid ratio, the volume reactant per volume catalyst, and, consequently, the conversion can be very different in different channels.

5. *Effectiveness Factor*

The thickness of the catalytic layer deposited on channel walls is very small: The average varies typically from 10 to 150 μm. The first approximation concerning the deposit distribution is that the layer is distributed uniformly around the channel periphery. This may be true if circular channels are considered. The typical cross section of the monolith channels is, however, square. In this case a significant nonuniformity in the washcoat thickness is often encountered (Fig. 6). The reason is that the liquid from which the

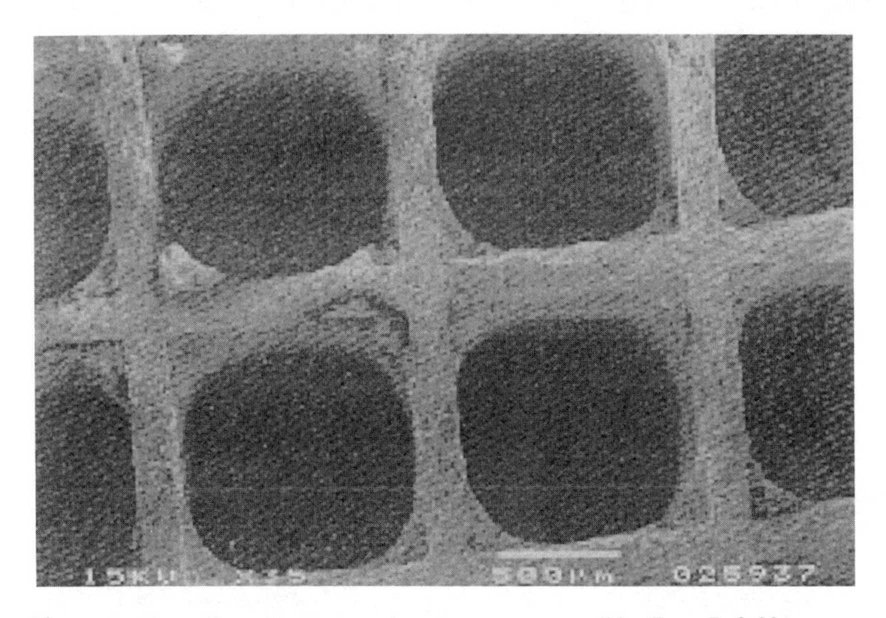

Figure 6 Nonuniform distribution of washcoat on a monolith. (From Ref. 22.)

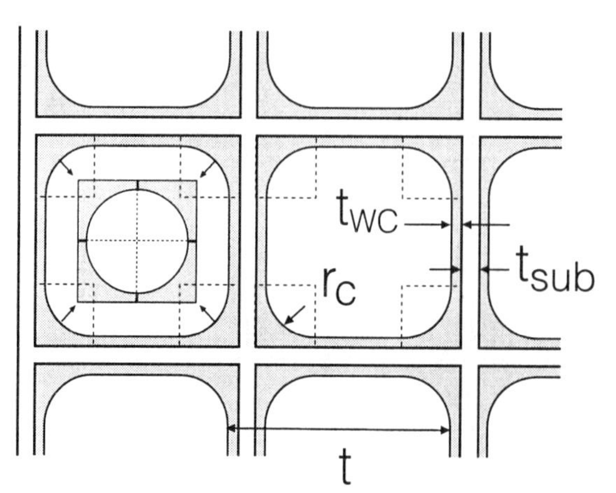

Figure 7 A monolithic structure with square channels with a nonuniform washcoat. The left cell illustrates how the washcoat conceptually is divided into slab regions along the sides and "circle-in-square" geometry in the corners, for calculation of the effectiveness factor. (From Ref. 14.)

washcoat is formed tends to settle in the corners due to surface tension forces. This maldistribution is fixed at the stage of thermal treatment after the layer is deposited. The problem of nonuniform deposition was a subject of theoretical and experimental analysis by Camp et al. [23], Kolb and Cerro [3,24], and Kolb et al. [25]. The significant maldistribution of the catalytic material may influence the effectiveness factor.

The problem of reaction diffusion in square channels with a nonuniform washcoat distribution in two spatial dimensions can be solved numerically using, e.g., the finite element method, as illustrated by Hayes and Kolaczkowski [22].

Edvinsson and Cybulski [14] presented a simpler method for evaluation of the overall effectiveness factor using the concept of a "circle-in-square" layer. According to this concept the washcoat layer can be divided into two sections (Fig. 7): (1) one of uniform thickness along the side of the channels wall, and (2) a circular section at the

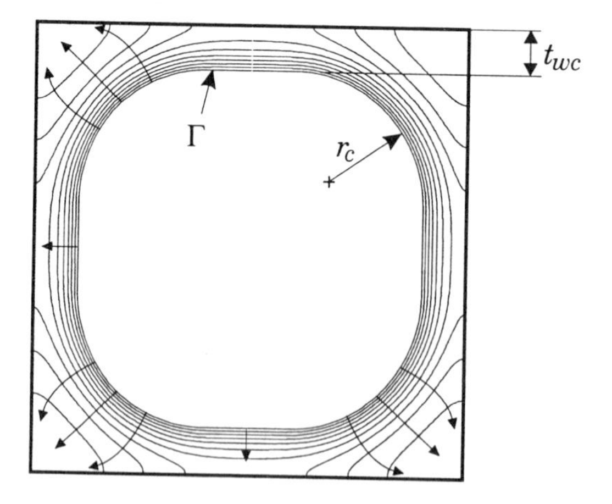

Figure 8 Concentration profiles in a washcoated monolith for a first-order reaction. (From Ref. 1.)

corners. The direction of diffusive flux is assumed to be normal to the curves of constant concentration (Fig. 8). Axial and peripheral mass dispersion inside the washcoat layer has been neglected. The method has been applied to a consecutive reaction scheme:

$$A \rightarrow B \rightarrow C \tag{44}$$

with both reactions being of the first order [1].

Exemplary results of modeling processes inside the catalytic layer are presented in Fig. 9. The solid lines show the dependency of the overall effectiveness factor on the relative distribution of the catalyst between the corners and the side regions. The two cases represent two levels of the first-order rate constants, with the faster reaction in case (b). As expected, the effectiveness factor of the first reaction drops as more catalyst is deposited in the corners. The effectiveness factor for the second reaction increases in case (a) but decreases in case (b). The latter behavior is caused by depletion of B deep inside the catalytic layer. What might be surprising is the rather modest dependency of the effectiveness factor on the washcoat distribution. The explanation is that internal diffusion is not important for slow reactions, while for fast reactions the available external surface area becomes the key quantity, and this depends only slightly on the washcoat distribution for thin layers. The dependence of the effectiveness factor on the distribution becomes more pronounced for consecutive reactions described by Langmuir–Hinshelwood–Hougen–Watson kinetics [26].

Cybulski and Moulijn [27] proposed an experimental method for simultaneous determination of kinetic parameters and mass transfer coefficients in washcoated square channels. The model parameters are estimated by nonlinear regression, where the objective function is calculated by numerical solution of balance equations. However, the method is applicable only if the structure of the mathematical model has been identified (e.g., based on literature data) and the model parameters to be estimated are not too numerous. Otherwise the estimates might have a limited physical meaning. The method was tested for the catalytic oxidation of CO. The estimate of effective diffusivity falls into the range that is typical for the washcoat material (γ-alumina) and reacting species. The Sherwood number estimated was in between those theoretically predicted for square and circular ducts, and this clearly indicates the influence of rounding the corners on the external mass transfer.

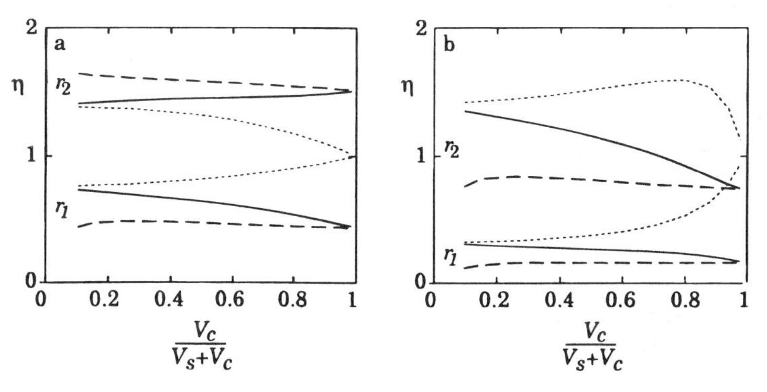

Figure 9 Effect of catalyst maldistribution for a consecutive reaction. $x_{cat} = 0.2$, $k_2/k_1 = 0.2$, $c_{b,2}/c_{b,1} = 0.5$. (a) $\phi_{A,1}^2 = 1$; (b) $\phi_{A,1}^2 = 10$. (From Ref. 1.)

D. Heat Transfer

Little has been published about heat transfer in narrow channels. Some attempts have been made to model heat transfer for segmented flow in small tubes (not capillaries) by Hughmark [28] and Oliver and Young Hoon [29,30]. The concept adopted by these authors is that heat transfer in a two-phase system may be approximated by heat transfer to a single fluid (the liquid phase) contained in a series of shorter tubes with some form of intermediate mixing. However, these studies have been carried out for larger tubes (2.54-cm ID) in which turbulent flow also occurs. Thus, they are not directly applicable to heat transfer monoliths.

Usually, an adiabatic operation of MR is assumed in modeling. This seems to be a reasonable assumption, especially for ceramic monoliths of low thermal conductivity and in the absence of significant radial temperature gradients. In practice, there is no convective radial heat transfer because of barriers between the adjacent channels. The radial heat transfer occurs only by conduction through the walls and, to some extent, by convection in liquid plugs.

III. MODELING OF MONOLITH REACTORS

A. General Description

In monolith reactor modeling, the surface reactions, the mass transfer within the channels, the residence-time distribution, and the external recirculation of gas and liquid must be considered.

The flow in the monolith reactor is distributed over the cross section. This flow can be uneven, giving different gas holdup in different channels. The liquid flow in a monolith channel is separated into a fast-moving liquid slug and a slow-moving liquid film close to the wall (Fig. 10). The film moves due to gravity when exposed to the gas bubble and is pushed by the liquid slug when it passes. Due to the surface tension and the pushing of the liquid slug, the liquid film in the liquid slug is somewhat thicker than the film exposed to gas.

The liquid flow is driven by gravity, and a small pressure increase is obtained at the bottom of the reactor. The pressure difference can be used to recirculate the gas. This recirculated gas is mixed with the fresh gas at the inlet. Some of the recirculated gas is also entrained by the liquid in the lower monoliths (Fig. 11).

The liquid is separated from the gas and recirculated by an external pump. This liquid recirculation flow is mixed with fresh reactants and added at the top of the monolith.

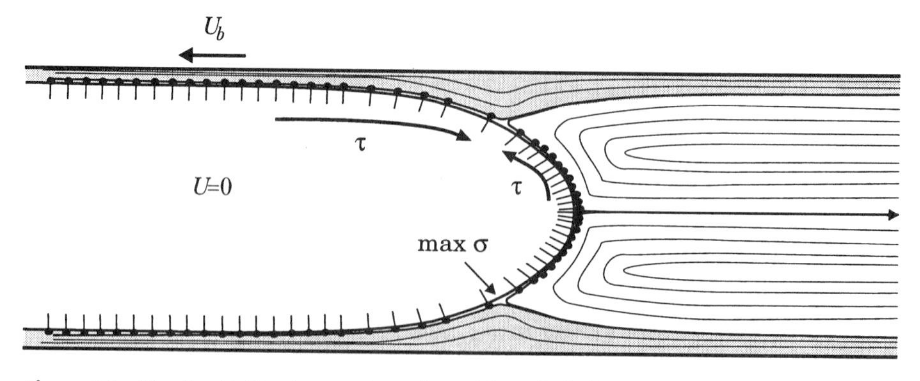

Figure 10 The liquid film in two-phase flow. (From Ref. 1.)

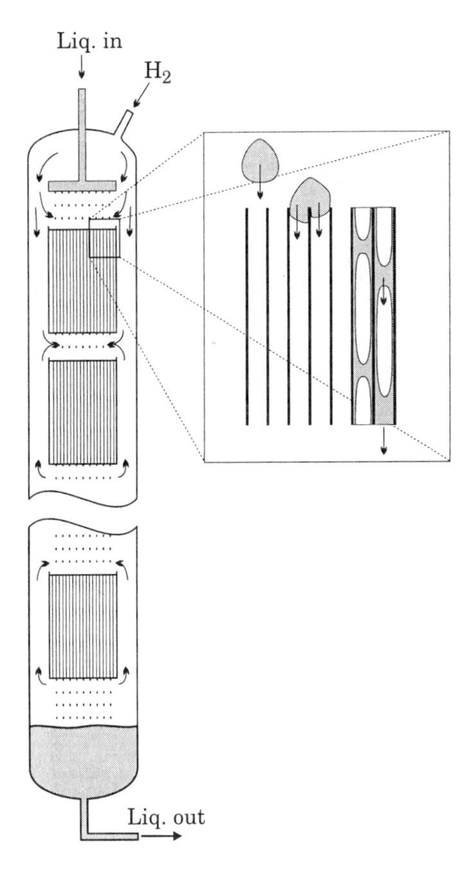

Figure 11 The monolith reactor. (From Ref. 26.)

B. Conversion in a Monolith Channel

The reaction occurs on the catalytic walls, and the transport of reactants is by diffusion and convection. There is a recirculation within the liquid slug, giving enhanced mass transfer. The liquid-phase component is transported to the film both by convection and by diffusion, and it is transported through the film due only to diffusion. The gas-phase component is transported by diffusion to the catalytic wall through the thin liquid film surrounding the bubble, as well as by diffusion into the liquid slug and by convection and diffusion through the recirculating liquid and finally by diffusion through the film.

The liquid film has a varying thickness and is alternately exposed to the gas and to the liquid with different concentrations. However, the film damps the effect of varying concentration, and the concentration at the wall is almost constant. The time constant for diffusion in the liquid film is $\delta_f^2/2D \approx 0.1$ sec. (Eq. 32), and the contact time for the gas bubble and the liquid slug is $L_g/U_m \approx 0.02$ sec. Thus the wall concentration will be almost constant, and the mass transfers directly from the gas bubble and through the liquid slug can be added using the same driving potential.

A balance of the gas-phase component A over an element containing a bubble and a liquid slug (Fig. 12) is given by

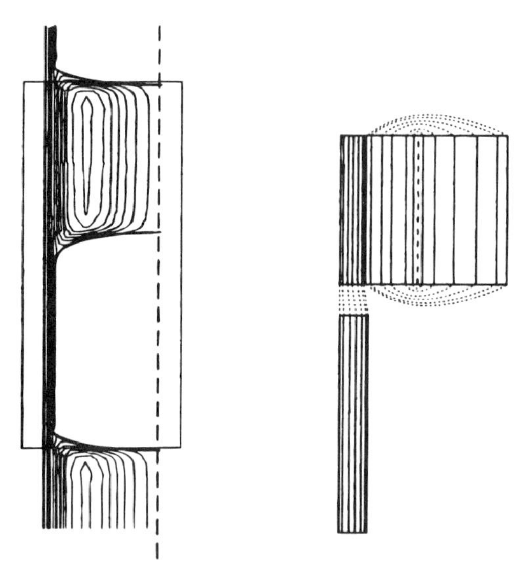

Figure 12 Modeling element in two-phase flow. (From Ref. 1.)

$$U_m \frac{d(V_g c_{A,g})}{dz} + U_m \frac{d(V_l c_{A,l})}{dz} = -N_A \cdot A \tag{45}$$

with $N_A \cdot A$ from Eq. (29). For the liquid component B, the corresponding equation is given by

$$U_m \frac{d(V_l c_{B,l})}{dz} = -k_{LS}(c_{B,l} - c_{B,\text{wall}})A_{LS} \tag{46}$$

The wall concentration is then given by the balance with the surface reaction:

$$\frac{N_A}{\nu_A} = \frac{N_B}{\nu_B} = \eta r(c_{A,\text{wall}}, c_{B,\text{wall}})t_{\text{wc}} \tag{47}$$

where ν_A and ν_B are the stoichiometric coefficients in the reaction and η is the effectiveness factor.

C. Recirculation of Gas and Liquid

The liquid is recirculated with an external pump that gives a controlled volumetric flow rate q_l. The recirculated gas flow is determined by the amount that can be entrained by the liquid into the monolith. The total linear velocity in the channels is controlled by gravity. The gas entrained by the liquid is given by the difference between the total linear velocity and the linear velocity of the liquid in the channels and can be calculated from the relationship

$$q_g + q_l = A_R \epsilon_c U_m \tag{48}$$

with the reactor cross section A_R and open area fraction ϵ_c.

D. Residence-Time Distribution

The axial dispersion in a single channel is low due to the very thin film surrounding the bubbles. For the low conversion that is usually obtained in a single pass through the monolith reactor, the residence-time distribution within the channels will have an insignificant effect on conversion. However, the difference between the channels can be important. In downflow where the velocity is controlled by gravity, the linear velocity will be almost the same in all channels, but the gas hold-up will be different in the channels due to uneven liquid distribution over the cross section.

IV. MONOLITH REACTORS VS. CONVENTIONAL REACTORS

A. General Description

A number of industrially important reactions are carried out in three-phase reactors. Typically, a gas and a liquid-phase reactant are converted to products in the presence of a solid catalyst. There are two basic concepts for designing such reactors, the fixed-bed reactor and the slurry reactor [31].

In a fixed-bed the reactor vessel is filled with catalyst particles having sizes in the range 1–3 mm. The catalyst particles may be spherical, cylindrical, or of more sophisticated forms, including eggshell catalysts, with active species concentrated near the outer surface of the particle. The gas and liquid phases are passed through the bed.

Several flow arrangements are possible, though cocurrent downflow is often preferred. One important flow regime is trickling flow and hence the term TBR. The main advantages are the high catalyst loading and the low investment cost. A drawback is the high pressure drop caused by using small catalyst particles, which are needed for processes controlled by internal diffusion. Fines may also be formed and can then plug the bed, thereby increasing pressure drop. A remedy is to use larger particles, which in the case of fast reactions can result in low catalyst utilization due to the relatively long diffusion distance. In addition, this may have an adverse effect on selectivity. Special care is also required to prevent flow maldistribution, which can cause incomplete wetting in some parts of the bed, resulting in lower overall production rates and poorer selectivity. For strongly exothermic reactions, more severe consequences—such as hot-spot formation and possibly even runaways—must be considered. The relatively low radial heat transfer rate in a large reactor implies that operation is essentially adiabatic.

In an SR, the catalyst is present as finely divided particles, typically in the range 1–200 μm. A mechanical stirrer, or the gas flow itself, provides the agitation power required to keep the catalytic particles in suspension. One advantage is the high catalyst utilization: Not only is the diffusion distance short, it is also possible to obtain high mass transfer rates with proper mixing. It is also reasonably simple to control the temperature. An important advantage in the case of a rapidly deactivating catalyst is the ease with which the catalyst can be replaced. There are, however, a number of problems associated with handling fine catalyst particles. They have to be separated from the products, and this is usually troublesome; plugging of lines and valves can occur; pyroforicity of the catalyst may also require special procedures. Moreover, the catalyst load is limited to what can be kept in suspension with a reasonable power input. Back-mixing is significant and may necessitate the staging of reactors, and this increases the cost.

The MR is an alternative to the two reactor types [32,33], which in some respects can be considered as a compromise between the two. Several processes are in the development stage, and one, the hydrogenation step in the production of H_2O_2 using the alkyl-anthraquinone process, has reached full scale, with several plants in operation [34]. The scale-up of MR can be expected to be straightforward in most other respects, since the conditions within the individual channels are scale-invariant. Scaling-up of MRs has proved to be very simple indeed: just multiplication of the number of channels in the full-scale reactors. This procedure was checked for the hydrogenation of alkyl-anthraquinones in a hydrogen peroxide plant [12]. An adiabatic operation is not any serious restriction for two-phase processes realized in MRs. Due to the high heat capacity of the liquid phase, the temperature rise is much smaller than that in gas-phase processes.

B. Comparison Between Trickle-Bed and Monolith Reactors

1. *Monolith vs. Trickle-Bed Reactor: A Qualitative Comparison*

TBRs are operated at comparable superficial liquid velocities and much higher superficial gas velocities than the monoliths. The gas velocities for the latter are restricted by the Taylor flow requirements; for higher velocities the transition to annular flow would occur. Hence, in this respect TBRs are superior to monoliths for processes characterized by high consumption of gaseous reactant. However, this superiority is reached at the cost of a much higher pressure drop in TBRs, almost two orders of magnitude higher than that in monoliths. It should also be emphasized that random nonuniformities of various kinds, including flow maldistributions arising inside a packed bed, are inherent features of TBRs. These can cause undesirable and sometimes dangerous hot spots in such reactors. In monoliths, no maldistributions are formed inside individual channels if flow is uniform at the inlet. Further, external mass transfer in monoliths is better than that in trickle beds. Thus, for external mass transfer-limited processes, monoliths could be favored. For internal diffusion-controlled processes a detailed comparison for each process has to be made. However, the short diffusion path in monoliths implies an easier manipulation of the selectivity than in trickle beds. Therefore, changes in flow, temperature, and/or concentrations can more easily influence rates and selectivity in MRs, and periodic operation (an alternate flow of gas and liquid plugs in the Taylor flow regime) is an inherent feature of the monolithic pseudo-steady-state processes.

The most important differences between the TBR and the MR are summarized in Table 2. Obviously, less experience has been accumulated with monoliths, in particular for three-phase applications. They are also more expensive. Therefore, the use of monoliths can be economically justified only for three-phase processes, in which it offers a distinct advantage, like higher yield (improved selectivity), increased throughput of a plant, or lower investment or running costs. Of particular interest are situations in which an MR substantially simplifies the design or operation of a unit. It should be possible to reduce the cost of monoliths when manufacturers design monoliths for the chemical industry. The high cost of monoliths is in part explained by the fact that they are designed mostly to meet the demands of high-temperature transient operation in vehicles.

For simple reactions occurring fully in the kinetic regime, the TBR will outperform the MR, due to the higher load and the lower cost of the catalyst. In processes involving faster reactions, especially if there are selectivity concerns (like the intermediate products being the desired one), it is not obvious which reactor is better suited for the process.

Table 2 Characteristics of Trickle-Bed and Monolith Reactors for Three-Phase Processes

	TBR	MR
Catalyst:		
Manufacture	Well-established	Established for gas phase; techniques for liquid phase at developmental stage
Cost	Low	Moderate-high
Volumetric catalyst load	0.55–0.6 for conventional packings; <0.3 for shell catalysts	0.05–0.25
Handling	Replacement using well-established procedures	Monolithic blocks are assembled in frames that can be stacked on top of one another
Size	Pellets, 1–5 mm	Channels, 1–4 mm, possibly with a washcoat, 50–150 μm
External surface area	1000–3000 m^2/m^3	1500–2500 m^2/m^3; external mass transfer better than in TBR
Diffusion length	0.1(shell)–2.5 mm	50–150 μm
Operating conditions:		
Superficial velocities	Liq. 0.005–0.05 m/sec Gas 0.05–1.5 m/sec	Liq. 0.03–0.15 m/sec Gas 0.05–1.0 m/sec; the sum of superficial velocities should be ~0.4 m/sec
Pressure drop	High for small particles	Very low
Mode of operation	Pseudo-steady-state	Inherently unsteady
Design:		
Experience	Many units in operation	Only one full scale process in operation
Gas recirculation	External, needs compression	Internal, no pump needed
Scale-up	Gas/liquid maldistributions can appear	Simple
Inlet distribution	Good distribution required; liquid tends to flow toward the wall	Very good distribution needed

Note: From Ref. 26.

The diffusion length in a particulate catalyst is longer than in a monolithic catalyst, and hence the effectiveness factor is typically lower for the TBR [14]. As a consequence, the conversion per unit volume of catalyst is lower in the TBR. Moreover, the diffusion length can influence the selectivity of the reaction if the rate is affected by diffusion. Two ways to reduce the diffusion length in TBRs are: (1) to use smaller catalyst particles, and (2) to use a catalyst with the catalytically active species located near only the outer surface of the particle (an eggshell). However, the first remedy will increase the pressure drop until it becomes unacceptable, and the second one reduces the catalyst load in the reaction zone, making the loads of TBR and MR comparable. For instance, the volumetric catalyst load for a bed of 1-mm spherical particles with a 0.1-mm-thick layer of active material is 0.16. The corresponding load for a monolithic catalyst made from a commercial cordierite structure (square cells, 62 channels/cm^2, wall thickness 0.15 mm), also with a 0.1-mm-thick layer of active material, is 0.26.

The frictional pressure drop is always lower in the MR, by up to two orders of magnitude, for all relevant sizes of catalyst particles (shell catalysts of a greater size included). For an MR operating in downflow mode, it is possible to balance the frictional pressure drop with the hydrostatic pressure of the liquid inside the channels. The essentially zero net pressure drop provides an opportunity to operate the MR with internal recirculation of gas. Since the gas does not need to be recompressed, an open passageway from the bottom of the reactor to the top is all that is needed (Fig. 11).

2. *A Comparison of Monolith and Trickle-Bed Reactors: Modeling*

The differences between the TBR and the MR originate from the differences in catalyst geometry, which affect catalyst load, internal and external mass transfer resistance, contact areas, as well as pressure drop. These effects have been analyzed by Edvinsson and Cybulski [14,26] via computer simulations based on relatively simple mathematical models of the MR and TBR. They considered catalytic consecutive hydrogenation reactions carried out in a plug-flow reactor with cocurrent downflow of both phases, operated isothermally in a pseudo-steady state; all fluctuations were modeled by a corresponding time average:

$$A \underset{r_1}{\overset{H_2}{\rightarrow}} B \underset{r_2}{\overset{H_2}{\rightarrow}} C \tag{49}$$

The reaction rate was assumed to obey a Langmuir–Hinshelwood–Hougen–Watson (LHHW) type of rate expression:

$$\begin{aligned} r_1 &= k_1 \frac{K_{H_2}c_{H_2}}{1 + K_{H_2}c_{H_2}} \frac{K_A c_A}{1 + K_A c_A + K_B c_B} \\ r_2 &= k_2 \frac{K_{H_2}c_{H_2}}{1 + K_{H_2}c_{H_2}} \frac{K_B c_B}{1 + K_A c_A + K_B c_B} \end{aligned} \tag{50}$$

These kinetic expressions can be useful in many situations, since they capture two key aspects of heterogeneous catalysis: the rate of the reaction, and the saturation of the surface by the reactants. The values assigned to the various kinetic and adsorption parameters in this work produce rates that agree well with those reported in the literature. The liquid-phase components were considered nonvolatile. The saturation concentration of H_2 was evaluated using Henry's law. All physical parameters were treated as constants. The catalyst properties were representable for a supported noble metal hydrogenation catalyst.

For the TBR, spherical catalyst particles of uniform size with the catalytically active material either uniformly distributed throughout the catalyst or present in a shell were considered. For the MR, channels of square cross section were assumed to have walls covered by the washcoat distributed in such a way that the corners are approximated by the "circle-in-square" geometry, while the sides are approximated by a planar slab geometry. The volumetric load of catalytic material was a function of the washcoat thickness and the radius of the washcoat in the corner. An extruded monolith with catalytic species incorporated into the walls was also taken into account. All catalytic material was assumed to be uniformly distributed in a washcoat only. For details of modeling, the reader is referred to the original papers of Edvinsson and Cybulski [14,26].

The criteria chosen for more detailed comparison of the performance of the MR and the TBR are the space time yield, STY_v, in moles of the intermediate product B formed per cubic meter reactor and second, the selectivity, S, in net moles of B formed per mole of A consumed, and the pressure drop, in bars.

3. *Results of Modeling: General Remarks*

First, consider a crude design of the TBR. For a selective consecutive hydrogenation of the type assumed, it is desirable to use as small particles as possible, since this improves both the effectiveness factor and selectivity. The limitation is the acceptable pressure drop. If only these three performance criteria are considered, the optimal design is a reactor that is very short and with a very large diameter. There are obviously practical limits on the ratio between the length and the diameter of the reactor (L_R/d_R). For highly packed beds, the pressure drop becomes a limiting factor, while for shallow beds significant maldistributions can appear. Moreover, there exists an optimal economic ratio L_R/d_R with respect to investment costs. The main operating parameters affecting pressure drop are the particle diameter and liquid flow rate.

With respect to pressure drop, the situation is quite different for the MR. As the liquid phase enters the top of the monolith it accelerates/decelerates until it reaches the linear velocity at which the hydrostatic pressure drop balances the frictional force. This velocity is approximately

$$U_l = \frac{\rho_l g d_t^2}{32\mu_l} \tag{51}$$

For the fluids and the monoliths considered in this comparison, this limit is approximately 0.50 m/sec (depending on the catalyst load), which is in good agreement with the value found experimentally by Irandoust et al. [12]. If the liquid load is less than this limit, gas will be sucked in as well. Hence the sum of the linear velocities will tend to be close to the maximum flow rate of liquid alone. Frictional pressure drop in the MR is up to two orders of magnitude lower than in the TBR. Consequently, for the MR we may consider very high flow rates and high columns, higher than appears to be of practical interest, before the pressure drop becomes a restriction with the physical properties considered. For practical reasons an upper limit of 20 m for L_R was taken. In order to make a comparison between the MR and the TBR, some more restrictions were imposed. In a consecutive-reaction system like the one considered, selectivities must be compared at the same conversion, and results discussed below are for a 50% conversion of reactant A.

a. Catalyst Geometry. As mentioned earlier, the most obvious difference between the MR and the TBR lies in the shape of the catalyst support and in the distribution of active material. Figure 13 illustrates the relationship between the catalyst geometry, the productivity (or space time yield per volume, STY_v), and the selectivity (S). The comparison is made for the same superficial flow rates and conversion (x_A = 10%). A low conversion level was chosen, so the effect of different pressure drops should not considerably affect the comparison. For each curve the thickness of the catalytic layer, and hence the catalyst load, decreases as we move to the right. Selectivity increases with decreasing thickness

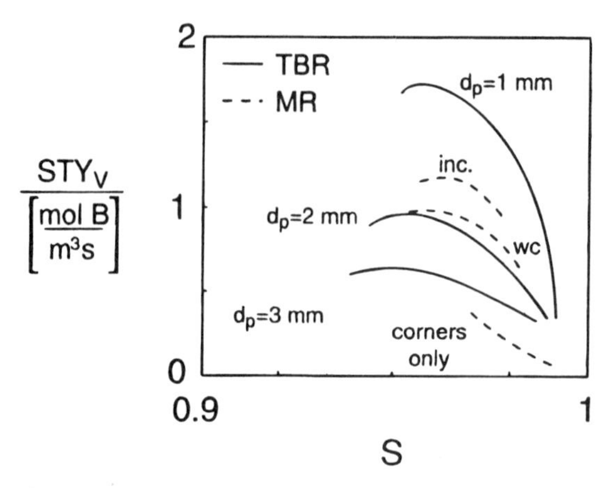

Figure 13 Attainable productivity and selectivity with various catalyst geometries. On each curve the thickness and load of the catalyst decrease from left to right. $k_1 = 40$ mol/m^3 sec, $k_2 = 0.1\ k_1$. Volumetric catalyst loads: TBR, 4.0–55%; MR: incorporated type (inc) 15.1–35.5%, washcoated (wc) 8.6–28.6%, and washcoated with all active material in the corners (corners only) 0.6–6.6%. (From Ref. 26.)

while the productivity goes through a maximum. For the TBR this can be explained by the lower effectiveness factor associated with thicker catalyst layers. For the MR there is an additional effect in that the external surface area decreases as the washcoat becomes thicker since it grows toward the center of the channel. It can be concluded that for a reasonably fast reaction the best performance in terms of STY_v and S can be attained using low-diameter shell catalysts. The performance of a monolithic catalyst is comparable to that of 2-mm particles. Moreover the incorporated type is somewhat better than the washcoated one. The performance of an MR with a very poorly distributed washcoat, all active material in the corners, has been added to illustrate the negative effect of nonuniform distribution.

b. Pressure Drop. The advantage of small particles in the TBR is limited by the acceptable pressure drop. As the size of a TBR is increased, it is necessary to increase both the reactor length and the diameter, since the ratio between them must be kept reasonable. As a consequence, the acceptable pressure drop per unit length decreases, and this necessitates the use of larger particles. This also causes the external surface area to decrease and the diffusion distance to increase. In Fig. 14, productivity and selectivity of a TBR are plotted against the length of the reactor, which is filled with the smallest particles producing an acceptable pressure drop (2 bar) for a fixed conversion ($x_A = 0.50$). The lines for an MR of the same length operated with zero pressure drop (i.e., with balance between the frictional and the hydrostatic pressure drop) are also plotted. The productivity of a TBR for the range of length considered is higher than that of a MR, obviously at the cost of a much higher pressure drop. This is due to the higher volumetric catalyst load in the TBR. Contrary to this, selectivity is rather higher for an MR, except for shell particles of the very thin catalytic layer in the TBR. The MR is almost insensitive to variations in bed depth. The importance of this difference will depend on the acceptable pressure drop, the required space velocity, and the viscosity of the liquid.

c. Design and Operating Variables. At first a comparison is made for a reference case where the only parameters varied are the superficial liquid flow rate at the inlet,

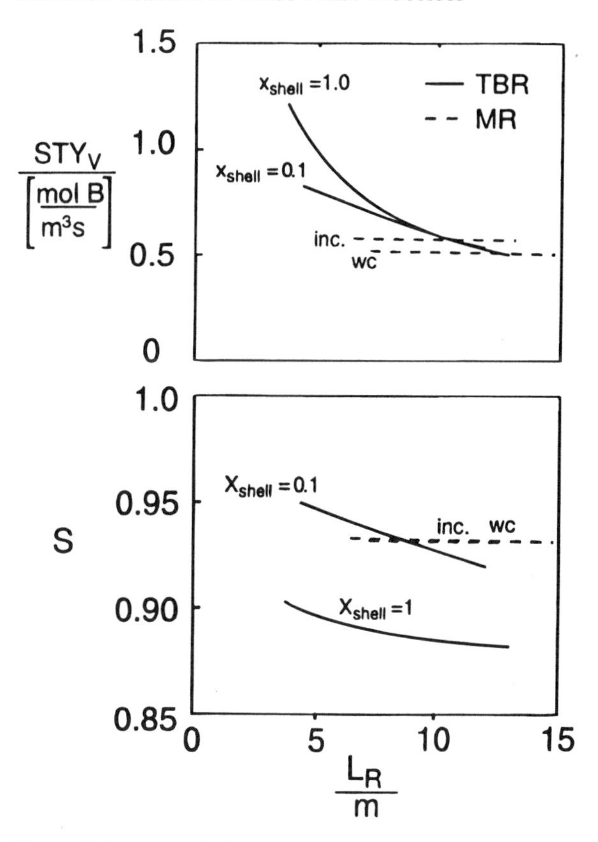

Figure 14 Productivity and selectivity as a function of the reactor length. MR: $u_{L,\text{sup}} = 0.02\text{–}0.04$ m/sec, (wc) washcoated $t_{\text{wc}} = 50$ μm, $r_c = 150$ μm, (inc) incorporated $t_{\text{wc}} = 50$ μm, $t_{\text{sub}} = 50$ μm, $r_c = 0$ μm. TBR: $d_p = 1.0\text{–}2.5$ mm, $u_{L,\text{sup}} = 0.019\text{–}0.037$ m/sec. (From Ref. 26.)

$u_L^{(0)}$, the reactor length L_R, and the depth of the catalyst layer. In the case of the TBR the particle diameter is varied, while for the MR both the thickness of the washcoat and the corner radius are varied. Hence, for the MR the catalyst load varies in addition to the diffusion distance. The relation between the conversion of A, x_A, and $u_L^{(0)}$ is illustrated in Fig. 15. The calculations are limited by an allowed pressure drop of 5 bar, indicated by the dotted line in Fig. 15a. For the MR (Fig. 15b) no further increase in conversion is obtained as the catalyst load increases beyond the intermediate level of 17%. This is better illustrated in Fig. 16, where it can be seen that conversion for the MR actually passes through an optimum as the catalyst load is increased. The decrease at high loads can be explained by the reduction of surface area that follows with increasing thickness of the washcoat. In addition, the selectivity decreases as the catalytic layer becomes thicker. The substrate A is consumed faster in the TBR because of the higher catalyst load.

Figure 16 is constructed for $L_R = 10$ m and $x_A = 0.50$. The highest STY_v is obtained for the TBR using the smallest particles possible. The pressure drop is approximately 5 bar, which is close to what can be accepted. Both pressure drop and STY_v decrease fast with increasing particle size, and for the 2-mm particles the STY_v has dropped below the highest STY_v that can be reached with the washcoated MR. For both reactors the selectivity decreases with increasing catalyst thickness. Selectivity is higher for the MR. The diffusion

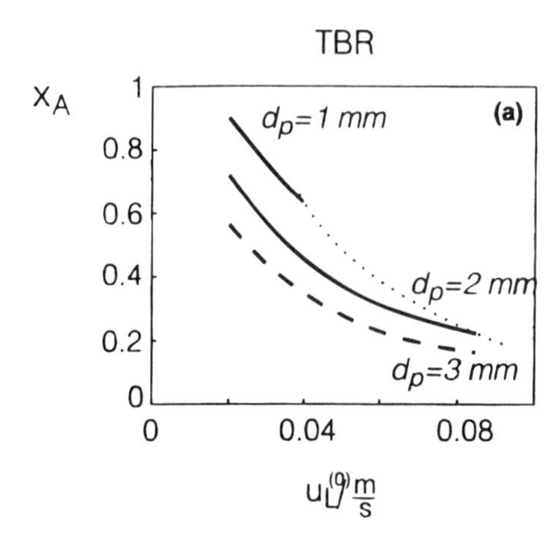

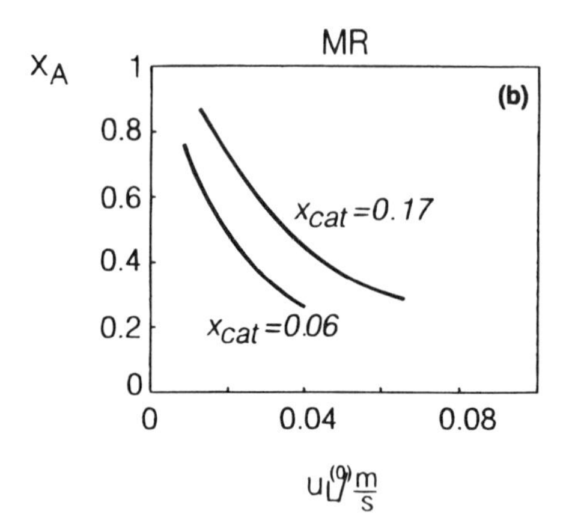

Figure 15 Conversion vs. superficial liquid flow rate. $u_G^{(0)} = 0.3$ m/sec; $L_R = 10$ m; $k_1 = 40$ mol/m^3 sec; $k_2 = 0.1\ k_1$; $K_A = K_B = 0.0025$ m^3/mol; $K_{H_2} = 0.0296$ m^3/mol. TBR: dashed line is shell catalyst with $t_{shell} = 0.05\ d_p$, solid line is uniform activity ($t_{shell} = 0.5\ d_p$). The dotted line indicates a pressure drop of 5 bar. (From Ref. 14.)

distance in a TBR can be reduced by using a shell catalyst, which also reduces catalyst loads. A curve for a shell catalyst with active material in only the outer 10% of the radius is shown in Fig. 16 as well.

A less restrictive comparison can be made at a fixed conversion but allowing the liquid velocity, and hence L_R, to vary. In Fig. 17, the L_R required to reach $x_A = 0.50$ is shown with the resulting productivity and selectivity. STY_v and S are relatively insensitive to $u_L^{(0)}$ and hence to L_R for the MR. It should also be noted that although STY_v for the two washcoat loads (17 and 33%) is virtually the same, the selectivity is predicted to differ substantially. In these cases, the increase in catalyst load causes an increase in conversion that is almost perfectly balanced by a decrease in selectivity. The similarity of the shell catalyst and the MR is also apparent, though the pressure drop is greater for the TBR. The trends in relative performance for superficial gas velocity within the range tested ($u_G^{(0)} = 0.1$–0.8 m/sec) were the same for both reactors.

d. Reaction Kinetics. For slow reactions without side reactions, the TBR is favored simply because higher catalyst loads can be used. Fast reactions and/or competing side

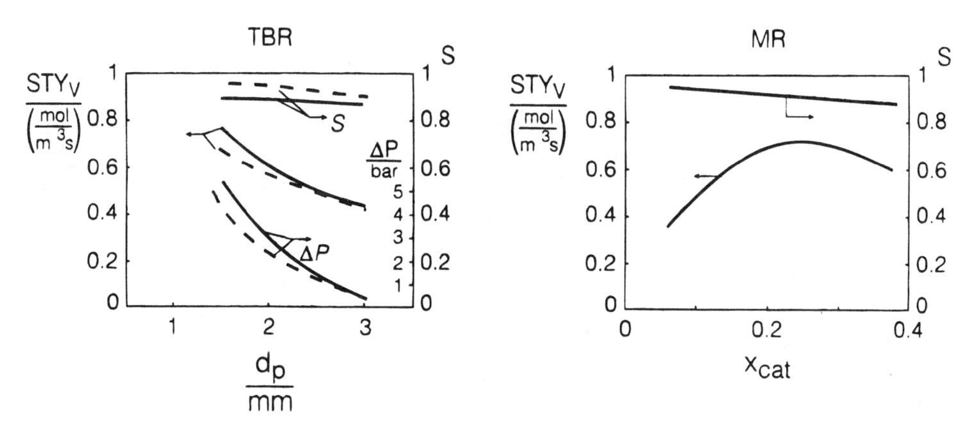

Figure 16 Dependence of productivity, selectivity, and pressure drop (TBR) on the catalyst dimensions. $u_G^{(0)}$ = 0.3 m/sec; x_A = 0.50; L_R = 10 m; k_1 = 40 mol/m^3 sec; k_2 = 0.1 k_1; K_A = K_B = 0.0025 m^3/mol; K_{H_2} = 0.0296 m^3/mol. TBR: dashed line is shell catalyst with t_{shell} = 0.05 d_p, solid line is uniform activity (t_{shell} = 0.5 d_p). (From Ref. 14.)

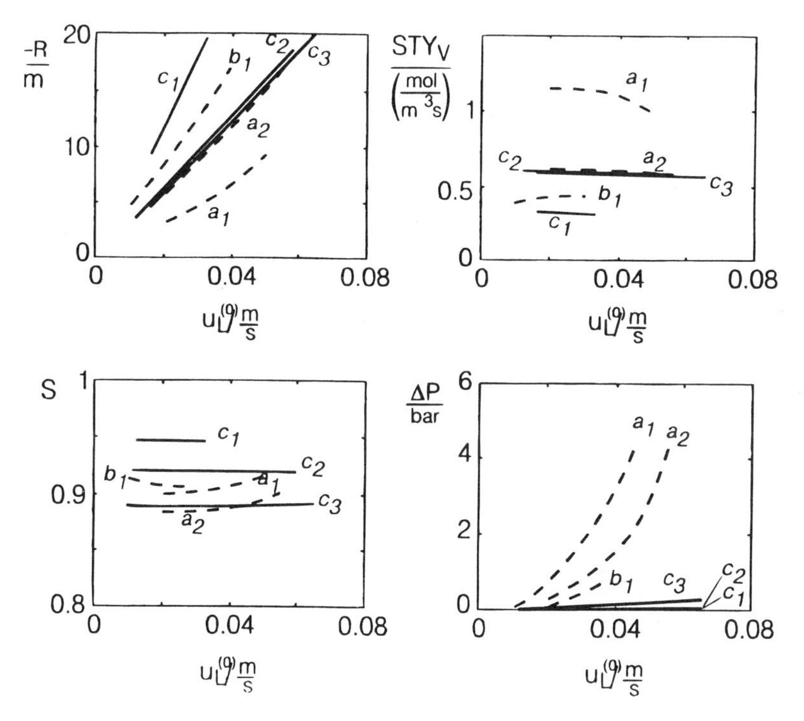

Figure 17 Dependence of reactor length, productivity, selectivity, and pressure drop on the superficial liquid rate. $u_G^{(0)}$ = 0.3 m/sec; x_A = 0.50; k_1 = 40 mol/m^3 sec; k_2 = 0.1 k_1; K_A = K_B = 0.0025 m^3/mol; K_{H_2} = 0.0296 m^3/mol. Catalyst geometries: a_1: d_p = 1 mm; a_2: d_p = 2 mm; b_1: d_p = 3 mm, t_{shell} = 150 μm; c_1: x_{cat} = 6%; c_2: x_{cat} = 17%; c_3: x_{cat} = 33% (solid line = washcoated monolith, dashed line = TBR). (From Ref. 14.)

reactions make large particles unsuitable, so smaller particles or shell catalysts must be used. Figure 18 illustrates the trade-off between catalyst volume on the one hand and surface area and diffusion length on the other. The curves show lines of equal STY_v in the MR and the TBR for the catalyst geometry indicated. The rate of the first reaction (in the absence of mass transfer) at the reactor inlet (r_{1s}) is used as the abscissa, and the ratio of rate constants k_2/k_1 is used as the ordinate. The lower left corner in the figure corresponds to a slow reaction with small selectivity problems, while the upper right corner represents the reverse situation. In general, an MR can perform better than a TBR if the initial rate exceeds 10 mole m^{-3} sec^{-1} for model parameters and design/operating variables considered. The area of the higher performance of an MR shifts to the left with an increase of particle size and/or shell thickness in a TBR. This is related strictly to the increasing diffusion path inside the particles. The border lines of equal performance for a monolithic catalyst of incorporated type (Fig. 18b) are shifted more to the left than for washcoated monoliths. This can be attributed to the higher volumetric catalyst load for the "incorporated" monolithic catalysts.

In Fig. 19, the relationship between the performance and the ratio k_2/k_1 is illustrated. For each set of kinetic constants, simulations were carried out corresponding to each combination of $u_L^{(0)} = 2$ cm/sec and $d_p = 1$ or 2 mm (TBR), $x_{cat} = 0.06$ or 0.168 (MR). As might be expected, both STY_v and selectivity decrease with the increase of the ratio k_2/k_1. STY_v is higher for the TBR, while selectivity is better for the MR. The relative performance appears to change only weakly.

The adsorption equilibrium constants in the reference case are set so that the equilibrium coverage would be 0.5 if the compound were present alone at the inlet concentration. The effect on STY_v and S of varying this equilibrium coverage is summarized in Fig. 20. The adsorption strength of the liquid-phase components affects the performance very weakly. The influence of adsorption strength of hydrogen is greater, and this can be understood by considering the impact it has on the effectiveness factors. For stronger adsorption the reaction rate drops less rapidly as we penetrate deeper into the catalytic material, and hence the catalyst utilization increases. Moreover, the tendency for the

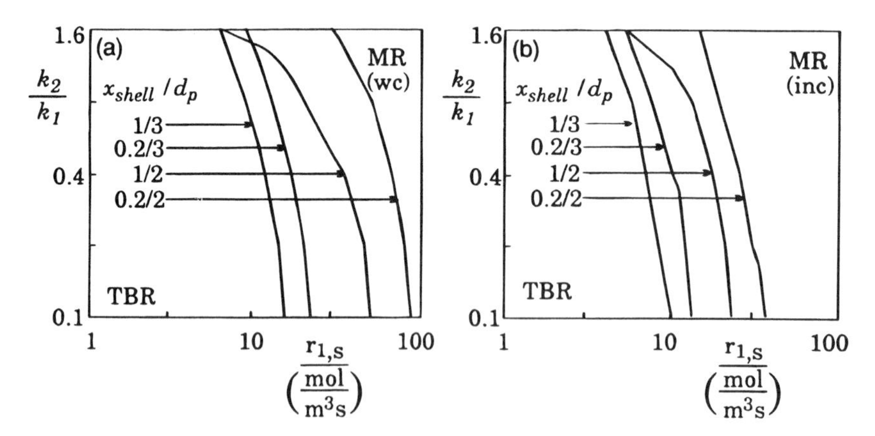

Figure 18 Contours indicating equal productivity for the TBR and the MR as a function of the rate constants. Monolith geometries: (a) washcoated, $t_{wc} = 50$ μm, $t_{sub} = 150$ μm, $r_c = 150$ μm; (b) incorporated, $t_{wc} = 50$ μm, $t_{sub} = 50$ μm, $r_c = 0$ μm. TBR geometries are given in the figure. (From Ref. 26.)

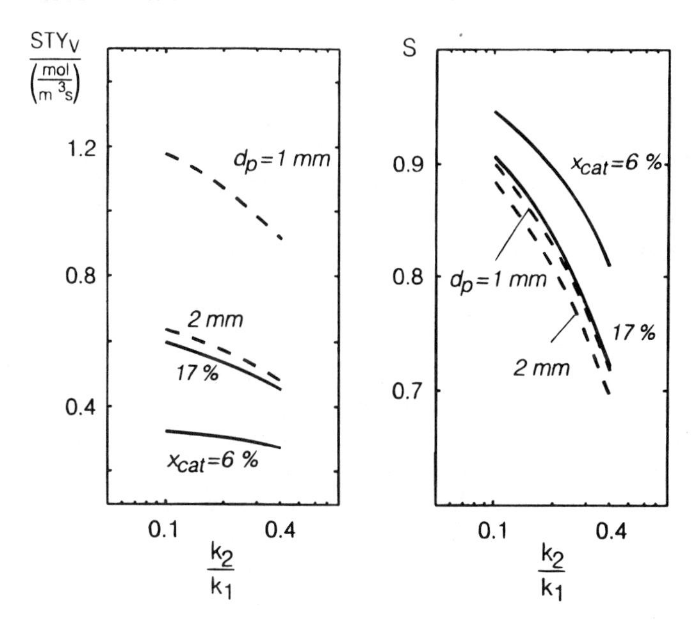

Figure 19 Dependence of productivity and selectivity on the ratio of k_2/k_1. $u_G^{(0)} = 0.3$ m/sec; $x_A = 0.50$; $k_1 = 40$ mol/m^3 sec; $k_2 = 4$–16 mol/m$_3$ sec; $K_A = K_B = 0.0025$ m^3/mol; $K_{H_2} = 0.0296$ m^3/mol (solid line = washcoated monolith, dashed line = TBR). (From Ref. 14.)

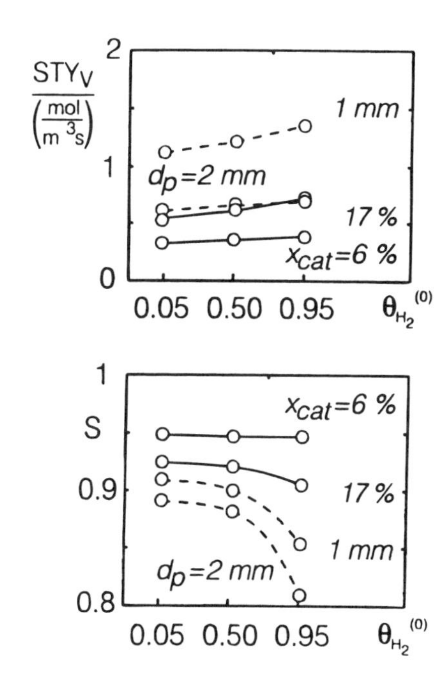

Figure 20 Dependence of productivity and selectivity on the adsorption constants. $u_G^{(0)} = 0.3$ m/sec; $x_A = 0.50$; (solid line = washcoated monolith, dashed line = TBR) $K_A = K_B = 0.0025$ m^3/mol; $\theta_{H_2}^{(0)} = 0.05$: $k_1 = 400$ mol/m^3 sec, $k_2 = 0.1\ k_1$, $K_{H_2} = 0.0016$ m^3/mol; $\theta_{H_2}^{(0)} = 0.50$: $k_1 = 40$ mol/m^3 sec, $k_2 = 0.1\ k_1$, $K_{H_2} = 0.0296$ m^3/mol; $\theta_{H_2}^{(0)} = 0.05$: $k_1 = 21.1$ mol/m^3 sec, $k_2 = 0.1\ k_1$, $K_{H_2} = 0.5625$ m^3/mol. (From Ref. 14.)

reaction to proceed deeper inside the catalyst causes the selectivity to decrease. This explains the moderate effect on the MR relative to the TBR.

The relative dependence of performance on the rate constant k_1 is illustrated in Fig. 21. In all cases STY_v increases and S falls with increasing rate constant. Again it should be noted that the relative performance depends only weakly on the rate constant.

In the results just discussed, the attempt was to determine conditions for equal performance, from which regions of relative superiority of either reactor could be identified. Moreover, the relationship between the location of this boundary and some key parameters has been studied. It is difficult to show the location of the boundary using two- or three-dimensional plots. The most practical method would be to optimize a process for a defined set of model parameters with respect to design and operating conditions, and to compare optima for the MR and the TBR to find which of the two is superior.

In spite of difficulties in simple comparisons, some general conclusions can be drawn. The highest catalyst load and, consequently, conversion of the substrate is reached in the TBR. Use of a monolith with a catalytically active substrate allows reasonably high catalyst loads, approximately half that of the TBR, to be reached. Hence, for noncompeting slow reactions, this determines the superiority of the TBR. The picture is not so obvious for competing reactions with an intermediate as the product desired. Then the diffusion length may be a factor determining yield and/or productivity. Even then, STY_v is higher for the TBR for the process conditions studied here if particles of diameter less than 2 mm are used. It becomes, however, commensurate for particles with diameter about 2 mm. The diffusion lengths are generally shorter in the MR compared to the TBR. This is the reason why selectivity for a network of consecutive reactions and the kinetic model

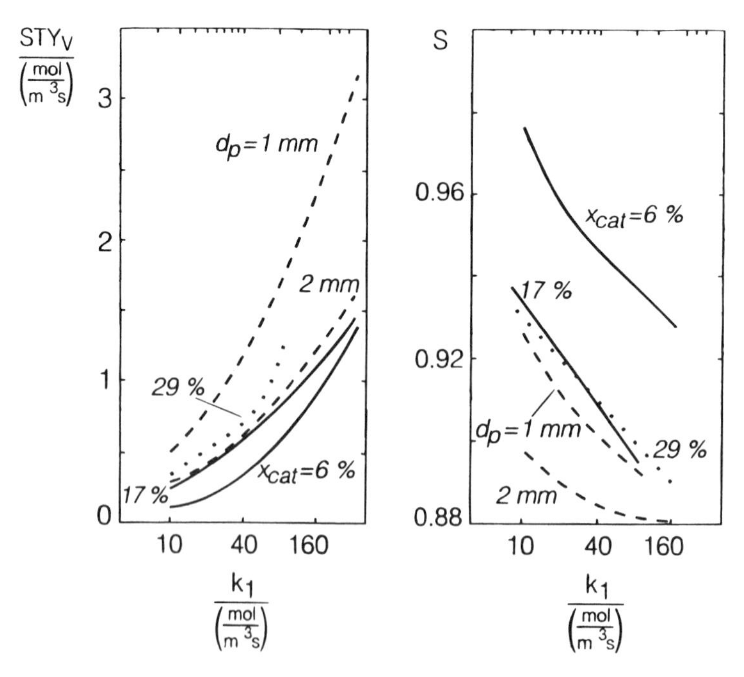

Figure 21 Dependence of productivity and selectivity on the rate constants k_1 and k_2. $u_G^{(0)}$ = 0.3 m/sec; x_A = 0.50; k_2 = 0.1 k_1; $K_A = K_B$ = 0.0025 m^3/mol; K_{H_2} = 0.0296 m^3/mol (solid line = washcoated monolith, dotted line = extruded monolith, dashed line = TBR). (From Ref. 14.)

considered here is better for the MR within the whole range of design and operating variables studied. The use of a shell catalyst enables similar diffusion lengths to be reached in the TBR. This, of course, reduces the catalyst load in the TBR. The TBR with shell catalyst and the MR can be compared only for a specific process after optimization for both reactors is made. If a process is limited by external mass transfer, both reactors would perform similarly. The highest surface area is obtained for the TBR with very small particles, which are characterized by high pressure drop and poor flooding. If particles with diameter 2 mm are used, the surface area of the MR with the cell density of 62 channels/cm^2 becomes commensurate.

A higher productivity of the TBR can be achieved only for small particles. This is, however, at the cost of relatively high pressure drop in the TBR. The use of small particles in the TBR causes a pressure drop up to 5 bar per reactor unit, which becomes limiting for long reactors and high fluid velocities. This can be a problem if large capacity is needed. The frictional pressure drop for MR is much smaller and in practice never limiting. Indeed, gravity alone can give too high velocities. To reach lower velocity, it would be desirable to use monoliths with still higher cell densities (e.g., 90 channels/cm^2 or more). If the fluid contains fines, the straight channels of the MR can offer an advantage, since the tendency to clog the reactor is expected be smaller.

e. Considerations of Economics. An economic evaluation of the two process alternatives requires detailed knowledge of price and cost factors. But some general observations can be made. The cost of monoliths is higher than for the conventional catalyst shapes, and, when large volumes of catalyst are required for a relatively slow reaction, this is likely to be a decisive factor. If only the outer portion of the catalyst can be utilized effectively (selectivity or mass transfer constraints), the MR becomes competitive. An important factor in this respect is the acceptable pressure drop, since it determines how small a catalyst can be used in the TBR but has little effect on the performance of the MR. This can be expected to be more important when only small pressure drops are accepted, e.g., when the system pressure is low. Most of the progress made in monolith technology is related to gas-phase applications. For processes in which slight modifications of such a catalyst suffice, the cost difference should be moderate.

The design of the reactor vessel and fluid distributors is perhaps slightly more complex and more expensive for the MR. Much more experience is available in the design of TBRs compared to MRs, in which only one process is operated at full scale. TBRs are, however, liable to experience problems with fluid distribution throughout the bed. The scale-up of MRs is in principle more straightforward, though little help is available in the literature on the design of liquid distributors. A possibly unique feature of the MR operated at zero net pressure drop, hydrostatic pressure balancing the frictional one, is the simple way in which internal gas recirculation can be achieved. Since the gas does not need to be recompressed, an open passageway from the bottom of the reactor to the top is all that is needed. This can simplify the process and hence reduce operation and investment costs.

f. Conclusions. An MR can be an attractive alternative to a TBR for hydrogenations proceeding in a consecutive-reaction scheme with an intermediate as the product desired. For the model parameters and design/operating variables considered, an MR can perform better for fast reactions characterized by initial rates higher than 10 mole m^{-3} sec^{-1}. For such fast reactions, performance indices are better than those for a TBR with particles of size greater than 2 mm. Selectivity is better for the MR over almost the whole range of design and operating variables studied. Space time yield (STY_v) is rather higher for TBR if particles of diameter less than 2 mm are used, although MR might become competitive at greater catalyst load, especially if monoliths with catalytically active substrate are considered.

The selectivity of the TBR can be increased for this type of reaction if a shell catalyst is used. The use of shell catalysts leads to reduction in catalyst load, which then becomes comparable to that of a MR. The higher productivity of the TBR is reached at the cost of relatively high pressure drop, up to 5 bar per reactor unit, which limits the size of the reactor. A TBR is characterized by much higher frictional pressure drop, which is negligible in an MR and is balanced by hydrostatic pressure with zero net pressure drop. This creates a unique possibility of operation with an internal hydrogen, pumpless recirculation.

C. Comparison Between Slurry and Monolith Reactors

Mass transfer–limited processes favor SRs over monoliths as far as the overall process rates are concerned. Moreover, SRs are more versatile and less sensitive to gas flow rates. However, the productivity per unit volume is not necessarily higher for SRs because of the low concentration of catalyst in such reactors. There is also no simple answer to the selectivity problem, and again each process should be compared in detail for both reactors. For a kinetic regime, monoliths can be more advantageous due to their easier operation. The catalyst does not disintegrate due to the stirrer action, and catalyst separation is avoided. Catalysts are often pyrophoric materials, handling of which is usually a hazardous operation. The benefits of MRs can be achieved only for stable catalysts. For quickly deactivating catalysts, SRs are easier to operate, since replacement of decayed catalysts is simpler.

V. REACTOR DESIGN

Industrial experience with the three-phase MR is limited, since only few large-scale industrial plants are running in the world today. Very little has been published about these industrial reactors and reactor scale-up. Also, most modeling has been done on cylindrical channel geometry, while most industrial reactors use sinusoidal or square geometry. Hence this chapter mainly summarizes our own experience with three-phase monolith reactors, with limited reference to the literature.

Cocurrent downflow with slug or Taylor flow has been most widely used. Other possible designs, e.g., cocurrent upflow and froth flow, have to our knowledge been tested only in laboratory and pilot plant reactors. Consequently, we will focus on downward slug flow, and the main areas of interest are scale-up, liquid distribution, space velocity, stacking of monoliths, gas–liquid separation, recirculation, and temperature control.

The modeling of single channels in Section II of this chapter is accurate for circular channels, but most reactors contain square or sinusoidal channels. There are very few measurements of film thickness and mass transfer in square and sinusoidal channels. In reactor design we have to rely on reaction rate measurements in laboratory and pilot plant reactors, and scale up the results to industrial size.

A. Reactor Scale-Up

Scaling up three-phase monolith reactors from pilot plant to industrial size is easy in some areas and more difficult in others. Since there is no interaction between the channels, the behavior within the monolith channels is independent of scale. Adding more parallel channels will not affect the flow, the mass and heat transfer, or reactions in each channel, as long as the flow distribution is uniform. Also, both the pilot plant reactors and the industrial reactors are adiabatic due to the absence of radial mixing.

The difficult part is the distribution of gas and liquid, both at the top of the first monolith and between the monoliths. In a small pilot plant reactor with one single monolith, there are minor problems with the liquid distribution. To experience the problems in industrial scale, we need a pilot plant reactor with a diameter of at least 1 m and several monoliths stacked above each other.

B. Liquid Distribution

In cocurrent downflow the liquid distribution is the most sensitive part, since there is no redistribution of liquid within the monolith. The liquid must be equally distributed to every channel at the top. On the other hand, when the liquid is evenly distributed, there is no maldistribution further down in the monolith. Nonuniform liquid distribution gives nonuniform conversion, resulting in larger reactors and lower selectivities.

Several different methods to distribute the liquid have been tested. Flooding the monolith with liquid for a short time gives distribution to all channels. However, the linear velocity in the monolith is very high, 0.2–0.5 m/sec. An optimal plug length of gas and liquid is 1–2 cm, giving a pulsing frequency of 10–50 Hz. Furthermore, this method does not provide an even liquid distribution unless the liquid is added very fast or very uniformly over the monolith.

Spraying the liquid in small drops has been shown to provide the simplest solution. Small liquid drops, much smaller than the channel diameter, are sprayed uniformly over the monolith inlet. The monolith is wetted by these drops, and the liquid starts to flow down at low velocity in a thin film on the channel wall. When more liquid is added, a meniscus is formed and the liquid starts to flow at a much higher velocity, taking up the annular liquid and forming a much larger liquid plug. A sieve plate that forms larger drops can also work, provided that the distance between the holes is of the same order as the channel diameter. The drops will break up when they hit the monolith surface, but more liquid will flow in channels that are directly below the sieve holes. The advantage of a sieve plate is that it is simple and requires less pressure drop. The disadvantage of both spray and sieve plate is that it is difficult to control the liquid and gas plug length.

Cocurrent upflow is unstable: The areas with high gas holdup will accelerate, and there will be a downflow in channels with low gas holdup. The gas must be evenly distributed in very small bubbles over the whole cross section to get an acceptable performance. It is much easier to distribute small drops in a continuous gas phase than small bubbles in a continuous liquid phase. There are some advantages of cocurrent upflow, e.g., control of space velocity, but so far the experiments have shown much better performance in cocurrent downflow.

Crynes et al. [35] have developed what they call a "monolith froth reactor." They introduce the gas through a porous glass frit just below the monolith, forming a froth that is fed into the reactor. Oxidation of phenol was studied in a 5-cm-wide and 33-cm-long

MR with a porous glass frit (145–175-μm pore diameter). They obtained very good mass transfer properties, but the residence-time distribution of the liquid phase corresponded to a stirred tank reactor.

C. Space Velocity

The main problem with cocurrent downflow is the very high linear velocity. Due to the very low pressure drop, gravity will give the liquid a velocity of 0.2–0.5 m/sec for liquids with waterlike viscosity in monoliths with 60 ch/cm^2. This low residence time gives too low a conversion for most reactions. Adding a back pressure will create an unstable system, where gas will flow upward in some channels and liquid downward in others. Using monoliths with smaller channels can solve this problem. Since $U \propto d_t^2$, a decrease in diameter by a factor $\sqrt{10}$ will decrease the velocity by a factor 10 and at the same time increase the mass transfer area by a factor $\sqrt{10}$. Unfortunately, monoliths with 186 ch/cm^2 are the smallest available at the moment. Until the manufacturers can make monoliths with smaller channels, we have to recirculate the liquid to get the desired conversion. The recirculation will also smoothen the differences due to the nonuniform flow distribution in the monolith.

The space velocity in cocurrent upflow, e.g., in the froth reactor, can be controlled within large areas by the pumping rate. There is an upper velocity limit for formation of small bubbles in the glass frit, and the very high back-mixing in the monolith indicates that draining of the monolith down to the inlet area can be a problem at low velocities. The residence-time distribution in the monolith froth reactor has been studied by Patrick et al. [36] and Thulasidas et al. [37].

D. Arrangement of Monoliths

Large extruded monoliths are usually hexagonal, with a diameter of about 30 cm and about 20 cm in length. These monoliths can be bunched to form desired reactor diameters up to several meters. Corrugated monoliths can be made up to several meters in diameter, but they are also only about 20 cm in length. Building longer reactors by stacking the monoliths above one another can be difficult. When the liquid leaves the monolith it is accelerated due to gravity, forming a liquid beam with decreasing cross section. Even 5 cm below the monolith, the liquid beam has decreased its cross section to less than half that of the monolith. The second monolith must be placed very close to the first so that meniscuses are formed, thereby directing the flow from the first to the second monolith.

A packing material can also be used between the monoliths. It must direct the flow from the upper to the lower monolith without further segregation of the two phases. With a proper choice of packing material, even a redistribution of the liquid to compensate for nonuniform flow at the top of the monolith could occur.

E. Gas–Liquid Separation

Gas and liquid must be separated within the reactor so that the liquid can be pumped back to the spraying nozzle. The gas–liquid flow at high velocity that hits a liquid surface below the monolith bed will in many cases cause a foam problem. Lowering the velocity and providing a large contact area with the gas bulk, by directing the liquid from the monolith to the bottom with a cyclonlike device, is in most cases adequate to avoid this problem.

F. Recirculation

Since the residence time in cocurrent downflow is very short, it is necessary to recirculate the liquid and the gas. Also, the best performance from a mass transfer point of view is when the gas and liquid volume flow rates are about equal, and with a bubble/slug length of about 0.5–2 cm. In these cases the molar flow of gas is much less than that of liquid, and the gas component will be consumed before the liquid component reaches complete conversion. Without recirculation, new gas must be added to the liquid further down in the reactor.

The liquid is usually pumped externally back to the sieve plate or spray nozzle. The liquid is saturated with gas and will probably contain small bubbles. The pump must be able to handle this kind of flow. The liquid may also contain particulates from the catalyst. Precautions against a pressure buildup in the spraying device or the sieve plate must be taken as well.

When the liquid flows down through the monolith by gravity, it decreases the pressure above the monolith. The pressure above the monolith will be lower than that below the monolith. This pressure difference is enough to recirculate the gas in a wide internal channel without pumping. The gas/liquid ratio can be controlled by the liquid flow rate alone.

G. Temperature Control

Since there is no radial flow, the reactor will be very close to adiabatic, even with small reactor diameters. There are also limited possibilities to introduce a heat exchanger between the monolith beds. The two-phase flow is sensitive to disturbances. However, in the existing plants temperature control is no problem, due to the high heat capacity of the liquid and the low conversion in each passage. An external heat exchanger in the liquid flow is sufficient to control the reactor temperature.

VI. FUTURE WORKS

For accurate design calculations we need more data from working industrial reactors. The actual flow distribution in and between stacked monoliths in large reactors is an unknown area at the moment. In laboratory reactors, we need more data on mass transfer with other geometries than cylindrical.

The present monoliths that are developed for emissions control in cars are not optimal for chemical three-phase reactors. Development of new monoliths, both metallic and ceramic, with higher cell densities and other geometries at much lower prices, can be expected when the market starts to grow.

NOTATION

$a_0 a_1$	coefficients in Eq. (16)	—
A	mass transfer surface; reactant A	m^2; —

B	reactant B	—
Bo	Bond number	—
c	concentration	mol m^{-3}
Ca	capillary number	—
d	diameter	m
D	diffusivity; dispersion coefficient	m^2 sec^{-1}
f_l	volume fraction of fluid consisting of liquid	—
g	acceleration due to gravity	m sec^{-2}
k	mass transfer coefficient	m sec^{-1}
k_1, k_2	rate constants	mol m^{-3} sec^{-1}
K	adsorption equilibrium constant	m^3 mol^{-1}
L	length	m
MR	monolith reactor	—
N	molar flux	mol sec^{-1} m^{-2}
N_c	parameter in Eq. (23)	—
p	pressure	Pa
p_g	pressure inside the gas plug	Pa
p_l	pressure in the liquid at the interface	Pa
ΔP	pressure drop	Pa
q	volumetric flow rate	m^3 sec^{-1}
r	radial coordinate; reaction rate	m; mol m^{-3} sec^{-1}
r_c	radius at channel corner	m
R_1	principal radius of curvature	m
R_2	principal radius of curvature	m
Re	Reynolds number	—
S	selectivity	—
Sc	Schmidt number	—
Sh	Sherwood number	—
SR	slurry reactor	—
$STYv$	space time yield	mol(B) m^{-3} sec^{-1}
t	time	sec
t_{wc}	thickness of the washcoat layer	m
t_{shell}	thickness of catalytic shell	m
t_{sub}	thickness of monolith wall	m
TBR	trickle-bed reactor	—
u	velocity in the axial direction	m sec^{-1}
	superficial velocity (in fig. 5, 14–17, and 19–21)	m sec^{-1}
U	linear velocity	m sec^{-1}
v	velocity in the radial direction	m sec^{-1}
V	volume	m^3
W	ratio introduced in Eq. (13)	—
We	Weber number	—
x	conversion	—
x_{cat}	catalyst load (catalyst volume/reactor volume)	—
x_{shell}	thickness of catalytic layer/particle radius	—
z	axial coordinate	m

Greek Symbols

α	parameter introduced in Eq. (43)	—
Γ	boundary	—
δ_f	film thickness	m

η	effectiveness factor	—
θ	coverage	—
μ	liquid viscosity	pa sec
ν	stoichiometric coefficient	—
ϵ	open area fraction	—
ρ	liquid density	kg m^{-3}
σ	surface tension	N m^{-1}
τ	characteristic time; residence time	sec
ϕ	Thiele modulus; frequency of slugs	—; sec^{-1}

Subscripts

A	reactant A
b	bubble
B	reactant B
c	capillary; channel; corner
cat	catalyst
D	diffusive
fr	frictional
g	gas; gas plug
G	gas
GL	gas–liquid
GS	gas–solid
l	liquid; liquid plug
L	liquid
LS	liquid–solid
m	mean
orif	orifice
p	particle
R	reactor
s	side region; reactor inlet
st	hydrostatic
t	tube
tot	total
w	wake

Superscripts

*	interface
(0)	at the inlet of the reactor

REFERENCES

1. R. Edvinsson, *Monolith Reactors in Three-Phase Processes*, Ph.D. dissertation Chalmers University of Technology, Göteborg, Sweden, 1994.
2. T.C. Thulasidas, M.A. Abraham, and R.L. Cerro, Bubble-Train Flow in Capillaries of Circular and Square Cross Section, *Chem. Eng. Sci. 50(2)*:183 (1995).
3. W.B. Kolb and R.L. Cerro, The Motion of Long Bubbles in Tubes of Square Cross Section, *Phys. Fluids A* 5:1549 (1993).

4. S. Irandoust and B. Andersson, Simulation of Flow and Mass Transfer in Taylor Flow Through a Capillary, *Computers Chem. Engng. 13(4/5)*:519 (1989).
5. F. Bretherton, The Motion of Bubbles in Tubes, *J. Fluid Mechanics 10*:166 (1961).
6. P. Concus, On the Film Remaining in a Draining Circular Cylindrical Vessel, *J. Phys. Chem. 74*:1818 (1970).
7. J. Snyder and H. Adler, Dispersion in Segmented Flow through Glass Tubing in Continuous Flow Analysis: The Ideal Model, *Anal. Chem. 48*:1017 (1976).
8. F. Fairbrother and A. Stubbs, Studies in Electroendosmosis. Part IV. The "Bubble Tube" Method of Measurement, *J. Chem. Soc. 1*:527 (1935).
9. G. Taylor, Deposition of a Viscous Fluid on the Wall of a Tube, *J. Fluid Mechanics 10*:161 (1961).
10. R. Marchessault and S. Mason, Flow of Entrapped Bubbles through a Capillary, *Ind. Eng. Chem. 52*:79 (1960).
11. S. Irandoust and B. Andersson, Liquid Film in Taylor Flow through a Capillary, *Ind. Eng. Chem. Res. 28*:1684 (1989).
12. S. Irandoust, B. Andersson, E. Bengtsson, and M. Siverström, Scaling up of a Monolithic Catalyst Reactor with Two-Phase Flow, *Ind. Eng. Chem. Res. 28*:1489 (1989).
13. C.N. Satterfield and F. Özel, Some Characteristics of Two-Phase Flow in Monolithic Catalyst Structures, *Ind. Eng. Chem. Fundam. 16(1)*:61 (1977).
14. R.K. Edvinsson and A. Cybulski, A Comparative Analysis of the Trickle-Bed and the Monolithic Reactor for Three-Phase Hydrogenations, *Chem. Eng. Sci. 49 (24B)*:5653 (1994).
15. R. Clift, J.R. Grace, and M.E. Weber, *Bubbles, Drops, and Particles*, Academic Press, New York, 1978.
16. S. Irandoust and B. Andersson, Mass Transfer and Liquid-Phase Reactions in a Segmented Two-Phase Monolithic Catalyst Reactor, *Chem. Eng. Sci. 43*:1983 (1988).
17. S. Irandoust, S. Ertlé, and B. Andersson, Gas–Liquid Mass Transfer in Taylor Flow through a Capillary, *Can. J. Chem. Eng. 70*:115 (1992).
18. K. Kawakami, K. Kawasaki, F. Shiraishi, and K. Kusunoki, Performance in a Honeycomb Monolithic Bioreactor in a Gas–Solid–Liquid Three-Phase System, *Ind. Eng. Chem. Res. 28*:394 (1989).
19. V. Hatziantoniou and B. Andersson, Solid–Liquid Mass Transfer in Segmented Gas–Liquid Flow Through a Capillary, *Ind. Eng. Chem. Fundam. 21*:451 (1982).
20. J.R.F. Guedes de Carvalho, S.S.S. Cardoso, and J.A.S. Teixeira, Axial Mixing in Slug Flow. The Use of Injected Air to Reduce Taylor Dispersion in a Flowing Liquid, *Trans. Inst. Chem. Eng. 71(A)*:28 (1993).
21. H. Pedersen and C. Horvath, Axial Dispersion of a Segmented Gas–Liquid Flow, *Ind. Eng. Chem. Fundam. 20*:181 (1981).
22. R.E. Hayes and S.T. Kolaczkowski, Mass and Heat Transfer Effects in Catalytic Monolith Reactors, *Chem. Eng. Sci. 49(21)*:3587 (1994).
23. C.E. Camp, W.B. Kolb, K.L. Sublette, and R.L. Cerro, The Measurement of Square Channel Velocity Profiles Using a Microcomputer-Based Image Analysis System, *Experiments in Fluids 10*:87 (1990).
24. W.B. Kolb and R.L. Cerro, Film Flow in the Space between a Circular Bubble and a Square Tube, *J. Colloid and Interface Sci. 159*:302 (1993).
25. W.B. Kolb, A.A. Papadimitriou, R.L. Cerro, D.D. Leavitt, and J.C. Summers, The Ins and Outs of Coating Monolithic Structures, *Chem. Eng. Progr. February*:61 (1993).
26. R.K. Edvinsson and A. Cybulski, A Comparison between the Monolithic Reactor and the Trickle-Bed Reactor for Liquid Phase Hydrogenations, *Catalysis Today 24*:173 (1995).
27. A. Cybulski and J.A. Moulijn, *Mass Transfer in a Monolithic Catalyst Under Reacting Conditions: Oxidation of Carbon Monoxide*. Unpublished works of the Delft University of Technology, Delft, The Netherlands, (1994).

28. G.A. Hughmark, Holdup and Heat Transfer in Horizontal Slug Gas–Liquid Flow, *Chem. Eng. Sci. 20*:1007 (1965).
29. D.R. Oliver and A. Young Hoon, Two-Phase Non-Newtonian Flow, Part I. Pressure Drop and Hold-up, *Trans. Inst. Chem. Eng. 46*:T107 (1968).
30. D.R. Oliver and A. Young Hoon, Two-Phase Non-Newtonian Flow, Part II. Heat Transfer, *Trans. Inst. Chem. Eng. 46*:T116 (1968).
31. Y.T. Shah, *Gas–Liquid–Solid Reactor Design*, McGraw-Hill, New York, 1979.
32. A. Cybulski and J.A. Moulijn, Monoliths in Heterogenous Catalysis, *Catal. Rev. Sci. Eng. 36(2)*:179 (1994).
33. S. Irandoust and B. Andersson, Monolithic Catalysts for Nonautomobile Applications, *Catal. Rev. Sci. Eng. 30(3)*:341 (1988).
34. N.-H. Schöön, Recent Progress in Liquid-Phase Hydrogenation: With Aspects from Microkinetics to Reactor Design. Proceedings of the 6th Natl. Sympos. Chem. React. Eng., Warsaw, Poland, and the Second Nordic Symposium on Catalysis, Lyngby, Denmark, 1989.
35. L.L. Crynes, R.L. Cerro, and M.A. Abraham, Monolith Froth Reactor: Development of a Novel Three-Phase Catalytic System, *AIChE J. 41(2)*:337 (1995).
36. R.H. Patrick, Jr., T. Klindera, L.L. Crynes, R.L. Cerro, and M.A. Abraham, Residence Time Distribution in Three-Phase Monolith Reactor, *AIChE J. 41(3)*:649 (1995).
37. T.C. Thulasidas, R.L. Cerro, and M.A. Abraham, The Monolith Froth Reactor: Residence Time Modelling and Analysis, *Trans IChemE. 73 (part A)*:314 (1995).

11

Monolithic Reactors for Countercurrent Gas–Liquid Operation

Swan Tiong Sie and Paul J. M. Lebens
Delft University of Technology, Delft, The Netherlands

I. INTRODUCTION

Three-phase fixed-bed reactors, in which a solid catalyst is contacted with a flow of gas and liquid, are widely applied in the industry for processes where the reactants are in the gas as well as in the liquid phase, due to the incompatibility of these two phases under process conditions. In the oil industry in particular, hydrotreating of heavy oils is at present carried out mostly in fixed-bed reactors in which oil and a hydrogen containing gas pass together over a solid catalyst.

In these reactors the fixed bed of catalyst is generally a more or less random packing of catalyst particles with a diameter in the range of 1–3 mm. Both oil and gas flow through this bed in the downward direction, i.e., cocurrently. These so-called trickle-flow or trickle-bed reactors are applied on a very large scale: Their estimated total capacity is well above 1 billion tons per annum.

From a theoretical point of view, the cocurrent mode of operation is in many cases not the optimal one, and countercurrent operation, in which gas and liquid flow in opposite directions over the fixed catalyst, should be preferable, as will be discussed below. Countercurrent flow of gas and liquid through a random packing of catalyst particles is difficult or hardly obtainable under industrial conditions, however. One possibility for obtaining the desired countercurrent flow is to use fixed catalysts with a well-defined specific geometry instead of a random packing. An example of a reactor with such a structured catalyst is the internally finned monolithic reactor (IFMR), to be discussed below.

II. INCENTIVES FOR COUNTERCURRENT FLOW IN GAS–LIQUID–SOLID REACTORS

Countercurrent operation can have advantages over cocurrent operation in situations where the desired conversion is suppressed by (by-)products generated in the process that act as a catalyst poison (product inhibition) or when the conversion is limited by thermodynamic equilibria. When the catalytic poisons are taken up in the gas phase, their concentration will decrease in the direction of the liquid flow due to the sweeping effect of the gas, whereas in the cocurrent mode of operation their concentration increases as they accumulate in the gas. Hence, the reaction will be less suppressed with increasing conversion in the countercurrent case, which is particularly important if high conversions are aimed at.

In an equilibrium-limited conversion process, the barrier to further conversion will be lifted if a reaction product that limits the conversion of the liquid is taken up in the gas phase and swept away rather than allowed to accumulate in the gas stream in the direction of the flowing liquid.

In most applications of trickle-flow reactors, the conversions generate heat that causes a temperature rise of the reactants, since the industrial reactors are generally operated adiabatically. In the cocurrent mode of operation, both the gas and the liquid rise in temperature as they accumulate heat, so there is a significant temperature profile in the axial direction, with the highest temperature at the exit end. When the total adiabatic temperature rise exceeds the allowable temperature span for the reaction, the total catalyst volume is generally split up between several adiabatic beds, with interbed cooling of the reactants. In the countercurrent mode of operation, heat is transported by gas and liquid in both directions, rather than in one direction only, and this may increase the possibility of obtaining a more desirable temperature profile over the reactor.

An example of a product-inhibited conversion is the hydrodesulfurization of medium-to-heavy petroleum fractions. The removal of sulfur from such oils can generally be described as a second-order reaction in total sulfur [1]. This high apparent order is a reflection of the presence of a variety of sulfur-containing compounds that have widely differing reactivities for hydrodesulfurization and implies that a relatively large proportion of sulfur is removed from the oil by conversion of the bulk of more reactive compounds in an early stage of the reaction. The conversion of the more refractive sulfur compounds occurs far more slowly in a later stage.

Hydrogen sulfide, which is a product of the hydrodesulfurization reaction, is therefore generated in large quantities already in the oil inlet part of the bed, and in the cocurrent operation this initially generated hydrogen sulfide will pass through the remaining part of the bed. Thus, the main part of the bed downstream operates under a relatively high hydrogen sulfide partial pressure, as can be seen from Fig. 1.

Since hydrogen sulfide suppresses the activity of hydrodesulfurization catalysts, the downstream part of the bed operates under suboptimal conditions, which is particularly disadvantageous since the conversion of the more refractive sulfur-containing compounds in this part of the bed calls for the highest catalytic activity. The suppression of hydrodesulfurization activity by hydrogen sulfide is illustrated by the experimental data in Fig. 2.

Figure 1 also shows the hydrogen sulfide pressure profile in the case of countercurrent operation. It can be seen that the main part of the bed where high catalyst activity is needed now operates under H_2S-lean conditions. Only a relatively small part of the bed operates under H_2S-rich conditions, and in this part suppression of catalytic activity is less serious, since here the conversion of relatively reactive compounds takes place. Therefore, the countercurrent mode of operation will be clearly superior, and this is the more true the deeper the desulfurization target.

Another example of product inhibition is the suppression of hydrocracking catalyst activity by ammonia in the hydrogen gas. Since nitrogen-containing compounds are generally present in petroleum fractions used as feed for hydrocracking, conversion of these compounds by hydrodenitrogenation generates ammonia as by-product. Ammonia

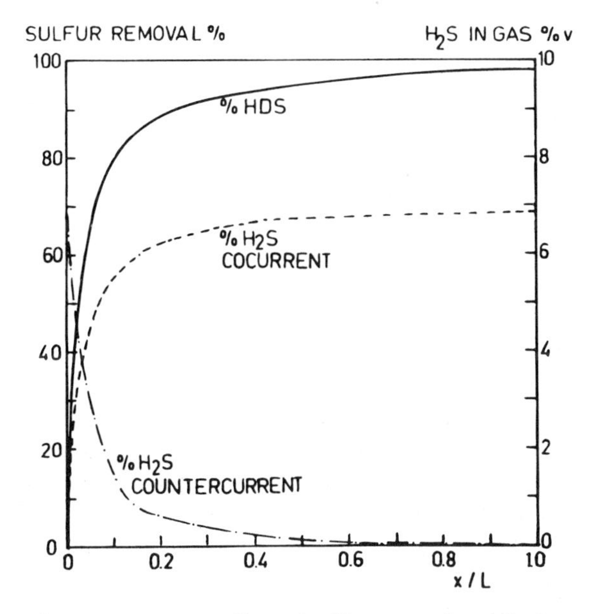

Figure 1 Axial profiles of sulfur removal and hydrogen sulfide concentration in the gas for hydrodesulfurization of oil following second-order kinetics in total sulfur.

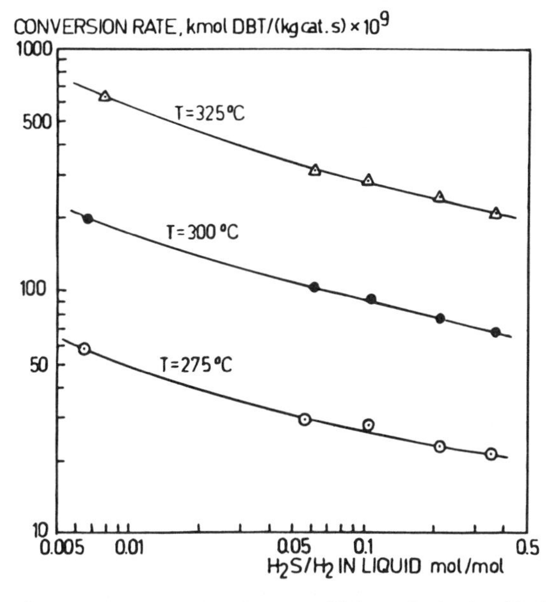

Figure 2 Effect of hydrogen sulfide on hydrodesulfurization of dibenzothiophene (DBT) over a Co/Mo/Alumina catalyst. (Based on experimental data from Ref. 2.)

Table 1 Effect of Ammonia on the Activity of Hydrocracking Catalysts (From Ref. 3)

Feed	ppm NH_3 in H_2	k, for Ni/W/silica-alumina	k, for Ni/W/Y-zeolite
n-Heptane	0	7.0	13.6
n-Heptane	270	0.07	0.9
Naphtha	0	5.0	11.2
Naphtha	300	0.07	1.0

Note: k = first-order rate constant for cracking to C_6 and lighter, in L/(L · h).
T = 380°C; P = 75 bar; hydrogen/feed = 2000 NL/L.

strongly suppresses the activity of the acidic catalyst used in the hydrocracking process for the desired cracking reaction, as evidenced by the data given in Table 1. In the cocurrent mode of operation this inhibiting effect of ammonia results in a substantial loss of conversion capability. When very high conversions are required so that this loss is unacceptable it is customary to carry out the hydrocracking process in two stages: The first stage is a hydrodenitrogenation stage where the nitrogen-containing compounds are converted; after removal of ammonia from the effluent stream, the cracking reaction is carried out in a second-stage reactor. It will be clear that countercurrent operation may obviate the need to operate in two separate stages while permitting high cracking conversions.

An example of an equilibrium-limited reaction is the hydrogenation of aromatics in petroleum fractions. This is illustrated by Fig. 3, which shows that under the prevailing conditions, saturation of aromatics is kinetically limited at temperatures below about 370°C, but that above this temperature, thermodynamic limitation occurs. In cocurrent operation, the hydrogen partial pressure will be lowest at the reactor outlet, due to the

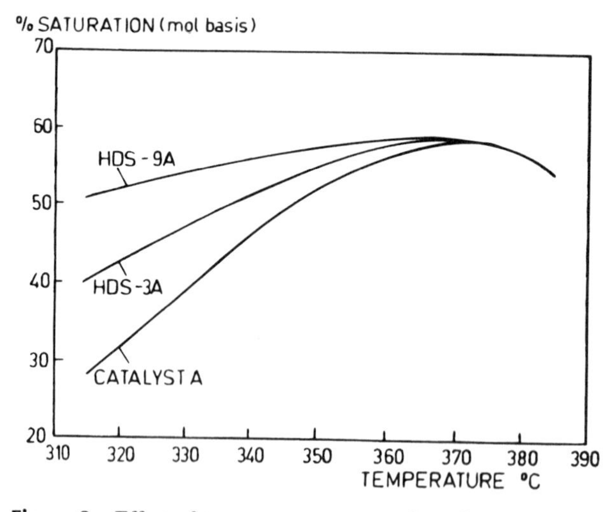

Figure 3 Effect of temperature on saturation of total polyaromatics in a gas oil over sulfided Ni/Mo/Alumina catalysts. LHSV = 2 L/(L · h); P = 750 psig; hydrogen/oil = 2000 scf/bbl. (From Ref. 4.)

combined effects of hydrogen consumption, pressure drop over the bed, and the buildup of impurities (low molecular weight cracked fragments, hydrogen sulfide, ammonia, water, etc.). Due to the heat of the reaction, the temperature at the exit of an adiabatic reactor will be highest. Hence, aside from inhibition of the hydrogenation activity of the catalyst in the downstream part of the bed by hydrogen sulfide and ammonia, the outlet situation is most unfavorable for deep conversion of aromatics when operating cocurrently. In this case too, countercurrent operation should be more favorable, since it alleviates the thermodynamic equilibrium constraint as well as the inhibiting effects of hydrogen sulfide and ammonia.

The three examples just discussed represent cases of considerable industrial importance and show that there is a significant incentive for countercurrent operation.

III. POSSIBILITIES OF COUNTERCURRENT OPERATION IN FIXED-BED REACTORS

A. Conventional Packed Catalyst Beds

Countercurrent flow of gas and liquid, with gas flowing upward and liquid flowing downward over a packing of solid particles, is quite common in distillation and absorption columns. However, in these packings the particles are of a relatively large size, typically 1–2 in. in diameter. In addition, the particles are generally of a shape that leads to a high voidage (e.g., Raschig rings, Berl saddles, or Pall rings giving packing voidages of about 0.9). In catalytic beds, however, much smaller particles are used, with typical diameters between 1 and 3 mm. Moreover, the voidage is much lower, typically around 0.4.

Because of the latter reasons, the permeability of a typical bed of catalyst is much lower than in a packed distillation column. This makes countercurrent operation in catalytic beds possible only at very low gas and liquid velocities, which are of little industrial interest. At practical velocities, the downward flow of liquid is impeded by the upward flow of gas, giving rise to flooding. Figure 4 shows the well-known flooding correlation of Sherwood et al. [5], and it can be seen that a typical hydrotreating case would be far above the flooding limit.

Enlargement of the catalyst particles so as to alleviate the flooding constraint leads to a poor utilization of the catalyst particle as a consequence of pore diffusion limitation. The occurrence of pore diffusion limitation in hydrodesulfurization of a gasoil is demonstrated in Fig. 5. Particles with diameters in the centimeter range, as required to avoid flooding at the desired liquid and gas velocities, will have a very low effectiveness factor, implying that poor use is made of reactor space, which is a significant cost factor for a high-pressure process. For the same reason, a very loose packing with a very high voidage will be unacceptable in such a process.

For the above reasons, countercurrent operation in a packed catalyst bed is seldom applied in practice. One example is the Arosat process [7], where a relatively small catalyst bed is operated in the countercurrent mode at the tail end of a conventional trickle-bed reactor to allow deeper hydrogenation of aromatics in kerosine. Since the absolute amounts of aromatics to be converted is small, little hydrogen is required, and consequently gas velocities can be low, while catalyst utilization is not so important, since the conversion is largely equilibrium limited. A more recent example is the Synsat process [8], which

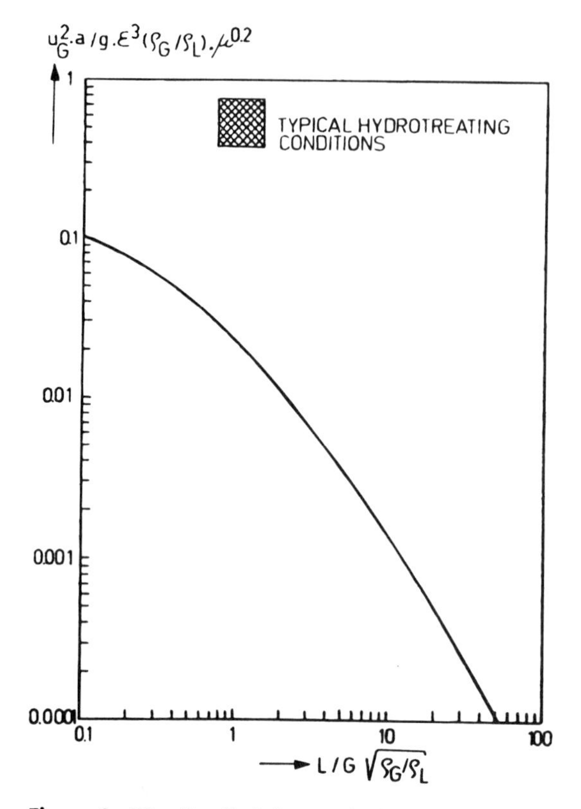

Figure 4 Flooding limit in a packed bed. (From Ref. 5.)

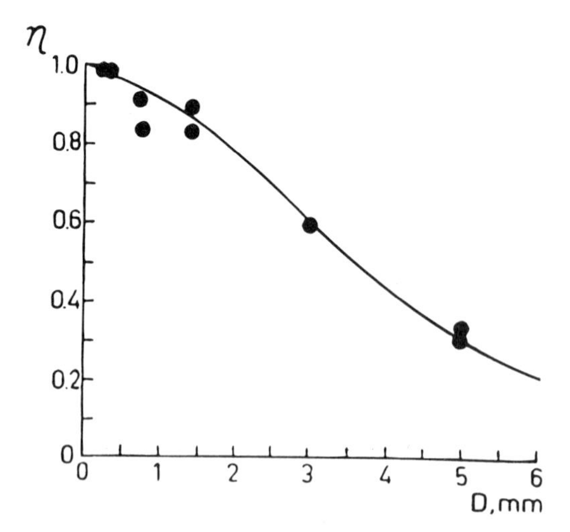

Figure 5 Catalyst effectiveness factor as a function of particle diameter in gasoil hydrodesulfurization. (From Ref. 6.)

combines hydrodesulfurization of gasoil with aromatics saturation in a similar process configuration.

B. Comparison of Packed Beds with Conventional Monoliths

As stated before, the volume of catalyst per unit volume of reactor space, which is to be termed the fractional catalyst volume (equal to 1 minus the voidage) and the degree of utilization of this catalytic material are important factors in a high-pressure, relatively slow catalytic process, such as the hydroprocessing of oils. The effectiveness factor of catalyst particles of arbitrary shape can be correlated with a generalized Thiele modulus Φ_{gen}, defined by

$$\Phi_{gen} = \frac{V}{SA}\left(\frac{k_m \rho_p}{D_{eff}}\right)^{0.5} \tag{1}$$

in which V is the volume and SA the external surface of the particle, k_m is the reaction rate constant per unit mass of catalyst, ρ_p is the particle density, and D_{eff} is the effective intraparticle diffusivity. For a given reaction over a given catalyst material, the degree of pore diffusion limitation is thus determined by the ratio V/SA, which may be considered as a characteristic dimension of the catalyst particle. For a proper comparison between monoliths and packed beds, the ratio SA/V as well as the fractional catalyst volume should both be of the same order of magnitude for the two systems.

Flow through monoliths having a large number of straight, parallel channels requires a much lower pressure difference than flow through a comparable packed bed of particles. This is because in the latter case the interstitial channels are tortuous and have repeated constrictions, and the turbulent character of the flow gives rise to effective momentum transfer from the moving fluid to the stationary solid. In the case of two-phase flow through the bed, there is also a very intensive momentum transfer between the flowing gas and liquid.

In the less turbulent flow through the straight channel of a monolith, momentum transfer from the fluid to the wall is less effective; and in the case of two-phase, countercurrent annular flow, momentum transfer between gas and liquid will also be less than in the interstitial channels of a packed bed. The lower rates of momentum transfer, which is the reason for the higher permeability of monoliths, should in principle improve the possibility for achieving countercurrent flow of gas and liquid at realistic velocities.

However, when considering monoliths having comparable fractional catalyst volumes and SA/V ratios as typical catalyst particles in fixed beds, countercurrent flow of gas and liquid is still very problematic. At the small channel diameter of about 1 mm (see Table 2) and at realistic velocities of gas and liquid, the liquid, which should flow downward as a film along the wall, will easily bridge the channel and form a slug, which will be transported upward by the gas. Thus, instead of the desired annular countercurrent flow, a segmented flow, or Taylor flow, in the upward direction will be obtained. This phenomenon is akin to the flooding in packed beds.

C. Monoliths with Internally Finned Channels

To reduce the tendency for slug formation and to enlarge the window for countercurrent operation, the diameter of the channel can be enlarged. But this is accompanied by a

Table 2 Comparison of Packed Beds of Particulate Catalysts with a Monolith Catalyst

Catalyst Shape	Spheres	Cylindrical extrudates	Monolith (incorporated)
Diameter, mm	2.5	1.5	—
Length, mm	—	6.0	—
Channel diameter, mm	—	—	1
Wall thickness, mm	—	—	0.6
Cells per sq. inch	—	—	250
Voidage	0.35	0.40	0.39
Fractional catalyst volume	0.65	0.60	0.61
Surface area			
per unit volume of reactor, mm^{-1}	1.6	1.8	1.6
per unit volume of catalyst (*SA*/*V*), mm^{-1}	2.4	3.0	2.6

reduction of the *SA*/*V* ratio. This ratio can be maintained by the incorporation of longitudinal fins in the channel that increase the external surface. In the case of monoliths prepared from catalytic material (so-called incorporated monoliths as contrasted to monoliths with an active washcoat layer), the fins also contribute to the fractional catalyst volume. Thus, the fractional catalyst volume can be enhanced for monoliths having a given wall thickness, or conversely the wall thickness may be decreased while maintaining the fractional catalyst volume.

Another advantage of the fins is that a liquid that wets the solid surface (as is generally the case with oil and a porous catalyst) tends to occupy the grooves between the fins, as depicted in Fig. 6. Thus, an organized division of the available channel cross section for liquid flow and gas flow is obtained, with a relatively large part available for liquid flow in combination with a relatively large liquid/solid contact area. In the case of countercurrent flow, the fins may stabilize the relatively thick liquid layer against formation

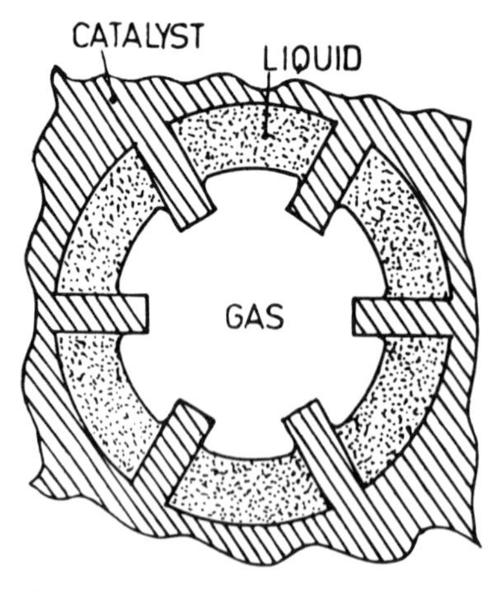

Figure 6 Schematic representation of the cross section of an internally finned channel.

of ripples by the upward drag of the gas, which may eventually bridge the channel and initiate slug formation.

IV. GEOMETRIC ASPECTS OF INTERNALLY FINNED MONOLITHS

The geometry of the internally finned monolith can be varied in a wide range. A number of possible geometric configurations are described in a patent on the application of the internally finned monolith as an element in a gas–liquid–solid reactor [9]. Geometric variables include channel size and shape, number of fins, fin height and thickness, channel array, pitch, and wall thickness.

Figure 7 shows a part of a monolith with internally finned channels of circular cross section. Figure 8 shows cross sections of possible monoliths with circular, triangular, square, or hexagonal channels.

Table 3 lists some data on internally finned monoliths of the types depicted in Fig. 8, with relative instead of absolute values for the dimensions. It can be seen that for the relative wall thickness chosen, the fractional catalyst volumes (in the case of incorporated catalytic monoliths) are of the same order of magnitude (about 0.6) as in conventional packings of particulate catalysts.

Figure 9 shows how the SA/V ratio of members of these geometric types varies with changing absolute dimensions, as represented by the pitch p. For the purpose of comparison, the figure also shows the SA/V values of some particulate catalysts of common shape and size, viz., cylindrical extrudates of $\frac{1}{16}$-in. and $\frac{1}{8}$-in. diameter (length-over-diameter ratio 4). It can be inferred that for the geometric types considered, pitch values between 3 and 8 mm give similar SA/V ratios as the above packings of extrudates. This corresponds with channel diameters between approximately 2.5 and 5.5 mm.

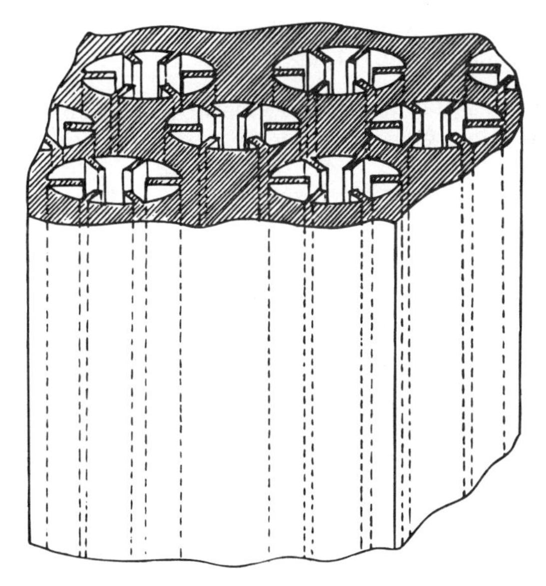

Figure 7 Drawing of a part of an internally finned monolith with channels of circular cross section.

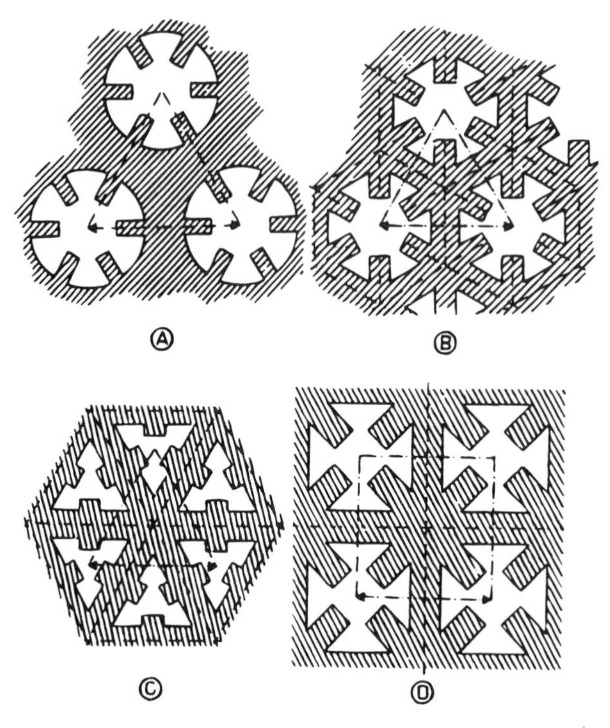

Figure 8 Cross sections of some internally finned monoliths with different channel shapes. A: Circular channels; B: hexagonal channels; C: triangular channels; D: square channels. The line between the two arrows is the pitch p.

V. HYDRODYNAMICS AND MASS TRANSPORT IN INTERNALLY FINNED MONOLITHS

A. Hydrodynamic Studies with an Internally Finned Tube

Since in the envisaged monolith the channels will be of identical geometry, the hydrodynamic situation in every channel will be the same if gas and liquid are evenly distributed.

Table 3 Some Possible Internally Finned Monolith Geometries

Basic channel cross section	Circle	Equilateral triangle	Square	Hexagon
Channel arrangement	T	T	S	T
Wall thickness	$0.2p$ (min.)	$0.2p$	$0.2p$	$0.2p$
Number of fins	6	3	4	6
Fin height	$0.2p$	$0.1p$	$0.2p$	$0.2p$
Fin thickness	$0.1p$	$0.2p$	$0.2p$	$0.2p$
Fractional catalyst volume	0.57	0.55	0.60	0.68
SA/V	$10.0/p$	$10.8/p$	$8.1/p$	$8.8/p$

Note: T = Triangular; S = Square; p = pitch.

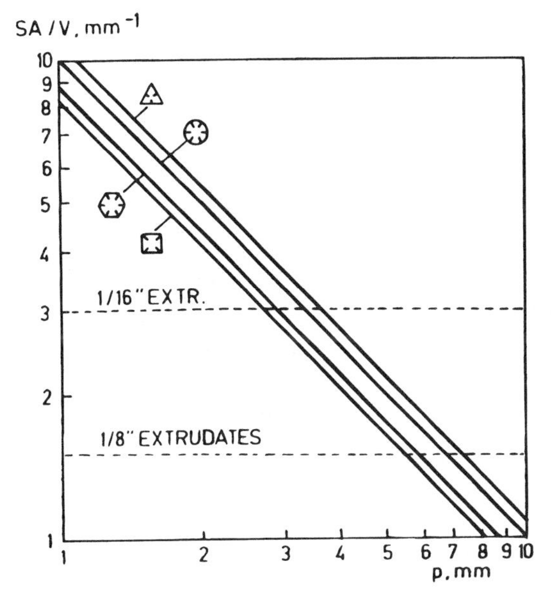

Figure 9 Surface area/volume ratio of the monolith types of Table 3, as a function of the pitch *p*.

Therefore, the hydrodynamic behavior of gas and liquid in countercurrent flow has been studied in single tubes provided with internal fins. The tubes were made of transparent polycarbonate to allow visual inspection and video recording of the phenomena observed. The tubes were typically 1 m long and were mounted in a vertical position. Liquid was fed to the top end of the tube, while gas was fed in at the bottom end.

1. Limitations Caused by Liquid at the Bottom End

In initial studies with liquids such as water, ethylene glycol, *n*-decane, and ethanol and with air of room temperature and atmospheric pressure it turned out that countercurrent annular flow in a channel of a few millimeters' width was possible, but only in a relatively narrow window of velocities much below the desired range corresponding with the operation of large industrial trickle-flow reactors. The main limitation proved to be caused by the behavior of the liquid exiting at the bottom end of the tube. With a small-diameter tube cut perpendicular to the axis, a droplet may form that temporarily closes off the entry to gas. If the gas has a sufficiently high overpressure, the hanging drop of liquid may be forced back into the tube, causing the formation of a slug of liquid that moves in the upward direction. While moving upward, this slug may be reduced in volume as it loses liquid to the liquid film already present, so that the total downward flow in the bottom part of the channel has to increase. The overall result is a fluctuating flow at the bottom that excerts its influence further up the tube, up to a certain height. The affected zone will be longer with increasing liquid and gas velocities. Ultimately, a point will be reached where hardly any liquid can flow out at the bottom, liquid in the tube being effectively blown upward. This situation is akin to total flooding.

Figure 10(B) illustrates the closure of the bottom end of a narrow tube by a pendent drop of liquid. But even in the case of a somewhat wider tube, where such a total closure does not necessarily occur, pendent liquid can still form an obstruction to entering gas,

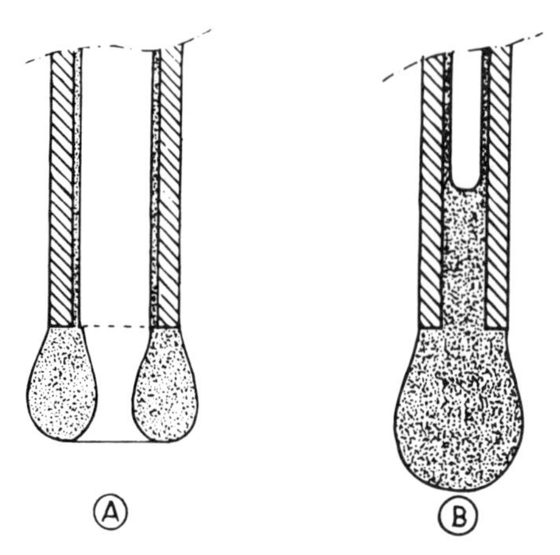

Figure 10 Constriction (A) and closure (B) of bottom channel opening by pendent liquid.

as illustrated by Fig. 10(A). The constriction of the tube entrance will cause a higher local velocity of the gas, which leads to a lower local pressure according to the law of Bernoulli. At sufficiently high gas velocities, pendent liquid may be sucked inward, resulting in slug formation, as described earlier for narrower tubes completely closed off by liquid drops.

2. *Tubes with a Slanting Bottom End*

This problem at the bottom outlet can be alleviated by cutting the tube at an oblique angle at the bottom end, as depicted in Fig. 11. Tubes with such a beveled end have been shown by English et al. to have a lower flooding tendency when installed vertically in dephlegmaters in which vapors enter at the bottom [10]. In line with their findings, beveling of the internally finned tube was found to shift significantly the onset of slugging. The cutoff angle proved to be important: Small angles have little effect, and significant reduction of the slugging tendency is obtained only at cutoff angles greater than approximately 45°.

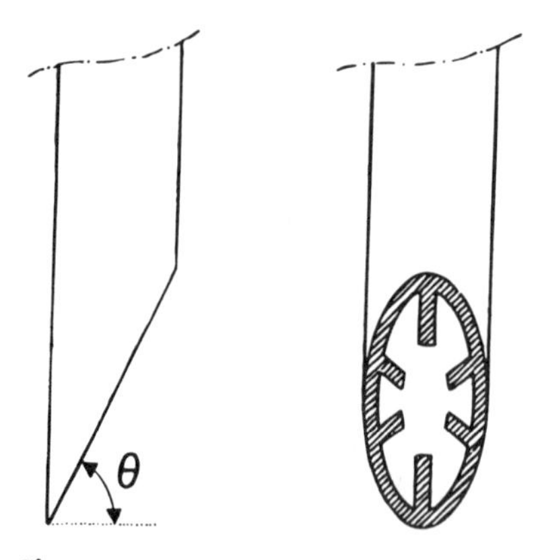

Figure 11 Internally finned tube cut off at an angle at the lower end. Left: side view. Right: front view.

Figure 12 shows some results of pressure drop measurements over a 1-m-long internally finned round tube (4-mm internal diameter, six fins, fin height 1 mm, fin thickness 0.5 mm) with a cutoff angle at the bottom end of 60°. With *n*-decane as the liquid and air at ambient temperature and pressure as the gas, the pressure drop increases steadily with increasing gas velocity until a certain critical gas velocity is reached. Below this critical velocity, the pressure drop is low, viz., orders of magnitude lower than in a fixed bed of catalyst under comparable conditions. It can also be seen that under these conditions the superficial velocity of the liquid in the internally finned tube has little effect on the pressure drop.

Above the critical velocity, the pressure drop increases dramatically. Visual observations have indicated that this abrupt rise in pressure drop corresponds with the onset of slugging. It can be inferred from Fig. 12 that the liquid velocity has a significant effect on the value of the critical gas velocity.

Figure 13 shows these transitions in a plot of the superficial velocities of *n*-decane as the liquid against that of air as the gas for three internally finned tubes cut off at the bottom at different angles. The lines through the points mark the division between the regime of annular flow with no or only little fluctuations at the tube bottom and the regime of serious slugging. The transition points in this graph were determined from pressure drop measurements as exemplified by Fig. 12, but essentially the same results were obtained with transitions determined visually. The cutoff angle can be seen to have an important effect: In the range of 60–70°, the highest angle allows the highest velocities of liquid and/or gas before slugging occurs.

Figure 13 also shows the range of liquid and gas velocities corresponding with the operation of industrial trickle-flow reactors. It follows that an internally finned tube of the present geometry with an outlet angle of 70° allows counterflow operation in the

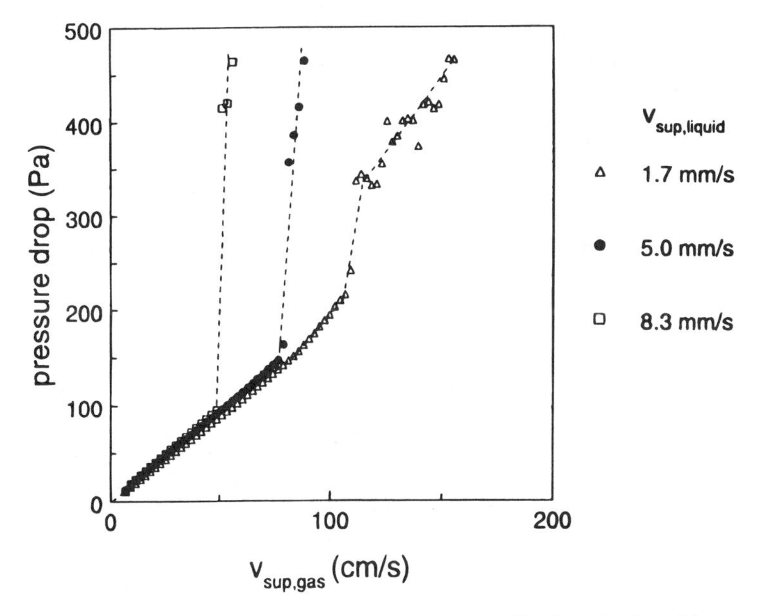

Figure 12 Pressure drop over a 1-m-long internally finned tube with countercurrent flow of *n*-decane and air at ambient conditions. Circular channel of 4-mm i.d. with six fins of 1-mm height and 0.5-mm thickness. Tube end cut off at an angle of 60°.

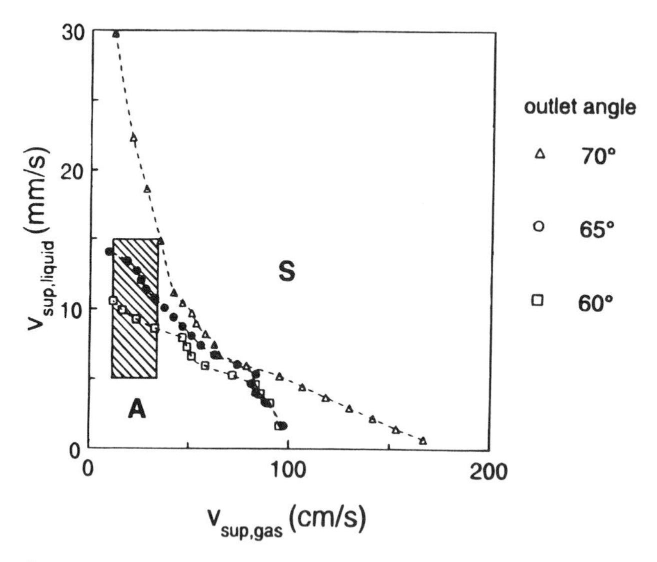

Figure 13 Map of flow regimes for *n*-decane and air flow at ambient conditions. A = countercurrent annular flow; B = slugging flow. The hatched area represents the range of velocities corresponding with the operation of large industrial trickle-bed reactors. Channel cross section is the same as in Fig. 12.

annular flow regime at industrially relevant velocities, but only marginally so for the most demanding cases.

3. *Flow Regime Transitions with a Specially Adapted Tube Outlet*

From the observable effect of the cutoff angle on the flow regime transition it can be deduced that notwithstanding the favorable results obtained at an angle of 70°, the transition is still determined by the outlet geometry to a large extent. This implies that the limits to counterflow operation in the internally finned tube proper have not yet been reached in the previously discussed experiments.

To obtain a better idea of the latter limits, a special measure was taken to avoid the interaction of liquid and gas at the channel outlet; viz., gas was introduced through a capillary inserted in the bottom end of the channel, and the liquid was drained through a wad of mineral wool, as shown in Fig. 14. Although this solution may not be of industrial interest, it is helpful in understanding the hydrodynamic behavior of countercurrent flow in the internally finned channel in the absence of inlet and outlet effects.

Figure 15 shows a flow regime map for an internally finned channel of the same cross section as described before, but with gas introduction through an inserted capillary. It can be seen that the transition from annular countercurrent flow to slugging (largely cocurrent) flow has shifted dramatically. As can be inferred from this figure, operation at the desired liquid and gas velocities corresponding with those in large industrial trickle-bed reactors is now comfortably within the stable annular flow region.

B. Mass Transfer Considerations

The reduced interaction between gas and liquid during annular flow in the internally finned channel as compared with two-phase flow in a packed bed not only affects transfer

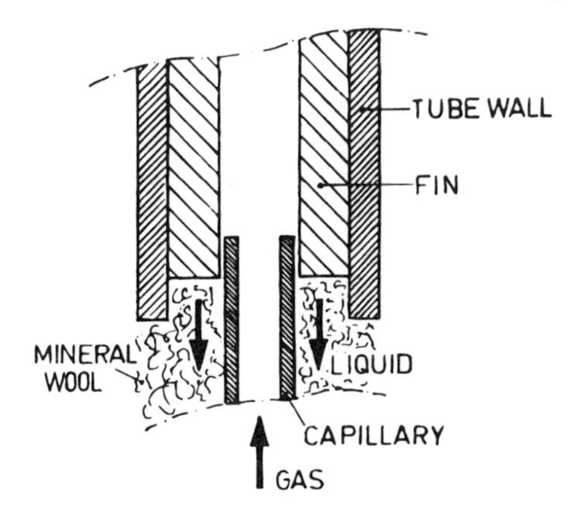

Figure 14 Outlet of internally finned tube with gas introduction through an inserted capillary.

of momentum (reduction of momentum transfer being beneficial for the lowering of the flooding tendency in countercurrent flow), but equally affects the rate of mass transfer for the reaction. Compared with trickle flow through a packed bed, gas-to-liquid mass transfer is likely to be considerably less efficient in an internally finned monolith operating in the annular flow regime.

The consequences of this reduction in mass transfer may not be too serious, however, in applications such as hydroprocessing of oils. Under the usual trickle-flow conditions at which these processes are now operated, the contribution of gas/liquid mass transfer to the overall mass transfer resistance is generally negligible and considerably smaller

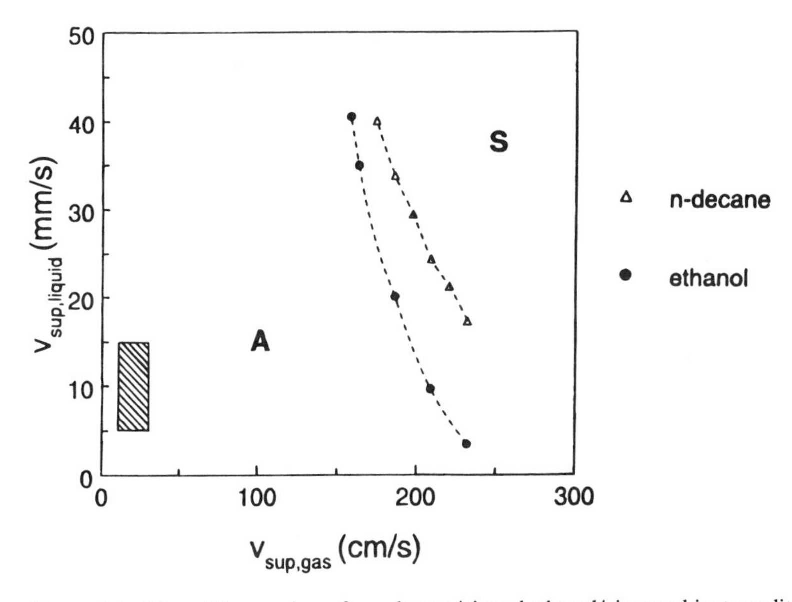

Figure 15 Map of flow regimes for *n*-decane/air and ethanol/air at ambient conditions for internally finned tube with gas introduction through inserted capillary. A = countercurrent annular flow; B = slugging flow. The hatched area represents the range of velocities corresponding with the operation of large industrial trickle-bed reactors. Channel cross section is the same as in Fig. 12.

than that of intraparticle mass transfer. As stated before, the SA/V ratio may be considered to be a measure of the latter resistance, and the concept of the internally finned monolith aims at maintaining similar ratios as in usual packed beds of catalyst.

VI. POTENTIAL APPLICATIONS

Potential applications of internally finned monoliths as elements in gas–liquid–solid reactors may be found in the area of hydroprocessing of petroleum, which is now dominated by the application of trickle-bed reactors and which represents a large-volume business. Incentives for countercurrent operation in this field have already been discussed in Section II.

The relatively high surface area that can be realized in fairly wide channels may also be of interest in other applications with cocurrent two-phase flow or with single-phase flow. In the latter applications, exploitation of the relatively low pressure drop can be the main reason for choosing this type of monolith.

VII. SUMMARY AND CONCLUSIONS

The internally finned monolith as catalyst in a fixed-bed reactor can be designed for similar volumes of active catalyst material per unit of reactor space and similar surface-area-to-volume ratios as common particulate catalysts in fixed beds.

In contrast to packed catalyst beds, however, countercurrent flow of gas and liquid is in principle possible in internally finned monoliths at realistic fluid velocities that are of interest for large-scale industrial applications. The main limitation to countercurrent flow at high velocities is at the outlet of a channel, rather than in the channel itself. With a suitable design of the outlet geometry, however, this problem can be alleviated so that countercurrent operation becomes possible in the velocity range of interest.

Countercurrent flow of gas and liquid, in contrast to the cocurrent trickle flow now in use, can have important advantages. The incentives for countercurrent operation are illustrated by several examples in the field of oil hydroprocessing.

NOTATION

a	surface area per unit volume of bed
D	diameter
D_{eff}	effective diffusivity
G	gas loading, mass flow per unit cross section of bed
g	gravity constant
k	reaction rate constant
k_m	reaction rate constant based on catalyst mass
L	liquid loading, mass flow per unit cross section of bed
P	pressure
p	pitch
SA	external surface area of catalyst particle
T	temperature
u_G	superficial velocity of gas in packing, based on empty reactor or column

u_L superficial velocity of liquid in packing, based on empty reactor or column
V volume of catalyst particle
$v_{sup,gas}$ superficial velocity of gas in channel, based on empty channel
$v_{sup,liquid}$ superficial velocity of liquid in channel, based on empty channel

Greek Letters

ϵ voidage
η catalyst effectiveness factor
μ dynamic viscosity of liquid
ρ_G density of gas
ρ_P density of catalyst particle
ρ_L density of liquid

REFERENCES

1. H. Beuther and B.K. Schmid, Reaction Mechanisms and Rates in Residue Hydrodesulphurization, Proc. 6th World Petroleum Congress, Sect. III, Verein zur Förderung des 6 Welt Erdoel Congresses, Hamburg, 1963, pp. 297–387.
2. D.H. Broderick and B.C. Gates, Hydrogenolysis and Hydrogenation of Dibenzothiophene Catalyzed by Sulfided $CoO/MoO_3/Gamma\text{-}Al_2O_3$: The Reaction Kinetics, *AIChE J. 27*:553–673 (1981).
3. S.T. Sie, Past, Present and Future Role of Microporous Catalysts in the Petroleum Industry. In *Advanced Zeolite Science and Applications* (J. Jansen, M. Stöcker, H. Karge, and J. Weitkamp, eds.), Studies in Surface Science and Catalysis no. 85. Elsevier, Amsterdam, 1994, pp. 587–631.
4. D.C. McCulloch, *Advantages of Cat Feed Hydrotreating*. Paper presented at the 1975 NPRA Annual Meeting, San Antonio, Texas, March 23–25, 1975.
5. T.K. Sherwood, G.H. Shipley, and F.A.L. Holloway, Flooding Velocities in Packed Columns, *Ind. Eng. Chem. 30 (7)*:765–769 (1939).
6. J.W. LeNobel and J.H. Choufour, Development in Treating Processes for the Petroleum Industry, Proc. 5th World Petroleum Congress, Sect. III., 5th WPC Inc., New York, 1959, pp. 233–243.
7. J.W. Reilly, M.C. Sze, U. Saranto, and U. Schmidt, Aromatics Reduction Process is Commercialized, *Oil and Gas J. 17 (Sept)*:66–68 (1973).
8. G.L. Hamilton and A.J. Suchanek, Flexible New Process Converts Aromatics in a Variety of Diesel Feedstocks, *Oil and Gas J. 89(26)*:56–59 (1991).
9. S.T. Sie, A. Cybulski, and J.A. Moulijn, Internally finned channel reactor, Netherlands Pat. Appl. 92.01923, 4 Nov. 1992.
10. K.G. English, W.T. Jones, R.C. Spillers, and V. Orr, Flooding in a Vertical Updraft Partial Condensor, *Chem. Eng. Progress, 59(7)*:51–53 (1963).

12

Parallel-Passage and Lateral-Flow Reactors

Swan Tiong Sie and Hans Peter Calis
Delft University of Technology, Delft, The Netherlands

I. INTRODUCTION

The parallel-passage reactor (PPR) and the lateral-flow reactor (LFR) are fixed-bed reactors suitable for the treatment of large volumes of gas at relatively low pressure, as are typical for end-of-pipe cleaning of combustion gases and other stack gases. In such applications, a low pressure drop, e.g., below 10 mbar, is generally required, and this demand can be met by these reactors. In addition, resistance to fouling by dust particles in the gas is important in a number of cases, and the PPR is particularly suitable for such cases.

In the treatment of flue gases, the PPR and LFR provide an alternative to the industrially applied monoliths (honeycombs), which also feature low pressure drop and high dust tolerance. However, since the PPR and LFR can use catalysts in the shape and size as used in conventional fixed beds, no dedicated catalyst manufacturing plants are generally required to fulfill the catalyst needs, and there are no special requirements for catalyst handling beyond those for traditional fixed-bed catalysts. Another advantage of the use of conventionally shaped, relatively small catalyst particles over monoliths is that they can easily withstand thermal stresses in applications where rapid temperature rises or drops occur during start-up and shutdown, so special devices for preheating to avoid cracks or detachment of washcoat layers in monoliths are unnecessary.

With respect to catalyst morphology, the PPR and LFR are in principle not different from normal fixed-bed reactors, but they owe their characteristics to the structured arrangement of the catalyst particles in the reactor space.

The parallel-passage reactor was conceived and patented in the late 1960s [1–3] and saw its first application in the Shell flue gas desulfurization process in the early 1970s [4]. The lateral-flow reactor was conceived as a constructional modification of the PPR, and its first application was for NO_x removal from flue gas of a gas-fired furnace in the early 1990s [5].

II. PRINCIPLES AND FEATURES OF PARALLEL-PASSAGE AND LATERAL-FLOW REACTORS

In contrast with a traditional fixed-bed reactor, where the catalyst particles are present in a single bed or a small number of beds of unstructured packing, the catalyst in the parallel-passage reactor is confined between wire gauze screens that devide the total reactor space in a regular array of a large number of catalyst layers with empty passages in between. The gas flows through these passages *along* the catalyst layers, instead of through the bed as in a traditional fixed-bed reactor. This principle of the PPR is illustrated in Fig. 1. The thickness of the catalyst slabs and the gas passages between them are typically in the range of 4–15 mm.

Because the gas flows through straight channels that are much wider than the tortuous interstitial channels of a normal packing, the pressure drop over the PPR is substantially lower than over a traditional fixed bed, the difference amounting to several orders of magnitude. The reacting molecules are transferred from the flowing gas to the stationary catalyst inside the screens mainly by diffusion, and the reaction products have to diffuse from the catalyst through the catalyst layers into the gas streams. The straightness of the gas passages also prevents particulates present in the gas from being caught by impingement upon obstacles, and thus the PPR can be used for treating dust-containing gases, similarly to monolithic (honeycomb)-type reactors, which are also applied in treating flue gas.

In the lateral-flow reactor the catalyst is also contained in structures of gauze screens that form alternating layers of catalyst and empty passages for gas. However, in contrast to the PPR, where all the gas passages connect the inlet directly with the outlet by being open at both ends, the gas passages of the LFR are each closed off at one end, neighboring passages being open and closed at different ends, as shown in Fig. 2. Thus, the gas is forced to flow *through* the layers of catalyst, instead of alongside them as in the PPR.

In principle, the LFR is a fixed-bed reactor with a very low aspect ratio, i.e., the ratio of bed height to bed diameter. Typically, the thickness of the catalyst layers is in the range of 15–75 mm. Hence, the reactor can be considered as a "pancake" reactor, in which the "pancake" has been folded for convenient accommodation in the reactor space. Because of the shallowness of the bed and its very large cross section, the pressure drop is much lower than in the case of a fixed bed of more conventional dimensions.

In operation with particulates containing gases, the LFR is more prone to fouling than is the PPR. Similarly, as in a normal fixed bed, dust particles may be caught by impingement on solid surfaces, and this may give rise to plugging of the bed. However,

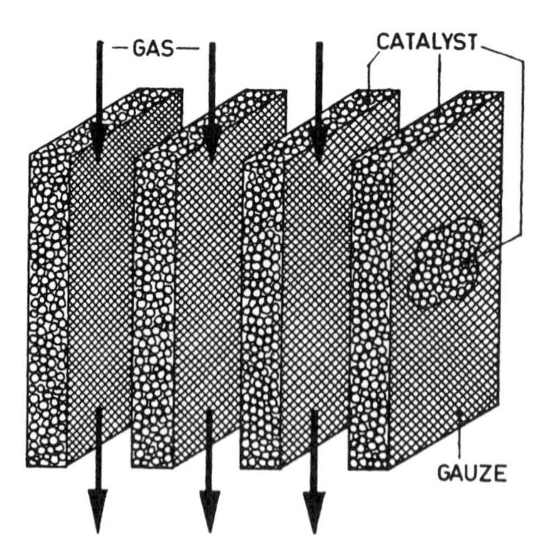

Figure 1 Schematic drawing of a parallel-passage reactor.

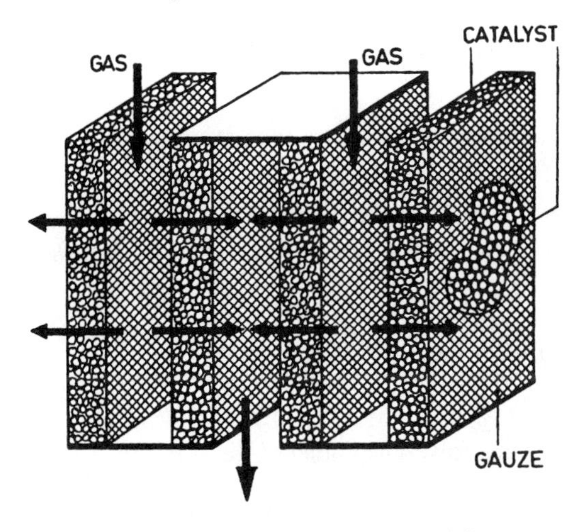

Figure 2 Schematic drawing of a lateral-flow reactor.

because of the much larger cross section of the bed, the plugging time of an LFR is likely to be substantially longer than that of a more traditional fixed bed. Thus it may be feasible to design the LFR for operation with particulates containing gas by incorporation of facilities for continuous or periodic withdrawal of dust-containing catalyst and recycling the catalyst after removal of the trapped dust.

In the PPR and LFR geometries shown in Figs. 1 and 2, the catalyst is contained between two flat, parallel screens, which form so-called envelopes. The PPR and LFR principle can, however, also be embodied in different geometries. For instance, it is possible to use a corrugated screen, as shown in Fig. 3, to form gas passages in the form of channels rather than slits.

The performance of the PPR and LFR as a reactor depends upon several factors, the most important being mass transfer and uniformity of flow. These aspects, as well as the fouling in operation with dust-containing gases, will be discussed in more detail below.

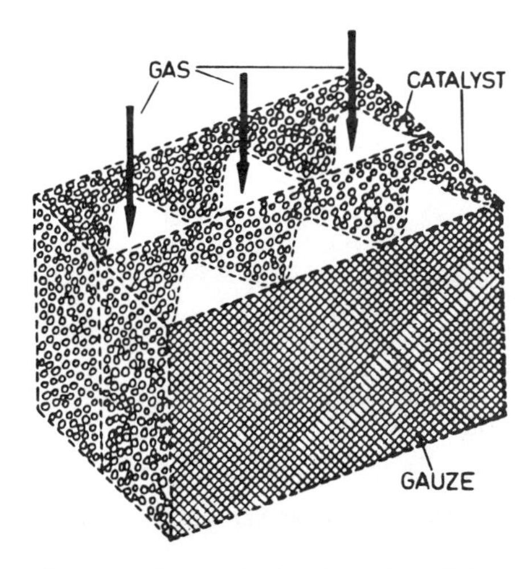

Figure 3 Schematic drawing of parallel-passage reactor with corrugated screen.

III. FLOW AND TRANSPORT PHENOMENA IN PARALLEL-PASSAGE REACTORS

A. Pressure Drop

The pressure drop for flow through the gas passages resulting from momentum transfer from the flowing gas to the stationary screen-enclosed catalyst slabs follows from the general Fanning equation

$$\Delta P = 4f \cdot \frac{L}{d_h} \cdot \frac{1}{2} \rho v^2 \tag{1}$$

in which the friction factor f is a function of the Reynolds number $\left(\mathrm{Re} = d_h \cdot \rho \cdot \frac{v}{\eta}\right)$, d_h is the hydraulic diameter, ρ is the density, η is the dynamic viscosity, and v is the average velocity in the channel.

At low Reynolds number, where the flow through the gas passage is laminar, the friction factor is inversely proportional to the gas velocity:

$$4f = \frac{\text{Const.}}{\mathrm{Re}} \tag{2}$$

whereas for fully turbulent flow, f is independent of Re.

Figure 4 shows the experimentally determined friction factor as a function of the Reynolds number for a laboratory PPR module with six 4-mm-thick catalyst slabs of 68-mm width and 500-mm height, spaced apart with a pitch of 11 mm, and made up from 2.2-mm-diameter glass spheres enclosed in 0.5-mm gauze mesh [6]. It can be seen that the transition of laminar to turbulent flow occurs already at a low Reynolds number (approximately 1000), which is attributable to the roughness of the channel walls caused by the wire gauze.

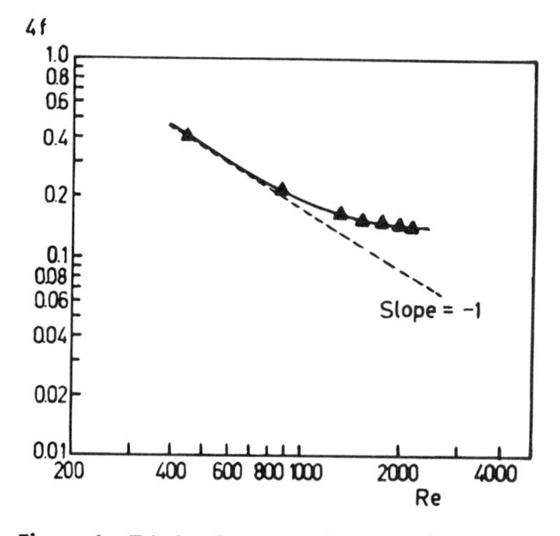

Figure 4 Friction factor as a function of the Reynolds number for flow through the gas passages. Air under ambient conditions, module A. (From Ref. 6.)

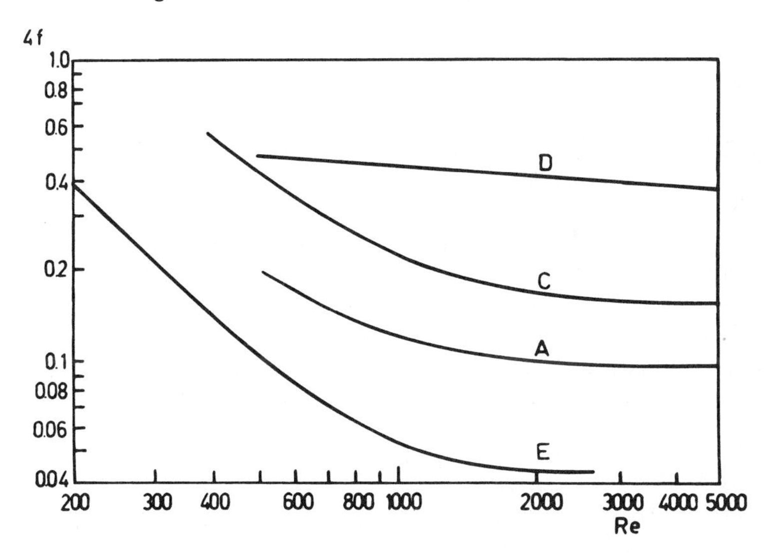

Figure 5 Friction factor for different PPR geometries (see Table 1). Simulated flue gas at approximately 1 bar and 160°C. (Adapted from Ref. 7.)

Figure 5 shows friction factors for different parallel-passage reactor modules tested by Calis et al. [7]. Particulars on the geometry of these modules are given in Table 1.

B. Mass Transfer

1. *Mass Transfer Resistances in a PPR*

In the PPR concept, the gas flows *along* the catalyst bed, so the catalyst particles are essentially surrounded by a stagnant gas and are not in direct contact with the gas stream.

Table 1 Characteristics of Some Laboratory PPR Modules

	Module				
	A	C	D	E	F
PPR type	Currugated screen	Corrugated screen	Flat screen	Flat screen	Wavy PPR
Hydraulic:					
Bed diameter, mm	12.1	12.1	14.5	9.5	12.0
Channel diameter, mm	11.8	11.8	12.4	5.1	9.4
Wire mesh: Thickness, mm	0.30	0.20	0.30	0.30	0.30
Length, mm	500	500	500	500	500
Cross section, mm	160 × 60	160 × 60	160 × 60	160 × 60	160 × 60
Number of catalyst beds	20	16	11	21	14
Number of channels	19	18	11	21	14

Note: From Ref. 7.

The catalytic conversion is therefore dependent upon mass transfer from the gas flowing through the stagnant gas surrounding the catalyst particles and through the intraparticle pores to the catalytic surface.

As discussed before, the transition from laminar to turbulent flow in the PPR channels already occurs at relatively low Reynolds number as a consequence of the roughness of the channel walls. Under typical operating conditions in practice, flow through the channels is quite turbulent, in contrast to the situation generally prevailing in monoliths as used in exhaust convertors, where due to the much smaller channel diameter and smoothness of the wall, flow is generally laminar. Therefore, in a PPR mass transfer in the gas inside the channel is generally relatively fast.

Neglecting the mass transfer resistance in the gas channel, the total mass transfer resistance in the PPR is made up from the following contributions (see Fig. 6):

1. From the flowing gas to a stationary gas film at the outer surface of the screens
2. Through the screen
3. Through a stationary film at the inner surface of the screen
4. Through the interstitial channels of the bed
5. Through a stationary film around the catalyst particles
6. Through the intraparticle pores to the catalytic surface.

According to Hoebink et al. [8], steps 2 and 3 are much faster than the other steps. Step 1 is not limiting either, except for channel geometries where the channel cross section has a very sharp corner leading to a relatively thick stagnant zone of gas inside the channel. Except for catalyst particles in which the reactants penetrate only in a very shallow outer layer (very low catalyst effectiveness factor), step 5 is generally also negligible as compared with step 6, since the stationary film around the catalyst particles is generally much thinner than the particle radius, and free diffusion through this thin film is consequently much faster than the effective diffusion through the intraparticle pores toward the heart of the particle. Consequently, step 4 (intrabed/extraparticle diffusion) and step 5 (intraparticle diffusion) are the main factors that determine the performance of the PPR.

2. *Intraparticle Mass Transfer*

Intraparticle diffusion limitation (step 6) is a well-documented phenomenon in catalysis in general, and its effect on catalyst utilization is given by the Thiele modulus ϕ, defined by

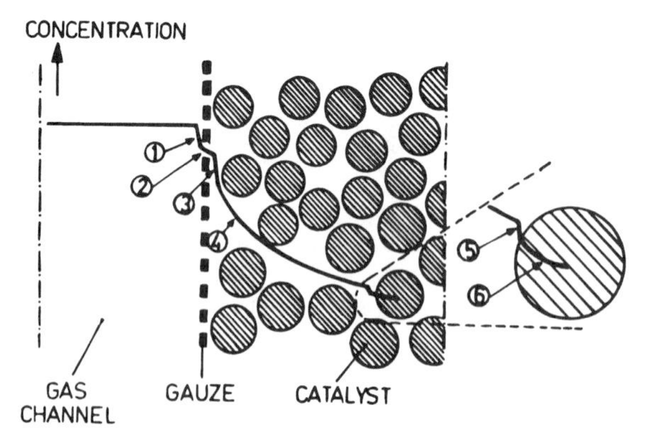

Figure 6 Mass transfer resistances in a PPR.

$$\phi = R\left(\frac{k}{D_{\text{eff.,}p}}\right)^{0.5} \tag{3}$$

in which R is a characteristic dimension of the catalyst particle, k is the reaction rate constant, and $D_{\text{eff.,}p}$ is the effective intraparticle diffusivity. The relationship between the effectiveness factor for catalyst utilization and the Thiele modulus is shown in Fig. 7 for a first-order reaction [9].

3. *Effective Intrabed Diffusivity with Stagnant Gas*

The intrabed, extraparticle mass transfer can be described by a lateral effective bed dispersion coefficient $D_{\text{eff.,}b}$, which in the absence of flow through the bed is given by

$$D_{\text{eff.,}b}(\text{stagnant}) = \frac{\epsilon}{\tau} D_g \tag{4}$$

in which ϵ is the bed voidage, τ is a tortuosity factor, and D_g is the diffusivity in bulk gas. With typical values of 0.4 and 2 for the voidage and tortuosity factor, respectively, the effective diffusivity in the bed is about 0.2 times the diffusivity in bulk gas.

4. *Enhancement of Effective Intrabed Diffusivity by Axial Flow of Gas*

Although in the conceptual PPR the gas flows *along* the catalyst layers instead of through the beds, the gas in the bed is not completely stagnant. The pressure gradient in the channel causes a small parallel flow in the axial direction through the bed. This flow would occur even when the screens are impermeable to gas, as long as the ends of the catalyst layer are in open connection with the gas inlet and outlet. This parallel flow through the bed, which is depicted in Fig. 8(A), gives rise to a convective contribution, so

$$D_{\text{eff.,}b}\ (\text{nonstagnant}) = \frac{\epsilon}{\tau} D_g + x_m u \tag{5}$$

in which x_m is a mixing length, which depends on the size and shape of the particle, and

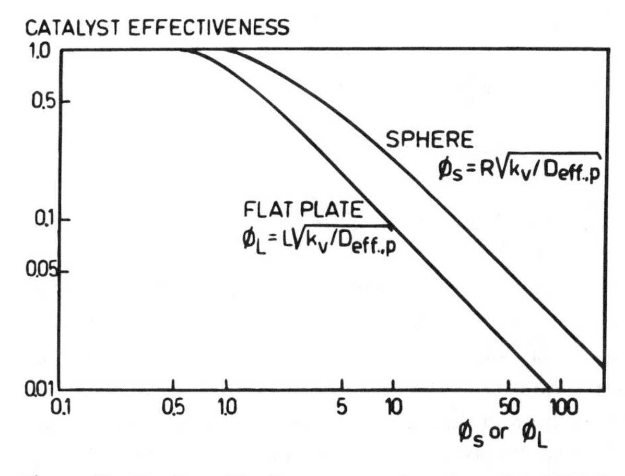

Figure 7 Catalyst effectiveness as a function of the Thiele modulus.

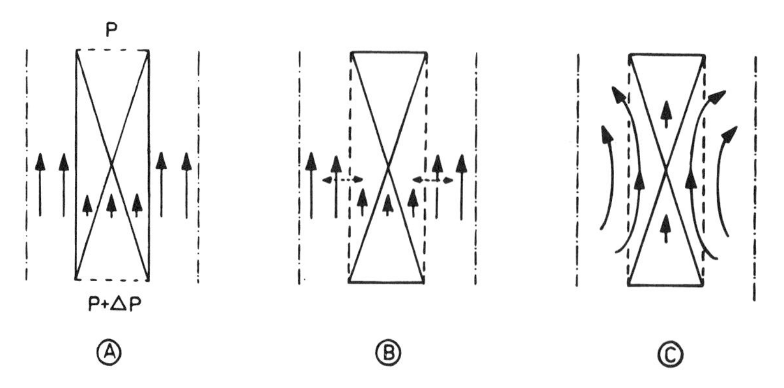

Figure 8 Enhancement of intrabed mass transfer by axial flow through the bed. A: Parallel flow driven by the pressure difference between inlet and outlet; B: momentum transport by diffusing molecules; C: convective momentum transport by penetrating eddies.

u is the superficial velocity of the gas through the bed. The superficial velocity is given by the well-known Ergun equation describing pressure drop for flow through porous media:

$$\frac{\Delta P}{L} = 150\,\frac{\eta(1-\epsilon)^2}{d_p^2\epsilon^3}\,u + 1.75\,\frac{\rho u^2}{d_p}\,\frac{1-\epsilon}{\epsilon^3} \tag{6}$$

Although the velocity of this parallel flow is very small compared with the velocity in the gas passages of the PPR, its contribution to the intrabed mass transfer can be significant. Typically, this velocity through the bed is some three orders of magnitude smaller than in the gas channels, which attests to the much lower pressure drop of a PPR compared with a fixed-bed reactor.

For thin beds, the actual average velocity of the gas in the bed is higher than the average velocity in the interior of the bed as calculated by the Ergun equation. This is because the local voidage close to the wall is higher than the average voidage in the packing. The effect of a finite layer thickness can be accounted for by a factor K, which is given by

$$K = 16 - 8\left(1 - \frac{2}{d_b/d_p}\right)^2 \tag{7}$$

in which d_b and d_p are diameters of the bed and the particle, respectively. Thus, the convective contribution to mass transfer in the bed can be written as [10,11].

$$D_{\text{conv}} = \frac{x_m}{K}\,u = \frac{F d_{\text{char}}}{K}\,u \tag{8}$$

in which F is a shape factor and d_{char} is a characteristic dimension of the particle. For example, for spherical particles, $d_{\text{char}} = d_p$ and $F = 1.15$.

The above mechanism of mass transport enhancement through lateral dispersion as a consequence of axial flow induced by the pressure gradient in the channel is also operative when the catalyst layers would be bounded by solid walls, as mentioned before. In fact, however, the screens enclosing the catalyst beds are permeable to gas molecules. This gives rise to momentum transport through the gauze screens. This phenomenon has been discussed by Hoebink et al. [8], who found that the gas velocity in the bed close to the screen was 5–10 times higher than far away from the gauze, which is too large a difference to be explained by the effect of the wall on the local voidage.

Momentum transport through the bed, which is an additional factor for enhanced intrabed mass transport, is operative also under laminar flow condition, as is illustrated by Fig. 8(B). When the flow through the channels is turbulent, there is also a convective transport of mass and momentum through the gauze screen as a consequence of eddies penetrating through the gauze screens, as depicted in Fig. 8(C). These mechanisms of momentum transport considerably enhance the effective intrabed diffusivity in PPRs with relatively thin catalyst layers.

Figure 9 shows some results of estimates by Calis et al. [7] of the intrabed effective diffusivities in a laboratory PPR module (module A of Table 1) filled with different catalysts in either granular or extrudate form and applied in experiments on removal of NO_x from a simulated flue gas. At low velocities in the gas channels the effective diffusivity is between 0.1 and 0.3 times the diffusivity in bulk gas, as is to be expected for stagnant gas in the bed. The higher effective intrabed diffusivity found for extruded catalysts as compared with granular catalyst can be largely ascribed to the higher voidage of the packing of extrudates.

At higher gas velocities in the channels, the effective intrabed diffusivity increases substantially, reaching values as high as or even higher than the diffusivity in the bulk gas. This enhancement of intrabed diffusivity cannot be ascribed solely to parallel flow as calculated by the Ergun equation, since a 2.5–9-times-higher axial gas velocity is needed to account for the experimentally found enhancement of the intrabed diffusivity. The substantial enhancement must be attributed to momentum transport through the screen [6].

5. *Mass Transport Contribution by Lateral Flow*

In a perfectly regular PPR, the channels are straight and of constant cross section. At any axial position the pressure in adjacent channels is therefore the same. In a real PPR, however, there can be variations in channel cross section due to tolerances in manufactur-

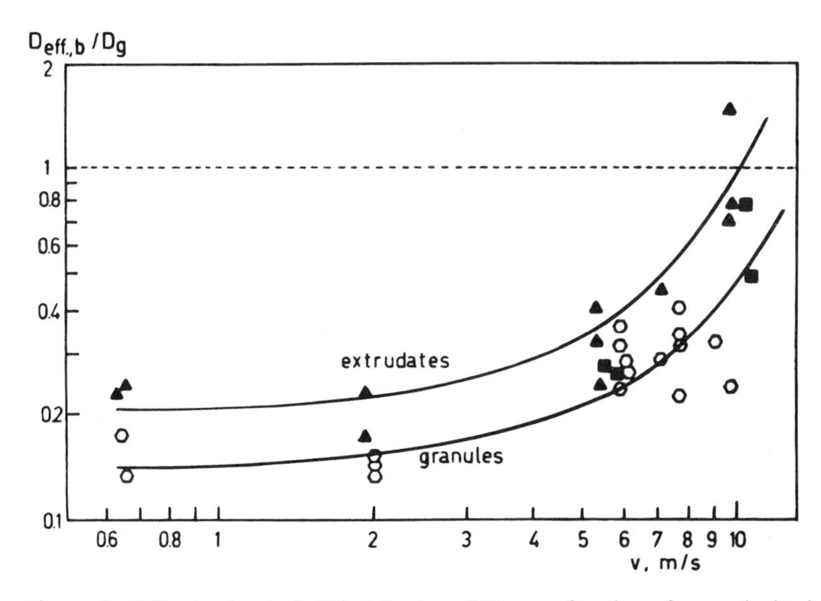

Figure 9 Effective intrabed diffusivity in a PPR as a function of gas velocity in the channels during NO_x removal with ammonia. (From Ref. 7.)

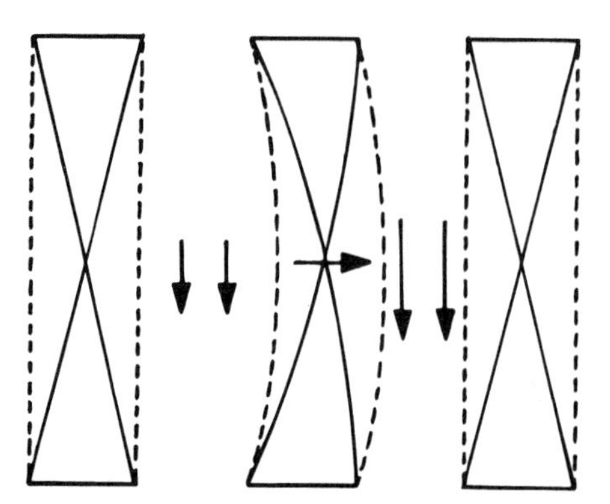

Figure 10 Lateral flow induced by the Bernoulli effect.

ing, bending, or bulging of the screens. This gives rise to differences in gas velocities in different channels at a given axial position, which in turn affects the local pressure P according to the well-known Bernoulli's law:

$$P + \rho g h + \tfrac{1}{2} \rho v^2 = \text{constant} \tag{9}$$

in which g is the height above a reference level and g is the gravity constant.

The pressure difference in adjacent channels provides a driving force for a lateral flow of gas through the bed, as is illustrated in Fig. 10. This cross flow of gas provides an additional contribution to mass transport from the gas in the channels to the catalyst in the bed.

This phenomenon of cross flow can be exploited in a PPR in which the screens are not perfectly flat but have a slight sinusoidal or zigzag shape. Such a "wavy" PPR is depicted in Fig. 11.

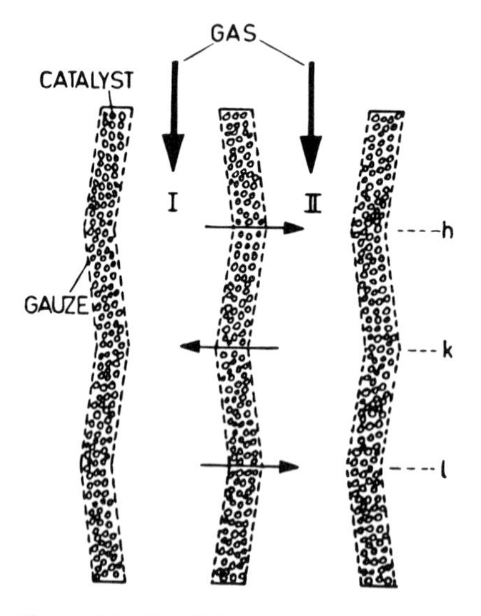

Figure 11 Parallel-passage reactor with "wavy" catalyst slabs.

C. Catalyst Bed Utilization

The limited rate of mass transport through the bed can give rise to incomplete utilization of the catalyst in the bed, in analogy with the incomplete utilization of a catalyst particle due to pore diffusion limitation.

Data on bed utilization have been collected by Calis et al. [7] for the laboratory PPR modules listed in Table 1 in experiments on NO_x removal from a simulated flue gas by reaction with ammonia. The bed effectiveness was determined by comparison of the NO_x conversion achieved in the PPR with that in a fixed-bed reactor with the same catalyst, taking proper account of the reaction kinetics. The fixed-bed reactor can be considered as a plug-flow reactor with a bed utilization of 100%.

Figure 12 shows results obtained with a PPR with corrugated screen (module A of Table 1) with different catalysts. It can be seen that the degree of utilization of the bed is not complete, but can reach reasonably high values (up to 70%) by operating at high gas velocities. Bed effectiveness is lowest for the most active catalyst, A, which is in line with expectations. For catalyst C, which has an extrudate shape, the effectiveness of the bed is higher than for catalyst B, which has a similar activity but is in a granular form. The higher bed effectiveness in the case of catalyst C is due to a higher intrabed diffusivity as a consequence of a higher voidage and greater mixing length in a packing of extrudates.

Figure 13 compares bed effectiveness for module A, in which the ends of the catalyst beds have been closed off with the effectiveness of module B, which has the same geometry as A but in which the ends of the beds allow passage of gas. It can be inferred that B has a slightly higher bed effectiveness, which attests to the contribution of parallel flow through the bed to mass transfer.

Figure 14 compares bed effectiveness for PPR modules with flat screens of different catalyst layer thickness. Module D, with 7-mm-thick catalyst slabs, has a lower bed

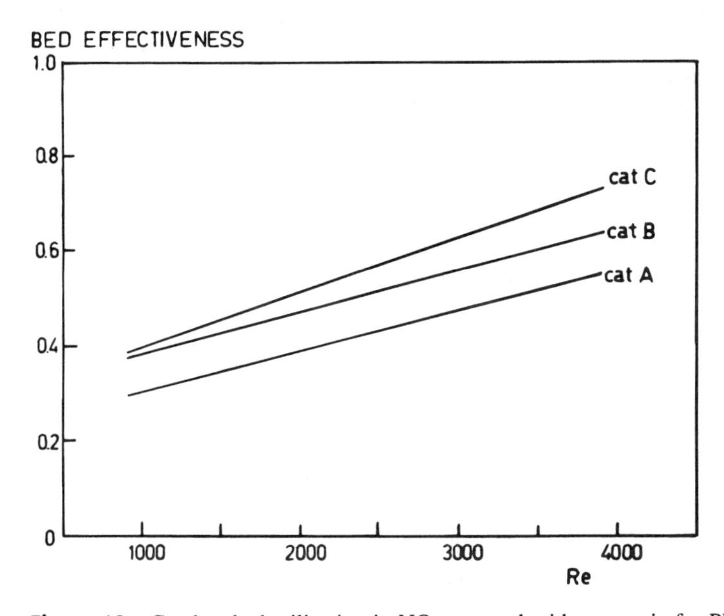

Figure 12 Catalyst bed utilization in NO_x removal with ammonia for PPR module A filled with different catalysts. Catalyst A: granules, high activity; catalyst B: granules, lower activity; catalyst C: extrudates, same activity as B. (From Ref. 7.)

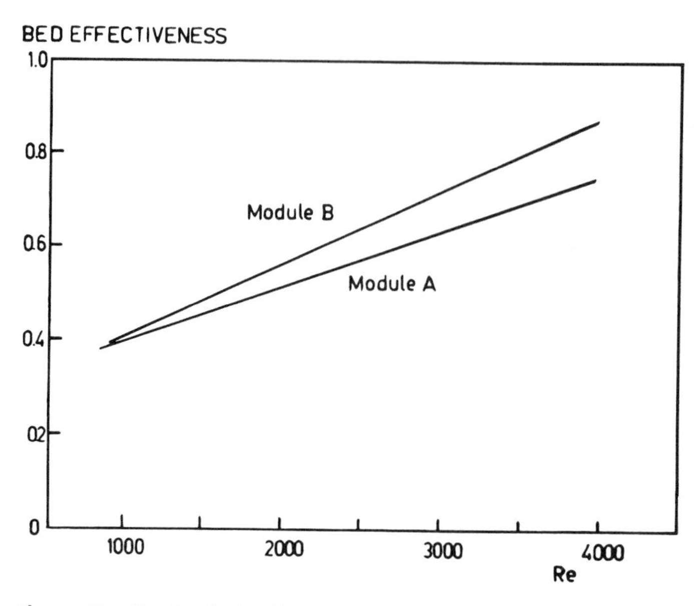

Figure 13 Catalyst bed utilization in NO_x removal with ammonia. Module A: PPR with corrugated screen, bed ends closed; Module B: same, but bed ends open to gas. (From Ref. 7.)

effectiveness than module E, with 4-mm-thick slabs, as is to be expected. The figure also shows the bed utilization of a "wavy" PPR (module F) having 6-mm-thick catalyst layers with a geometry as shown in Fig. 11. At low velocities, where the convective contributions to intrabed mass transport are low, the bed effectiveness of module F, with 6-mm-thick catalyst layers, is in between that of modules D and E, with 7- and 4-mm-thick catalyst slabs, respectively, as is to be expected. With increasing gas velocity in the gas passages, however, the effectiveness of module F increases rapidly and becomes higher than that of module E, which can be attributed to the cross flow occurring in the "wavy" PPR.

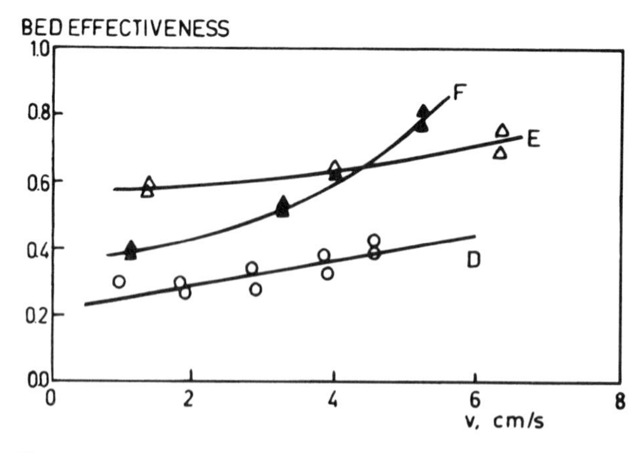

Figure 14 Catalyst bed utilization in NO_x removal with ammonia for PPR modules of different geometry. Module D: 7-mm-thick flat catalyst slabs; module E: 4-mm-thick flat catalyst slabs; Module F: 6-mm-thick "wavy" catalyst slabs. (From Ref. 7.)

D. PPR Design Considerations

The bed effectiveness is an important factor for an optimal design of a PPR for a given application. A lower bed effectiveness means that more catalyst and a larger reactor is needed for the same process duty. Another important factor is pressure drop. As discussed above, a PPR with thin catalyst layers will have a high degree of catalyst utilization, so the amounts of catalyst and reactor space required are low. However, with thinner catalyst layers the construction cost of the PPR will be higher.

As shown above, a high gas velocity in the channels is in general beneficial for obtaining a high catalyst bed effectiveness but will increase the pressure drop. Figure 15 compares the PPR modules listed in Table 1 with respect to the relationship between catalyst bed effectiveness and pressure drop. The PPRs with corrugated screens, A and C, and the PPR D with flat screens, which all have similar hydraulic diameters for the cross sections of the gas passages and catalyst beds, show a similar effectiveness at the same pressure drop.

PPR modules E and F, which have thinner catalyst layers and will be more costly to construct, show a better relationship between bed effectiveness and pressure drop. The "wavy" PPR, F, is more advantageous at higher gas velocities, which, however, cause higher pressure drops.

For a PPR of a given geometric type, the gas velocity in the channels is an important design factor. Figure 16 shows how the costs related to pressure drop and reactor construction vary with the choice of gas velocity and that there is in general an optimum situation at a certain gas velocity. The precise location of the optimum will depend on the relative importance of the various cost factors, which will vary for different applications.

IV. FLOW AND TRANSPORT PHENOMENA IN LATERAL-FLOW REACTORS

A. Pressure Drop

As stated before, the LFR can be regarded as a fixed-bed reactor with a very low ratio of bed length over cross section. The pressure drop over the bed can be calculated from the Ergun equation for flow through porous media (Eq. 6).

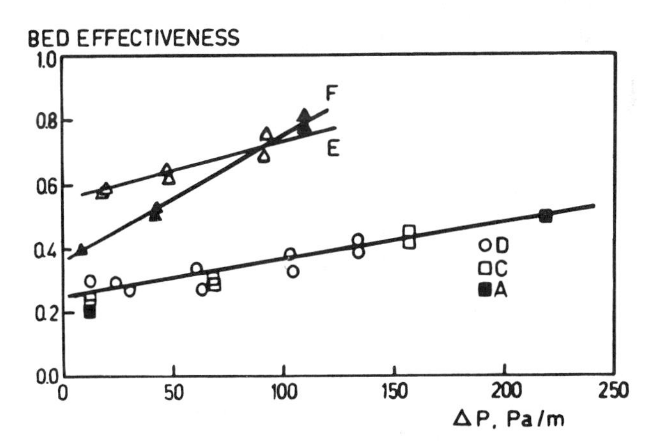

Figure 15 Relationship between catalyst bed utilization and pressure drop for different PPR modules in NO_x removal with ammonia. For details on PPR geometry, see Table 1. (From Ref. 7.)

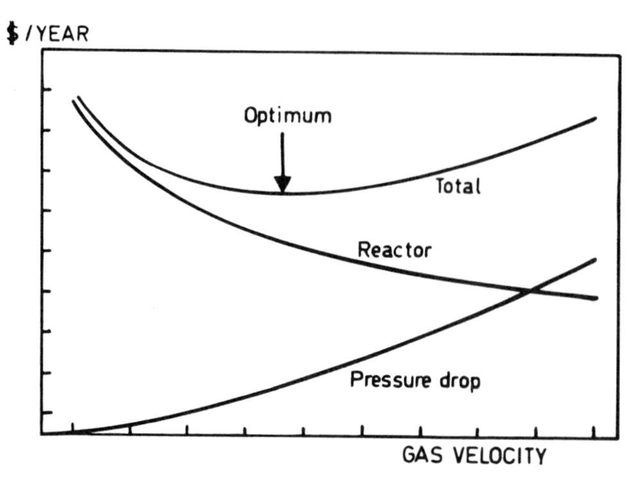

Figure 16 Effect of gas velocity in a PPR on process costs.

At a constant gas hourly space velocity, the superficial velocity of gas flowing through the bed is directly proportional to the bed length, i.e., to the thickness of the catalyst layers in the LFR. For laminar flow, the pressure drop per unit length is proportional to the velocity, while in the fully turbulent regime it is proportional to the square of the velocity. Hence the pressure drop over the total bed is proportional to the second power of the bed length for flow in the laminar region, and to the third power of the bed length in the fully turbulent regime. Hence, there is a very strong dependence of pressure drop on bed length (depth), as demonstrated by Fig. 17. As shown in this figure, the pressure drop over the catalyst layer in an LFR (thickness typically less than 75 mm) is several orders of magnitude lower than the pressure drop over a conventional fixed bed (bed length on the order of meters).

B. Deviations from Plug Flow

In contrast with the PPR, bed utilization in a fixed-bed reactor is essentially complete. Mass transfer outside the bed is generally not a limiting factor, for the main resistance is in most cases the intraparticle diffusion, which gives rise to incomplete utilization of the catalyst particle (see Section III.B.2).

Whereas a conventional fixed-bed reactor with flow of gas closely approaches an ideal plug-flow reactor in the sense that the residence-time distribution of the gas is very narrow, this may no longer be the case in very shallow beds, as in the LFR. For reactions with a positive order, extra catalyst is therefore required to achieve the same conversion as in an ideal plug-flow reactor.

Causes for deviations from ideal plug flow are molecular diffusion in the gas and dispersion caused by flow in the interstitial channels of the bed, and uneveness of flow over the cross section of the bed.

1. *Longitudinal Diffusion and Dispersion*

Molecular diffusion in the gas and dispersion resulting from flow through the interstitial channels cause a spread in residence time that can be described by an apparent diffusivity in longitudinal direction (i.e., the direction of the gas stream in the bed), $D_{ap.,l}$.

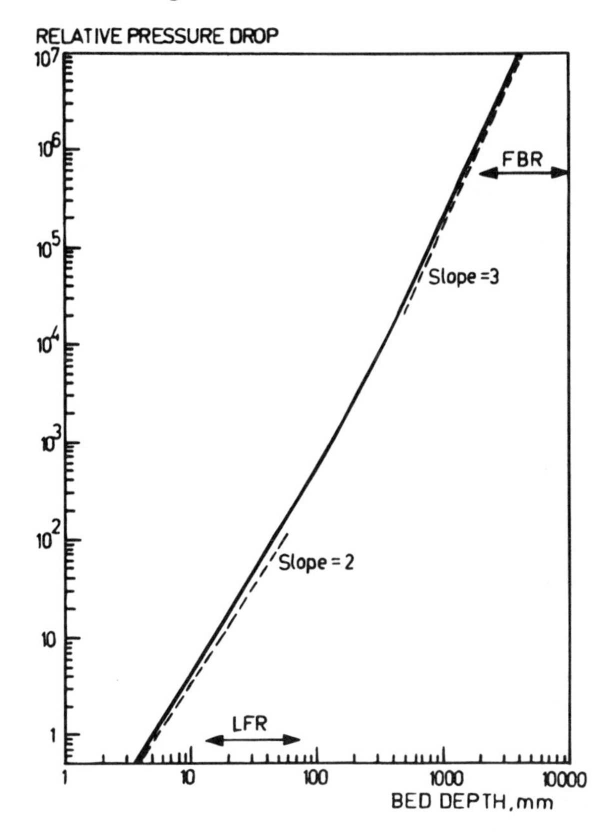

Figure 17 Effect of bed depth of fixed-bed reactors on pressure drop at constant space velocity. Air at 1 bar and 200°C; GHSV = 10,000 $Nm^3/(m^3 \cdot hr)$. $d_p = 2$ mm, $\epsilon = 0.4$. FBR = conventional fixed-bed reactor; LFR = lateral flow reactor.

A criterion for an acceptably small deviation from an ideal plug-flow reactor has been proposed by Gierman [12], based on the argument that the temperatures required for a given conversion in the real reactor and the ideal plug-flow reactor should not differ by more than 1°C, which is approximately the attainable accuracy of temperature definition in practice. This criterion is given by the expression

$$\text{Pé} = \frac{Lu}{D_{\text{ap.},l}} > 8n \ln \frac{1}{1 - X} \tag{10}$$

in which Pé is the axial bed Péclet number, L is the bed length or bed depth (i.e., the layer thickness of the catalyst layer in the LFR), u is the superficial gas velocity, n is the reaction order, and X is the conversion.

At low gas velocities, low pressures, and high temperatures, molecular diffusion in the gas can be the governing factor for $D_{\text{ap.},l}$. To satisfy the criterion of Gierman, the catalyst layers in the LFR should be thicker than a minimum value of L given by

$$L_{\min} = 8 \frac{\epsilon/\tau}{u} D_g \cdot n \ln \frac{1}{1 - X} \tag{11}$$

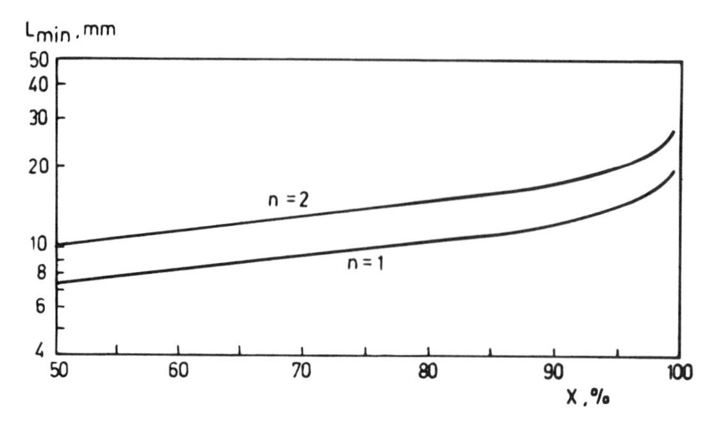

Figure 18 Minimum bed depth for acceptable deviation from plug flow as determined by gas diffusion. CO in air at 1 bar and 200°C; GHSV = 2000 $Nm^3/(m^3 \cdot hr)$. $\epsilon = 0.4$, $\tau = 2$.

This minimum layer thickness as a function of conversion is shown in Fig. 18 as a function of conversion and reaction order for representative conditions for flue gas treatment. It can be seen that a layer thickness in the range 15–75 mm is generally adequate.

At high velocity and with gases of low diffusivity, convective dispersion in the bed will be the dominating factor for $D_{ap,l}$, particularly with relatively large catalyst particles.

The longitudinal dispersion for flow through a packed bed is correlated with the dimensionless axial Bodenstein number Bo, defined as Bo $= d_p u / D_{ap.,l}$. At the low linear velocities typical for the operation of the LFR, Bo tends to approach a constant value of approximately 0.4, as found by Gierman [12]. Hence, Eq. (10) can be written as

$$\frac{L}{d_p} > 20n \ln \frac{1}{1 - X} \tag{12}$$

Figure 19 shows the minimum value of the ratio between layer thickness and particle diameter as a function of conversion for a first- and second-order reaction. It can be

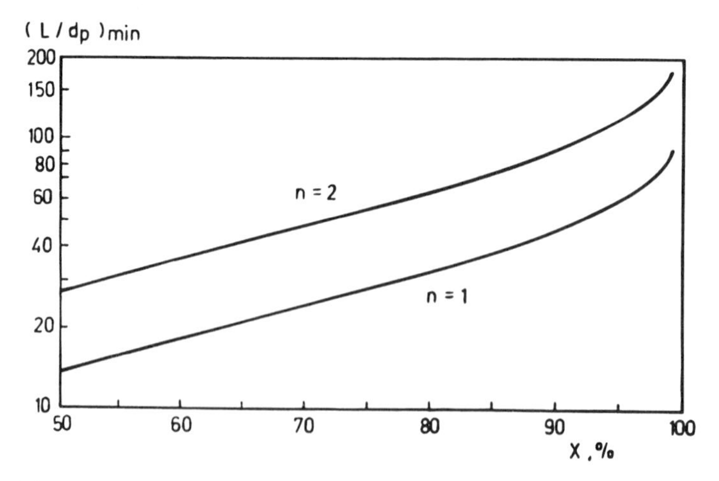

Figure 19 Minimum ratio between bed depth and particle diameter for acceptable deviation from plug flow, as determined by longitudinal dispersion in the bed.

inferred that for particle diameters larger than 1 mm, dispersion in the bed is a more important factor than molecular diffusion under the above conditions. For example, with particles of 1-mm diameter, a layer thickness of about 40 mm is required to satisfy the criterion proposed by Gierman for achieving 90% conversion in a first-order reaction.

2. *Velocity Variations*

Another factor that gives rise to a spread of gas residence time in the bed is an uneven velocity over the cross section. For reactions with a positive order, the velocity variation results in a loss of conversion, which must also be compensated by the use of an extra amount of catalyst over that required in an ideal reactor.

Possible causes for uneven flow through a thin catalyst slab are local variations in porosity and/or average particle size, which can occur due to improper filling of the LFR. A more intrinsic reason for uneven flow through the catalyst layers is the occurrence of pressure gradients in the channels. Since the LFR is designed for a low pressure drop over a catalyst layer, pressure gradients in the gas channels can have a significant effect on the local driving force for lateral flow.

Pressure gradients in the gas channels are caused by friction to flow in the channels and by variations of gas velocity in the channels. Friction in the channels causes a reduction in pressure in the flow direction, in both the inlet and outlet channels. Since these channels generally have their openings at the opposite ends of a catalyst slab, the pressure gradients in the inlet and outlet channels are in the same direction, and therefore there is some compensating effect in the local driving force for lateral flow through the bed.

A decrease of gas velocity in the inlet channel occurs as a result of loss of gas by lateral flow through the bed. Conversely, in the outlet channel the gas velocity increases as a result of the accumulation of gas that has passed through the bed. When the openings of the inlet and outlet channels are at the opposite ends of the catalyst slabs, the velocity gradients in the channels are in opposite directions. Thus, in two adjacent channels the gas velocities at a given position relative to the channel entrance are in general not the same. The differences in velocity give rise to pressure differences according to Bernoulli's law (Eq. 9) that affect the local driving forces for lateral flow through the bed.

Figure 20 is a schematic picture of the flow distribution in a typical LFR reactor. It can be inferred that the pressure difference is not constant over the catalyst slab and consequently that the lateral flow through the catalyst slab is not uniform.

In applications of the LFR for flue gas treatment, the volume of gas is not substantially altered by the treatment since the component to be converted is generally present in minor concentrations. Consequently, there is a simple relation between the velocity profiles in the channels and the velocity distribution in the bed, as shown in Fig. 21.

As mentioned before, the uneven flow through the catalyst slabs gives rise to a loss in conversion, which should be compensated by an extra volume of catalyst. The role of pressure gradients in the channels can be reduced by adopting thicker catalyst layers in the PPR, but this will give rise to a higher pressure drop of the LFR. An optimal design of a LFR for a given application therefore requires a careful balancing of the various design parameters.

V. BED-FOULING BEHAVIOR WITH SOLIDS-LADEN GAS STREAMS

A. Fouling of the Parallel-Passage Reactor

In a conventional fixed bed, where the gas flows through the packing of particles, particulates present in the gas deposit in the bed due to impingement on the solid surfaces. This

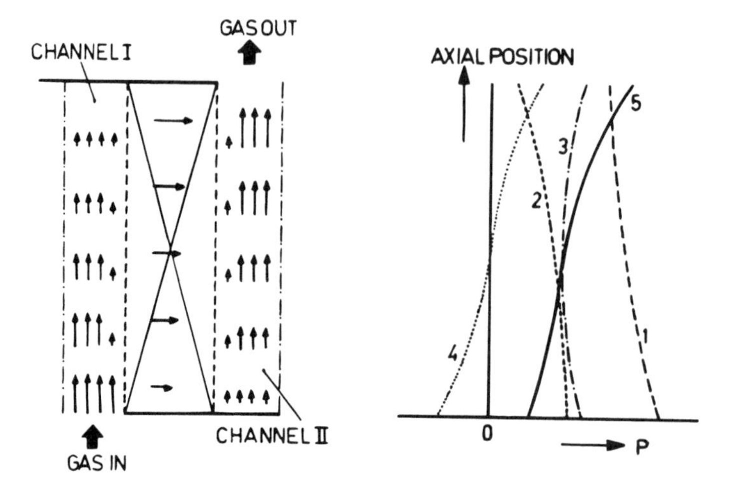

Figure 20 Patterns of velocity and pressure in a typical LFR: 1. Pressure profile in channel I as caused by friction. 2. Pressure profile in channel II as caused by friction. 3. Difference between 1 and 2. 4. Pressure profile as caused by Bernoulli effect. 5. Total pressure difference across catalyst slab.

generally occurs in a rather shallow layer, so the bed is very rapidly plugged by fly ash or other dust particles. In the PPR, particles in the gas can pass through the straight gas passages, and most of them will leave the reactor with the exiting gas stream.

Although the PPR can be operated with dust-containing gases, this does not mean that the reactor performance is totally unaffected by dust. Dust particles from the gas stream can enter the bed by diffusion (Brownian movement) and can also be transported by eddies penetrating into the bed when the flow in the channel is turbulent, as discussed before in the context of mass transfer. The occurrence of lateral flow through the catalyst

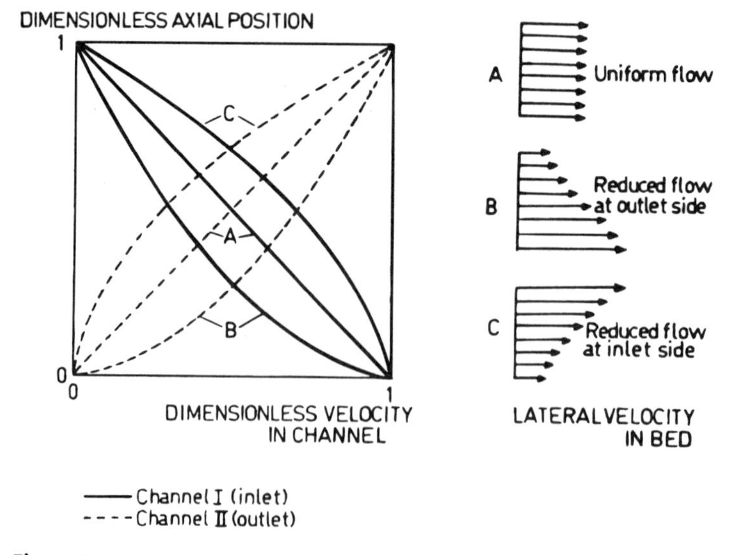

Figure 21 Relationships between velocity variations in channels and velocity profile in the bed.

layers as a consequence of deviations from perfect regularity of practical reactors is another cause for transport of dust particles from the gas stream into the catalyst bed.

The dust particles can deposit on the surfaces of the screens and the solid particles in the bed and thus reduce the permeability of the screen ("screen blinding") and the voidage of the packing. This manner of PPR fouling is depicted in Fig. 22.

Measurements on dust accumulation in a laboratory PPR module, carried out by Calis [6] with a silica powder and a fly ash (median particle diameter between 20 and 30 μm) at a solids concentration between 5 and 10 g/m^3 in air at ambient conditions and a gas velocity of about 1 m/sec showed that about 10% of the dust introduced is trapped, the major part of 90% being entrained by the exit gas.

At the above rate of dust accumulation, it can be estimated that the void space in the catalyst beds of the PPR will be completely filled with fluffy dust in a period on the order of 1 day. Assuming that the dust accumulation rate is linearly dependent upon the dust concentration in the feed gas (i.e., constant trapping efficiency), it follows that the fouling time of the PPR with a gas containing between 50 and 300 mg/m^3 of fly ash (0.02–0.12 gr/SCF, typical values for the offgas of oil-fired furnaces) would be on the order of a month to half a year.

Even with completely fouled catalyst beds, the PPR remains operable, in the sense that gas can still pass through the reactor (unless deposition of dust particles in the gas passages occurs without re-entrainment, which can eventually lead to bridging of deposited layers and blocking of the passages). Aside from a possible chemical poisoning of the catalyst by components of dust, the main effect of accumulated fluffy dust in the catalyst bed is an increase in intrabed mass transfer resistance. Apart from a reduction in bed voidage, which can reduce the static component in the intrabed diffusivity by a factor of approximately 2 to 4, the convective contribution is most strongly affected. This is because the decrease in voidage and average particle size lowers the parallel flow and the mixing length substantially.

The overall effect of fouling on reactor performance and whether the reactor performance in the fouled state remains acceptable depend on the design premises. For a PPR in which the bed effectiveness in the clean state is largely determined by the chemical activity or intraparticle mass transfer, the loss of reactor performance by fouling need not be too great and may still be acceptable even if the process has not been designed on the basis of a (partially) fouled reactor.

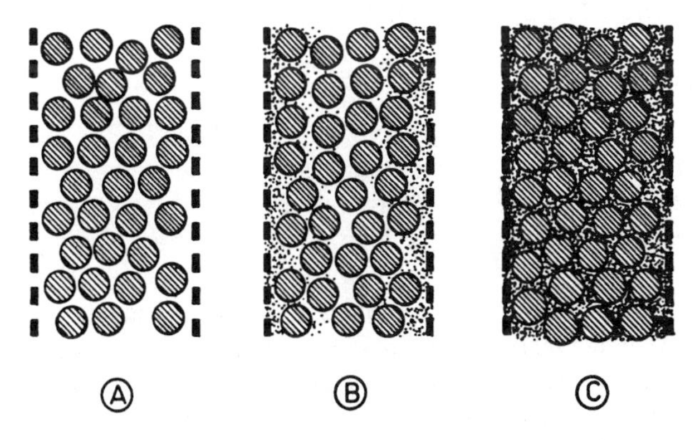

Figure 22 Schematic representation of the fouling of a catalyst slab in a PPR by dust. A: clean; B: partially fouled; C: completely fouled.

For operations with high dust-containing gases, e.g., gases from coal-fired boilers, which can contain several g/Nm^3 fly ash, there is the option to install the PPR downstream of the electrostatic precipitator that is generally installed to avoid unduly large emissions of particulates to the atmosphere. Another option to combat the fouling of the PPR is periodically to apply steam blowing, similar to the soot blowing customarily applied in the operation of boilers to remove deposited particulates from heat exchange surfaces. At high steam velocities in the channel, turbulence and vibrations of the catalyst slabs can cause detachment of particulates from the surfaces and their re-entrainment by the gas stream.

B. Fouling of the Lateral-Flow Reactor

The LFR traps particulates in the feed gas more efficiently than the PPR as a consequence of the flow of gas through the bed. The LFR is therefore more prone to fouling, and in contrast to the PPR the fouling also affects its operability, since the pressure drop increases significantly by fouling. The LFR is therefore primarily conceived for application in low fouling situations, e.g., the treatment of off-gases from gas-fired furnaces.

Compared to conventional fixed beds, the fouling rate of the LFR may be two orders of magnitude slower due to the much larger cross section of the bed. Under low fouling conditions, in which a normal fixed bed would be plugged within a week, the LFR may therefore be operated for a year or longer.

For operations under more strongly fouling conditions, options have been developed to overcome the fouling problem, such as periodic or continuous withdrawal of fouled catalyst from the reactor and reintroduction of the catalyst after removal of trapped dust. The removal and reintroduction of catalyst from the catalyst slabs is easily achieved when the gas flows horizontally through vertically oriented catalyst slabs. Fouled catalyst can be removed at the bottom end of the slab by gravity and clean catalyst added at the top of the slumped catalyst bed.

For vertical gas flow, which may be preferred in certain furnace configurations, vertical catalyst slabs can still be used when the bottom and top ends are tilted at an angle greater than the angle of repose of the solid material. Thus, gravity discharge of fouled catalyst material is possible, as illustrated in Fig. 23.

VI. CONSTRUCTIONAL ASPECTS AND SCALING UP

A. Construction of Reactor Modules

For industrial reactors of large volume, the structured packing of the LFR and PPR is for practical reasons best made up from standard modules ("unit cells") that are arranged in the reactor space by stacking. These modules are shop fabricated and filled before being installed in the reactor. Thus time can be saved in the startup of a plant and during catalyst replacement.

LFR and PPR modules with flat screens may be constructed by spot welding of metal screens to metal rods that serve to maintain the required distance between the screen surfaces and to provide rigidity to the construction. The screen envelopes may be packed with catalyst while the entrances of the gas passages are temporarily blinded. Alternatively, the whole module may be filled with catalyst, and subsequently excess catalyst in the gas passages is allowed to flow out.

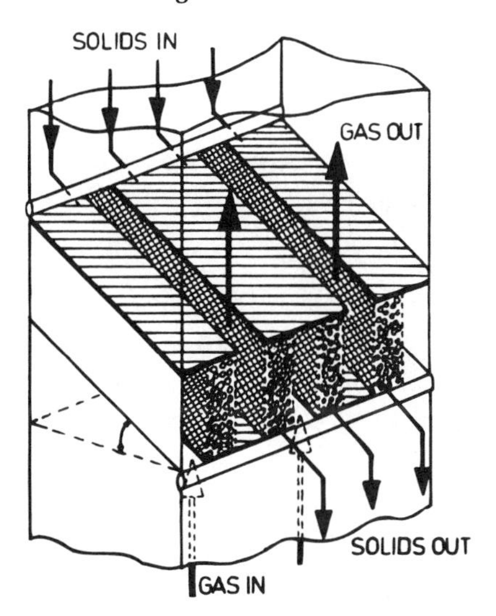

Figure 23 LFR with vertical gas flow, allowing on-stream withdrawal and addition of catalyst. (Adapted from Ref. 5.)

Parallel-passage reactors with corrugated screens were developed later [13]. Corrugated screens allow cheaper manufacture since the corrugations can serve to hold the screens apart from each other. An additional advantage is that filling with catalyst can be more easy. This can be illustrated by the PPR construction shown in Fig. 24, featuring zigzag folded screens and screens with folded ridges, the directions of the folds in alternate screens differing by 90°. It can be inferred that the gas passages indicated by B in this figure are separated from the catalyst space and that the latter space can be filled simply by pouring catalyst down the vertical direction A.

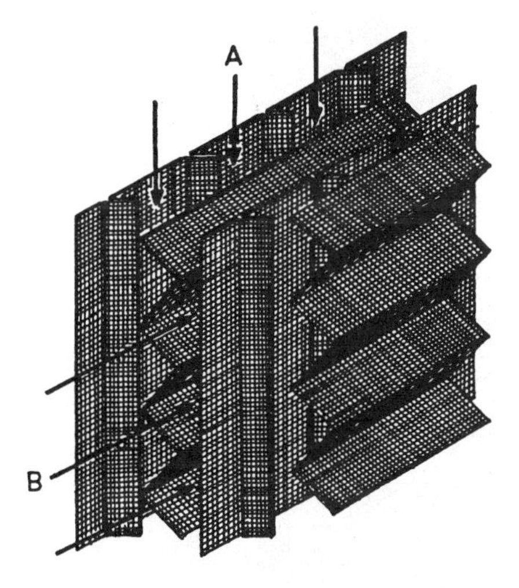

Figure 24 Construction of a PPR with a zigzag screen and a screen with folded ridges. A: direction of catalyst filling; B: direction of gas flow.

A photograph of PPR modules for an industrial reactor is shown in Figure 25.

To avoid the use of gauze screens in a PPR, the catalyst material can have the form of porous plates, similar to ceramic tiles [14]. These plates are positioned in racks. Such a PPR bridges the gap between the PPR with granular catalyst enclosed by screens and the honeycombs or monoliths that are used in off-gas treatment.

B. Scale-Up Aspects

The regular structure of the PPR and LFR is very favorable with respect to the scaling up of laboratory test reactors and demonstration units to industrial-size reactors. Since the basic geometric elements of the gas passages and catalyst beds remain the same, scaling up involves merely a multiplication of these elements. Data on the hydrodynamics and chemical conversions obtained in an adiabatic pilot unit consisting of a series of single unit cells are in principle identical to those of a large industrial reactor, which can be considered as a large number of identical cell stacks operated in parallel.

Because of the low pressure drop of the reactor, the distribution of gas at the reactor inlet and outlet has to receive proper attention.

The precisely defined geometry of the regular packing in the LFR and PPR facilitates computational modeling approaches. Both basic modeling techniques such as computational fluid dynamics and kinetic modeling and more empirical modeling based on correlations of laboratory and field data can be applied as useful tools in design and scaling up.

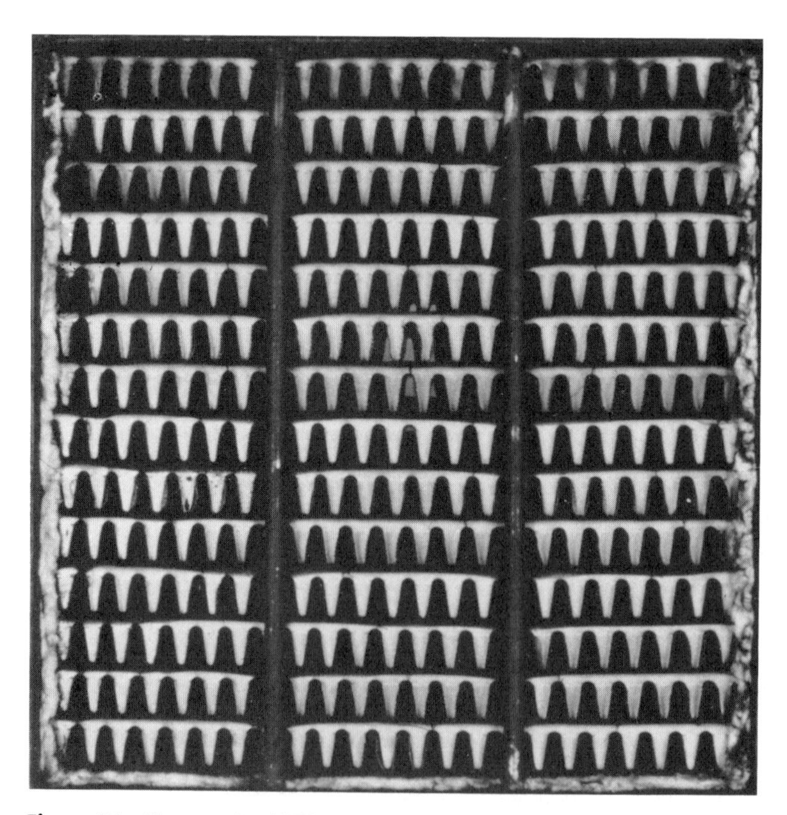

Figure 25 Photograph of PPR modules. (Courtesy Shell Research.)

VII. INDUSTRIAL APPLICATIONS

A. The Shell Flue Gas Desulfurization Process

The Shell flue gas desulfurization (SFGD) process described in 1971 [4] removes sulfur oxides from flue gas in a PPR using a regenerable solid adsorbent (acceptor) containing finely dispersed copper oxide. At a temperature of about 400°C, sulfur dioxide reacts with copper oxide to form copper sulfate according to the reaction.

$$CuO + \tfrac{1}{2}O_2 + SO_2 \rightarrow CuSO_4 \qquad \text{(i)}$$

In situ regeneration of the adsorbent is carried out by reaction with a reducing gas, e.g., hydrogen, at essentially the same temperature:

$$CuSO_4 + 2H_2 \rightarrow Cu + SO_2 + 2H_2O \qquad \text{(ii)}$$

which releases sulfur as a concentrated stream of pure sulfur dioxide. In the subsequent adsorption (acceptance) cycle, oxidation of copper occurs rapidly by the oxygen present in the flue gas:

$$Cu + \tfrac{1}{2}O_2 \rightarrow CuO \qquad \text{(iii)}$$

so that copper is returned to the active state for reaction with sulfur oxides.

The net result of reactions (i), (ii), and (iii) is the oxidation of hydrogen to water, which provides the driving force for transforming the diluted sulfur oxides in the flue gas to a stream of pure sulfur dioxide (after condensation of water), which can be converted to elemental sulfur in the Claus process.

The SFGD process is in principle an isothermal process, in that acceptance and regeneration are carried out at the same temperature level. This temperature level is favorable with respect to integration with a boiler or furnace, since the reactors can be installed between the economizer and air preheater, thus avoiding the need for reheating the stack gas.

Because the PPR is operated as an adiabatic reactor, the strongly exothermic oxidation reaction (iii) causes a temperature wave traveling through the bed, giving rise to a peak outlet temperature in the initial period of the acceptance cycle, as shown in Fig. 26. In this figure, the temperature profile predicted by a mathematical model developed at the Shell laboratory in Amsterdam is compared with the profile measured in an industrial reactor to be described later. During the initial oxidation period, the copper is not yet active for reaction with sulfur oxides, so there is a slip of sulfur in the initial period, as can be seen in Fig. 27. It can be inferred that the sulfur dioxide concentration profile of the effluent of the industrial reactor is in close agreement with the profile predicted on the basis of a kinetic model developed at the Shell laboratory in Amsterdam.

The essential elements in the development of the SFGD process are the development of a mechanically and chemically stable active acceptor that can withstand thousands of acceptance/regeneration cycles [15,16] and the parallel-passage reactor as a dust-tolerant system. Following an intensive development program, including the operation of a demonstration unit at Shell's refinery in Pernis for about 20,000 operating hours, an industrial unit was built at the Yokkaichi refinery of Showa Yokkaichi Sekiyu in Japan [17]. The unit, which was designed to effect 90% desulfurization of 125,000 Nm^3/hr of flue gas (mainly from an oil-fired boiler), was successfully started up in 1973 and has operated for many years. Data on the sulfur removal performance of the unit are listed in Table 2.

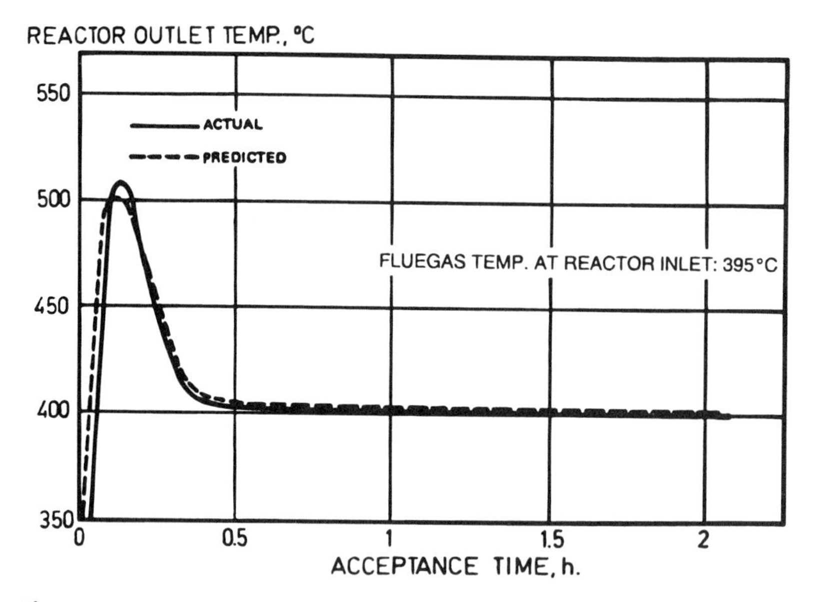

Figure 26 Reactor outlet temperature during acceptance cycle of the SFGD process. (From Ref. 17.)

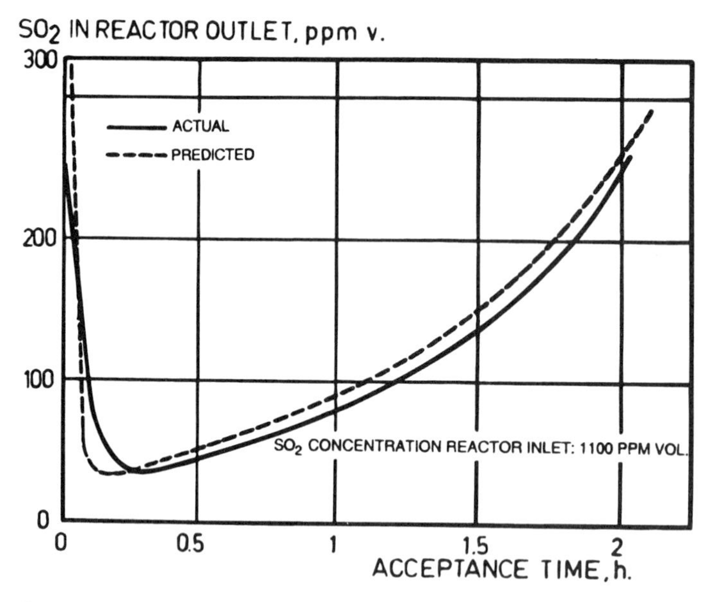

Figure 27 Sulfur dioxide concentration in the reactor effluent during the acceptance cycle of the SFGD process. (From Ref. 17.)

Table 2 Sulfur Dioxide Removal Performance of SFGD plant at Showa Yokkaichi Sekiyu Refinery

	Actual	Design
SO_2 removal efficiency, %	90	90
H_2 consumption, wt H_2/wt S	0.20	0.19
Absorber efficiency, %	99.5	99.9
SO_2 concentration of stripped water, ppm w. S	10	5
Total S conc. of stripped water, ppm w. S	—	20

NOTE: From Ref. 17.

A flow scheme of the unit, which features two parallel-passage reactors operated in the swing mode with automatic sequence control and sulfur dioxide absorption/stripping system to smoothen the fluctuating sulfur dioxide stream, is shown in Fig. 28. Figure 29 shows a photograph of the unit.

B. Simultaneous Removal of Sulfur and Nitrogen Oxides

A modification of the SFGD process just described is the Shell flue gas treating process, which not only removes sulfur oxides from flue gas, but can also effect a substantial reduction of the nitrogen oxides content. This is based on the activity of copper, whether in the oxidic or sulfate form, to catalyze the reaction of nitrogen oxides with ammonia according to the reactions

$$NO + NH_3 + \tfrac{1}{2}O_2 \rightarrow N_2 + \tfrac{3}{2}H_2O \qquad \text{(iv)}$$

$$NO_2 + \tfrac{4}{3}NH_3 \rightarrow \tfrac{7}{6}N_2 + 2H_2O \qquad \text{(v)}$$

Hence, by dosing ammonia during the acceptance cycle of the SFGD process, sulfur oxides as well as nitrogen oxides are removed from the flue gas. This variant of the SFGD process has also been applied on an industrial scale in the unit at Showa Yokkaichi refinery in Japan.

C. The Shell Low-Temperature NO_x Reduction Process

The PPR and LFR are also applied in a more recently developed dedicated process for NO_x removal from off-gases. The Shell low-temperature NO_x reduction process is based on the reaction of nitrogen oxides with ammonia (reactions iv and v), catalyzed by a highly active and selective catalyst, consisting of vanadium and titania on a silica carrier [18]. The high activity of this catalyst allows the reaction of NO_x with ammonia (known as selective catalytic reduction) to be carried out not only at the usual temperatures around 300°C, but at substantially lower temperatures down to 130°C. The catalyst is commercially manufactured and applied in the form of spheres (S-995) or as granules (S-095) [19].

The low temperature allows the Shell process to be applied as an "add-on" process to existing furnaces and boilers so that major modifications in the heat recovery sections are not necessary. This is particularly true in the case of low-sulfur gases, such as the off-gases of gas-fired furnaces. Higher sulfur contents can cause deposition of ammonium

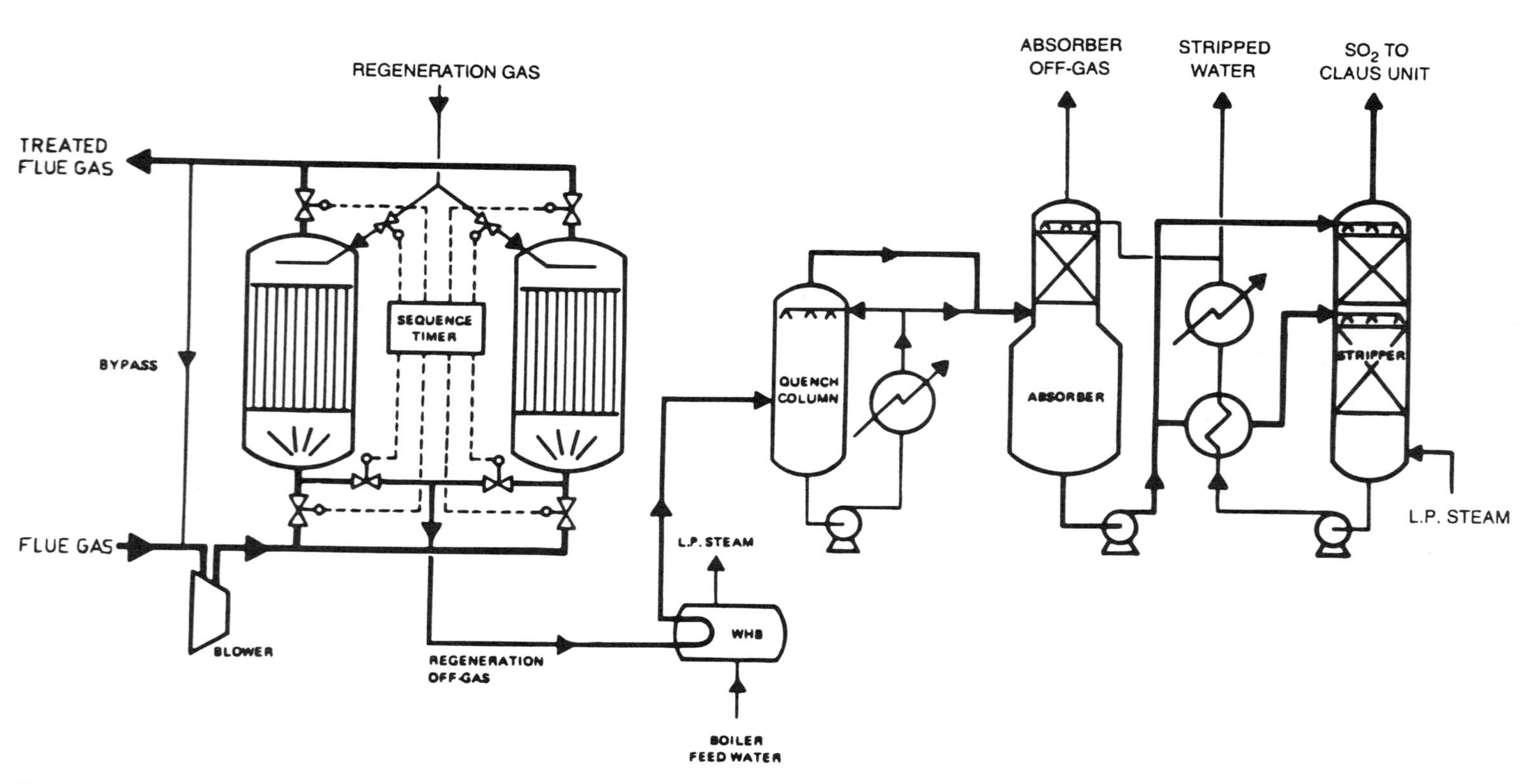

Figure 28 Flow scheme of the SFGD process as applied for sulfur oxides removal from refinery furnace off-gas. (From Ref. 17.)

Figure 29 Photograph of the SFGD plant at Showa Yokkaichi refinery, Japan. (Courtesy Shell Research.)

(hydro) sulfate, which causes a loss of catalyst activity. Therefore, there is a minimum operating temperature that depends on the sulfur content of the gas, ranging from about 150 to 260°C for sulfur dioxide concentrations in the flue gas between 10 and 1000 ppm v. [19]. Typical performance data of the Shell $DeNO_x$ process obtained in a semiindustrial fixed-bed test facility are listed in Table 3.

The performance of the PPR for NO_x removal by the Shell low-temperature NO_x reduction has been investigated extensively [20]. In the first commercial application of the Shell process with parallel-passage reactors, flue gases of six ethylene cracker furnaces at Rheinische Olefin Werke at Wesseling, Germany, are treated in a PPR system with 120-m^3 catalyst in total to reduce the nitrogen oxide emissions to about 40 ppm v. Since its successful start-up in April 1990, the unit has performed according to expectations

Table 3 DeNO$_x$ Performance of V/Ti/Silica Catalyst

Temperature, °C	135–155
GHSV, $Nm^3/(m^3 \cdot hr)$	2500–3500
NH_3/NO_x ratio	0.6–0.8
NO_x conversion, %	60–80
NH_3 slip, ppm v. (dry basis)	<5
Run length, months	6
Catalyst deactivation	not detectable

Note: S-995, 3-mm spheres, in a semiindustrial fixed-bed reactor treating a slip stream of flue gas from a commercial ethylene cracker furnace. (From Ref. 19.)

and without noticable catalyst deactivation [21]. A photograph of the unit is shown in Fig. 30.

The first application of the LFR in the Shell low-temperature NO_x reduction process was to treat the flue gas of a gas-fired furnace in a California refinery. The unit has been designed to treat flue gas containing 5 ppm v. of sulfur dioxide at 190°C and a space velocity of 5000 $Nm^3/(m^3 \cdot hr)$. The unit was started up successfully in 1991 [21].

D. Future Perspectives

Although the SFGD process has been proven as an industrial process for removing sulfur dioxide from flue gas (with optional reduction of NO_x), since its successful start up in 1973 it has found no further application so far. The main reason is that the advent of fuel oil hydrodesulfurization, which occurred in the period shortly thereafter, has provided a competitive and more convenient route for users of oil-fired furnaces to meet legislative requirements on sulfur oxide emissions. For coal-fired boilers, where sulfur oxides removal from flue gas remains necessary, the technology of scrubbing with lime or limestone slurries has since then been established as a cheaper way to remove sulfur oxides from the flue gas. Nevertheless, the SFGD process, which produces pure salable sulfur as an end product, is in principle an environmentally more friendly option. Possible future restrictions on the disposal of spent limestone slurry and limitations on the marketability of gypsum as a construction material may revive interest in regenerable adsorption processes such as SFGD.

The Shell DeNO$_x$ process as an "add-on" process is of interest for a wider range of applications. In addition to the treatment of gases from combustion sources such as furnaces and boilers, we may also consider NO_x removal from heaters, gas turbines, stationary reciprocating gas engines, etc. The modular construction of the PPR and LFR makes these types of reactor suitable for a wide range of reactor sizes, down to relatively small ones. We may also foresee applications in the treatment of NO_x-containing waste gases from the chemical industry, e.g., in nitric acid and caprolactam production or in catalyst manufacture.

The PPR and LFR, as dust-tolerant reactor systems with low pressure drop, have potential for many end-of-pipe catalytic processes for the cleaning of waste gases to reduce emissions that are increasingly the subject of environmental concern. The cleaning of waste gases includes, besides removal of sulfur and nitrogen oxides, the removal of volatile organic compounds, halogen-containing compounds, ammonia, and compounds with offensive odors.

Figure 30 Photograph of parallel-passage reactors for NO_x removal at Rheinische Olefin Werke at Wesseling, Germany. (Courtesy Shell Research.)

VIII. SUMMARY AND CONCLUSIONS

The parallel-passage and lateral-flow reactors are reactors that feature a low pressure drop and the ability to handle gases containing dust. They owe these characteristics to the specific arrangement of the catalyst in regular structures: Catalyst particles of a similar morphology as in traditional fixed beds are enclosed in geometric structures made of screens.

An advantage of the PPR and LFR over monoliths is that the catalyst does not need to be manufactured in a dedicated plant and that no special facilities for handling are required. The PPR and LFR can withstand thermal shocks very well since there is no danger of cracking of the catalyst or detachment of washcoat layers.

The effectiveness of the PPR and LFR as catalytic reactors or adsorbers can be high if they are designed with due consideration given to the flow and mass transfer characteristics. Scale up and reactor modeling benefit from the modular construction and the well-defined geometry of these reactors.

The applicability of the PPR and LFR as industrial reactors has been proven in industrial processes, viz., the Shell flue gas desulfurization process and the Shell low-temperature NO_x removal process.

The low pressure drop and dust tolerance of the PPR and LFR are of potential interest in many end-of-pipe treatments of waste gases to reduce emissions that meet with increasing environmental concern.

NOTATION

Bo	Bodenstein number
$D_{ap.,l}$	apparent longitudinal diffusivity in bed
$D_{eff.,b}$	effective intrabed diffusivity
$D_{eff.,p}$	effective intraparticle diffusivity
D_g	molecular diffusivity of gas
d_b	bed diameter
d_{char}	characteristic dimension of particle
d_h	hydraulic diameter
d_p	diameter of particle
F	particle shape factor
f	Fanning friction factor
g	gravity constant
h	height
K	correction factor for finite bed width
k	reaction rate constant
L	bed length or thickness of catalyst layer
m_x	mixing length in packing
n	reaction order
P	pressure
Pé	bed Péclet number
R	radius of catalyst particle
u	superficial velocity of flow through packing
v	average velocity in channel
X	conversion

Greek Letters

ϵ	voidage
η	dynamic viscosity
ρ	density
τ	tortuosity factor
ϕ	Thiele modulus

ACKNOWLEDGMENT

The authors wish to thank Ir J. E. Naber of the Koninklijke/Shell-Laboratorium Amsterdam for critically reviewing the manuscript.

REFERENCES

1. H.J.A. van Helden, J.E. Naber, J. Zuiderweg, and H. Voetter (to Shell Oil Co.), Removal of sulfur oxides from gas mixtures, U.S. Pat. 3,501,897 (1970).
2. J.E. Naber and C.W.J. Verweij (Shell Internationale Research Mij. N.V.), An apparatus for gas phase catalytic conversion, German Offenl. 1,907,027 (1969).
3. F.M. Dautzenberg, J.E. Naber, and C.W.J. Verweij (Shell Internationale Research Mij. N.V.), Device for contacting gases with a solid, German Offenl. 2,030,677 (1970).
4. F.M. Dautzenberg, J.E. Naber, and A.J.J. van Ginneken, Shell's Flue Gas Desulfurization Process, *Chem. Eng. Progr. 67* 8:86–91 (1971).
5. R. Samson, F. Goudriaan, O. Maaskant, and T. Gilmore, The Design and Installation of a Low-Temperature Catalytic NO_x Reduction System for Fired Heaters and Boilers, paper presented at the 1990 Fall Int. Symposium of the American Flame Research Committee, Oct. 8–10 1990, San Francisco, Calif.
6. H.P.A. Calis, *Development of Dustproof, Low Pressure Drop Reactors with Structured Catalyst Packings—The Bead String Reactor and the Zeolite-Covered Screen Reactor*, Ph.D. dissertation, Delft University, Delft, The Netherlands, 1995.
7. H.P. Calis, T.S. Everwijn, A.W. Gerritsen, C.M. van den Bleek, F.G. van Dongen, and F. Goudriaan, Mass Transfer in a Parallel Passage Reactor, *Chem. Eng. Sci. 49*:4289–4297 (1994).
8. J.H.B.J. Hoebink, E.P.J. Mallens, K.A. Vonkeman, and G.B. Marin, Transport Phenomena in a Parallel Passage Reactor. Paper presented at the A.I.Ch.E. 1993 Spring National Meeting, Houston, Texas, March 28–April 1, 1993.
9. C.N. Satterfield and T.K. Sherwood, The Role of Diffusion in Catalysis, Addison-Wesley, Reading, MA, 1963.
10. O. Kalthoff and D. Vortmeyer, Ignition/Extinction Phenomena in a Wall-Cooled Fixed-Bed Reactor, *Chem. Eng. Sci. 35*:1637–1643 (1980).
11. E.U. Schlünder and E. Tsotsas, Wärmeübertragung in Festbetten, durchmischten Schüttgütern und Wirbelschichten, Georg Thieme Verlag, Stuttgart, Germany, 1988.
12. H. Gierman, Design of Laboratory Hydrotreating Reactors. Scaling Down of Trickle-flow Reactors, *Appl. Catal. 43*:277–286 (1988).
13. P.L. Zuideveld and M.J. Groeneveld (Shell Internationale Research Mij. B.V.), Contacting device for gas and solid particles, Eur. Pat. Appl. 293,985 (1988).
14. F.M. Dautzenberg, H.W. Kouwenhoven, J.E. Naber, and C.W.J. Verweij (Shell Internationale Research Mij. N.V.), Apparatus and molded articles as catalysts and acceptors for removing impurities, especially sulfur dioxide, from stack gases, Ger. Offenl. 2,037,194 (1971).
15. H.W. Kouwenhoven, F.W. Pijpers, and N. van Lookeren Campagne (Shell Internationale Research Mij. N.V.), Removal of sulfur dioxide from oxygen containing gases, Brit. Pat. 1,089,716 (1964).
16. H.J.A. van Helden and J.E. Naber (Shell Internationale Research Mij. N.V.), Removal of sulfur dioxide from an oxygen-containing gas, Brit. Pat. 1,160,662 (1969).
17. W. Groenendaal, J.E. Naber, and J.B. Pohlenz, The Shell Flue Gas Desulfurization Process: Demonstration on Oil- and Coal-Fired Boilers, *A.I.Ch.E. Symposium Series 72 (156)*:12–22 (1976).

18. M.J. Groeneveld, G. Boxhoorn, H.P.C.E. Kuipers, P.F.A. van Grinsven, R. Gierman, and P.L. Zuideveld, Preparation, Characterization, and Testing of New V/Ti/SiO_2 Catalyst for Denoxing and Evaluation of Shell Catalyst S-995, Proc. 9th Int. Congr. on Catalysis, M.J. Phillips and M. Ternan, eds., Vol. 4:1743–1749. The Chemical Institute of Canada, Ottawa, 1988.
19. F. Goudriaan, C.M.A.M. Mesters, and R. Samson, Shell Process for Low-Temperature NO_x control, Proc. of the joint EPA-EPRI Symposium on Stationary Combustion NO_x Control, San Francisco, March 6–9, 1989, Vol. 2, pp. 8–39.
20. F. Goudriaan, H.P. Calis, F.G. van Dongen, and M.J. Groeneveld, Parallel-Passage Reactor for Catalytic Denoxing, Paper presented at the 4th World Congress of Chemical Engineering, Karlsruhe, Germany, June 16–21, 1991.
21. A. Woldhuis, F. Goudriaan, M.J. Groeneveld, and R. Samson, Process for Catalytic Flue Gas Denoxing, Paper presented at the Society of Petroleum Engineers Symposium on Health, Safety and Environment in Oil and Gas Exploration and Production, The Hague, November 11–14, 1991.

13

Bead-String Reactor

Hans Peter Calis, Kálmán Takács, Albert W. Gerritsen, and C. M. van den Bleek
Delft University of Technology, Delft, The Netherlands

I. INTRODUCTION

In addition to and as an alternative for the existing concepts of low-pressure-drop reactors with structured catalyst packings, discussed in the previous chapters, a new concept is proposed in this chapter: the bead-string reactor (BSR). The BSR was invented [1] as an alternative for a parallel-passage reactor (PPR) with extremely thin catalyst beds, viz. beds of only one catalyst-particle-diameter width.

The BSR is characterized by conventional particulate catalyst material that is fixed on parallel strings that are arranged, in the reactor, parallel (or cross) to the flow; see Fig. 1. Important in the definition of this concept is that the catalyst material is fixed to the strings due to its shape, without being glued. The catalyst material may consist of commercially available hollow extrudates or ring-shaped pellets that are mechanically stringed on wires or stacked on rods. Alternatives for mechanical stringing of conventional particles are to extrude silica or alumina paste *around* a carrier fiber (like the insulation material of electric cords) or to pill powder onto wire structures. In the latter cases, rods of catalyst material supported by wires are obtained, rather than strings of particulate catalyst material.

A. Advantages of the Bead-String Reactor (BSR) and Potential Applications

Below are listed several attractive features of the BSR concept and ways in which these features can be exploited.

Low pressure drop. Because of the shape of the void space through which the fluid flows, i.e., noncircular channels that are straight in the direction of the flow, the pressure drop across a BSR is comparable to that of a monolithic reactor. This feature is most profitable in processes operating at low pressure and high space velocities, such as catalytic removal of NO_x, SO_x, or volatile organic compounds (VOCs) from flue gases. The pressure drop can be manipulated by means of the voidage (see next item).

Voidage can be varied between 100% and as low as 10%. By choosing the pitch of the strings in a BSR, the voidage can be optimized with respect to pressure drop on one hand and reactor size on the other. The lower voidage limit of 10% is obtained by arranging the strings in a regular triangular array, at such a pitch that the strings (almost)

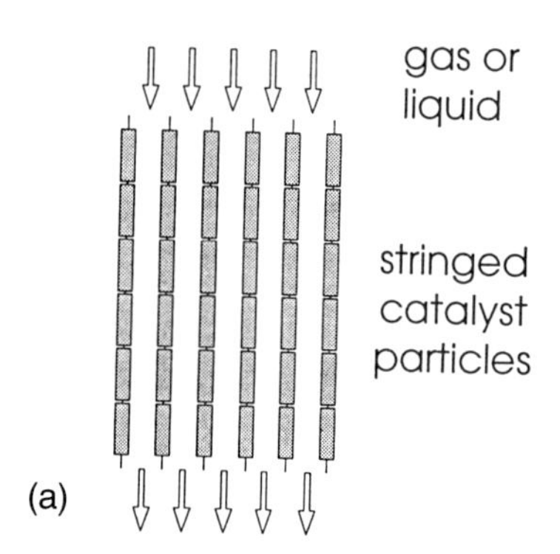

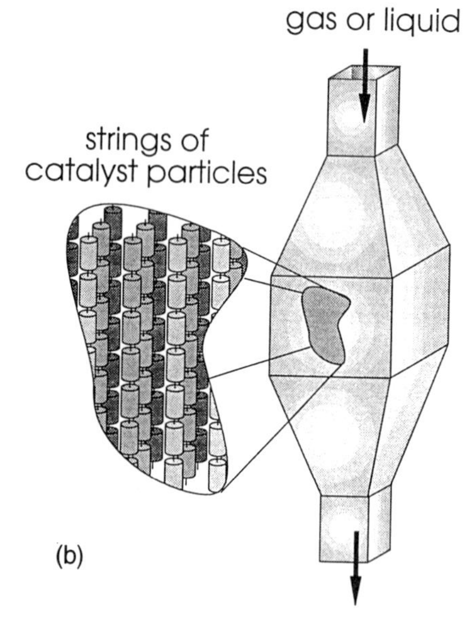

Figure 1 Schematic of the bead-string reactor. (a) Principle of operation; (b) artist's impression.

touch each other. In a traditional, randomly packed catalyst bed the voidage is typically ca. 45%. In lateral-flow reactors, parallel-passage reactors, and incorporated monolithic reactors it is often higher than 60%(!). In washcoated monolithic reactors the fraction of the reactor volume that is occupied by the active catalyst material (i.e., the washcoat) is even less than suggested by the voidage, because the inactive ceramic body usually occupies a considerable amount of space. The low voidage that can be attained with the BSR concept allows the design of very compact reactors, with a yet relatively low pressure drop.

Lateral transport of heat and reactants across the diameter of the reactor. This feature is important when the reaction has a high heat effect or when insufficient mixing of the reactants would have a strong adverse effect on the performance. Examples are selective oxidation processes (highly exothermic), where lateral or radial temperature gradients would decrease the selectivity of the conversion. Another example is the selective catalytic reduction process of nitric oxide with ammonia, where mixing of ammonia with the flue gas is often a point of great concern. Because the void space in a BSR is continuous not only in the axial direction but also in the lateral direction, lateral dispersion of heat and reactants is much better than in a monolithic reactor. An interesting point is that lateral transport in longitudinal flow through rod bundles (as in a nuclear reactor) can be significantly enhanced by spontaneous pulsating flows in the lateral plane, at highly turbulent flows [2].

Effective mass and heat transfer between gas and catalyst phase. Because the BSR geometry induces transition to turbulent flow at relatively low Reynolds numbers (as low as 500), mass and heat transfer between gas and catalyst phase is faster than in monolithic reactors. This is because the flow in a monolithic reactor is usually laminar. The advantage of the BSR can be exploited in processes with a large heat effect, where heat transfer through the film layer is generally more important than intraparticle heat and mass transfer.

Possibility to use "standard" catalyst particles. When hollow extrudates or ring-shaped pellets are used as the catalyst material for the BSR, the catalyst can be manufactured according to standard procedures. Consequently, no additional catalyst development is necessary to apply an existing catalyst in a BSR. This is an advantage over the monolithic reactors, because with those reactors one has to deal with the peculiarities of the washcoat or the ceramic carrier body serving as the support of the active sites.

Virtually dustproof operation. This feature results from the continuous channel structure of the BSR: Small "piles" of dust collected between strings can collapse in several directions and can be entrained by gas flowing along it. This effect can be increased by a gentle swinging movement of the strings, induced by the gas flow, when the strings are connected only at the top end and not at the bottom end of the reactor. In combination with a suitable choice of the voidage or pitch of the strings, this feature enables treatment of flue gases with a high dust content that would cause rapid fouling of other reactor types.

No maldistribution of gas or liquid in three-phase processes. Regarding application of the BSR concept to gas/liquid/solid processes, an important advantage of the BSR is that adjacent strings do not (necessarily) touch. Because of the liquid surface tension, liquid will not spill over from one BSR string to another. Consequently, the initial liquid distribution is maintained throughout the BSR module. This feature is especially advantageous when incomplete catalyst wetting (which results from liquid maldistribution in traditional, randomly packed trickle-flow reactors) would lead to hot spots and decreased selectivity.

Possibility to operate three-phase processes not only in cocurrent but also in countercurrent mode at high gas and liquid flow rates. Because the liquid flow is "guided" by the BSR strings, flooding in countercurrent operation is prevented up to high loads, unlike the case in traditional trickle-flow reactors. In this application, catalyst particle shapes other than simple rings may prove advantageous, such as particles with fins that guide the liquid flow. The possibility to operate in countercurrent operation can be exploited, especially in three-phase processes in which reaction products inhibit the reaction. An example is deep hydrodesulfurization of gas oil. An additional advantage of the small gas-phase pressure drop across the reactor is the comparatively high partial pressure of gas-phase reactants (e.g., hydrogen) at the reactor outlet, from which the reaction rate may benefit.

Possibility to supply heat directly to the catalyst by using the strings onto which the catalyst material is fixed as electrical heating elements. Clearly this feature can be exploited for quick startup of reactors and for highly endothermic reactions. This enables, for example, the steam reforming of methane in a very compact, electrically heated BSR.

Steam reforming is nowadays carried out in large numbers (often more than a thousand) of parallel tubular reactors placed in large furnaces. The high number of tubes is necessary because of the limited diameter (ca. 10 cm) of the tubes, set by the high heat requirement of the reaction. Using the BSR concept, all catalyst could be contained in one reactor, the heat being supplied from "within" the catalyst. Because of the low pressure drop, much smaller catalyst particles could be used (the standard reforming catalyst has an efficiency factor of a few percent). The size of a reformer could thus be decreased by at least a factor of a hundred. Especially in small-scale applications, these cost savings may weigh up to the extra costs of electrical heating and the more complicated construction of a BSR. The usefulness of electrically heating a catalyst on a metal structure to enable very fast warming up (i.e., starting up) of the reactor was recently illustrated with respect to automotive exhaust gas cleaning [3].

B. Status of Development of the BSR

The BSR is still in an early stage of development, compared to the monolithic reactors and the parallel-passage reactor. To the best knowledge of the authors, no other groups have reported on reactor engineering aspects of reactors containing strings of catalyst particles. The work by the present authors was focused on the modeling of gas/solid applications; the same models are applicable to liquid/solid BSRs. The development of models for three-phase BSRs (and analogues; see Section V.C) is the subject of current research at Delft University of Technology. The fact that the BSR concept is new to the field of catalytic reactor engineering does not imply that the development of the BSR had to start from scratch. Momentum and heat transport phenomena of longitudinal flow through regular assemblies of cylindrical rods (commonly called "rod bundles") have been studied quite extensively: It is estimated that over 500 articles have appeared on this subject since 1959. The main incentive for that work has been the growing importance of compact heat exchangers and nuclear reactors several decades ago. Important literature reviews in this field are by Kakaç and Spalding [4], Johannsen [5], and Rehme [6]. However, since in a BSR the (metal) rods are substituted by catalytically active strings of particles, additional research is necessary.

C. Overview of This Chapter

To enable assessment of the feasibility of the BSR concept for a particular application, it is necessary to have an adequate mathematical model that describes the reactor performance in terms of pressure drop and reactant conversion. Ingredients of such mathematical models are relations that describe transport of momentum, mass, and heat in the reactor and relations that describe the kinetics of the reaction that is carried out. Furthermore, an estimate of the costs of structuring the catalyst is needed.

The information presented in this chapter allows the construction of such a mathematical model for a G/S or L/S BSR, provided the kinetics of the reaction are known. In most instances in this chapter, the fluid phase is referred to as a gas; unless stated otherwise, the relations presented are expected to be also valid for liquids. As was mentioned in the previous section, the development of the BSR concept for G/L/S applications is in a very early stage, and no suitable engineering relations have been developed yet.

In Section II, relations are given to predict momentum transport phenomena in a BSR (i.e., pressure drop and flow distribution). In Section III, mass transport phenomena are discussed. Due to the analogy between mass transport and heat transport, the relations presented in that section are also applicable to heat transport. In Section IV, five models for a G/S BSR are discussed, which have been validated experimentally and found to yield satisfying results. Finally, in Section V several possible manufacturing techniques for the BSR or analogues are proposed, to allow an estimate (or rather guesstimate) of the cost of structuring a catalyst according to the BSR principle.

II. MOMENTUM TRANSPORT IN A G/S BSR (PRESSURE DROP AND FLOW DISTRIBUTION)

A. Introduction

An engineering equation that is often used to calculate the pressure drop across straight conduits or channels with a constant cross section, as in the BSR, is

$$\Delta p = 4f \frac{1}{2} \rho \langle u \rangle^2 \frac{L}{d_h} \tag{1}$$

This relation follows directly from a force balance across the channel and the definition of the friction factor:

$$\frac{4f}{4} \equiv f \equiv \frac{\tau_{f,w}}{(1/2)\rho \langle u \rangle^2} \tag{2}$$

Consequently, the problem of predicting the pressure drop across a BSR can be rephrased as the problem of predicting $4f$.

The purpose of this chapter is to present experimentally verified relations predicting $4f$ as a function of the BSR geometry, i.e., pitch, particle size, and reactor size (wall effect), and as a function of the Reynolds number and the roughness of the strings. Although various types of string arrays are possible (e.g., square, triangular, and circular), only one type, the regular square array, will be discussed. The analysis of the other types of array is analogous to that of the square array.

Only steady-state flow is considered, because the time scale on which the flow reaches a steady state is usually small compared to the typical time scale of process changes. Furthermore it is assumed that there is only convective flow in the longitudinal direction, parallel to the strings. Although it has been shown (e.g., Ref. 2) that lateral secondary flows and flow pulsations may occur in longitudinal flow through rod bundles, the effect of these lateral flows on the axial pressure gradient proves to be negligible.

Three flow regimes are distinguished, based on the prevailing momentum transport mechanism: laminar, turbulent, and transition flow. An important question is in which flow regime(s) a BSR is expected to be operated. Consider the case that: catalyst particles with a diameter ranging from 3 to 17 mm are used; the catalyst efficiency factor is 0.75; the pitch is such that the reactor voidage is 40%; the space velocity is such that the reactant conversion is 0.9; the reactor length over diameter ranges from 1 to 5; and the reactor diameter ranges from 1 to 3 m. When these specifications are met, the Reynolds numbers will be below 2000. The Reynolds number is based on the average linear gas velocity and the hydraulic diameter of the channels "between" the strings. Although for circular tubes this would imply laminar flow only, it will be shown in this chapter that in a BSR, turbulent flow may occur at Reynolds numbers as low as 500. For this reason all flow regimes will be discussed in this chapter.

Another important question is whether hydrodynamic inlet effects have to be taken into account under typical conditions. It was shown [7] that these effects are insignificant under these conditions, and therefore they will not be discussed here.

In Sections II.B and II.C relations for laminar and turbulent momentum transport in BSR-like structures are discussed, based on the literature and on original work of the present authors. An extensive literature review was given fairly recently by Rehme [6]. Many references to that review will be given in this chapter because it presents many graphs and correlations that combine all available literature data. Furthermore, a considerable part of the literature quoted in there consists of internal reports and conference proceedings that are difficult to obtain. In Sections II.B and II.C some attention will be given to the effect of the reactor wall on the flow distribution and the pressure drop. Although in reactors with few rods or strings (such as nuclear reactors or lab-scale BSR modules) the wall effect is considerable, for all larger-scale applications the wall effect will be negligible.

In Section II.D the validity of the presented relations is illustrated using some experimental data obtained with lab-scale BSR modules. Finally, in Section II.E the pressure drop across a BSR will be compared to that across a monolithic reactor and a packed-bed reactor.

B. Relations for Laminar-Flow Momentum Transport

1. *Governing Equation and Domains*

The momentum balance over a small element of fluid for incompressible, steady-state fully developed flow in the z-direction results in the following partial differential equation (PDE):

$$\nabla \cdot (\mu \nabla u_z) = \frac{dp}{dz} \tag{3}$$

The Nabla operator (∇) is in fact two-dimensional in this equation. The pressure gradient dp/dz is a constant, i.e., place independent. Obviously, the assumption of constant density is very reasonable even for gases, since we are dealing with low-pressure-drop reactors.

In laminar flow, μ is the molecular viscosity; in that case Eq. (3) is a second-order linear two-dimensional PDE of the Poisson type. In turbulent flow, μ also depends on the velocity gradients (hence on the lateral position), and Eq. (3) then is quasilinear or nonlinear.

The procedure to find analytical or numerical relations for $4f$ is quite straightforward: (1) Solve the momentum balance over a unit cell of the channel for a given pressure gradient, either analytically or numerically, to find the velocity distribution; (2) integrate the velocity profile over the cross-sectional area of the channel, to obtain the average velocity; (3) combine the obtained average velocity and the given pressure gradient with Eq. (1), to obtain $4f$.

For the simplest geometry of all, the circular channel, this procedure can be carried out analytically and yields

$$4f = \frac{64}{\text{Re}} \quad \text{or} \quad 4f \cdot \text{Re} = 64 \tag{4}$$

The product of $4f$ and Re is usually called the *laminar geometry parameter*, since for

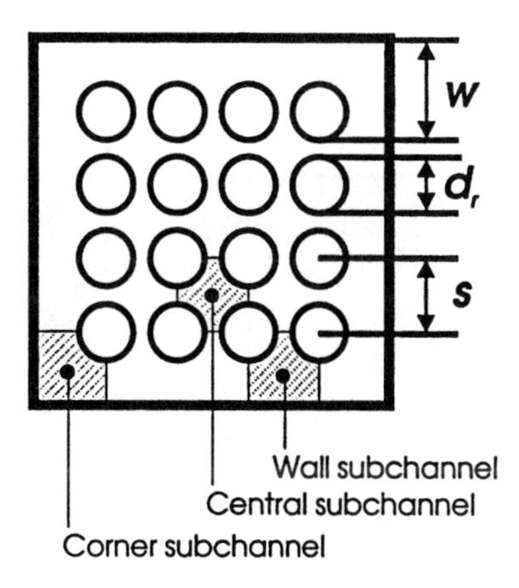

Figure 2 Geometrical definitions for a BSR or a rod bundle. s is the pitch, w the wall distance, s/d_r the relative pitch, and w/d_r the relative wall distance.

laminar flow its value depends only on the geometry of the channel and not on Re. To see in which domains Eq. (3) should be solved for a BSR, first some geometric definitions have to be given. In Fig. 2 a basic type of array is shown, the regular square array (cross section perpendicular to the direction of flow). It is characterized by three types of subchannels: central, wall and corner subchannels.

Under the important assumption (to be discussed in Section II.B.3) that the flow through central subchannels is influenced negligibly by the walls, and that the flow through wall and corner subchannels is influenced only by the nearest wall, there is a high degree of symmetry within the channels. In Fig. 3 all symmetry lines are shown that result from these assumptions. A symmetry line is defined here as a line connecting points at which the derivative of the longitudinal velocity, with respect to the normal vector of the line, equals zero (i.e., $\partial u_z/\partial \underline{n} = 0$).

From Fig. 3 it is seen that there are only three unit cells in this array, which are shown in Fig. 4; one subchannel contains 2–8 unit cells. The unit cells are the domains for which Eq. (3) should be solved to obtain the flow profile in these arrays. The boundary conditions are that (1) at the surface of the cylinder and the reactor wall, the fluid velocity equals zero, and (2) on the symmetry lines the derivative of u_z with respect to the normal vector of the symmetry line equals zero.

Figure 3 Symmetry lines in rod bundles. On the symmetry lines, $\partial u_z/\partial \underline{n} = 0$, where $\underline{n}$ is the normal vector of the symmetry line.

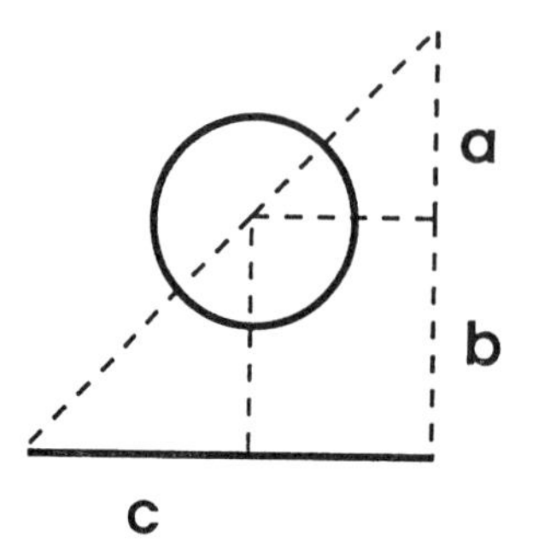

Figure 4 The unit cells that make up the subchannels in rod bundles. (a) Central subchannels; (b) wall subchannel; (c) corner subchannel. The dashed boundaries of the unit cells indicate symmetry.

2. *Solutions for the Central Subchannels*

In one of the first articles on this subject [8], the general analytical solution of Eq. (3) was derived. This general solution is easy to find, but it contains infinite series and (integration) constants that depend on the boundary conditions. Those were determined for the central cells of square and triangular arrays, using the boundary collocation method [8]. More recent publications on this subject are based mostly on complete numerical solution using finite-element methods.

Since, for a given type of array, the geometry of a rod bundle is characterized by the relative pitch (or the voidage), $4f \cdot \mathrm{Re}$ is a function of the relative pitch. For the range of relative pitches that is of greatest practical interest, literature data on the laminar central cell geometry parameter were correlated [6]. For regular square arrays the following correlation is accurate within 2%:

$$4f \cdot \mathrm{Re} = 162.8\left(\frac{s}{d_r} - 1\right)^{0.434} = 162.8\left(\sqrt{\frac{\pi}{4(1-\varepsilon)}} - 1\right)^{0.434} \tag{5}$$

for $1.05 \leq s/d_r \leq 2.0$, or $0.29 \leq \varepsilon \leq 0.80$.

When the spacing of the rods is increased, the flow along a rod is less influenced by the surrounding rods. As a result the flow will become more symmetric with respect to the longitudinal axis of each rod. At high relative pitches the central subchannel in Fig. 2 can be approximated by an annular channel, for which $\partial u_z/\partial r = 0$ at a certain radius r^*; see Fig. 5. The radius r^* is chosen such that the cross-sectional area of the annulus equals that of the actual subchannel. The analytical solution of Eq. (3) for this so-called *equivalent annular ring zone* approximation is easily obtained, and the resulting geometry parameter can be most conveniently expressed as a function of the voidage:

$$4f \cdot \mathrm{Re} = \frac{64\varepsilon^3}{(\varepsilon - 1)(\varepsilon^2 + 2\ln(1-\varepsilon) + 2\varepsilon)} \tag{6}$$

This relation is accurate within 1.5% for relative pitches higher than 2.8 ($\varepsilon > 0.90$) in a square array [6].

3. *Solutions for the Wall and Corner Subchannels; Wall Effects*

By solving Eq. (3) for rod bundles with different numbers of rods, it was shown [9] that an accurate estimation of the flow rate through wall, corner, and central subchannels is

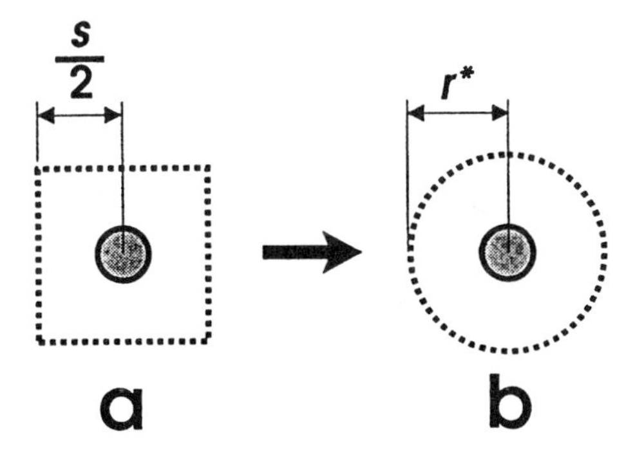

Figure 5 Approximation of a central subchannel in (a) a square array by (b) an annular channel.

obtained when symmetry on the boundary with the neighboring subchannels is assumed. This is the justification for the symmetry lines drawn in Fig. 3. Among other things, it was shown that the error in the flow rate through a central subchannel neighboring a wall subchannel, due to the assumption of symmetry, is less than 1% for relative pitches (i.e., s/d_r) smaller than 2.5 and relative wall distances (i.e., w/d_r) between 0.6 and 1.3. These and other results [9] show that the pressure drop characteristics of all subchannels that make up a finite rod bundle are closely approximated by those of the unit cells indicated in Fig. 4.

It can also be shown [7] that the assumption of symmetry at the boundaries between the subchannels leads not only to an accurate approximation of the flow rate through the individual subchannels (which is an average over the subchannel), but also to an accurate flow profile.

It is easily understood that when the distance between the reactor wall and the first strings (i.e., the wall distance) is much larger than the pitch of the strings, the fluid flows preferentially through the wall subchannels, effectively bypassing the catalyst material. This effect becomes especially pronounced when the reactor contains only a small number of strings or rods, say, less than 20 on a row. However, when the wall distance is approximately equal to the pitch of the strings, this wall effect is negligible in a real-size BSR (at least 50 strings on a row). This implies that the pressure drop characteristics of a BSR can be evaluated sufficiently accurately by considering only a central subchannel of the structure, just like a monolithic reactor is often modeled on the basis of only one channel. This subject is discussed in more detail in Ref. 7.

C. Relations for Turbulent-Flow Momentum Transport

1. *Critical Reynolds Numbers for Onset and Completion of Turbulence*

Whereas a circular channel has only one characteristic size (its diameter), a rod bundle has two: Δ_1 and Δ_2, defined in Fig. 6. When the flow rate through a rod bundle is increased, the flow in the narrow part of a subchannel (at Δ_1) will remain laminar longer than in the wide part (at Δ_2). This explains the observation often reported in the literature that in a rod bundle the transition from laminar to turbulent flow takes place over a much wider Reynolds range than in circular channels.

The onset of turbulence is defined as the Reynolds number at which $4f \cdot \mathrm{Re}$ starts to deviate from the laminar geometry parameter. The completion of turbulence is defined

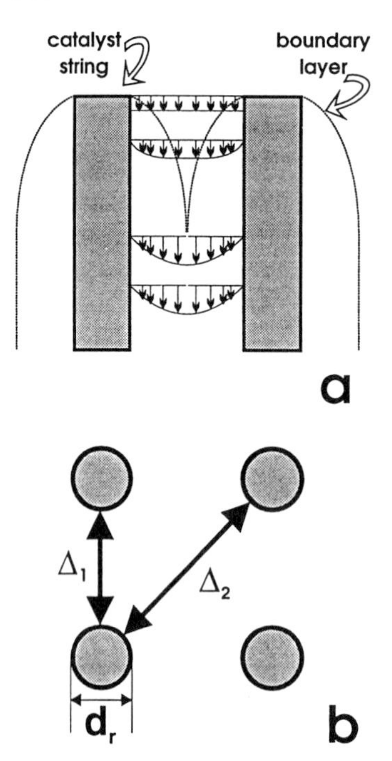

Figure 6 (a) Development of laminar flow in a rod bundle (side view); (b) definition of wall-to-wall distances in a rod bundle (top view).

as the Reynolds number at which the friction factor has become proportional to $Re^{-0.25 \text{ to } 0}$. The order in Re of the turbulent friction depends on the roughness of the rods: For smooth rods the order will be approximately -0.25, for very rough rods 0. The dependence of the friction factor on the surface roughness can be explained from models for turbulent flow. For an introduction to this subject, see, e.g., Ref. 10.

In Ref. 6 a graph is presented containing all experimental results on the onset and completion of turbulence in smooth rod bundles. There appears to be a wide spread in the experimental data, attributed to differences between the test sections with which the data were obtained, notably differences in relative wall distance and number of rods. There is no clear correlation between the onset and completion Reynolds numbers and the relative pitch, despite the curves that are suggested in the graph. For regular square arrays the onset Reynolds numbers range from 200 to 2000. The completion of turbulence takes place between $Re = 3000$ and $Re = 10^4$.

An important point is that these results have been obtained with bundles of smooth rods, as used in heat exchangers and nuclear reactors. In circular tubes, especially the completion Reynolds number depends strongly on the relative wall roughness. Consequently, for a BSR containing strings of catalyst particles, at least the completion Reynolds numbers and probably also the onset Reynolds numbers are lower than mentioned above.

2. *Semiempirical Relations for Smooth Rod Bundles*

In Ref. 11 the friction factor for turbulent flow in channels with smooth circular or noncircular cross sections is related to the geometry parameter for laminar flow. The laminar geometry parameter determines the values of the parameters A and G^* in the following expression for the turbulent friction factor:

$$\sqrt{\frac{2}{f}} = A\left(2.5 \ln \mathrm{Re}\sqrt{\frac{f}{2}} + 5.5\right) - G^* \tag{7}$$

Equation (7) is based on the universal flow profile for turbulent flow. Values of A and G^* were obtained from the theoretical analysis of the laminar and turbulent pressure drop characteristics of assemblies of circular channels with different diameters. These values also proved to give accurate friction factor predictions for noncircular channels, such as symmetric and asymmetric annuli and rod bundles.

Correlations of A and G^* with the laminar geometry parameter are given by Ref. 6. These relations predict that author's own experimental data within 6% and also adequately describe the other experimental data.

Based on the observation that the friction factors for turbulent flow in rod bundles differ only little from those of circular tubes, a very simple empirical correlation was proposed [12]:

$$4f = \left[0.57 + 0.18\left(\frac{s}{d_r} - 1\right) + 0.53(1 - e^{-a})\right]4f_t \tag{8}$$

where $a = 0.58 + 9.2(s/d_r - 1)$ and $4f_t$ is the friction factor for turbulent flow in hydraulically smooth circular channels. The latter may be calculated with the widely used Prandtl/Karman/Nikuradse relation.

For relative pitches between 1.08 and 2.7, the friction factors predicted by both methods described above deviate less than 10%, the latter always giving the higher prediction.

3. *Influence of Surface Roughness*

Using Prandtl's boundary layer equations [13], it can be shown that in a BSR with stringed catalyst particles, the roughness of the strings is expected to be of the same order of magnitude as the laminar boundary layer thickness [7]. Though the roughness of the catalyst material itself is normally small, because two catalyst particles on a string will never be perfectly aligned there is a considerable increase of the apparent roughness of the strings. Consequently, it can be expected that in nonlaminar flow the friction factor depends on both the roughness of the strings and the Reynolds number.

A model to account for roughness in a BSR may be derived from Eq. (7). The part of Eq. (7) within parentheses represents the universal velocity profile for turbulent flow along hydraulically smooth surfaces. For nonsmooth surfaces, the same expression for the velocity profile has been proved experimentally to be adequate, but then the second constant is smaller than 5.5; the first constant, 2.5, appears not to depend on the surface roughness. From this it can be made plausible that the roughness in rod bundles could be described with an empirical roughness function, $R(h^+)$, implemented in Eq. (7):

$$\sqrt{\frac{2}{f}} = A\left(2.5 \ln \mathrm{Re}\sqrt{\frac{f}{2}} + 5.5 + R(h^+)\right) - G^* \tag{9}$$

By analogy with the universal velocity profile for nonsmooth surfaces, $R(h^+)$ would be expected to be between −5.5 and 0. In Section II.D it will be shown that this procedure indeed allows an adequate description of pressure drop data in rough rod bundles.

4. *CFD Techniques for Flow in Rod Bundles*

Using the methods discussed in Section II.C, it is possible to predict friction factors and distribution of the flow over the subchannels in a BSR for hydraulically smooth turbulent flow. In most cases this will be sufficient for the design of a BSR. For a more detailed analysis, e.g., investigation of secondary flows in the lateral plane within subchannels or the lateral mixing between subchannels, computational fluid dynamics (CFD) techniques are necessary. An introduction to the application of CFD techniques for flow in rod bundles is given by Ref. 4; due to the enormous progress in the field of CFD during the last 16 years, a large number of publications have appeared since. Unfortunately, the standard versions of commercial CFD programs, like FLUENT and PHOENICS, are not suited for rod bundles, because of the combination of Cartesian and cylindrical domains. Consequently, if detailed calculations are necessary, the best option is to seek access to the software used by research institutes in the field of nuclear reactors or heat exchangers. Because the relations presented or referred to in this chapter provide sufficient information for preliminary reactor design purposes, the use of CFD techniques is not further discussed here.

D. Some Experimental Pressure Drop Data

To give an impression of the validity of the relations presented in this chapter, two graphs of the friction factor vs. the Reynolds number for two lab-scale G/S BSR modules are taken from Ref. 7. The characteristic size in the calculation of Re, as well as the channel diameter in the calculation of $4f$, is the hydraulic diameter of the reactor, i.e., four times the open area over the perimeter of the strings and the reactor wall.

Figure 7 pertains to a BSR with a square cross section of 70 mm, filled with a regular square array of strings of hollow cylindrical extrudates with a diameter of 1.6 mm and a particle length of ca. 6 mm. The relative pitch of the strings was 3.1, corresponding to a reactor voidage of 93%. The solid line represents the friction factor predicted from

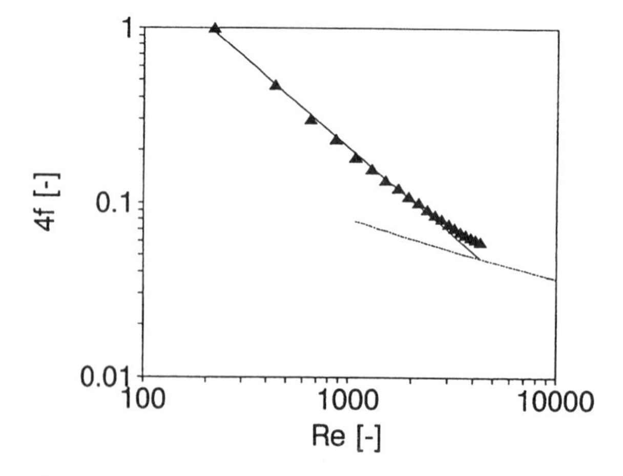

Figure 7 Experimental friction factors obtained with strings of extruded catalyst particles.

numerical solution of Eq. (3), taking into account the (small) wall effect. The dashed line represents the prediction based on Eq. (7) for turbulent flow in smooth rod bundles. It is seen that at the lowest Reynolds numbers, the experimental friction factors match the predicted values; at Re = 2500 the experimental values start to deviate from the prediction for laminar flow, indicating the transition to turbulent flow.

Figure 8 pertains to a BSR with a square cross section of 70 mm, filled with a regular square array of strings of ceramic pills with a particle diameter and length of 8 mm. The relative pitch of the strings was 1.1, corresponding to a reactor voidage of 34%. The solid line again represents the prediction for laminar flow. The lower dashed line represents the prediction based on Eq. (7) for turbulent flow in smooth rod bundles, whereas the upper dashed line represents the description with Eq. (9), containing an empirical roughness parameter fitted from these experiments. It is seen that the deviation from laminar flow starts already at Re = 100, which is in agreement with other experiments; in general, at low voidages (≤35%) the transition started at Reynolds numbers below 400, whereas at high voidages (≥80%) the flow remained laminar up to Reynolds numbers of 3000. It is further seen from this graph that the friction factor for turbulent flow can be adequately described with Eq. (9) on the basis of one empirical parameter, $R(h^+)$. This implies that only one pressure drop experiment in the turbulent-flow regime already provides sufficient information to predict the pressure drop under all conditions. Other experiments, in which smooth glass rods were used rather than bead strings, showed that in that case the friction factor for turbulent flow could be adequately predicted with Eq. (7), as expected.

E. Comparison of the BSR with the Monolithic Reactor and the Packed-Bed Reactor

Based on the discussion in the previous sections, it is now possible to compare the pressure drop across a BSR to that across an incorporated monolithic reactor and a randomly

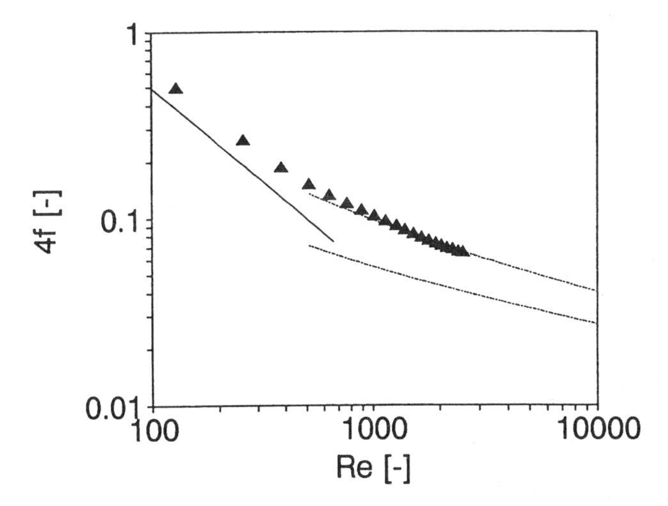

Figure 8 Fit of the roughness function to the experimental friction factors for strings of steatite beads. Solid line: theoretical, laminar flow; lower dashed line: theoretical, turbulent flow with smooth rods; upper dashed line: fit for turbulent flow with fitted roughness function [$R(h^+) = -3.5$]; symbols: experimental.

packed bed of spherical particles. This was done in Fig. 9, where the pressure drop across a certain length of reactor is plotted as a function of the reactor voidage. A constant superficial gas velocity was used in this graph. Furthermore, laminar flow was assumed for the BSR and the monolithic reactor. In this comparison the particle diameter in the BSR and the packed bed, and the wall thickness in the monolith, were chosen such that at a given voidage, the ratio of catalyst volume to external surface area was the same in each of the three reactors. This implies that for a given intrinsic catalyst activity, the catalyst effectiveness factor in each of the three reactors was (approximately) the same at a given voidage, resulting in a fair comparison.

It is seen that the pressure drop across the BSR is close to that across the monolithic reactor, and both are about two orders of magnitude lower than that across the randomly packed-bed reactor. Obviously, for a washcoated monolithic reactor the pressure drop will be slightly higher than for the incorporated monolith used in this comparison, depending on the ratio between the thickness of the washcoat and the ceramic or metal support.

From this comparison it is concluded that the BSR is indeed comparable to the monolithic reactors with respect to the pressure drop behavior, at least in the case of laminar flow (which normally prevails in monolithic reactors). A direct implication is that the mass and heat transfer characteristics of the two reactors will also be approximately the same in the case of laminar flow.

F. Conclusions

The description of momentum transport in a BSR is analogous to that in nuclear reactors and certain types of heat exchangers, and suitable engineering relations can be obtained from those fields. Unlike in nuclear reactors and heat exchangers, the wall effect will normally be insignificant in a real-size BSR, due to the large number of elements (i.e., strings) in the reactor. Consequently, the pressure drop across a BSR can normally be evaluated on the basis of a unit cell containing one string.

Unlike in a monolithic reactor, the flow in a BSR changes gradually from laminar to turbulent, over a relatively large range of Reynolds numbers. The transition Reynolds

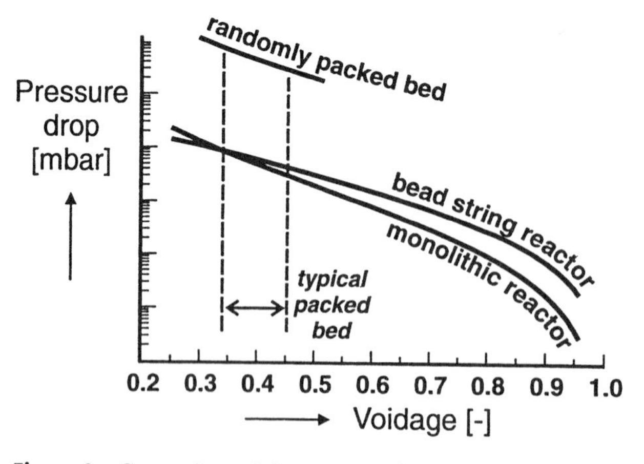

Figure 9 Comparison of the pressure-drop across a unit length of BSR, monolith, and randomly packed bed.

number ranges from below 400 (voidage $\leq$ 35%) to 3000 (voidage $\geq$ 80%). Using the relations presented or referred to in this chapter, the pressure drop for laminar flow in a BSR can be predicted with an accuracy of a few percent. For turbulent flow the pressure drop can be predicted with the same accuracy when the bundle is hydraulically smooth, to be judged from comparison of the string roughness with the thickness of the laminar boundary layer. When the bundle is hydraulically rough, an empirical roughness parameter should be introduced, which then allows an adequate description of the pressure drop.

Finally, it is concluded that the BSR is indeed comparable to the monolithic reactors with respect to the pressure drop behavior, at least in the case of laminar flow.

III. MASS TRANSPORT AND HEAT TRANSPORT IN A G/S BSR

A. Introduction

As was mentioned in Section I, heat transport phenomena of longitudinal flow through rod bundles (i.e., transport from the rods to the fluid v.v.) have been studied quite extensively in the past and are still receiving constant attention, due to the importance of compact heat exchangers and nuclear reactors. This has led to many theoretical and empirical relations to predict Nusselt numbers as a function of the relative pitch and (for nonlaminar flow) the Reynolds number; a relatively recent review of those relations that pertain to smooth rod bundles is by Rehme [6].

Knowing this, we might expect that a study of the existing literature would be sufficient to allow modeling of mass transfer in a bead-string reactor, because heat transfer relations usually can be translated directly into mass transfer relations, since the governing differential equations are identical. Unfortunately, this is not completely true, for two reasons.

The first reason is that in the translation of heat transfer relations, special attention has to be given to the boundary conditions. Most texts on heat transfer deal only with the standard boundary conditions of constant wall temperature or constant heat flux, whereas in a chemical reactor constant surface concentration or constant mass flux can be assumed only in special cases.

The second reason is that no relations have been found that predict heat (or mass) transfer in rod bundles with the unusual type of roughness that is encountered in the BSR. The known relations on smooth rods will obviously underestimate the heat (or mass) transfer rate in a BSR in turbulent flow.

Consequently, the goal of this section is to:

Give a brief overview of the most convenient heat transfer relations that are proposed in the literature; an extensive literature review was given fairly recently by Rehme [6].

Discuss the consequences of the nonstandard boundary conditions arising in the description of mass transfer in a plug-flow reactor with catalytically active walls.

Illustrate the presented relations with experimental data on mass transfer for turbulent flow in a BSR with hydraulically rough strings of beads.

The same assumptions are made as in the discussion of momentum transport, in the previous section.

Under typical conditions, the effect of enhanced mass transfer in the entrance zone of the reactor on the overall reactor performance, due to the development of the concentration profile, is negligible. For more details, see Ref. 7.

In Section III.B, theoretical and empirical mass transfer and heat transfer relations are discussed. The theoretical relations pertain to laminar flow, because for that flow regime the governing equations can be relatively easily solved numerically. The empirical relations pertain to turbulent flow in smooth rod assemblies. In this section the implications of the nonstandard boundary conditions in a BSR are also discussed. In Section III.C, the validity of the presented relations will be illustrated using experimental data obtained for turbulent flow in a lab-scale BSR with hydraulically rough strings of beads.

B. Theoretical and Empirical Mass Transfer and Heat Transfer Relations

1. *Governing Equations and Domains*

The discussion of mass transfer and heat transfer in rod bundles pertains to the same geometrical and mathematical domains as the discussion of momentum transfer; see Fig. 10. In the previous section it was stated that in a real-size BSR, the pressure drop and flow distribution are influenced only negligibly by the reactor wall; the same holds true for the mass transfer and heat transfer characteristics. Consequently, only mass (and heat) transfer in the central subchannel will be discussed here.

The aim of this discussion is to obtain relations to predict the value of the mass transfer coefficient k_g, which is defined by

$$\phi''_{\mathrm{mol},A} \equiv k_g(c_{A,b} - c_{A,s}) \tag{10}$$

In this equation the bulk concentration $c_{A,b}$ is defined as the mixing cup average of the concentration in the fluid domain; the surface concentration $c_{A,s}$ is defined as $c_A|_{r=R}$. By using concentrations in this equation, it is implied that the system is isothermal and ideal in the thermodynamic sense. If this is not the case, partial pressures or even fugacities should be used instead.

According to Fick's law, the flux to the catalyst surface is given by

$$\phi''_{\mathrm{mol},A} = -D_{\mathrm{free}} \left.\frac{\partial c_A}{\partial r}\right|_{\substack{r=R \\ \text{fluid domain}}} \tag{11}$$

Combining Eqs. (10) and (11), it is seen that k_g can be obtained from

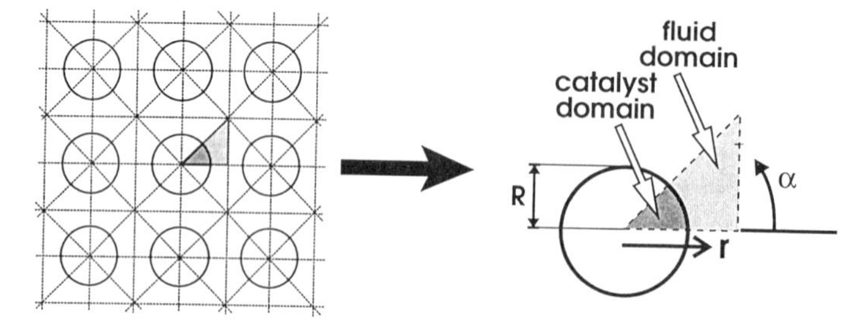

Figure 10 Unit cell of a central subchannel of a regular square array.

$$k_g = \frac{-D_{\text{free}} \left. \dfrac{\partial c_A}{\partial r} \right|_{\substack{r=R \\ \text{fluid domain}}}}{c_{A,b} - c_{A,s}} \tag{12}$$

The procedure that leads to relations for k_g as a function of the reactor geometry and process conditions is to obtain values for the right-hand side of Eq. (12) by analytically or numerically solving the partial differential equations (PDEs) that govern mass transfer in this system, for various reactor geometries and process conditions.

The governing equation for mass transfer that should be solved to find values of k_g follows from the component balance over a small element of the fluid domain. For steady-state, fully developed, incompressible flow in the z-direction, the component balance results in

$$\nabla \cdot (D \nabla c_A) = u_z \frac{\partial c_A}{\partial z} \tag{13}$$

For turbulent flow the diffusivity D is a function of the velocity gradients, hence of the lateral position; for laminar flow it is the constant free molecular diffusivity D_{free}. The axial gas velocity u_z follows from the momentum balance over a small element of the fluid domain, as discussed in the previous section.

At the symmetry lines that bound the fluid domain (see Fig. 10), the boundary conditions of Eq. (13) are simple: The derivative of c_A with respect to the normal vector should equal zero. The boundary condition at the external catalyst surface is the relation that couples the fluid domain with the catalyst domain:

$$D_e \left. \frac{\partial c_A}{\partial r} \right|_{\substack{r=R \\ \text{catalyst domain}}} = D_{\text{free}} \left. \frac{\partial c_A}{\partial r} \right|_{\substack{r=R \\ \text{fluid domain}}} \tag{14}$$

Consequently, an equation describing the concentration in the catalyst domain is also needed. This equation follows from a component balance over a small element of the catalyst domain; for steady state it yields

$$\nabla^2 c_A = \frac{1}{D_e} \Re_A \tag{15}$$

The boundary condition of Eq. (15) at the external catalyst surface is given by Eq. (14); the boundary condition at the symmetry lines is that the derivative of c_A with respect to the normal vector should equal zero.

These equations are identical to the equations describing mass transfer in monolithic reactors. For monolithic reactors it was shown [14] that when the reaction rate is very fast compared to the mass transfer rate in the fluid domain, the boundary condition of Eq. (15) becomes identical to the standard heat transfer boundary condition of constant wall temperature; when the reaction rate is very slow compared to the mass transfer rate in the fluid domain, the boundary condition of Eq. (15) becomes identical to the standard heat transfer boundary condition of constant heat flux. The influence of the boundary conditions on the mass transfer coefficient in case of laminar flow is discussed in the following section.

An interesting point is that unlike in a circular channel, the mass transfer rate in a BSR varies along the perimeter of a rod or string. In circular ducts (i.e., normal pipes) there is axial symmetry, which implies that $(\partial c_A/\partial r)|_{r=R}$ and $c_{A,s}$ in Eq. (12) are constant along the perimeter of the wall, i.e., circumferentially. In a rod bundle, however, symmetry with respect to the axis of each rod is approximated only at high relative pitches (>2.8 for a regular square array); cf. the discussion of the annular ring-zone approximation in the previous section. At lower relative pitches, $(\partial c_A/\partial r)|_{r=R}$ and/or $c_{A,s}$ will vary circumferentially. Consequently, k_g and Sh ($\equiv k_g d_h/D_{free}$) will then be local variables that depend on the angular coordinate α. In the case of circumferential variations of the mass transfer rate, a suitable mass transfer coefficient that is of engineering use is the circumferential average of k_g. Unless stated otherwise, the values of k_g presented here are circumferential averages. The importance of the circumferential variations in case of laminar flow will be discussed in the following section.

2. *Relations for Laminar Flow*

a. Angular Dependence of Sh. Because in a rod bundle axial symmetry with respect to each rod is approximated only at high relative pitches, circumferential variations of the concentration and/or mass flux can arise along the perimeter of the rods, because of the nonsymmetrical velocity field around the rod. The momentum transfer equivalent of this phenomenon was discussed by Ref. 8; in momentum transfer, the obvious boundary condition of the governing PDE is constant axial and circumferential velocity (namely, 0 everywhere!). It was shown that in a regular square array, the wall shear stress has its minimum at an angle α (see Fig. 10) of 0 and its maximum at an angle of 45°, as is easily understood. The circumferential variations increase with a decreasing relative pitch: For a relative pitch of 1.1 the local wall shear stress at $\alpha = 0$ is ca. 58% lower than the circumferential average, whereas at a relative pitch of 1.5 the value at $\alpha = 0$ is only ca. 12% lower than the circumferential average.

Whereas the circumferential variations of the local wall shear stress (i.e., the momentum flux) in itself are not of interest in the study of the BSR, the analogous variations in mass flux or surface concentration are indeed. In Ref. 15 a graph is presented of the local heat flux relative to the circumferential average, for the constant-temperature boundary condition, as a function of α and s/d_p. These data are based on a semianalytical solution of the governing PDE, following the procedure described by Ref. 8 (see Section II.B.2). At a relative pitch of 1.2 the local flux at $\alpha = 0$ is ca. 64% lower than the circumferential average; at a relative pitch of 1.5 the flux at $\alpha = 0$ is still ca. 20% lower than the circumferential average. In the case of a constant surface temperature, the local heat fluxes are directly proportional to the local Nusselt (or Sherwood) numbers.

For the boundary condition of constant mass flux, it was shown [16] that significant circumferential variations in the local surface concentration will arise at low relative pitches. As expected, the surface concentration at $\alpha = 0$ then is lower than the circumferential average.

For a BSR built up of cylindrical catalyst rods (i.e., "infinitely" long catalyst cylinders), the boundary condition of Eq. (14) will result in circumferential variation of both the mass flux and the surface concentration. Because of the varying surface concentration, the concentration profile in the rod will not be axisymmetric, which will influence the effectiveness factor of the rod. Fortunately, the influence can be expected to be limited. This is because the strongest (relative to the flux) circumferential variations of the surface concentration are obtained for the boundary condition of constant flux, but this boundary condition corresponds to a relatively low reaction rate, which will usually correspond to small intraparticle concentration gradients.

b. Influence of Relative Pitch on Sh. Following the procedure described in Section III.B.1, circumferentially averaged Sherwood numbers were calculated for a regular square array, as a function of the relative pitch, for both extreme boundary condition [16]. The results are displayed in Fig. 11. It is seen that for relative pitches larger than 1.1 (corresponding to a reactor voidage of 0.35 or higher), the effect of the boundary condition on the calculated Sherwood number is smaller than 30%.

Because the values of Sh calculated for those two boundary conditions form the limiting values for a BSR, the maximum error in Sh is limited to 30% (for $s/d_p > 1.1$) when either of the limiting values is used for predicting the mass transfer rate. In predicting the performance of a BSR, the Sherwood number is of interest only when the reaction rate is relatively fast. Since in that situation the boundary condition resembles the constant-temperature condition, it is recommended to use that limiting Sherwood number for simple engineering calculations. For very precise predictions of the mass transfer rate, however, or when the relative pitch is chosen smaller than 1.1, it will be necessary to find the mass transfer rate by solving the governing equations discussed in Section III.B.1.

3. Relations for Turbulent Flow

Unfortunately, no empirical mass transfer or heat transfer relations were found that consider the type of artificial surface roughness that is encountered in a BSR. The relations that are discussed in the first part of this section pertain to hydraulically smooth surfaces and are therefore expected to predict the lower limit of Sherwood numbers in a BSR.

Because of the relatively flat velocity profile in turbulent flow, the channel geometry has only a small influence on the friction factor (as discussed in the previous section) and the Sherwood and Nusselt numbers. The turbulent Sherwood and Nusselt numbers of rod bundles can therefore be related to those of circular tubes. The few experimental data, as compiled by Ref. 6, suggest that for relative pitches between 1.1 and 2.0 (which

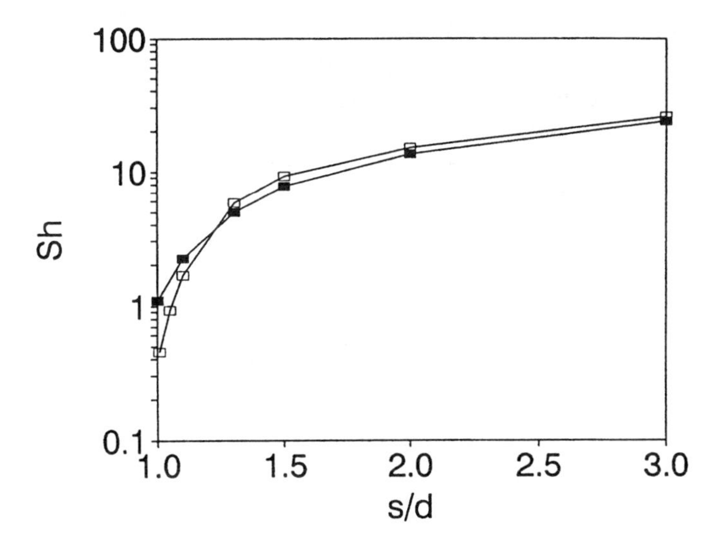

Figure 11 Theoretical Sherwood numbers of laminar flow in a BSR with a regular square array, as a function of the relative pitch. ■ = Sh for boundary condition of constant wall concentration; □ = Sh for boundary condition of constant flux.

is expected to be the interesting range for the BSR), the deviation of the rod bundle Nusselt number from the circular tube Nusselt number is less than 20%. Often this accuracy will be sufficient. In compiling the experimental data, Ref. 6 used the following correlation for the circular tube Nusselt number:

$$\mathrm{Nu}_t = 0.02087\mathrm{Re}^{0.7878} \quad \text{for Pr} = 0.7 \tag{16}$$

For other Prandtl numbers, Nu_t can be calculated from Eq. (16) using the theoretical proportionality of Nu with $\mathrm{Pr}^{1/3}$.

A more accurate approximation is obtained by using a correlation obtained from the equivalent ring-zone theory. As discussed in the previous section, axial symmetry is approximated for each rod in a bundle when the relative pitch becomes "large." Then the angular coordinate drops out of the governing equations, and they can be relatively easily solved. The equivalent ring-zone solution, as compiled by Ref. 6, predicts $\mathrm{Nu}/\mathrm{Nu}_t$ ratios increasing from 1.03 for a relative pitch of 1.0, to 1.26 for a relative pitch of 2.0. The values just mentioned pertain to a Pr value of 0.7, but the influence of Pr is very weak. A fitted correlation of the equivalent ring-zone solution (for Pr = 0.7) was presented by Ref. 6:

$$\frac{\mathrm{Nu}}{\mathrm{Nu}_t} = 0.855\left(\sqrt{\frac{4}{\pi}\frac{s}{d_r}}\right)^{0.1688}\left(\sqrt{\frac{4}{\pi}\frac{s}{d_r}} + 1\right)^{0.2122} \tag{17}$$

Although the assumption of axial symmetry is obviously wrong for relative pitches close to 1.0, the equivalent ring-zone solution adequately fits the experimental data with some 15% error, even at the lowest relative pitches. Combination of Eqs. (16) and (17) therefore seems sufficiently accurate for engineering purposes.

An interesting point is the dependence of the turbulent Nu numbers on the boundary condition. For laminar flow it was shown that the dependence of Nu on the boundary condition rapidly fades away when the relative pitch is increased: The difference in Nu between the two limiting boundary conditions is less than 30% for relative pitches larger than 1.1. Because of the flatter velocity profile in turbulent flow, the dependence on the boundary conditions is weak in turbulent flow, except for small relative pitches. An estimate of the maximum influence of the boundary condition on the turbulent Nu number can be obtained from the respective values for laminar flow.

Because pressure drop measurements are much faster and cheaper than mass transfer or heat transfer measurements, it is tempting to try to relate the Sherwood and Nusselt numbers to the friction factor. A relation that has proved successful for smooth circular tubes is obtained from a plausible assumption that is known as the film layer model. The assumption is that for turbulent flow the lateral velocity, temperature, and concentration gradients are located in thin films at the wall of the channel; the thickness of the films is indicated with δ_h, δ_T, and δ_c, respectively. According to the film model, the lateral velocity gradient at the channel surface equals $\langle u\rangle/\delta_h$, the lateral temperature gradient equals $(T_b - T_s)/\delta_T$, and the lateral concentration gradient equals $(c_{A,b} - c_{A,s})/\delta_c$. From these assumptions, and the theoretical knowledge that $\delta_h/\delta_T \approx \mathrm{Pr}^{1/3}$ and $\delta_h/\delta_c \approx \mathrm{Sc}^{1/3}$ (for Pr and Sc $\geq$ 1), the following relation is easily derived:

$$\mathrm{Nu} \approx \frac{f}{2}\mathrm{Re}\mathrm{Pr}^{1/3}; \qquad \mathrm{Sh} \approx \frac{f}{2}\mathrm{Re}\mathrm{Sc}^{1/3} \tag{18}$$

This relation is usually called after Chilton and Colburn. For smooth circular channels, the Nusselt numbers predicted with Eq. (18) agree with experimental data within 25% for $Re > 5 \times 10^3$, which supports the assumptions of the film layer model. Unfortunately, Eq. (18) overestimates Nu and Sh for channels that are not hydraulically smooth. The error increases rapidly with increasing roughness, because the hydrodynamic boundary layer thickness is more strongly affected by wall roughness than is the thermal or concentration boundary layer thickness; cf. Ref. 17.

From this the following procedure arises to predict the limits of Sherwood and Nusselt numbers in the BSR from simple pressure drop measurements:

1. Measure the pressure drop characteristics, and compare the experimentally determined friction factors with the predictions for smooth bundles, according to the relations discussed in the previous section. This gives an impression of the roughness of the strings.
2. For the Reynolds number of interest, calculate Sh or Nu with Eq. (18), using (a) the experimentally determined friction factor and (b) the predicted friction factor for smooth bundles. The true value of Sh or Nu is then expected to lie between the predicted (a) and (b) values.

The usefulness of this procedure will be illustrated in the following section.

C. Some Experimental Data on Mass Transfer in Turbulent Flow

To give an impression of the validity of the mass transfer relations for turbulent flow in a BSR presented in this section, two graphs of experimentally determined Sherwood numbers vs. the Reynolds number for two lab-scale G/S BSR modules are taken from Ref. 7. The data were obtained by inserting naphthalene beads into strings of ceramic beads arranged in a regular square array and determining the rate of naphthalene mass decrease. The data were corrected for the enhancement of the mass transfer rate obtained when only one naphthalene particle is inserted in a string, due to the developing concentration profile.

Figure 12a pertains to a BSR with a square cross section of 70 mm, filled with a regular square array of strings of ceramic beads with a particle diameter and length of 6.5 mm. The relative pitch of the strings was 1.1, corresponding to a reactor voidage of 45%. Pressure drop experiments showed that in these BSR configurations the flow was turbulent at Reynolds numbers higher than 1000 and that the strings had a considerable roughness. The dotted line represents the Sherwood numbers predicted with Eqs. (16) and (17) under the assumption that the strings were hydraulically smooth. The dashed line represents the Chilton–Colburn prediction [Eq. (18)] based on the predicted friction factor for smooth strings. The solid line represents the Chilton–Colburn prediction [Eq. (18)] based on the experimentally determined friction factor. Figure 12b pertains to an analogous BSR module, but with a relative pitch of 1.4, corresponding to a reactor voidage of 67%.

In both BSR modules, the Sherwood number lies between the two Chilton–Colburn predictions, as expected. The most important conclusion to be drawn from these graphs, is that the Sherwood number for turbulent flow in a BSR can be predicted with an accuracy of ca. 30% (which is usually acceptable) on the basis of one single pressure drop experiment in the turbulent-flow regime. From this pressure experiment the empirical roughness function can be fitted, with which the friction factor can be adequately predicted as a function of Re, as discussed in the previous section; from these an upper estimate of Sh

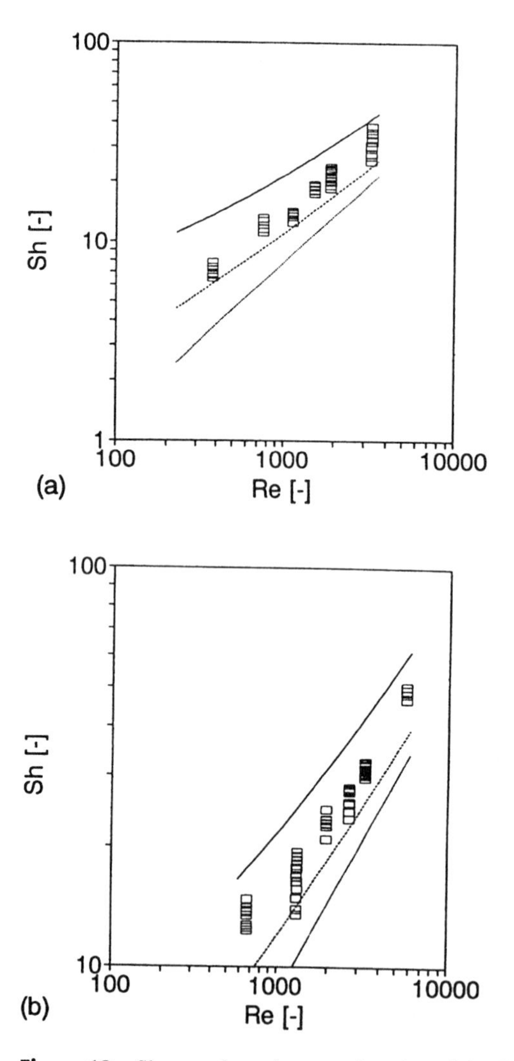

Figure 12 Sherwood number as a function of the Reynolds number. (a) $s/d_r = 1.1$, $d_r = 6.5$ mm; (b) $s/d_r = 1.4$, $d_r = 6.5$ mm. □ = experimental data; lower line = prediction from ring-zone approximation; middle line = Chilton–Colburn prediction based on predicted friction factor for smooth rods; upper line = Chilton–Colburn prediction based on experimental friction factor.

as a function of Re is obtained, using the Chilton–Colburn analogy. A lower estimate is also obtained from the Chilton–Colburn analogy, this time using the friction factor predicted for smooth rods.

D. Conclusions

The large amount of literature on heat transfer in rod bundles allows a reasonably accurate prediction of the mass transfer characteristics of a BSR. Deviations from predictions based on literature correlations may occur because of two facts:

1. The boundary condition of the governing partial differential equation that describes mass transfer in a reactor with catalytic walls, differs from the standard boundary conditions assumed in most texts on heat transfer in rod bundles (either constant wall temperature/concentration or constant heat/mass flux). However, the Sherwood number of a reactor with catalytic walls will lie between the values obtained for these two standard boundary conditions, which deviate less than 30% for relative pitches higher than 1.1.
2. The literature usually deals with hydraulically smooth rod bundles; the type of artificial roughness of the BSR strings seems unique.

Based on the experimental study it is concluded that for any Reynolds number in the turbulent-flow regime, the experimental Sherwood number falls between the two values predicted with the Chilton–Colburn analogy, from (a) the experimentally determined friction factor for the tested reactor and (b) the theoretical friction factor for a smooth rod bundle with the same relative pitch. Using these two predictions as the limiting values, the experimental Sherwood numbers can be predicted with an accuracy of ca. 30%, which is acceptable in most cases.

IV. MODELING OF THE BSR FOR G/S APPLICATIONS

A. Introduction

To assess the feasibility of the BSR as a competitor of the monolithic reactor, the parallel-passage reactor, and the lateral-flow reactor, it is necessary to do case studies in which the performance and price of these reactors are compared, for certain applications. To allow such case studies, two tools are needed: (1) mathematical models of the reactors that predict the reactor performance, and (2) an optimization routine that, given a mathematical reactor model and a set of process specifications, finds the optimum reactor configuration. Furthermore, data are needed on costs, safety, availability, etc. In this section, five mathematical models of different complexity for the bead-string reactor (BSR) are presented that can be numerically solved on a personal computer within a few hours down to a few minutes. The implementation of the reactor models in an optimization routine, as well as detailed cost analyses of the reactor, are beyond the scope of this text.

In Section IV.B the five mathematical BSR models will be discussed. This includes a discussion of the general assumptions or restrictions made in the development of the models and a discussion of the additional assumptions that lead to each of the separate models. The relations that were used to describe momentum and mass transfer have already been discussed in the previous two sections, and will therefore not be repeated here. Furthermore the kinetic model to be implemented in a BSR model is considered to be known. In Section IV.C the adequacy of the models will be illustrated based on the results of validation experiments. For those experiments, the selective catalytic reduction (SCR) of nitric oxide with excess ammonia served as the test reaction, using a BSR filled with strings of a commercial deNO$_x$ catalyst shaped as hollow extrudates. The kinetics of this reaction had been studied separately in a recycle reactor.

B. Description of the BSR Models

1. General Assumptions and Restrictions

The mathematical BSR models presented in this section are restricted to simple cases, specified by the following assumptions:

1. The flow is incompressible.
2. There is a steady-state situation.
3. The flow is hydrodynamically fully developed at the entrance of the reactor. As was stated in Section II, this assumption results in a negligible error in the prediction of the pressure drop, and therefore the influence of the flow development on the mass transfer rate is also negligible. Because the flow is fully developed, the gas velocity will only have one component, namely, in the axial (z-) direction, u_z.
4. The pressure gradient is in the z-direction; in other words, there are no lateral pressure gradients. Because the flow is fully developed, the pressure gradient is independent of the axial position (z) in the reactor.
5. There is no interaction between the various subchannels, i.e., no net exchange of momentum or reactants. This assumption will usually be valid in laminar flow but not in turbulent flow, where lateral flows are known to occur between neighboring subchannels. However, no significant effect of lateral flows on the pressure drop has been observed (see Section II).
6. The enhancement of the mass transfer rate in the first part of the reactor, resulting from the developing concentration profile, is negligible. It was stated in Section III that for relative pitches larger than 1.1 this assumption is valid, and that even for smaller relative pitches the effect of the enhanced mass transfer rate is small, except when film layer mass transfer is the limiting factor in the reactor performance.
7. Axial dispersion is negligible. Whether this assumption is valid, can be seen from the Péclet number (u_zL/D); for typical conditions (u_z = 1 m/sec, L = 0.5 m) and laminar flow, the Péclet number is larger than 1000, and even for turbulent flow it will be much larger than 10. In laminar flow, the radial flow profile within a subchannel will also result in deviation from plug flow. The effect of this deviation can be estimated by comparing the predictions from different mathematical models, one of which takes the flow profile into account and the other of which assumes plug flow.
8. The system is isothermal. This assumption is valid in an important projected BSR application, namely, catalytic deNOxing.
9. The chemical reaction has no effect on the flow profile. This assumption is plausible in many processes, unless very strong heat effects are involved.
10. Diffusion in the gas phase (in laminar flow) can be described with the Fick diffusion model; in the catalyst phase it can be described with a combination of free molecular and Knudsen diffusion. These assumptions are valid unless very high reactant concentrations are involved, the size of the reactant molecules is relatively large, or the affinity of the reactant toward the catalyst material is very large.
11. The diameter of the wire (or "kernel") on which the catalyst particles are stringed exactly equals the inner diameter of the hollow catalyst particles (or

"rings"). In practice it will normally be attempted to approximate this situation closely, in order to have the particles nicely aligned on the strings.

As was stated in Section II, it is not difficult to obtain a negligible influence of the reactor wall on the flow distribution in a BSR, and this situation will normally be desired. Therefore, in the simple mathematical models presented in this chapter, it is assumed that all subchannels are identical. To study the possible influence of the flow distribution, it was accounted for in one of the models.

2. *Five Models*

In spite of the eleven assumptions we have made, numerical solution of the mathematical system comprising the governing equations discussed in the previous sections is still time-consuming. Various simplifications can be made to obtain simpler mathematical systems or models. Next, five models are presented that were found to allow adequate description of the BSR in a limited range of conditions (that will be indicated for each model). To illustrate the different models, their distinguishing features are explained graphically in Fig. 13.

Because the pressure drop across the BSR can be calculated independently from the concentration profile, the modeling of the momentum transfer could be the same for all five models and will therefore not be discussed any further; see Section II for details on the modeling of the momentum transfer.

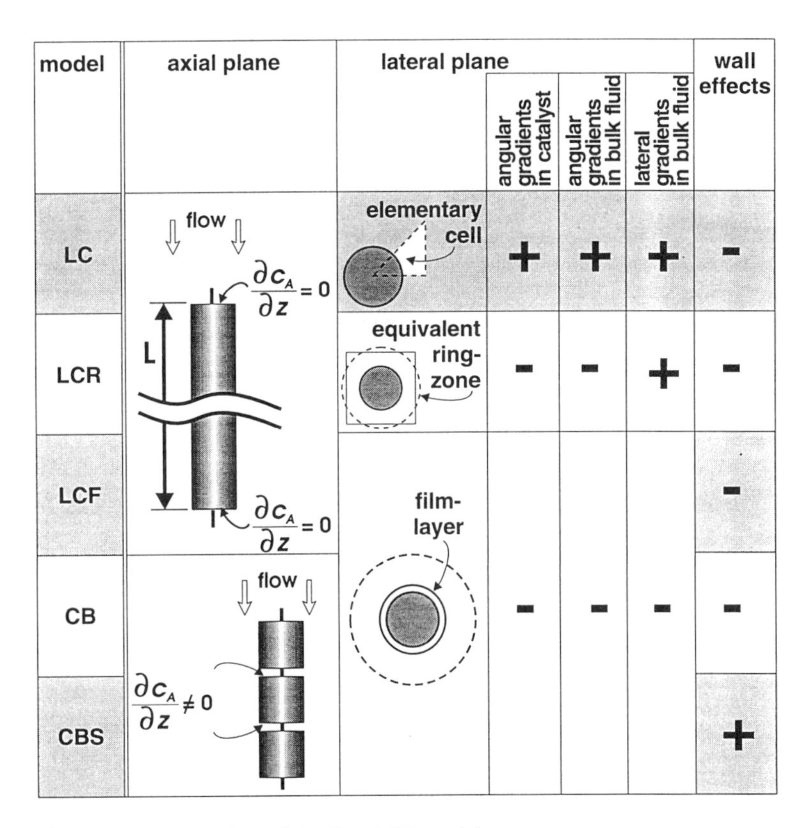

Figure 13 Illustration of the five BSR models.

a. The "Long Cylinder" (LC) Model. If the (axial) length of the catalyst beads is much larger than the outer particle radius, the molar flow of reactants entering a particle through the flat ends (i.e., the top and bottom of the particle) is small compared to the flow entering through the cylindrical surface. A string of catalyst particles can then be modeled as a long "catalyst rod" with blocked top and bottom ends; see Fig. 14. This is a useful simplification, because it allows reduction of the three-dimensional PDE of Eq. (15) to two dimensions. The justification for this reduction is that in a "long" catalyst cylinder (where "long" should be interpreted as the length of the reactor), the axial concentration gradients will be small compared to the radial gradients; hence the axial reactant flux is negligible compared to the radial flux. (The situation where the angular gradients are also small compared to the radial gradients is treated in Section IV.B.2.b).

The "long cylinder" approximation could be valid if the catalyst particles are tightly packed on the string (i.e., with only a very narrow gap between two consecutive particles); see Fig. 15. This is illustrated in Fig. 16, in which four contour plots are presented of the reactant concentration in and around cylindrical catalyst particles on a string, with different axial spacings of the particles. It is assumed that outside (and inside) the particles there is only reactant transport due to diffusion, and the concentration is taken constant at the outer side boundary of the annularly shaped fluid domain. It can be seen that even when the gap between the particles is small relative to the length of the particles, the concentration profile deviates considerably from the profile in an infinitely long cylinder. In practice the deviation will be even larger, because of extra convective reactant transport in the fluid phase. Consequently, the "long cylinder" approximation can usually not be justified by assuming that the transport of reactants is negligible in the "gap" between consecutive particles on a string. The approximation is therefore justified only for particles with a relatively large length-over-diameter ratio.

The model that results from the relations presented in Section IV.B.1 and the extra assumption of negligible axial reactant flux within the catalyst particles and negligible wall effects will be called the "long cylinder" (LC) model. In this model the governing equations are solved for the elementary cell depicted in Fig. 10.

When the length-over-diameter ratio of the catalyst beads is *not* much larger than 0.5 (and the particles are not very tightly packed on the strings), the LC model is obviously expected to underestimate the reactant conversion in the BSR if the performance is limited by intraparticle pore diffusion.

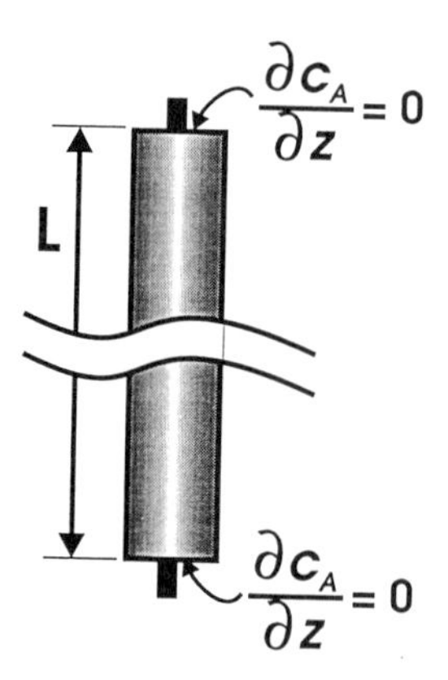

Figure 14 Model of a string of catalyst particles, with the "long cylinder" approximation.

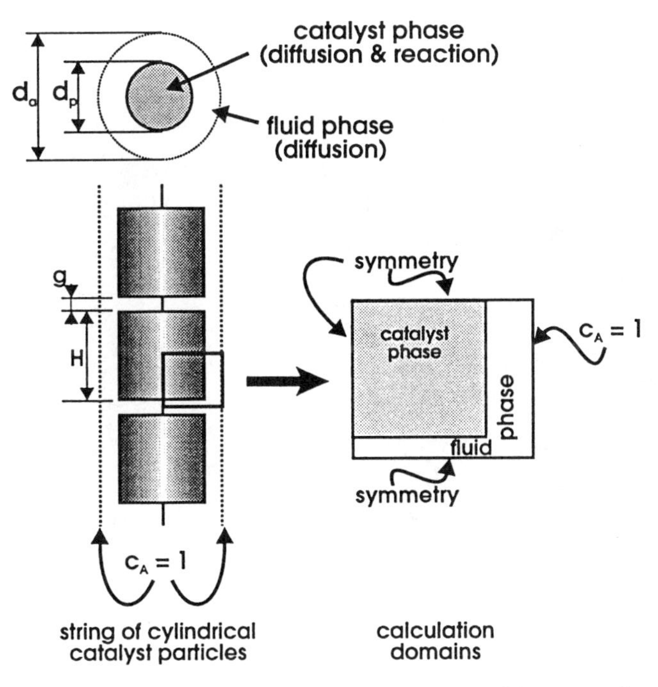

Figure 15 Model for studying the effect of the "gap" between stringed particles on the concentration profile in the particles. *Left*: top view and side view of a string in an annular space; right: mathematical domains.

b. The "Long Cylinder Ring-Zone" (LCR) Model. Upon increasing the relative pitch, the velocity and concentration profiles around each rod become more axisymmetric. For relative pitches higher than ca. 2.5, only a negligible error is introduced when the lateral plane is described by one coordinate (r) instead of two (r and α); see the discussion on the so-called "equivalent ring-zone approximation" in Section II.B.2. The LCR model is derived from the LC model by assuming axial symmetry in and around each rod. This assumption reduces Eq. (3) to a second-order, linear or (in turbulent flow) nonlinear, ordinary differential equation (ODE), Eq. (13) to a second-order, linear, quasilinear or nonlinear, two-dimensional PDE, and Eq. (15) to a second order, linear or nonlinear (depending on the reaction rate expression) ODE:

$$\frac{1}{r}\frac{\mathrm{d}}{\mathrm{d}r}\left(\mu r\frac{\mathrm{d}u_z}{\mathrm{d}r}\right) = \frac{\mathrm{d}p}{\mathrm{d}z} \tag{19}$$

$$\frac{1}{r}\frac{\partial}{\partial r}\left(Dr\frac{\partial c_A}{\partial r}\right) = u_z\frac{\partial c_A}{\partial z} \tag{20}$$

$$\frac{\mathrm{d}^2 c_A}{\mathrm{d}r^2} + \frac{1}{r}\frac{\mathrm{d}c_A}{\mathrm{d}r} = \frac{1}{D_e}\Re_A \tag{21}$$

Obviously, this drastically reduces the computation time needed to solve the model numerically.

It is expected that for relative pitches smaller than 2.5, the LCR predicts higher conversions than the LC model. This is because the equivalent ring-zone approximation then overestimates the transfer of momentum and mass (see Sections II and III).

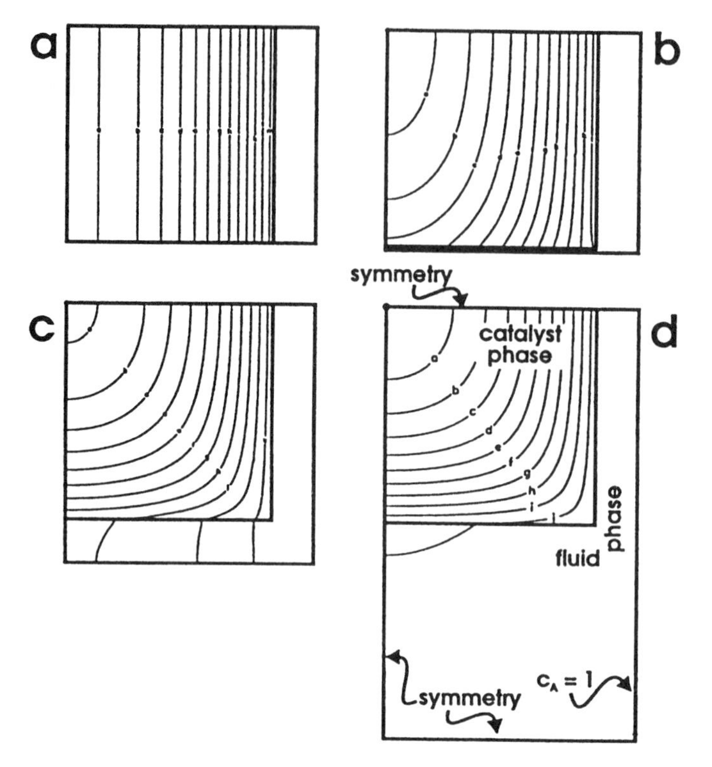

Figure 16 Concentration profiles in the calculation domains indicated in Fig. 15, for different values of g/H. At the east boundaries $c_A = 1$; the concentration difference between consecutive contour lines is 0.05. *Conditions*: (a) $g/H = 0$ (infinitely long cylinder); (b) $g/H = 1/50$; (c) $g/H = 1/5$; (d) $g/H = 1$. In all cases: $d_a/d_p = 1.2$; $H/d_p = 1$; $D_e/D_{\text{free}} = 1/20$; first-order reaction; $d_p^2k/D_e = 12$.

c. The "Long Cylinder Film" (LCF) Model. The calculation of the reactant concentrations in the fluid phase can be simplified significantly by adopting the film model, according to which the fluid phase is perfectly mixed in the lateral plane, except for a small layer at the surface of the catalyst. The model obtained by combining the "long cylinder" approximation discussed in the previous two sections and the film model is called the "long cylinder film" (LCF) model.

According to the film model, the rate of mass transfer from the bulk fluid phase to the catalyst surface is described by a mass transfer coefficient, and boundary condition (14) changes into

$$D_e \left.\frac{\partial c_A}{\partial r}\right|_{\substack{r=R \\ \text{catalyst domain}}} = k_g(c_{A,b} - c_{A,s}) \tag{22}$$

Because the bulk fluid phase is assumed to be perfectly laterally mixed, a mass balance over a thin lateral slice of a unit cell leads to the following ODE:

$$\frac{\mathrm{d}c_{A,b}}{\mathrm{d}z} = -\frac{4}{\langle u_z \rangle d_h} k_g(c_{A,b} - c_{A,s}) \tag{23}$$

This ODE replaces the PDE of Eq. (13).

The LCF model allows another (trivial) simplification. Because the bulk fluid phase is assumed to be perfectly mixed, it will also be axisymmetric; consequently, the concentration profile in the catalyst particles will also be axisymmetric. As a result, the three-dimensional PDE of Eq. (15), which could be reduced to a two-dimensional PDE due to the "long cylinder" approximation, can be further reduced to the second-order linear or nonlinear ODE of Eq. (21), since the angular coordinate drops out.

From this it follows that the LCF model consists (besides the momentum transfer model) of one first-order and one second-order ODE, and the numerical solution of this model therefore consumes considerably less time than does that of the previous two models. However, there is a complication in the LCF model. As was discussed in Section III.B.1, the value of the mass transfer coefficient depends on the ratio of the reaction rate to the diffusion rate in the fluid phase. Fortunately, the dependence is only weak (at least for relative pitches larger than 1.1), and going from a very small to a very large value of this ratio, the Sherwood number varies less than 30% ($s/d > 1.1$). It was also shown in Section III that, depending on the ratio, the Sherwood number may vary significantly along the perimeter of a particle or "rod," which can be accounted for by using a circumferentially averaged Sherwood number. A suitable value for the Sherwood number to be used in the LCF model can therefore be obtained from Fig. 11; whether to use the value for constant surface concentration or for constant flux can be assessed from the estimated ratio of the reaction rate to the diffusion rate in the fluid phase.

The LCF model will tend to underestimate the conversion obtained in a BSR, because of the "long cylinder" approximation.

d. The "Catalyst Bead" (CB) Model. If the length-over-diameter ratio of the catalyst particles is not much larger than 0.5 (and the particles are not very tightly packed on the strings), the reactant flow entering through flat ends of the cylindrical particles is no longer negligible. According to Eq. (14), the concentration profile in the particle is necessary to calculate the reactant flux from the fluid phase into the particle. Rather than solving a three-dimensional PDE [Eq. (15)] to calculate the reaction rate in each catalyst particle, we could use an approximative, algebraic method to estimate the *average* reaction rate in each catalyst particle (e.g., the well-known method named after Thiele or the Aris method developed by Ref. 18).

The approximative methods to calculate a catalyst particle effectiveness factor are always based on the assumption that the reactant concentration at the particle surface has one and the same value over the complete surface (see Fig. 17). Therefore, the use of these methods for predicting the BSR performance is—strictly spoken—allowed only if there is no external mass transfer limitation. The BSR can then be described with only one ODE:

$$\frac{dc_{A,b}}{dz} = -\frac{1}{\langle u_z \rangle} \frac{d_p}{d_h} \eta \Re_A \big|_{c_{A,s}} \tag{24}$$

in which the concentration at the particle surface equals the bulk concentration.

To account for a limited external mass transfer rate, the concentration at the particle surface is calculated with

$$A_p k_g (c_{A,b} - c_{A,s}) = V_p \eta \Re_A \big|_{c_{A,s}} \tag{25}$$

The mass transfer coefficient used in this equation should be the average value over the total external particle surface area A_p, including the flat ends of each particle. HQwever,

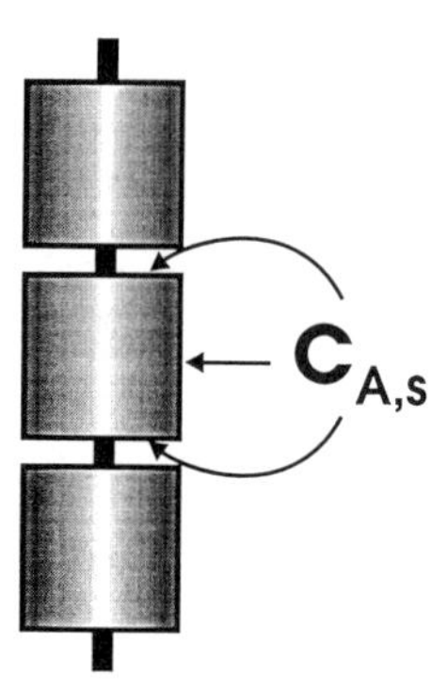

Figure 17 Assumption made in the "catalyst bead" models: The surface concentration is the same everywhere on one bead.

because suitable relations for the average mass transfer coefficient have not been developed yet, the relations discussed in Section III (pertaining to the cylindrical surface of rods) are used as an approximation.

The model that is formed by Eqs. (24) and (25), with the extra assumption that wall effects can be neglected (i.e., that the reactor is built up solely of central subchannel elementary cells such as the one depicted in Fig. 10), is called the "catalyst bead" (CB) model. Because of the assumption in the CB model that the value of the mass transfer coefficient at the flat ends of the particle equals the value at the cylindrical surface, it can be expected that this model overestimates the conversion obtained in a BSR, especially when the particle length-over-diameter ratio is much smaller than 0.5 and when the "gap" between consecutive particles on a string is small compared to the particle diameter. However, whether the CB model over- or underestimates the BSR performance also depends on the error made in the calculation of the particle effectiveness factor.

In the present study the particle effectiveness factor was estimated with the Aris method presented by Ref. 18. In the estimation, the particles were modeled as normal cylindrical catalyst particles, i.e., finite cylinders with no inert kernel. This was necessary because approximative methods to estimate an effectiveness factor cannot account for an inert kernel in a catalyst particle such as the metal wire in the axial hole of the BSR beads. The cylinder diameter used in the calculations was $d_{p,s}$, defined in Fig. 18 (the

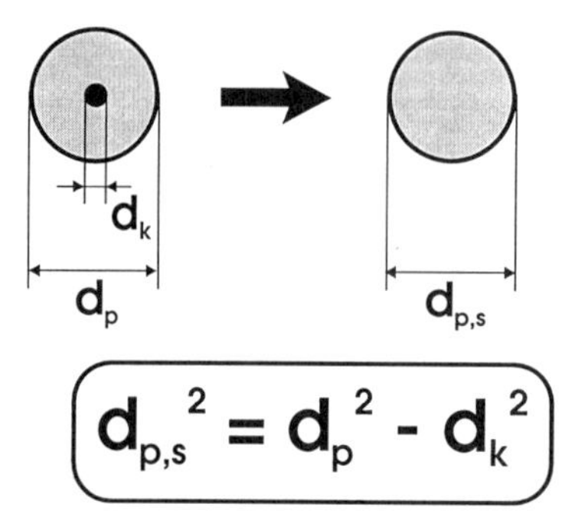

Figure 18 Transformation of a cylindrical particle with an inert kernel to an equivalent particle with no inert kernel.

subscript s stands for "shrunk cylinder"). In this way the active catalyst volume of the "shrunk" cylinder equals that of the real catalyst particles, and the particle volume-to-surface-area ratio of the "shrunk" cylinder approximately equals that of the real, stringed particles. The error that is made in the estimation of the effectiveness factor by using the Aris method with the "shrunk particle" modeling was less than 7% in the present study [7].

e. The "Catalyst Bead Subchannel" (CBS) Model. The "catalyst bead subchannel" (CBS) model is a refinement of the CB model, in the sense that it does not contain the assumption that the reactor contains solely central subchannels. Instead, in the CBS model the flow distribution over the central, wall, and corner subchannels is calculated following the procedures outlined in Section II; in other words, the wall effect is taken into account in this model. The reactant conversion is then calculated for each type of subchannel, under the assumption that there is no exchange of reactants between the subchannels. The overall outlet concentration is calculated as the mixing cup average of the outlet concentrations of all subchannels.

Obviously, the assumption of no lateral exchange between the subchannels is, strictly spoken, only (approximately) valid for a relative pitch of 1; however, for higher relative pitches it might be valid if the ratio of the residence time in the reactor to the time constant for lateral diffusion/dispersion is much smaller than 1. For lateral diffusion/dispersion over a distance equal to the pitch (s) or the width of the reactor (d_r), these ratios or Fourier numbers are, respectively,

$$\mathrm{Fo}_s = \frac{D_{\mathrm{lat}} L}{u_z s^2}; \qquad \mathrm{Fo}_d = \frac{D_{\mathrm{lat}} L}{u_z d_r^2} \tag{26}$$

In this equation, D_{lat} represents the effective lateral diffusion/dispersion constant; for laminar flow a value on the order of $\varepsilon D_{\mathrm{free}}$ is suitable, and for turbulent flow the molecular diffusion coefficient in this expression should be replaced by the turbulent diffusion coefficient. Based on these simple relations it can be calculated that under typical conditions, lateral reactant transport takes place only over distances of a few subchannels. In other words, if the gas velocity through the wall channels differs much from the velocity through the central subchannels, the nonuniform flow profile can have a significant effect on the overall reactant conversion. In these situations the CBS model can be expected to give a better estimate of the reactor performance than the CB model.

C. Illustration of the Validity of the Models

To illustrate the validity of the models presented in the previous section, results of validation experiments using lab-scale BSR modules are taken from Ref. 7. For those experiments, the selective catalytic reduction (SCR) of nitric oxide with excess ammonia served as the test reaction, using a BSR filled with strings of a commercial deNO_x catalyst shaped as hollow extrudates (particle diameter 1.6 or 3.2 mm). The lab-scale BSR modules had square cross sections of 35 or 70 mm. The kinetics of the model reaction had been studied separately in a recycle reactor. All parameters in the BSR models were based on theory or independent experiments on pressure drop, mass transfer, or kinetics; *none* of the models was later fitted to the validation experiments. The PDEs of the various models were solved using a finite-difference method, with centered differencing discretization in the lateral direction and backward differencing in the axial direction; the ODEs were solved mostly with a Runge–Kutta method [16]. The numerical error of the solutions was checked by changing the step size, and the step size was chosen small enough to keep the numerical error below half a percent.

Unfortunately, for the LC model even the mild tolerance of half a percent led to computation times of several days per run (on an Intel 486, 66-MHz processor). Therefore the adequacy of this model could not be fully tested. A few runs were calculated with the LC model for the experiments with a relative pitch of 1.0, where the strongest angular gradients in the catalyst particle are expected. The conversions calculated with the LC model appeared practically to coincide with the values calculated with the LCF model.

The computation time needed to solve the LCF, LCR, CB, and CBS models is between 15 and 45 seconds (486/66 MHz). The performance of these models is presented in the form of parity plots in Figs. 19a to 19d. The plots show a satisfactory agreement between the predicted NO conversion and the experimental value, especially when it is taken into account that the models have *not* been fitted in any way to the results of the validation experiments. The deviations between the model predictions and the experimental observations are more or less in agreement with the expected deviations discussed in Section IV.B. The models based on the "long cylinder" approximation (LCF and LCR) underestimate the conversion for most runs. Of the two models based on the "catalyst bead" approximation (CB and CBS), the first one generally overestimates the conversion, as expected. The second one, however, predicts the conversion correctly on the average, whereas a slight overprediction was expected. This implies that the negative effect of the flow distribution on the conversion is smaller than predicted by the model; this is plausible, because the CBS gives a prediction for the extreme situation that there is no lateral exchange of reactants at all between the subchannels, which is obviously only approximately true. For a more detailed statistical treatment of these results, see Ref. 7.

D. Conclusions

The performance of lab-scale BSR modules in the SCR of NO can be predicted with an error of ca. 10% by four relatively simple mathematical models, whose parameters were determined via independent experiments. As expected, the LCF and the LCR model generally underpredict the conversion, whereas the CB model generally overestimates it. The CBS model gives the most accurate prediction of the NO conversion: On the average, the predicted conversion deviates ca. 5% (relative) from the experimentally determined value. Such deviations can be attributed to stochastic variations in the experiments.

V. MANUFACTURING THE BSR AND BSR ANALOGUES

The subject of manufacturing catalyst packings for a BSR is touched upon only briefly in this chapter, since it is in a very early stage of development. As in the development of manufacturing techniques for ceramic monoliths, the first steps in the development of BSR manufacturing techniques are influenced more by art and common sense than by technology.

It is beyond doubt that it is very well possible to manufacture BSR catalyst packings. Valid questions are, however, how the manufacturing costs of a BSR compare to those of other reactors with structured catalyst packings, and which reactor engineering advantages the BSR concept offers in specific applications.

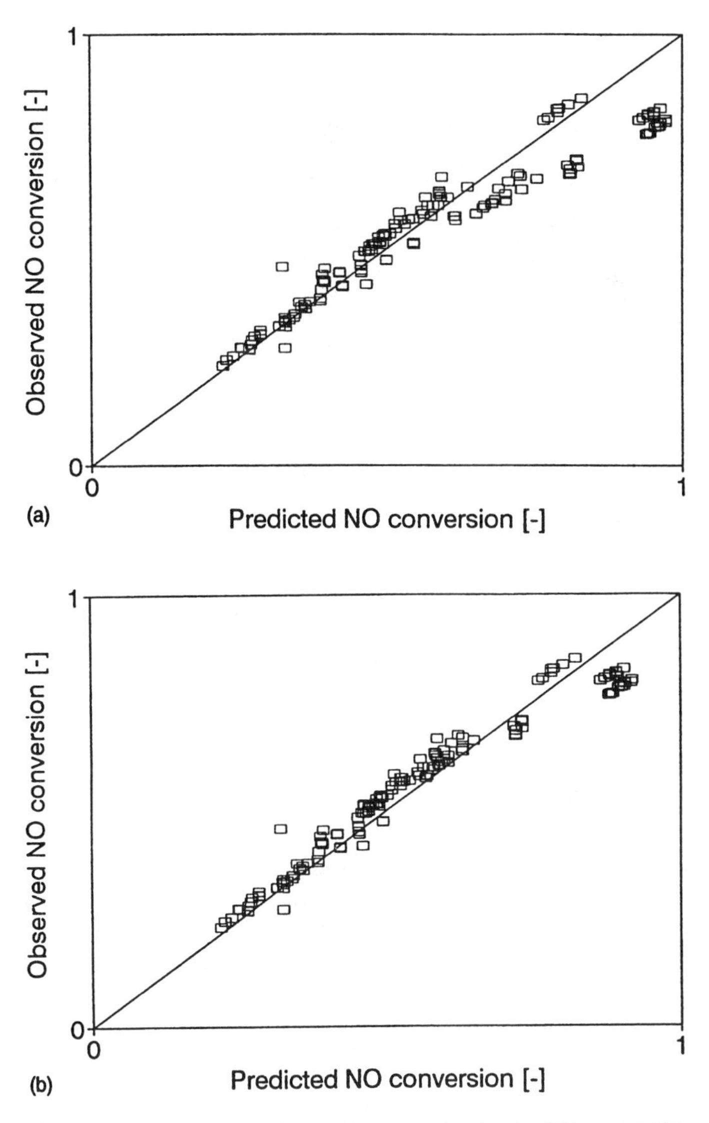

Figure 19 Parity plots of the NO conversion in the BSR. (a) LCR model; (b) LCF model; (c) CB model; (d) CBS model.

A. Mechanical Stringing of Particulate Catalyst Material

The simplest way to prepare a BSR packing is mechanically to string commercially available hollow extrudates or ring-shaped pellets on wires, or mechanically to stack them on metal rods. This is fairly simple, provided that the particles have constant, well-defined dimensions. The manufacturing costs are not prohibitive (estimated at several dollars per kg for particles 5 mm in diameter), though they increase rapidly with decreasing particle size. The expected feasible lower limit is 2 mm. This lower limit implies that the BSR concept is less suitable for processes with very high reaction rates. A typical borderline case is a first-order reaction rate constant of 15 sec^{-1} (reaction rate expressed per unit

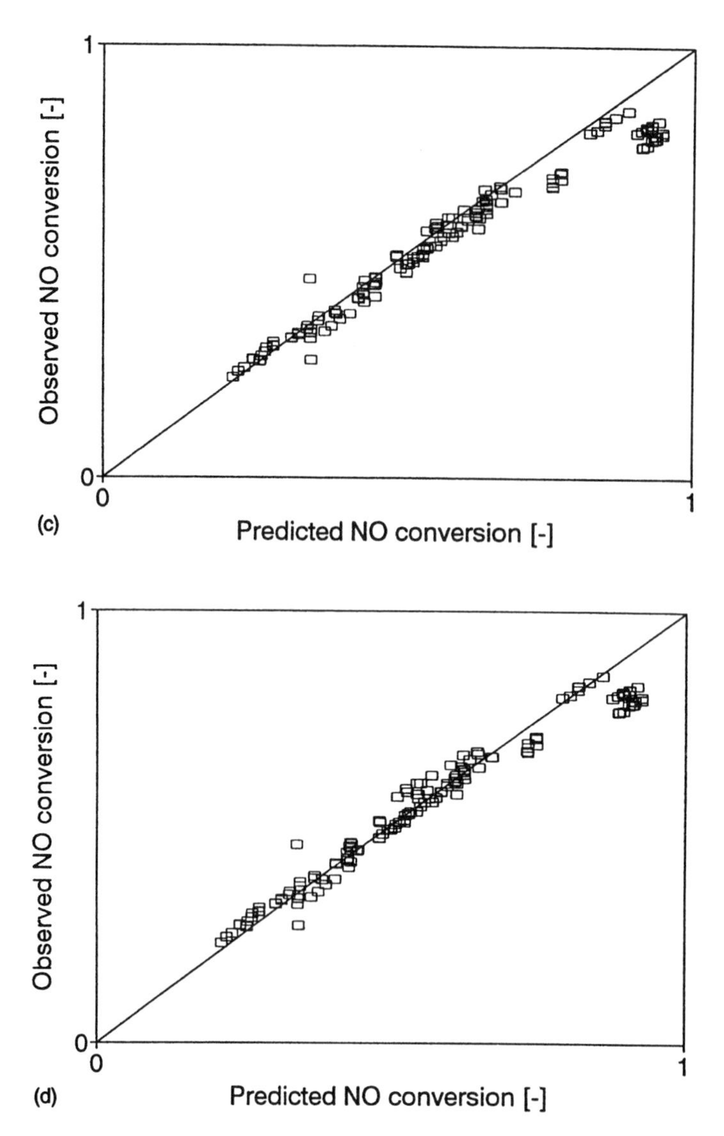

Figure 19 *continued*

catalyst particle volume), at which cylindrical particles or rods with a diameter of 2 mm have an efficiency factor of 75–85%. However, the upper limit on the rate constant may be increased by using catalyst particles with a nonuniform distribution of the active material ("egg-shell catalysts").

B. Extrusion of Paste Around a Wire or Fiber

It is also possible to extrude silica or alumina paste *around* a carrier wire or fiber (like the insulation material of electric cords), although concessions may have to be made with respect to the formulation of the paste, to adapt its rheological behavior. An interesting point is that in a BSR the catalyst material itself doesn't need to have a large crushing strength (unlike normal particulate catalysts), since it is supported by a central wire or fiber and also doesn't have to carry the weight of the catalyst material on top of it. This allows optimization of the catalyst structure with respect to activity, which for normal extrudates is limited by strength limits. Even though in the extrusion technique the manufacturing costs for a BSR would be less dependent on the catalyst "rod" diameter than in mechanical stringing of standard particles, the estimated feasible lower limit is 2 mm.

C. Pilling of Powder onto Wire Structures

The last option for preparing catalyst packings for a BSR discussed here is the pilling of silica or alumina powder (mixed with lubricants and binders) onto wire structures. Rather than using the normal, rotating pilling machine with a series of vertical, cylindrical holes, a flat mold with parallel horizontal grooves could be used. By inserting a wire frame in the grooves of the mold, filling the grooves with powder, and compressing the powder with a punch, large elements of a structured catalyst packing could be made with one punch. As was mentioned in the previous section, the catalyst material itself doesn't need to be very strong, granting extra freedom in the formulation of the powder.

A point of special interest is that following this technique, a reactor with the same mixing properties as the well-known static mixers can be made, as indicated in Fig. 20. For this BSR analogue, the name "polylith reactor" [19] is proposed; the development of models for both G/S and countercurrent G/L/S mixing-rod reactors is the subject of current research at Delft University of Technology.

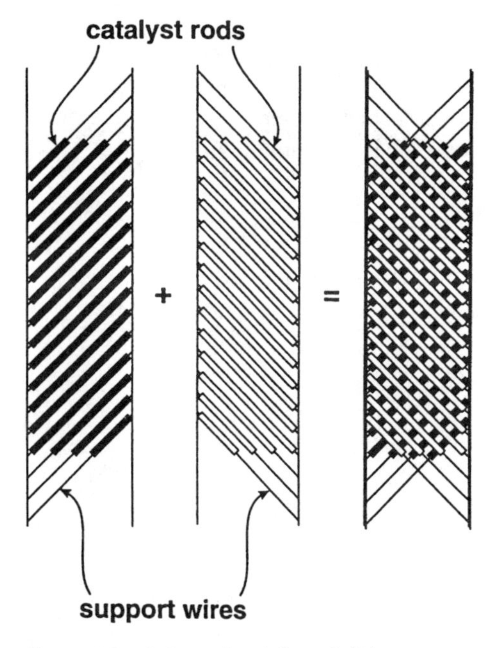

Figure 20 Schematic of the polylith reactor.

LIST OF SYMBOLS

A	dimensionless parameter in Eqs. (7) and (9)	—
A_p	external particle surface area	m^2
c_A	concentration of reactant A	mol_A/m^3_{fluid}
$c_{A,b}$	concentration of A in the bulk phase (mixing cup average of the fluid domain)	mol_A/m^3_{fluid}
$c_{A,s}$	concentration of A at the external catalyst particle surface	mol_A/m^3_{fluid}
d_h	diameter of hydraulic channel	m
d_p	particle diameter	m
$d_{p,s}$	equivalent diameter of "shrunk" particle	m
d_r	rod or string diameter	m
D	diffusivity (free molecular or turbulent)	m^2/sec
D_e	effective molecular pore diffusion constant	m^2/sec
D_{free}	free molecular diffusion constant	m^2/sec
D_{lat}	(effective) lateral dispersion coefficient	m^2/sec
f	(Fanning) friction factor ($\equiv 4f/4$)	—
$4f$	(Darcy–Weisbach) friction factor	—
$4f \cdot Re$	laminar geometry parameter	—
$4f_t$	(Darcy–Weisbach) friction factor for turbulent flow in smooth channels	—
G^*	dimensionless parameter in Eqs. (7) and (9)	—
k	first-order reaction rate constant	1/sec
k_g	mass transfer coefficient	m/sec
L	length of a channel	m
$\underline{n}$	normal vector	m
Nu_t	Nusselt number for a circular tube	—
p	pressure	Pa
r	radial coordinate	m
R	(outer) radius of a cylindrical or ring-shaped particle	m
$\Re_A$	reaction rate of A expressed per unit catalyst volume	$mol_A/(m^3_{cat} \cdot sec)$
$R(h^+)$	roughness function	—
s	pitch of rod or strings in a bundle	m
T_b	temperature of the bulk phase	K
T_s	temperature of the external catalyst particle surface	K
$\langle u \rangle$	average fluid velocity in a channel	m/sec
u_s	"undisturbed" fluid velocity in developing flow	m/sec
u_z	axial fluid velocity	m/sec
V_p	particle volume	m^3
w	wall distance	m
z	axial coordinate	m

Greek Letters

α	angular coordinate	rad
δ_c, δ_h, δ_T	concentration, hydrodynamic, and temperature boundary layer thickness	m

Δ_1, Δ_2	shortest and longest surface-to-surface distance in a BSR	m
Δp	pressure drop across a channel	Pa
ε	voidage of a bundle of rods or strings	$m^3_{void}/m^3_{reactor}$
η	catalyst particle effectiveness factor	—
μ	dynamic fluid viscosity, molecular or turbulent	Pa·sec
ν	kinematic fluid viscosity	m^2/sec
ρ	fluid density	kg_{fluid}/m^3_{fluid}
$\tau_{f,w}$	friction stress exerted by the fluid on the wall	N/m^2
$\phi''_{mol,A}$	molar flux of component A	$mol_A/(m^2 \cdot sec)$
∇p	pressure gradient across a channel	Pa/m

REFERENCES

1. A.W. Gerritsen, *Kralenreactor*, Dutch Patent Application Nr. 9000454, 26 February 1989.
2. K. Rehme, The structure of turbulence in rod bundles and the implications on natural mixing between the subchannels, *Int. J. Heat Mass Transfer 35(2)*:567 (1992).
3. A.T. Bell, L.E. Manzer, N.Y. Chen, V.W. Weekman, L.L. Hegedus, and C.J. Pereira, Protecting the environment through catalysis, *Chem. Eng. Prog.* (February 1995):26 (1995).
4. S. Kakaç and D.B. Spalding (eds.), *Turbulent Forced Convection in Channels and Bundles*, vol. 1, Hemisphere, Washington, 1979.
5. K. Johannsen, Longitudinal flow over tube bundles, *Low Reynolds Number Flow Heat Exchangers* (S. Kakaç, R.K. Shah, and A.E. Bergles, eds.), Hemisphere, Washington, 1983, p. 229.
6. K. Rehme, Convective heat transfer over rod bundles, *Handbook of Single-Phase Convective Heat Transfer* (S. Kakaç, R.K. Shah, and W. Aung, eds.), Wiley, New York, 1987, p. 7–1.
7. H.P. Calis, *Development of Dustproof, Low-Pressure-Drop Reactors with Structured Catalyst Packings—The Bead String Reactor and the Zeolite-Covered Screen Reactor*, Eburon P&L Press, Delft, The Netherlands, 1995.
8. E.M. Sparrow and A.L. Loeffler Jr., Longitudinal laminar flow between cylinders arranged in regular array, *AIChE J. 5(3)*:325 (1959).
9. J. Schmid, Longitudinal laminar flow in an array of circular cylinders, *Int. J. Heat Mass Transfer 9*:925 (1966).
10. J.M. Coulson and J.F. Richardson, *Chemical engineering* 3rd ed., vol. 1, Pergamon Press, Oxford, 1979, Ch. 9.
11. K. Rehme, Simple method of predicting friction factors of turbulent flow in non-circular channels, *Int. J. Heat Mass Transfer 16*:933 (1973).
12. P.N. Pustyl'nik, B.F. Balunov, and Blagoveshchenskii, A.Y., Heat transfer with forced longitudinal flow of air through a bundle of tubes with boundary condition Tw = const, *Thermal Engineering 37*:135 (1990).
13. N. de Nevers, *Fluid Mechanics for Chemical Engineers*, McGraw-Hill, New York, 1991, p. 386.
14. E. Tronconi and P. Forzatti, Adequacy of lumped parameter models for SCR reactors with monolith structure, *AIChE J. 38(2)*:201 (1992).
15. J.H. Kim, Heat transfer in longitudinal laminar flow along circular cylinders in a square array, *Fluid Flow and Heat Transfer over Rod or Tube Bundles* (S.C. Yao and P.A. Pfund, eds.), Winter annual meeting of the American Society of Mechanical Engineers, December 2–7, New York, 1979, p. 155.
16. K. Takács, *Mathematical Modelling of the Bead String Reactor*, Faculty of Chemical Engineering and Materials Science, Delft University of Technology, Delft, The Netherlands, 1994.

17. R.B. Bird, W.E. Stewart, and E.N. Lightfoot, *Transport Phenomena*, Wiley, New York, 1960, p. 401 and p. 647.
18. R.J. Wijngaarden and K.R. Westerterp, *Generalized Formulae for the Calculation of the Effectiveness Factor*, University of Twente, Enschede, The Netherlands, 1994.
19. H.P. Calis, A.W. Gerritsen, and C.M. van den Bleek, *Reactorpakking*, Dutch Patent Application No. 10.00176, 19 April 1995.

14

Open Cross-Flow-Channel Catalysts and Catalyst Supports

Jean-Paul Stringaro, Peter Collins, and Oliver Bailer
Sulzer Chemtech Ltd., Winterthur, Switzerland

I. INTRODUCTION

A. Limitations of Conventional Catalyst Structures

While the chemical aspects of catalyst performance (resistance to poisoning, activity, and selectivity) are being continually improved, only little has been done to date to improve the hydrodynamic aspects (heat transfer and mass transfer rates, pressure drop, and uniformity of distribution of concentration, temperature, and velocity) of heterogeneous contacting.

Today's catalysts and catalyst supports come in two basic forms, as dumped packings and as monoliths or honeycomb structures with parallel channels. Dumped packings do not perform too well in hydrodynamic terms. Transverse mixing is limited, and fluid motion through them is usually characterized by channeling and stagnant zones. Furthermore, the high pressure drop usually associated with chemical reactors filled with a dumped catalyst can be very costly. Monoliths or honeycomb structures are used in catalytic reactors when pressure drop is a problem, as is mainly the case in tail-end pollution-control reactors. The flow inside the honeycomb channels is usually laminar; mass transfer thus proceeds by molecular diffusion only and is substantially lower than under turbulent conditions, as shown in Fig. 1. This leads to poor mass transfer from the bulk stream to the catalyst walls. For heterogeneous catalytic reaction processes with fast kinetics, the rate of mass transfer of reactants from the carrying fluid to the catalyst surface determines the overall conversion rate. Thus for reactors equipped with monoliths, very large volumes are required to achieve an acceptable conversion level.

In addition, reactors equipped with monolithic bodies show a total absence of radial stream components. Inhomogeneities in concentration, temperature, and velocity profiles over the reactor cross section are thus propagated throughout the length of the reactor.

B. Open Cross-Flow Structures

Structures consisting of superimposed individual corrugated sheets, with the corrugations in opposed orientation such that the resulting unit is characterized by an open cross-flow

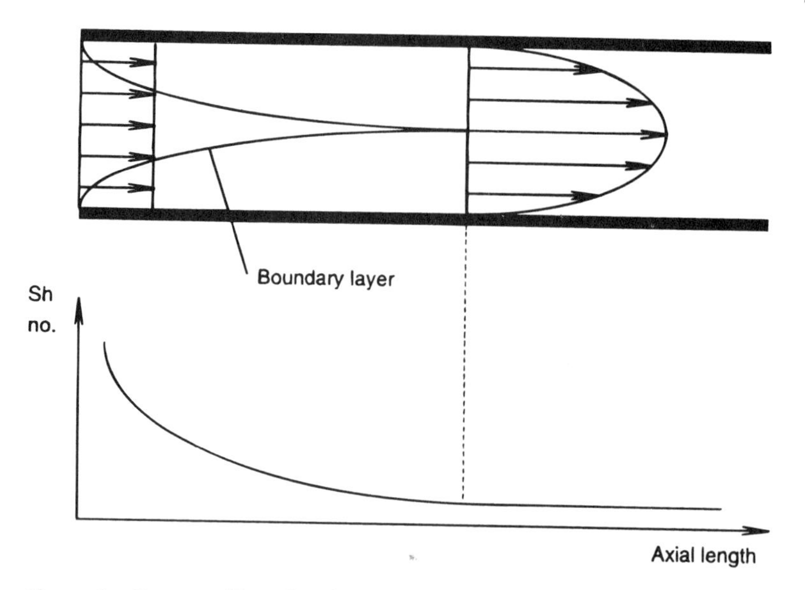

Figure 1 Contour of boundary layer with laminar velocity profile and schematic of mass transfer expressed as Sherwood number.

structure (OCFS) pattern, are well proven as distillation column packings and as static mixers. In distillation columns such packings permit optimal gas–liquid contact and, thereby, high separation performance at low pressure drop, while static mixers prove ideal where good macro-level mixing, improved heat transfer and mass transfer, and narrow residence-time distribution are required in continuous processing or in batch processing systems with recycling. With catalytic systems, however, mixing has to date been carried out upstream of the reaction zone, rather than within it.

Figure 2 shows an example of a static mixer and a schematic representation of how such structures operate—see, for example, [1] and [2]. The open intersecting channels divide the main fluid stream into a number of substreams. In addition to the lateral displacement caused by the obliquity of the channels, a fraction of each substream shears off into the adjacent channel at every intersection. This continuous division and recombination of the substreams causes transition from laminar to turbulent flow at Reynolds numbers (based on channel hydraulic diameter) as low as 200–300 and results in

Intensive radial mixing and thus uniform velocity, concentration, and temperature profiles, resulting in turn in

Narrow residence-time distribution, i.e., very little back-mixing (reverse flow in the axial direction) and virtually no stagnant zones. The Bodenstein number Bo (ratio of fluid velocity times reactor length to axial dispersion coefficient) is a measure of back-mixing: $Bo = \infty$ corresponds to plug flow, whereas $Bo = 0$ represents total back-mixing. With OCFS, $Bo \geq 50$ per meter of mixer length are obtained. This is equivalent to the residence-time behavior of a series of 25 or more ideally stirred tanks. In consequence, the residence-time distribution is narrow; i.e., the flow pattern approaches that of a plug-flow reactor,

High mass transfer rates, i.e., good utilization of the available catalyst surface.

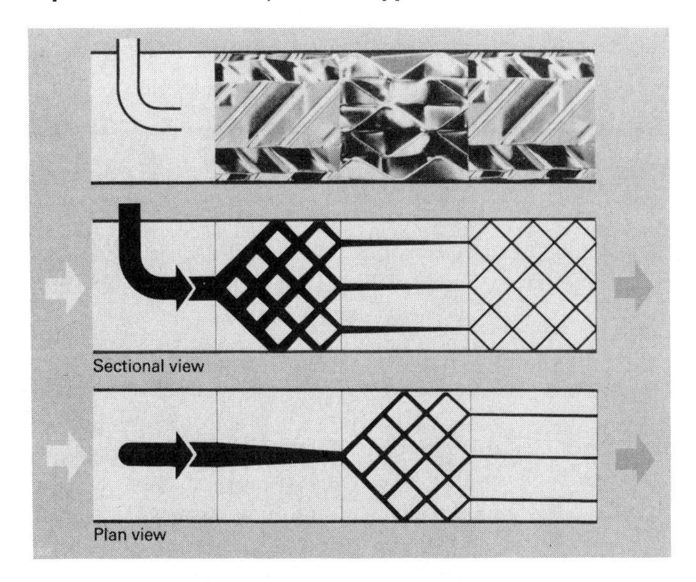

Figure 2 Principle of operation of a static mixer.

High heat transfer rates, making these supports ideal for use in exo- or endothermic applications.

The radial mixing in any one element takes place in one plane. In order to achieve homogeneous distributions of concentration, temperature, and velocity over the entire flow cross section, consecutive elements (along the reactor axis) are installed rotated through 90° on the axis—see Fig. 2.

Figure 3 illustrates typical flow patterns in OCFS, in a representative cross section, obtained using the techniques of computational fluid dynamics, i.e., by numerical solution

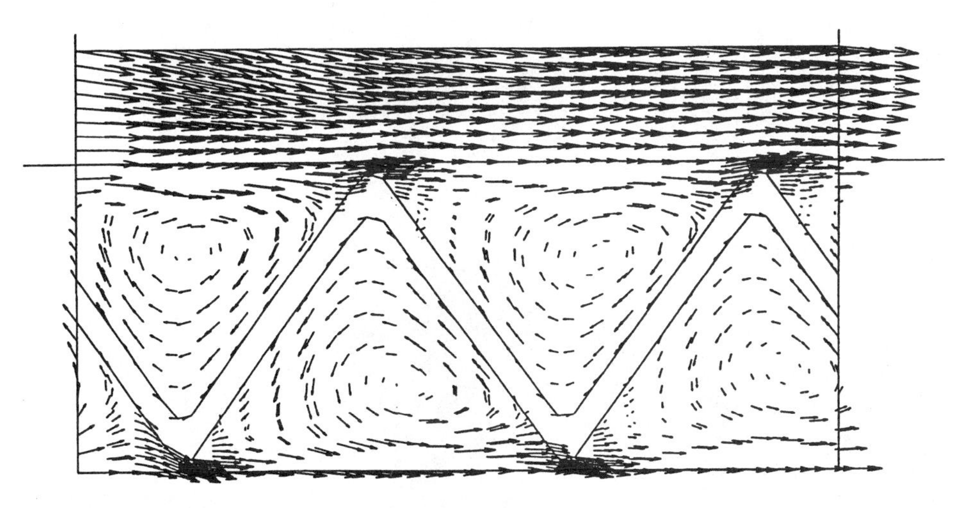

Figure 3 Flow pattern in OCFS, showing unimpeded streamlines in the channels and vorticity at the channel intersections. (Courtesy of Sulzer Chemtech Ltd.)

of the three-dimensional Navier–Stokes equations describing fluid flow. Clearly visible are the unimpeded vectors in the channels and the vorticity at the channel intersections.

II. REALIZATIONS OF OPEN CROSS-FLOW STRUCTURES

The OCFS principle as a catalyst support can be realized in several different ways, in response to the widely varying requirements placed by various processes on reactor internals.

A. Washcoated Metal

Metal substrates coated with a layer of controlled porosity and impregnated with catalytically active material—Fig. 4—offer high surface area and high void fraction and are therefore particularly attractive for applications where pressure drop is a parameter of primary significance.

The geometric surface area of such supports per unit of reactor volume is typically in the range 300–1800 m^2/m^3. Void fractions are in excess of 90%, which, together with the unimpeded flow paths offered by the open channels, result in values of pressure drop very much lower than those associated with conventional catalyst supports. The second consequence of the exceptionally high void fraction is a lengthening of residence time (by a factor of ca. 2) compared to dumped packings. The angle of the channels to the axis of flow is in principle variable between 0 and 90°; the optimum value is usually found in practice to lie between 30 and 45°. Broadly speaking, the efficiency of educt–catalyst contact increases with increasing channel angle, at the expense of increasing pressure drop.

The choice of alloy as substrate material is primarily determined by the necessity for an adequate key for the washcoat. Suitable alloys have been developed and proven for three-way automotive catalytic converters, in which the operating conditions place

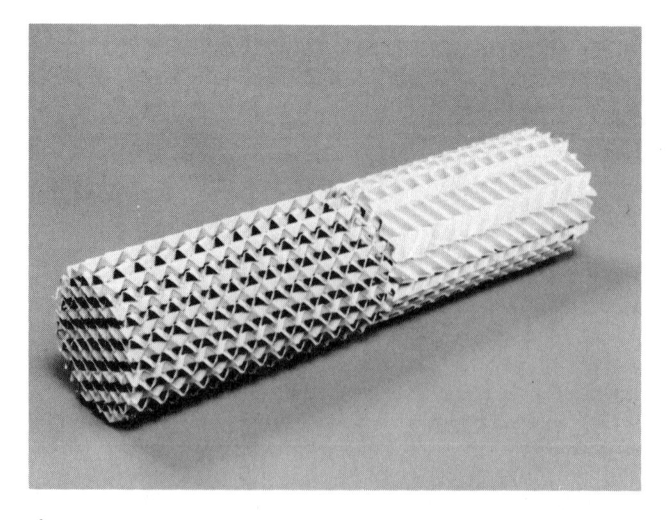

Figure 4 Catalyst support on the OCFS principle: Metal substrate with ceramic washcoat.

considerably higher demands on the catalyst support than those usually associated with process industry reactors. Similar technologies are employed in the washcoating and impregnation of these structures as for other metal catalyst supports, such as Raschig rings and monoliths. Washcoat materials are typically transition metal oxides, e.g., Al_2O_3; washcoat thickness and surface area are typically in the range 5–100 μm and 10–200 m^2 BET area/g washcoat material, respectively.

B. Granulate-Filled Sandwich

Granulate catalyst—e.g., ion-exchange resins, precious metals on alumina, or activated charcoal supports—may be embedded between two corrugated permeable screens to form a sandwich, and the sandwiches in turn stacked such as to give the characteristic OCFS structure—Fig. 5. This variant is of particular interest for heterogeneous gas–liquid systems such as trickle beds and bubble columns—see Section IV—and catalytic distillation—see Section V.

Where nonwashcoated supports have been in use, advantage can be taken of the properties of OCFS described in Section I without having to modify the catalyst recipe and/or seek a washcoat with similar properties of porosity etc. This shortens the duration of development and minimizes the number of process parameters to be simultaneously modified.

The washcoat stability problems associated with some liquid systems can be avoided.

If the catalyst life is short, it may prove more economical to refill sandwiches with regenerated/replacement catalyst than to dispose of the substrate (with or without the catalyst).

The volume of granulate is typically 25–30% of the volume of a reactor equipped with such a structure. This is significantly lower than the 60% volume fraction associated with fixed beds, but the difference is compensated for—when the rate-limiting step is either gas-film or pore diffusion resistance (Thiele modulus >5)—by the smaller particle size that can be employed in filling the sandwiches. That is, the geometric surface area of the granulate per unit of reactor volume will usually be the same or higher using the OCFS principle as in a conventional fixed-bed reactor.

A certain number of corrugated layers (sandwiches) is required in the plane perpendicular to the flow axis in order to achieve the mixing effect described in section I.B, which places an upper limit on the thickness of the individual sandwiches and hence on the particle size, particularly for vessels of small cross section. It will then be appreciated that the choice of particle size is, for small vessels, partly dependent on the cross-sectional dimensions of the reactor. The lower limit on particle size is given by the necessity to ensure satisfactory convection of reactants and products within the sandwich itself. As a guideline, the particle size is generally in the range 0.5–2.5 mm.

Other concepts for fixing catalyst particles in reactors do exist—see, for example, Stadig [3] and Sie [4], but these do not exhibit the advantages of OCFS—cross-mixing, high heat transfer and mass transfer rates, low pressure drop—described above.

C. Ceramic Extrudate

The technology now exists to extrude catalyst supports directly in the OCFS form from any extrudable matrix. Thus in addition to standard support ceramics—cordierite, steatite,

Figure 5 Catalyst support on the OCFS principle: sandwich structure with granulate catalyst embedded between two screens.

mullite, α-alumina, etc.—matrices including such additives as the vanadium/titanium mixtures widely used in $DeNO_x$ processes working with Japanese technology can also be manufactured—Fig. 6.

III. TWO-PHASE PROCESSES

The following discussion is concerned with processes for which reaction educts and products are in the same phase; the second phase referred to is that of the catalyst itself.

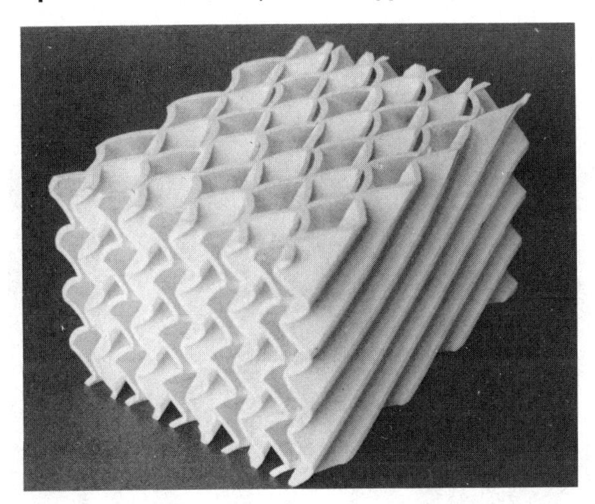

Figure 6 Catalyst/catalyst support on the OCFS principle: Ceramic matrix extrudate.

Mass transfer of reactants and products to and from the catalytically active surface in a reactor equipped with an OCFS is a function of the geometry of the structure and of the flow regime within it alone. Heat transfer to and from the reactor, on the other hand, is additionally a function of the flow regime between the structure and the reactor wall. For this reason mass transfer and heat transfer are treated separately in the following.

A. Mass Transfer

Gaiser and Kottke [5] have used a method similar to that developed by Marcinkowski and Zielinski [6] to visualize heat transfer and mass transfer improvements achieved by OCFS in comparison to parallel-channel devices. Their physicochemical method is based on convective mass transfer and produces colorization of the body wall as gas is passed through the support structures and reacting molecules are transferred from the gas bulk to the prepared wall material. The intensity of the colorization is a visual representation of the amount reacted and, since the surface reaction is very fast, of the mass transfer. Figure 7 shows the result of a typical such experiment.

A cross-flow and a parallel-channel structure are prepared in such a way that colorization of the surface takes place upon instantaneous chemical reaction with ammonia, which is fed as a pulse to an air flow passing over the investigated structures. It can be observed with the parallel-channel structure that there is strong colorization at the inlet, due to the flow phenomena associated with the entry region. The colorization decreases rapidly very soon thereafter, due to the establishment of a laminar boundary layer. Mass transfer is by molecular diffusion only, and the reactor dimensions necessary to transfer all of the ammonia from the bulk gas to the surface are considerably greater than those of the body examined.

In the cross-flow structure, mass transfer rates remain high throughout the length of the structure, and the ammonia is rapidly transferred from the bulk gas to the surface (intense colorization). The rapid reduction in concentration of ammonia results in a decrease in colorization of the surface, to practically zero within one-half of the "reactor" length: All of the added ammonia has been converted.

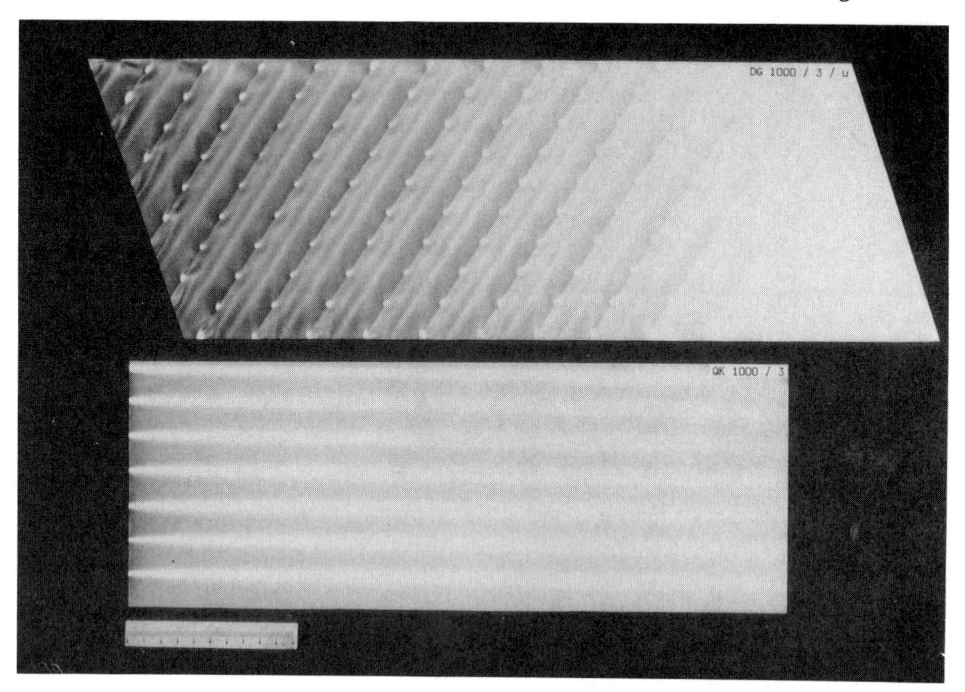

Figure 7 Mass transfer visualized in cross-flow and parallel-channel structures (flow direction from left to right). (From Ref. 5.)

Proof that the low intensity of colorization downstream of the entry region in the parallel-channel structure is due to formation of a laminar boundary layer, rather than to the lack of further ammonia for transfer, is offered by a second experiment conducted by Gaiser and illustrated in Fig. 8. Here a cross-flow structure is placed immediately downstream of the parallel-channel structure, and the mass transfer rate rises dramatically, indicating that much ammonia remained in the gas at the parallel-channel structure outlet.

Experiments show that with OCFS, mass (or heat) transfer from the bulk gas to the catalyst surface can be improved by factors of 5 to 10 compared to parallel-channel devices. For catalytic reaction processes *in which the rate of reaction is not the limiting step*, every such improvement of mass transfer leads to a higher overall reaction rate.

Kambing [7] has quantified the comparative mass transfer in OCFS and monoliths for the oxidation of propylene. With equal residence times, virtually 100% conversion was achieved in the OCFS at a throughput for which only 79% conversion was achieved in the monolith, although the temperature in the latter was 100°C higher and the surface area 1.4 times that of the OCFS. The mass transfer data, in the form of the Sherwood number, Sh, can be correlated with the Reynolds and Schmidt numbers, Re and Sc, respectively, in the form

$$\mathrm{Sh} = A \cdot \mathrm{Re}^b \cdot \mathrm{Sc}^{1/3}$$

but with values of A and b that are dependent on the geometry of the structure, i.e., that are functions of the corrugation amplitude and the angle of the channel to the reactor axis. For geometries of practical interest, the values of A and b lie in the ranges 0.8–2.5 and 0.3–0.6, respectively.

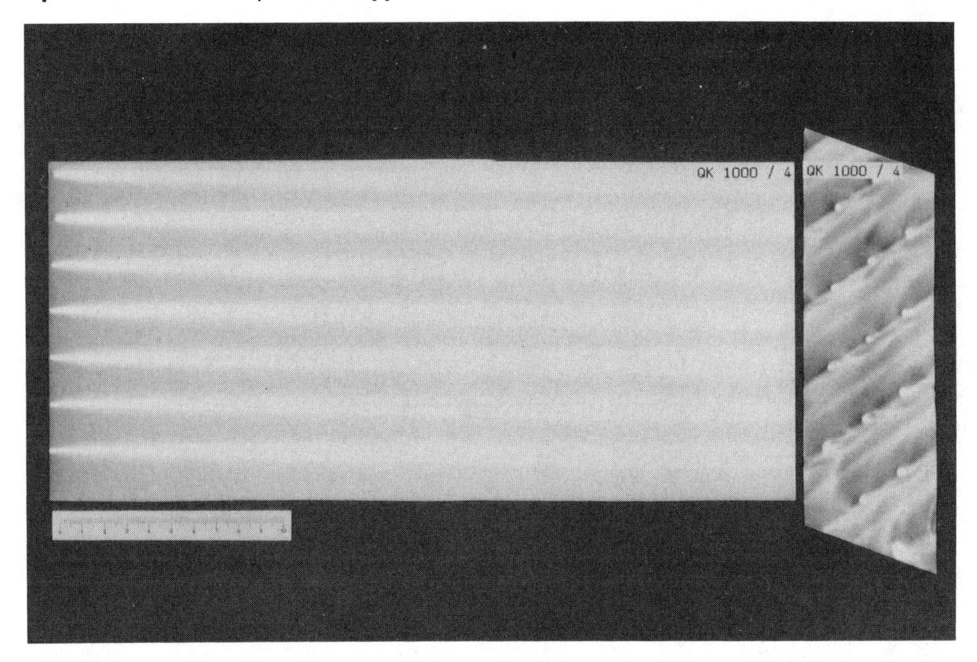

Figure 8 Mass transfer rate in cross-flow structure downstream of parallel-channel structure is substantially higher, indicating that reaction in monolith is far from complete (flow direction from left to right). (From Ref. 5.)

B. Heat Transfer

Strongly exo- or endothermic processes are generally carried out in multitube configurations, to permit heat removal/supply via a fluid medium. Within the tubes heat must be transferred at right angles to the flow axis; this transfer can be described by a radial heat transfer coefficient in the structure itself, λ_r, and a wall heat transfer coefficient, α_w. The overall heat transfer coefficient α_{tot} is then given, where the radial temperature profile is parabolic (see [9]):

$$\frac{1}{\alpha_{tot}} = \frac{1}{\alpha_w} + \frac{R}{4\lambda_r}$$

where R is the tube radius. Figure 9 illustrates the effect of varying λ_r and α_w on radial temperature profiles.

The experimental program of the University of Stuttgart [8–10] in particular permits the following guidelines to be given regarding the heat transfer behavior of OCFS:

In all but the densest structures (i.e., those of smallest corrugation amplitude and highest specific geometric surface area), convection is a significantly more important transfer mechanism than molecular diffusion or conduction in the substrate. It is thus important that these structures be installed in the manner illustrated in Fig. 2, since convective transfer takes place in one plane only in any one element length.

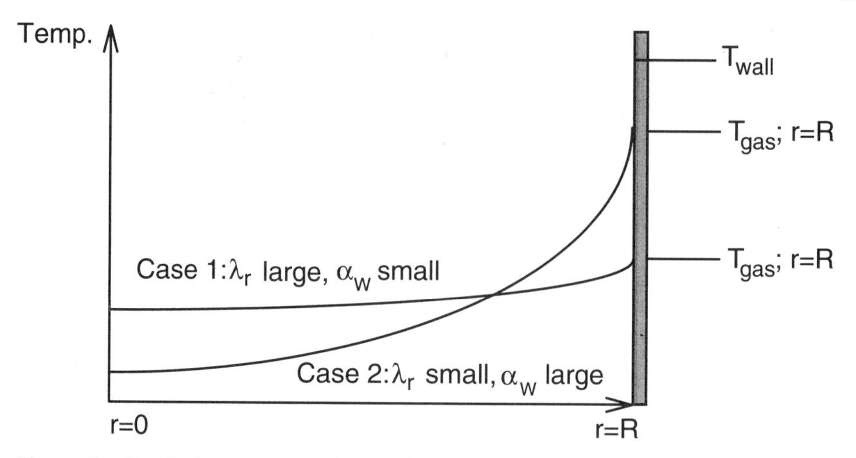

Figure 9 Qualitative representation of the effect on radial temperature profiles of varying heat transfer coefficients. (After Ref. 9.)

The value of λ_r increases with increasing angle between the channels of the structure and the axis of flow, since the flow path from axis to reactor wall is correspondingly shorter (see Fig. 10).

The value of λ_r increases with increasing amplitude of the corrugations of the structure, since the number of channel intersections in the axis-to-wall flow path decreases and the effectiveness of heat transfer through convection accordingly increases (see Fig. 10).

As a consequence of the foregoing, the increase in the value of λ_r with the gas flow rate is more pronounced in the less dense structures (see Fig. 11).

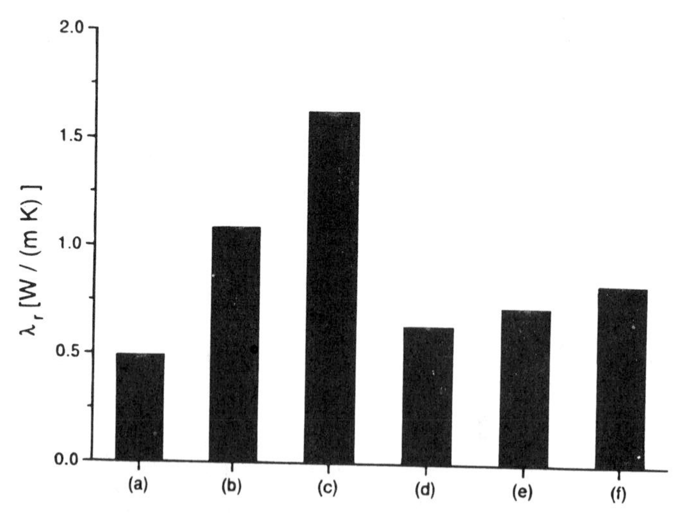

Figure 10 Radial heat transfer coefficient in OCFS and dumped packing (a) Corrugation amplitude 1.7 mm and angle of channel to flow axis 30°; (b) corrugation amplitude 4 mm and angle of channel to flow axis 30°; (c) corrugation amplitude 4 mm and angle of channel to flow axis 45°; (d) 5 mm spheres; (e) 2 mm spheres; (f) 5 mm cylinders.

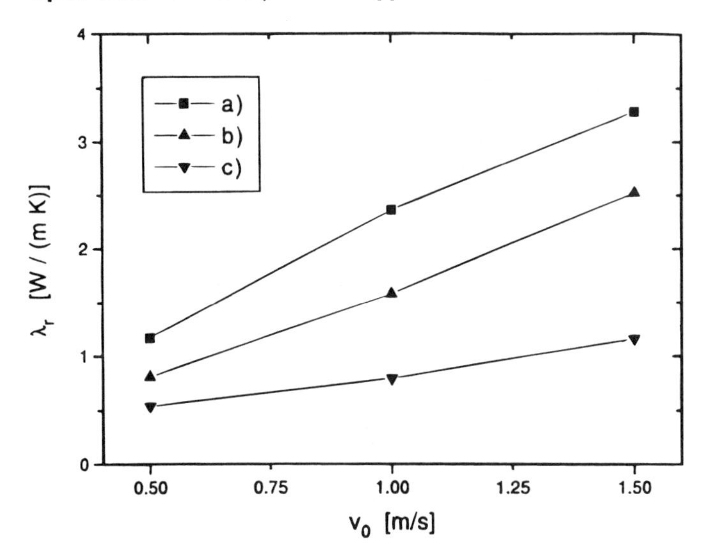

Figure 11 Radial heat transfer coefficient as function of gas flow rate. (a) Corrugation amplitude 4 mm and angle of channel to flow axis 45°; (b) corrugation amplitude 4 mm and angle of channel to flow axis 30°; (c) corrugation amplitude 1.7 mm and angle of channel to flow axis 30°.

The value of λ_r with these structures is higher than that in conventional fixed-bed structures, i.e., rings, cylinders, etc.—see Fig. 10. The primary significance of this is the resulting reduction of temperature gradients (see Fig. 12).

The value of α_w depends both on the fraction of the gas flow that bypasses the structure in the gap between the structure and the tube wall and on the flow regime therein. The larger the ratio of the gap to the hydraulic diameter of the channels of the structure, the higher the bypass stream velocity and the less effective the interaction with the well-mixed, highly turbulent flow in the structure. This in turn leads to substantially reduced heat transfer.

Figures 13 and 14 illustrate the importance of taking this gap into consideration in designing reactors to be equipped with OCFS, with two representative cases from the flow visualization experiments of Bey et al. [10]. By means of injection of an ink pulse into a water stream flowing through an OCFS, at two radial locations—in the structure and in the gap—the quality of the radial distribution/mixing in the structure itself and in the gap can be followed. Figure 13 shows flow at successively increasing flow rates (corresponding to Reynolds numbers representative of reactor operating conditions) with a large (undesirable) wall gap, Fig. 14 for the same flow rates with a small (desirable) gap.

In the case of a large gap it will be apparent that ink injected into the gap remains within it, even at the highest flow rates (turbulence intensities). And while radial mixing of the ink and water within the structure is good, little finds its way into the wall zone. This lack of convective transport impedes heat transfer from the reactant/product mixture into the surroundings. What these pictures also illustrate, of course, is bypass in the wall region, i.e., failure of a nonnegligible proportion of the reactants to contact the catalyst. The residence time of the ink in the wall region is less than half that of the ink in the OCFS.

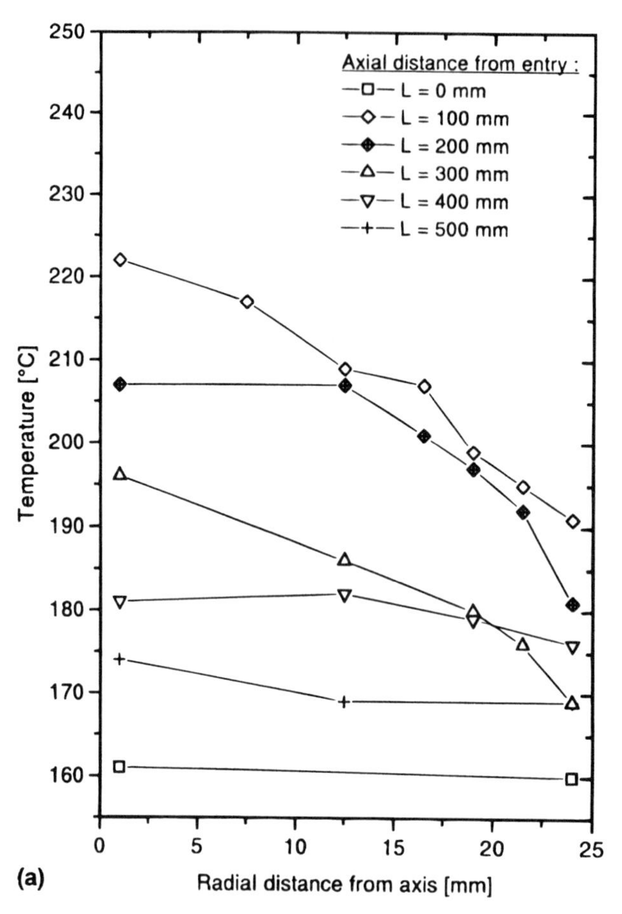

Figure 12 Comparison of radial temperature profiles in an OCFS (a) and in a fixed bed (b), for the test reaction catalytic combustion of propene, at entry to the reactor and at successive axial intervals of 100 mm; entry and coolant (outside wall) temperature 160°C; gas velocity 0.8 m/sec; adiabatic temperature rise 83°C. Note the flatter profiles in the OCFS and the difference in on-axis temperature between the OCFS and the fixed bed at the axial station $L = 100$ mm.

By contrast, when the wall gap is modest, convective exchange between the wall region and the channels of the OCFS is satisfactory, ensuring equalization of temperature between the flowing mixture and the wall and effectively eliminating bypass (as illustrated by equal rate of axial progress of the ink from the axis through the OCFS to the wall).

The temperature profile and absolute value in each structure must be compared before conclusions can be drawn for a specific process.

Replacement of a conventional support with an OCFS will result in a flatter temperature profile—beneficial in terms of selectivity and catalyst life—with values lying between the maximum and minimum obtaining in the fixed bed, due to the higher λ_r value. Should $\alpha_{w,OCFS}$ be equal to $\alpha_{w,FB}$, i.e., $\alpha_{tot,OCFS} > \alpha_{tot,FB}$, then the average temperature in the OCFS will be lower than that in the fixed bed.

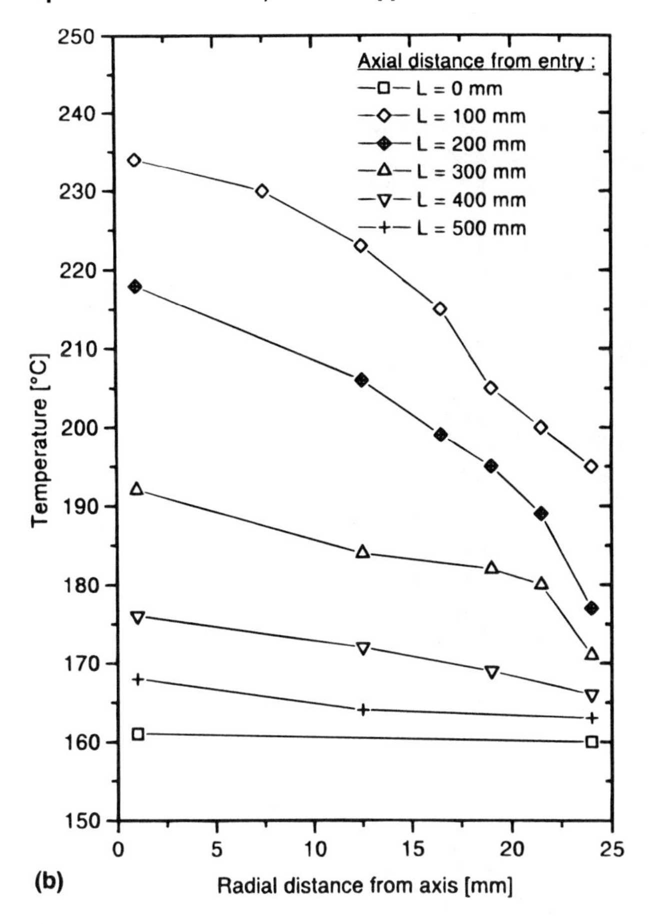

Figure 12 *Continued*

It will therefore be apparent that an optimization of the geometry, such as to achieve optimum λ_r and α_w values, requires experimental testing and/or simulation with accurate kinetic values for the process in question.

Finally, the intensive mixing in OCFS inhibits the formation of hot spots, thereby raising selectivity and extending catalyst life.

C. Pressure Drop

Pressure drop in gas flow in OCFS, in washcoated metal, is compared here to that in conventional supports for representative support surface areas per unit of reactor volume and representative gas flow rates. Figure 15 shows the comparison with cylinders and pellets, at the same throughputs (superficial velocities)—pressure drop is a factor of 10–100 higher in fixed beds, while the slope of the curves is comparable, indicating that the flow regime for both fixed beds and cross-flow structures is the same (either transitional or turbulent). The lower pressure drop in OCFS results from the higher void fraction—and consequently lower effective gas velocities—together with the unimpeded flow paths afforded by the structure. In the second comparison—Fig. 16—the slope of the pressure

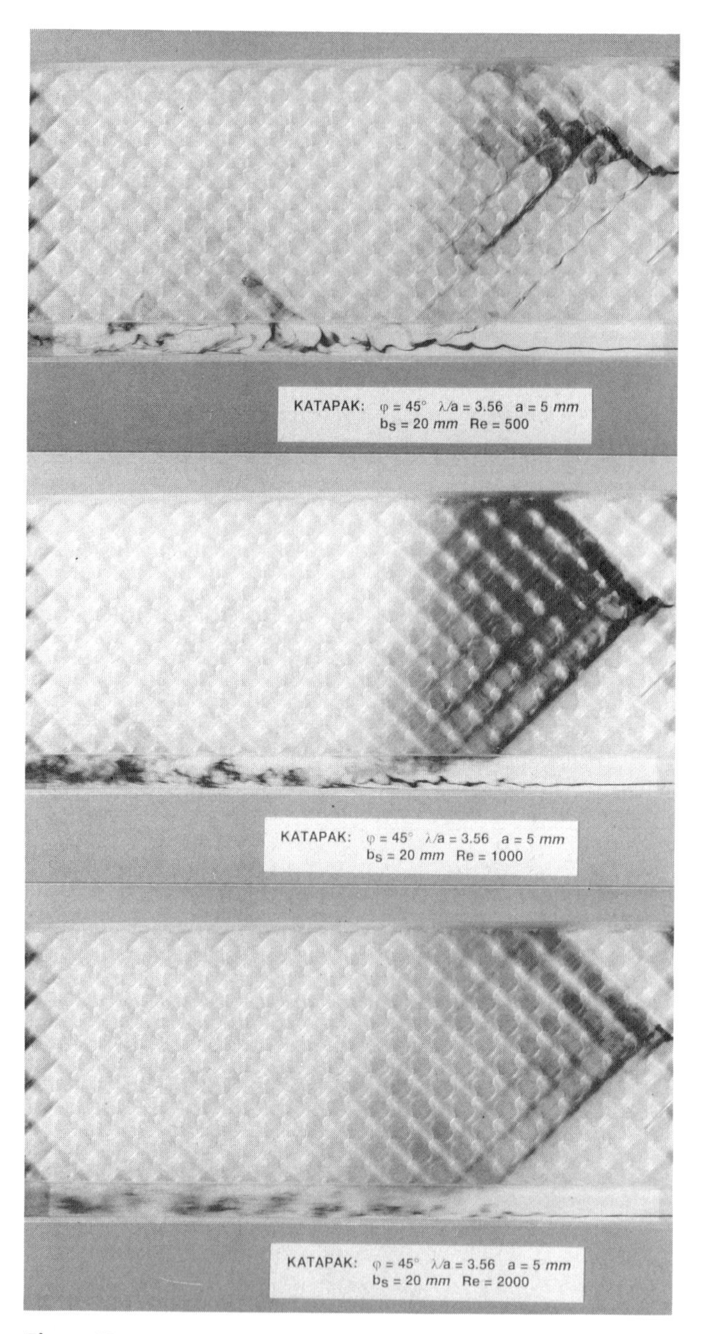

Figure 13 Flow visualization experiments showing effect of varying gap between OCFS and wall. Ink pulse injected into a water stream flowing (right to left) through an OCFS, at two radial locations, in the structure and in the gap (bottom of each picture). From top to bottom: successively increasing flow rates (corresponding to Reynolds numbers representative of reactor operating conditions) in a structure with a *large (undesirable) wall gap*. (From Ref. 10.)

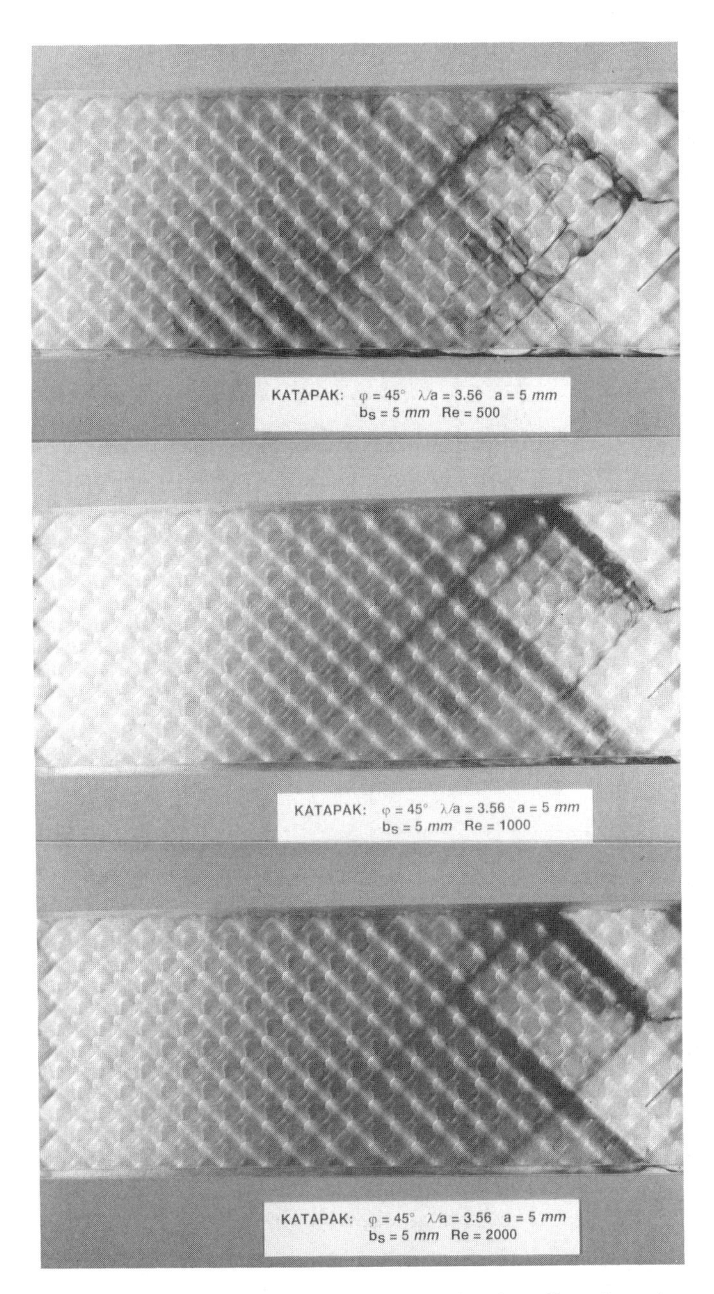

Figure 14 Flow visualization experiments showing effect of varying gap between OCFS and wall. Ink pulse injected into a water stream flowing (right to left) through an OCFS, at two radial locations, in the structure and in the gap (bottom of each picture). From top to bottom: successively increasing flow rates (corresponding to Reynolds numbers representative of reactor operating conditions) in a structure with a *small (desirable) wall gap*. (From Ref. 10.)

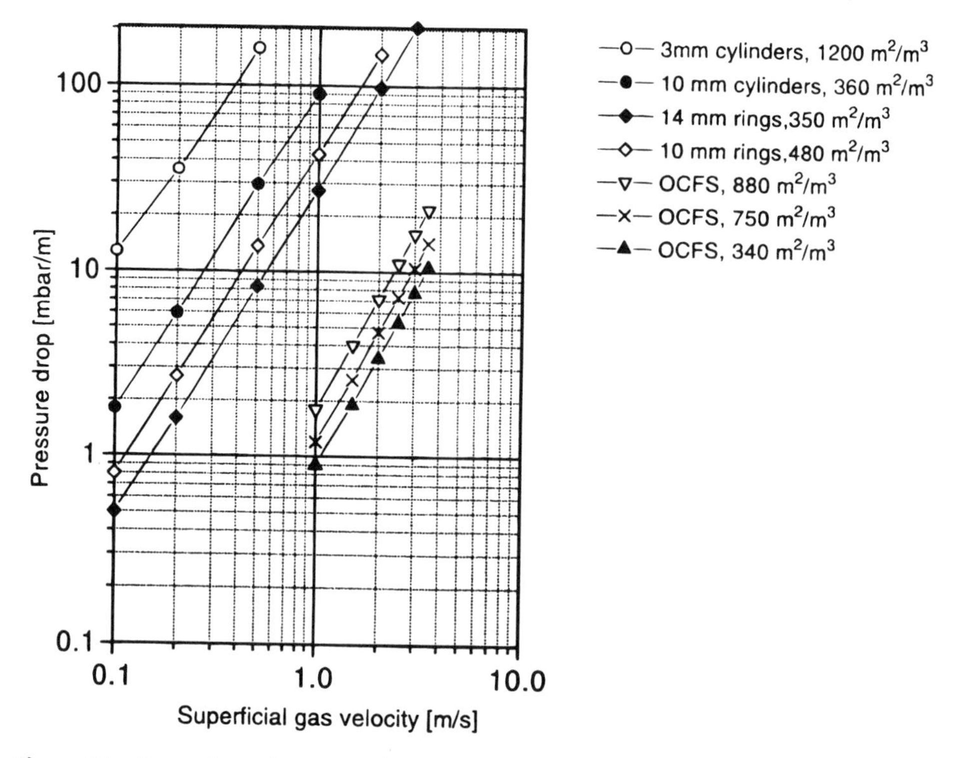

Figure 15 Comparison of pressure drop in OCFS structures and in dumped packings (rings and cylinders).

drop curves for monoliths is flatter, indicating that the flow regime in these is essentially laminar; pressure drop is of the same order of magnitude in both types of structure for the considered gas velocities.

It will be apparent from the foregoing that the optimum geometry, in terms of corrugation amplitude and angle of channel to flow axis, is process-specific, since it is dependent on the relative importance of a number of parameters: required specific geometric surface area, quantity of heat to be removed/supplied, and acceptable pressure drop. Thus, pilot tests must be conducted in connection with every proposed change to catalyst, support structure, and/or operating conditions. Fortunately, OCFS permit scale-up from such tests to be made with confidence, due to the regularity of their geometry and to the several decades of experience with these structures as static mixers.

D. Applications

Reactions that are strongly exothermic, such as selective oxidations, or those that are strongly endothermic, such as oxychlorinations, are usually carried out in multitube reactors. The catalyst is dumped into tubes of limited diameter—typically to 1 in. (25 mm)—such as to permit adequate radial heat transfer to/from the liquid bath surrounding them; as a result of their limited cross section, the overall number of such tubes is very large (typically on the order of 25,000). Minimization of pressure drop and maximization of

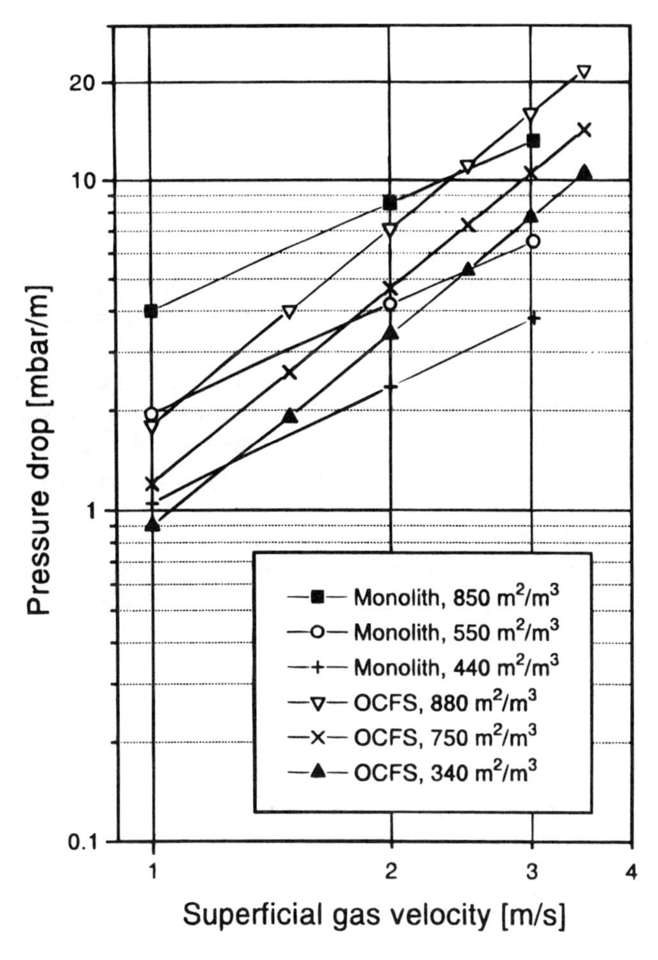

Figure 16 Comparison of pressure drop in OCFS structures and in monoliths.

selectivity—by means of precise temperature control and equalization of residence time—are key goals in reactor and process development.

With honeycombs (monoliths), pressure drop could be reduced by a factor of 100 or more, with corresponding reductions in equipment and operating costs. The total absence of radial flow in such structures, however, precludes their use in multitube reactors; the bulk heat generated by the reaction would not be transported to the tube walls, the selectivity of the process would decrease dramatically, and prevention of reaction runaway would prove difficult. Furthermore, the lack of radial vectors means that inhomogeneities in radial velocity profiles would be maintained; these inequalities in residence time would reduce selectivity and result in poor utilization of the catalyst in many of the channels.

With OCFS, the lower pressure drop in the catalyst bed results in reduction in the energy costs associated with recirculation of gas streams and—in new plants—lower investment costs due to the possibility of using boosters rather than compressors. Further potential for savings lies in the reduction of the number of reactor tubes, due to the increased tube diameter made possible by more efficient radial heat transfer. Of greatest significance, however, for processes such as the oxidation of *o*-xylene to phthalic anhydride or of ethylene to ethylene oxide, is the expected increased selectivity due to equalization of residence times across each individual reactor tube and from tube to tube and due to the elimination of hot spots.

IV. THREE-PHASE PROCESSES

The following discussion is concerned with processes in which both gas and liquid phases are present; the third phase referred to is that of the catalyst itself. In trickle-bed processes the gas phase is continuous; in bubble columns and loop reactors the gas phase is discrete.

The granulate-filled sandwich structure described in Section II.B permits continuously operated loop systems to be developed as replacements for batch processes. Loop reactors so equipped also exhibit advantages by comparison to conventional fixed-bed reactors: better exploitation of the catalytically active component due to smaller granulate size, lower pressure drop, and better gas–liquid mass transfer.

Operating experience with the sandwich structure in trickle-bed mode is as yet limited, but its behavior is sufficiently similar to that when employed as a catalytic distillation packing, in terms of liquid holdup/residence time, mass transfer, and pressure drop, to allow design of pilot plant for these parameters to be made with confidence.

Mazzarino and Sergi [11] have examined the liquid-phase catalytic hydrogenation of styrene in *n*-butanol in gas–liquid cocurrent flow, using a Pd-on-Al_2O_3 catalyst, and they summarize their findings as follows.

For the operating conditions listed below, the conversion rate is controlled by gas–liquid mass transfer of the gaseous reactant (hydrogen).

Temperature	273–313 K	Pressure	120 kPa
Styrene concentration	20–3000 ppm	H_2 partial pressure	5–120 kPa
Gas superficial velocity	10^{-2}–0.3 m/sec	Liquid superficial velocity	6×10^{-4}–8×10^{-3} m/sec
Reactor diameter	50 mm	Reactor length	1 m
Catalyst particle size	0.5–1 mm	Packing void fraction	0.75

The mass transfer coefficient is dependent primarily on the superficial velocity of the gas phase.

The use of an OCFS leads to a significant decrease in resistance to mass transfer with respect to conventional random packings.

Mass transfer of the key reactants in the liquid phase is not hindered by the permeable screen retaining the catalytic granulate.

Axial dispersion in an OCFS is higher than in a conventional packed bed; the extent to which this is of significance will depend on the process in question.

V. CATALYTIC DISTILLATION

For a number of the most important groups of catalytically promoted reactive processes in the chemical industries—e.g., etherifications and esterifications—conversion is equilibrium-limited, according to Le Châtelier's principle. It is thus desirable to conduct reaction and separation in the one vessel—by immediate removal of the products, conversion can be raised substantially above the equilibrium level associated with the operating conditions. Conversion of up to 100% is in principle possible with this approach. The advantages of catalytic distillation have been extensively documented in the literature in recent years—see, for example, Doherty and Buzad [12]—and can be briefly summarized as follows:

Minimization of the reverse reaction and fewer side reactions, resulting in higher selectivity than in a fixed-bed reactor at equivalent conversion levels.
Simplified temperature control—the temperature is the boiling point of the component mixture and is easily controlled by the system pressure.
Reduction of the number/volume of recycle streams (possibly none required).
Heat of reaction as boil-up for fractionation (exothermic reactions).
Less complex separation (no azeotropes with the reactants) and reduction of separation effort (fewer components), resulting in lower energy costs.
Fewer vessels and less control instrumentation.

In short, all of these advantages permit significantly lower investment and operating costs than those associated with classical processes.

A. Mass Transfer

The separation performance of the reaction section is key to realizing the full potential of catalytic distillation as described above. Efficient gas–liquid mass transfer ensures immediate removal of the products from the reaction zone, effectively shifting the equilibrium such that reactants are completely converted. Poor separation, on the other hand, will result in the reaction's proceeding little beyond the equilibrium value associated with a fixed bed operating under the same conditions. The remaining reactants will then have to be separated from the products downstream. The potential of catalytic distillation to avoid difficulties in separation due to azeotropes with the reactants is then lost. (It should be kept in mind, however, that while this is one of the major attractions of catalytic distillation, the process cannot help avoid all azeotropes that may form.)

The advantages of OCFS in distillation service have been proven over several decades—see, for example, Darton and Weve [13]; they may be summarized as follows:

High separation efficiency
Low pressure drop per unit of mass transfer/high capacity
Minimal liquid entrainment
Insensitivity to foaming and/or fouling

In addition, their regular structure permits scale-up from pilot units to be made with confidence.

The mass transfer performance of OCFS in rectification/stripping (structure in sheet metal) and in catalytic distillation (sandwich structure) is compared in Fig. 17. The basis of comparison is equivalent throughput capacity, since catalytic distillation columns are typically of constant diameter. It will be apparent that the behavior of both is qualitatively the same: Mass transfer (typically quantified in terms of the number of theoretical stages per unit height) remains sensibly constant over the greater part of the capacity range (good turndown capability), rises somewhat as the capacity limit is neared (attributable to increased gas–liquid friction, maximizing the gas–liquid contact area), before falling off

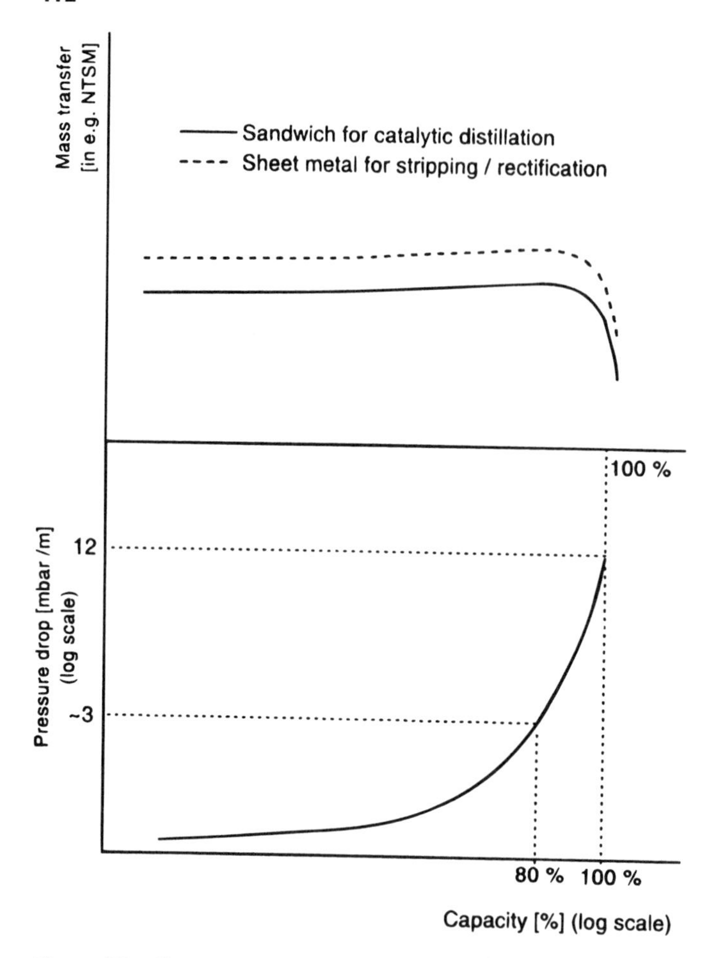

Figure 17 Characteristic performance curve (upper) and pressure drop curve (lower) for OCFS in distillation and catalytic distillation service, as function of throughput capacity.

sharply as the column approaches flooding. The absolute level of mass transfer achieved in a given height is somewhat higher in a sheet metal OCFS, since both sides of each corrugated layer are available for film formation and hence for gas–liquid contact; in the catalyst-filled sandwich variant, only one face of each layer provides for gas–liquid mass transfer because the other is in contact with the granulate.

Calculation of the height of the reaction section can be made in the same manner as for the rectification and stripping sections, by solving for the mass and enthalpy balances, having regard to the phase equilibrium. The effects of the chemical reaction (changes in the concentration, reaction enthalpy) must, however, be accounted for in the derivation of the balance equations. The accuracy of simulations of catalytic distillation processes that can be made with commercially available software packages is now approaching that necessary for column design. The complex phase equilibrium behavior in the reaction zone is, for many gas–liquid systems, within the capacity of such packages to compute—measurement of the kinetics for the particular catalyst and reaction remains in most cases necessary, however. The comparisons of Krafczyk and Gmehling [14] of simulation with experiments conducted with a catalytic distillation column equipped with an OCFS packing, for the production of methyl acetate, are an indication of what is currently achievable.

B. Capacity

Figure 17 also illustrates pressure drop in the gas stream in OCFS employed as distillation packings, as a function of capacity (gas throughput). The form of the curve is the same, irrespective of whether the structure is used for rectification/stripping or catalytic distillation. It approaches the vertical as flooding is neared; the value of pressure drop for the "pure" distillation and catalytic distillation packings at this condition is comparable—on the order of 12 mbar/m of packed height. This provides a quantifiable basis for the definition of 100% capacity. Distillation columns equipped with OCFS are then usually operated at 70–80% capacity, at which throughput the pressure drop is in the range 2–3 mbar/m of packed height.

C. Other Design Issues

The performance of OCFS as mass transfer devices is heavily dependent on the quality of the distribution of the gas and liquid phases across the column upon entry to the packed section—irrespective of whether its function is purely rectification/stripping or chemical conversion also. Optimal liquid distribution is, however, of additional importance in catalytic distillation, in ensuring contacting of reactants with the catalyst.

The parameters to which regard must be paid in the selection of a liquid distributor are flow rate, turndown ratio, bed height, and liquid properties. For packed beds exceeding a given height it may be necessary to collect and redistribute the liquid. The type of gas inlet system (simple inlet nozzle, nozzle with baffle plate, radial or tangential inlet tube) required will again depend on flow rate, turndown ratio, and gas properties and in addition on the ratio of column diameter to the height between the inlet and the bottom of the packed section.

Finally, the same care must be taken as with nonreactive distillation packings in ensuring that the gap between the structure and the column wall is sufficient to prevent flow of liquid onto the wall (on which it will then remain) while not so large as to promote gas bypass. This compromise is usually achieved with the aid of annular collars fitted to the structure itself.

D. Applications

1. *Etherification*

The catalytically promoted liquid-phase etherification of isobutene with an excess of methanol to produce MTBE has been carried out on a commercial scale in conventional fixed-bed reactors since the 1970s; isobutene conversion is on the order of 90–97%. MTBE can be subsequently separated from the inerts and excess methanol by distillation—although this is complicated by the presence of minimum boiling azeotropes between MTBE and methanol and between isobutene and methanol. Unreacted isobutene is, however, difficult to separate from *n*-butanes and *n*-butenes because of their low relative volatilities, and a fraction of the isobutene is thus usually not converted but leaves the MTBE plant with the C_4 inerts.

The catalytic distillation process of Smith [15], by providing for the fixing of the catalyst in a reactive section of a column between nonreactive stripping and rectification sections, and thereby for the continuous removal of MTBE from the reactants, boosts the conversion of isobutene to well in excess of 99%. The concept is still more economically attractive when OCFS are employed to secure the catalyst in the reactive section—DeGarmo et al. [16]—due to their significantly higher mass transfer efficiency.

Etherifications of various olefins can be conducted on the catalytic distillation principle, using methanol, ethanol, propanol, butanol etc., and the additional advantages of employing OCFS are equally relevant for many of these.

2. *Esterification*

One of the first commercial-scale catalytic distillation processes was that for the production of methyl acetate [17]. In this esterification process the liquid acid catalyst is continuously added to the reactant mixture; it finishes up in the bottoms and, following neutralization, in the wastewater from the plant.

Krafczyk and Gmehling [14] have shown that very high purity methyl acetate may be produced by a catalytic distillation process using an ion-exchange resin catalyst fixed in an OCFS packing. Figure 18 gives a schematic representation of their apparatus. This process operates at atmospheric pressure, with virtually complete conversion of the reactants (methanol/acetic acid, in equimolar proportions). The height of the reaction zone is extremely modest, for two reasons: Only ten theoretical stages are required, and the

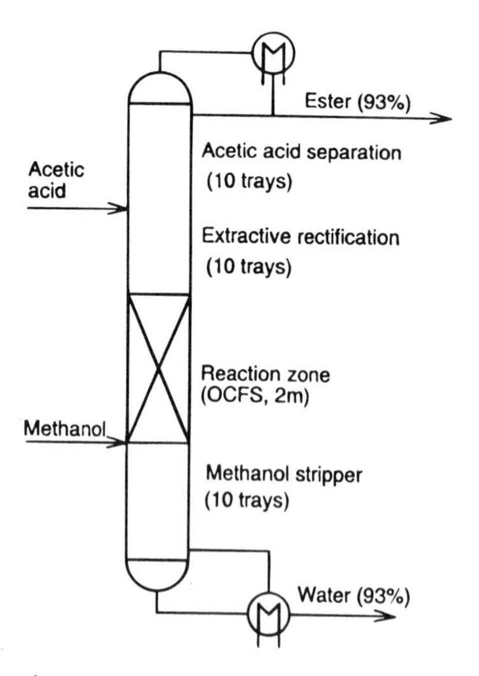

Figure 18 Configuration of catalytic distillation apparatus for production of methyl acetate. Reflux ratio = 2; pressure drop 6 kPa. (From Ref. 14.)

reaction section of a catalytic distillation column, equipped with an ion-exchange catalyst, provides some twenty times more proton sites per unit volume of reacting liquid than a homogeneous reactor, dramatically reducing the residence time required to achieve a given level of conversion. Finally, the heterogeneous approach involves no loss of catalyst and eliminates the need for downstream treatment of liquid acid, which is becoming less and less acceptable from an environmental standpoint.

VI. CONCLUSION

Catalyst supports in the OCFS are now making significant improvements in reactor design and operation possible. The newness of the concept means that only a small part of its potential has yet been realized, but success to date in fundamentally revising several process technologies is reason for belief that the palette of applications will expand significantly in the next few years. Much work remains to be done in quantifying heat transfer and mass transfer for the various process classes, and in characterizing flow regimes, before design becomes a matter of routine; as operating experience is gained, optimization of such design will follow.

REFERENCES

1. Mutsakis, M., Streiff, F.A., and Schneider, G., Advances in Static Mixing Technologies, *Chemical Engineering Progress* (July 1986), pp. 42–48.
2. Tauscher, W., and Schütz, G., Static Mixing Elements, *Sulzer Technical Review*, 2/1973.
3. Stadig, W.P., Catalytic Distillation: Combining Chemical Reaction with Product Separation, *Chemical Processing*, February 1987.
4. Sie, S.T., Fixed-Bed Reactors on the Basis of Structured Elements: How and Why, *NPT Procestechnologie*, April 1995 (in Dutch).
5. Gaiser, G., and Kottke, V., *Chem.-Ing.-Technik 61* (1989) pp. 729–731 (in German).
6. Marcinkowski, and Zielinski, *Chemia Stosowana* (*Applied Chemistry*, Polish), VII (4B), 515–523 (1970); *Przemysl Chemiczny* (*Industrial Chemistry*, Polish), 49 (5), 283–287 (1970).
7. Kambing, C., *Charakterisierung und Vergleich des Stoffaustausches an Katalysatorträgern mit Parallel- und Kreuzkanalstruktur*, Diplomarbeit der Universität Dortmund, 1994 (in German).
8. Eigenberger, G. Fixed-Bed Reactors, in *Ullmann's Encyclopedia of Industrial Chemistry*, 5th ed. (1992), Vol. B4, pp. 199–238.
9. Eigenberger, G., Kottke, V., Daszkowski, T., Gaiser, G., and Kern, H.-J., *Regelmäßige Katalysatorformkörper für technische Synthesen*, Fortschritt-Berichte VDI, Reihe 15, Nr. 112 (1993) (in German).
10. Bey, O., Gaiser, G., and Eigenberger G. *Studie zum radialen Transportverhalten in Kreuzkanal-Katalysatorstrukturen*, and Müller, F.-U., Holzer, J., Bey, O., and Gaiser, G., *Randgängigkeit der Strömung in Kreuzkanal-Katalysatorstrukturen und ihr Einfluß auf den wandseitigen Wärmeübergang*. Unpublished reports of the Institut für Chemische Verfahrenstechnik, University of Stuttgart (in German).
11. Mazzarino, I., and Sergi, M., *Mass Transfer in a Sandwich Packing Multiphase Reactor*, unpublished reports of the Dipartimento di Scienza dei Nateriali e Ingegneria Chimica, Politecnico di Torino.
12. Doherty, M.F., and Buzad, G., *Reactive Distillation by Design*, IChemE Symposium Series No. 128 (1992).

13. Darton, R., and Weve, D., Distillation—The Importance of Equipment Choice, *The Chemical Engineer*, 14 November 1991, pp. 19–24.
14. Krafczyk, J., and Gmehling, J., Einsatz von Katalysatorpackungen für die Herstellung von Methylacetat durch reaktive Rektifikation, *Chem.-Ing.-Technik 66* (1994), pp. 1372–1375 (in German).
15. Smith, L.A., Catalytic distillation process, U.S. Patent No. 4,307,254, December 22, 1981.
16. DeGarmo, J.L., Parulekar, V.N., and Pinjala, V., Consider Reactive Distillation, *Chemical Engineering Progress* (1992), pp. 43–50.
17. Agreda, V.H., Partin, L.R., and Heise, W.H., High-Purity Methyl Acetate via Reactive Distillation, *Chemical Engineering Progress* (1990), pp. 40–46.

15

Catalytic Filters for Flue Gas Cleaning

Guido Saracco and Vito Specchia
Politecnico di Torino, Torino, Italy

I. INTRODUCTION

A. Multifunctional Reactors

The growing need for energy and space savings has forced chemical engineers to work out new reactors capable of carrying out, besides the chemical reaction, other functions, such as separation, heat exchange, momentum transfer, and a secondary reaction. Agar and Ruppel [1] and Westerterp [2] recently reviewed a number of these apparatuses, calling them *multifunctional reactors*. Typical members of this class are, e.g., membrane reactors (combining a catalytically promoted reaction and a separation allowed by the membrane itself; see Chapters 16 and 17 of this book), reactive distillation columns (where separation between reactants and products is accomplished by distillation [3,4]), and catalytic reactors with periodic flow reversal (in which higher than adiabatic temperatures can be kept in the central part of the reactors, thus allowing complete combustion of VOCs [5]).

A common feature to all multifunctional reactors is that they allow substitution of at least two process units with a single reactor, where all the operations of interest are carried out simultaneously. A likely consequence is the reduction of investment costs, which is often combined with significant energy recovery and/or saving. Consider, for instance, the case of reactive distillation, where the heat of reaction allows reduction of the heat required at the boiler of the distillation column.

Furthermore, space saving is often achieved. Consider, for example, the catalytic abatement of nitrogen oxides from stationary sources (e.g., power plants) by selective reduction with ammonia. Normally, considerable volumes of honeycomb catalyst are required for this. To solve the related space requirement problems, several researchers proposed certain multifunctional reactors. Lintz and co-workers [6,7], for instance, tested the use of the Ljungstroem air preheater of the power plant as a chemical reactor for NO_x removal, by covering the air heater elements with a suitable catalyst. Conversely, Harris [8] coupled NO reduction catalyst with the silencer of large-scale natural gas engines. A further possibility explored was the coupling of the NO_x reduction with fly ash separation in high-temperature-resistant filters on which the catalyst had been deposited [9–11].

This last one has probably been the most interesting application of a particular multifunctional reactor, the *catalytic filter*, to which this chapter is dedicated.

B. Catalytic Filters: The Basic Concept

Figure 1 shows a schematic of a catalytic filter. Such devices are capable of removing particulates from flue gases (e.g., from waste incinerators, pressurized fluidized-bed coal combustors, diesel engines, boilers) and simultaneously abating chemical pollutants (e.g., nitrogen oxides, volatile organic compounds) by catalytic reaction. The catalyst is applied, in the form of a thin layer, directly onto the constituent material of the filter, which can be either rigid (filter tubes made of sintered granules) or flexible (a tissue of ceramic or metallic fibers).

The potential advantages of catalytic filters are those typical of multifunctional reactors (reduction of process units, space and energy savings, cost reduction, etc.). Later on some examples will elucidate these points (see Section IV). However, the following properties should be possessed by catalytic filters so that these opportunities actually can be exploited:

Thermochemical and mechanical stability.

High dust separation efficiency: Dust should not markedly penetrate the filter structure since this would lead to pore obstruction and/or to catalyst deactivation.

High catalytic activity so as to attain nearly complete catalytic abatement for conveniently high superficial velocities, i.e., those employed industrially for dust filtration: 10–80 m/hr.

Low pressure drops (though a certain head loss increase, compared with the virgin filters, has to be envisaged owing to the presence of the catalyst).

Low cost.

The present chapter will show how, and to what extent, these properties can be exploited and the limitations overcome.

II. THE CURRENT MARKET FOR HIGH-TEMPERATURE INORGANIC FILTERS

The use of porous inorganic filters for high-temperature applications is meeting with increasing interest [12,13]. Almost an entire book has recently been dedicated to this topic

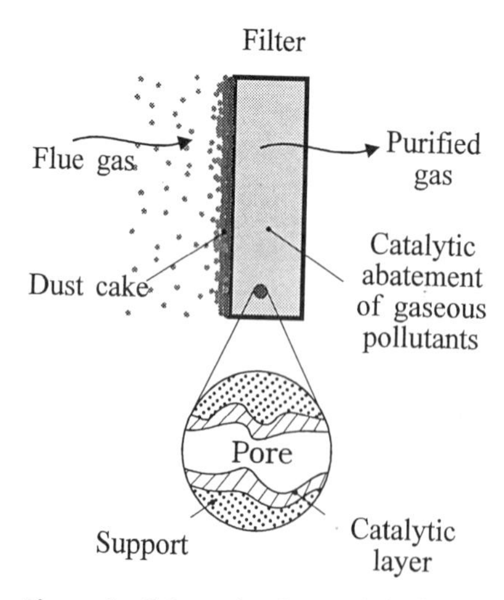

Figure 1 Schematic of a catalytic filter.

by Clift and Seville [14]. High-temperature particulate removal allows the performance of any heat recovery on clean flue gases, to keep temperature high enough so that chemical pollutants can be catalytically destroyed (which is very important for catalytic filter applications), to withstand temperatures (above 250°C) at which conventional polymer-based fabric filters would rapidly be damaged.

Most of the development of inorganic filter media has occurred in the last 10–15 years, along with the improvements achieved in filter manufacturing, which allowed for obtained products suitable for a wide range of high-temperature applications (catalyst and precious metal recovery, fly ash filtration from coal gasification or combustion effluents, soot filtration from diesel engines, etc.). To gain an idea of the current rate of development of these apparatuses, consider that in 1992 the world market for apparatuses for removing hot gas dust equaled $70–75 million, whereas estimates for 1996 are around $170–180 million [15]. Filter media may represent about 20–30% of this total.

Different applications generally require different filter properties (temperature resistance, pressure drop, filtration efficiency, etc.), and therefore a number of different types of filters, based on either ceramic (SiC, mullite, cordierite, etc.) or metallic (stainless steel, Hastelloy, Fecralloy, etc.) materials, are currently produced and commercialized. Table 1 summarizes such filter types, their producers, and their major application fields.

Most filters have a tubular or candle structure (whose length ranges mostly from 1 to 2 m) and are assembled into high-temperature baghouses. Figure 2 shows the exterior of such a baghouse; Fig. 3 shows the internal assembly of candles in the baghouse. Filtration occurs on the outer surface of the candles on which a dust cake progressively grows. Whenever the related pressure drop becomes higher than a limiting value, the dust cake is removed through a jet-pulse technique (a strong pulse of air or of an inert gas is fed at the inner side of the filter thus causing detachment of the dust cake). A similar technique would also be adopted for ceramic cordierite monoliths produced with the so-called wall-flow deadend-channel configuration [16].

Table 1 Commercial Inorganic Filters: Types and Application Fields

Filter type	Producers	Main applications
Rigid ceramic sintered filters	Cerel, Universal Porosics, Industrial Filters and Pumps, among others	Coal gasification, fluidized-bed coal combustion, waste incineration
Pulp-type $SiO_2 \cdot Al_2O_3$ fiber candle filters	BWF, Cerel	Separation of metal dust, fluidized-bed coal combustion, waste incineration
Ceramic woven-fabric filters	3M	Catalyst recovery, coal-fired boilers, metal smelting, soot filtration
Ceramic cross-flow filters	Coors	Applications up to 1500°C
Ceramic cordierite monoliths	Corning, Ceramem, NGK insulators, among others	Coal gasification, fluidized-bed coal combustion, waste incineration, soot filtration
Sintered porous-metal-powder filters	Pall, Mott, Newmet, Krebsöge, Fuji, among others	Catalyst and precious metal recovery
Sintered, stainless steel, semi rigid fiber filters	Bekaert, Memtec, among others	Catalyst and metal dust recovery, soot filtration

Figure 2 External view of a high-temperature baghouse. (Courtesy of Schumacher GMBH, Crailsheim, Germany.)

The constituent material can be in either a granular or a fibrous form. In the latter case, fibers can be either arranged in a tissue (e.g., Nextel filters by 3M [17]) or randomly dispersed and held together by a binder (e.g., KE-85 cartridge filters by BWF [18]). Fibrous-type filters, whether the fibers are metallic or ceramic, generally allow comparatively lower pressure drops, due to their high porosity (up to 80%), and are homogeneous. Granular filters, which can also be either ceramic or metallic, generally have a higher mechanical resistance but much higher pressure drops, due mostly to the lower porosity of the products (<50%). However, recent developments allowed the production of granular asymmetric filters made of two layers: the outer layer (about 100 μm thick with pores of a few tenths of a micron) is the true filter media, while the inner layer (15–20 mm thick with pores

Figure 3 A set of SiC Dia-Schumalith candle filters in a high-temperature baghouse. (Courtesy of Schumacher GMBH, Crailsheim, Germany.)

of a much wider size) acts as a support, giving mechanical resistance to the filter [13]. Such filters allow much higher permeability than their homogeneous counterparts. A SEM micrograph of the cross section of a SiC double-layer asymmetric filter (Dia-Schumalith by Schumacher, Crailsheim, Germany) is given in Fig. 4.

The main applications for high-temperature-resistant filters lie at present in the following fields:

Catalyst/metal recovery: e.g., catalysts from fluidized catalytic reactors, nickel, platinum, pharmaceutical products, silicon; several hundred plants are currently operating worldwide for such purposes.

Waste incineration: for either medical or municipal wastes, a few units, employing rigid candle filters, have recently been installed, especially in the Netherlands and the USA.

Pressurized fluidized-bed combustion (PFBC) of coal: This is probably the most promising application field for inorganic filters, despite the fact that only a few large-scale examples of application have been successfully tested in very recent years [14]. Dust is removed just at the outlet of the boiler, allowing heat recovery from cleaned gases.

Figure 4 SEM picture of the cross section of a double-layer Dia-Schumalith filter. (Courtesy of Schumacher GMBH, Crailsheim, Germany.)

Coal gasification: fly slag from the gasification process is removed at high temperature, allowing direct utilization of the produced syngas in a gas turbine cycle for power generation purposes [19].
Soot entrapping: e.g., from diesel engine exhausts.
Dust recovery from calcination processes: e.g., magnesium oxide production.

For deeper insight into the applications of high-temperature filter media, the book by Clift and Seville [14] is recommended, and particularly the review by Bergmann, from Filter Media Consulting Inc. (La Grange, Georgia), therein included [15]. Section V of this chapter will discuss the new application opportunities allowed by catalytic activation of the above inorganic filters.

III. THE PREPARATION OF CATALYTIC FILTERS

The deposition of a suitable catalyst in the intimate body of the above-described filters is controlled primarily by the structure of the filter itself, but it is also influenced by the nature of its constituent material. In fact, shear stresses may arise at the interface between this material and the deposited catalyst, owing to thermal expansion mismatch between the two phases. Since most catalyst supports are based on inorganic oxides, this problem would be particularly serious for metal-based filters, owing to their much higher thermal expansion coefficients. However, in some metal alloys, such as the FeCrAlloy, a thin surface layer of a metal oxide (e.g., Al_2O_3) is formed at high temperatures, which improves their thermal resistance and allows a proper basis for catalyst anchoring.

In the case of fibrous nonsintered filters, the catalyst deposition can be performed before the filter itself is assembled. For instance, Morrison and Federer [20] developed a sol-gel technique for lining the alumina-based Nextel fibers by 3M with a vanadia catalyst, to be used in a catalytically active diesel particulate trap. Further, some researchers at Babcock and Wilcox [9] prepared catalytic filters in which a preformed zeolitic catalyst (NC-300 by Norton) was incorporated, once again to promote NO_x reduction with ammonia. The techniques employed in this case are those listed in a series of patents [21–23]. A suspension of finely divided catalyst may be sprayed on the fibrous filter before thermal treatment. Alternatively, the finely divided catalyst may be suspended in a gas stream, which is then passed through the bags to coat the filtering side uniformly; according to the inventors, the catalyst should lodge in the interstices of the weave pattern and remain there during operation, held by not-better-specified adhesive forces. Another imaginative solution, suggested by Pirsh [21] for the case of metal catalyst, is that of drawing filaments of such metals and interspersing them with ceramic fibers in the production of catalytic fabric filters. Where the metal oxide is to be the desired catalyst (as in the case of noble metals), exposure of the combined fabric/filament material to air or more severe oxidizing conditions will oxidize the surface of the filaments.

Kalinowsky and Nishioka [24] underlined how, in the case of catalytic fiber filters, the catalytic materials should be not only self-supporting but also flexible, capable of withstanding elevated temperatures, and highly resistant to abrasion, particularly to self-abrasion among the fibers. These authors developed a special sol-gel process that they claim to be the only one capable of producing catalytic fibers with the desired properties. Metal alkoxides are dissolved in the desired proportions in an organic solvent, such as methanol or ethanol. The fabric filter is then impregnated with the solution. The product is then dried in an atmosphere containing moisture, added in controlled amount to promote hydrolysis-condensation reactions. The hydrolysis reaction results in the replacement of organic groups by hydroxyl groups; in the condensation reaction, hydroxyl groups condense by slitting off water, resulting in gel formation. The gel thus formed is then heated to temperatures from 250 to 500°C to consolidate the structure. When substrates having numerous hydroxyl groups on the surface (e.g., glass fibers) are used, the coating remains strongly bound to them, owing to the formation of covalent bonds as a result of the condensation of the alkoxides and such surface hydroxyl groups. Finally, as concerns self-abrasion resistance, the best results were obtained by building up the catalyst coating by the sequential application of multiple thin layers from a relatively dilute alkoxide coating solution.

In case of sintered-type filters, the intrusion of any preformed catalyst particle in the intimate structure would obviously be hampered by the relatively small pore size of the filters themselves. However, the problem of self-abrasion, typical of the fibrous filter, is not present at all. In such cases, means to promote in situ catalyst formation and deposition have to be worked out. Montanaro and Saracco [25] tested some techniques for depositing a thin layer of a γ-Al_2O_3 catalyst support layer on the pore walls of an α-Al_2O_3 granular-type porous filter, through vacuum impregnation with suitable precursors and subsequent thermal treatments. The most promising techniques were the so-called sol-gel and nitrate-urea ones. In the former, the filter was impregnated with an alumina sol, then mildly dried and treated at 400–500°C to promote the formation of the γ-Al_2O_3 layer. In the latter, a concentrated aluminum nitrate–urea solution was used to impregnate the filter, then urea was hydrolyzed at 95°C to promote in situ $Al(OH)_3$ precipitation, which was finally calcined at 500°C. Both methods have been successfully employed to prepare catalytic filters to test the abatement of gaseous pollutants [11,26–28]. In particular, after deposition of V_2O_5 onto the γ-Al_2O_3 support layer, NO_x abatement tests were carried out, with positive results [11]. Two main conclusions were drawn from these experiments concerning catalytic filter preparation: (1) by repeating the deposition cycle, higher and higher catalyst loads can be achieved at the price of higher pressure drops, which above certain loads become unacceptable due to severe pore plugging; (2) the complete absence of defects in the deposited layer could not be avoided, especially at the grain boundaries of the filter, where cracks were prone to form.

Similar results were obtained more recently in the deposition of a $V_2O_5 \cdot TiO_2$ catalyst (see Fig. 5), more active and selective for the SCR of NO_x with NH_3 than $V_2O_5 \cdot Al_2O_3$ in the same filter media. However, different method was followed in this case, one similar to that proposed by Kalinowski and Nishioka [24] for fibrous filters. The filter was impregnated with a 20% b.w. tetrapropyl-ortotitanate solution in propanol; afterwards, the impregnated filter was kept in distilled water for 2 days, thus allowing water to diffuse into the filter structure, to hydrolyze the metal–organic compound, and thus to promote $Ti(OH)_4$ precipitation. Subsequent $Ti(OH)_4$ calcination at 500°C and V_2O_5 deposition were necessary to obtain the final product.

IV. SOME APPLICATION OPPORTUNITIES FOR CATALYTIC FILTERS

Several flue gases (e.g., from coal-fired boilers, incinerators, diesel engines) are characterized by high loads of both particulates (e.g., fly ashes, soot) and gaseous pollutants (NO_x, SO_2, VOCs, CO, etc.), which need to be removed for environmental purposes. In this context, several possible applications of catalytic filters can be envisaged, some of which

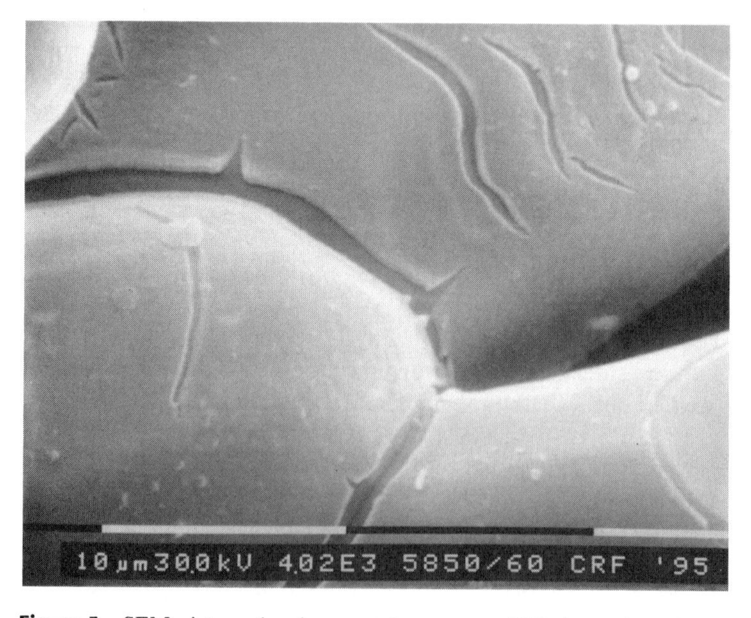

Figure 5 SEM picture showing a catalyst-support TiO_2 layer deposited on the pore walls of an α-Al_2O_3 granular filter.

have already been successfully tested experimentally, even at an industrial scale. Following is a summary of the prevalent such applications.

A. Coupling NO_x Reduction and Fly Ash Filtration

The aim in this case is to treat combustion flue gases, separating the fly ash by filtration and abating the nitrogen oxides by selective catalytic reduction with ammonia within the filter itself. Figure 6 shows, for comparison, two treatment routes for flue gases from PFBC coal boilers (NO_x: 200–1000 ppmv; SO_2: 500–2500 ppmv; O_2: 3–4%; fly ashes: 1000–10,000 ppmv), the first (Fig. 6a) based on conventional technologies (including conventional fabric filter, a wet scrubber for SO_2 removal with an alkaline solution, and

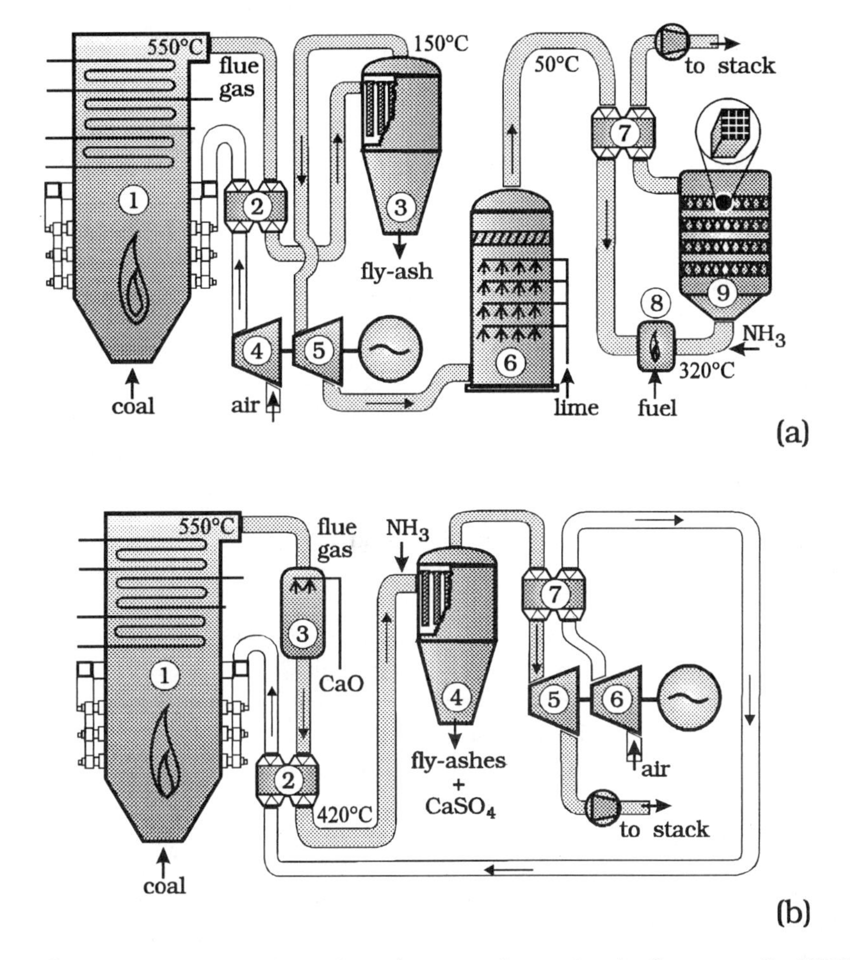

Figure 6 Schematics of two alternative routes for treating the flue gases of a PFBC coal boiler. (a) Process based on conventional technologies; (b) process employing catalytic filters. *Legend for scheme (a)*: 1. PFBC boiler; 2. air preheater; 3. fabric filter; 4. air compressor; 5. turbine; 6. SO_2 wet scrubber; 7. economizer; 8. postheater; 9. DeNOx catalytic unit. *Legend for scheme (b)*: 1. PFBC boiler; 2. air preheater; 3. SO_2 dry-scrubber; 4. catalytic filter unit; 5. turbine; 6. air compressor; 7. air preheater.

a DeNOx honeycomb reactor), the second (Fig. 6b) employing catalytic filters (SO_2 is removed by a preliminary dry-scrubbing with lime [29], leading to formation of $CaSO_4$ particles, which are then filtered together with the fly ashes by the catalytic filter).

Conventional filter bags based on polymer materials cannot withstand temperatures higher than about 200°C. This implies considerable energy consumption for the reheating of flue gases up to temperatures suitable for the catalytic converters (e.g., selective catalytic reduction of NO_x with ammonia, on dust-free flue gases, is typically performed on honeycomb structures at 320°C [30]). As already discussed, considerable space savings can be achieved if filters are catalytically activated, allowing the elimination, or at least the reduction, of the catalytic converter section of the plant.

Based on the results presented in Ref. 11, a thorough economic comparison was made for the two treatment alternatives shown in Fig. 6, by varying the operating pressure (from atmospheric to 10 bar), the boiler capacity (from 1 to 500 MWe), and pollutant concentrations (in the above-listed ranges). The overall results of such investigations are reported in Ref. 31; following are the most important conclusions.

Even considering very high engineering and contingency costs for the catalytic filters, the overall treatment costs (running plus investment) of the flue gases from PFBC coal boilers can likely be reduced by 15–30%, by varying the above process parameters.

The above margins were due primarily to the reduction of operating costs rather than of investment costs, due mostly to the absence of any energy consumption for preheating the flue gases before the SCR unit.

Other prevalent issues concern the fact that most of the heat recovery for air preheating is performed on clean flue gases, which allows for higher heat exchange coefficients and lower heat exchange surfaces.

Contrary to early expectations, the cost of the catalytic filter apparatus was higher than the sum of the fabric filters and SCR reactor costs, primarily because the cost of the baghouse (which has to work at more than 400°C and is not produced routinely, at present, but only for tailored processes) is higher than that of the catalytic filters.

On the basis of this last observation, it is easy to predict that if the catalytic filter technology can penetrate the market more deeply, investment cost will likely lower, due to higher productions, thus further improving the above economic advantages. Moreover, even better results could be accomplished if higher lime utilization efficiency were achieved in the dry-scrubbing unit, a point addressed by several research programs all over the world [32–35].

Patents were filed by Babcock & Wilcox [21–23] concerning the so-called SO_x–NO_x–Rox Box process, according to which, in line with Fig. 6b, contemporary SO_2 and NO_x removal (the former by adsorption on lime, the latter by catalytic reduction with ammonia) is accomplished by the use of catalytic filters, prepared as described in Section III. A schematic of the catalytic baghouse assembly is presented in Fig. 7. The results of the application of such technology to the treatment of a lab-scale atmospheric fluidized-bed coal boiler (capacity: 0.5 MWe) were reported in Refs. 9 and 29. The achieved abatement efficiencies were 70–80% for SO_2, 90% for NO_x(NH_3/NO ratio = 1; ammonia slippage = 10–15%), and 99% for particulate. Since March 1992 a 5-MWe demonstration project, funded by the U.S. Department of Energy and by the Ohio Coal Development Office,

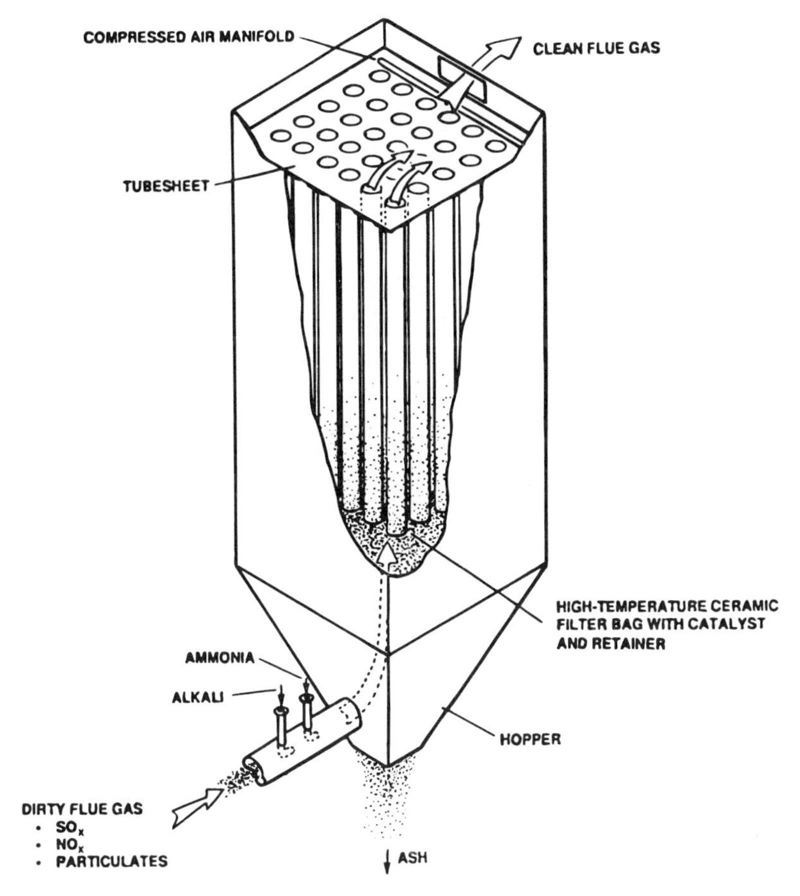

Figure 7 Schematic of the catalytic filter baghouse by Babcock & Wilcox for the SO_x–NO_x Rox Box process. (Reprinted from Ref. 9, with permission of the American Institute of Chemical Engineers. Copyright © 1992, all rights reserved.)

has been operating at the R.E. Bruger Plant of Ohio Power. Early results from this large-scale testing demonstrates that the above limits can even be exceeded.

Finally, the Owens-Corning Fiberglass Co. and the Energy & Environmental Research Center (Grand Forks, North Dakota) have been testing a similar application with a glass fiber filter, activated with a V–Ti catalyst, according to the earlier-discussed procedures of Kalinowski and Nishioka [24]. The extent of basic NO_x reduction achieved in small-scale pilot plants (about 0.2 MWe) was nearly 72% for an NH_3/NO ratio of 0.78 and an ammonia slippage of 5.6% [10]. A very interesting point treated in the experimental runs was the effect of various types of poisons (fly ashes of acidic or alkaline nature, SO_2, HCl, H_2SO_4, etc.), eventually present in the flue gases from coal burners or waste incinerators on the catalyst activity and durability. In particular, it was demonstrated that fly ashes (especially when characterized by high Na contents) can markedly deteriorate the performance of the catalytic filter, by reacting with the catalytically active principles.

A solution to this problem might eventually be found with filters allowing a modest penetration of the dust inside the filter structure. Double-layered filters such as those shown in Fig. 4 might indeed offer a solution by keeping the thin filtration layer inert and activating only the large pore support. Provided suitable preparation techniques can be found for such a product, the small-pore filtration layer would easily prevent fine fly ashes from reaching the catalytic layer. Dust penetration data in double-layered filters, reported in Ref. 13, strengthen this hypothesis.

B. Other Emerging Potential Applications

Among the other emerging opportunities for the application of catalytic filters, the treatment of diesel engine emissions is by far the most studied field. Diesel engines produce exhaust gases containing carbonaceous particulate (soot), nitrogen oxides, unburned hydrocarbons, etc. The traditional approach to soot removal is based on the use of ceramic traps. Once the pressure drop across the trap becomes unacceptable, regeneration is operated by raising the trap temperature so that the entrapped soot spontaneously burns out. However, the very high temperatures reached during the regeneration step (generally higher than 1000°C) often lead to short-term trap deterioration. A solution to the problem might be the use of filters onto which a soot-combustion catalyst is applied. The aim is to filter the soot and simultaneously promote its catalytic combustion at the same exhaust temperatures (200–400°C), thereby avoiding any periodic regeneration. Since all of Chapter 18 of this book is dedicated to this specific topic, no further details will be given here.

Another possible application lies in the treatment of the syngas produced by biomass or peat gasification processes. Such a process is the focus of intense research and development in Finland, as recently pointed out in Ref. 36. In the power range of 50–150 MWe, the integrated, gasification combined cycle (IGCC) seems to be the most attractive way to exploit gasification for power production purposes. Figure 8 shows a schematic of the IGCC cycle in which a catalytic filter unit has been included.

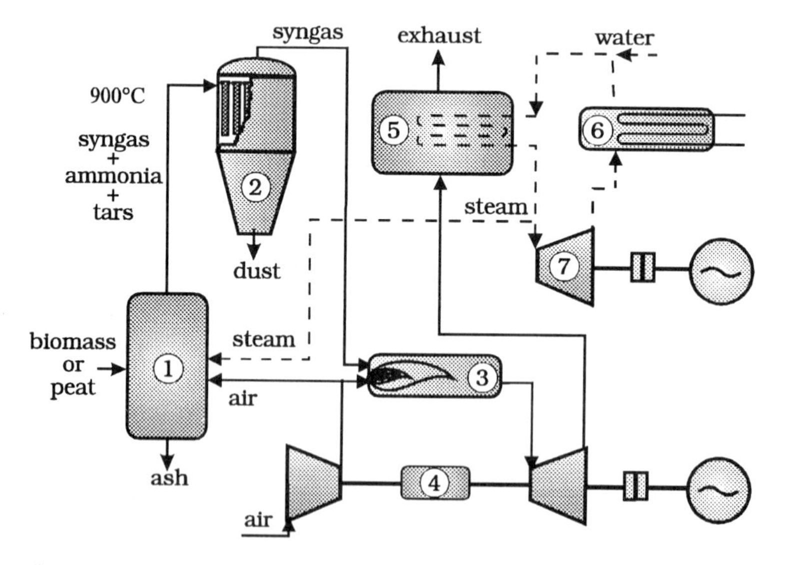

Figure 8 Potential application of catalytic filters in the IGCC cycle. *Legend*: 1. gasifier; 2. catalytic filter unit; 3. combustion chamber; 4. gas turbine-compressor setup; 5. boiler for heat recovery from exhaust gases; 6. condenser; 7. steam turbine.

In addition to the main components (N_2, CO, CH_4, CO_2, H_2O, H_2), the outlet gas stream from the fluidized-bed gasification unit contains impurities, such as dust, tar, and ammonia. Normally, dust is satisfactorily removed by means of high-temperature-resistant ceramic filters. Conversely, tar remains harmful, owing to its capability of depositing in the filter and other downstream units in the form of soot. On the other hand, ammonia produces nitrogen oxides when the gas is burned. Simell and co-workers [36] recently worked out a Ni catalyst capable of decomposing both ammonia and tar to nonharmful gases (H_2, N_2, lower-size hydrocarbons, etc.) at about 900°C (i.e., the outlet temperature of the gasifier). By applying such a catalyst onto the pore walls of the ceramic filters, a multifunctional catalytic filter could easily be attained as a more promising alternative to that which separates filtering and catalytic treatment units.

Another application could be in the treatment of the emissions from small-scale wood burners, containing pollutants such as particulate, CO, and unburned hydrocarbons (methane, naphthalene, etc.). Järås and co-workers [37] are actually studying catalysts for the total combustion of such hydrocarbons. Such a catalyst might be coupled with high-temperature ceramic filters for solving at once the entire range of pollution problems entailed by the considered emissions.

Besides the cited application opportunities of catalytic filters, surely many others can be proposed for solving specific problems of flue gas treatment.

V. SOME ENGINEERING AND MODELING ISSUES

The modeling of mass transfer and reaction in catalytic filters can be compared, in a first approximation, with the twin problem concerning honeycomb catalysts. The pores of the filters will have as counterparts the channels of the monolith, whereas the catalyst layer deposited on the pore walls of the filter will be related to the wall separating the honeycomb channels, which in general are made exclusively of catalytic material. Considering, for example, the DeNOx reaction, Fig. 9 shows schematically the NO concentration profiles within the channels/pores and the catalyst wall/layer of the two reactor configurations.

Owing to the comparatively small size of the pores (up to 100 μm, compared to a pitch of a few millimeters for the honeycomb channels) and the small thickness of the catalyst layer (a few microns, compared to some tenths of a millimeter for the catalytic wall of the honeycomb channels), both internal and external mass transfer limitations to NO conversion in catalytic filters can easily be neglected. An efficiency factor equal to unity can thus be assumed with confidence for NO reduction, contrary to honeycomb catalysts, for which this parameter is hardly higher than 0.05% at the conventional operating temperatures (320–380°C).

A second advantage of catalytic filters over honeycomb converters for the DeNOx reaction lies in the comparatively small degree of SO_2 oxidation they should allow. This last reaction has to be kept to a minimum, since the formed SO_3 would react with the ammonia slip to form ammonium sulphate deposits in the pipeline and apparatuses downstream of the NO_x converter, causing their obstruction in a relatively short time. Oxidation of SO_2 on V–Ti catalysts is a rather slow reaction compared with NO reduction, so the efficiency factor of honeycomb catalysts for this reaction is practically 1, despite the relatively thick catalyst walls. Figure 10 shows, for kinetics and operating parameters given in Ref. 38, the conversion attained for the NO reduction and the SO_2 reaction as a function of the wall's thickness of a typical DeNOx honeycomb catalyst. As expected,

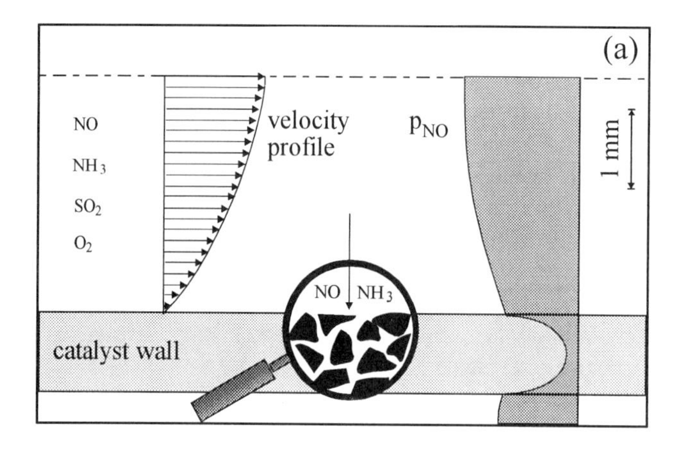

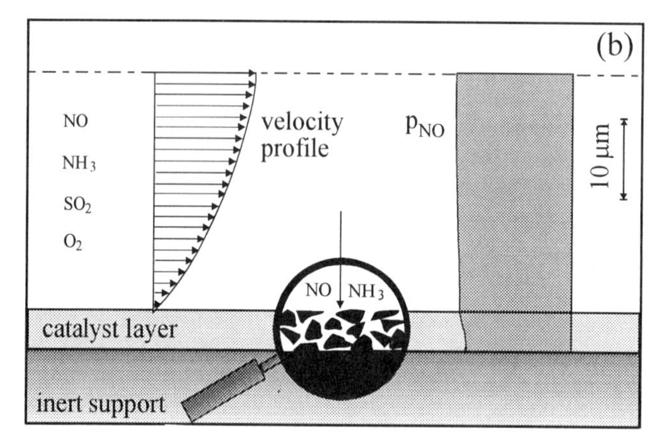

Figure 9 Mass transfer and reaction in (a) honeycomb catalysts and in (b) catalytic filters for the selective catalytic reduction of NO_x with NH_3.

SO_3 production is proportional to the wall thickness, whereas NO reduction is practically not affected above a critical, rather small wall thickness. As a consequence, one should produce honeycombs with very thin walls so as to reduce SO_3 formation without affecting NO conversion. However, the honeycomb wall thickness is limited by either mechanical resistance or manufacturing problems. The thinnest channel walls ever manufactured for the considered application are in the range of 0.5–0.6 mm [39].

In catalytic filters, since the catalyst layer is much thinner than the catalyst wall of honeycomb, SO_3 formation is lowered proportionately. However, drawbacks for catalytic filters can be found in their more complex preparation procedure, in the probably lower long-term stability of the layer of the deposited catalyst, and in their higher pressure drop. However, concerning this last point, it has to be admitted that the catalytic filter on its own does not represent an alternative to the honeycomb converter alone, but to the combination of a traditional dedusting device and the honeycomb itself. In this context, the problems related to comparatively high pressure drops might be minor, if any, and in any case more or less critical, depending on the particular application of interest.

The only modeling study on catalytic filters that has appeared in the literature [27–28] concerned a γ-Al_2O_3-deposited α-Al_2O_3 granular filter on which a model reaction

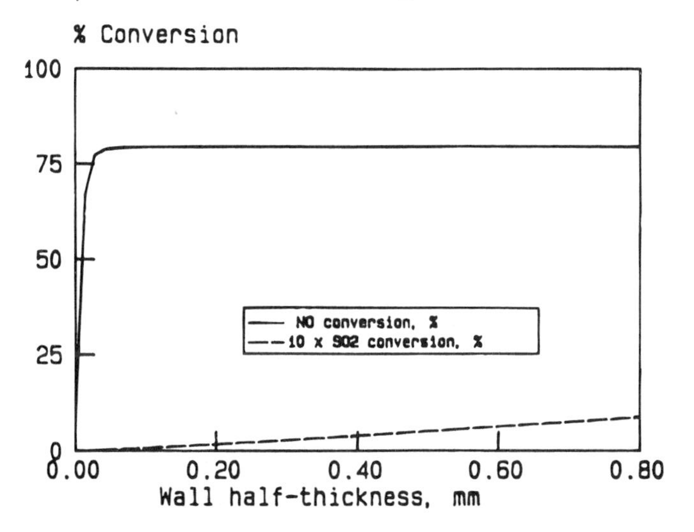

Figure 10 Typical trend of NO reduction and SO_2 oxidation in a honeycomb DeNOx catalyst as a function of the wall thickness, as emerging from calculations in Ref. 38. (Reprinted from Ref. 38, with kind permission from Elsevier Science Ltd., The Boulevard, Langford Lane, Kidlington OX5 1GB, UK. Copyright © 1994, all rights reserved.)

(2-propanol dehydration, catalyzed by the γ-Al_2O_3 itself) was performed. The major conclusion drawn by the authors was that, for those filters that underwent a single γ-Al_2O_3 deposition cycle through the aforementioned nitrate-urea method, a certain degree of catalyst bypassing should be present. In other words, they concluded that a single catalyst deposition step allows a catalyst uneven distribution throughout the filter, letting some pores have a higher catalyst load compared with others. This forced them to adopt for these filters a pseudo-homogeneous model based on a bimodal pore size distribution so as to account for the fact that those pores, which are less catalytically active, are also more permeable, thus implying lower overall conversions throughout the filter. However, a second deposition cycle already seemed to recover the above uneven-distribution defects, thus enabling the use of a simpler model based on a monomodal pore size distribution.

The above modeling study showed therefore the importance of having a proper knowledge of the pore texture and catalyst distribution of the catalytic filter, since they can seriously affect its performance. This suggests, in line with Ref. 40, the need of a proper characterization of the porous structure of the catalytic filters, concerning pore connectivity, pore size distribution, presence of deadend pores, etc., since each of these features might play a primary role in reactor performance. On the basis of such characterization work, valuable information could be drawn in order to choose or optimize the preparation routes.

VI. CONCLUSIONS

The basic properties and potentials of catalytic filters for simultaneous abatement of dust and gaseous pollutants have been reviewed. On the basis of the discussed literature, and of the growing interest arising in the producers of high-temperature filters for the considered reactor (some work is being carried out under secrecy agreement by some of such producers

[41]), it seems reasonable to predict that market penetration will probably be gained in a few years.

The extent of such penetration will largely depend on their long-term durability and on the initial investment cost of these rather new products. The work already done on catalytic filters has demonstrated that it is possible to achieve practically complete catalytic conversion of gaseous pollutants (e.g., NO_x) for superficial feed velocities of industrial interest, thus rendering the coupling of filtration and catalytic abatement convenient. The point now is to assess how the catalytic filters will resist long-term exposure to relatively harsh environments containing potential poisons for the catalyst itself (e.g., fly ash, sulfur, and/or chlorinated compounds, steam). From this viewpoint, the above-discussed recently published work [10] has probably set a path for forthcoming research. Further, preparation techniques could also be improved, always taking into account process economics, so as to get catalytic layers well stuck on the pore walls of the filters and capable of a good resistance to the mechanical stresses that arise from thermal fatigue and from the jet-pulse cleaning technique. The key to this future research lies in the hands of materials scientists.

REFERENCES

1. Agar D.W. and Ruppel W., Multifunktionale Reaktoren für die Heterogene Katalyse, *Chemie Ing. Tech. 60:*731 (1988).
2. Westerterp K.R., Multifunctional reactors, *Chem. Eng. Sci. 47* (9/11):2195 (1992).
3. Shoemaker J.D. and Jones E.M., Cumene by catalytic distillation, *Hydroc. Process* (June issue):57 (1987).
4. Smith L.A. and Huddleston M.N., New MTBE design now commercial, *Hydroc. Process* (March issue):121 (1982).
5. Matros Yu. Sh., *Catalytic Processes Under Unsteady-State Conditions.* Elsevier, Amsterdam (1989).
6. Kotter M., Lintz H.-G. and Turek T., Selective catalytic reduction of nitrogen oxide by use of the Ljungstroe air heater as reactor: A case study, *Chem. Eng. Sci. 47* (9/11):2763 (1992).
7. Lintz H.-G. and Turek T., The selective catalytic reduction of nitrogen oxides with ammonia in a catalytically active Ljungstroem heat exchanger, in *New Frontiers in Catalysis,* Guczi et al., eds, Elsevier, Amsterdam (1993).
8. Harris H.L., The S.I. natural gas engine exhaust and the converter/silencer as a system, Presented at the 12th Annual Energy-Sources Technology Conference, Houston, Texas (1989).
9. Kudlac G.A., Farthing G A., Szymasky T. and Corbett R., SNRB Catalytic baghouse laboratory pilot testing, *Environmental Progress 11:*33 (1992).
10. Ness S.R., Dunham G.E., Weber G.F. and Ludlow D.K., SCR catalyst-coated fabric filters for simultaneous NO_x and high-temperature particulate control, *Environmental Progress 14:*69 (1995).
11. Saracco G., *Gas-solid reactors based on catalytically active porous barriers.* PhD Dissertation, Turin Polytechnic, Turin, Italy (1995), in Italian.
12. Alvin M.A., Lippert T.E. and Lane J.A., Assessment of porous ceramic materials for hot gas filtration applications, *Ceramic Bulletin 70:*1491 (1991).
13. Zievers J.F., Eggerstedt P. and Zievers E.C., Porous ceramics for gas filtration, *Ceramic Bulletin 70:*108 (1991).
14. Clift R. and Seville J.P.K., eds., *Gas Cleaning at High Temperatures,* Chapman & Hall, London (1993).

15. Bergman L., The world market for hot gas media filtration: Current status and state of the art, in *Gas Cleaning at High Temperatures,* Clift R. and Seville J.P.K., eds., Chapman & Hall, London (1993), p. 294.
16. Abrams R.F. and Goldsmith R.L., Compact ceramic membrane gas filter, in *Gas Cleaning at High Temperatures,* Clift R. and Seville J.P.K., eds., Chapman & Hall, London, (1993), p. 346.
17. Gennrich T.J., High temperature ceramic fiber filter bags, in *Gas Cleaning at High Temperatures,* Clift R. and Seville J.P.K., eds., Chapman & Hall, London, (1993), p. 307.
18. Skroch R., Mayer-Schwinning G., Morgenstern U. and Weber E., Investigation into the cleaning of fiber ceramic filter elements in a high pressure hot gas dedusting pilot plant, in *Gas Cleaning at High Temperatures,* Clift R. and Seville J.P.K., eds., Chapman & Hall, London, (1993), p. 280.
19. Phillips J.N. and Dries H.W.A., Filtration of flyslag from the Shell coal gasification process using porous ceramic candles, in *Gas Cleaning at High Temperatures,* Clift R. and Seville J.P.K., eds., Chapman & Hall, London, (1993), p. 127.
20. Morrison E.D. and Federer W.D., Sol-gel derived diesel soot combustion catalysts on Nextel ceramic fiber filters. Presented at the 5th A.I.Ch.E. Meeting, Miami, FL (1992).
21. Pirsh E.A. (Babcock and Wilcox Co.), Filter house and method for simultaneously removing NO_x and particulate matter from a gas stream, U.S. Patent 4,220,633 (1980).
22. Pirsh E.A. (Babcock and Wilcox Co.), Filter house having catalytic filter bags for simultaneously removing NO_x and particulate matter from a gas stream, U.S. Patent 4,309,386 (1982).
23. Doyle J.B., Prish E.A. and Downs W. (Babcock and Wilcox Co.), Integrated injection and bag filter house system for SO_x-NO_x-particulate control with reagent/catalyst regeneration, U.S. Patent 4,793,981 (1988).
24. Kalinowski M.R. and Nishioka G.M., Method for applying porous metal oxide coatings to relatively non-porous fibrous substrates, U.S. Patent 4,732,879 (1988).
25. Montanaro L. and Saracco G., Influence of some precursors on the physico-chemical characteristics of transition aluminas for the preparation of ceramic catalytic filters, *Ceramics International 21:*43 (1995).
26. Saracco G. and Montanaro L., Catalytic ceramic filters for flue gas cleaning. I. Preparation and characterization, *Ind. Eng. Chem. Res. 34:*1471 (1995).
27. Saracco G. and Specchia V., Catalytic ceramic filters for flue gas cleaning. II. Performance and modelling thereof, *Ind. Eng. Chem. Res. 34:*1480 (1995).
28. Saracco G. and Specchia V., Studies on Sol-Gel Derived Catalytic Filters, *Chem. Eng. Sci. 50:*3385 (1995).
29. Chu P., Downs B. and Holmes B., Sorbent and ammonia injection at economiser temperatures upstream of a high-temperature baghouse, *Environmental Progress 9:*149 (1990).
30. Bosch H. and Janssen F., Catalytic reduction of nitrogen oxides. A review of the fundamentals and technology, *Catalysis Today 2:*369 (1987).
31. Chiosso F., *Treatment of flue gases by means of catalytic filters: A technical-economical analysis.* M.S. Thesis, Turin, Polytechnic, Turin, Italy (1995), in Italian.
32. Sakadata M., Shinbo T., Harano A., Yamamoto H. and Kim H.J., Removal of SO_2 from flue gas using ultrafine CaO particles, *J. Chem. Eng. Jpn. 27:*550 (1994).
33. O'Dowd W.J., Markussen J.M., Pennline H.W. and Resnik K.P., Characterisation of NO_2 and SO_2 removals in a spray dryer/baghouse system, *Ind. Eng. Chem. Res. 33:*2749 (1994).
34. Sanders J.F., Keener T.C. and Wang J., Heated fly ash/hydrated lime slurries for SO_2 removal in spray dryer absorbers, *Ind. Eng. Chem. Res. 34:*302 (1995).
35. Tsuchiai H., Ishizuka T., Ueno T., Hattori H. and Kita H., Highly active absorbent for SO_2 removal prepared from coal fly ash, *Ind. Eng. Chem. Res. 34:*1404 (1995).
36. Simell P., Kurkela E., Ståhlberg P. and Hepola J., Catalytic hot gas cleaning of gasification gas, Proceedings of the 1st World Congress on Environmental Catalysis (Pisa, Italy), Centi G. et al., eds., (1995), p. 41.

37. Carnö J., Berg M. and Järås S., Catalytic abatement of emissions from small-scale combustion of wood. A comparison of the catalytic effect in model and real flue gases, *Fuel 75:*959 (1996).
38. Tronconi E., Beretta A., Elmi A.S., Forzatti P., Malloggi S. and Baldacci A., A complete model of SCR monolith reactors for the analysis of interacting NO_x reduction and SO_2 oxidation reactions, *Chem. Eng. Sci. 49:*4277 (1994).
39. Binder-Begsteiger I., Improved emission control due to a new generation of high-void-fraction SCR-catalysts, Proceedings of the 1st World Congress on Environmental Catalysis (Pisa, Italy), Centi G. et al., eds., (1995), p. 5.
40. Mc Greavy C., Draper L. and Kam E.K.T., Methodologies for the design of reactors using structured catalysts: modelling and experimental study of diffusion and reaction in structured catalysts, *Chem. Engng Sci.* 49:5413 (1995).
41. Mader H.G. (BWF GMBH, Offingen, Germany), private communication (1994).

16

Reactors with Metal and Metal-Containing Membranes

Vladimir M. Gryaznov
Russian University of People's Friendship, Moscow, Russia

Natalia V. Orekhova
Topchiev Institute of Petrochemical Synthesis, Russian Academy of Sciences, Moscow, Russia

I. INTRODUCTION: THE ADVANTAGES OF CATALYST-MEMBRANE SYSTEMS

Catalyst-membrane systems are promising structured catalysts. The perspectives for control of heterogeneous catalytic reactions by the combination of a catalyst and a membrane selectively permeable for one of the reactants have been discussed [1]. Catalyst-membrane systems enhance reaction rate and selectivity due to the directed transfer of reactants and energy.

The simplest catalyst-membrane system is shown in Fig. 1. The membrane cross section is shaded, and the catalyst is marked by black spots. The reaction $A + B \rightarrow C$

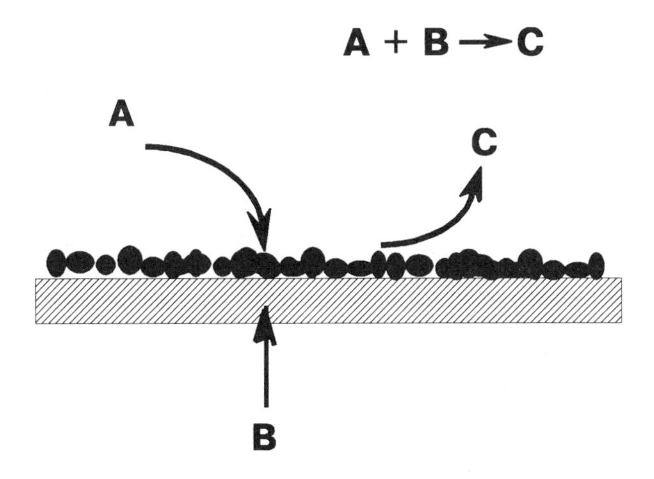

Figure 1 Schematic of the catalyst-membrane system dividing the reactor into two compartments.

may be performed in the following way: The reactant A is supplied into the upper compartment of the reactor. The second reactant, B, comes from the lower compartment, through the membrane. It provides independent control of the surface concentration of the two reactants and suppression of competing adsorption of A and B, which is inevitable on conventional catalyst and decreases the reaction rate.

A. Independent Tuning of the Surface Concentrations of Two Initial Substances

The other advantage of a catalyst-membrane system is that it maintains the desirable concentration of one reagent, for example, hydrogen, along the whole catalyst length. A small surface concentration of hydrogen is favorable to reach incomplete hydrogenation products without transformation of the initial substance into a less valuable, saturated compound. Figure 2 depicts the product $B = \eta x$, where η is the selectivity towards cyclopentadiene (CPD) into cyclopentene (CPE) hydrogenation and x is the CPD conversion degree as a function of x. Curve 1 represents the series of the runs in which the mixture of CPD vapors and hydrogen was contacted with palladium alloy foil as a conventional catalyst. Much higher B values at $x > 0.7$ were found when hydrogen was supplied through the mentioned foil to the other side of its surface, where CPD vapors were present (curve 2).

The removal of one of the reaction products through the catalyst-membrane system is a powerful tool for increasing the reaction rate and the degree of conversion of reversible reaction [2]. Table 1 shows the equilibrium extent of dehydrogenation of some hydrocarbons at atmospheric pressure and the one performed after the removal of various amounts of the hydrogen formed.

For a comparison with known commercial processes used for increasing the yield of butadiene, it is enough to indicate that the removal of 90% of the hydrogen formed in the course of butene dehydrogenation at 800 K and atmospheric pressure is equivalent to

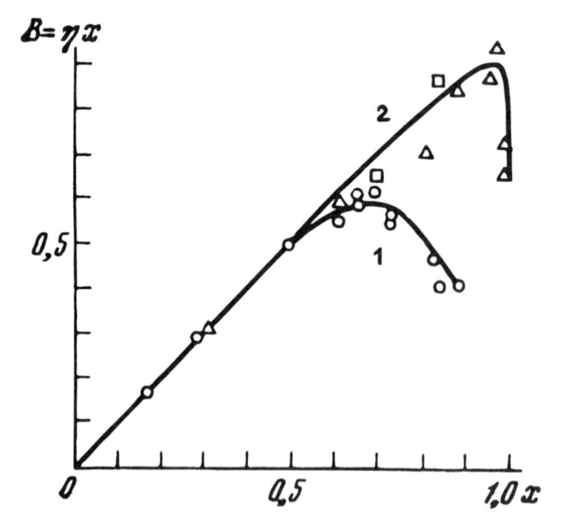

Figure 2 Dependence of cyclopentene yield B on CPD conversion degree without (curve 1) and by (curve 2) hydrogen transfer through the membrane catalyst at 343 K (○), 358 K (□), and 373 K (△).

Table 1 Equilibrium Degrees of Dehydrogenation of Some Hydrocarbons at an Overall Pressure of 1 atm and After the Removal of Various Amounts of the Hydrogen Formed

		Amount of hydrogen removed		
Reaction	Temperature, K	0	0.90	0.98
$C_2H_6 = C_2H_4 + H_2$	900	0.22	0.51	0.76
$C_3H_8 = C_3H_6 + H_2$	900	0.59	0.87	0.95
trans-butene-2 = butadiene + H_2	800	0.15	0.38	0.73
isopentene = isoprene + H_2	800	0.17	0.42	0.69
ethylbenzene = styrene + H_2	800	0.21	0.50	0.75

a 10-fold dilution of the butene with steam or to carrying out the process at an overall pressure of 0.1 atm. However, the removal of the hydrogen through a membrane catalyst does not require such energetic costs as for the generation of vacuum or butene dilution with superheated steam.

B. Coupling of Hydrogen Evolution and Consumption Reactions on the Monolithic Membrane Catalysts

Hydrogen transfer from the zone of its formation increases not only the equilibrium yields of this reaction, but its selectivity as well. For example, if, together with dehydrogenation, cracking of the initial hydrocarbon takes place, then removal of the hydrogen formed through a membrane catalyst facilitates dehydrogenation, but the cracking products remain in the same reaction space and retard the side reaction.

The catalytically active material may be present on both sides of the membrane. If the substance penetrating through the membrane is being formed on the catalyst adjacent to one surface of the membrane and is being consumed at its other surface on the second catalyst, these two reactions are coupled. Experimental evidence of such reaction coupling was found independently in the USSR [3] and in the USA [4]. Hydrogen formed during cyclohexane dehydrogenation on the inner surface of the palladium tube diffused through its wall to an outer surface, where it participated in toluene or xylene hydrodemethylation [3]. The patent [4] claimed ethane dehydrogenation on one surface of the palladium–silver alloy tube and the diffused hydrogen oxidation on its other surface. In both cases, hydrogen-porous membrane was the catalyst for two coupled reactions as well. Catalytic membranes of such type were elaborated as a result of the systematic study of hydrogen permeability and catalytic activity of different binary and ternary palladium-based alloys (see Refs. 5–11).

C. Co-current and Countercurrent Regimes

The additional tool for controlling the hydrogen concentration in a catalytic membrane reactor is a direction of the stream flows along adverse surfaces of the membrane. Figure 3 shows the effect of the directions of flow of hydrogen and CPD vapors on the degree of CPD hydrogenation occurring on a palladium alloy containing 4 wt.% of indium [12]. The rate of hydrogen transfer to the hydrogenation chamber was higher with a countercurrent flow of hydrogen–nitrogen mixture and CPD vapor–nitrogen mixture than

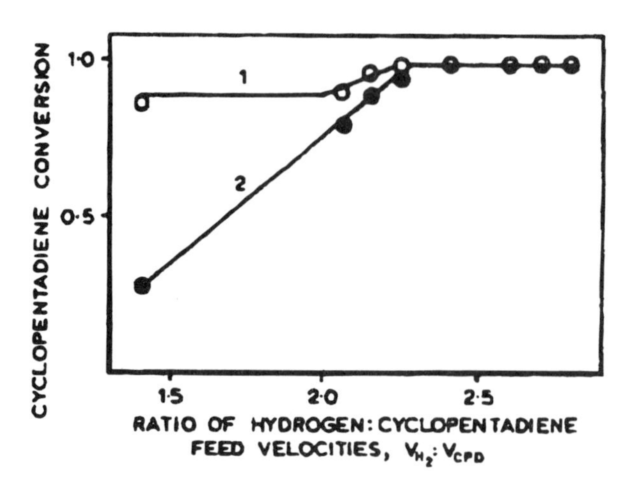

Figure 3 Dependence of cyclopentadiene conversion on the H_2/CPD ratio in the feed velocities for countercurrent (curve 1) and cocurrent (curve 2) flows.

for cocurrent flow. The CPD conversion for the countercurrent regime decreased only slightly with a decrease in the ratio of hydrogen to CPD (curve 1). At cocurrent regime, the CPD conversion drastically decreased with a decrease in this ratio (curve 2).

For the coupling reactions of terpene alcohol: borneol dehydrogenation into camphor on a copper catalyst and cyclopentadiene hydrogenation on palladium–ruthenium alloy foil, countercurrent flow again proved to be more effective than cocurrent flow [13]. As Fig. 4 depicts, the cyclopentene concentration (curve 1 in Fig. 4a) in CPD hydrogenation products is much higher for countercurrent flow and decreased more slowly in the course of the experiment than for cocurrent flow (curve 1 in Fig. 4b). The same increase in hydrogenation selectivity at countercurrent regime was observed by coupling cyclohexane dehydrogenation with 1,3-pentadiene hydrogenation [14].

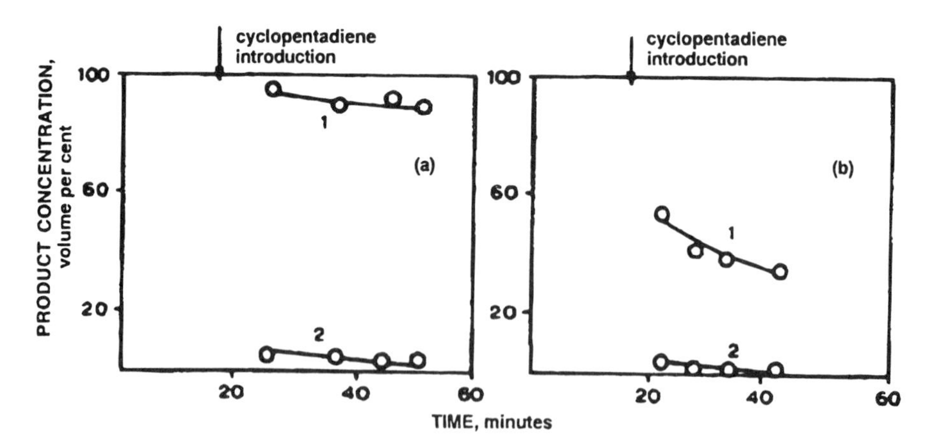

Figure 4 Time dependence of concentrations of cyclopentene (curve 1) and cyclopentane (curve 2) during coupling of CPD hydrogenation with borneol dehydrogenation on the other surface of the foil (a) countercurrent flows and (b) cocurrent flows.

II. REACTORS WITH METALLIC MEMBRANES IN THE FORM OF FOILS AND TUBES

A. Properties of Metallic Membranes

The first step in the application of nonporous metal membranes was made by T. Graham [15], who discovered palladium permeability for hydrogen only. He mentioned that

> an excellent opportunity of observing the penetration by hydrogen of a compact plate of palladium, 1 millim. in thickness, was afforded by a tube of that metal constructed by Mr. Matthey. This tube was said to have been welded from palladium near the point of fusion of the metal. The length of the tube was 115 millims., its internal diameter 12 millims., thickness 1 millim., and external surface 0.0053 of a square metre. It was closed by thick plates of platinum soldered at both ends, by which the cavity of the palladium tube could be exhausted of air.

Catalytic activation of hydrogen condensed either in the palladium sponge or in the foil was disclosed in the same paper [15]. At the ambient temperature, persalt of iron became protosalt and chlorine-water was transformed into hydrochloric acid.

The mechanism of hydrogen transfer through palladium was understood much later. It is well known now that this process starts with molecular hydrogen chemisorption on palladium surface, which is accompanied by the molecules' dissociation on atoms. The next step is the diffusion of hydrogen atoms into the palladium bulk, with the formation of a palladium–hydrogen solid solution. These atoms reach the opposite surface of the palladium membrane, diffuse to the surface, and finally recombine and desorb to the surrounding gas or liquid phase. Each one of these steps may be rate limiting for the permeation process. The main characteristic of palladium membrane productivity for hydrogen flux is the permeability constant (P), which is a product of the values of the hydrogen diffusion (D) and solubility (S) constants. Each of these constants depends on temperature.

The unique property of palladium is the opportunity to dissolve large amounts of hydrogen. The intensive study of hydrogen absorption by palladium during the last decade [16–23] revealed that besides hydrogen's occupying the interstitial sites in the bulk of the palladium lattice, there are hydrogen atoms bonded with the subsurface sites of the palladium lattice, that is, the sites between the second and third layers of the palladium lattice. According to Ref 24, hydrogen binds to subsurface sites easier than to surface ones. The calculated enthalpy of subsurface desorption for polycrystalline palladium is 32 kJ/mol [19], which is much closer to bulk hydrogen desorption enthalpy (19 kJ/mol) than to enthalpy of chemisorbed hydrogen desorption (80 kJ/mol). The palladium–silver alloy proposed by Graham [15] has some advantages in comparison with pure palladium in hydrogen permeability. This alloy, in the form of foil and thin-walled tubes, was used industrially for ultrapure hydrogen production. Now several branches of industry, such as petrochemistry, metallurgy, electronics, aviation, and astronautics, use the palladium alloy membrane for hydrogen purification because of the possibility of obtaining hydrogen of ultrapurity. The first diffusion unit with hydrogen productivity about 100,000 m^3/day was commercialized in the USA over 25 years ago and worked at a temperature of 623 K and at a pressure of 3.4 MPa. Such severe conditions were used because of the coexistence at lower temperatures of α- and β-hydride phases [25]. Both phases have the face-centered cubic texture of palladium, but the lattice constant increases at 297 K from 0.3890 nm for Pd to 0.3894 nm for α-phase and 0.4025 nm for β-phase. The maximal

H/Pd atomic ratio (n) for α-phase equals 0.008, and the minimal n for β-phase equals 0.607. The interval between these n values corresponds to the biphase field in which the increasing in n does not cause an increase in the hydrogen equilibrium pressure. With an increasing temperature, hydrogen solubility in palladium decreases and the biphase field cuts down. The mutual transformations of the two phases create strains in the material and may result in a splitting of the membrane.

The alloying of palladium with some other metals permits one to overcome the disadvantages of pure palladium and to prepare the materials with a hydrogen permeability above that of palladium. The insertion of a second and a third component into the palladium membrane may increase its mechanical strength, the hydrogen solubility, and catalytic activity of the membrane toward hydrogen dissociation. This was discussed in many original papers and reviews [26–36].

The most usable binary palladium alloy disclosed up to now is that of palladium with 25 at.% silver. Some other metals of groups IB, IV, and VIII of the periodical table and the rare earth elements have been studied as a second component of palladium alloys for hydrogen separation. Table 2 [6] lists the hydrogen permeabilities of miscellaneous binary palladium alloys at 623 K as compared with that of palladium.

Thus the palladium alloy with 53% copper proved to be more permeable than palladium [37]. However, the maximal operating temperature for membranes of this alloy is 623 K. Palladium–ruthenium alloys are more thermostable and may be used up to 823 K. At the increase in ruthenium content from 1 to 9.4 at.%, the hydrogen permeability of the alloys attained a maximum at a ruthenium content of about 4.5%. The long-term strength of this alloy at 823 K after service for 1000hr was greater by a factor of almost 5 than that of pure palladium [35].

B. Palladium and Palladium Alloys as Hydrogen-Permeable Membrane Catalysts of Hydrogenation and Dehydrogenation

The earliest applications of palladium-based membranes in catalysis were connected with the study of catalytic and electrocatalytic reaction mechanisms [38–43]. Then B. Wood [44] used a thimble fabricated from palladium/23 wt.% silver for cyclohexane dehydrogenation at 398 K to cyclohexene. This unusual product was obtained at very low hydrogen atom concentration on the membrane catalyst surface. In the complete absence of hydrogen, the dehydrogenation was not detected. Benzene was formed with longer contact of the reactants with the catalyst. Wood's conclusion that the most active surface for dehydrogenation of a hydrocarbon must sustain a concentration of sorbed hydrogen atoms that is very low, but not zero, was confirmed in isopentenes dehydrogenation [45] on the palladium/10 wt.% nickel tube at 723 K. Figure 5 shows that dehydrogenation rate raises with the increase of the hydrogen/isopentenes partial pressures ratio to 1. Further addition of hydrogen depressed the dehydrogenation rate.

During the last two decades, a number of dehydrogenation reactions with hydrogen removal through the palladium-based membranes was studied. Thus, membranes of palladium–ruthenium alloys were used for isopropanol [46] and cyclohexanol [47] dehydrogenation; the cyclohexanediol-1,2 was dehydrogenated into pyrocatechol on the membrane of palladium–7% rhodium [48]. The systems, consisting of an industrial catalyst and hydrogen-permeable palladium foil or tube, have been used for butane and butenes [49] or cyclohexane [50] dehydrogenation. In all these cases, the membranes gave advantages

Table 2 Hydrogen Permeability of Miscellaneous Binary Palladium Alloys as Compared with That of Palladium at 623 K

Material, at.%	P_{all}/P_{Pd}	Ref.	Material, at.%	P_{all}/P_{Pd}	Ref.
80.0 Pd–20.0 Ag	1.72	27	92.0 Pd–8.0 Y	5.0	34
75.0 Pd–25.0 Ag	1.73	27	98.5 Pd–1.5 La	1.41	29
70.0 Pd–30.0 Ag	1.02	27	94.3 Pd–5.7 Ce	2.06	34
48.0 Pd–52.0 Ag	0.09	27	90.0 Pd–10.0 Ce	0.89	35
75.0 Pd–25.0 Ag	2.0	32	98.5 Pd–1.5 Nd	2.15	33
97.0 Pd–3.0 Au	1.06	27	96.6 Pd–1.4 Sm	2.18	33
88.2 Pd–11.8 Au	0.96	27	92.0 Pd–1.4 Gd	4.20	33
73.5 Pd–26.5 Au	0.42	27	99.1 Pd–0.9 Sn	1.15	33
60.3 Pd–39.7 Au	0.09	27	96.2 Pd–1.8 Sn	0.61	33
84.4 Pd–15.6 Cu	0.48	27	95.5 Pd–4.5 Sn	0.35	33
56.8 Pd–41.2 Cu	0.08	27	99.5 Pd–0.5 Pb	1.69	33
41.3 Pd–52.7 Cu	1.06	27	99.0 Pd–1.0 Pb	1.85	33
42.2 Pd–57.8 Cu	0.18	27	97.4 Pd–2.6 Pb	2.28	33
32.8 Pd–67.2 Cu	0.01	27	97.1 Pd–2.9 Re	0.94	33
96.7 Pd–3.3 Cu	2.18	33	83.2 Pd–16.8 Ni	0.19	27
93.5 Pd–6.5 Cu	1.28	33	99.0 Pd–1.0 Ru	1.12	35
90.4 Pd–9.6 Cu	0.87	33	96.0 Pd–2.0 Ru	1.13	35
87.3 Pd–12.7 Cu	1.08	33	95.6 Pd–4.4 Ru	1.22	35
84.3 Pd–15.7 Cu	0.64	33	93.7 Pd–6.3 Ru	1.09	35
81.4 Pd–18.6 Cu	0.46	33	91.1 Pd–8.9 Ru	0.87	35
50.4 Pd–49.6 Cu	0.13	33	90.6 Pd–9.4 Ru	0.60	35
48.3 Pd–51.7 Cu	0.31	37	96.5 Pd–4.5 Ru	1.74	36
45.2 Pd–54.8 Cu	0.79	33	95.0 Pd–5.0 Ru	0.33	27
44.2 Pd–55.8 Cu	0.46	33	98.0 Pd–2.0 Rh	1.13	33
42.2 Pd–57.0 Cu	0.26	33	95.0 Pd–5.0 Rh	1.05	33
95.3 Pd–4.7 B	0.94	27	93.0 Pd–7.0 Rh	0.82	33
97.6 Pd–2.4 Y	1.97	33	90.0 Pd–10.0 Rh	0.59	33
94.0 Pd–6.0 Y	3.50	32	85.0 Pd–15.0 Rh	0.38	33
96.0 Pd–12.0 Y	3.76	32	80.0 Pd–20.0 Rh	0.18	33

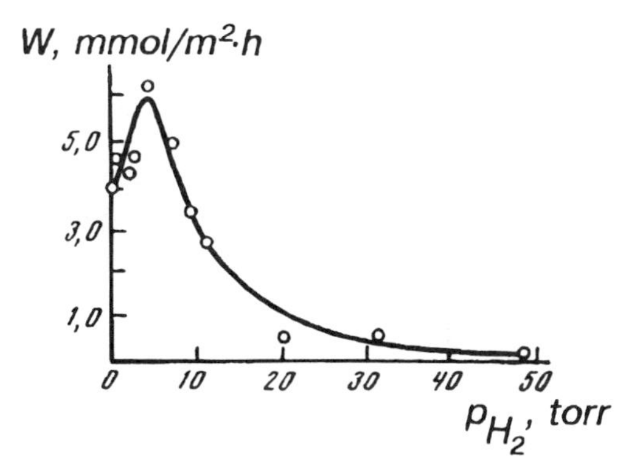

Figure 5 Dependence of isopentenes dehydrogenation rate on the ratio of hydrogen/isopentenes partial pressures P_H/P_i.

for dehydrogenation in productivity or selectivity as compared with conventional forms of catalysts.

Liquid-phase hydrogenation on membrane catalysts may be performed at low hydrogen pressure, unlike the case of conventional catalysts. The hydroquinone yield was equal to 98% when hydrogenating quinone on palladium–ruthenium alloy foils [51]. The threefold increase in the reaction rate for foil of 0.02-mm thickness in comparison with five-times-thicker foil suggests that in the second case the rate-determining step was hydrogen diffusion through the catalytic foil. A. Farkas [52] obtained similar results when comparing the para-hydrogen conversion rate on palladium disc with the rate of hydrogen diffusion through the disc.

It was found [53] that the rate of pentadiene-1,3 (PD) hydrogenation by hydrogen feeding through the palladium–ruthenium membrane catalyst was twice as larger as that for the mixture of PD and hydrogen. Butyndiol-1,4 was hydrogenated [54] into ethylenic alcohol, with the rate 2.5 mol/(m^2hr). Acetylenic alcohol with 10 C atoms was converted to corresponding ethylenic alcohol even more swiftly (4 mol/(m^2hr)). The selectivity of both processes was 99% on palladium/6 wt.% ruthenium tube. These hydrogenation rates surpass those known for the conventional catalyst and have no diffusion limitation.

The products of complete hydrogenation were found [54] above a conversion of 85% of the initial substance. Using the conventional catalyst, such as Raney nickel or palladium on carbon, the products of complete hydrogenation were formed from the very beginning of hydrogen addition. Obviously, direct hydrogenation of a $C \equiv C$ triple bond to a single bond does not take place during hydrogen introduction through the membrane catalyst when the atomic hydrogen is dominated on the catalyst surface. The addition of molecular hydrogen increased the rate of single C—C bond formation from triple bonds. An important advantage of monolithic palladium-based membrane catalysts is high selectivity and reaction rate in hydrogenation and hydrogen evolution reactions. The list of these reaction in chronological order is presented in Table 3 [38–103].

The data of Tables 2 and 3 show that palladium–ruthenium alloys with mass % of ruthenium from 4 to 7 have high hydrogen permeability, catalytic activity toward many reactions with hydrogen evolution or consumption, and good mechanical strength [35]. Seamless tubes with a wall thickness of 100 and 60 μm, as well as foils of 50-μm thickness made of the mentioned alloy, are commercially available in Russia. The tube of outer diameter of 1 mm and wall thickness of 0.1 mm is stable at a pressure drop of up to 100 atm and a temperature up to 900 K. The application of such tubes for membrane reactor will be discussed in next part of this section.

Palladium-based alloys are important, but not unique, examples of membrane catalysts. Silver is permeable to oxygen and has catalytic activity for oxidation reactions. Some of these reactions have been investigated [104,105] on silver seamless tubes. For example, oxygen introduction through this membrane catalyst permits one to obtain the transformation of 85% of methanol into formaldehyde. At the same conditions 23% of formaldehyde was produced from the mixture of methanol vapors and oxygen by using the conventional process. The products of complete oxidation were detected on the membrane catalyst at much higher temperature only. Some other examples of reactions on oxygen-porous membranes are listed in Ref. 11.

C. Reactors with Monolithic Palladium-Based Membranes

The first pilot reactor with 196 1-m-long tubes made of palladium–nickel alloy according to the patent [106] is shown on Figs. 6 and 7. Figure 6 shows the soldering of the tube

Table 3 Applications of Monolithic Membrane Catalysts Based on Palladium and Palladium Alloys

Reaction	Catalyst	Temperature, K	Ref.
Hydrogenation of quinone	Pd, foil	375	38
H_2–D_2–exchange, hydrogenation of ethylene	Pd, foil	423–573	39
Hydrogenation of ethylene	Pd, tube	293–448	40
Hydrogenation of ethylene	Pd–35 Ag, tube	327	41
Hydrogenation of acetylene	Pd, tube	373–473	42
Hydrogenation of cyclohexene	Pd–23 Ag, thimble with Au layer	343–473	43
Coupling of dehydrogenation of ethane and oxidation of hydrogen	Pd–25 Ag	725	4
Dehydrogenation of cyclohexane to cyclohexene	Pd–23 Ag	398	44
Isomerization of butene-1 and *trans*-butene-2	Pd–23 Ag, Pd–60 Au	573–603	55
Coupling of dehydrogenation of cyclohexane and demethylation of Ō-xylene	Pd, tube	703	3
Coupling of dehydrogenation of *trans*-butene-2 and demethylation of toluene or hydrogenation of benzene	Pd, Pd–20 Ag, tube	653–713	56
Hydrogenation of isoprene to 2-methyl-butene-1, 2,4-hexadiene, and 1,5-hexadiene to hexene, styrene to ethylbenzene, acrolein to propion aldehyde, and methyl vinylketone to methylethyl ketone	Pd–25 Ag, capillary	523	57
Dehydrogenation of cyclohexanol with formation of cyclohexanone and phenol	Pd–25 Ag	625	57
Hydrogenation of butene-1 and cyclohexene	Pd–23 Ag, tube with thin layer of Au	383	58
Dehydrogenation of isopentane to isopentene and isoprene	Pd–5, 5 Ni, Pd–10 Rh, Pd–10 Ru	743–870	59
Dehydrogenation of *n*-butane preferentially to 1-butene, and isobutane to isobutene	Pd–25 Ag	603	60
Hydrogenation of 1,3-butadiene and 1-butene to butane, methyl methacrylate to methyl isobutyrate, and di-*tert*-butyl ethylene to di-*tert*-butyl ethane	Pd–25 Ag, capillary tube	298–573	61
Dehydrogenation of cyclohexane	Pd–Ru (4.5;5;6;7;7.5;8.5;9)	623	35
Hydrogenation of benzene to cyclohexane and cyclohexene	Pd–5.9 Ni	373–473	62
Hydrogenation of quinone	Pd, Pd–10 Ru	403	51
Coupling of dehydrocyclization of undecane and hydrodemethylation of dimethylnaphtalene	Pd–5.9 Ni	860	63

Table 3 *Continued*

Reaction	Catalyst	Temperature, K	Ref.
Hydrogenation of cyclopentadiene	Pd–Ru (4.4;9.8), Pd–Rh (2;5)	300–510	64
Hydrogenation of furan, 2,3-dihydrofuran, silvan, and furfural	Pd–5.9 Ni	325–573	65–67
Methanol steam reforming	Pd–23 Ag, tube	500–600	31
Coupling of dehydrogenation of isoamylenes and hydrodemethylation of toluene or oxidation of hydrogen*	Pd–5.9 Ni	723	68
Hydrogenation of 2-methyl-1,4-naphtoquinone in acetic anhydride solution with subsequent etherification of 2-methyl-1,4-naphtohydroquinone to vitamin K_4^*	Pd–5.5 Ni	405–408	69
Dehydrogenation of 1,2-cyclohexanediol	Pd–5 Ti; Pd–Rh (7;15); Pd–Cu (37;39;42)	503–773	48
Dehydrogenation and hydrogenation of cyclohexene	Pd–5.9 Ni	433–573	70
Hydrogenation of butadiene	Pd	321–373	71
Dehydrogenation of isopropyl alcohol	Pd–5.5 Ni; Pd–Ru (6;8;10)	473	46
Coupling of dehydrogenation of isopropyl alcohol and hydrogenation of cyclopentadiene	Pd–10 Ru	493	72
Hydrogenation of acetylene and ethylene	Pd–5.9 Ni	293–463	73
Hydrogenation of cyclopentadiene	Pd–9.8 Ru	343–393	74
Hydrogenation of nitrobenzene	Pd–Ru (6;10)	303–473	75
Hydrogenation of naphthalene to tetraline	Pd–15 Rh	353–423	76
Decomposition of hydroiodic acid	Pd–23 Ag	>873	77
Hydrogenation of 1,3-pentadiene, isoprene, cyclopentadiene to corresponding olefines	Pd–9.8 Ru	353–473	78
Hydrogenation of ethylene	Pd	373	79
Hydrogenation of 2,4-dinitrophenol to 2,4-diaminophenol	Pd–5.5 Ru	390	80
Coupling of borneol dehydrogenation and cyclopentadiene hydrogenation	Pd–5.9 Ni; Pd–10 Ru; Pd–15 Rh	473–543	81
Dehydrocyclization of *n*-hexane	Pd–5.9 Ni, tube	793	82
Hydrogenation of carbon dioxide	Pd–Ru, foil	563–663	83
Coupling of butane dehydrogenation and hydrogen oxidation	Pd–9.8 Ru; Pd–5.5 Sn; Pd–23 Ag, foils	753–823	84
Hydrogenation of *cis,trans*-butene-1,4-diol to *cis,trans*-butanediol	Pd–Ru, foil	363	85
Hydrogenation of 2-butyne-1,4-diol to *cis,trans*-butenediol	Pd–Ru, foil	333	85

Table 3 *Continued*

Reaction	Catalyst	Temperature, K	Ref.
Hydrogenation propylene to propane and 1-butene to butane	Pd, foil	373	86
Hydrogenation of carbon monoxide	Pd–Ru;	523–673	87
Hydrogenation of acetylenic and ethylenic alcohols	Pd–5.9 Ni, Pd–Ru (4;6;8;10)	323–473	88
Hydrogenation of dehydrolinaool*	Pd–6 Ru	410	88
Hydrogenation of 1,3-cyclooctadiene, 1,5-cyclooctadiene, and cyclooctatetraene to cyclooctene	Pd–9.8 Ru, foil	353–473	89
Coupling of cyclohexanol dehydrogenation to cyclohexanone and cyclopentadiene hydrogenation to cyclopentene	Pd–9.6 Ru, foil	500–550	90
Hydrogenation of phenol to cyclohexanone	Pd–9.8 Ru, foil	400–500	90
Hydrogenation of dicyclopentadiene	Pd–9.8 Ru; Pd–15 Rh	400–410	91
Hydrogenation of benzoquinone	Pd	295	92
Hydrogenation of 1,3-butadiene	Pd	373	93
Hydrogenation of ethylene	Pd–23 Ag; Pd–7.8 Y	373–573	94,95
Cyclization of 1,3-pentadiene to cyclopentene and cyclopentane	Pd–Ru; Pd–Rh	423	96
Hydrogenation of α-methylstyrene to cumene	Pd–6 Ru; Pd–15 Rh, Pd–5.9 Ni	293–400	97
Hydrogenolysis of propane to methane and ethane	Pd–6 Ru, tube	433–533	98
Aromatization of propane	Pd	823	99
Coupling of cyclohexane dehydrogenation and 1,3-pentadiene hydrogenation	Pd–6 Ru	490	14
Photolysis of water	Pd coated with TiO_2	800	100
Hydrogenation of cyclohexene to cyclohexane; hydrogenation of cyclodecene to cyclodecane	Pd–6 Ru, tube	343	101
Hydrodesulfurization of thiophene	Pd	650	102
Dehydrogenation of cyclohexane	Amorphous Pd–Si (15;17.5;20)	423–498	103

*Reactions in pilot plants.

ends to a nickel manifold disc. It separates the total amount of the tubes into two parts for feeding the reagents down and up inside the tubes. The hydrogen evolved during a dehydrogenation or needed for hydrodemethylation goes through the tubes walls. The drawback of this model was that the high gasodynamical resistance of the lower part permits the reagents to reverse flow inside the tubes. This lower part may be omitted by means of U-shaped tubes. Such a version of the catalytic reactor was presented in the patent [106]. The constructional arrangements of the apparatus according to the invention [107] eliminates stagnant zones and provides uniform distribution of the velocity of the

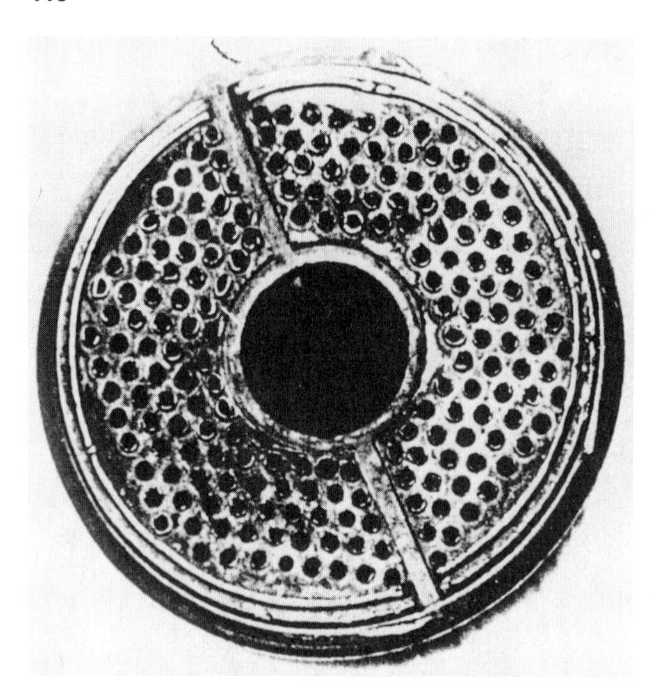

Figure 6 Nickel manifold of pilot plant reactor with welded ends of 196 palladium–nickel tubes.

Figure 7 General view of the bank of 196 palladium–nickel tubes with upper and lower manifolds according to the patent. (From Ref. 106.)

flow of reactants throughout the cross-sectional area of the apparatus, as well as uniform exposure of the external surfaces of the tubes to the substances taking part in the reaction. Figure 8 shows that the apparatus housing and the tube banks are made in the shape of a right prism, the tube banks being arranged with respect to each other so that the outlet holes of the tubes are disposed in opposite directions and staggered. This provides a snug arrangement of the tubes in the apparatus and enlarges the active surface of the tubes per unit volume of the housing.

However, this reactor is disadvantageous in that the surface area of the tubes is commensurate with the surface area of the material of the reactor shell, which may lead to undesirable side processes. A reactor that is subdivided into two compartments, A and B (Fig. 9), by a partition in the form of a thin plate coiled in a double spiral was proposed [108]. Each compartment serves for carrying out one of the reactions to be conjugated. The thin plate is made of a material that is selectively permeable to a reactant common to the reactions being conjugated and catalytically active toward both reactions. The edges

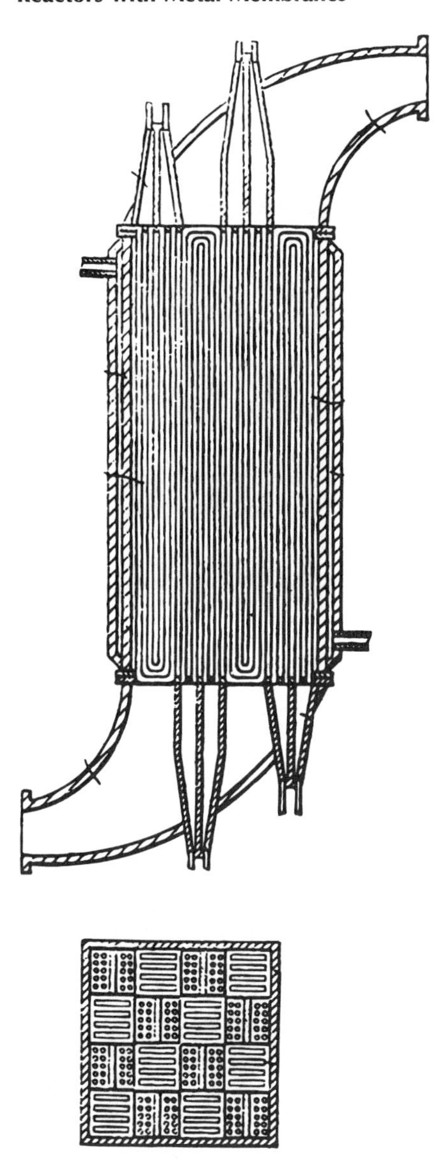

Figure 8 Vertical and horizontal section views of the catalytic membrane reactor according to the patent. (From Ref. 107.)

of this plate are built into the reactor walls. For preventing the deformation of the thin plate, corrugated sheets are inserted into both compartments of the reactor.

The other reactor for carrying out simultaneously reactions involving evolution and consumption of hydrogen [109] consists of at least two cellular foils made of palladium alloy. Each foil has alternating and oppositely directed projections arranged in rows. The ratio of the height of the projections to the foil thickness is within the range of 10:1 to 200:1. The foils are arranged so that the projections of one foil oppose the projections of the neighboring foil, and a gap is defined between the foils for passage of the starting material and for discharge of the reaction products.

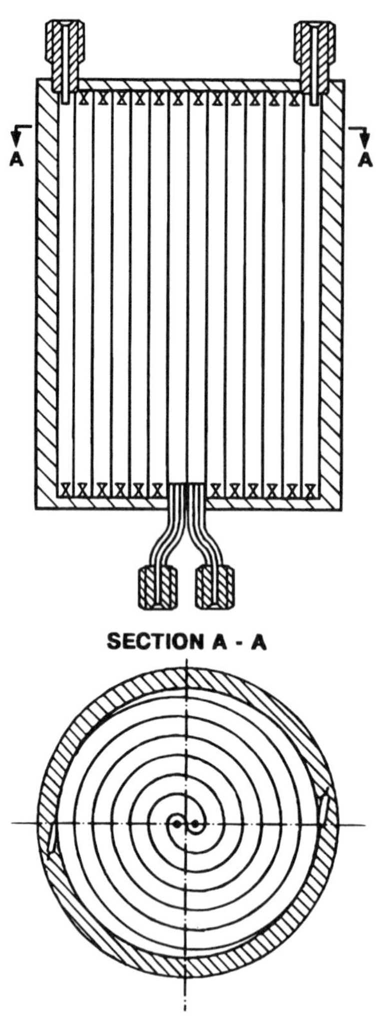

Figure 9 Longitudinal section view of the catalytic membrane reactor according to the patent and a section taken along line A–A. (From Ref. 108.)

Into the body (1) (Fig. 10) and the covers of the reactor [110], the edges of hydrogen-permeable foils (2) possessing catalytic activity with regard to both of the reactions being conjugated are hermetically sealed. The reactor space is separated into two groups of chambers (3 and 4) with corresponding inlet tubes (5 and 6). The chambers of each group are interconnected in an alternate pattern by V-shaped channels (7) formed in the side walls of the body.

These reactors were employed for coupling of isopentenes dehydrogenation into isoprene with toluene hydrodemethylation. Hydrogen evolved during the first reaction in one compartment dissolved in palladium alloy plates and penetrated through them to the other compartment, where the second reaction took place.

The increase in membrane catalyst surface per unit volume of the reactor shell was achieved [111] by means of thin-walled palladium alloy tubes in the form of plane double-

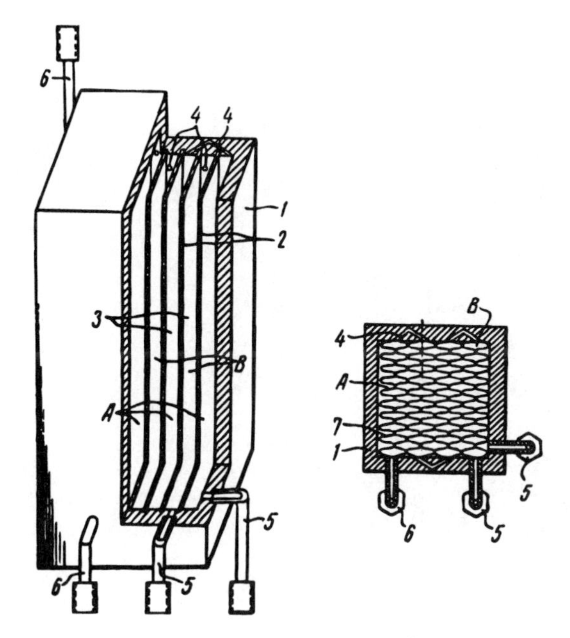

Figure 10 Catalytic membrane reactor according to the patent. (From Ref. 110.)

start spirals (Fig. 11a). The spirals are stacked one on the other, with the inlet and outlet ends of the tubes being secured in tubular headers positioned perpendicular to the plane of the spiral (Fig. 11b). This constructional arrangement enables the apparatus volume to be filled to a maximum with tubes. The spirals was compressed by two crosses (Fig. 11c) to prevent vibration of the spirals, which can damage its soldered junctions to manifolds. The blocks of spirals were mounted within the reactor shell and were joined in parallel to diminish hydraulic resistance to flow inside the tubes. A stainless steel reactor with 200 spirals with a total tubes length of 400 m was used for laboratory experiments on liquid-phase hydrogenation [10].

III. REACTORS WITH METAL-CONTAINING MEMBRANE CATALYSTS ON DIFFERENT SUPPORTS

A. Thin Films of Palladium and Palladium Alloys on Dense Other Metals and on Porous Supports

Besides the compact membrane catalysts described in Section II, there are two types of composite membrane catalyst: porous and nonporous. Composite catalyst consists of at least two layers. The first bilayered catalyst was prepared by N. Zelinsky [112], who covered zinc granules with a porous layer of palladium sponge. The sponge became saturated with the hydrogen evolved during hydrochloric acid reaction with zinc and at room temperature actively converted hydrocarbon iodates into corresponding hydrocarbons.

Modern bilayered catalysts have a mechanically strong and gas-permeable support. It may be a sheet of sintered metal powder or a tube made from a heat- and corrosion-

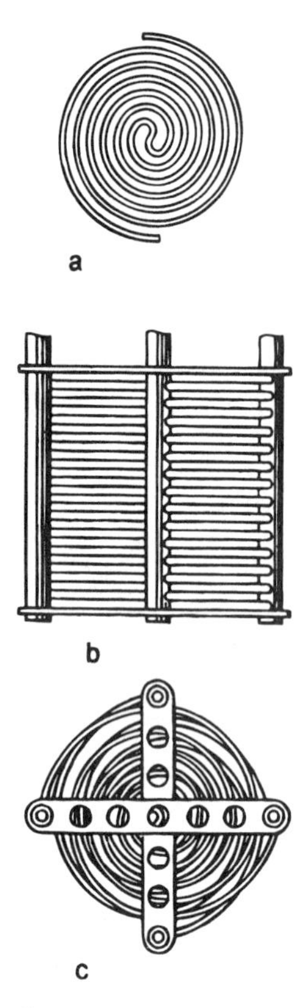

Figure 11 Thin-walled tubular membrane catalyst in the shape of (a) a double-start flat spiral, (b) a side view, and (c) a top view of the spiral block.

resistant oxide. This microporous material may not be a catalyst, but it is covered by very thin film of the catalyst, for example, palladium alloy or silver. An intermediate layer is usually used to obtain a nonporous thin film of catalytic active metal and to depress the diffusion of the catalyst and support elements.

Composite membrane with a nonporous catalyst layer has all the advantages of the compact membrane catalyst and two more. The hydrogen flow is 50 times higher, and the amount of precious metal on the unit of surface 100 times less, than for membrane catalyst in the form of foil or tube.

The methods for preparation of nonporous composite membrane catalyst are discussed in Ref. 10. The porous stainless steel sheets were covered with a dense palladium alloy film by magnetron sputtering [113] or by corolling of palladium alloy foil and porous steel sheet. The electroless plating of palladium or palladium alloy on stainless steel [114] or on porous alumina ceramic [115,116] gives the composite membranes with an ultrathin, dense palladium top layer.

Composite membrane catalysts can also be assembled with polymeric supports or intermediate layers [117–119]. These membranes were tested as membrane catalysts for selective hydrogenation of some dienic hydrocarbons and proved to be as selective as monolithic palladium alloy membranes [117]. The use of polyarilyde has been proposed in order to widen the temperature range of polymer-supported membrane application

[120]. Polyarilyde is resistant in air up to 623 K, and its hydrogen-from-nitrogen separation factor is about 100. On the base of polyarylide, asymmetric membranes have been created [120]. When such polyarylide membranes are covered with a 1-micron palladium alloy layer, they possess higher permeability for hydrogen than nonmetallized membranes at temperatures above 373 K, and are not permeable for other gases.

Composite membranes consisting of two or more different metals were claimed in the patent [121]. The thick layer of inexpensive and highly hydrogen-permeable metal (e.g., vanadium, niobium, or titanium) was coated with thin layers of chemically stable and hydrogen-permeable metal (e.g., palladium) after covering both surfaces with a thick layer with intermetallic-diffusion barriers [122]. A series of such membranes was being developed at Bend Research [123,124] for specific applications. The composite-metal membranes were made from metal foils by a hot-pressing. Prior to lamination, the oxide barriers were applied to both surfaces of the vanadium or niobium foil. The highest hydrogen flux was obtained with palladium coating layers, silicon dioxide as barriers, and vanadium as the support ($Pd/SiO_2/V$). At a temperature of 700°C, hydrogen flux was permanent after 8 days. Control composite membrane without any intermetallic-diffusion barrier reduced hydrogen permeability three times after 2 days at 700°C.

The composite membrane $Pt/SiO_2/V$ proved to be resistant to irreversible poisoning by hydrogen sulfide after 8 hours' exposure at temperatures between 700 and 800°C. Platinum coating layers were 25 μm thick. Similar palladium layers were damaged within 15 sec [124].

B. Clusters of Catalytically Active Metals in Pores of Membranes

Porous composite membrane catalysts have higher, but less selective, gas permeability than nonporous ones. Thus the membrane prepared by the metal–organic chemical vapor deposition method [125] have a hydrogen productivity as high as 0.1 mol/(m^2 sec) at temperatures of 600–800 K and a pressure of 0.1 MPa. The hydrogen-to-nitrogen selectivity factor of this membrane is about 1000 at the same conditions. The alumina mesoporous membranes with dispersed metal particles have high productivity and selectivity on hydrogen [126]. Methane conversion by steam-reforming on such membrane is two times higher than for a system with a granular catalyst. The method suggested [127] for preparation of microporous palladium membrane was to sputter palladium on the surface of a porous glass membrane, insert polymethylmethacrylate into the pores of the glass, and subsequently dissolve the glass in 20% HF.

Porous composite membrane catalysts for liquid-phase hydrogenation were prepared [128] by criochemical technology. Ultradispersed particles of binary palladium alloys were imbedded into the porous stainless steel sheet under ultrasonic treatment. The catalysts contain less than 1% of palladium. The flow reactor was divided into two chambers by the catalyst sheet. The stream of liquid 2,6-dimethylocten-2-yn-7-ol-6 was introduced into one chamber. Hydrogen under atmospheric pressure passed into the other chamber and diffused through the catalyst. The productivity of acetylenic alcohol hydrogenation into ethylenic one, i.e., linalool, was 30 times higher than by palladium alloy foil with respect to palladium weight. The most selective (96%) were palladium–manganese clusters. The reactions on different types of metal-containing composite membrane catalysts are listed in Table 4.

Table 4 Composite Metal-Containing Membrane Catalysts

Reaction	Catalyst	Temperature, K	Ref.
Hydrogenation of cyclohexene	Pd–23 Ag, thimble with Au layer	343–473	43
Hydrogenation of butene-1 and cyclohexene	Pd–23 Ag, tube with thin Au layer	383	58
Dehydrogenation and hydrogenation of cyclohexene	Pd–Ni tube, covered by thin layer of Au	473–573	70
Hydrotreating of carbon to liquid fuel*	Porous tubes from Ni and Mo powders, sulfided by H_2S	673	129
Hydrogenation of cyclopentadiene	Polydimethybiloxane with Pd complexes or films	403	130
Dehydrogenation of cyclohexane	Porous glass tube, impregnated by Pt	480	131
CO oxidation to CO_2	Anodic Al_2O_3 with Pt layer	465	132
Dehydrogenation, hydrogenolysis, and hydrogenation of ethane	Anodic Al_2O_3 with Pt layer	473	132
Water–gas shift reaction	Porous glass tube with 20-micron Pd layer and Fe–Cr oxide catalyst	673	133
Hydrogenation of cyclohexene	Pt or Pd on composite PTFE/Nafion	290–350	134
Hydrogenation of dehydrolinalool*	Pd clusters in porous stainless steel	423	128,135
Dehydrogenation of cyclohexane	Pd-impregnated Vycor glass tube	288–345	136
Hydrogenation of ethylene and decomposition of H_2S	Pt-SiO_2-V-SiO_2-Pd	1000	122–124
Methanol dehydrogenation to formaldehyde	Pd-impregnated porous Al_2O_3 with P and Cu in pores	550	137

*Reactions in pilot plants.

C. Reactors with Metal-Containing Composite Membrane Catalysts

The main advantages of reactors with composite membrane catalysts are the higher hydrogen permeability and smaller amount of precious metals in comparison with those presented in Section II. All constructions of the reactors with plane membrane catalyst may be used for composites of thin palladium alloy film and porous metal sheet. The design of reactors with composite membranes on polymeric support may be the same as for diffusion apparatus with polymeric membranes (see, for example, Ref. 138). A very promising support for the composite membrane catalysts is hollow carbon fiber [139], once properly thermostable adhesives are found.

Ceramic plates with palladium alloy may be joined to a stainless steel reactor shell by special welding. Anodized alumina plate 0.4 mm thick covered with palladium–ruthenium alloy by cathodic sputtering was sealed to the reactor body with phosphate adhesive [140]. Tubular ceramic supports may be joined with reactor modules through a multiple brazing technique [141]. Until now the junction of composite membrane catalysts to the reactor shell was not as easy and durable as for palladium alloy foils or tubes.

D. Systems of Metal-Containing Membrane and Granular Catalyst

The most general case of catalyst-membrane systems are systems containing a conventional granulated catalyst and a membrane catalyst. Two varieties of such systems are possible: (1) a pellet catalyst with a monolithic membrane or (2) a pellet catalyst with a porous (sometimes composite) membrane. The inorganic membrane reactors with or without selective permeability are discussed in Chapter 17 of this book. Examples of applications of systems of selective metal-containing membrane and granulated catalyst are presented in Table 5.

IV. CURRENT AND POTENTIAL APPLICATION OF METAL AND METAL-CONTAINING MEMBRANES FOR CATALYSIS AND SEPARATION

Metal- and alloy-containing membranes are currently applied mainly in ultrapure hydrogen production. Pilot plants with palladium alloy tubular membrane catalyst were used in Moscow for hydrogenation of acetylenic alcohols into ethylenic ones. In the Topchiev Institute of Petrochemical Synthesis, a laboratory-scale reactor of the same type was tested

Table 5 Systems of Metal-Containing Membrane and Granulated Catalysts

Reaction	Membrane	Catalyst	Temperature, K	Ref.
Dehydrogenation of butane to butadiene	Ag, tubes	Cr_2O_3–Al_2O_3	780	142
Methane steam-reforming	Pd–23 Ag, tubes	Ni	723	143
Borneol dehydrogenation to camphor	Pd–5.9 Ni, Pd–10 Ru, Pd–10 Rh, foils	Cu, wire	520	81
Butane dehydrogenation to butadiene	Pd–9.8 Ru	Cr_2O_3–Al_2O_3	723–823	49,84
Dehydrogenation of cyclohexane to benzene	Pd, tube	Pt/Al_2O_3	473	144
Dehydrocyclization of alkanes	Pd–Ag, foil	Zeolite ZSM	723–823	145
Methane steam-reforming*	Pd, foil or tubes	Ni spheres	1123	146,147
Methane steam-reforming	Pd, tube	Pt/Sn, Rh or Ni	400	148
Methane steam-reforming	Pd layer on Vycor glass tube	Ni	350–500	149
Dehydrogenation of methylcyclohexane to toluene	Pd–23 Ag, tube	Sulfided Pt	573–673	150
Ethane dehydrogenation to ethene	Pd–23 Ag tube	Pd/Al_2O_3	660	151

*Reaction in pilot plant.

for selective hydrogenation of phenylacetylene into styrene. Structured metal catalyst for the Du Pont process of directly combining hydrogen and oxygen to form hydrogen peroxide was prepared by the attachment of palladium and platinum to a stainless steel mesh using washcoating techniques [152].

Potential applications of membrane reactor catalysis were mentioned during the discussion "The future of catalysis" that was held 27–29 June 1993 at Heemskerk. J. F. Roth [153] noted that there is a very strong incentive to develop a new selective route to ethylene via selective dehydrogenation, perhaps using membrane reactor catalysis. Membrane reactor dehydrogenation could be also useful for improving production of propene or styrene. The importance of developing new materials possessing very high hydrogen transport rates was stressed. The use of 0.2-μm palladium alloy dense film supported on polycrystalline alumina [140] may solve this problem. In R. A. Bader's [154] opinion, an unfavorable ratio of the amount of wastes per unit of bioactive product may be improved by combining subsequent catalytic and/or noncatalytic steps of a given synthesis to a "one-pot reaction." Selective hydrogenation of 2-methylnaphtoquinone-1,4 into 2-methylnaphthohydroquinone-1,4 was combined with etherification of the latter with acetic anhydride because of the high corrosion resistance of palladium alloy membrane catalyst. It makes vitamin K_4 production a one-stage process, and it increases its yield to 95% [69].

J. A. Cusumano [155] forecast that by the year 2000, membrane catalytic reactors will be used on a large scale. I. Pasquon [156] said: "An important breakthrough in the field of catalysis could be originated by the development of process based on the use of membranes." B. Delmon [157] appreciated removing the product through membrane catalyst but saw no reason for applying membranes in methane oxidative coupling or methane functionalization. I. Pasquon [158] pointed out that the introduction of oxygen through a membrane can increase the selectivity but that we do not have a membrane that can resist the high temperature of methane oxidative coupling. The creation of refractory and selective composite membranes is one of the important tasks for the future.

An important advantage of metal-containing membrane catalysts is the high selectivity in the hydrogenation of diolefines and acetylenic compounds as well in hydrogen evolution reactions.

V. CONCLUSIONS

The foregoing analysis of monolithic metal-containing membrane catalysts and reactors shows that composites with very thin palladium alloy film on refractory porous support are more selective than common catalysts for partial hydrogenation reactions important for the production of vitamins, drugs, fragrances, hard fats, and other valuable substances of high purity. Some hurdles in hermetically sealing composite membrane catalysts into the reactor shell have been overcome, especially for the systems on the base of porous stainless steel sheet. The dense coating of these supports with a catalytically active film of metal or alloy was prepared at laboratory conditions, but there are possibilities for scaling up. Very durable and easily prepared monolithic reactors with tubular palladium alloy membrane catalysts of 50-μm wall thickness can be used to produce small amounts of special chemicals. Porous composite membrane catalysts obtained by ultrasonic introduction of clusters or ultradispersed metal powders into the pores of stainless steel sheets have many interesting applications. Such systems are less selective towards hydrogen flow, but the catalytic selectivity proved to be sufficiently good and the hydrogen flux high. The manufacture of the mentioned types of high-temperature membrane reactors are under development.

REFERENCES

1. V.M. Gryaznov, Catalysis by selectively permeable membranes, *Dokl. Akad. Nauk SSSR. 189*:794 (1969).
2. V.M. Gryaznov, Reactions coupling by membrane catalysts, *Kinetika i Kataliz. 12*:640 (1971).
3. V.M. Gryaznov, Sov. Pat. 274,092 (1969).
4. W.C. Pfefferle, U.S. Pat. 3,290,406 (1966).
5. V.M. Gryaznov, Hydrogen-permeable palladium membrane catalysts, *Plat. Met. Rev. 30*:68 (1986).
6. V.M. Gryaznov and N.V. Orekhova, *Catalysis by Noble Metals. Dynamic Features* (in Russian). Moscow, Nauka, 1989.
7. J.N. Armor, Catalysis with permselective inorganic membranes, *Appl. Catal. 69*:1 (1989).
8. H.P. Hsieh, Inorganic membrane reactors, *Catal. Rev.-Sci. Eng. 33*:1 (1991).
9. J. Shu, B.P.A. Grandjean, A. Van Neste, and S. Kaliaguine, Catalytic palladium-based membrane reactors. A review, *Can. J. Chem. Eng. 69*:1036 (1991).
10. V.M. Gryaznov, Platinum metals as components of catalyst-membrane systems, *Plat. Met. Rev. 36*:70 (1992).
11. G. Saracco and V. Specchia, Catalytic inorganic-membrane reactors: Present experience and future opportunities, *Catal. Rev.-Sci. Eng. 36*:305 (1994).
12. N.N. Mikhalenko, E.V. Khrapova, and V.M. Gryaznov, The effect of hydrogen and cyclopentadiene vapors flow directions along opposite surfaces of membrane catalyst on its hydrogen permeability and conversion depth of hydrogenation, *Zhurn. Fiz. Khim. 60*:511 (1986).
13. V.M. Gryaznov, M.M. Ermilova, L.S. Morozova, et al., Palladium alloys as hydrogen permeable catalysts in hydrogenation and dehydrogenation reactions, *J. Less-Comm. Met. 89*:529 (1983).
14. N.V. Orekhova, M.M. Ermilova, and V.M. Gryaznov, Combination of cyclohexane dehydrogenation and pentadiene-1,3 hydrogenation in granulated and monolithic membrane catalyst systems, *Proc. Acad. Sci. USSR. Phys. Chem. Sec. 321*:789 (1991).
15. T. Graham, On the absorption and dialytic separation of gases by colloid septa, *Phyl. Trans. Roy. Soc. 156*:399 (1866).
16. W.B. Carter, Investigation of desorption kinetics of hydrogen from the palladium (110) surface. Report, 1982. DOE/NBM-1046; *Chem. Abstr. 98*:204945v (1983).
17. M.G. Cattania, W. Penka, and R.J. Behm, Interaction of hydrogen with palladium (110) surface, *Surf. Sci. 126*:382 (1983).
18. R.J. Behm, W. Penka, and M.G. Cattania, Evidence for "subsurface" hydrogen on Pd(110). An intermediate between chemisorbed and dissolved species, *J. Chem. Phys. 78*:7486 (1983).
19. E. Wicke, Some present and future aspects of metal-hydrogen systems, *Z. Phys. Chem. 143*:1 (1985).
20. K.H. Rieder, M. Baumberger, and W. Stocker, Selectiver Ubergang von Wasserstoff von spezifischen Chemisorptionplatzen auf Pd(110) ins Innere, *Helv. phys. Acta. 57*:214 (1984).
21. M. Baumgarten, K.H. Rieder, and W. Stocker, Selektiver Ubergang von Wasserstoff von spezifischen Chemisorptionplatzen auf Pd(110) ins Innere, *Helv. phys. Acta. 59*:110 (1986).
22. K.H. Rieder and W. Stocker, Hydrogen chemisorption on Pd(100) studied with He scattering, *Surf. Sci. 148*:139 (1984).

23. B. Tardy and J.C. Bertolini, Site d'adsorption de l'hydrogene sur Pd(100) a 200 K, *C.R. Acad. sci. Ser. 2. 302*:813 (1986).
24. M. Lagos, Subsurface bonding of hydrogen to metallic surfaces, *Surf. Sci. 122*:L601(1982).
25. G.L. Holleck, Diffusion and solubility of hydrogen in palladium and palladium-silver alloys, *J. Phys. Chem. 74*:503 (1970).
26. E. Wicke and K. Meyer, Uber den Einfluss von Grenzflachenvorgangen bei der Permeation von Wasserstoff durch Palladium, *Z. Phys. Chem. N.F. 64*:225 (1969).
27. A.G. Knapton, Palladium alloys for hydrogen diffusion membranes. A review of high permeability materials, *Plat. Met. Rev. 21*:44 (1977).
28. F.A. Lewis, The palladium-hydrogen system. Pt. III. Alloy systems and hydrogen permeation, *Plat. Met. Rev. 26*:121 (1982).
29. G.J. Grashoff, C.E. Pilkington, and C.W. Corti, The purification of hydrogen, *Plat. Met. Rev. 27*:157 (1983).
30. J.E. Philpott, Hydrogen diffusion technology, *Plat. Met. Rev. 29*:12 (1985).
31. W.D. Gunter, J. Myers, and S. Girsperger, *Hydrogen: Metal Membranes, Hydrothermal Experimental Technique.* New York, 1987, p. 100.
32. D.T. Hughes and J.R. Harris, A comparative study of hydrogen permeability and solubility in some palladium solid solution alloys, *J. Less- Comm. Met. 61*:9 (1978).
33. A.P. Mischenko, *The Influence of Composition, Thermal Treating and Other Activation Methods on the Hydrogen Permeability of Some Palladium Alloys. The Metals and Alloys as Membrane Catalysts* (in Russian). Moscow, Nauka, 1981, p. 56.
34. D.T. Hughes and J.R. Harris, Hydrogen diffusion membranes based on some palladium–rare earth solution alloys, *Z. Phys. Chem. N.F. 117*:185 (1979).
35. V.M. Gryaznov, A.P. Mischenko, V.P. Polyakova et al., Palladium-ruthenium alloys as the membrane catalysts, *Dokl. Akad. Nauk SSSR 211*:624 (1973).
36. J. Cohn, U.S. Pat. 2,819,162 (1958).
37. D.L. MacKinley, U.S. Pat. 3,439,474 (1969).
38. A.R. Ubellohde, Septum hydrogenation, *J. Chem. Soc.*:No. 8, 2008 (1949).
39. V.B. Kazansky and V.V. Voevodsky, *The Study of Mechanism of Catalytic Reactions on Metallic Palladium. Physics and Physical Chemistry of Catalysis. (Problems of kinetics and catalysis. V. 10)*, Moscow. Acad. Sci. USSR. 1960, p. 398.
40. M.I. Temkin and L.O. Apelbaum, *The Application of Semipermeable Membrane for Study of Link Features of Surface Reactions. Physics and Physical Chemistry of Catalysis. (Problems of kinetics and catalysis. V. 10)*, Moscow. Acad. Sci. USSR, 1960, p. 392.
41. M. Kowaka, *J. Japan Insts. Metals 23*:655 (1959).
42. M. Kowaka and M.J. Joncich, Effect of diffusing hydrogen on the reaction over palladium, *Mem. Inst. Scient. and Industr. Res. Osaka Univ. 16*:113 (1959).
43. B.J. Wood and H. Wise, The role of adsorbed hydrogen in the catalytic hydrogenation of cyclohexene, *J. Catal. 5*:655 (1966).
44. B.J. Wood, Dehydrogenation of cyclohexane on a hydrogen-porous membrane, *J. Catal. 11*:30 (1968).
45. N.V. Orekhova, M.M. Ermilova, V.S. Smirnov, and V.M. Gryaznov, Influence of the reaction products on the rate of isoamilenes dehydrogenation by the membrane catalyst (in Russian), *Izv. Akad. Nauk SSSR. Ser. khim.*:No. 11, 2602 (1976).
46. N.N. Mikhalenko, E.N. Khrapova, and V.M. Gryaznov, Dehydrogenation of isopropanol on the membrane catalysts of binary alloys of palladium with ruthenium and nickel, *Neftekhimia 18*:189 (1978).
47. N.L. Basov, V.M. Gryaznov, and M.M. Ermilova, Dehydrogenation of cyclohexanol with removal of hydrogen through a membrane catalyst, *Russ. J. Phys. Chem. 67*:2185 (1993).

48. M.E. Sarylova, A.P. Mischenko, V.M. Gryaznov, and V.S. Smirnov, The influence of binary palladium alloys on the route of cyclohexanediol-1,2 transformations, *Izv. Akad. Nauk SSSR. Ser. khim.*:No. 2, 430 (1977).
49. N.V. Orekhova and N.A. Makhota, *The Effect of Hydrogen Separation Through the Membranes from Palladium Alloys on Dehydrogenation of Butane, Membrane Catalysts Permeable for Hydrogen or Oxygen* (in Russian), Moscow, Nauka, 1985, p. 49.
50. N. Itoh, A membrane reactor using palladium, *AIChE J. 33*:1576 (1987).
51. A.P. Maganjuk and V.M. Gryaznov, *Peculiarities of Liquid-Phase Hydrogenation by Membrane Catalysts Permeable for Hydrogen Only (in Russian), Analis Sovremennikh Zadach v Tochnikh Naukakh.* People's Friendship University Press, Moscow, 1973, p. 176.
52. A. Farkas, On the rate determing step in the diffusion of hydrogen through palladium, *Trans. Faraday Soc. 32*:1667 (1936).
53. V.M. Gryaznov, V.S. Smirnov, and M.G. Slin'ko, The development of catalysis by hydrogen-porous membranes, Proceedings of the 7th Int. Congr. on Catalysis, Tokyo, 1980, p. 224.
54. A.N. Karavanov and V.M. Gryaznov, The liquid-phase hydrogenation of acetylenic and ethylenic alcohols on the membrane catalysts from the palladium-nickel and palladium-ruthenium binary alloys, *Kinetika i Kataliz 25*:69 (1984).
55. S.H. Inami, B.J. Wood, and H. Wise, Isomerization and dehydrogenation of butene catalyzed by noble metals, *J. Catal. 13*:397 (1969).
56. V.M. Gryaznov, V.S. Smirnov, L.K. Ivanova, and A.P. Mischenko, Reactions coupling by hydrogen transfer through the catalyst (in Russian), *Dokl. Akad. Nauk SSSR. 190*:144 (1970).
57. P.G. Simmonds, G.R. Shoemake, and J.E. Lovelock, Palladium hydrogen system: Efficient interface for gas chromatography-mass-spectrometry, *Anal. Chem. 42*:881 (1970).
58. R.S. Yolles, B.J. Wood, and H. Wise, Hydrogenation of alkenes on gold, *J. Catal. 21*:66 (1971).
59. V.M. Gryaznov, V.S. Smirnov, A.P. Mischenko, and A.A. Rodina, U.S. Pat. 3,562,346 (1971).
60. B.J. Wood, U.S. Pat. 3,702,876 (1971).
61. P.G. Simmonds and C.F. Smith, Novel type of hydrogenator, *Anal. Chem. 44*:1548 (1972).
62. V.M. Gryaznov, L.F. Pavlova, and V.B. Khlebnikov, *The study of surface concentration with the help of membrane catalyst*, Mechanisms of hydrocarbon reactions symposium (1973). Budapest, Academiai Kiado, 1975, p. 107.
63. V.M. Gryaznov, V.S. Smirnov, and M.G. Slin'ko, Heterogeneous catalysis with reagent transfer through the selectively permeable catalyst, Proceedings V Intern. Congr. Catal. (J.W. Hightower, ed.). Amsterdam: North Holland, 1973. V.2, p. 1139.
64. Smirnov V.S., M.M. Ermilova, N.V. Kokoreva, and V.M. Gryaznov, Selective hydrogenation of cyclopentadiene over membrane catalysts, *Dokl. Akad. Nauk SSSR 220*:647 (1975).
65. S.A. Giller, V.M. Gryaznov, L.F. Pavlova, and L.F. Bulenkova, *Khimia Heteorcycl. Soed.*:No. 5, 599 (1975).
66. L.F. Bulenkova, V.M. Gryaznov, and Ya. F. Oshis, *Izv. Latv. Akad. Nauk. Ser. Khim.*:No. 6, 696 (1975).
67. L.F. Bulenkova, V.M. Gryaznov, and L.F. Pavlova, *Izv. Latv. Akad. Nauk. Ser. Khim.*:No. 6, 701 (1975).
68. V.S. Smirnov, V.M. Gryaznov, M.M. Ermilova, and N.V. Orekhova, The study of coupling of dehydrogenation of isoamylenes with reactions of hydrogen consumption on palladium-nickel membrane catalyst, *Dokl. Akad. Nauk SSSR 224*:391 (1975).
69. A.P. Maganyuk, V.M. Gryaznov, P.V. Kostoglodov, R.P. Evstigneeva, and I.K. Sarycheva, Sov. Pat. 540,859 (1976).
70. H. Augilar, V.M. Gryaznov, L.F. Pavlova, and V.D. Yagodovsky, Conversion of cyclohexene on goldplated palladium-nickel membrane catalyst, *React. Kinet. Catal. Lett. 7*:181 (1977).
71. H. Inoue, H. Nagamoto, and M. Shinkai, Reactor with catalytic membrane, *Asahi Garasu Kogyo Gijutsu Shokai Kenkyi Hokoku 31*:277 (1977).

72. N.N. Michalenko, E.V. Khrapova, and V.M. Gryaznov, Effect of hydrogen transfer through the membrane catalyst from the Pd–Ru alloy on the rate of isopropanol dehydrogenation and cyclopentadiene hydrogenation, *Neftekhimiya 18*:354 (1978).
73. V.M. Gryaznov, E.A. Zelyaeva, S.G. Gul'yanova, and A.P. Filippov, The conversion of acetylene and ethylene on the palladium-nickel alloy, *Izv. Vuzov. Ser. Khimiya. Khim. Tekhnol. 22*:911 (1979).
74. M.M. Ermilova, N.L. Basov, V.S. Smirnov, V.M. Gryaznov, and A.N. Rumyantsev, Dicyclopentadiene monomerization and hydrogenation of cyclopentadiene in methane flow over Pd–Ru alloy, *Izv. Akad. Nauk SSSR, Ser. khim.*: No. 8, 1773 (1979).
75. A.P. Mischenko, V.M. Gryaznov, and V.S. Smirnov, Sov. Pat. 685,661 (1979).
76. V.M. Gryaznov, V.S. Smirnov, K.M. Dyumaev, M.M. Ermilova, and N.V. Fedorova, Sov. Pat. 704,936 (1979).
77. J. Yeheskel, D. Leger, and P. Courvoisier, Thermal decomposition of hydroiodic acid and hydrogen separation, *Adv. Hydrogen Energy 1*:569 (1979).
78. V.M. Gryaznov, M.M. Ermilova, L.D. Gogua, et al., Selective hydrogenation of dienic hydrocarbons C5 on membrane catalyst from Pd–Ru alloy, *Izv. Akad. Nauk SSSR, Ser. khim.*: No. 12, 2694 (1980).
79. H. Nagamoto, H. Inoue, Mechanism of ethylene hydrogenation by hydrogen permeable palladium membrane, *J. Chem. Eng. Japan 14*:377 (1981).
80. A.P. Mischenko and V.M. Gryaznov, Ger. Pat. 3,013,799 (1981).
81. V.S. Smirnov, V.M. Gryaznov, M.M. Ermilova, N.V. Orekhova, et al., Ger. Pat. 3,003,993 (1981).
82. V.I. Lebedeva and V.M. Gryaznov, Effect of hydrogen removing through the membrane catalyst on dehydrocyclization of *n*-hexane, *Izv. Akad. Nauk SSSR, Ser. khim.*: No. 3, 611 (1981).
83. V.M. Gryaznov, S.G. Gul'yanova, Yu. M. Serov, and V.D. Yagodovski, Some features of carbon dioxide hydrogenation over palladium-ruthenium membrane catalyst with nickel coating, *Zhurn. Fiz. Khim. 55*:815 (1981).
84. N.V. Orekhova, N.A. Makhota, Dehydrogenation of alkanes and alkenes in presence of membranes from palladium alloys. *Metals and Alloys as Membrane Catalysts*, Moscow, Nauka, 1981, p. 168.
85. V.M. Gryaznov, T.M. Belosljudova, A.P. Maganjuk, A.N. Karavanov, et al, Br. Pat. Appl. 2,096,595 (1982).
86. H. Nagamoto, and H. Inoue H., *3rd Pacific Chem. Eng. Congr. 3*:205 (1983).
87. Yu. M. Serov, O.S. Gur'yanova, S.G. Gul'yanova, and V.M. Gryaznov, Hydrogenation of CO over hydrogen permeable membrane catalysts at atmospheric pressure, Chem. Synthesis Based on C1 Molecules. Abstracts of papers of Vsesouzn. Conf. Moscow, Topchiev Inst. of Petrochemical Synthesis, 1984, p. 28.
88. A.N. Karavanov and V.M. Gryaznov, Effect of hydrogen content in membrane catalyst on hydrogenation selectivity of dehydrolinalool, *Kinetika i Kataliz 25*:74 (1984).
89. M.M. Ermilova, N.V. Orekhova, L.S. Morozova, and E.V. Skakunova, Selective hydrogenation of cyclopolyolefines on membrane catalyst from palladium-ruthenium alloy, Membrane Catalysts Permeable to Hydrogen and Oxygen, Moscow, Topchiev Inst. of Petrochemical Synthesis, 1985, p. 70.
90. N.L. Basov and V.M. Gryaznov, Dehydrogenation of cyclohexanol and hydrogenation of phenol to cyclohexanon on membrane catalyst from palladium-ruthenium alloy, Membrane Catalysts Permeable to Hydrogen and Oxygen, Moscow, Topchiev Inst. of Petrochemical Synthesis, 1985, p. 117.
91. V.M. Gryaznov, M.M. Ermilova, S.I. Zavodchenko, and M.A. Gordeeva, Hydrogenation of dicyclopentadiene (tricyclo[5.2.1.02,6]-deca-3,8-diene) on membrane catalysts from palla-

dium binary alloys of ruthenium and rhodium, Materials of 8th Conf. Young Scientists of University of Peoples Friendship: Moscow, 1985, p. 167.

92. D.V. Sokolsky, B. Yu. Nogerbekov, and L.A. Fogel, Hydrogenation of benzoquinone on palladium membrane catalyst, *Izv. Akad Nauk Kaz. SSR, Ser. Khim.*: No. 6, 16 (1985).
93. H. Nagamoto and H. Inoue, The hydrogenation of 1,3-butadiene over a palladium membrane, *Bull. Chem. Soc. Jap. 59*:3935 (1986).
94. I.T. Caga, J.M. Winterbottom, and I.R. Harris, Pd-diffused membranes as ethylene hydrogenation catalyst, *Inorg. Chim. Acta 140*:53 (1987).
95. A.F. Al-Shammary, I.T. Caga, J.M. Winterbottom, A.Y. Tata, and I.R. Harris, Palladium-based diffusion membranes as catalysts in ethylene hydrogenation, *J. Chem. Technol. Biotechnol. 52*:571 (1991).
96. V.M. Gryaznov, A.P. Mischenko, and M.E. Sarylova Br. Pat. Appl. 2,187,758A (1987).
97. V.I. Lebedeva and V.M. Gryaznov, Hydrogenation of α-methylstyrene on membrane catalysts, *Izv. Akad. Nauk SSSR. Ser. khim.*: No. 5, 1018 (1988).
98. E.V. Skakunova, M.M. Ermilova, and V.M. Gryaznov, Hydrogenolysis of propane on palladium-ruthenium membrane catalysts. *Izv. Akad. Nauk SSSR, Ser. khim.*: No. 5, 858 (1988).
99. S. Uemiya, T. Matsuda, and E. Kikuchi, Aromatization of propane assisted by palladium membrane reactor, *Chem. Lett.*: No. 8, 1335 (1990).
100. M. Arai, K. Yamada, and Y. Nishiyama, Evolution and separation of hydrogen in the photolysis of water using titania-coated catalytic palladium membrane reactor, *J. Chem. Eng. Jap. 25*:761 (1992).
101. J.N. Armor and T.S. Farris, Membrane catalysis over palladium and its alloys, Proceedings of 10th Int. Congr. on Catalysis, paper O-94, 1993.
102. M. Arai, Y. Wada, and Y. Nishiyama, Thiophene hydrodesulfurization by catalytic palladium membrane systems, *Sekiyu Gakkaishi 36*:44 (1993).
103. N. Itoh, T. Machida, W.-C. Xu, and H. Kimura, Amorphous Pd–Si alloys for hydrogen-permeable and catalytically active membranes, *Catal. Today 25*:241 (1995).
104. V.M. Gryaznov, V.I. Vedernikov, and S.G. Gul'yanova, The role of oxygen diffused through the silver membrane catalyst in heterogenous oxidation, *Kinetika i Kataliz 27*:142 (1986).
105. A.G. Anshits, A.N. Shigapov, S.N. Vereshchagin, and V.N. Shevnin, C2 hydrocarbons formation from methane on silver membrane, *Catal. Today 6*:593 (1990).
106. V.M. Gryaznov et al., U.S. Pat. 3,779,711 (1973).
107. N.D. Fomin, V.M. Gryaznov, A.P. Mischenko, et al., Br. Pat. 2,056,043 (1983).
108. V.M. Gryaznov et al., U.S. Pat. 3,849,076 (1974).
109. V.M. Gryaznov et al., Br. Pat. 1,342,869 (1974).
110. V.M. Gryaznov et al., U.S. Pat. 4,014,657 (1977).
111. V.M. Gryaznov, Surface catalytic properties and hydrogen diffusion in palladium alloy membranes, *Z. Phys. Chem. N.F. 147*:123 (1986).
112. N.D. Zelinsky, Uber Reductionvorgange in Gegenwart von Palladium, *Ber. Dtsch. chem. Ges. 31*:3203 (1898).
113. V.M. Gryaznov, O.S. Serebryannikova, Yu. M. Serov, M.M. Ermilova, A.N. Karavanov, A.P. Mischenko, and N.V. Orekhova, Preparation and catalysis over palladium composite membranes, *Appl. Catal. A: General 96*:15 (1993).
114. J. Shu, B.P.A. Granjean, E. Ghali, and S. Kaliaguine, Simultaneous deposition of Pd and Ag on porous stainless steel by electroless plating, *J. Membr. Sci. 77*:181 (1993).
115. G. Yan and Q. Yuan, The preparation of ultrathin palladium membrane, Inorganic Membranes. Proceedings 2nd Int. Conf. on Inorganic Membranes, Montpellier, France, July 1991, Key Engineering Materials, 1991, v. 61–62, p. 437.
116. E. Kikuchi and S. Uemiya, Preparation of supported thin palladium–silver alloy membranes and their characteristics for hydrogen separation, *Gas Sep. Purif. 5*:261 (1991).

117. V.M. Gryaznov, V.S. Smirnov, V.M. Vdovin, et al., Br. Pat. 1,528,710 (1978).
118. R.V. Bucur and V. Mecea, The duffusivity and solubility of hydrogen in metallized polymer membranes measured by the non-equilibrium stripping potentiostatic method, *Surf. Coat. Technol. 28*:387 (1986).
119. P. Mercea, L. Muresan, V. Mecea, D. Silipas, and I. Ursu, Permeation of gases through poly(ethyleneterephthalate) membranes metallized with palladium, *J. Membr. Sci. 35*:19 (1988).
120. V.M. Gryaznov, M.M. Ermilova, S.I. Zavodchenko, and N.V. Orekhova, Hydrogen permeability of some metallopolymer membranes, *Polymer Science (Russia) 35*:365 (1993).
121. A.C. Makrides, M.A. Wright, and D.N. Jewett, U.S. Pat. 3,350,846 (1967).
122. D.J. Edlund, U.S. Pat. 5,139,541 (1992).
123. D.J. Edlund and W.A. Pledger, Thermolysis of hydrogen sulfide in a metal-membrane reactor, *J. Membr. Sci. 77*:255 (1992).
124. D.J. Edlund, D. Friesen, B. Johnson, and W.A. Pledger, Hydrogen-permeable membranes for high-temperature gas separation, *Gas Sep. Purif. 8*:131 (1994).
125. S. Yan, H. Maeda, K. Kusakabe, and S. Moro-Oka, Thin palladium membrane formed in support pores by metal–organic chemical vapor deposition method and application to hydrogen separation, *Ind. Eng. Chem. Res. 33*:616 (1994).
126. M. Chai, M. Machida, K. Eguchi, and H. Arai, Promotion of hydrogen permeation on metal-dispersed alumina membranes and its application to a membrane reactor for methane steam reforming, *Appl. Catal. A: General. 110*:239 (1994).
127. H. Masuda, K. Nishio, and N. Babe, Preparation of microporous metal membrane using two-step replication of interconnected structure of porous glass, *J. Mater. Sci. Lett. 13*:338 (1994).
128. A.N. Karavanov, V.M. Gryaznov, V.I. Lebedeva, A. Yu. Vasil'kov, and A. Yu. Olenin, Porous membrane catalysts with Pd-Me-clusters for liquid-phase hydrogenation of dehydrolinalool. Abstracts of First Int. Workshop on Catalytic Membranes. IWCM, Lyon-Villeurbanne (France), 1994, C23.
129. C. Karr and K.B. McCaskill, U.S. Pat. 4,128,473(1978).
130. V.M. Gryaznov, V.S. Smirnov, M.M. Ermilova, et al., Catalytic properties and hydrogen permeability of polydimethylsilohaxane with palladium complexes or films, Abstracts of III Intern. Conf. Surface and Colloid Sci., Stockholm, 1979, p. 369.
131. Y.-M. Sun and S.-J. Khang, Catalytic membrane for simultaneous chemical reaction and separation applied to a dehydrogenation reaction, *Ind. Eng. Chem. Res. 27*:1136 (1988).
132. R.C. Furneaux, A.P. Davidson, and M.D. Ball, Eur. Pat. Appl. 244,970 (1987).
133. E. Kikuchi, S. Uemiya, N. Sato, H. Inoue, K. Ando, and T. Matsuda, Membrane reactor for the water-gas shift reaction, *Chem. Lett.*: No. 2, 489 (1989).
134. A.M. Hodges, M. Linfon, A. W.-H. Mau, K.J. Cavell, J.A. Hey, and A.J. Seen, Perfluorated membranes as catalyst supports, *Appl. Organomet. Chem. 4*:465 (1990).
135. A. Yu. Vasil'kov, A. Yu. Olenin, V.A. Sergeev, A.N. Karavanov, E.G. Olenina. and V.M. Gryaznov, Membrane catalysts for hydrogenation with cryochemically synthesized palladium clusters, *J. Cluster Sci. 2*:117 (1991).
136. K.C. Canon and J.J. Hacskaylo, Evaluation of palladium-impregnation on the performance of Vycor glass catalytic membrane reactor, *J. Membr. Sci. 65*:259 (1992).
137. J. Deng and J. Wu, Formaldehyde production by catalytic dehydrogenation of methanol on inorganic membrane reactors, *Appl. Catal. A. 109*:63 (1994).
138. S.-T. Hhwang and K. Kammermeyer, *Membranes in Separation*, Wiley, New York, 1975.
139. V.M. Linkov and R.D. Sanderson, Carbon membrane-based catalysts for hydrogenation of CO, *Catal. Lett. 27*:97 (1994).
140. P.P. Mardilovich, P.V. Kurman, A.N. Govyadinov et. al., The gas permeability of anodic alumina membranes with palladium-ruthenium alloy layer, *Russian J. Phys. Chem.*, in press.

141. F.M. Velterop, U.S. Pat. 5,139,191 (1992).
142. A.I. De Rosset and C. Hills, U.S. Pat. 3,375,288 (1968).
143. H.J. Setzer and A.C.W. Eggen, U.S. Pat. 3,450,500 (1969).
144. N. Itoh, Membrane reactor for effective performance of reversible reactions, *Kagaku kogaku 50*:808 (1986).
145. D.M. Clayson and P. Howard, Br. Pat. Appl. 2,190,397 (1987).
146. M. Oertel, J. Schmitz, W. Weinrich, D. Jendryssek-Neuman, and R. Schulten, Hydrogen preparation by natural gas steam reforming with integrated hydrogen separation, *Chem. Eng. Technol. 10*:248 (1987).
147. J. Schmitz and H. Gerke, Membrane technology. Less hydrocarbons, more hydrogen, *Chem. Ind. (Dusseldorf) 111*:58,60 (1988).
148. A. Andersen, J.M. Dahl, K.J. Jens, E. Rytter, A. Slagtern, and A. Solbakken, Hydrogen acceptor and membrane concept for direct methane conversion, *Catal. Today 4*:389 (1989).
149. S. Uemiya, N. Sato, H. Ando, T. Matsuda, and E. Kikuchi, Promotion of methane steam reforming by use of palladium membrane, *Sekiyu Gakkaishi 33*:418 (1990).
150. K. Ali Jawad, E.J. Newson, and D.W.T. Rippin, Exceeding equilibrium conversion with a catalytic membrane reactor for the dehydrogenation of methylcyclohexane, *Chem. Eng. Sci. 49*:2129 (1994).
151. E. Gobina and R. Hughes, Ethane dehydrogenation using a high-temperature catalytic membrane reactor, *J. Membr. Sci. 90*:11 (1994).
152. M.J.H.R., *Plat. Met. Rev. 39*:65 (1995).
153. J.F. Roth, Evolving nature of industrial catalysis, *Appl. Catal. A: General 113*:131 (1994).
154. R.A. Bader, Catalysis in agrochemicals and pharmaceuticals industry. Frontiers and future developments, *Appl. Catal. A: General 113*:141 (1994).
155. J.A. Cusumano, Role of catalysis in achieving environmentally sustainable growth in 21st century, *Appl. Catal. A: General 113*:181.
156. I. Pasquon, What are the important trends in catalysis for the future? *Appl. Catal. A: General 113*:193.
157. B. Delmon, *Appl. Catal. A: General 113*:197.
158. I. Pasquon, *Appl. Catal. A: General 113*:198.

17

Inorganic-Membrane Reactors

Guido Saracco and Vito Specchia
Politecnico di Torino, Torino, Italy

I. INTRODUCTION

A membrane reactor is a particular type of multifunctional reactor where one or more chemical reactions, generally catalytically promoted, are carried out in the presence of a membrane; this last, thanks to its permselectivity, affects the course of the reactions, allowing improvements of either the achievable conversion (e.g., equilibrium reactions) or the selectivity toward intermediate products (e.g., consecutive reaction schemes).

The original idea of coupling catalysts and membranes dates back to the 1960s. Michaels [1], among the first, suggested that considerable increase in the conversion of thermodynamically limited reactions could be achieved by the use of membranes capable of being selectively permeated by one of the reaction products. Polymer membranes, which were facing rapid development in those years, thanks to the discovery of the phase-inversion preparation technique, were considered, at first, as the most promising candidates for such an application. Due to the modest temperature resistance of polymeric membranes, only low-temperature membrane reactors were considered, mostly in the biotechnological field. This research branch of membrane reactors has never stopped since then. Recent reviews are Refs. 2 and 3.

In the middle '80s, though, a significant increase of interest in membrane reactors was kindled by the opportunity to produce inorganic membranes, either metallic or ceramic, capable of withstanding as high temperatures as those typical of gas–solid catalytic processes (200–600°C). Techniques such as the sol-gel deposition of thin layers, the track-etching, the electroless plating, and the chemical vapor deposition process, among many others, reached precision and reproducibility levels that were not predictable a decade earlier. This allowed significant expansion of the potential application fields of membrane reactors, since new processes were proposed and tested for the chemical, petrochemical, and pharmaceutical industries, as well as for environmental protection purposes. As a consequence, the literature in the field has grown fast, as enlightened by many former reviews [4–12]; however, no important industrial applications of inorganic-membrane reactors (IMRs) have met with success.

Despite the unique properties of inorganic membranes vs. the rather well-established polymeric ones (see Table 1 for a comparison), issues such as membrane instability, insufficient permeability or permselectivity, or simply the unbearable costs implied still hamper the application of inorganic-membrane reactors in the process industry.

However, in view of the significant potential advantages that IMRs can guarantee, several research programs (funded by, e.g., the U.S. Department of Energy and CEE BRITE programs) are in progress nowadays, with the goal of filling the gaps remaining in practical viability. Besides, a European Science Foundation Network on Inorganic Membrane Reactors has recently been constituted by several research groups in Europe. As a consequence, thinner and defect-free permselective inorganic layers with comparatively high thermal and chemical stability and with relatively small and homogeneously dispersed pores are under constant development.

Since membrane reactors based on metal membranes have been thoroughly addressed in the previous chapter, attention hereafter is paid mostly to ceramic-membrane reactors, elucidating their basic features and application opportunities, and updating to June 1995 the literature information given in other recent reviews [11,12].

II. BASIC FEATURES OF INORGANIC-MEMBRANE REACTORS

The basic features of IMRs are summarized next, before getting deeper into details concerning specific applications.

A. Membrane Structure and Shape

First of all, inorganic membranes can be divided into two main categories: unsupported and supported (see Fig. 1). The former are also called symmetric, the latter asymmetric.

Symmetric membranes were the first ones to be produced. Typical symmetric membranes are Vycor glass or solid-electrolyte ones, whereas, in general, an asymmetric structure is preferred for any other material so as to get a proper balance among membrane permselectivity, permeability (the lower the permeability, the lower is the transmembrane flux at a given pressure difference) and mechanical strength [8]. In fact, an inorganic

Table 1 Ceramic Membranes vs. Polymeric Membranes

Pros	Cons
· Long-term stability at high temperatures	· High capital costs
· Resistance to harsh environments	· Brittleness
· Resistance to high pressure drops (>30 bar)	· Low membrane surface per unit volume in modules
· Inertness to microbial degradation	· Difficulty achieving high selectivities on large-scale membranes
· Easy cleanability after fouling	· Generally low permeability of these selective membranes
· Easy catalytic activation	· Difficult membrane-to-module sealing at high temperatures

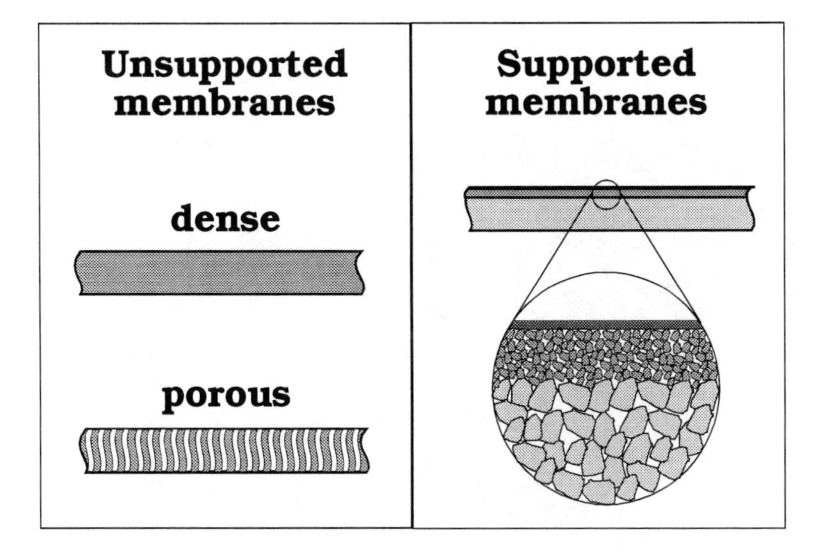

Figure 1 Basic inorganic-membrane structures.

membrane should possess a certain mechanical resistance for practical application; in other words, it should be self-supporting. As described later, this last need entails a certain membrane thickness that, for symmetric membranes, simply implies too low permeabilities. Hence, the idea of assigning to a porous support the structural resistance of the product, and to a thin permselective membrane layer its separation properties.

The structure of supported membranes is generally made of two or three supporting porous layers plus a permselective top layer. As represented in Fig. 1b, the supporting layers possess a decreasing average pore size as long as the permselective layer is approached. This is done to minimize the overall pressure drops, with the obvious constraint that thin permselective layers cannot be supported directly on large-pore-size supports; otherwise, massive formation of defects such as cracks or pinholes would take place [13]. Figure 2, for example, shows a SEM micrograph of a γ-Al_2O_3 membrane deposited on a double-layer α-Al_2O_3 support structure.

The membrane geometry can be either flat or tubular; this would obviously entail different reactor configurations at an industrial scale (see Fig. 3). Flat membranes can easily be stacked onto one another (Fig. 3a) by interposition of corrugated plates. However, modules based on flat membranes cannot guarantee, at present, a membrane surface per unit volume higher than 30 m^2/m^3, which is quite a low value compared to those attainable with polymer-membrane modules (up to 1000 m^2/m^3), thanks to their capability of being assembled into spiral-wound or hollow-fiber modules [13]. Dealing with inorganic membranes, these last module configurations have been attempted only for Pd alloy self-supporting structures [14], at the price of a very low membrane permeability. A flat membrane system was employed at the Jet Propulsion Laboratory (California Institute of Technology, Pasadena) for the purification of O_2 via solid-electrolyte membranes ($Y_2O_3 \cdot ZrO_2$) at high temperatures [15]. However, oxygen losses due to imperfect sealing (obtained through precision grinding) between membranes and module was the main cause for the failure of this O_2 production system compared with other technologies. Sealing

Figure 2 SEM micrograph of a double-layered α-Al_2O_3 inorganic-membrane support with a γ-Al_2O_3 top layer. (Courtesy of S.C.T., Tarbes, France.)

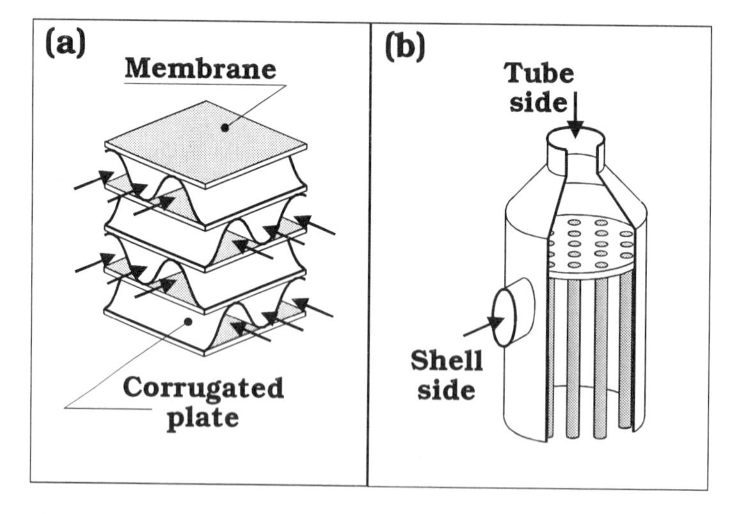

Figure 3 Possible module configurations for IMRs. (a) Flat membranes; (b) tubular membranes.

ceramic membranes into high-temperature-resistant modules is indeed a major technological problem, owing to the thermal mismatch between the two counterparts to be joined. This topic has been thoroughly addressed in Ref. 11.

Shell-and-tube modules (Fig. 3b) seem to be more promising than flat-membrane ones, since they can develop up to 250 m^2/m^3 [13]. Indeed, most of the recent literature on membrane reactors concerned tubular membranes. The lower the tube diameter, the higher the specific surface areas attainable. However attempts to manufacture hollow-fiber supported ceramic membranes were not completely satisfactory owing to the unacceptable brittleness of the obtained membranes from the practical application viewpoint [16–18].

B. Flow Patterns

Figure 4 shows the different ways an IMR can be operated concerning the flow patterns at opposite membrane sides. Countercurrent or cocurrent flows are typical of tubular membranes, which, as previously emphasized, likely represent the best shape for industrial-scale applications. A comparative study of these two flow patterns was done by Mohan and Govind [19–21] in an experimental and modeling study concerning the conversion increase of equilibrium-limited reactions in membrane reactors. In particular, they pointed out that, despite the fact that countercurrent flow gives a better distribution along the membrane of the driving force for reactant transport, it can be disadvantageous in comparison to cocurrent flow when the residence time of the gases flowing at opposite sides of the membrane is rather low. In such conditions, a certain back diffusion of the products can take place at the membrane end where reactants are fed (product concentration practically zero) and the sweep gas is discharged (high product concentration). The same authors stress how the choice of the correct flow pattern is nonetheless not a simple matter, which strongly depends on the particular reaction of interest. To strengthen this concept they report that if cyclohexane dehydrogenation could be driven to higher conversions with a countercurrent flow pattern, a cocurrently operated reactor gave better results for propylene disproportionation. We will get into greater detail concerning this point in the modeling section of this chapter (Section VI).

The stirred-chambers setup, adopted almost exclusively with flat membranes and operated either in a batch or in a continuous way, is particularly suitable for diffusion

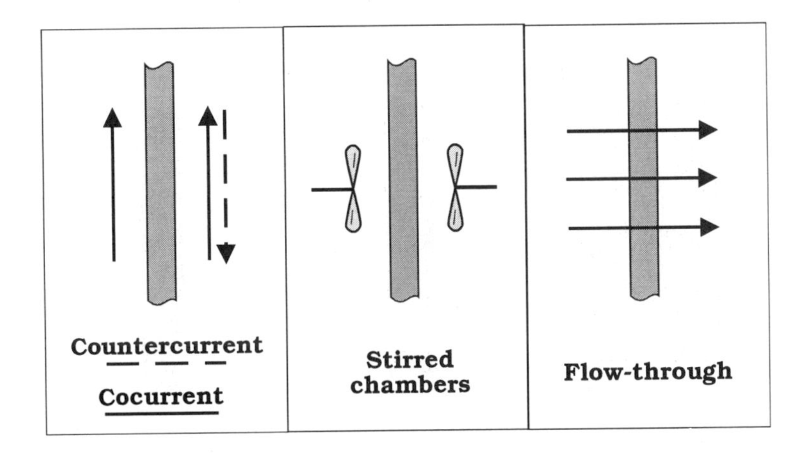

Figure 4 Flow patterns for IMRs.

and reaction tests [22], due to the very controlled boundary conditions it ensures, but it is probably of minor interest for practical membrane reactor applications. Finally, the flowthrough setup has sometimes been proposed for certain nonseparative applications [23].

C. Coupling Catalysts and Membranes

In a catalytic-membrane reactor, membranes and catalysts have to be combined. Figure 5 shows how this combination can be done. Adris and co-workers [24] proposed and tested a fluidized-bed membrane reactor for the steam-reforming of methane (Fig. 5a). The membrane, permselective toward hydrogen, is immersed in a fluidized bed of catalyst pellets. This setup allows the coupling of the most typical properties of fluidized-bed reactors (good degree of mixing, high heat transfer coefficients, etc.) with the separation properties of the membrane. Vacuum is applied to extract the permeating compound throughout the membrane. A possible drawback to this system might lie in the strong abrasion that the membrane has to face.

The coupling of a permselective membrane with a packed bed of catalyst pellets (Fig. 5b) has been one of the most widely studied membrane reactor setups. Generally, the catalyst fixed bed is enclosed on the tube side of a porous membrane, although several cases can be found in the literature in which permselective tubular membranes have been inserted at regularly spaced intervals into the packed bed of catalyst pellets (e.g., Ref. 25). The most interesting property of this membrane reactor type is that the amount of catalyst and the membrane surface area can be varied almost independently within wide ranges, so as to optimize the coupling of reaction and separation.

According to the third catalyst-membrane coupling possibility, represented in Fig. 5c, the surface of the membrane is deposited with some catalytic material. This setup is typical of solid-electrolyte membranes, where the catalyst is also playing the role of the electrode, necessary to drive the permeation of ions throughout the membrane at a desired rate. Problems may arise here concerning the fact that the catalyst per unit membrane surface is limited to some extent, and that several catalytic materials (e.g., metal oxides) are poor electricity conductors [26].

In the last type, the membrane itself, supported or unsupported, is catalytically active (Fig. 5d). This can be either due to the fact that the constituent material is intrinsically

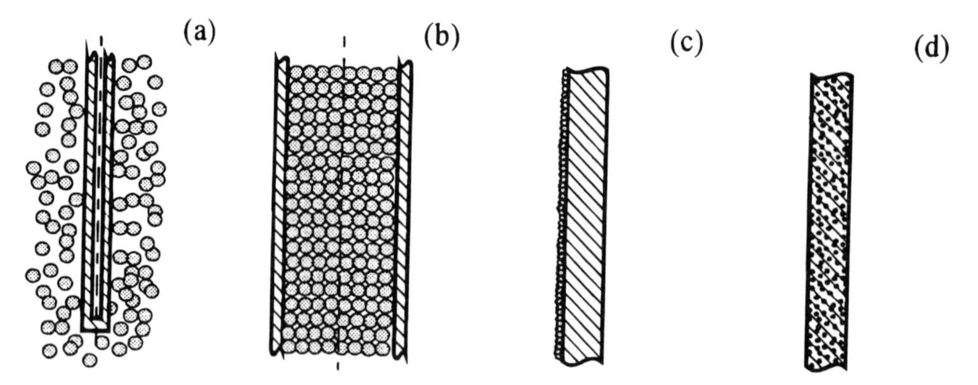

Figure 5 Coupling catalysts and inorganic membranes. (a) Fluidized-bed IMR; (b) packed-bed IMR; (c) catalyst-deposited membrane reactor; (d) catalytically active IMR.

active (e.g., Pd-alloy membranes [14]; perovskite membranes [27]) or because it has been deposited, through suitable techniques, with catalytic materials [28]. The limited amount of catalyst present inside the membrane generally corresponds to rather low overall catalyst loads in the reactor and to low specific productivity, also considering the previously quoted difficulty in getting high surface areas per unit volume in inorganic-membrane modules. However, for those applications, described later on, in which the reaction has to be confined within the membrane (e.g., separate feed of reactants [29–31]), the above disadvantages can be tolerated in view of other peculiar properties of these membrane reactors (e.g., high selectivity toward intermediate reaction products and high heat removal efficiency).

D. Major Application Opportunities

Two major application fields can be envisaged for inorganic-membrane reactors, depending on the fact that membrane permselectivity is essential (separative applications) or not (nonseparative applications). Separative applications are those in which membrane permselectivity is an essential property of the reactor. Figure 6 presents the most interesting applications belonging to this category. The most widely studied application by far is the selective removal of one of the reaction products, which permeates through the membrane, so that an *increase of the per-pass conversion of equilibrium-limited reactions* can be attained. The equilibrium constant is obviously not affected, but the product is simply removed from further contact with the catalyst, thus hindering the reverse reaction. The conversion enhancement is limited though by the permeability of the reactants, unless a very selective membrane is employed. A combined advantage of such desired high permselectivity lies in the fact that the reaction product passing through the membrane can be recovered in pure form.

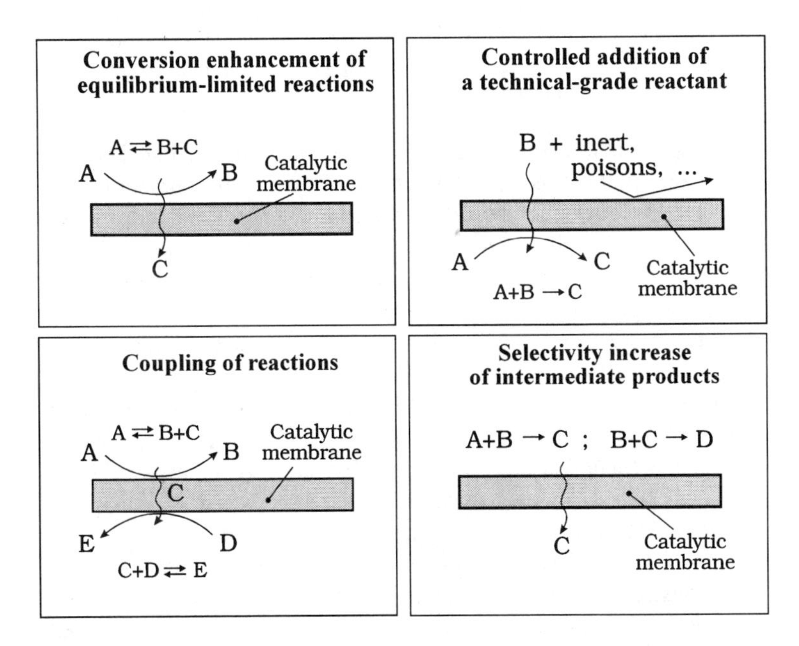

Figure 6 Potential application opportunities of IMRs with permselective membranes.

Dehydrogenations have been the most widely studied reactions in this field. The most likely reason why is the small dimension of the hydrogen molecule compared with that of the hydrocarbons (e.g., cyclohexane, ethylbenzene, propane) to be dehydrogenated, allowing certain permselectivities throughout porous membranes. Further, hydrogen selectively permeates also through some interesting materials for dense membrane (Pd alloys, SiO_2, proton-conductive solid electrolytes, etc.).

Consider, for instance, ethylbenzene dehydrogenation to styrene. The traditional plant used in the process industry [32] is based on an fixed-bed catalytic reactor to which a preheated mixture of ethylbenzene and steam, which prevents coke formation, is fed. The reaction products then normally undergo a rather complex separation scheme, mostly based on distillation columns, aimed at recovering styrene (the desired product), benzene, toluene and H_2 (by products), and a certain amount of unconverted ethylbenzene, which has to be recycled. The overall conversion per pass is typically around 60%, whereas selectivity is close to 90%.

If a membrane reactor were used employing a completely permselective membrane, total conversion might be achieved, thus eliminating the need for ethylbenzene recycle and the associated operating costs, recovering pure hydrogen at one side of the membrane. From a different viewpoint, equal conversions would be obtained at lower temperatures, with a potential benefit for the reaction selectivity (parasite reactions such as coke formation are more and more severe at increasing temperature), or at higher mean pressures, thus reducing the reactor volumes.

Abdalla and Elnashaie [33] stated that styrene yield could easily be 20% higher than in traditional plants if permselective membrane were to be used. From an economic viewpoint, these authors also calculated that

> assuming 330 working days a year and the value of $600 for a metric ton of styrene (*Chemical Marketing Report*, 1992), 1% improvement in the styrene production corresponds to a dollar value of about $376,200 per year for Polymer Corporation, Sarnia, Ontario, Canada (production rate of 190 MTPD) and $1,683,000 per year for the Saudi Petrochemical Company (SADAF) of the Saudi Basic Industries Corporation (SABIC), Saudi Arabia (production rate of 850 MTPD).

However, if the membrane is not permselective enough, all the above-listed advantages would be markedly reduced, the most upsetting drawbacks being the persisting need for reactant recycle (conversion could not be driven to completeness) and the missed simplification of the separation section. It appears clear therefore that, at least for the circumvention of chemical equilibria with inorganic-membrane reactors, membrane permselectivity is a prevalent property, which has to be coupled with thermochemical stability and a sufficiently high permeability. The synthesis of such membranes is the most difficult challenge left for materials scientists, as discussed later.

A second interesting application field of permselective membrane reactors is that of the so-called coupling of reactions, proposed first by Gryaznov [34], for the contemporary handling of a dehydrogenation (endothermic) and a hydrogenation (exothermic) at the two sides of a Pd membrane permeated by hydrogen. This author stated that coupling can take place at three levels: *energetic* (the heat generated by the exothermic reaction supports the endothermic one), *thermodynamic* (both reactions are driven to higher conversions than the equilibrium ones), and *kinetic* (typical only of Pd membranes, which enhance the reaction kinetics owing to the monatomic form of the hydrogen transferred by the membrane). In some cases a primary reaction is coupled to another one, whose major aim is that of reacting away as soon as possible the permeating species, thereby increasing the permeation flux (e.g., $H_2 + {}^1/_2O_2 \rightarrow H_2O$; [35]).

It has to be underlined, though, that reaction coupling reduces the number of degrees of freedom on which one can play for controlling the operation. In some cases one may have to feed an additional quantity of the key reactant (e.g., H_2) directly to the gas phase at one side of the membrane (e.g., the hydrogenation one) so as to offset thermal (heat of reaction) or stoichiometric deficiencies of the system [36].

Another interesting application opportunity for membrane reactors is supplying gradually along the reactor, through the membrane, one of the reactants. If the membrane is permselective toward the key reactant, this one can be fed at a technical grade. The inert gases or the potential catalyst poisons eventually present in the feed would be rejected by the membrane, thus preventing them from reaching the location where the reaction actually takes place. By this means pure oxygen can be driven to the catalyst of, e.g., partial-oxidation-reaction reactors [37,38] starting from air.

Moreover, the *controlled addition of a reactant* along the reactor length can have favorable effects on reaction selectivity. Some reactions, such as partial oxidations or hydrogenations, are conveniently driven to high selectivity by keeping rather low reactant concentrations in the reacting mixture. This can be accomplished through the membrane, which can be used to dose the reactant at the desired rate in all parts of the reactor at once. Further, by simply keeping the bulk of the two reactants separated, any premixing of them can be prevented, together with the consequently promoted side reactions (deep oxidations or hydrogenations) and safety problems (formation of explosive mixtures).

A final application that can be envisaged for permselective IMRs concerns the *enhancement of reaction selectivity toward intermediate products* of consecutive reaction pathways. Such a goal could be attained by developing a membrane capable of separating the intermediate product from the reaction mixture [39,40]. The most critical point in this regard is that intermediate product molecules (e.g., partially oxidized hydrocarbons) are often larger in size than the complete reaction products (e.g., CO_2) or the reactants themselves (e.g., O_2). This seriously complicates the separation process, limiting the number of selective transport mechanisms that can be utilized for the purpose of capillary condensation, surface diffusion, or multilayer diffusion (described later in this chapter).

In a less investigated concept, membranes can be used in some rather innovative reactor configurations to control the way the reactants come into contact. In some of these applications, membrane permselectivity is not important. It is a rather obvious consequence that only porous inorganic membranes are of interest in this context. Figure 7 shows a couple of such applications. In the first case, the membrane, as formerly described for separative applications, is used to feed one of the reactants along the reactor, thus enabling the above-described advantages concerning selectivity [41,42]. Since, in this case, the membrane is not permselective, only pure reactants can be employed. In the second sketch in Fig. 7, a non-permselective catalytic membrane reactor with separate feed of reactants is represented. Such a reactor setup has recently been proposed by van Swaaij and co-workers [29,30,43,44]. According to this reactor setup, two key reactants are fed to a microporous catalytic membrane from opposite membrane sides. Provided the kinetics are fast enough compared to the transport of reactants, the reaction takes place in a limited zone inside the membrane (practically a surface, for infinitely fast reactions), reached by reactants in proportion to their stoichiometric coefficients. Any change in reactant concentration in the gas feeds results in a shift of the reaction zone inside the membrane

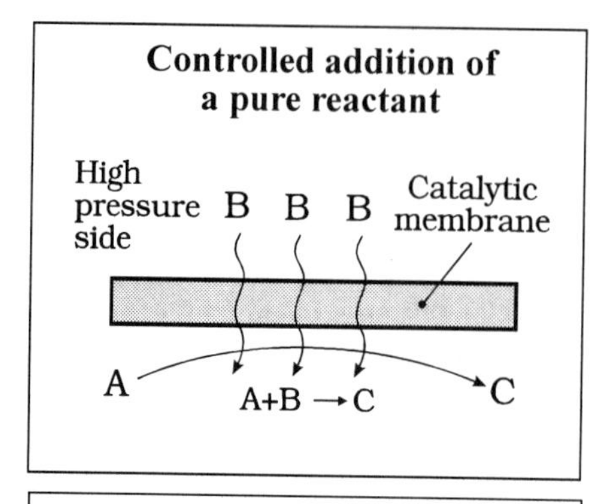

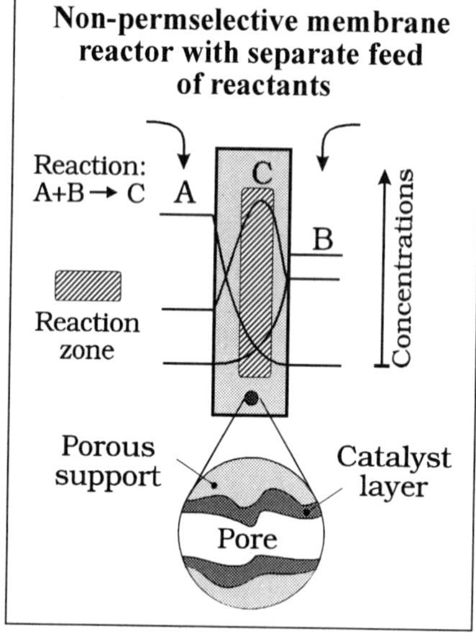

Figure 7 Potential application opportunities for IMRs with nonpermselective membranes.

without losing the above property (see Fig. 8). Any slip of reactant toward the opposite side of the membrane is prevented as well.

Other properties of this reactor can possibly render it attractive for hydrocarbon combustion processes. In fact, for markedly fast kinetics, the overall attainable conversion will become controlled almost exclusively by transport phenomena, which are much less temperature sensitive than kinetics, allowing operation with lower risks of thermal runaways and consequent catalyst damage. Moreover, a certain flexibility in controlling the reactor is given by the possibility to vary independently the flow rates, the concentrations, and the pressures of the two separate reactant feeds. Further, the formation of explosive mixtures is hampered by avoiding any premixing of the reactants. Finally, by applying a pressure difference over the membrane, the products can be shifted preferentially toward the low-pressure chamber, allowing one to: keep one of the reactants (i.e., the hydrocarbon) pure enough to be recycled; increase the overall conversion, limited only by the flux of the reactant that diffuses against the pressure gradient (see Fig. 8); reduce the residence

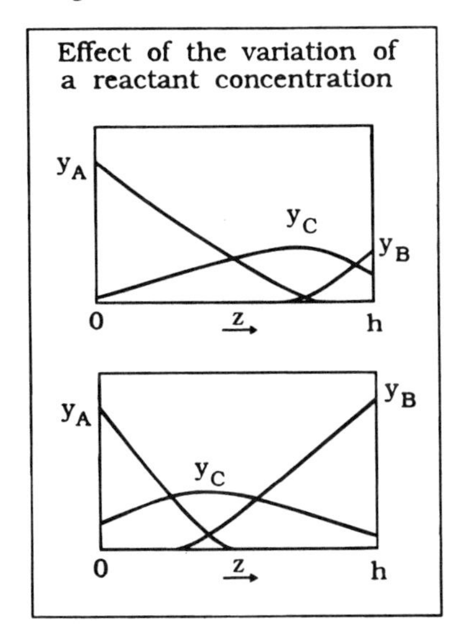

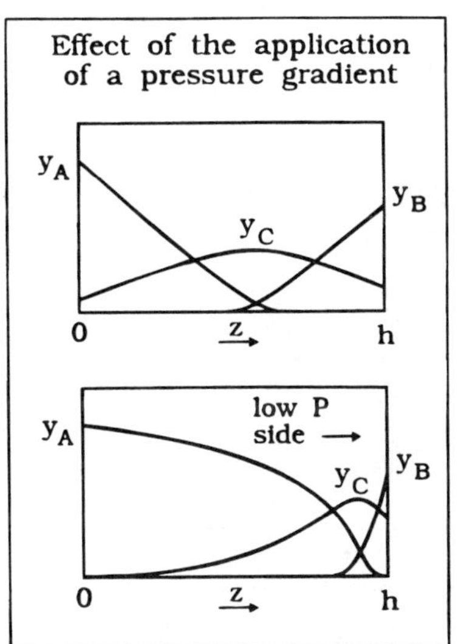

Figure 8 Effect of the variation of the feed mole fraction of reactant B or of the application of a pressure difference over the membrane in a nonpermselective catalytic IMR in which a reaction $A + B \rightarrow C$ is carried out.

time of the products in the catalytic membrane, resulting in a higher selectivity for, e.g., partial oxidation products that are removed from the membrane by the convective flow, preventing deeper oxidation.

III. MECHANISMS OF (SELECTIVE) TRANSPORT THROUGH INORGANIC MEMBRANES

Before getting into details concerning the most widely used preparation routes and applications of permselective inorganic membranes, it seems convenient to describe briefly the major transport mechanisms that can govern the selective permeation of gases throughout porous or dense ceramic membranes. At the end of this section, conclusions will be drawn concerning the type of mechanisms and the desired membrane structures that need to be synthesized in order to achieve a possible breakthrough for membrane reactors in the process industry. Figure 9 summarizes the most important permeation mechanisms through inorganic membranes.

Considering at first transport mechanisms in porous membranes, *viscous flow* (Fig. 9), also called Poiseuille flow, takes place when the mean pore diameter is larger than the mean free path of gas molecules (pore diameter higher than a few microns), so that collisions between different molecules are much more frequent than those between molecules and pore walls. In such conditions, no separation between different molecules can be attained [45].

As the pore dimension decreases (down to fractions of a micron) or the mean free path of molecules increases, which can be achieved by lowering the pressure or raising the temperature, the molecules collide more frequently with the pore walls of the membrane than with one another. When the so-called *Knudsen flow* (Fig. 9) is achieved,

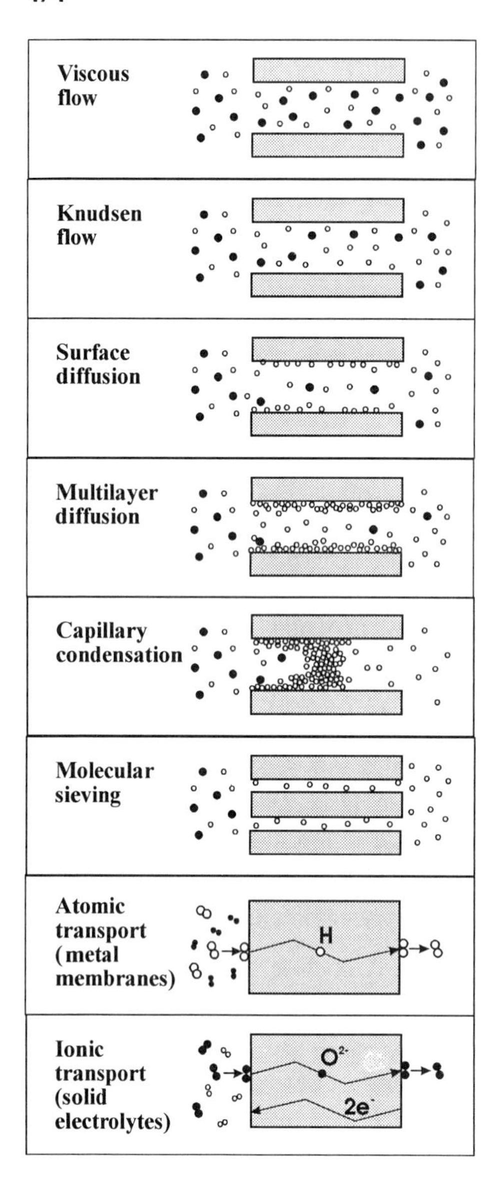

Figure 9 Transport mechanisms through inorganic membranes.

the permeating species flow throughout the membrane, almost independent of one another. Transmembrane fluxes are then proportional to the square root of the molecular weight of the different gaseous compounds [45]. Therefore, the highest achievable separation factor between two different molecules becomes equal to the square root of the ratio of the two molecular weights. As a consequence, membranes operating in the above regime have higher permeability for small molecules (e.g., H_2) than for big ones (e.g., hydrocarbons), but the separation factors in most cases remain markedly below the values needed for practical applications. Hence, in equilibrium circumvention experiments (see Fig.

6) the attainable conversion increase remains seriously limited by permeation of the reactants [46].

Poiseuille and Knudsen flow generally govern the mass transfer through the membrane in the above-defined nonseparative applications of membrane reactors. In this context, with special reference to the transition region from Poiseuille to the Knudsen flow (pore size = 0.1 ÷ 1 μm), the so-called *Dusty-Gas Model* [47], which combines a Stefan–Maxwell expression of diffusive fluxes and a Darcy expression of convective ones, was frequently employed, especially for the IMR with separate feed of reactants [30,48,49]. Such studies clearly indicated that the use of the Stefan–Maxwell approach to diffusion has to be preferred over a simple Fick one, especially when large pressure differences are imposed across the membrane.

Higher permselectivities than with Knudsen flow can in principle be achieved with *surface flow* (Fig. 9). According to this transport mechanism, one of the permeating molecules can preferentially physisorb on the pore walls [50]. Selective transport of the adsorbed molecules is thus enabled even though rather low selectivities can be achieved, unless pores are sufficiently small (i.e., the specific surface area is high). As a side effect, the adsorbed molecules reduce the effective diameter through which the other molecules can migrate, thus further reducing the permeability of those molecules that move within the gas phase inside the pores themselves. Pore sizes as small as a few nanometers, though, are required to emphasize this effect.

As concerns the modeling of surface diffusion, Ulhorn and co-workers [51] worked out a quantitative expression for the calculation of the surface-flow contribution to permeation fluxes, but its general applicability is questionable. In a more recent study, Jaguste and Bhatia [52] describe the combination of surface flow and viscous flow in a γ-Al_2O_3 porous media.

A major drawback of surface diffusion for high-temperature membrane reactor application lies in the fact that adsorptive bonds between molecules and surfaces become less and less strong as long as temperature increases, thus lowering the separation factors achievable.

For molecule–surface interactions that are particularly strong, Ulhorn and co-workers [53] proposed the existence of another flow mechanism, the *multilayer diffusion* (Fig. 9), a sort of intermediate regime between surface flow and capillary condensation.

When one of the components can condense within the pores, the *capillary condensation* mechanism is enabled (Fig. 9). The condensate fills the pores and then evaporates at the permeate side, where a low pressure is imposed [54]. Moreover, the transfer of rather big molecules is generally favored by this mechanism over rather small ones. Provided the pore dimension is small and homogeneous enough, and the pores themselves uniformly dispersed over the membrane, this mechanism allows for very high selectivities (separation factors between 80 and 1000, as reported in [55]) limited only by the solubility of noncondensable molecules in the condensate. However, capillary forces are strong enough to promote this mechanism only with small pore sizes at relatively low temperatures. Hence, as for surface or multilayer diffusion, the practical chances of application appear poor in inorganic-membrane reactors.

The last mechanism indicated in Fig. 9 for porous inorganic membranes is *molecular sieving*. This is achieved when pore diameters are small enough to let only smaller molecules permeate while mechanically preventing the bigger ones from getting in. Provided the pores are monodispersed in dimension, selectivity may reach, in this case, very high values in a temperature range limited only by membrane stability. Therefore, as described, this mechanism is probably the most attractive among those governing transport in porous membranes, and a considerable part of the research efforts of materials scientists working in the field of inorganic membranes is addressed to the synthesis of mol-sieve membranes (e.g., zeolite membranes).

A theory of gas diffusion and permeation has recently been proposed [56] for the interpretation of experimental data concerning molecular-sieve porous glass membranes. Other researchers [57,58], on the basis of experimental evidences, pointed out that a Stefan–Maxwell approach has to be preferred over a simple Fick one for the modeling of mass transfer through zeolite membranes.

Shifting to dense membranes, some metal membranes, namely, Pd alloys or Ag membranes (to which the preceding chapter was entirely dedicated), are capable of being selectively permeated by hydrogen or oxygen, respectively. The overall transport mechanism is based on the following sequence of steps. Gas molecules chemisorb and dissociate on one side of the membrane, then atoms dissolve in the metal matrix and diffuse toward the opposite side, on the grounds of a concentration gradient kept across the membrane; at the permeate side atoms combine and desorb as molecules. Each one of these consecutive steps may become rate controlling depending on temperature, pressure, and gas mixture composition. A deeper analysis of transport mechanisms in Pd-alloy membranes is given in Ref. 9.

Some ceramic dense materials, such as SiO_2 [59], are also capable of being selectively permeated by hydrogen according to a solution-diffusion-desorption mechanism. The only difference with Pd-alloy membranes lies in the fact that hydrogen diffuses across silica membranes in a molecular form.

A final transport mechanism through dense inorganic membranes is typical of solid electrolytes: After a dissociative chemisorption, atoms are ionized and then transported through the crystalline lattice until they lose their charge, combine, and desorb as molecules at the permeate side. The most typical gases that can undergo such a mechanism are oxygen and hydrogen. In any case, the driving force behind the diffusive transfer can be provided by an electrical potential gradient, imposed through suitable electrodes, deposited on both sides of the membrane (*electrochemical pumping*). In other cases, such as *fuel-cell or sensor applications*, a chemical reaction promotes the transfer of the permeating species, setting an electrical potential difference between two electrodes placed at opposite sides of the membrane. A deep insight into transport mechanisms in solid-electrolyte membranes is given in Ref. 26. Finally, it is worth mentioning that some materials exhibit both electronic and ionic conductivity. Considering oxygen conductivity, such materials (i.e., perovskites [38,60]) transport not only oxygen ions (thus functioning as oxygen separators) but also, countercurrently, electrons. No external electrodes are required in this case, with the oxygen partial pressure difference across the membrane acting as the only driving force.

IV. INORGANIC-MEMBRANE REACTORS FOR SEPARATIVE APPLICATIONS

In the following two subsections the preparation and some representative application case studies of permselective inorganic membranes will be addressed.

A. Membrane Preparation

As discussed in the former section, high permselectivities can be achieved for porous inorganic membranes only if comparatively small pore dimensions can be obtained all over the membrane, enabling transport mechanisms such as molecular sieving. Armor stated in a recent review [6] that "to achieve a size-selective separation of reactants and products, we need inorganic membranes with pores <8 Å." Moreover, this goal has to be reached on relatively small membrane thicknesses not to compromise unacceptably the permeability of the membrane itself. Unfortunately, most membrane reactor experiments up to now were performed with porous membranes operating in the less selective regimes (Knudsen or surface flow); hence, though indicative of promising opportunities, the results obtained have not yet been satisfactory enough to allow a breakthrough in the market.

As concerns dense membranes, permselectivity (often limited only to hydrogen or oxygen) is not a major concern. The challenge is once again to combine the very high selectivity these membranes allow with reasonable permeability for practical application. Concerning this last point, Dixon and co-workers [61], in their modelistic study on the application of O_2-permeable-membrane reactors to waste reduction and recovery, conclude that higher permeabilities than those currently available have to be achieved to gain competitiveness. This means either that thinner membranes have to be produced or that new materials with higher oxygen permeabilities have to be found.

A list of permeabilities and permselectivities of inorganic membranes is given in Table 2.

Following are some details about the preparation of permselective inorganic membranes and future perspectives. The book by Bahve [13] is recommended for deeper insight into these topics.

A serious breakthrough in the synthesis of inorganic porous membranes occurred in the past decade with the development of slip casting techniques, based on sol-gel processes, for the deposition of microporous layers on porous supports [8]. The best results obtained by these means at both laboratory and industrial scale were though defect-free γ-Al_2O_3 top layers having a pore dimension of 3–5 nm (i.e., Membralox membranes from SCT-France; see Fig. 2) and supported on α-Al_2O_3 porous tubes. Although fair permeabilities could be attained as a consequence of the asymmetric multilayer structure, permselectivity was still not sufficient (see Table 2), being controlled almost exclusively by Knudsen diffusion. Similar results were also obtained with the so-called Vycor glass membranes prepared according to a phase-separation and leaching technique described in Ref. 8. Moreover, serious stability problems arise during long-time operation at high temperatures (γ-Al_2O_3 is a metastable phase whose layer undergoes a certain pore-size increase at temperatures higher than about 450–500°C).

Some techniques were developed to stabilize such membranes by deposition of dopants (e.g., La or Y salts). Lin and co-workers [74] developed a technique to coat the grain surfaces of nanostructured alumina, titania, and zirconia membranes with a suitable dopant. By these means, they obtained a rise of 200°C for the phase transformation γ-Al_2O_3/α-Al_2O_3, of 150°C for TiO_2(anatase)/TiO_2(rutile), and of 300°C for ZrO_2(tetragonal)/ZrO_2(monoclinic). Doping also allowed them to retard the surface area loss and the pore growth of the three membranes.

Further, concerning the point of selectivity, several attempts were made to decrease the pore size of sol-gel synthesized membranes by depositing some metal oxides (e.g., SiO_2) within the membrane pores through techniques such as chemical vapor deposition,

Table 2 Permeabilities and Permselectivities of Some Inorganic Membranes

Membrane material	Temperature (K)	Permeability (mol/m·Pa·sec)	Gas	Separation factor	Ref.
Porous Membranes					
γ-Al_2O_3 (5 nm)	1000	3×10^{-12}	O_2	$O_2/N_2 = 1.1$	62
γ-Al_2O_3 (3 nm)	293	1.5×10^{-11}	H_2	—	13
γ-Al_2O_3 (4 nm)	295	2×10^{-11}	H_2	$H_2/N_2 = 4.7$	62
Vycor glass	293	1.1×10^{-10}	H_2	$H_2/N_2 = 3.3$	63
Vycor glass	293	3.1×10^{-11}	O_2	—	63
SiO_2-modified (<2 nm)	343	9.5×10^{-12}	H_2	$H_2/N_2 = 163$	64
SiO_2-modified (≈1 nm)	473	1×10^{-13}	H_2	$H_2/N_2 = 2000$	65
carbon (mol-sieve)	773	2.4×10^{-13}	H_2	—	66
carbon (mol-sieve)	292	3.5×10^{-13}	O_2	$O_2/N_2 = 13$	67
silicalite-1	292	1.44×10^{-11}	CH_4	$CH_4/C_3H_6 = 10$	57
zeolite ZSM-5	303	6.2×10^{-14}	H_2	$H_2/C_4H_{10} = 151$	68
zeolite ZSM-5	458	1×10^{-13}	H_2	$H_2/C_4H_{10} = 54$	69
Dense Membranes					
Ag	675	5.4×10^{-16}	O_2	$O_2/N_2 = \infty$	45
Ag	1075	2.0×10^{-13}	O_2	$O_2/N_2 = \infty$	45
Pd	293	1.2×10^{-12}	H_2	$H_2/N_2 = \infty$	13
Pd	673	5.7×10^{-12}	H_2	$H_2/N_2 = \infty$	6
Pd–Ag (23%)	673	1.7×10^{-11}	H_2	$H_2/N_2 = \infty$	6
Pd–Y (7.8%)	573	8.9×10^{-11}	H_2	$H_2/N_2 = \infty$	70
SiO_2	723	1.3×10^{-15}	H_2	$H_2/N_2 = 3100$	71
SiO_2	973	5.4×10^{-13}	H_2	$H_2/N_2 = 2500$	59
SiO_2	873	3.5×10^{-14}	H_2	$H_2/N_2 > 1000$	72
Y_2O_3 (8%)–ZrO_2	1073	$5\text{–}50 \times 10^{-18}$	O_2	$O_2/N_2 = \infty$	62
Y_2O_3 (25%)–Bi_2O_3	923	1.7×10^{-13}	O_2	$O_2/N_2 = \infty$	62
$SrCo_{0.8}Fe_{0.2}O_3$	923	3.6×10^{-12}	O_2	$O_2/N_2 = \infty$	73
$La_{0.79}Sr_{0.2}MnO_{3-\alpha}$	973	5×10^{-15}	O_2	$O_2/N_2 = \infty$	27
$La_{0.79}Sr_{0.2}MnO_{3-\alpha}$	1133	5×10^{-13}	O_2	$O_2/N_2 = \infty$	27

chemical vapor infiltration, electrochemical vapor infiltration, repeated sol-gel depositions, and pyrolysis of impregnating polymer precursors [65,75–81]. A recent review of this topic is Ref. 82. Reasonably good results in terms of permselectivity increase were obtained in most of the above cases, but always at the price of a decrease of membrane permeability (see data in Table 2 concerning SiO_2-modified membranes), apart from some stability problems detected for silica-modified membranes under harsh hydrothermal conditions. In fact, a serious drawback of the above technique lies in the fact that the increase in selectivity is gained by decreasing not only the pore size but also the overall porosity of the membrane.

The best answer to the permeability/permselectivity optimization would be to synthesize very thin layers of materials having a comparatively high porosity and pore sizes in the range 5–8 Å so as to achieve molecular sieving effects. Instead of the modification of already available membranes, the synthesis of new membranes seems more appropriate to reach the above goal. The two most promising candidates in this context are carbon and zeolite membranes.

Carbon membranes, synthesized mostly by pyrolysis of polymeric membrane precursors [66,67,83,84], are excellent separators and in most cases outperform polymer membranes. Moreover, they can easily be formed in a multilayer hollow-fiber structure that is resistant to crack formation and rather flexible. However, their stability in oxygen-containing environments is obviously limited, especially at high temperatures, which limits the potential of these membranes in IMRs. Further, due to the strong adsorption efficiency of carbon, these membranes are prone to get progressively plugged by adsorption of organic contaminants eventually present in the feed, thus needing periodic regenerative cleaning with suitable solvents or stripping agents [84].

The synthesis and application of zeolite membranes therefore remains probably the most interesting research field for inorganic porous-membrane applications. An entire chapter, later on in this book, is dedicated to these promising membranes. Zeolites are crystalline; their pores arise from the lattice spacings (a few angstroms) of their molecular structure and should be stable until the crystalline structure itself remains unchanged. Moreover they possess a fair thermal resistance, chemical inertness, and mechanical strength. Another interesting issue concerning these materials is that their physicochemical properties can be varied by ion exchange, thus modifying also their adsorptive properties.

Since zeolites were formerly synthesized in powder form, new methods for the preparation of zeolite membranes were attempted in recent years. Most of the early attempts regarded the use of conventional hydrothermal synthetic methods to prepare thin zeolitic layers. Matsushita Electric Industries [85] first attempted to produce zeolite membranes by reacting sodium silicate with a caustic hydroxide directly on the surface of a porous alumina support, followed by hydrothermal treatment. Sano and co-workers [86–88] prepared thin zeolite membranes on Teflon, filter paper, or stainless steel substrate by immersing these supports in an aqueous alkali solution containing aluminum, silicon, and some organic templating reagents, followed by hydrothermal treatment. The problem was that defects or pinholes were unavoidable. Others [89] produced zeolitic layers strongly bound to aluminosilicate supports; however they admitted that transmembrane transport was controlled by the intercrystal spacings (about 0.1 μm) rather than by the pores of the zeolite itself. Geus and co-workers [90–91] came to nearly the same conclusion.

From the technical standpoint, it has to be emphasized how a defect, such as a crack, on these very selective membranes might compromise the performance of the entire membrane due to the very high transport rate it allows compared to that through the regular pores. This becomes quite critical, especially when a pressure difference is imposed over the selective layer favoring the appearance of convective flows. As a consequence, an almost complete absence of defects has to be reached during the preparation of these membranes, as well as a sufficient mechanical strength so as to prevent defect formation under operating conditions. This is a real crucial point that holds also for dense membranes (addressed below) and that has to be reached not only on small lab-scale membranes but also on industrial-scale ones.

Since a zeolite-membrane surface based on a single crystal width appears an almost unreachable goal, a step toward the synthesis of defect-free zeolite membranes was made by letting crystals grow closely to one another on a mesoporous support. Some researchers at Mobil [92] have recently synthesized unsupported thin and, unfortunately, very fragile zeolite membranes, showing a densely packed structure composed of ZSM-5 crystals grown together. After their experience the need for a suitable porous support to allow high-temperature separation processes was universally recognized.

Since then, layers of grown-together zeolite crystals have been prepared on porous supports of stainless steel [93] or of porous alumina [69,72,94], showing very promising results (see Table 2). However, major steps still have to be taken in order to render these highly selective porous membranes reliable and cheap enough to be produced at an industrial scale. If these problems are solved, the porous IMR technology will probably make its way toward practical success.

Now considering dense membranes, attention will be focused only on ceramic membranes, since a detailed description of the preparation and properties of the interesting and promising metal membranes have been described in detail in the preceding chapter of this book. Data concerning the permeability of Ag and Pd-alloy membranes, though, are listed in Table 2 for comparison.

As already discussed, some dense ceramic materials are permeable to gas molecules rather than to ions, in accordance with a solution-diffusion model [95]. For instance, some researchers [71,96] first succeeded in making a 0.1-μm-thick silica membrane on a Vycor glass support by means of a modified chemical vapor deposition technique. $SiCl_4$ and water vapor were fed together at the same side of a porous tubular membrane at 600–800°C, thus promoting the formation of SiO_2 inside the Vycor glass tube. This very thin membrane showed a high selectivity to hydrogen, at the price of a modest permeability (see Table 2). Ioannides and Gavalas [97] tested this dense silica membrane for isobutane dehydrogenation to isobutene at 500°C. The membrane retained its permselectivity and permeability during several days of operation. Similar membranes with a SiO_2/C/Vycor structure have recently been produced by Megiris and Glezer [98] by low-pressure oxidation of triisopropylsilane.

Recently, Kim and Gavalas [59] improved the above technique based on $SiCl_4$ hydrolysis, by feeding the two key reactants to the support tube, not simultaneously but alternatively. By these means, thinner (5–10 μm) SiO_2 membranes were obtained, thus increasing the overall permeability of the membrane without affecting seriously its permselectivity (see Table 2). However, even if the chemical vapor deposition technique is less prone to crack formation of the deposited layer as compared to "wet" techniques, which need drying and calcination steps, wide margins for the improvement of the permselectivity of currently produced dense SiO_2 membranes still seem to be left, since the H_2/N_2 permeability ratio of dense amorphous silica should be higher than 10^5 [59].

Further interesting materials for dense inorganic membranes are solid electrolytes, which allow very high selectivities (almost unlimited) but generally show low permeabilities. Permeating fluxes become significant only at high temperatures (>600°C), where, on the other hand, they can work properly due to their generally good thermal stability.

Typical solid electrolytes are stabilized-ZrO_2, -ThO_2, or -CeO_2, solid solutions of Bi_2O_3 in alkali, $SrCeO_3$, etc. These materials can selectively transfer oxygen or hydrogen. However, novel solid electrolytes capable of transferring different species (i.e., F, C, N, S, . . .) could in principle be prepared [4]. For instance, β-Al_2O_3 can selectively transport Na^+ ions.

New materials are being developed having higher permeabilities than the above conventional solid electrolytes. For instance, Gür and co-workers [27] developed perovskite membranes, based on La, Sr, Mn oxides, capable of transporting oxygen through a vacancy diffusion mechanism at a rate 1000-fold higher than the conventional Y_2O_3(8%)–ZrO_2. Such a difference in permeability can also be appreciated from data in Table 2. Similar results were also obtained more recently with $La_{1-x}Sr_xCoO_3$ membranes [99].

Concerning proton conductors, Govind and Zaho [100] stated that metal-based membranes could be outperformed by solid electrolyte membranes based on materials such as $SrCe_{0.95}Yb_{0.05}O_{3-\alpha}$, owing to the very good resistance of these materials in harsh environments or at high temperatures.

In any case, the fundamental problem to be solved also for this kind of membrane is to develop technologies capable of producing, with relatively low costs, very thin membranes in line with the work presented in Ref. 27. This would allow reduced power consumption in electrochemical pumping applications, or increased energy conversion efficiency in fuel-cell applications [76,101].

On the other hand, concerning the electrodes deposited whenever necessary on the membrane surfaces, there is, as already stated, the need to combine electrical conductivity with good catalytic activity. This appears, though, to be a task for most catalytic materials based on metal oxides, suitable, for instance, for methane oxidative coupling. In order to combine both catalytic and electric properties on a single, tailor-made electrode, two main routes can be envisaged [26]: manufacturing electrodes made of both a metal and a ceramic material in the form of simple macroscopic mixtures or, better, of the so-called *cermets* [102,103]; synthesising mixed-conduction ceramic electrodes, capable of transferring both positive and negative charges and of carrying an imposed potential while enhancing the available area for the captation of oxygen or hydrogen [104].

B. Some Application Case Studies

Some of the most interesting application opportunities that have been tested on separative inorganic-membrane reactors are listed in Table 3, where recent literature references are cited as well. For more information concerning the huge number of reactions ever tested on such reactors, see Ref. 11. Some representative cases are discussed next.

Most of the studies on IMRs focused on equilibrium-restricted reactions, where selective permeation of reactants (mostly H_2, in some cases O_2) led in any case to improvements compared to conventional fixed-bed reactors. However, it has to be admitted

Table 3 Most Interesting Reactions for the Process Industry or for Environmental Protection Performed on Separative Inorganic-Membrane Reactors

	References
Methane steam-reforming	105,106
Water–gas shift reaction	107,108
Ethane dehydrogenation	109
Propane dehydrogenation	106
Cyclohexane dehydrogenation	18,28
Ethylbenzene dehydrogenation	110,111
Methane oxidation to syngas	38
Oxidative coupling of methane	112
CO hydrogenation to hydrocarbons	113,114
Partial oxidation of ethylene	115
Partial oxidation of propane to acrolein	115
SO_2 removal	116,117
H_2S removal	118,119
NO_x decomposition	117
NH_3 decomposition	120

that almost every membrane reactor pilot plant study ended with promising results, indicating wide potential for this technology, though on no occasion were the results enough to ensure commercial success, as a consequence, e.g., of membrane instability, too low permeability, and insufficient permselectivity.

Concerning dehydrogenation, perhaps the most interesting latest results are those coming from the use of zeolite membranes in membrane reactors. A group working at the IRC (Institute des Recherches sur la Catalyse, Lyon, France) recently studied the application of a tubular zeolite membrane containing a fixed bed of catalyst for isobutane dehydrogenation to isobutene, getting 50% isobutene yield increase due to equilibrium displacement [121]. Literature on less innovative membranes is still flourishing. For instance, some researchers recently applied γ-Al_2O_3 membranes modified by deposition of metals such as Ru, Pd, Rh, and Pt. The permeability to hydrogen of such membranes exceeded the limitations of the Knudsen diffusion mechanism. This allowed them to achieve, in the temperature range 300–500°C, conversions of methane steam-reforming twice as high as the equilibrium value [105]. Similar results were recently also achieved by Deng and Wu [122] concerning methanol dehydrogenation on sol-gel-derived, catalytically active, γ-Al_2O_3 membranes.

It is also worth it mentioning that some Dutch researchers at the National Energy Research Foundation (ECN, Petten) are developing a process for hydrogen recovery from coal-derived gases [107]. Conventional approaches for CO_2 control in integrated gasification combined-cycle (IGCC) power plants conventionally consist of two separate units, the former for CO reaction with steam (water–gas shift reaction) to obtain CO_2 and H_2, the latter for low-temperature CO_2 separation. By using membranes highly selective to hydrogen, combined with a proper catalyst for the water–gas shift reaction, pure H_2 can be recovered directly from the gaseous reacting mixture, with further advantages concerning the improved CO conversion and the reduced steam consumption. For the same reaction, Seok and Hwang [108], operating with a Vycor glass membrane and a $RuCl_3$ catalyst, observed high conversions (up to 85%) per pass through their reactor.

Although most studies on permselective IMRs considered the circumvention of chemical equilibria, more recently, catalytic membranes have been looked at as tools to control reactions taking place at the membrane so as to drive them to higher yields in, e.g., intermediate oxidation products.

Indirect routes for converting methane into valuable products, such as methanol and formaldehyde, require partial oxidation of methane to form syngas ($CO + H_2$), either by steam-reforming or by direct oxidation, and the subsequent conversion into upgraded products (Fisher–Tropsch or methanol synthesis). Steam-reforming, though, is a rather expensive process, energy and capital expensive due to the endothermicity of the reaction. Although direct partial oxidation of methane is a potential alternative, air cannot be used as the oxygen source because downstream processing requirements cannot tolerate nitrogen, and recycling with cryogenic separation is required [38]. However, dense inorganic membranes selectively permeable to oxygen can solve the problem, allowing the feed of air from which oxygen is selectively separated and dosed, in a controlled way, to the CH_4 feed side, thus promoting its partial oxidation to syngas with good yields. Balachandran and co-workers [38] actually demonstrated that this technology is feasible, operating methane partial oxidation to CO and H_2 with 99% CH_4 conversion and about 90% selectivity toward CO in La-Sr-Fe-Co-O perovskite-membrane reactors. In some cases, operation for 500 hr was achieved with no damage to the membranes.

Conversely, Nozaki and Fujimoto [112] developed a supported PbO dense membrane (doped with K_2O—an oxidative coupling promoter—and supported on a porous SiO_2–Al_2O_3 tube) through which oxygen could selectively permeate, promoting methane oxidative coupling at the opposite membrane side. Selectivity toward C_2 hydrocarbons reached 90% at 800°C; however, the specific reaction rate per unit membrane area remained rather low, due mainly to slow oxygen permeation.

Some final issues concern the use of permselective IMRs in the treatment of flue gases for environmental protection purposes. Special concern was devoted to the treatment of coal-derived flue gases or in coal gasification systems.

Winnick and co-workers at the Georgia Institute of Technology (Atlanta, PA) proposed and tested a couple of electro-driven IMR processes. The first process, for SO_2 removal from flue gases from coal-fired boilers, is based on a composite membrane made of a molten salt solution ($K_2V_2O_7$ + V_2O_5) entrapped in a porous matrix, which is sandwiched between an anode and a cathode. The membrane is permselective toward SO_2. Therefore, by applying current, SO_2 can be pumped out of the flue gases, easily achieving 90% removal with an almost 100% current efficiency [116]. The second process, ment for H_2S removal from natural gas streams, is based on a membrane similar to the former one, in which carbonates are used as the molten electrolyte. Reduction of H_2S to sulphide ion and H_2 gas takes place at the cathode, and the sulfide ion migrates in a molten electrolyte away from the reaction zone. Up to 90% H_2S removal could easily be achieved by these means [118].

Finally, Cicero and Jarr [117] reported on an IMR process carried out on behalf of the U.S. Department of Energy for NO_x abatement from flue gases. The membrane was made of yttria-stabilized zirconia sandwiched between two electrodes. By applying an electrical potential difference across the membrane, oxygen, originating from NO_x decomposition on the cathode, could be driven out of the reaction site, thus enhancing conversion. NO_x per-pass conversion as high as 91% was easily achieved at operating temperatures ranging from 650 to 1050°C.

V. INORGANIC-MEMBRANE REACTORS FOR NONSEPARATIVE APPLICATIONS

A. Membrane Preparation

Making nonpermselective membranes for nonseparative applications is not a problem at all. Depending on the particular application, different properties are required for the membrane. First of all, the nonpermselective membrane can be either inert or catalytically activated.

As discussed later, some researchers [123] have recently developed a membrane reactor concept according to which an inert porous membrane was used to supply oxygen in a controlled way to a fixed bed of catalyst so as to drive to higher selectivities the oxidative coupling of methane (see Fig. 7). The membrane they developed was based on a commercially available microporous alumina membrane having an average pore size of about 10 μm. Since a critical property of the membrane for the above application is its absolute permeability, in order to optimize this parameter the above researchers deposited silica in the membrane pores by dipping the membrane into silica sols, followed by calcination at 800°C. In a similar way they developed tubular membranes with a nonuniform permeation pattern along their axial length [42] so as to modulate the oxygen feed in each reactor section according to an optimum value.

Contrary to the just-described case, in a nonpermselective-membrane reactor with separate feed of reactants (Fig. 7), the membrane has to be catalytically active. In this cases, the work done until now by van Swaaij and co-workers [29,30,48,124] clearly indicated that a pore size (0.1–5 μm), corresponding to the so-called transition zone between the Knudsen and the Poiseuille regime, has to be chosen so as to get a proper balance between membrane permeability and the actual possibility of keeping the reaction zone completely inside the short thickness of the membrane. The above membranes develop rather low surface areas (up to a few $m^2 \cdot g^{-1}$) which are not suitable for a direct catalyst supporting. For such a reason, the above group developed a technique for depositing a thin layer of a catalyst support material (a transition alumina) on the pore walls of an α-Al_2O_3 basic membrane (Fig. 7). This is at first impregnated with a concentrated solution of aluminum nitrate and urea, then kept at 95°C for about 12 hr (thus promoting urea decomposition and aluminum hydroxide precipitation), followed by drying and calcination at 500°C. By these means the specific surface area was enhanced almost 10-fold without serious variation of membrane permeability. On this modified membrane Pt was then deposited by impregnation/calcination techniques in a satisfactorily dispersed way [29].

Similar techniques have been used by different researchers for the manufacture of catalytically active porous barriers for use as catalytic filters [125–127] (see Chapter 15 of the present book) and as catalytic burners [128,129].

Finally, some authors [31,130,131] employed γ-Al_2O_3 thin supported layers (pore size: 4 nm) for ethylene partial oxidation in membrane reactors with separate feed of reactants. In such cases the membrane material had a specific surface area high enough to guarantee a direct catalyst support.

B. Some Application Case Studies

Table 4 lists some of the most interesting reactions tested until now on nonseparative inorganic-membrane reactors. As for separative applications, we refer the reader to a recent review of ours [11] for a more complete list. A few case studies are discussed next in some detail.

In the last decade many research programs started concerning oxidative coupling of methane. Santamaria and co-workers [41,42,123] recently demonstrated that nonperm-

Table 4 Most Interesting Reactions for the Process Industry or for Environmental Protection Performed on Nonseparative Membrane Reactors

	Refs.
Hydrocarbon catalytic combustion	29,30
Liquid-phase hydrogenations	132,133
Oxidative coupling of methane	41,42,123,134
Methanol oxidation to formaldehyde	31
Partial oxidation of ethylene	131,135
SO_2/H_2S Claus reaction	124
CO oxidation	44,48
NO_x reduction with NH_3	130,136

selective membrane reactors based on a porous inert membrane enclosing a fixed bed of catalyst pellets can reach yields very close to the limits required to achieve commercialization (i.e., 25–30%, depending on the selectivity, when methane–oxygen mixtures are used). Their catalyst was Li/MgO and their membrane a SiO_2-modified commercial α-Al_2O_3 membrane, as discussed above. The basic concept, already explained before, is that of dosing oxygen (fed in pure form) in a controlled way along the reactor so as to keep its concentration low enough to avoid parasite reactions (methane oxidation to CO and CO_2). A fair temperature control is also achieved, allowing the avoidance of hot spots, compared to conventional fixed-bed reactors with premixed feed. The best result they obtained is a 23% yield in oxidative coupling products.

Coming to nonseparative membrane reactors with a catalytic membrane, most of the properties of this reactor setup for gas-phase applications were outlined a few pages earlier (Fig. 7 and related comments).

This reactor concept was first demonstrated to be promising for those reactions that require strict stoichiometric feed of reactants (i.e., selective catalytic reduction of NO_x with NH_3, SO_2 abatement to elemental gaseous sulfur with H_2S [124,136]). These studies showed some promising features of this reactor setup, though the reactor was not amenable from the economic viewpoint due to the very low specific conversions per unit membrane surface it guaranteed.

Probably more chances to succeed lie in the use of this reactor for catalytic combustion considering either enhancement of intermediate-product yield [22] or CO complete oxidation to CO_2 [48] or low-NO_x hydrocarbon catalytic combustion for heat production purposes [29,30]. For example, these last authors, in their experimental work concerning propane catalytic combustion on a tubular Pt/γ-Al_2O_3 activated porous membrane whose preparation was already discussed, succeeded in finding conditions in which the membrane reactor remained ignited, with no need for any external heating device, exchanging heat mostly by radiation, to heat exchange surfaces placed on the tube side of the membrane (a cooling oil pipe) and on the shell side of it (the outer wall of the shell-and-tube module).

Recent modeling and experimental studies by Harold and co-workers [131,135] demonstrated that catalytic-membrane reactors with separate feed of reactants can be used in partial oxidation systems, where important enhancements of intermediate-product yield can be achieved owing to the effect of some mass transfer limitations. In their concept the porous support, to which a thin catalytic membrane is anchored (Fig. 1), provides a mass transfer resistance that lowers the oxygen partial pressure when air is fed at the support side of the membrane reactor. The presence of lower oxygen partial pressures on the catalysts favors partial oxidations vs. total oxidations, thus increasing the yields of partially oxidized products compared with the case in which both sides of the membrane were exposed to the same feed mixture of the two reactants. Their theoretical results, however, were in qualitative agreement only with their experimental results concerning ethylene oxidation to acetaldehyde and carbon oxides on a V_2O_5-activated γ-Al_2O_3 membrane deposited on an α-Al_2O_3 porous support. Experiments with the separate feed of reactants led to higher acetaldehyde yields compared with the case in which all key reactants (i.e., C_2H_4 and O_2) were fed in a mixture at the membrane side.

Finally, a nonpermselective membrane can be used in multiphase applications. The original idea belongs to Harold and Cini [137]. Their reactor consisted of a supported catalytically active tubular membrane (Pd/γ-Al_2O_3 on a two-layer α-Al_2O_3 porous support) separating the two reactants: the gas flows at the tube side (membrane side), the liquid at the shell side (support side). Capillary forces let the liquid penetrate the pores of the support and those of the membrane itself where the reaction with the gas flowing along the membrane surface occurs. Apart from a well-controlled and -defined reaction interface and a good temperature control, this reactor setup allows much higher catalyst effectiveness factors compared with conventional trickle-bed reactors for those reactions that are limited by volatile reactant concentration. In fact, the entrance of the catalytic pores is always exposed to the gaseous reactant, thus eliminating the mass transfer resistance provided in conventional trickle-bed reactors by the liquid film covering at least in part the catalyst pellets. Their experimental evidences concerning α-methylstyrene hydrogenation to cumene [132] were in line with their modeling results. A further confirmation of the potentials was recently given in Ref. 133, using nitrobenzene hydrogenation as a model reaction on a Pt/γ-Al_2O_3 membrane. These authors demonstrated that hydrogen could not be the limiting reactants, as opposed to conventional multiphase reactors.

VI. MODELING OF INORGANIC-MEMBRANE REACTORS

Several investigators have faced the problem of modeling of membrane reactors either to achieve a proper interpretation of their experimental data or to assess the role of the various operating parameters (temperature, membrane permeability and permselectivity, feed flow rates, and concentrations, etc.) on the performance of membrane reactors. In some other cases [61,138] modeling studies helped to point the way toward future experimental work concerning, e.g., the need for thinner or more permeable or more stable membranes to outperform conventional technologies for given applications.

It has to be expected that most of these studies involved a rather simple reactor setup, meant for testing some opportunities at a laboratory scale rather than for large-scale industrial applications. When membrane reactors penetrate the process industry, new models, much more complicated than those assembled until now, will probably have to be solved. In this section we will rapidly go through some of the most representative modeling approaches proposed and solved by different researchers for different application fields. In particular, Table 5 highlights some of the modelistic studies on separative-membrane reactors (ranging from porous to dense membranes, from equilibrium circumvention studies to selectivity enhancement ones); Table 6 lists the most relevant features and conclusions of some modeling work dedicated to nonseparative-membrane reactor applications. The review of models of membrane reactors written by Tsotsis and co-workers [10] is recommended for deeper insight into these topics.

Starting with separative-membrane reactors, most of the early modeling studies aimed at assessing the role of fluid dynamics on the reactor performance (in Table 5, see the work on cyclohexane dehydrogenation membrane-enclosed reactors using porous Vycor glass or Pd membranes, performed by some Japanese researchers). First of all, a rather obvious conclusion was drawn: The selective permeation of hydrogen allows a noticeable increase (up to 200%) of the per-pass conversion throughout the reactor. A less trivial issue, noticed for porous-membrane reactors by Mohan and Govind [19] and Itoh and co-workers [139], is that the achievable conversion goes through a maximum at varying the membrane thickness. Mohan and Govind [20,21] extended this concept, demonstrating that an optimum membrane permeability can be found at which conversion is maximum. Both of the above observations can find a common explanation: For a given permselectivity, the more permeable (i.e., the thinner) the membrane, the higher is the amount of reactant that passes through the membrane to reach the permeate side and the

Table 5 Modeling Schemes for Separative Inorganic-Membrane Reactors

Reaction/membrane type	Basic model assumptions	Basic equations	Major conclusions	Refs.
Cyclohexane dehydrogenation/Vycor glass membrane, noncatalytic	Membrane-enclosed packed-bed reactor. Plug-flow regime at both membrane sides. Isothermal system. No axial or radial diffusion. Mass transfer rate constant all over the membrane. Negligible pressure drop at the catalyst side.	*Reaction side*: mass balance equation including diffusive, convective and reaction terms. *Permeate side*: as above, without reaction terms.	Equilibrium conversion is exceeded (up to twofold) by selective permeation of H_2. Good agreement with experimental results. An optimum membrane thickness (i.e., membrane permeability) maximizes conversion, which is increased by enhancing the sweep gas flow rate.	19,45,139
Cyclohexane dehydrogenation, HI decomposition, propylene disproportionation/ Vycor glass membrane, noncatalytic	Same as above.	Same as above. Effect of cocurrent or countercurrent flow patterns studied. Effect of recycle streams or intermediate feeds studied.	An optimum permeability is detected above which reactant loss in the permeate and product back-permeation reduce the attainable conversion. Reactant recycle or intermediate feed helps in overcoming the above limitations. The choice between cocurrent and countercurrent patterns depends on operating conditions.	20
Ethylbenzene dehydrogenation/Vycor glass membrane, noncatalytic	The system is adiabatic. Heat and mass transfer resistance outside the membrane are negligible. No axial or radial diffusion. Plug flow at both reactors sides. Gas permeabilities independent of concentrations.	Mass balance eqs., including diffusive, convective, and reaction terms at both sides. Convective heat transfer eqs. at both sides.	Considerable advantages vs. conventional reactors in terms of temperature lowering, conversion enhancing, steam (decocing agent) consumption. Reactant permeation limits the achievable conversion increase.	21

Table 5 *Continued*

Reaction/membrane type	Basic model assumptions	Basic equations	Major conclusions	Refs.
$CH_3OH + {}^1/_2\, O_2 \rightarrow CH_2O + H_2O$/Vycor glass membrane, noncatalytic, sealed at one end	Same as above. Catalyst pellets are at the shell side.	Same as above. Since one end of the tube side is closed, the feed flow rate is zero at that side.	Increasing the space time and the membrane area increases conversion. Higher permselectivities or permeabilities needed for further improvements.	138
Ethane dehydrogenation/asymmetric γ-Al_2O_3 membrane, noncatalytic	Membrane-enclosed packed-bed reactor. Plug-flow regime at both membrane sides. Isothermal system. No axial or radial diffusion. Negligible pressure drop at the shell side.	*Reaction side*: mass balance equation including diffusive, convective, and reaction terms. *Permeate side*: as above, without reaction terms. Cocurrent flow. Ergun law used for pressure drops at the catalyst side.	Increasing the sweep gas flow rate produces an increase in conversion due to the decreased partial pressure of hydrogen in the shell side and the corresponding increase of H_2 permeation out of the reaction zone.	140
$CH_4 + O_2 \rightarrow CH_2O + H_2O$/inert perselective membrane	Membrane-enclosed packed-bed reactor. Plug-flow regime at both membrane sides. Isothermal system. No axial or radial diffusion. Negligible pressure drop at both membrane sides.	*Reaction side*: mass balance equation, including diffusive, convective, and reaction terms. *Permeate side*: as above, without reaction terms. Cocurrent flow.	Selective permeation of the intermediate oxidation product (formaldehyde) increases selectivity, leaving the overall conversion almost unaffected.	39
Cyclohexane dehydrogenation/Pt-deposited Vycor glass membrane	Well-mixed conditions at both membrane sides. Isothermal system. Reactants are fed at the shell side.	*Feed and reactant sides*: mass balances with no reaction term. *Catalytic*: diffusion and reaction terms are considered. Results are compared with an inert-membrane fixed-bed setup for equal overall catalyst amount.	High conversion increase can be noticed for high residence times. In such conditions the catalytic-membrane reactor outperforms the inert-membrane-enclosed packed-bed reactor.	141,142

Table 5 *Continued*

Reaction/membrane type	Basic model assumptions	Basic equations	Major conclusions	Refs.
Ethane dehydrogenation/ Pt-deposited asymmetric γ-Al_2O_3 membrane	Cocurrent flow. Isothermal system	Same as above.	Conversion increases with operating temperature and sweep gas flow rate. A maximum in conversion is obtained at an optimum thickness.	143
Generic first-order reaction/catalytic membrane	Well-mixed conditions at both sides. Constant effective diffusivities within the membrane. Ideal gas assumption. No pressure difference across the membrane.	Heat and mass balances with no reaction terms outside the membrane, with reaction terms inside of it. Catalyst distribution effects are investigated.	The optimal catalyst distribution function is Dirac delta at the reactant side of the membrane. In case the catalyst loading is bounded, a multiple-step function turns out to be the optimal one.	144
1-Butene dehydrogenation coupled with hydrogen oxidation/ dense Pd membrane	Membrane-enclosed packed-bed reactor. Plug-flow regime at both membrane sides. Both isothermal and adiabatic conditions considered. No axial or radial diffusion. Negligible pressure drop at the catalyst side.	Mass balance eqs., including diffusive, convective, and reaction terms at both sides. Countercurrent and cocurrent flow patterns considered. Convective heat transfer eqs. at both sides in adiabatic conditions.	Coupling of reactions results in conversion increase. Countercurrent mode works better than cocurrent one. For a broad range of conditions the adiabatic reactor gives better results than the isothermal one.	145
Cyclohexane dehydrogenation/dense Pd membrane	Same as above. The effect of plug flow or of perfect mixing behavior is investigated at both membrane sides.	Same as above.	Whenever one of the reactor sides is mixed, conversion is decreased. Countercurrent flow generally outperforms cocurrent one, except for low feed rates and reaction rates.	146
Ethylbenzene dehydrogenation/dense Pd membrane	Membrane-enclosed packed-bed reactor. Plug-flow regime at both membrane sides.	Complete model, including heat and mass transfer differential balances in the packed bed. Constant hydrogen permeability across the membrane.	The yield of styrene is increased up to 20% by use of the membrane, as a result of higher conversion and selectivities. The higher the sweep gas flow rate, the higher the conversion achieved.	33

Table 5 *Continued*

Reaction/membrane type	Basic model assumptions	Basic equations	Major conclusions	Refs.
$2CO_2 = 2CO + O_2$ $2NO = N_2 + O_2$ *O*-xylene + $O_2 \rightarrow$ phthalic anhydride Dense O_2-permeable membranes	Membrane-enclosed packed-bed reactor. Plug-flow regime at both membrane sides. Isothermal conditions considered except for *O*-xylene partial oxidation. No axial or radial diffusion. Negligible pressure drop at the catalyst side.	*Reaction side*: mass balance equation, including diffusive, convective, and reaction terms. *Permeate side*: as above, without reaction terms. Cocurrent flow. Heat balances only in case of *O*-xylene partial oxidation.	Higher membrane permeabilities of those currently available are needed for NO decomposition. Stable materials at more than 2000°C are instead needed for CO_2 decomposition purposes. For *O*-xylene partial oxidation, the air stream fed at the shell side acts as a coolant and mitigates the temperature excursions, though higher permeabilities have to be achieved to reduce the required surface area.	61

more intense is the hydrogen back-permeation from the permeate to the feed side at the membrane end at which the fresh reactant is fed and the sweep gas is removed. Both these phenomena, which are not important with the extremely selective dense membranes, become prevalent above certain membrane permeabilities, thus lowering the achievable conversion. Reactant recycle or intermediate feed helps to some extent to overcome the problem of reactant losses in the permeate and consequent conversion lowering [20].

Another interesting point concerns the position of the membrane within the reactor, especially when fixed beds of catalysts are employed. In this case some researchers [25,111] noticed that the membrane should not be placed close to the inlet of the fixed bed because here the conversion is still markedly below the equilibrium limit and important reactant permeation can take place, making the use of membranes detrimental rather than beneficial.

Another interesting, though rather evident, conclusion of the above modeling works (see also Refs. 23 and 140) was that increasing the sweep gas flow rate, fed at the permeate side of the membrane to remove the permeating gases, also increases the conversion. There are several drawbacks to the use of large amounts of sweep gases: First of all, recovery of the permeate gases would be rather difficult from very dilute gas streams. Second, the use of any gas other than air or steam would likely be too expensive, but either oxygen or steam do interfere at high temperatures with some of the reactions involved or with several membrane materials. Third, it would be rather cost intensive to warm the sweep gas flow rate up to the reaction temperature. For such reasons, the coupling of reactions can be advantageous. Consider, for instance, the work of Itoh and Govind [145], which modeled 1-butene hydrogenation coupled with the oxidation of hydrogen at the permeate side of a Pd-membrane reactor. In this case the oxidation reaction removes hydrogen, thus enhancing its gradient across the membrane, its permeation flux,

Table 6 List of Modeling Schemes of Nonseparative Inorganic-Membrane Reactors

Reaction/membrane type	Basic model assumptions	Basic equations	Major conclusions	Refs.
A(g) + B(l) → Products(l) (e.g., α-methylstirene hydrogenation)/catalytic porous γ-Al_2O_3 membrane	Catalytic membrane pores are filled with the liquid reactant. No radial concentration variations in the gas phase. Isothermal system. Reactants are fed cocurrently from opposite sides.	Eqs. considered: momentum and material balances in the tube core and in the shell annulus; mass balances in the membrane include convective, diffusive, and reaction terms.	The membrane provides for reduced transport limitations, better temperature control, and a well-defined reactant interface. Higher catalyst effectiveness factors are achieved.	135,137
Partial oxidations (e.g., ethylene + O_2 → acetaldehyde)/catalytic microporous layer deposited on a macroporous inactive support	Isothermal and isobaric conditions. Fick law describes diffusion in the membrane. External mass transfer limitations are negligible. The model is solved for given reactant concentrations at the opposite membrane sides.	Simple differential mass balances accounting for only diffusive fluxes and reaction terms across the membrane.	For a range of conditions, higher selectivities can be found using separate reactant feeds compared with the case in which both sides of the membrane are exposed to the same reactant mixture.	131
$SO_2 + 2H_2S = {}^3/_8S_8 + 2H_2O$ γ-Al_2O_3-impregnated α-Al_2O_3 membrane	Isothermal system. The model is solved for given component fluxes at membrane interfaces. Reactants are fed from opposite sides.	Differential mass balances across the membrane combine Stefan–Maxwell-type diffusive fluxes, a surface diffusion term, the Darcy expression for convective fluxes, and the reaction terms.	Model fits the experimental data, though a certain underestimation of the converted H_2S is noticeable, which is attributed to uncertainty in the determination of surface diffusivities.	124
$C_3H_8 + 5O_2 \rightarrow 3CO_2 + 4H_2O$ Pt/γ-Al_2O_3 deposited in a macroporous α-Al_2O_3 membrane.	Same as above, but here either plug-flow or well-mixed approach is tested for both membrane sides	Same as above, with the exclusion of the surface flow contribution, owing to the high operating temperatures (about 500°C). An analytically solved model is also proposed under the hypothesis of very fast kinetics, which shrink the reaction zone to a surface. Cocurrent and countercurrent operation are considered	At high temperatures and in the absence of pressure differences over the membrane the simplified model is satisfactory, slips of reactants are prevented, and countercurrent operation slightly outperforms cocurrent one. The plug-flow assumption is preferable to the well-mixed one. Application of pressure gradients increases conversion and requires the complete set of differential equations for a proper modeling.	29,30

and thus the overall 1-butene conversion. Further, the heat produced at the permeate side sustains the endothermic dehydrogenation taking place at the opposite side.

Concerning the choice of the flow pattern scheme (countercurrent, cocurrent, well-mixed chambers, etc.) to be used in separative-membrane reactor applications. Mohan and Govind [20] concluded that it depends on the specific operating conditions needed. Itoh [146], studying cyclohexane dehydrogenation on a Pd-membrane reactor, observed that countercurrent mode generally outperformed cocurrent one, except for a low feed flow rate and slow reaction kinetics. This is due to the fact that, in a countercurrently fed membrane reactor working in the above conditions, back-permeation of hydrogen from the permeate to the feed side takes place at the entrance of the reactor, which tends to lower the conversion in this zone.

Considering now catalytically active porous membranes, Sun and Khang [141] showed that with the total amount of catalyst being equal, this reactor configuration outperforms the inert-membrane-enclosed fixed-bed reactor, provided the residence time in the reactor is sufficiently high, thus allowing the reactants to reach the catalytic membrane in convenient amount. Their modeling conclusions were confirmed by later experimental work [142]. More recently, some researchers [144] performed a modeling study concerning the optimization of the catalyst distribution in a catalytic-membrane reactor on which a first-order reaction takes place. Their conclusions were rather simple: A Dirac–delta function of the concentration of the catalyst in the membrane placed at the feed side allows the highest conversions. In other words, it is better to promote the reaction as close to the membrane as possible (on its surface), letting the rest of the membrane work as a mere separator of some of the reaction products. In case the local catalyst load of the membrane cannot overpass a given limit (as in all real cases), the optimal catalyst distribution turns to be a multiple-step function, which tends to the Dirac–delta function as long as the above limit is increased.

Coming finally to the modeling of nonseparative-membrane reactor applications, almost exclusively dedicated to membrane reactors with a separate feed of reactants, little has to be added to the conclusions listed in Table 6. An interesting issue is that, as with separative applications, most of the models proposed are isothermal. This a reasonable approximation for those systems in which a liquid phase is involved [137] or whenever very low reactant concentrations are considered [124], but may become more severe when highly exothermic reactions take place. Saracco and co-workers [29,30], who studied propane catalytic oxidation on a membrane reactor with a separate feed of reactants, noticed that an isothermal model was accurate enough to predict their experimental data when operating in the transport-controlled regime, when high temperatures and reaction kinetics reduce the reaction zone almost to a very thin layer inside the catalytic membrane. However, at low temperatures a steady-state multiplicity phenomenon takes place: Reactor ignition and extinction take place at different membrane temperature, which cannot be predicted by simple isothermal models. Heat balances are currently being assembled in the modeling so as to get a proper fit of the experimental data in the low-temperature operating regime of the reactor.

The need to govern heat balances properly in membrane reactors will certainly become a major task if large-scale industrial units are ever to be put into operation. Whether the performed reaction is endothermic (dehydrogenation) or exothermic (oxidation), innovative means to supply or remove heat from large-scale membrane reactor modules will have to be designed. The isothermicity assumption valid for several lab-scale membrane reactors will not hold anymore, and much more complex modeling will certainly have to be developed.

While waiting for industrial-scale membrane reactors, the most intriguing field in which modeling work still has to be done is that of transport through molecular-sieve membranes [56].

VII. CONCLUSIONS

The major features of and application opportunities for inorganic-membrane reactors have been described in some detail. We can conclude that inorganic-membrane reactors actually show promise for improving either conversion of equilibrium-limited reactions (e.g., dehydrogenations) or selectivity toward some intermediates of consecutive reaction pathways (e.g., partial oxidations).

However, at least for separative applications, most hopes to find consistent application of inorganic-membrane reactors lie in the development of inorganic membranes having pores of molecular dimensions (<10 Å, e.g., zeolitic membranes). Such membranes should moreover be: thin enough to allow reasonable permeability, defect-free, resilient, and stable from the thermal, mechanical, and chemical standpoints. Such results should not be achieved only at a lab scale (a lot of promising literature has recently appeared in this context), but should also be reproducible at a large, industrial scale. Last, but not least, such membranes should not be unacceptably expensive, in both their initial and their replacement costs.

The primary role in developing such membranes will belong to materials scientists. Meanwhile, chemical engineers should devote part of their work to gaining a better understanding of highly selective transport mechanisms, designing modules with high specific membrane areas and with suitable heat supply/removal systems, and developing more complex modeling for such unconventional reactors.

When inorganic-membrane reactors will achieve commercialization? Opinions differ in the wide and still-growing population of researchers working in this area. At a recent conference devoted entirely to inorganic-membrane reactors (1st International Workshop on Catalytic Membranes; Lyon, France, 1993), Dr. J. N. Armor, a specialist in the field from Air Products & Chemicals (Allentown, PA) stated that commercialization for most membrane reactors "will not occur until the next century. It takes time to develop any invention. Then there's scale up, which for the first catalytic membrane reactors will take longer than for the later versions" [121]. However, at the same conference Dr. R. Govind, a chemical engineering professor at Cincinnati University who has been working in the field since the early '80s, said that a huge number of patents on membrane reactor manufacturing and application has been filed all over the world, "making me believe that commercialization may only be a few years away." We share Dr. Armor's cautionary opinion, in the hope of being surprised soon. After all, the next century is just around the corner.

REFERENCES

1. Michaels A.S., New separation technique for the CPI, *Chem. Eng. Progr. 64:*31 (1968).
2. Belfort G., Membranes and bioreactors: A technical challenge in biotechnology, *Biotechnol. Bioeng. 33:*1047 (1989).
3. Drioli E., Iorio G. and Catapano G., Enzyme membrane reactors and membrane fermentors, in *Handbook of Industrial Membrane Technology,* M.C. Porter, ed., Noyes, Park Ridge, NJ, (1990), p. 401.

4. Catalytica Study Division, *Catalytic Membrane Reactors: Concepts and Applications,* Catalytica Study No. 4187, Mountain View, California (1988).
5. Armor J.N., Catalysis with permselective inorganic membranes, *Appl. Catal. 49:*1 (1989).
6. Armor J.N., Challenges in membrane catalysis, *Chemtech 22:*557 (1992).
7. Armor J.N., Membrane catalysis: Where is it now, what needs to be done?, *Catalysis Today 25:*199 (1995).
8. Hsieh H.P., Inorganic membrane reactors—A review, *Catal. Rev.—Sci. Eng. 33:*1 (1991).
9. Shu J., Grandjean B.P.A., van Neste A. and Kaliaguine S., Catalytic palladium-based membrane reactors: A review, *Can. J. Chem. Eng. 69:*1036 (1991).
10. Tsostsis T.T., Champagnie A.M., Minet R.G. and Liu P.K.T., Catalytic Membrane Reactors, Chapter 12 in *Computer-Aided Design of Catalysts,* E.R. Becker and C. Pereira, eds; Dekker, New York, (1993), Chap. 12.
11. Saracco G. and Specchia V., Catalytic inorganic membrane reactors: Present experience and future opportunities, *Catal. Rev.—Sci. Eng. 36:*305 (1994).
12. Saracco G., Versteeg G.F. and van Swaaij W.P.M., Current hurdles to the success of high temperature membrane reactors, *J. Membr. Sci. 95:*105 (1994).
13. Bahve R.R., *Inorganic Membranes: Synthesis and Applications,* Van Nostrand-Reinhold, New York (1991).
14. Gryaznov V.M., Platinum metals as components of catalyst-membrane systems, *Plat. Met. Rev. 36:*70 (1992).
15. Clark D.J., Losey R.W. and Suitor J.W., Separation of oxygen by using zirconia solid electrolyte membranes, *Gas Separation and Purification 6:*201 (1992).
16. Anzaj H. and Yanagimoto T., Jap. Pat. Pend. 62-52185 (1987).
17. Lee K.H. and Kim Y.M., Asymmetric hollow inorganic membranes. Presented at ICIM '91, Montpellier, France, (1991), p. 17.
18. Okubo T., Haruta K., Kusakabe K., Morooka S., Anzai H. and Akiyama S., Equilibrium shift of dehydrogenation at short space-time with hollow fiber ceramic membrane, *Ind. Eng. Chem. Res. 30:*614 (1991).
19. Mohan K. and Govind R., Analysis of a cocurrent membrane reactor, *AIChE J. 32:*2083 (1986).
20. Mohan K. and Govind R., Analysis of equilibrium-shift in isothermal reactors with a permselective wall, *AIChE J. 34:*1493 (1988).
21. Mohan K. and Govind R., Studies on a membrane reactor, *Sep. Sci. Technol. 23:*1715 (1988).
22. Veldsink J.W., *A Catalytically Active, Non-permselective Membrane Reactor for Kinetically Fast, Strongly Exothermic Heterogeneous Reactions.* Ph.D. dissertation, Twente University of Technology, Enschede, the Netherlands (1993).
23. Michaels A.S., New membrane processes: Evaluation and prospects. Presented at the 7th ESMST Summer School, Twente University, Enschade, The Netherlands (1989).
24. Adris A.M., Lim C.J. and Grace J.R., A fluidized-bed membrane reactor for the steam reforming of methane, *Chem. Eng. Sci. 49:*5833 (1994).
25. Oertel M., Schmitz J., Weirich W., Jendryssek-Neumann D. and Schulten R., Steam reforming of natural gas with integrated hydrogen separation for hydrogen production, *Chem. Eng. Technol. 10:*248 (1987).
26. Gellings P.J., Koopmans H.J.A. and Burggraaf A.J., Electrocatalytic phenomena in gas phase reactions in solid electrolyte electrochemical cells, *Appl. Catal. 39:*1 (1988).
27. Gür T.M., Belzner A. and Huggins R.A., A new class of oxygen selective chemically driven nonporous ceramic membranes. Part I. A-site doped perovskites, *J. Membr. Sci. 75:*151 (1992).
28. Cannon K.C. and Hacskaylo J.J., Evaluation of palladium-impregnation on the performance of a Vycor glass catalytic membrane reactor, *J. Membr. Sci. 65:*259 (1992).
29. Saracco G., Veldsink J.W., Versteeg G.F. and van Swaaij W.P.M., Catalytic combustion of propane in a membrane reactor with separate feed of reactants. I. Operation in absence of trans-membrane pressure gradients, *Chem. Eng. Sci., 50:*2005 (1995).

30. Saracco G., Veldsink J.W., Versteeg G.F. and van Swaaij W.P.M., Catalytic combustion of propane in a membrane reactor with separate feed of reactants. II. Operation in presence of trans-membrane pressure gradients, *Chem. Eng. Sci., 50:*2833 (1995).
31. Zaspalis V.T., van Praag W., Keizer K., van Ommen J.G., Ross J.H.R. and Burggraaf A.J., Reactions of methanol over catalytically active alumina membranes, *Appl. Catal. 74:*205 (1991).
32. Voge H.H., Dehydrogenation, in *Encyclopedia of Chemical Processing and Design,* Dekker, New York, (1982), Vol. 14, p. 276.
33. Abdalla B.K. and Elnashaie S.S.E.H., Catalytic dehydrogenation of ethylbenzene to styrene in membrane reactors, *AIChE J. 40:*2055 (1994).
34. Grayznov V.M., Sov. Pat. 274,092 (1964).
35. Cales B. and Baumard J.F., Production of hydrogen by direct thermal decomposition of water with the aid of a semipermeable membrane, *High Temp.-High Press. 14:*681 (1982).
36. Basov N.L. and Gryaznov V.M., Membrane catalysts permeable for hydrogen or oxygen, *Membr. Katal.:*117 (1985), in Russian.
37. Omata K., Hashimoto S., Tominaga H. and Fujimoto K., Oxidative coupling of methane using a membrane reactor, *Appl. Catal. 52:*L1 (1989).
38. Balachandran U., Dusek J.T., Sweeney S.M., Poeppel R.B., Mieville R.L., Parampalli S.M., Kleefisch M.S., Pei S., Kobylinski T.P., Udovich C.A. and Bose A.C., Methane to syngas via ceramic membranes, *Ceramic Bulletin 74:*71 (1995).
39. Agarwalla S. and Lund C.R.F., Use of a membrane reactor to improve selectivity to intermediate products in consecutive catalytic reactions, *J. Membr. Sci. 70:*129 (1992).
40. Bernstein C.J. and Lund C.R.F., Membrane reactors for catalytic series and series-parallel reactions, *J. Membr. Sci. 77:*165 (1993).
41. Coronas J., Menendez M. and Santamaria J., Methane oxidative coupling using porous ceramic membrane reactors. Part II. Reaction studies, *Chem. Engng. Sci. 49:*2015 (1994).
42. Coronas J., Menendez M. and Santamaria J., Development of ceramic membrane reactors with non-uniform permeation pattern. Application to methane oxidative coupling, *Chem. Eng. Sci. 49:*4749 (1994).
43. Sloot H.J., Versteeg G.F. and van Swaaij W.P.M., A non-permselective membrane reactor for chemical processes normally requiring strict stoichiometric feed rates of reactants, *Chem. Eng. Sci. 45:*2415 (1990).
44. Veldsink J.W., van Damme R.M.J., Versteeg G.F. and van Swaaij W.P.M., A catalytically active membrane reactor for fast, heterogeneously catalyzed reactions, *Chem. Eng. Sci. 47:*2939 (1992).
45. Hwang S.-T. and Kammermeryer K., *Techniques in Chemistry: Membranes in Separation,* Wiley Interscience, New York, (1975).
46. Itoh N., Shindo Y., Haraya T., Obata K., Hakuta T. and Yoshitome H., Simulation of a reaction accompanied by separation, *Int. Chem. Eng. 25:*138 (1985).
47. Mason E.A. and Malinauskas A.P., *Gas transport in porous media: The Dusty-Gas Model, Chemical Engineering Monographs 17,* Elsevier, Amsterdam (1983).
48. Veldsink J.W., Versteeg G.F. and van Swaaij W.P.M., A catalytically active membrane reactor for fast, highly exothermic, heterogeneous gas reactions, *Ind. Eng. Chem. Res. 34:*763 (1995).
49. Veldsink J.W., van Damme R.M.J., Versteeg G.F. and van Swaaij W.P.M., The use of the dusty-gas model for the description of mass transport with chemical reaction in porous media, *Chem. Eng. J. 57:*115 (1995).
50. Kapoor A., Yang R.T. and Wong C., Surface diffusion, *Catal. Rev.—Sci. Eng. 31:*129 (1989).
51. Ulhorn R.J.R., Keizer K. and Burggraaf A.J., Formation and gas transport mechanisms in ceramic membranes, *ACS Symp. Ser.:*239 (1989).
52. Jaguste D.N. and Bhatia S.K., Combined surface and viscous flow of condensable vapor in porous media, *Chem. Eng. Sci. 50:*167 (1995).

53. Ulhorn R.J.R., Keizer K. and Burggraaf A.J., Gas transport and separation with ceramic membranes. Part I. Multilayer diffusion and capillary condensation, *J. Membr. Sci. 66:*259 (1992).
54. Kitao S., Ishizaki M. and Asaeda M., Permeation mechanism of water through fine porous ceramic membrane for separation of organic solvent/water mixtures. Presented at ICIM '91, Montpellier, France, (1991), p. 175.
55. Sperry D.P., Falconer J.L. and Noble R.D., Methanol-hydrogen separation by capillary condensation in inorganic membranes, *J. Membr. Sci. 60:*185 (1991).
56. Shelekin A.B., Dixon A.G. and Ma Y.H., Theory of gas diffusion and permeation in inorganic molecular-sieve membranes, *AIChE J. 41:*58 (1995).
57. Kapteijn F., Bakker W.J.W., Zheng G., Poppe J. and Moulijn J.A., Permeation and separation of light hydrocarbons through a silicalite-1 membrane: Application of the generalized Maxwell–Stefan equations, *Chem. Eng. J. 57:*145 (1995).
58. Krishna R. and van den Broeke L.J.P., The Maxwell–Stefan description of mass transfer across zeolite membranes, *Chem. Eng. J. 57:*155 (1995).
59. Kim S. and Gavalas G.R., Preparation of H_2-permselective silica membranes by alternating reactant vapor deposition, *Ind. Eng. Chem. Res. 34:*168 (1995).
60. Lin Y.S., Wang W. and Jan H., Oxygen permeation through thin mixed-conducting solid oxide membranes, *AIChE J. 40:*786 (1994).
61. Dixon A.G., Moser W.R. and Ma Y.H., Waste reduction and recovery using O_2-permeable membrane reactors, *Ind. Eng. Chem. Res. 33:*3015 (1994).
62. Lin Y.S., Porous and dense inorganic membranes for gas separations. Presented at the AIChE Annual Meeting, Miami, FL, paper 23c (1992).
63. Shindo Y., Obata K., Hakuta T., Yoshitome H., Todo N. and Kato J., Permeation of hydrogen through a porous vycor glass membrane, *Adv. Hydrogen Energy Progr. 2:*325 (1981).
64. Way J.D. and Roberts D.L., Hollow fiber inorganic membranes for gas separation, *Sep. Sci. Technol. 27:*219 (1992).
65. Ulhorn R.J.R., Huis In't Veld M.H.B.J., Keizer K. and Burggraaf A.J., Synthesis of ceramic membranes. Part I. synthesis of non-supported and supported γ-alumina membranes without defects, *J. Mater. Sci. 27:*527 (1992).
66. Koresh J.E. and Soffer A., Molecular sieve carbon permselective membrane. Part I. Presentation of a new device for gas-mixture separation, *Sep. Sci. Technol. 18:*723 (1983).
67. Jones C.W. and Koros W.J., Characterization of ultramicroporous carbon membranes with humidified feeds, *Ind. Eng. Chem. Res. 34:*158 (1995).
68. Yan S., Maeda H., Kusakabe K. and Morooka S., Thin palladium films formed in support pores by metal-organic chemical vapor deposition method and application to hydrogen separation, *Ind. Eng. Chem. Res. 33:*616 (1994).
69. Masuda T., Sato A., Hara H., Kouno M. and Hashimoto K., Preparation of a dense ZSM-5 zeolite film on the outer surface of an alumina ceramic filter, *Appl. Catal. A 111:*143 (1994).
70. Al-Shammary A.F.Y., Caga I.T., Winterbottom J.M., Tata A.Y. and Harris I.R., Palladium based diffusion membranes as catalysts in ethylene oxidation, *J. Chem. Technol. Biotechnol. 52:*571 (1991).
71. Gavalas G.R., Megiris C.E. and Nam S.W., Deposition of H_2-permeable SiO_2 films, *Chem. Eng. Sci. 44:*1829 (1989).
72. Yan Y., Davis M.E. and Gavalas G.R., Preparation of zeolite ZSM-5 membranes by in-situ crystallization on porous α-Al_2O_3, *Ind. Eng. Chem. Res. 34:*1652 (1995).
73. Teraoka Y., Zhang H., Furukawa S. and Yamazoc N., Oxygen permeation through perovskite-type oxides, *Chem. Lett.* 1743 (1985).
74. Lin Y.S., Chang C.H. and Gopalan R., Improvement of thermal stability of porous nanostructured ceramic membranes, *Ind. Eng. Chem. Res. 33:*860 (1994).

75. Lin Y.S. and Burggraaf A.J., CVD of solid oxides in porous substrates for ceramic membrane modification, *AIChE J. 38:*445 (1992).
76. Lin Y.S., de Vries K.J., Brinkman H.W. and Burggraaf A.J., Oxygen semipermeable solid oxide membrane composites prepared by electrochemical vapor deposition, *J. Membr. Sci. 66:*211 (1992).
77. Ulhorn R.J.R., Huis In't Veld M.H.B.J., Keizer K. and Burggraaf A.J., High permselectivities of microporous silica-modified γ-alumina membranes, *J. Matls. Sci. Lett. 8:*1135 (1989).
78. de Lange R.S.A., Hekkink J.H.A., Keizer K. and Burggraaf A.J., Microporous sol-gel modified membranes for hydrogen separation. Presented at ICIM '91, Montpellier, France (1991).
79. Asaeda M. and Du L.D., Separation of alcohol/water gaseous mixtures by thin ceramic membranes, *J. Chem. Eng. Jpn. 19:*72 (1986).
80. Asaeda M., Du L.D. and Fuji M., Separation of alcohol/water gaseous mixtures by improved ceramic membranes, *J. Chem. Eng. Jpn. 19:*84 (1986).
81. Okubo T. and Inoue H., Introduction of specific gas selectivity to porous glass membranes by treatment with tetraethoxysilane, *J. Membr. Sci. 42:*109 (1989).
82. Xomeritakis G. and Lin Y.S., Chemical vapor deposition of solid oxides in porous media for ceramic membrane preparation. Comparison of experimental results with semianalytical solutions, *Ind. Eng. Chem. Res. 33:*2607 (1994).
83. Jones C.W. and Koros W.J., Carbon molecular sieve gas separation membranes—I. Preparation and characterization based on polyimide precursors, *Carbon 32:*1419 (1994).
84. Jones C.W. and Koros W.J., Carbon molecular sieve gas separation membranes—II. Regeneration following organic exposure, *Carbon 32:*1427 (1994).
85. Matsushita Electric Ind., Jpn. Pat. Appl. 60–129, 119 (1985).
86. Sano T., Yangishita H., Kiyozumi K., Mizukami F. and Haraya K., Separation of ethanol-water mixture by silicalite membrane on pervaporation, *J. Membr. Sci. 95:*221 (1994).
87. Sano T., Ejiri S., Hasegawa M., Kawakami G., Enomoto N., Tamai Y. and Yangishita H., Silicalite membrane for separation of acetic acid-water mixture, *Chem. Lett.:*153 (1994).
88. Sano T., Kiyozumi Y., Kawamura M., Mizukami F., Takaya H., Mouri T., Inaoka W., Toida Y., Watanabe M. and Toyoda K., Preparation and characterisation of ZSM-5 zeolite film, *Zeolites 11:*842 (1991).
89. Suzuki K., Kiyozumi Y. and Sekine T., Preparation and characterization of a zeolite layer, *Chemistry Express 5:*793 (1990).
90. Geus E.R., Preparation and characterisation of composite inorganic zeolite membranes with moleculare sieve properties. Ph.D. Dissertation, Technical University of Delft, The Netherlands (1993).
91. Geus E.R., den Exter M.J. and van Bekkum H., Synthesis and characterization of zeolite (MFI) membranes on porous ceramic supports, *J. Chem. Soc. Faraday Trans. 88:*3101 (1992).
92. Tsikoyiannis J.G. and Haag W.O., Synthesis and characterization of a pure zeolitic membrane, *Zeolites 12:*126 (1992).
93. Bakker W.J.W., Metal-supported zeolite membranes, in *OSPT Procestechnologie,* A.C.M. Franken, ed., (1993), p. 65.
94. Matsukata M., Nishiyama N. and Ueyama K., Preparation of a thin zeolite membrane, *Studies on Surface Science and Catalysis 84:*1183 (1994).
95. Shelby J.E., Molecular solubility and diffusion, in *Treatise of Materials Science and Technology,* Academic Press, New York, (1979), Vol. 17.
96. Nam S.W. and Gavalas G.R., Stability of H_2-permselective SiO_2 films formed by chemical vapor deposition, *AIChE Symp. Ser. 85(268):*68 (1989).
97. Ioannides T. and Gavalas G.R., Catalytic isobutane dehydrogenation in a dense silica membrane reactor, *J. Membr. Sci. 77:*207 (1993).

98. Megiris C.E. and Glezer J.H.E., Synthesis of H_2-permselective membranes by modified chemical vapor deposition—Microstructure and permselectivity of SiO_2/C/Vycor membranes, *Ind. Eng. Chem. Res. 31:*1293 (1992).
99. Itoh N., Kato T., Uchida K. and Haraya K., Preparation of pore-free disk of $La_{1-x}Sr_xCoO_3$ mixed conductor and its oxygen permeability, *J. Membr. Sci. 92:*239 (1994).
100. Govind R. and Zaho R., Selective separation of hydrogen at high temperature using proton conductive membranes, presented at the AIChE Annual Meeting, Miami, FL, paper 23 (1992).
101. Meng G., Cao C., Yu W., Peng D., de Vries K. and Burggraaf A.J., Formation of ZrO_2 and YSZ layers by microwave plasma assisted MOCVD process. Presented at ICIM '91, Montpellier, France, 11 (1991).
102. Lee A.L., Zabransky R.F. and Huber W.J., Internal reforming development for solid oxide fuel cells, *Ind. Eng. Chem. Res. 29:*766 (1990).
103. Eng D. and Stoukides M., Catalytic and electrocatalytic methane oxidation with solid oxide membranes, *Catal. Rev.—Sci. Eng. 33:*375 (1991).
104. van Dijk M.P., de Vries K.J. and Burggraaf A.J., Study of oxygen electrode reaction using mixed conducting oxide surface layers. I. Experimental methods and current overvoltage experiments, *Solid State Ionics 21:*73 (1986).
105. Chai M., Machida M., Eguchi K. and Arai H., Promotion of hydrogen permeation on metal-dispersed alumina membranes and its application to a membrane reactor for methane steam reforming, *Appl. Catal. A 110:*239 (1994).
106. Tsotsis T.T., Champagnie A.M., Vasileidias S.P., Ziaka Z.D. and Minet R.G., The enhancement of reaction yield through the use of high temperature membrane reactors, *Sep. Sci. Technol. 28:*397 (1993).
107. Bracht M., Bos A., Pex P.P.A.C., van Veen H.M. and Alderlisten P.T., Water gas shift membrane reactor. Presented at the 1st International Workshop on Catalytic Membranes, Lyon, France (1994).
108. Seok D.R. and Hwang S.-T., Recent development in membrane reactors, *St. Surf. Sci. Catal. 54:*248 (1990).
109. Ziaka Z.D., Minet R.G. and Tsotsis T.T., A high temperature catalytic membrane reactor for propane dehydrogenation, *J. Membr. Sci. 77:*221 (1993).
110. Gallaher G.R., Gerdes T.E. and Liu P.K.T., Experimental evaluation of dehydrogenations using catalytic membrane processes, *Sep. Sci. Technol. 28:*309 (1993).
111. Tiscareno-Lechuga F., Hill Jr. G.C. and Anderson M.A., Experimental studies of the non-oxidative dehydrogenation of ethylbenzene using a membrane reactor, *Appl. Catal. 96:*33 (1993).
112. Nozaki T. and Fujimoto K, Oxide ion transport for selective oxidative coupling of methane with new membrane reactor, *AIChE J. 40:*870 (1994).
113. Hazbun E.A., U. S. Pat. 4,791,079 (1988).
114. Gür T.M. and Huggins R.A., Methane synthesis over transition metal electrodes in a solid state ionic cell, *J. Catal. 102:*443 (1986).
115. Chan K.K. and Brownstein A.M., Ceramic membranes—Growth prospects and opportunities, *Ceramic Bulletin 70:*703 (1991).
116. McHenry D.J. and Winnick J., Electrochemical membrane process for flue gas desulfurization, *AIChE J. 40:* 143 (1994).
117. Cicero D.C. and Jarr L.A., Application of ceramic membranes in advanced coal-based power generation systems, *Sep. Sci. Technol. 25:*1455 (1990).
118. Alexander S.R. and Winnick J., Removal of hydrogen sulfide from natural gas through an electrochemical membrane separator, *AIChE J. 40:*613 (1994).
119. Kameyama T., Dokiya M., Fujishige M., Yukokawa H. and Fukuda K., Posibility of effective production of hydrogen from hydrogen sulfide by means of a porous vycor glass membrane, *Ind. Eng. Chem. Fundam. 20:*97 (1981).

120. Collins J.P., Way J.D. and Kraisuwansarn N., A mathematical model of a catalytic membrane reactor for the decomposition of NH_3, *J. Membr. Sci. 77:*265 (1993).
121. Caruana C.M., Catalytic membranes beckon, *Chem. Eng. Progr.* (November issue):13 (1994).
122. Deng J. and Wu J., Formaldehyde production by catalytic dehydrogenation of methanol in inorganic membrane reactors, *Appl. Catal. A, 109:*63 (1994).
123. Lafarga D., Santamaria J. and Menendez M., Methane oxidative coupling using porous ceramic membrane reactors. Part I. Reactor development, *Chem. Eng. Sci. 49:*2005 (1994).
124. Sloot H.J., Smolders C.A., van Swaaij W.P.M. and Versteeg G.F., High-temperature membrane reactor for catalytic gas-solid reactions, *AIChE J. 38:*887 (1992).
125. Montanaro L. and Saracco G., Influence of some precursors on the physico-chemical characteristics of transition aluminas for the preparation of ceramic catalytic filters, *Ceramics International 21:*43 (1995).
126. Saracco G. and Montanaro L., Catalytic ceramic filters for flue gas cleaning. I. Preparation and characterisation, *Ind. Eng. Chem. Res. 34:*1471 (1995).
127. Saracco G. and Specchia V., Studies on sol-gel derived catalytic filters, *Chem. Eng. Sci. 50:*3385 (1995).
128. Bos A., Doesburg E.B.M. and Engelen C.W.R., Preparation of catalysts on a ceramic substrate by sol-gel technology, *Eurogel '91,* S. Vilminot, Nass R. and Schmidt H., eds., Elsevier, Amsterdam, (1992), p. 103.
129. Podyacheva O., Ketov A., Ismagilov Z., Ushakov V., Bos A. and Veringa H., Development of supported perovskite catalysts for high temperature combustion, in *Environmentl Catalysis,* Centi G. et al., eds, SCI Publ., Rome, (1995), p. 599.
130. Zaspalis V.T., van Praag W., Keizer K., van Ommen J.G., Ross J.H.R. and Burggraaf A.J., Reactor studies using alumina separation membranes for dehydrogenation of methanol and *n*-butane, *Appl. Catal. 74:*223 (1991).
131. Harold M.P., Zaspalis V.T., Keizer K. and Burggraaf A.J., Intermediate product yield enhancement with a catalytic inorganic membrane—I. Analytical model for the case of isothermal and differential operation, *Chem. Eng. Sci. 48:*2705 (1993).
132. Cini P. and Harold M.P., Experimental study of the tubular multiphase catalyst, *AIChE J. 37:*997 (1991).
133. Peureux J., Torres M., Mozzanega H., Giroir-Fendler A., and Dalmon J.A., Nitrobenzene liquid-phase hydrogenation in a membrane reactor, *Catalysis Today 25:*409 (1995).
134. Chanaud P., Julbe A., Larbot A., Guizard C., Cot L., Borges H., Giroir-Fendler A. and Mirodatos C., Catalytic membrane reactor for oxidative coupling of methane. Part 1: Preparation and characterization of LaOCl membranes, *Catalysis Today 25:*225 (1995).
135. Harold M.P., Zaspalis V.T., Keizer K. and Burggraaf A.J., Improving partial oxidation product yield with a catalytic inorganic membrane. Presented at the 5th Annual Meeting of the NAMS, Lexington, Kentucky, (1992), paper 11B.
136. Sloot H.J., *A non-permselective membrane reactor for catalytic gas-phase reactions,* Ph.D. Dissetation, Twente University of Technology, Enschede, The Netherlands (1991).
137. Harold M.P., Cini P., Patenaude B. and Venkataraman K., The catalytically impregnated ceramic tube: An alternative multiphase reactor, *AIChE Symposium Series 85 (268):*26 (1989).
138. Song J.Y. and Hwang S.-T., Formaldehyde production from methanol using a porous Vycor glass membrane. Proceedings of ICOM '90, Chicago, (30)540 (1990).
139. Itoh N., Shindo Y., Haraya T. and Hakuta T., A membrane reactor using microporous glass for shifting equilibrium of cyclohexane dehydrogenation, *J. Chem. Eng. Jpn. 21:*399 (1988).
140. Tsotsis T.T., Champagnie A.M., Vasileiadis S.P., Ziaka E.D. and Minet R.G., Packed bed catalytic membrane reactors, *Chem. Engng. Sci. 47:*2903 (1992).
141. Sun Y.M. and Khang S.-J. Catalytic membrane for simultaneous chemical reaction and separation applied to a dehydrogenation reaction, *Ind. Eng. Chem. Res. 27:*1136 (1988).

142. Sun Y.M. and Khang S.-J., A catalytic membrane reactor: Its performance in comparison with other types of reactors, *Ind. Eng. Chem. Res. 29:*232 (1990).
143. Champagnie A.M., Tsotsis T.T., Minet R.G. and Wagner E., The study of ethane dehydrogenation in a catalytic membrane reactor, *J. Catal. 134:*713 (1992).
144. Yeung K.L., Aravind A., Zawada R.J.X., Szegner J., Cao G. and Varma A., Nonuniform catalyst distribution for inorganic membrane reactors: Theoretical considerations and preparation techniques, *Chem. Eng. Sci. 49:*4823 (1994).
145. Itoh N. and Govind R., Development of a novel oxidative membrane reactor, *AIChE Symp. Ser. 85 (268):*10 (1989).
146. Itoh N., Dehydrogenation by membrane reactors, *Sekiyu Gakkaishi 33:*136 (1990).

18

Ceramic Catalyst Supports and Filters for Diesel Exhaust Aftertreatment

Suresh T. Gulati
Corning Incorporated, Corning, New York

I. HISTORICAL BACKGROUND

The diesel engine is the most fuel-efficient internal combustion engine, with superior durability and reliability as compared to other internal combustion engines. It is the workhorse for providing power to trucks, buses, cars, and locomotives in addition to marine, construction, agriculture, and mining equipment. The end user has treasured its fuel economy, durability, and performance plus minimal maintenance. However, the early engines were built and operated with less regard to exhaust emissions and subsequent air pollution. Emissions of smoke and odor, with their undesirable smell, were considered mere nuisance. Dr. Paul H. Schweitzer, professor of mechanical engineering at Pennsylvania State University, believed that there were cures for diesel's "bad breath" [1]. He testified before the U.S. Congress on diesel engine pollution, which led to the 1965 amendments to the Clean Air Act that required research into ways of controlling both the smoke and the odor. The initial studies of smoke and odor were sponsored by the National Air Pollution Control Administration of the U.S. Department of Health, Education and Welfare. These studies launched diesel technology in a new direction that was to open up new challenges and opportunities for the transportation industry. In 1970, the Clean Air Act was amended further and the Environmental Protection Agency (EPA) was created. Congressional edicts held the engine manufacturers accountable to an entirely new type of driving force, namely, the environment.

A. EPA Smoke Test

The chief transportation target of the 1970 Clean Air Act was the automobile; hence, stringent regulations were directed at automakers. Visible smoke from diesel engines used in heavy-duty vehicles was regulated first. There was a concern that smoky stacks would result in many more complaints as motorists had to pay more for cleaned-up cars. Thus, the first smoke opacity laws were based on the appearance of smoke to the eye. Although the health effect was debated at the time, there was little doubt that smoke obscures visibility. This alone was a sufficient basis for California and U.S. regulations. In 1970,

California required the roadside in-use visual inspection, while EPA mandated engine smoke certification using a new end-of-stack opacity meter. These regulations leveled the playing field for Japanese, European, and U.S. manufacturers. The 1970 opacity limits were set at 40% during acceleration and 20% during simulated hill climbing at full load. These limits were tightened in 1974 to 20% and 15%, respectively. A momentary peak opacity limit of 50% was added on.

B. Diesel Emissions Regulations

In 1974, due to health concerns, in addition to tightening the opacity limits, EPA introduced emissions regulations for hydrocarbons (HC), carbon monoxide (CO), and oxides of nitrogen (NO_x). The effect of CO poisoning by replacing oxygen in the blood as well as the photochemical formation of atmospheric ozone from reactive hydrocarbons and oxides of nitrogen were well established. In 1983, the 13-mode steady-state test was replaced by a transient emissions test that tightened these regulations further; see Table 1. In 1988, particulate regulations were introduced in response to chronic health impairment from fine particles, e.g., emphysema and bronchitis.

Today, the substances of biggest concern are particulates and oxides of nitrogen. Visible smoke has been reduced to a minimal level through improved combustion. Likewise, HC and some of the exhaust odorants have been reduced by means of improved combustion efficiency. Even CO emission has not been a major problem for the inherently low-CO diesel engine. However, with higher combustion efficiency, the NO_x emissions have increased. Thus, both NO_x and particulates have become a major concern. The use of turbochargers has helped not only in increasing the power output but also in improving the air/fuel ratio, which lowers the peak combustion temperature and NO_x emissions. Since the particulates can also arise from oil consumption, which is high relative to gasoline engine, design changes in piston liner, piston rings, turbo seals, and valve oil seals have been necessary to reduce oil consumption without compromising engine life. Durability requirements for diesel engines, as set forth by EPA, call for a minimum of 8 years or the mileage listed in Table 2.

II. DIESEL EMISSIONS AND CHARACTERIZATION

Although the on-highway diesel engine has received the greatest attention in terms of emissions, a large number of diesel engines serve to power ships, locomotives, off-

Table 1 U.S. EPA Emissions Limits for Heavy-Duty Diesel Engine (9/bhp hr)

Year	HC	CO	NO_x	Particulate material	Fuel sulfur (wt%)
1988	1.3	15.5	10.7	0.60	
1990	1.3	15.5	6.0	0.60	
1991	1.3	15.5	5.0	0.25	
1994	1.3	15.5	5.0	0.10	0.05
1996	1.3	15.5	4.0	0.05	0.02
1998	1.3	15.5	4.0	0.10	0.02

Table 2 EPA's Requirements for Engine Durability

Engine size	GVW (10^3 lb)	Durability (yr)	Durability (10^3 mi)
Light heavy-duty	10–19.5	8	110
Medium heavy-duty	19.5–33	8	185
Heavy heavy-duty	>33	8	290

road vehicles, and equipment used in mining, construction, and agriculture. Emissions regulations were expected to take effect for nonroad diesel engines over 50 hp in 1996. Similarly, new diesel-powered locomotives were also expected to comply with emissions legislation by the end of 1995.

The other significant segment of diesel application is the passenger car, notably in Europe and Japan. While the diesel car is uncompetitive in the United States relative to the gasoline engine in terms of emissions, performance, and price, its fuel economy in Europe and Japan is much more superior due to fuel price and tax policy. Consequently, passenger cars with diesel engine are quite popular in Europe and constitute a significant fraction of total passenger car fleet, as shown in Fig. 1 [2]. Of the total worldwide diesel production of approximately 15×10^6 units [2], passenger cars account for 31%, medium-duty trucks and buses account for 30%, heavy-duty trucks and buses account for 15%, stationary applications account for 15%, and off-road transportation accounts for 9% [2].

A. Exhaust Gas Composition

Diesel exhaust consists of gaseous, liquid, and solid emissions. Gaseous emissions, whose approximate size is shown in Fig. 2, comprise of N_2, CO_2, CO, H_2, NO/NO_2, SO_2/SO_3,

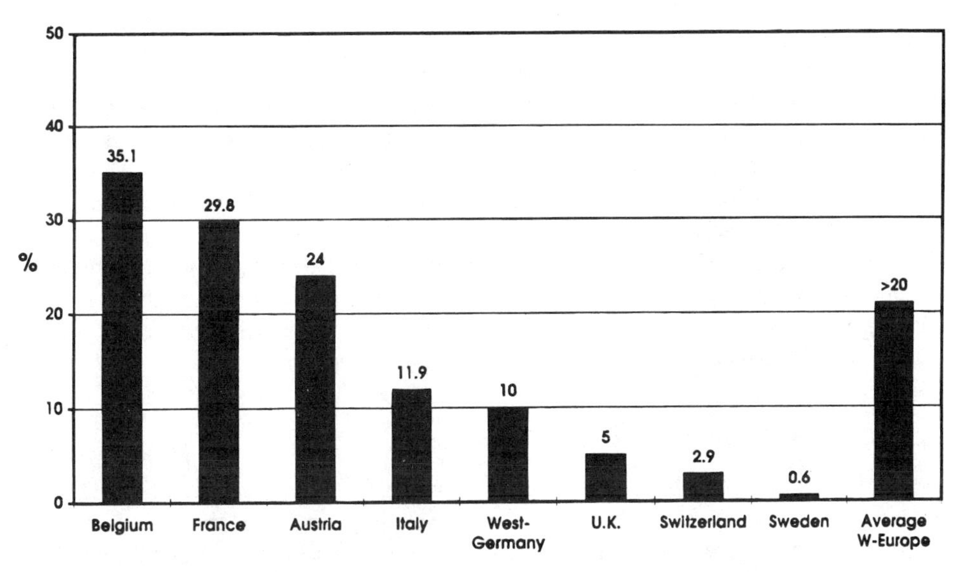

Figure 1 Relative importance of diesel-powered passenger cars in Europe during 1984. (Courtesy of Elsevier Science Publishers.)

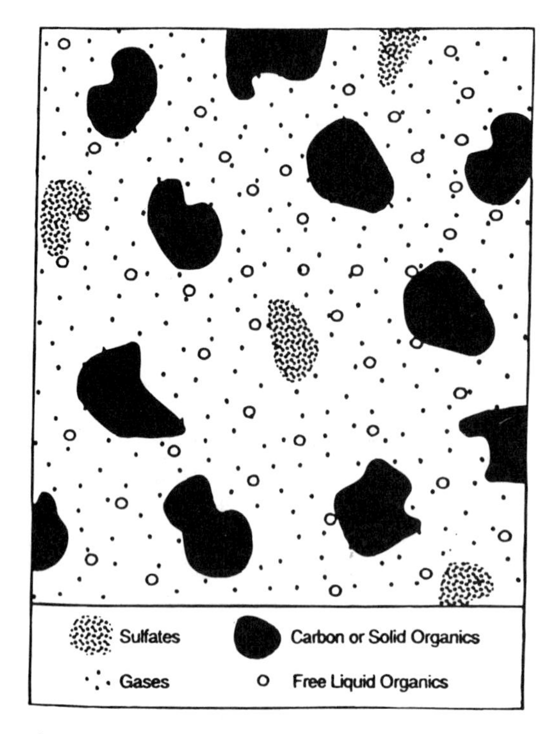

Figure 2 Schematic of diesel emissions showing approximate sizes. (Courtesy of Automotive Engineering.)

HC(C_2–C_{15}), oxygenates, and organic nitrogen and sulfur compounds. Liquid emissions include H_2O, H_2SO_4, HC(C_{15}–C_{40}), oxygenates, and polyaromatics. Solid emissions are made up of dry soot, metals, inorganic oxides, sulfates, and solid hydrocarbons. The physical and chemical processes responsible for soot formation are illustrated in Fig. 3, which also shows the soot size and production time. Similarly, the formation of diesel particulate, with soot carbon at the center surrounded by other organic and inorganic combustion products, is illustrated in Fig. 4. The World Health Organization has labeled the polar fraction of diesel particulate as potentially carcinogenic and hence a potential health hazard. The relative contribution of the above-listed emissions to total emissions in Europe, where diesels are most popular, has been estimated as follows [2]: CO at 2.2%, HC at 6.5%, NO_x at 23%, and particulates at 11.4%. Particulates arise from soot, sulfates, and hydrocarbons (from both fuel and oil). Of the three methods commonly employed for trapping these particulates, namely, wall-flow filter, wiremesh filter, and ceramic foams, we will focus on the wall-flow filter in this chapter.

B. Worldwide Legislation for Diesel Emissions

Since the diesel engine is the workhorse of urban buses and heavy-duty trucks, worldwide, and since diesel exhaust poses health concerns, including cancer, emphysema, and other respiratory diseases, the aftertreatment of diesel exhaust is receiving worldwide attention. Both wall-flow filters and catalytic converters are being tested, the former for trapping solid particulates and the latter for converting hydrocarbons by oxidizing the soluble

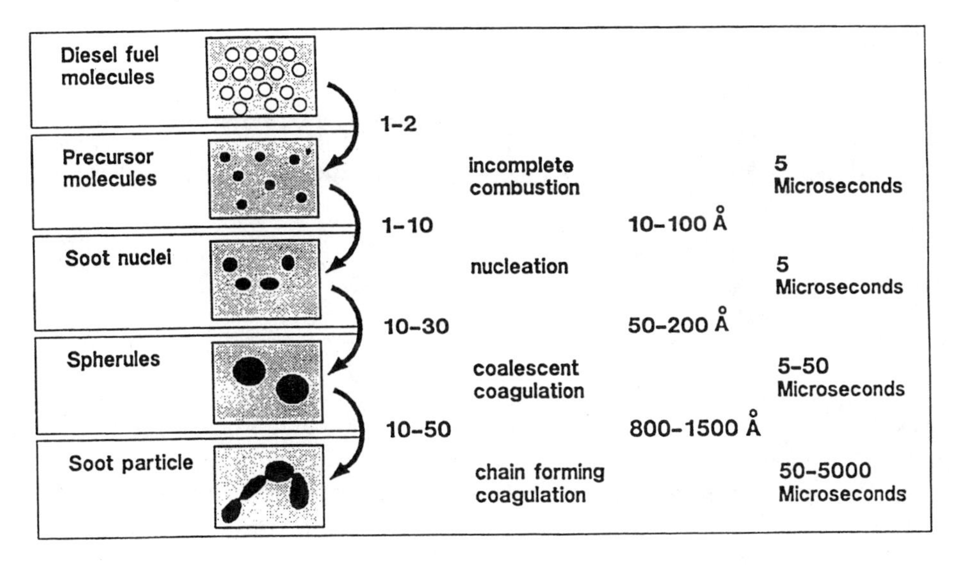

Figure 3 Physical and chemical processes leading to soot formation. (Courtesy of Elsevier Science Publishers.)

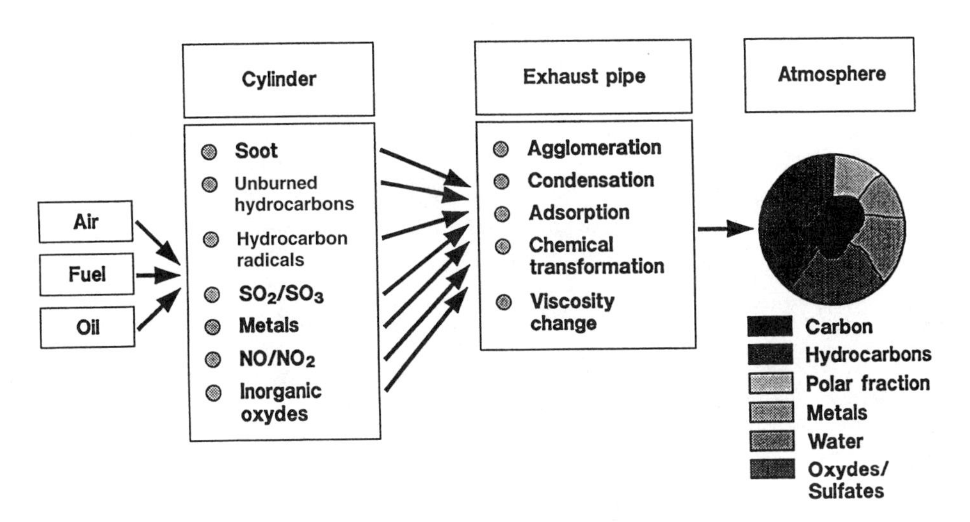

Figure 4 Physical and chemical processes responsible for diesel particulate formation. (Courtesy of Elsevier Science Publishers.)

organic fraction (SOF) portion of diesel exhaust [3]. Table 3 summarizes the worldwide legislation for particulate material (PM) and oxides of nitrogen (NO_x) through the end of this century.

Many field tests are under way, or have been completed, to prove the performance of filter and converter systems. Buses fitted with diesel filters are being, or have been, tested in New York, Philadelphia, Los Angeles, Munich, Athens, Taipei, Seoul, Tokyo, Mexico City, Sao Paulo, Santiago, and other urban areas of the world. With advances in

Table 3a Heavy-Duty Truck Engine Emission Standards (g/bhp/hr)

	North America		Japan		European Union	
Present	NO_x	5.0	NO_x	5.81 DI	NO_x	5.97
	PM	0.10		5.07 IDI	PM	0.27
			PM	0.71		
1996 MY					NO_x	5.22
					PM	0.11
1998 MY	NO_x	4.0				
	PM	0.10				
1999 MY			Proposed		Proposed	
			NO_x	3.36	NO_x	3.73
			PM	0.19	PM	0.075
2004 MY	NO_x	2.0				
Test Cycle	Transient Speed/Load		Japan 13—Mode		European 13—Mode	
Test Fuel	S .03–.05%		S .05% 1995?		S 0.2% 10/1/94	
	Cetane No. 40–48		Cetane No. 45 min.		0.05% 10/1/96	
					Cetane 49–53	

Table 3b The 2000 Values: The German Proposal for Exhaust Emission Standards for Passenger Cars (conformity of production) in the New European Driving Cycle (g/km)

	Stage 1 from 92/93 91/441/EEC	Stage 2 from 96/97 94/12/EEC	UBA proposal for 2000	UBA proposal for 2003
Gasoline:				
CO	3,16	2,2	2,2	
HC + NO_x	1,13	0,5		
NO_x			0,14	
NMHC			0,05	
Diesel:				
CO	3,16	1,0	1,0	1,0
HC + NO_x	1,13	0,7		
NO_x			0,4	0,14
NMHC			0,05	0,05
PM	0,18	0,08	0,05	0,025

engine design, low-sulfur fuel, sensor technology, and ceramic composition, the diesel filter is becoming one of the viable aftertreatment devices, notably for buses and trucks.

III. DESIGN/SIZING OF DIESEL PARTICULATE FILTER

The ceramic wall-flow filter is an innovative extension of the extruded honeycomb catalyst support described in Chapter 2. This filter concept, shown in Fig. 5, involves having the alternate cell openings on one end of the unit plugged in checkerboard fashion. The

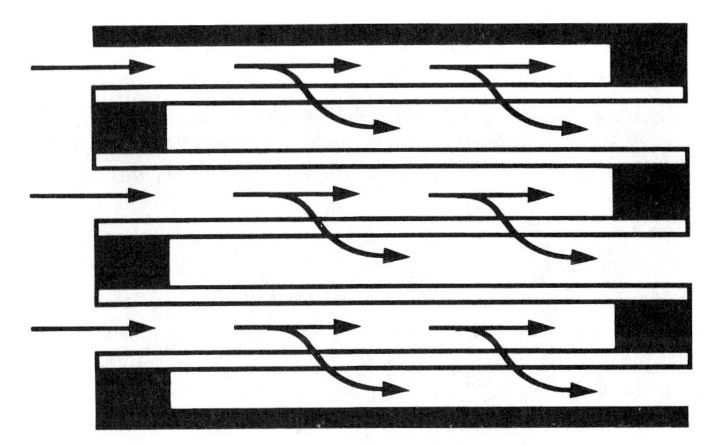

Figure 5 Wall-flow filter concept with alternate plugged cells. (Courtesy of the Society of Automotive Engineers.)

opposite end or face is plugged in a similar manner, but one cell displaced, allowing no direct path through the unit from one end to the other, as indicated in Fig. 6. The exhaust gas entering the upstream end is therefore forced through the wall separating the channels and exits through the opposite end by way of an adjacent channel. In this way, the walls of the honeycomb are the filter medium [4,5]. They can be made sufficiently porous to allow exhaust gas to pass through without excessive pressure drop. This wall-flow concept offers a large amount of filter surface area in a reasonably compact volume together with high filtration efficiency. Periodically, the soot that is collected is oxidized to CO_2—a process known as regeneration—which renders the filter clean. The fact that the filter is constructed of a special ceramic material results in its capability of withstanding high temperatures while being chemically inert. More precisely, it is a porous cordierite ceramic with magnesia/alumina/silica composition ($2MgO_2 \cdot Al_2O_3 \cdot 5SiO_2$). A key property of this

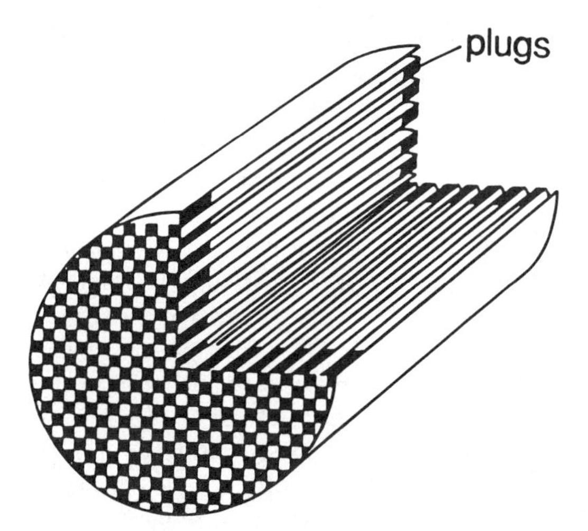

Figure 6 Schematic of diesel filter with checkerboard plug pattern. (Courtesy of the Society of Automotive Engineers.)

composition is a very low coefficient of thermal expansion. The material used to plug the cell openings in the faces is similar in nature to the body in composition and thermal characteristics. It is a high-temperature foaming cement that during firing seals to the cell walls and is impervious to gas flow. The walls contain a series of interconnected pores of a volume and size sufficient to enable the exhaust gas to flow completely through but restrain most of the particles.

The performance characteristics of wall-flow filter can be varied and managed. Its collection efficiency can be controlled to a large degree by the properties of the walls that form the channels. These include total pore volume, pore size distribution, and the thickness of the wall itself. The flow through the wall can be made more restrictive by adjusting the porosity of the wall. A smaller pore volume creates a highly efficient filter but at the same time restricts the flow and produces high back pressure. Conversely, with porosity adjusted in the opposite direction, low back pressure is achieved, but at the expense of reduction in collection efficiency.

A. Performance Requirements

The four basic requirements that the filter must meet are [6]:

1. Adequate filtration efficiency to satisfy particulate emissions legislation
2. Low pressure drop to minimize fuel penalty and conserve engine power
3. High thermal shock resistance to ensure filter integrity during soot regeneration
4. High surface area per unit volume for compact packaging

Although a high filtration efficiency would make the filter more effective, it must not be accomplished at the expense of high back pressure or low thermal integrity. Indeed, the microstructure and plugging pattern of the ceramic filter can be tailored to obtain filtration efficiencies ranging from 50% to 95% per engine manufacturers' specification. Furthermore, recent advances in ceramic composition have led to filters with high filtration efficiency, acceptable back pressure, and excellent thermal integrity [7].

As the particulate matter is trapped in the filter walls, it begins to build up on the surface of open cells, forming a soot layer that also acts as a membrane. With increasing thickness of the soot layer, the hydraulic diameter of the channel decreases, resulting in higher back pressure. Obviously, the initial and final channel sizes must be controlled via filter design and soot accumulation level to limit the back pressure to acceptable value. Again, this can be accomplished by designing the microstructure, the cell geometry, the plugging pattern, and the size of ceramic filter, which in turn are dictated by engine size, flow rate, and engine-out emissions.

The dominant component of trapped particulates is soot carbon, which is formed during the combustion of a fuel-rich mixture in the absence of adequate oxygen. Although some of the soot may be oxidized to CO_2 during the latter part of the power stroke, a major portion does not get oxidized due to slow process [8]. The other major component of particulate matter consists of heavy unburned hydrocarbons. Since the chemical energy of soot carbon and heavy hydrocarbons is high, once they are ignited during regeneration they release a great deal of heat that, if not dissipated continuously, can result in high temperature gradients within the filter [9]. Thermal stresses associated with such gradients must be kept below the fatigue threshold of the filter material to ensure its thermal integrity over its lifetime [10,11]. This is best accomplished by using a ceramic composition with ultralow thermal expansion and modestly high fatigue threshold [7]. Other approaches to improving thermal integrity include the use of fuel additives and/or catalysts to effect regeneration at lower temperature [9]. Alternatively, more frequent regenerations can also reduce the temperature gradients and enhance thermal integrity, but at the expense of fuel penalty if a burner is used for regeneration.

The honeycomb configuration of ceramic filters offers a high surface area per unit volume, thereby permitting a compact filter size [12]. The absolute filtration surface area depends on cell size, filter volume, and the plugging pattern, all of which are design parameters whose optimization, as will be shown shortly, calls for trade-offs in pressure drop, filtration efficiency, mechanical durability, thermal integrity, and space availability.

B. Composition and Microstructure

The filter composition that has performed successfully over the past decade is cordierite ceramic with the chemical formula of $2MgO \cdot 2Al_2O_3 \cdot 5SiO_2$. Its unique advantages include low thermal expansion, which is ideal for thermal shock resistance, and a tailorable microstructure, to meet filtration and pressure drop requirements. The extrusion technology for producing automotive catalyst supports also helps manufacture diesel filters. Consequently, the unit cell design can be achieved via die design, while the porosity and microstructure are best controlled by composition and process modifications.

The most common cell density employed for diesel filters is 100 cells/in.^{2}, with a 0.017-in.-thick cell wall. This choice offers the best compromise in terms of filtration area and back pressure. While a 200-cells/in.^{2} structure offers 41% larger filtration area and has been used for diesel filters, it may result in higher pressure drop. Similarly, thicker cell walls (0.025 in. thick) offer 50% higher strength but result in higher pressure drop. Another parameter that affects pressure drop is mean pore size, which can range from 12 μm to 35 μm. Although the pressure drop decreases with increasing pore size, so does filtration efficiency; hence, a compromise is necessary in tailoring the pore size. The wall porosity also affects pressure drop and mechanical strength. Both pressure drop and mechanical strength decrease as wall porosity increases, thus calling for a compromise in selecting the wall porosity. Most filter compositions and manufacturing processes are designed to yield a wall porosity of 48–50%. Table 4 summarizes the geometric and microstructural properties of six different extruded cordierite diesel filters [4,7,13,14]. Filters with low mean pore size, namely, EX-47 and DHC-221, are designed to offer high filtration efficiency (>90%); those with intermediate mean pore size, namely, EX-54 and

Table 4 Properties of Extruded Cordierite Diesel Filters with 100/17 Cell Structure and Checkerboard Plugging Pattern

Filter designation	Wall porosity	Mean pore size (μm)	Open frontal area	Specific filtration area (in.^{2}/in.^{3})
EX-47	50%	13.4	34.4%	16.6
EX-54	50%	24.4	34.4%	16.6
EX-66	50%	34.1	34.4%	16.6
EX-80	48%	13.4	34.4%	16.6
DHC-221	48%	13.0	34.4%	16.6
C-356E	50%	20.0	34.4%	16.6
DHC-141E	48%	35.0	34.4%	16.6

C-356E, are designed for medium filtration efficiency (80–90%); and those with large mean pore size, namely, EX-66 and DHC-141E, are designed for low filtration efficiency (60–75%).

C. Cell Configuration and Plugging Pattern

Figure 6 shows the wall-flow filter with square cell configuration and checkerboard plugging pattern. The open frontal area (OFA) and specific filtration area (SFA) for such a filter are defined in terms of cell spacing L and wall thickness t; see Figure 7b.

$$\text{OFA} = 0.5\left(\frac{L - t}{L}\right)^2 \tag{1}$$

$$\text{SFA} = \frac{2(L - t)}{L^2} \tag{2}$$

Since the cell density N for square cell structure is given by

$$N = \frac{1}{L^2} \tag{3}$$

it follows from Eq. (2) that the specific filtration area is directly proportional to cell density. On the other hand, as the cell density increases, the hydraulic diameter defined by

$$D_h = L - t \tag{4}$$

decreases. Hence a portion of the total pressure drop due to gas flow through the open channels of the filter, which depends inversely on the square of the hydraulic diameter, increases. Thus, care must be exercised in selecting the appropriate cell density [12].

Other factors that play a key role in designing the filter are its mechanical integrity and filtration capacity. The former is defined by mechanical integrity factor MIF, which, for a given wall porosity, depends on cell geometry via

$$\text{MIF} = \frac{t^2}{L(L - t)} \tag{5}$$

The filtration capacity is the total amount of soot that can be collected prior to safe regeneration. It is related directly to total filtration area TFA, defined by the product of specific filtration area and filter volume; i.e.,

$$\text{TFA} = \frac{2(L - t)}{L^2} V_f \tag{6}$$

where the filter volume V_f is given by

$$V_f = \frac{\pi}{4} d^2 \ell \tag{7}$$

in which d and ℓ denote filter diameter and length, respectively.

As noted earlier, most filter compositions enjoy 50% wall porosity to limit the pressure drop to acceptable levels. The mean pore size, which also has a bearing on pressure drop due to gas flow through the wall, is dictated primarily by the filtration efficiency requirement. As the emissions legislation becomes more stringent, filtration

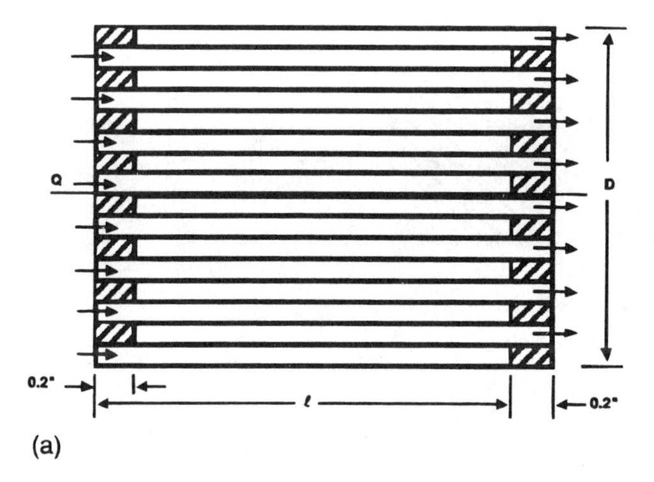

(a)

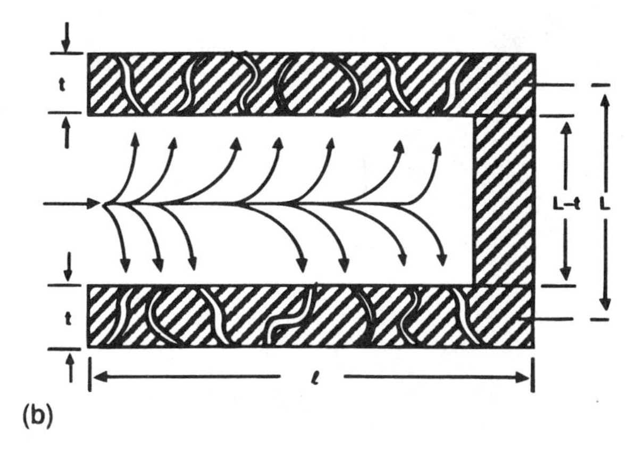

(b)

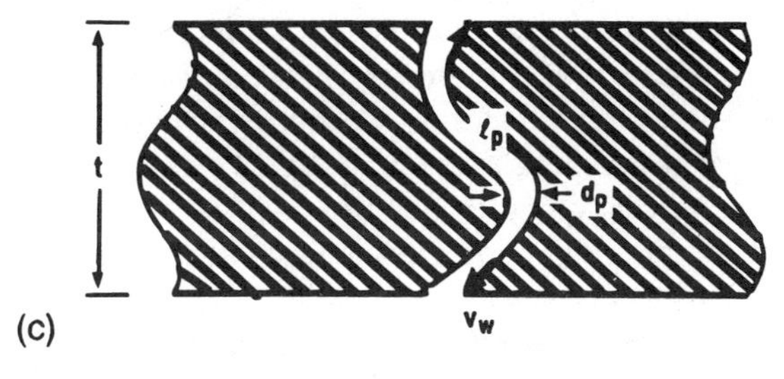

(c)

Figure 7 Flow model for pressure drop calculations. (a) Entry and exit losses; (b) pressure drop through clean channel; (c) pressure drop through cell wall; (d) pressure drop through sooted channel. (Courtesy of Society of the Automotive Engineers.)

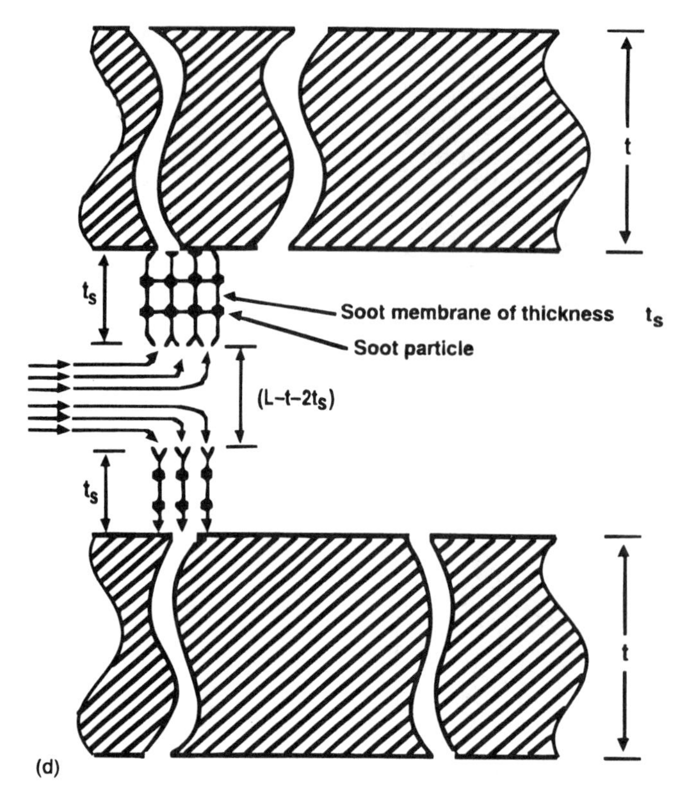

Figure 7 *Continued*

efficiencies equal to or greater than 90% become desirable, calling for a mean pore diameter of 12–14 μm. With microstructure fixed in this manner, the two common cell configurations for diesel filters that have been manufactured are 100/17 and 200/12. It may be verified that they have identical open frontal area and mechanical integrity factor. However, the specific filtration area of 200/12 is 41.5% greater than that of the 100/17 configuration, implying a lower* filter volume for the former, which may be desirable to meet space constraints. On the other hand, the hydraulic diameter of 200/12 is 30% smaller than that of the 100/17 configuration, implying a higher pressure drop for the former, which may not be acceptable. Furthermore, the 200/12 configuration may also experience fouling due to ash build-up following several regenerations.

The model for total pressure drop will be discussed in a later section; however, for a comparison of two different cell configurations we need to write the expression for the pressure drop due to gas flow through open channels, Δp_{ch}, namely,

$$\Delta p_{\mathrm{ch}} = \frac{C V_{\mathrm{ch}} \ell}{D_h^2} \tag{8}$$

where C is a constant and V_{ch} denotes the gas velocity through the channel, which is given by

* For constant total filtration area.

$$V_{\text{ch}} = \frac{Q}{A_o} \tag{9}$$

Here, Q is the flow rate through the filter and A_o is the open cross-sectional area, given by

$$A_o = \frac{\pi}{4} d^2 \times \text{OFA} \tag{10}$$

In view of identical open frontal areas, filters with the 100/17 and 200/12 cell configurations will have identical open cross-sectional areas and gas velocity through their respective channels under conditions of constant flow rate. Thus, the pressure drop Δp_{ch} will now be proportional to ℓ / D_h^2 according to Eq. (8). We call this ratio "back pressure index," or BPI:

$$\text{BPI} = \frac{\ell}{D_h^2} \tag{11}$$

Since the specific filtration area of the 200/12 configuration is 41.5% greater, the filter length with such a configuration can be 58.5% smaller than that of the filter with the 100/17 cell configuration for identical total filtration areas. In this manner, Eq. (11) helps estimate the back pressure penalty* due to smaller hydraulic diameter of 200/12 cell configuration.

The results of this exercise are summarized in Table 5, which compares the properties and performance parameters of filters with the two different cell configurations. It shows that despite the compact volume of the 200/12 filter, it will experience 17% higher back pressure than the 100/17 filter. Such a back pressure penalty, as will be shown later, may well exceed 17% as the soot membrane begins to build up on the surfaces of the open channel walls. It is clear from Table 5 that filter design often calls for trade-offs in performance parameters, which, in turn, require prioritization of durability and performance requirements on the part of the filter designer.

Table 5 Properties and Performance Parameters of Diesel Filters with Two Different Cell Configurations and Constant Total Filtration Areas

Property and performance parameter	100/17 Cell	200/12 Cell
L (in.)	0.100	0.071
t (in.)	0.017	0.012
n (cells/in.^{2})	100	200
OFA	0.345	0.345
MIF	0.035	0.035
D_h (in.)	0.083	0.059
SFA (in.^{2}/in.^{3})	16.6	23.5
BPI (in.$^{-1}$)	1.0	1.17
TFA (in.^{2})	X	X
ℓ (in.)	ℓ	0.585 ℓ

* The pressure drop due to channel flow is a significant fraction of the total pressure drop through the filter.

D. Filter Size and Contour

Both mechanical and thermal durability requirements favor a circular contour for the filter, since it lends itself to robust packaging and at the same time experiences less severe temperature gradients during regeneration. Furthermore, circular filters are easier to manufacture and control tolerances, making them more cost effective than noncircular contours. Indeed, the latter have also been manufactured for special applications where space constraint is the dominating factor.

Filter size is generally dictated by engine capacity and is normally equal to engine volume. This rule of thumb for designing filter size has worked well in both mobile and stationary applications, in that it helps control soot collection and regeneration without impairing filter durability and imposing a high back pressure penalty. We will illustrate these benefits with a realistic example.

Consider a 10-liter, 230-hp diesel engine for a medium- to heavy-duty truck for urban areas. We design the total filter volume to be 10 liters, with a microstructure commensurate with 90% filtration efficiency. Based on prior experience we limit the soot loading to 10 g/liter of filter volume to ensure safe regeneration at 2-hr intervals. Then,

$$\text{total soot collected} = \frac{10 \times 10}{2} = 50 \text{ g/hr}$$

$$\text{rate of soot emitted by engine} = \frac{50}{0.9} = 55.5 \text{ g/hr}$$

which in standard units works out to 0.24 g/bhp hr. This is a good representation of the soot output of new, modern-day diesel engines, as shown in Table 1 (post-1990 engine). Note that the filter will help reduce the soot emissions from 0.24 to 0.024 g/bhp hr due to its 90% collection efficiency.

We will next develop the pressure drop model in the next section and estimate the back pressure due to the above loading.

E. Pressure Drop Model

The pressure drop model is based on the following assumptions [15]:

1. Incompressible gas
2. Laminar flow
3. Constant density and viscosity at a given temperature
4. Circular pores in filter walls
5. No cross flow between pores

Referring to Fig. 7, we see that the total pressure drop across the filter is made up of five components, namely,

$$\Delta p_{\text{total}} = \Delta p_{\text{en}} + \Delta p_{\text{ch}} + \Delta p_w + \Delta p_s + \Delta p_{\text{ex}} \tag{12}$$

The entrance and exit losses, Δp_{en} and Δp_{ex}, are generally small compared with other losses, unless the cross-sectional area changes drastically. Hence they will be neglected. The remaining three losses can be estimated from the generic equation for a circular pipe:

$$\Delta p = \frac{32\mu v \ell}{g d^2} \tag{13}$$

where

$$\mu = \text{gas viscosity (lb/sec ft)}$$
$$v = \text{gas velocity through pipe (ft/sec)}$$
$$\ell = \text{effective length of pipe (ft)}$$
$$d = \text{effective diameter of pipe (ft)}$$
$$g = \text{gravitational acceleration (ft/sec}^2\text{)}$$

We will apply Eq. (13) to estimate each component of pressure drop through a 10-liter filter, 10.5 in. diameter $\times$ 7 in. long, with the 100/17 cell configuration. To this end, we assume

$$\text{engine size} = 10 \text{ liters} = 0.353 \text{ ft}^3$$
$$\text{engine speed} = 1500 \text{ RPM}$$
$$\text{gas temperature} = 325°\text{C}$$

Then

$$Q = \text{flow rate} = 0.353 \times 750 = 265 \text{ ft}^3\text{/min at } 325°\text{C}$$
$$= 531 \text{ SCFM}$$

$$\mu = 2 \times 10^{-5} \text{ lb/sec ft}$$

Estimate of Δp_{ch}. For the checkerboard plug pattern,*

$$A_{open} = \frac{\pi}{4}\left(\frac{10}{12}\right)^2\left(\frac{0.083}{0.100}\right)^2 0.5 = 0.188 \text{ ft}^2$$

$$v_{ch} = \frac{Q}{A_{open}} = \frac{531}{(0.188)(60)} = 47.2 \text{ ft/sec}$$

$$d_{ch} = L - t = 0.0069 \text{ ft}$$

$$\ell = 7 - 0.4 = 6.6 \text{ in.} = 0.55 \text{ ft}$$

$$\Delta p_{ch} = \frac{32\mu v_{ch}\ell}{g d_{ch}^2} = 10.83 \text{ lb/ft}^2 = 0.153 \text{ in. Hg} \tag{14}$$

Estimate of Δp_w. The effective length of pores in the filter wall depends on their tortuosity and mean pore diameter, which for the filter will be taken as 12.5 μm. The effective pore length is approximately $3t$, with t being the wall thickness [16]. The gas velocity through the pores is readily obtained by the continuity equation, namely,

$$v_{ch}(L - t)^2 = v_w \times 4P(L - t)\ell \tag{15}$$

where P denotes the fractional porosity of the filter walls, which will be taken as 0.5. Substituting $L = 0.1$ in., $t = 0.017$ in., $P = 0.5$, and $\ell = 6.6$ in. in Eq. (15), we obtain

$$v_w = 0.0063 v_{ch} = 0.297 \text{ ft/sec}$$

Substituting $\ell_p = 3t = 0.00425$ ft, $d_p = 12.5$ μm $= 0.000041$ ft, and $v_w = 0.297$ ft/sec in Eq. (13) we obtain

* We assume a diameter of 10 in. for the checkerboard region due to fully plugged peripheral region, 0.25 in. wide.

$$\Delta p_w = 14.91 \text{ lb/ft}^2 = 0.211 \text{ in. Hg}$$

Thus, the pressure drop through the wall is 38% higher than that through the channel. This estimate of Δp_w is based on clean and open pores. As these pores accumulate soot, their mean diameter will decrease, the flow velocity will increase, and Δp_w will go up. To re-estimate Δp_w, we can still use Eq. (13) once we know the amount of soot trapped in the pores.

Estimate of Δp_s. The pressure drop through the soot membrane is negligible due to both its open structure and its small thickness. However, as the membrane thickness increases with the continuous deposition of soot, the hydraulic diameter of the sooted channel decreases and the gas velocity increases, thereby contributing to Δp_{ch}. To estimate the incremental pressure drop due to the soot membrane, we must first study the kinetics of soot deposition.

Recall that the maximum allowable soot accumulation for safe regeneration is typically 10 g per liter of filter volume. For a filter volume of 10 liters, the total soot collected prior to regeneration is 100 g over a 2-hour filtration cycle. With a filtration efficiency of 90%, the soot output of 230-hp engine is given by:

$$\text{soot output} = \frac{100}{2 \times 230 \times 0.9} = 0.242 \text{ g/bhp hr}$$

$$\text{soot accumulation rate} = \frac{50}{60} = 0.833 \text{ g/min}$$

$$\text{active filter volume} = \frac{\pi}{4} \times 100 \times 6.6 = 518 \text{ in.}^3$$

$$\text{total filtration area} = \text{SFA} \times V_f = 8605 \text{ in.}^2$$

The soot density has been reported in the literature and is approximately 0.056 g/cm^3, or 0.917 g/in.^{3} [17]. Using this value we can estimate the rate at which the soot volume, hence the soot membrane thickness, builds up:

$$\text{rate of soot volume collected per filter} = \frac{0.833}{0.917} = 0.909 \text{ in.}^3\text{/min}$$

$$\text{rate of increase in soot membrane thickness} = \frac{0.909}{8605} = 0.00011 \text{ in./min}$$

$$\text{total thickness of soot membrane after 2 hr} = 0.013 \text{ in.}$$

$$d_h = 0.083 - 0.025 = 0.058 \text{ in.} = 0.00483 \text{ ft}$$

$$A_{open} = 0.188\left(\frac{0.058}{0.083}\right)^2 = 0.0918 \text{ ft}^2$$

$$v_{ch} = \frac{Q}{A_{open}} = \frac{531}{0.0918 \times 60} = 96.4 \text{ ft/sec}$$

Substituting into Eq. (13), we obtain

$$\Delta p_s = \frac{32 \times 2 \times 10^{-5} \times 96.4 \times 0.55}{32.2(0.00483)^2}$$

$$= 45.2 \text{ lb/ft}^2$$

$$= 0.639 \text{ in. Hg}$$

Thus, the pressure drop through a sooted channel (with 10 g/liter of soot loading) is three times as large as that through the wall and over four times as large as that through the clean channel.

The above computations were also carried out for a 10.5-in.-diameter × 5-in.-long filter with the 200/12 cell configuration (and an identical total filtration area as the 10.5-in.-diameter × 7-in.-long filter with the 100/17 cell configuration). Table 6 compares the individual pressure drop components for the two filters. It is clear from this table that the largest contribution comes from flow through the sooted channel. Furthermore, the small hydraulic diameter of the 200/12 cell results in nearly three times higher pressure drop than that for the 100/17 cell, which explains the popularity of the 100/17 cell configuration for filter applications.

The foregoing pressure drop model is only an approximation that helps quantify the effect of flow rate, open frontal area, and hydraulic diameter. It also provides the relative contributions of open and sooted pores in the wall as well as those of open and sooted channels to the total pressure drop. A more refined model is needed that must correlate with the experimental data.

IV. REGENERATION TECHNIQUES

In addition to high filtration efficiency, the pressure drop across the filter must be controlled to minimize the fuel economy penalty. Early tests by the Ford Motor Company showed that for a pressure drop of 1 in. Hg, the typical diesel-powered vehicle traveling at 40 mph would incur a 1% loss in fuel economy [17]. To prevent a further increase in pressure drop, the filter must be regenerated periodically by oxidizing the soot carbon. To facilitate oxidation of the soot carbon, the exhaust gas entering the filter must be above 540°C and contain sufficient oxygen to initiate the reaction. Since the exhaust temperature under normal driving conditions is under 540°C, this requirement can be relaxed downward either by catalyzing the filter or by using fuel additives. Next we provide a brief review of common regeneration techniques.

A. Throttling

The exhaust temperature of a diesel engine can be increased by throttling the air flow supplied to the engine. By reducing the air flow, the overall air/fuel ratio is decreased, which increases the average combustion temperature and hence the exhaust temperature.

Table 6 Comparison of Pressure Drop for Two Filters with Identical Total Filtration Area but Different Cell Configurations (in. Hg)

	10.5-in. diameter × 7-in.-long filter (100/17 cell)	10.5-in. diameter × 5-in.-long filter (200/12 cell)
Δp_{ch}	0.153	0.211
Δp_w	0.211	0.302
Δp_s	0.639	2.180
Total Δp	1.003	2.693

The effect of intake throttling on the exhaust temperature of a 5.7-liter engine is shown in Fig. 8. The exhaust temperature for the standard engine seldom exceeds 300°C (up to speeds approaching 55 mph), which is too low to incinerate particulate matter. Throttling not only reduces the heat loss to unused excess air, it lowers the manifold pressure, thereby increasing the pumping loss. This increases the fuel input by the driver to maintain power and results in a hotter exhaust temperature [18]. Figure 8, which is based on laboratory tests, shows two different particulate regeneration regions: one with a catalyst assist and one without. Note that regeneration cannot be achieved below 37 mph without the catalyst and below 21 mph with the catalyst. However, vehicle tests showed that regeneration by throttling was not achieved even in a catalyzed filter due, primarily, to limited oxygen availability. Throttling for a diesel engine is limited to an exhaust oxygen concentration of 2–5%. Under heavy loads and adequate oxygen availability, throttling can lead to a high enough temperature to effect regeneration. Thus, throttling regeneration cannot be applied at steady-state vehicle speeds and light loads. Furthermore, throttling can have an adverse effect on engine-out emissions and fuel consumption. It increases HC, CO, and smoke emissions while decreasing the NO_x emissions [17,19]. Fuel consumption increases due to a richer air/fuel ratio, higher heat losses, and negative pumping work.

In the absence of sufficient oxygen, overthrottling is sometimes employed to raise exhaust temperature to the regeneration level. However, with insufficient oxygen, soot combustion does not occur. And when the engine returns to the unthrottled condition defined by the lower engine speed and higher oxygen content, the overheated filter can experience "runaway" regeneration, with a peak temperature approaching 1400°C (due to insufficient exhaust flow), which can lead to melting and thermal cracking [17]. Failures of this nature can be prevented by reducing particulate loading and increasing the flow rate.

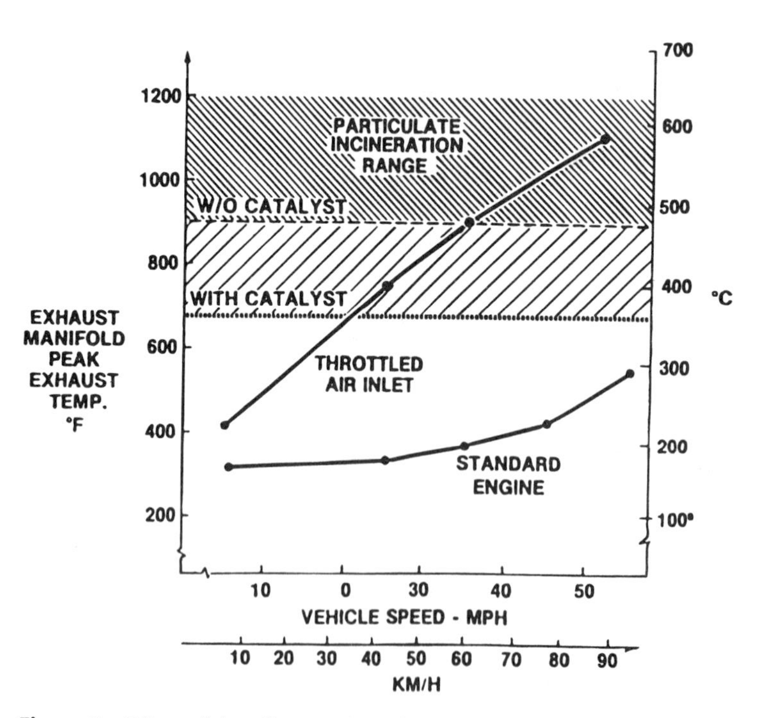

Figure 8 Effect of throttling on the exhaust temperature of a 5.7-liter engine. (Courtesy of the Society of Automotive Engineers.)

B. Burner Regeneration

A diesel-fueled burner placed in the exhaust system, in front of the filter, can effect its regeneration at nearly all engine speeds and load conditions [17–20]. However, such a system, shown schematically in Fig. 9a, is not only complex in terms of the sophisticated electronic controls needed to ensure filter reliability, but it requires high fuel consumption and a large air pump to heat up all of the exhaust gas to 540°C. Second, a modulated, high-pressure-burner fuel flow system is required to maintain the filter inlet temperature at a safe level. Third, air must be circulated continuously through the burner nozzle to minimize fouling by particulate deposits.

Isolating the filter from the engine exhaust during regeneration by using a bypass system, as shown in Fig. 9b, overcomes many of the above issues. First, the regeneration process is independent of the engine operating conditions; second, the electronic control system to effect regeneration is considerably simplified; and third, the energy required to heat the entry face of the filter to 540°C is reduced by nearly an order of magnitude due to the much reduced exhaust flow, as shown in Fig. 10. In the bypass mode, little or none of the exhaust flow is allowed to flow through the filter. Thus the air pump is considerably smaller and the outlet pressure requirement is considerably lower than those for the in-line burner, particularly when the filter is located after the muffler. The regeneration process is initiated when the back pressure across the filter reaches a specified level, e.g., 3 in. Hg. Alternatively or simultaneously, the exit temperature is monitored by a thermocouple. A sudden and significant increase in exit temperature is a good indicator of the onset of regeneration.

When the entry face of the filter approaches 650°C, soot oxidation begins with the combustion front propagating slowly toward the exit face. Consequently, the burner may be shut off halfway through the regeneration cycle and the increased oxygen concentration utilized to accelerate the remainder of the soot oxidation process. Care must be taken to limit the midbed temperature to less than 1400°C and the temperature gradients to 35°C/cm to prevent melting or cracking of the filter. Figure 11 shows temperature profiles of the combustion front during burner regeneration, with a peak value of 1260°C. As the combustion front moves toward the exit face, the peak temperature increases progressively due to preheating of the combustion air in the preceding region where regeneration had

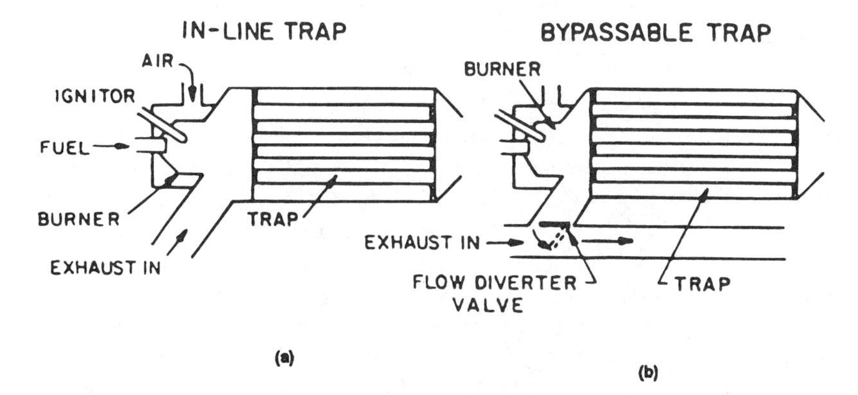

Figure 9 Schematic of diesel-fueled burner regeneration system. (a) In-line burner, (b) burner with bypass system. (Courtesy of the Society of Automotive Engineers.)

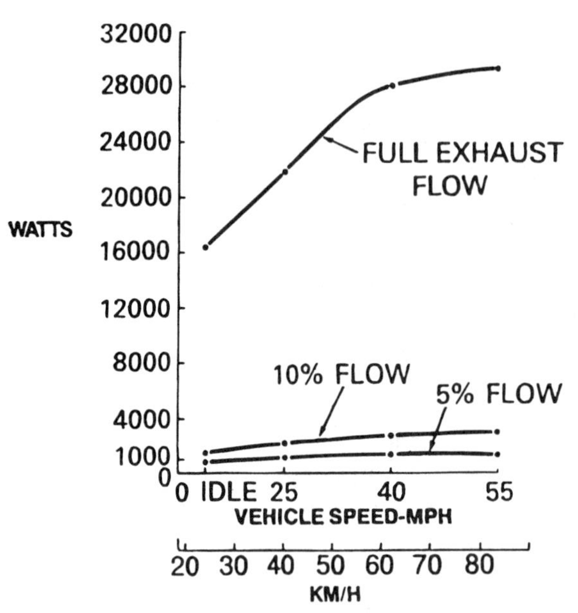

Figure 10 Energy required to raise the exhaust temperature of a 5.7-liter diesel engine to 540°C. (Courtesy of the Society of Automotive Engineers.)

been completed. The interval between regenerations is dictated not only by the fuel economy penalty, but also by the back pressure and critical soot mass for sustained self-regeneration.

C. Electrical Regeneration

Since the bypass system has demonstrated good feasibility, the complex burner with its electronic controls may be replaced by a simple resistance heater powered by the alternator on the vehicle; see Fig. 12. A small air pump (3–5 CFM) is used to transfer heat from the heater to the filter and provide sufficient oxygen for soot oxidation. Typical heaters are fabricated from two nichrome resistance elements contained in MgO powder insulation with a 0.260-in.-diameter stainless steel sheath. The energy requirement for regeneration is typically 3 kW and is readily achieved from a 90-amp alternator at a speed of 6400 rpm.

The heater is turned on when the back pressure across the filter reaches 5 in. Hg. Once the heater temperature reaches 760°C, the air pump is turned on to transfer heat and oxygen to the entry face of the filter. The combustion front moves slowly toward the exit face, much as in burner regeneration (see Fig. 11). At the end of regeneration, which may last 8–10 min, the air pump is shut off and the flow diverter valve returned to its original position to allow all of the engine exhaust gas to pass through the filter. The completion of the regeneration cycle is best indicated by the back pressure value's approaching that of a clean filter. In contrast to burner regeneration, where the air pump is on from the beginning, electrical regeneration does not trigger the air pump until the heating element reaches a temperature of 760°C. Consequently, both the peak temperature and temperature gradients are lower (100–200°C lower) in the case of electrical regeneration, which alleviates melting and cracking issues [17–20].

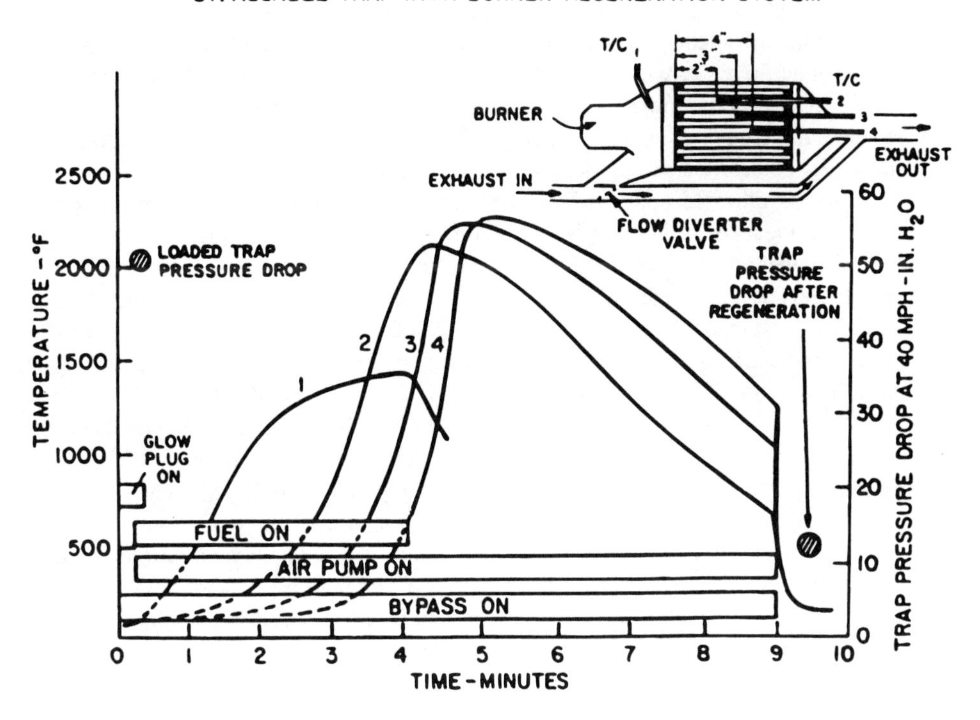

Figure 11 Temperature profiles of the combustion front during burner regeneration of a cordierite filter with a bypass system. (Courtesy of the Society of Automotive Engineers.)

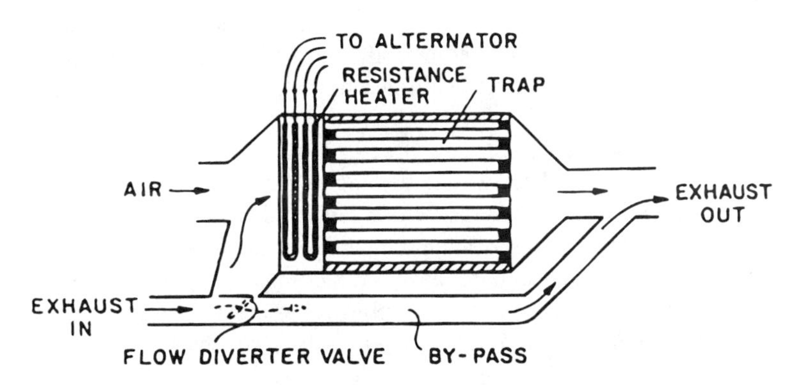

Figure 12 Schematic of a cordierite filter with a bypass and electrical regeneration system. (Courtesy of the Society of Automotive Engineers.)

D. Catalytic Regeneration

Catalytic additives in diesel fuel have also proven effective in reducing the regeneration temperature sufficiently to effect self-regeneration under normal driving conditions [17–20]. Fuel additives consisting of organometallic compounds of Cu, Ni, Ce, Mo, Mn, Zn, Ca, Fe, and Ba have been evaluated with respect to their effectiveness in lowering the regeneration temperature while remaining chemically stable in diesel fuel. Screening tests show that octoate-based compounds of Cu, Cu, and Ni, and Cu and Ce + Fe are very effective. At elevated temperatures, these compounds decompose into oxides of metals, which are ideal for promoting soot oxidation.

To evaluate the effectiveness of these additions, steady-state tests were conducted on a chassis dynamometer at an equivalent vehicle speed of 40 mph. Under normal load conditions, the exhaust temperature was only 200°C. After loading the filter with soot to a back pressure of 6 in. Hg, the chassis dynamometer was operated at successively higher loads at the constant vehicle speed of 40 mph to increase the filter inlet temperature in 30°C increments. Different additives in Phillips 2D diesel fuel helped complete the regeneration of the sooted filter at different temperatures. For example, 0.25 g/gal of Cu and 0.2 g/gal of Ce reduced the regeneration temperature to 280°C, nearly 230°C lower than that without the additive. The peak midbed temperature seldom exceeded 480°C with other combination of fuel additives, as shown in Fig. 13. The temperature gradients were also very low, with little or no concern over filter durability. Thus, regeneration with fuel additives is a viable option, although ash build-up can lead to plugging and high back pressure.

V. PHYSICAL PROPERTIES AND DURABILITY

Physical properties of cordierite ceramic diesel filters, which can be controlled independent of geometric properties, have a major impact on their performance and durability. These include microstructure (porosity, pore size distribution, and microcracking), coefficient of thermal expansion (CTE), strength (crush strength, isostatic strength, and modulus of rupture), structural modulus (also called E-modulus), and fatigue behavior (represented by the dynamic fatigue constant). These properties depend on both the ceramic composition and the manufacturing process, which can be controlled to yield optimum values for a given application.

The microstructure of diesel filters not only affects physical properties like CTE, strength, and structural modulus, but it has a strong bearing on filter/catalyst interaction, which, in turn, affects the performance and durability of catalyzed filter. The coefficient of thermal expansion, strength, fatigue, and structural modulus of the diesel filter, which also depend on cell orientation and temperature, have a direct impact on its mechanical and thermal durability [21–25]. Finally, since all of the physical properties are affected by washcoat formulation, washcoat loading, and washcoat processing, they must be evaluated before and after the application of washcoat to assess filter durability.

A. Physical Properties

The initial filter compositions, shown in Table 4,* were designed to offer different microstructures to meet different filtration efficiency and back pressure targets set by engine

* Ex-47, EX-54, and EX-66 diesel filters are no longer in production; they have been replaced by EX-80 filter.

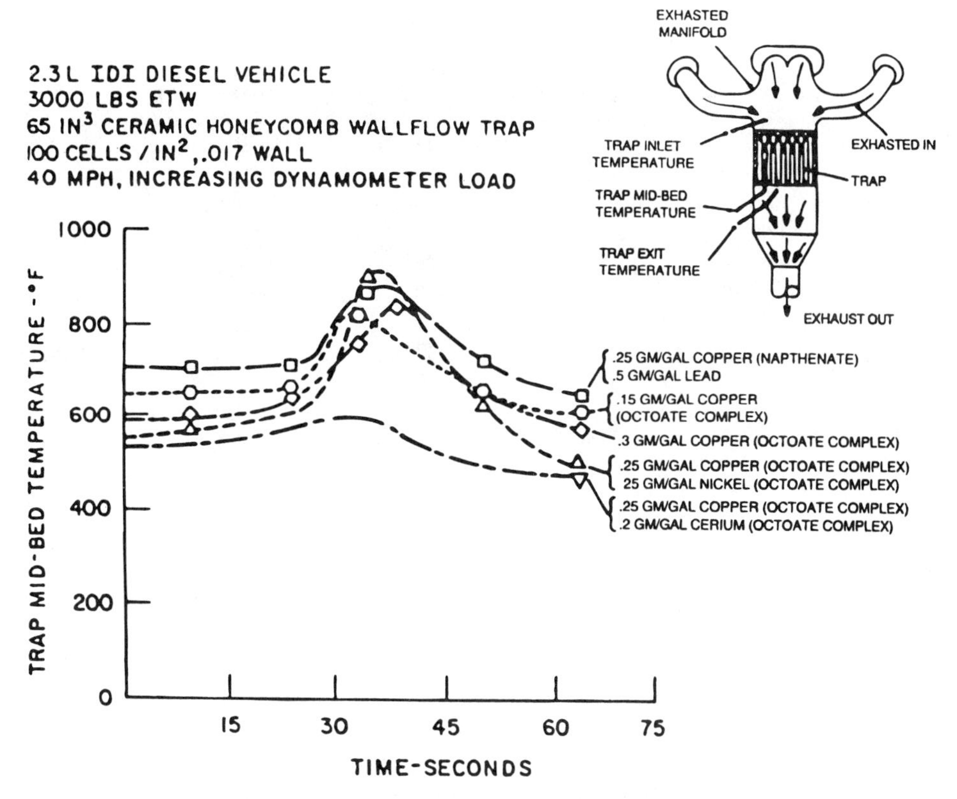

Figure 13 Peak midbed temperature vs. time during regeneration with different fuel additives. (Courtesy of the Society of Automotive Engineers.)

manufacturers [4]. However, they were not optimized with respect to thermal durability, which became a critical requirement to survive regeneration stresses. A secondary process, such as acid leaching, helped reduce the coefficient of thermal expansion, thereby improving thermal durability significantly, but had an adverse effect on mechanical durability and filter cost. A more advanced filter composition, EX-80, with superior performance was developed in 1992. This material is a stable cordierite composition with low CTE and has demonstrated improved long-term durability over a wide range of operating conditions. Moreover, it offers high filtration efficiency and low pressure drop. The low CTE reduces thermal stresses, thereby permitting numerous regeneration cycles without impairing the filter's durability. This composition is now the industry standard for diesel exhaust aftertreatment [7]. For the sake of completeness, however, we will summarize the properties of early filter compositions together with those of EX-80 composition. The specimen size, orientation, and test technique for measuring the physical properties are identical to those for ceramic catalyst supports for gasoline fuel, which are described in Chapter 2.

Table 7 compares the nominal physical properties of four different filter compositions, which differ in their mean pore diameter. The strength and modulus of elasticity data are those measured at room temperature. The axial coefficient of thermal expansion is the average value over the 25–800°C temperature range. It is clear from Table 7 that

Table 7 Physical Properties Data for Cordierite Ceramic Diesel Filters

	EX-47 100/17	EX-54 100/17	EX-66 100/25	EX-80 100/17
Wall porosity (%)	50	46	45	48
Mean pore size (μm)	12	25	35	13.4
Weight density (g/cm^3)	0.39	0.45	0.54	0.42
Crush A strength (psi)	1130	900	1300	1595
Crush B strength (psi)	380	300	430	325
3D-isostrength (psi)	585	460	660	500
2D-isostrength (psi)	500	390	560	425
Axial MOR (psi)	365	320	418	410
Tangential MOR (psi)	142	131	176	184
Axial MOE (10^6 psi)	0.81	0.83	1.06	0.75
Axial CTE 25°–800°C (10^{-7}/°C)	8.8	8.7	10.5	3.3

the EX-80 filter offers an optimum combination of properties, namely, small mean pore size, high strength, low modulus of elasticity (MOE), and low coefficient of thermal expansion (CTE), which together ensure superior performance compared with that of the other three filter compositions.

The values of the axial and tangential modulus of rupture (MOR) at high temperature, which impact the filter's thermal durability, are summarized in Table 8. Again, the EX-80 filter excels in this property relative to other filter compositions. Similarly, the axial* E-modulus data in Table 9 show that the EX-80 filter has similar stiffness to that of the EX-47 and EX-54 filters, notably at or below 600°C (representative of the filter's peripheral temperature during regeneration).

Finally, the axial CTE data for the four filter compositions are summarized in Table 10. These data also demonstrate that the EX-80 filter has the lowest expansion coefficient over the entire temperature range, which implies low thermal stresses during regeneration and excellent thermal durability for this filter.

B. Thermal Durability

Thermal durability refers to a filter's ability to withstand both axial and radial temperature gradients during regeneration. These gradients depend on soot distribution, soot loading,

Table 8 MOR Data for Cordierite Ceramic Diesel Filters with 100/17 Cell Structure

Temp.	Axial MOR (psi)				Tangential MOR (psi)			
(°C)	EX-47	EX-54	EX-66	EX-80	EX-47	EX-54	EX-66	EX-80
25	365	305	300	408	145	115	120	162
200	370	260	270	385	150	120	110	156
400	372	260	240	363	160	120	100	156
600	390	265	260	380	170	125	110	166

* The tangential E-modulus is exactly 50% of the axial E-modulus and hence is not included in Table 9.

Table 9 Axial E-Modulus Data for Cordierite Ceramic Diesel Filters (10^6 psi)

Temp. (°C)	EX-47 (100/17)	EX-54 (100/17)	EX-66 (100/25)	EX-80 (100/17)
100	0.63	0.61	1.02	0.63
200	0.63	0.61	1.03	0.64
300	0.63	0.62	1.04	0.64
400	0.64	0.62	1.05	0.65
500	0.65	0.62	1.06	0.66
600	0.66	0.65	1.08	0.68
700	0.68	0.68	1.11	0.71
800	0.70	0.70	1.16	0.75
900	0.75	0.77	1.31	0.83
1000	0.83	0.91	1.43	0.92

Table 10 Axial CTE Data for Cordierite Ceramic Diesel Filters (10^{-7} in./in./°C)

Temp. (°C)	EX-47 (100/17)	EX-54 (100/17)	EX-66 (100/25)	EX-80 (100/17)
100	−4.1	−9.4	−7.7	−1.33
200	−1.6	−5.1	−4.4	−9.6
300	0.8	−1.4	−1.5	−6.5
400	2.8	1.2	0.6	−4.0
500	4.4	3.2	2.4	−1.9
600	6.0	5.0	4.3	−0.2
700	7.8	6.8	6.3	1.2
800	9.4	8.5	8.4	2.9
900	10.9	9.7	10.5	4.3
1000	12.6	10.5	12.6	5.4

O_2 availability, and flow rate and give rise to thermal stresses that must be kept below the fatigue threshold of the filter material to prevent cracking. A detailed analysis of thermal stresses requires temperature distribution, which is readily measured with the aid of 0.5-mm-diameter Type-K chromel–alumel thermocouples during the regeneration cycle [12,21]. Several examples illustrating this technique are given in Section VI. To assess the relative thermal durability of different filter candidates we will compute the thermal shock parameter, TSP, using physical properties data and Eq. (16):

$$\mathrm{TSP} = \frac{(\mathrm{MOR}/E)T_p}{\alpha_c(T_c - 25) - \alpha_p(T_p - 25)} \tag{16}$$

In Eq. (16), MOR and E denote modulus of rupture and structural modulus, respectively, T_c and T_p denote the temperature of the center and peripheral regions, respectively, of the filter during regeneration, and α_c and α_p denote the corresponding CTE values. In view of plugging around the peripheral region, there is no gas flow in that region and the temperature T_p is typically 400°C. The center temperature, on the other hand, is higher, depending on soot loading, O_2 content, and flow distribution. We will assume T_c to range from 600°C (low soot loading) to 1000°C (high soot loading) and compute TSP values

Table 11 Axial Thermal Shock Parameter for Cordierite Ceramic Diesel Filters (for T_p = 400°C)

Temp. (°C)	EX-47 (100/17)	EX-54 (100/17)	EX-66 (100/25)	EX-80 (100/17)
600	2.42	1.32	1.49	4.20
700	1.38	0.77	0.83	2.40
800	0.93	0.52	0.53	1.50
900	0.68	0.40	0.37	1.10
1000	0.52	0.33	0.28	0.80

for each of these T_c values while keeping T_p = 400°C. The results of this exercise are summarized in Table 11. Note that the TSP values for EX-80 filter are nearly 50–250% higher than those for other filters, due primarily to its two- to threefold lower CTE values (see Table 10). The higher TSP value signifies improved thermal shock resistance and extended thermal durability. Alternatively, it permits higher regeneration stresses without impairing the filter's durability.

The power-law fatigue model [26,27] helps estimate the safe allowable regeneration stress for a specified filter life. Denoting the filter's short-term modulus of rupture by S_2, the safe allowable stress S_1 is given by

$$S_1 = S_2\left(\frac{t_2}{t_1}\right)^{1/n} \tag{17}$$

where t_1 denotes the specified filter life, t_2 denotes the equivalent static time for measuring the short-term modulus of rupture, and n denotes the dynamic fatigue constant of the filter composition. The last is obtained by measuring MOR as a function of the stress rate at temperature T_p. Table 12 summarizes n values for the four filters at T_p = 200°C and 400°C [7]. Both the mean value and the 95% confidence interval are listed for n. For a conservative estimate of S_1, the lowest value of n should be used in Eq. (17). The equivalent static time t_1 is defined as the actual test duration for measuring MOR divided by $(n + 1)$. Since the typical test duration is 30 sec and the lowest value of n is approximately 29 (with the exception of the EX-47 filter), $t_1 \cong 30/30 \cong 1$ sec.

Filter life is generally specified in terms of the number of regeneration cycles over the vehicle's lifetime. We will assume a filter life of 120,000 mi with a regeneration interval of 200 miles and a regeneration duration of 10 min. This translates to t_1 = 6000 min = 360,000 sec. Substituting these values into Eq. (17), we arrive at the safe allowable

Table 12 Dynamic Fatigue Data for Cordierite Ceramic Diesel Filters in the Axial Direction

Temp. (°C)	EX-47 (100/17)	EX-54 (100/17)	EX-66 (100/17)	EX-80 (100/17)
200	21	34	27	—
400	24	44	42	53
Combined data	23	38	38	—
95% Conf. Int'l.	17–30	27–62	28–60	32–92

stress in the axial direction, as shown in Table 13. Again, it is clear from this table that EX-80 filter can sustain higher regeneration stresses than the other filters. This superiority derives from its higher fatigue constant and MOR value, which, in turn, are related to its optimized microstructure.

C. Mechanical Durability

The mechanical durability of the ceramic filter depends not only on its tensile and compressive strengths but also on its packaging design [12]. In addition to mechanical stresses due to handling and processing, the filter package must be capable of withstanding in-service stresses induced by gas pulsation, chassis vibration, and road shocks. The design of a robust packaging system for catalyst supports for the gasoline engine discussed in Chapter 2 is equally applicable to filters. Tables 7 and 8 demonstrate more than adequate strength for tourniquet canning, which is recommended for long-term mechanical durability. In addition, pre–heat treatment of intumescent mat also promotes mechanical durability [28].

VI. APPLICATIONS

Ceramic wall-flow filters have performed successfully since their introduction in the 1980s in passenger cars. A substantial number of filters has been installed to date, and they continue to meet emissions, back pressure, and durability requirements. With new developments in filter materials, additives and catalyst technology, packaging designs, and regeneration techniques, ceramic wall-flow filters are being tested in new and more severe applications, including buses and trucks. The following examples help illustrate their design, performance, and durability.

A. Large Frontal Area Filter

As our first example, we analyze the thermal durability of a EX-54, 100/17 filter (10.63-in.-diameter × 12 in. long) during regeneration. The thermal stresses during regeneration, which control filter durability, are governed not only by the filter properties and geometry but also by the regeneration conditions, e.g., flow rate, flow distribution, soot loading, % oxygen, and burner temperature. This example illustrates the effects of flow rate, burner temperature, and filter geometry on regeneration stresses.

Table 13 Safe Allowable Stress in the Axial Direction for Long-Term Durability

Filter composition	S_2 (psi)	n	t_2 (sec)	S_1 (psi)
EX-47	372	17	2	183
EX-54	260	27	1	166
EX-66	240	28	1	156
EX-80	363	32	1	249

The regeneration process was simulated by thermal cycling a clean filter with an electronically controlled burner and a centrifugal blower (for combustion air). The typical cycle consisted of 5 min of heating (with burner on) and 3 min of cooling (with burner off). Several engine conditions, namely, full load, normal load, and idling, could be simulated by adjusting both the burner temperature and the flow rate. The radial temperature distribution during regeneration was obtained with the aid of 27 Type-K chromel–alumel thermocouples at 5-sec intervals (during the 8-min cycle), as shown in Fig. 14. With the burner temperature at 715°C, the maximum radial gradient at midsection L3 occurred at t = 185 sec with center and periphery temperatures approaching 930°C and 300°C; see Fig. 15 [24].

The thermal stresses were computed by the finite element code ANSYS [29], taking full advantage of the axial symmetry of the filter; see Fig. 16. Both the temperature-dependent physical properties of the EX-54 filter (Section V) and the time-dependent thermocouple data were used as inputs to stress analysis. The maximum stresses in the axial and tangential directions at the midsection are summarized in Table 14. It should be noted in Table 14 that the radial temperature gradient is the major contributor to thermal stresses; those due to axial gradient are less than 20%.

The axial stress approaches a value of 540 psi, far in excess of the high-temperature axial strength of 260 psi; see Table 8. Similarly, the tangential stress approaches 200 psi,

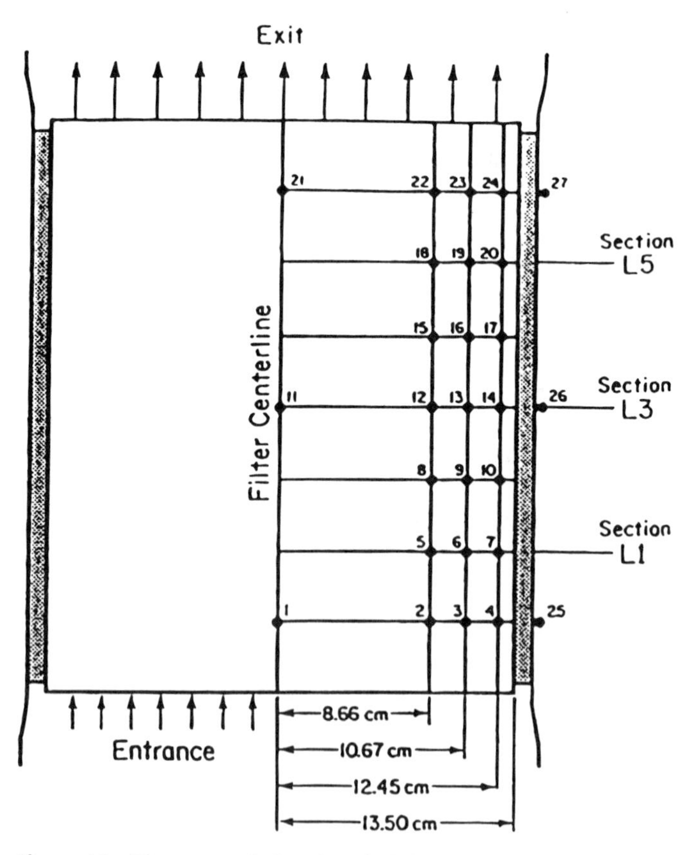

Figure 14 Thermocouple locations in LFA filter. (Courtesy of Deut. Keramische Gesellschaft.)

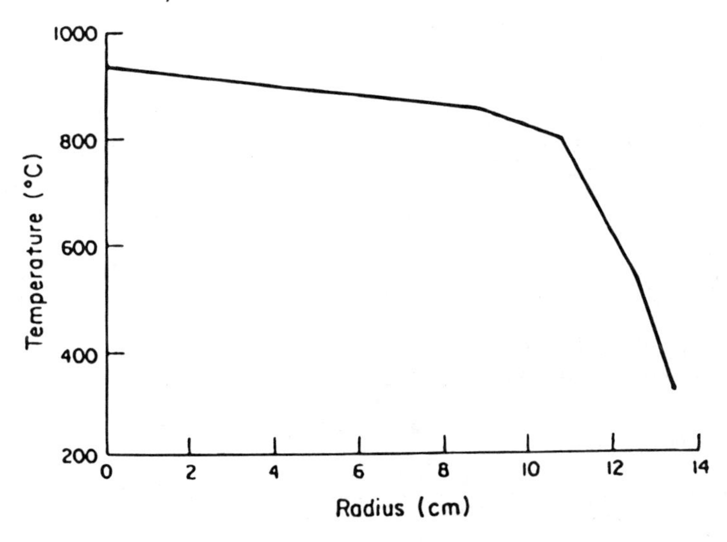

Figure 15 Regeneration temperature profile in the radial direction at midsection (T_{burner} = 715°C, t = 185 sec.) (Courtesy of Deut. Keramische Gesellschaft.)

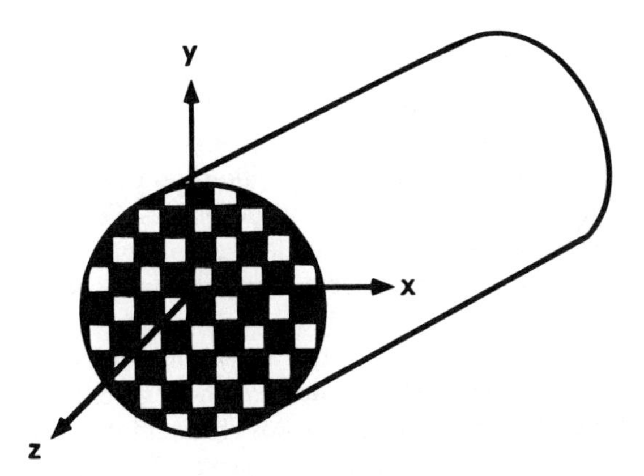

Figure 16 Coordinate system for computing thermal stresses in diesel filter.

Table 14 Maximum Tensile Stress at Midsection During Simulated Regeneration

	Axial Stress (psi)	Tangential Stress (psi)
Radial gradient	450	200
Axial gradient	90	0
Total stress	540	200

T_{burner} = 715°C

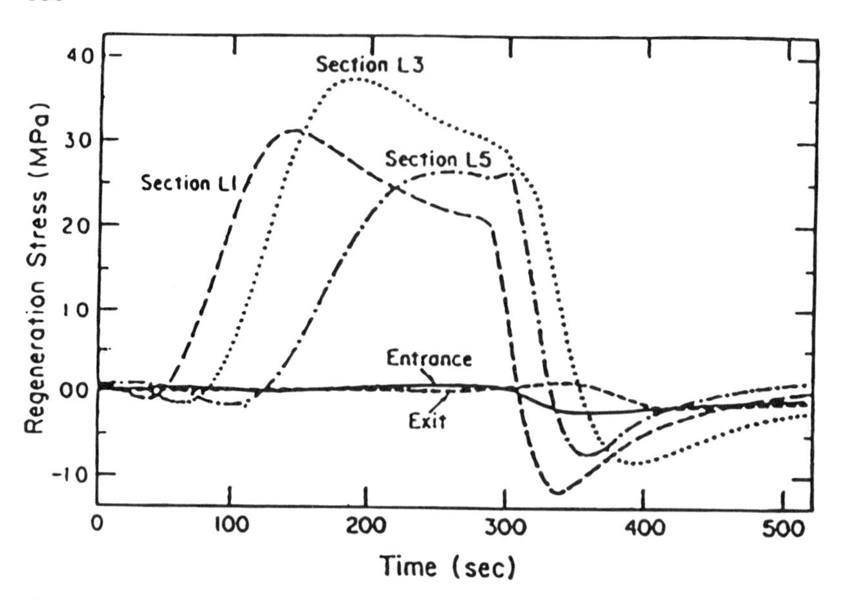

Figure 17 Dynamics of the axial stress-time history during regeneration at T_{burner} = 715°C. (Courtesy of Deut. Keramische Gesellschaft).

which also exceeds the tangential strength of 120 psi. Figure 17 shows the variation of maximum axial stress at five different filter sections with time. It is clear that as the regeneration front moves from entry to exit sections, so does the dynamic stress wave from sections L1 to L3 to L5. The maximum stress occurs at midsection and can result in ring-off failure if its magnitude and duration exceed the safe allowable values. Furthermore, the simulated regeneration with low flow rate and a burner temperature of 715°C is too severe for the specific LFA filter and hence is unacceptable from a durability point of view.

The above stresses may be reduced either by lowering the burner temperature or by using a smaller filter length. The former was verified by simulating the regeneration process with a burner temperature of 600°C. The maximum stress was reduced from 540 to 330 psi, and tangential stress was reduced from 200 to 125 psi. Thus, a 115°C-lower burner temperature reduced the maximum stresses by nearly 40%! Using the lower burner temperature, which proved highly beneficial, the filter length was varied from 12 to 6 in. stepwise fashion to reduce the axial stress further; see Table 15. It is clear from Table 15

Table 15 Effect of Filter Length on Maximum Axial Stress at Midsection of EX-54 Filter

Filter Length (in.)	Axial Stress (psi)
12.0	333
10.6	319
9.4	290
8.3	254
7.1	218
6.0	167

that the axial stress can be halved to a safe allowable value (see Table 13) by using a lower burner temperature and reducing the filter length to 6 in., thereby ensuring long-term durability.

Other ways to promote filter durability, which are currently being examined, include the modification of filter properties via composition and process research and the upgrading of packaging design via heavier mat, higher mount density, and stiffer shell.

B. Filter Durability for 6.2-Liter Light-Duty Diesel Engine

In this second example, we examine the durability of another LFA filter of EX-47, 100/17 composition (7.5-in. diameter × 8 in. long) for a 6.2-liter light-duty diesel engine with a burner bypass system; see Fig. 18. In addition to meeting the regulation limit of 0.13 g/mile particulate emissions, the filter must demonstrate a life durability of 120,000 miles. Based on the properties data alone (Tables 7–13), we see that the EX-47 filter offers low CTE, high strength, small mean pore diameter, and high filtration efficiency.

This example demonstrates how the filter properties are modified by the regeneration process, thereby limiting the regeneration stresses to below the threshold value and ensuring filter reliability. These modifications are attributed to successive but controlled microcracking that occurs during regeneration [25]. The reliability of all components of the regeneration system, i.e., exhaust metering valve, air blower, fuel supply, fuel pump, solenoid valve, burner, and flow distributor, was further ensured by careful control of inlet gas conditions via the air/fuel ratio, heating the gas mixture with an optimally designed fuel-powered burner, maintaining a certain minimum flow rate throughout the regeneration process, and limiting the soot loading to 20 g prior to regeneration.

The typical regeneration took 10 minutes, although shorter durations also met EPA specifications for an acceptable particulate level. The key advantages of shorter regeneration are significant fuel savings and minimal thermal fatigue of the ceramic filter, both of which are critical to the viability of a wall-flow particulate trap system. The soot filter

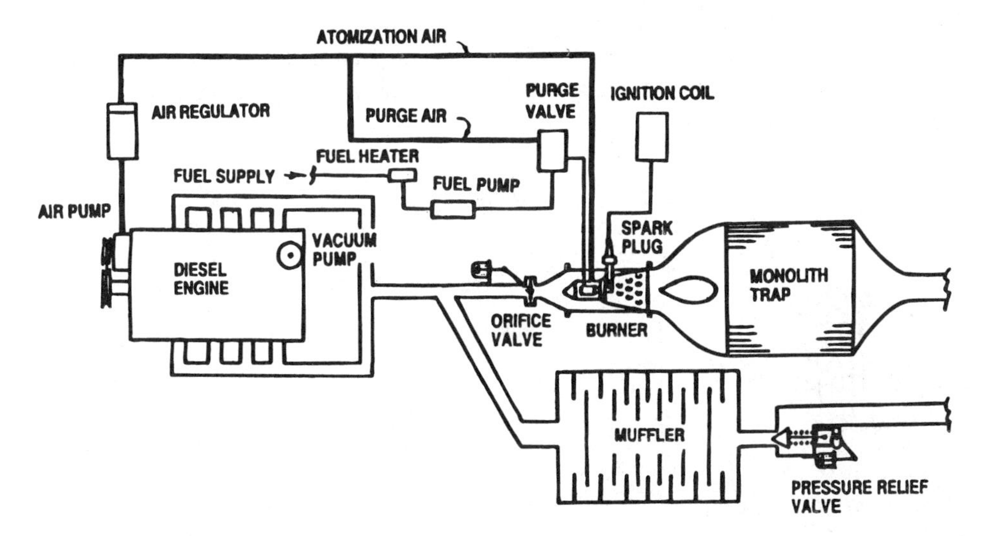

Figure 18 Burner bypass system for 6.2-liter light-duty diesel engine. (Courtesy of the Society of Automotive Engineers.)

Table 16 Operating Conditions During Regeneration of 7.5-in.-Diameter × 8-in.-Long EX-47 Filter

Engine RPM	Engine torque (ft lb)	Flow rate through filter (lb/hr)	% O_2 in inlet gas
650	idle	230	10.5%
1200	90	246	8.9%
2200	180	279	5.6%

Initial soot loading = 20 g

was regenerated in the test cell under three different engine conditions to simulate various driving conditions and terrains; see Table 16. Both the flow rate and the oxygen content of the inlet gas are also recorded in Table 16.

Of the three sources of stress, namely, mounting, vibrations, and regeneration, the latter two are most critical and are influenced by temperature gradients and physical properties. Figures 19 and 20 show thermocouple locations and the temperature–time history of the center and peripheral regions, defined by

$$T_c = \frac{1}{\pi r_o^2} \int_o^{2\pi} \int_o^{r^o} T(r) r \, d\theta \, dr \tag{18}$$

$$T_p = \frac{1}{\pi(R^2 - r_o^2)} \int_o^{2\pi} \int_{r_o}^{R} T(r) r \, d\theta \, dr \tag{19}$$

where r_o = 3.58 in. and R = 3.75 in. for this LFA filter. It is clear in Fig. 20 that a radial gradient of ~200°C is present for 5 min of the 8-minute cycle and produces tensile stresses in the peripheral region in the axial and tangential directions, which, if excessive, can lead to thermal fatigue and/or fracture. When the burner is turned off at the end of regeneration, the temperature profile is reversed, with the center becoming colder than the periphery. The thermal stresses also change sign and become tensile in the center region and compressive in the peripheral region. However, both the lower temperature and shorter duration of the cooling cycle will reduce the stress magnitude and pose little concern about internal fracture.

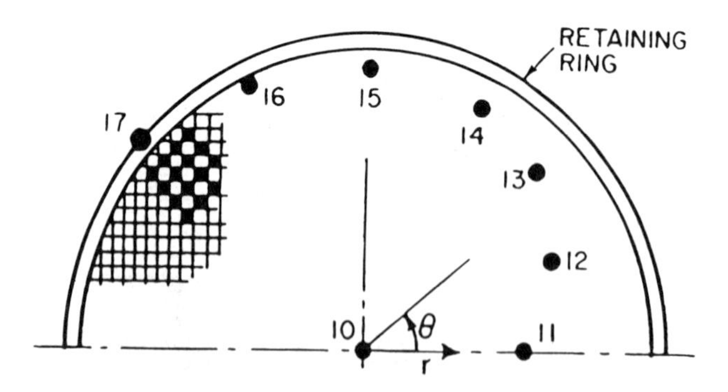

Figure 19 Thermocouple locations at the midsection of filter during regeneration. (Courtesy of the Society of Automotive Engineers.)

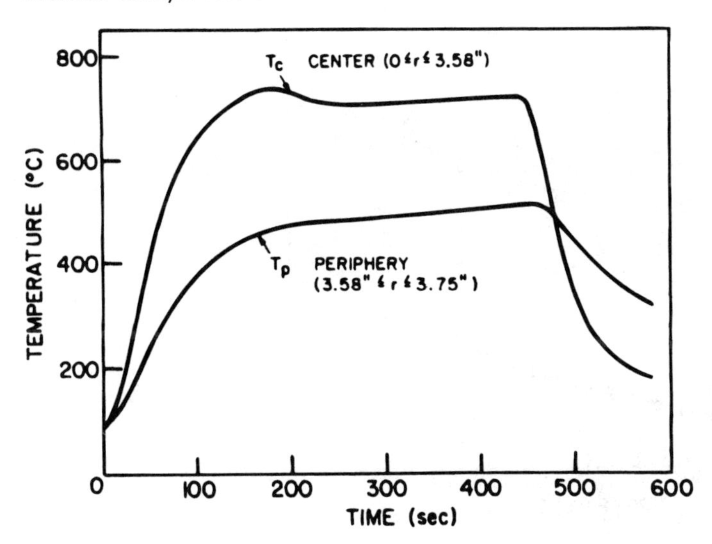

Figure 20 Temperature–time history at the center and the periphery of a filter's midsection during 8-min regeneration cycle at engine idle. (Courtesy of the Society of Automotive Engineers.)

Figure 21 shows the distribution of axial and tangential stresses in the peripheral region during regeneration. Both of these stresses exceed the MOR value in the region 3.6 in. $< r \leq$ 3.75 in. and produce macrocracking confined to that region. Consequently, the E-modulus of that region is reduced by nearly 60% with successive regenerations, as indicated in Fig. 22. Although the thermal expansion of this region is not affected, the regeneration stresses are reduced by 50% due to the lower E-modulus; see Fig. 23. As a result, the aged filter experiences stresses lower than its threshold limit and continues to perform reliably over its required lifetime.

Regeneration stresses at higher engine speeds and torque are 10–20% lower than those at engine idle; see Table 17. Thus, the likelihood of crack initiation and propagation

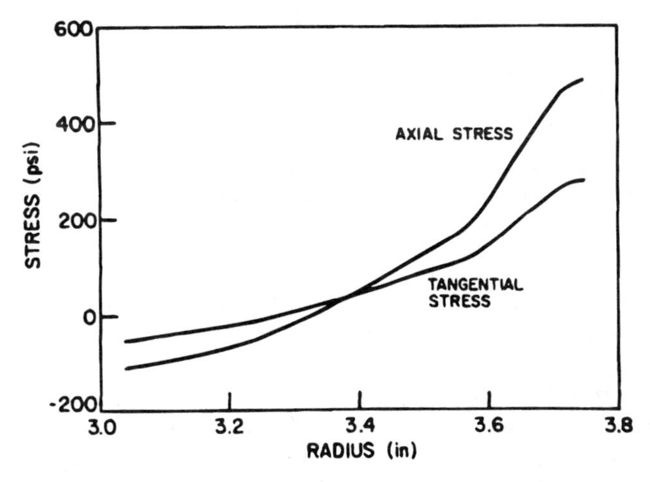

Figure 21 Radial variation of axial and hoop stresses at midsection at t = 200 sec during regeneration of a new filter at engine idle. (Courtesy of the Society of Automotive Engineers.)

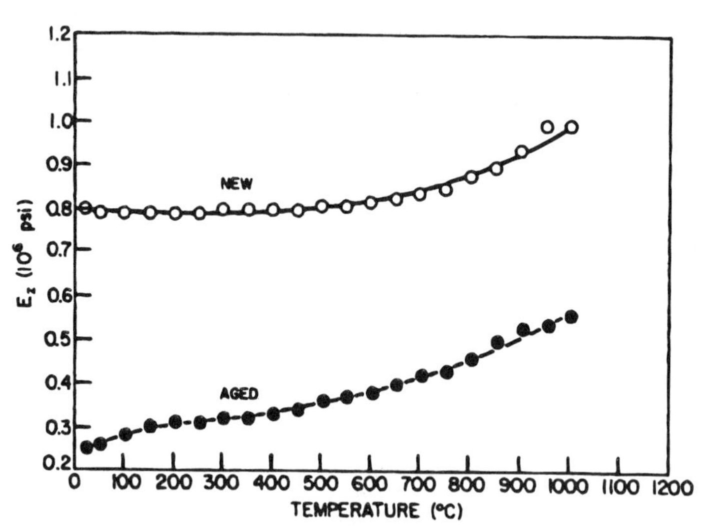

Figure 22 Variation of the axial modulus of new and aged filters with temperature. (Courtesy of the Society of Automotive Engineers.)

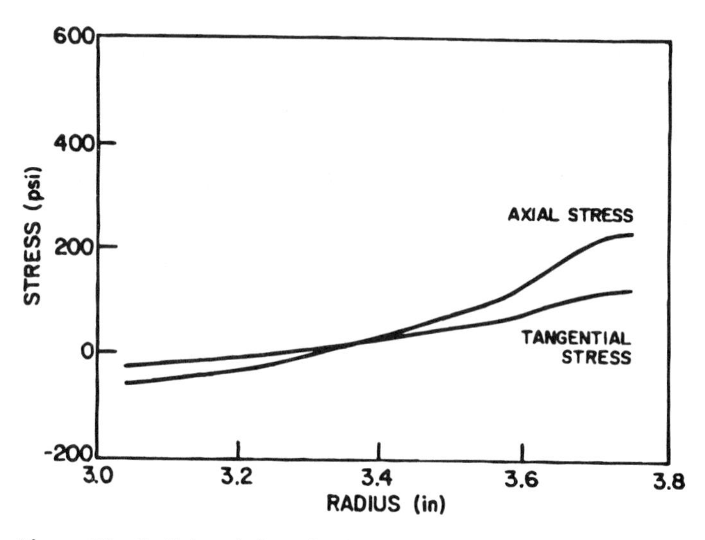

Figure 23 Radial variation of axial and hoop stresses at midsection at t = 200 sec during the regeneration of an aged filter at engine idle. (Courtesy of the Society of Automotive Engineers.)

is considerably lower under normal- and full-load conditions, which is good news from a total durability point of view since the majority of regenerations will occur under off-idle conditions.

Finally, the mounting design induces compressive stresses of 90 psi in the axial direction and 75 psi in the radial direction, which enhances the threshold limits of the EX-47 filter to 185 psi in the axial and 110 psi in the tangential direction. Comparing these to regeneration stresses in the aged filter, it is clear that the nature of macrocracking in the peripheral region will be stable and controlled due to low regeneration stresses.

Table 17 Maximum Thermal Stresses at Midbed of New and Aged EX-47 Filters Under Various Regeneration Conditions

Engine load		Maximum Stress (psi)			
		New filter		Aged filter	
Speed (RPM)	Torque (ft lb)	σ_θ	σ_z	σ_θ	σ_z
750	idle	295	515	130	240
1200	90	240	500	105	240
2200	180	215	460	100	225

This points out the critical role that mounting plays in ensuring both the mechanical and the thermal durabilities of ceramic wall-flow filters.

C. High-Tech Filter with Optimum Performance

In this final example, we focus on the performance and durability data for the EX-80 filter with optimized microstructure and physical properties. As noted in Section V, this high-tech composition offers a low mean pore size for high filtration efficiency and a low rate of pressure drop build-up at higher levels of soot accumulation. It has the lowest CTE values over the operating temperature range, which minimizes regeneration stresses. And it has the highest fatigue resistance, which permits the higher threshold stresses desirable for long-term thermal durability. The above combination of filter properties, achieved by optimizing both the raw materials and the manufacturing process, was necessitated by more stringent performance requirements and 290,000-mile lifetime durability [7].

The filtration performance of the 10.5-in.-diameter × 12-in.-long EX-80 filter, with a 100/17 cell structure, was measured by Ortech International (Ontario, Canada) using a 1989 Detroit Diesel 6V-92 TA DDEC II diesel engine and low-sulfur diesel fuel (D-1 with 0.1% S). A steady-state engine speed-load condition of 300 ft lb at 1600 RPM was selected to achieve an engine flow of approximately 500 SCFM and an exhaust temperature of 260°C. The engine soot output averaged 18 g/hr. All testing was performed with the engine's exhaust directed to, and diluted through, a dilution tunnel/PDP assembly of approximately 2000 SCFM. Forty-minute efficiency determinations were carried out until the engine's exhaust reached a back pressure of 5 in. Hg. The pressure drop across the filter was measured at 1-minute intervals during the test. Table 18 summarizes the collection

Table 18 Filtration Data for the LFA EX-80 Filter

Soot loading (g)	Collection efficiency (%)	Pressure drop (in. Hg)
5	87	1.6
15	88	2.0
50	92	3.1

efficiency and pressure drop data at three successive soot loadings; Fig. 24 captures the complete data.

Previous investigations have shown that filters with small pore size yield high efficiency and high back pressure while those with large pore size yield low efficiency and low back pressure [4,13,30]. While the EX-80 filter shows this trend at low soot levels, it soon shifts from conventional predictions, due to the formation of a soot membrane on the wall surface at higher loadings. During early stages of filtration, the intrinsic pore size distribution of the filter wall is altered by soot penetration, which is minimal for the EX-80 filter due to its small pore size. Consequently, the bulk of the soot ends up as a soot membrane that behaves as a filtration medium and reduces the rate of pressure drop build-up due to the absence of further alteration of the intrinsic pore structure of the filter wall. This dual-pore-size filtration mechanism sets in at 15 g of soot loading in the 10.5-in. × 12-in. filter tested, i.e., at 0.9 g/liter soot loading, and reduces the rate of pressure drop build-up from 0.38 to 0.02 in. Hg/g.

The effect of the low CTE and high fatigue resistance of the EX-80 filter (see Tables 10 and 12) is best demonstrated by the excellent thermal shock behavior of the 11.25-in.-diameter × 12-in.-long filter under the severe regeneration conditions produced by KHD Deutz Cologne, Germany [7]. Figure 25 shows the midbed temperature profile during the 10-minute regeneration cycle. It should be noted that the center temperature T_c is nearly 500°C higher than periphery temperature T_p for the bulk of the regeneration cycle. The peak axial thermal stress associated with these gradients was estimated by finite-element analysis to be 195 psi. The corresponding threshold strength of the EX-80 filter at T_p = 200°C is 250 psi, well above the peak thermal stress. This implies no-flaw growth during the 1450 regenerations, at 200-mile intervals, over the desired lifetime of 290,000 miles. Thus, the CTE, high MOR, and fatigue resistance ensure the physical durability of the EX-80 filter even under severe regeneration conditions.

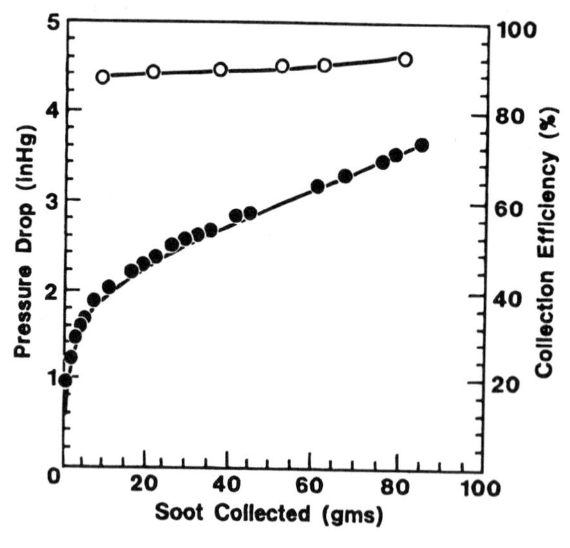

Figure 24 Pressure drop and filtration efficiency for the EX-80 filter as a function of soot loading. (Courtesy of the Society of Automotive Engineers.)

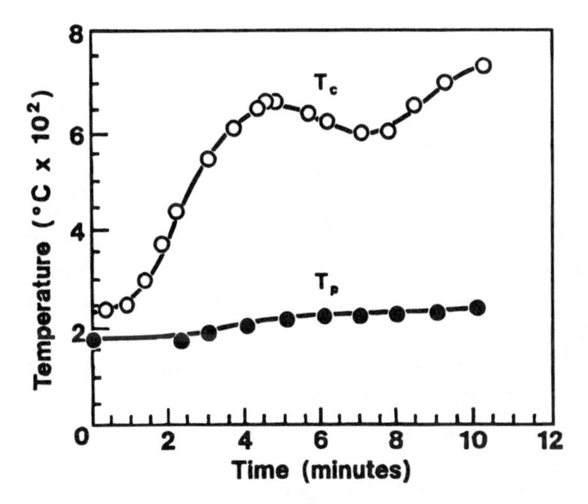

Figure 25 Thermal profile during severe regeneration as a function of regeneration time. (Courtesy of Society of the Automotive Engineers.)

D. Diesel Oxidation Catalysts

Recent advances in diesel technology have lowered exhaust particulates significantly such that diesel oxidation catalysts (DOCs) may provide the required incremental particle removal for MY 1994 + vehicles. The diesel oxidation catalysts use flow-through cordierite substrates with a large frontal area; see Fig. 26 [31]. Depending on the type of engine and its exhaust, they oxidize 30–80% of the gaseous HC and 40–90% of the CO present. They do not alter NO_x emissions. DOCs have been used in more than 60,000 diesel forklift trucks and mining vehicles since 1967 for HC and CO emissions [32].

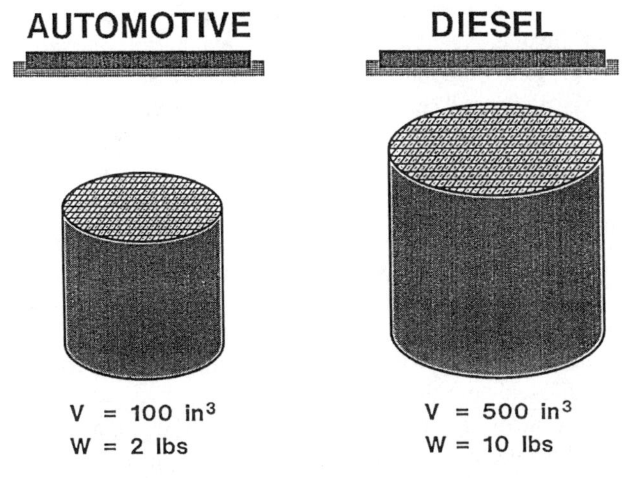

Figure 26 Relative size of automotive vs. diesel substrates. (Courtesy of the Society of Automotive Engineers.)

DOCs have little effect on dry soot (carbon), but engine tests show that they typically remove 30–50% of the total particulate load. This is achieved by oxidizing 50–80% of the SOFs present. DOCs are less effective with "dry" engines, in which particulates have a very low SOF content [32].

DOCs are similar to their gasoline engine counterparts in that they use a monolithic honeycomb support. Gases flow through the honeycomb with minimum pressure drop and react with the catalyst in the walls of the channels. The catalyst typically consists of platinum or palladium, or both, dispersed on a high-surface-area carrier such as aluminum oxide or silicon dioxide that maximizes contact with the gases and liquid organics. Research has focused on finding the best support/carrier/precious metal combination.

Although both metal and ceramic supports have been used, ceramic substrates offer stronger catalyst adhesion, less sensitivity to corrosion, and lower cost. The use of ceramics in automotive converters gives additional confidence in their performance.

Catalytic sites promote the reaction between HC gases, including those that would condense as SOFs downstream, and oxygen to form carbon dioxide and water. The sites also oxidize liquid SOFs, whether they are droplets that contact the catalyst or SOF gases that adsorb on them. SOFs adsorbed or condensed on the porous carrier are volatilized and then oxidized at the catalytic sites.

DOCs must function in a demanding environment. Although diesel exhaust temperatures are well below those in auto exhaust (200–600°C vs. 300–1000°C), diesel catalysts must contend with solids, liquids, and gases (not just gases) and deposits of noncombustible additives from lubricating oil. This last contains zinc, phosphorus, antimony, calcium, magnesium, and other contaminants that can shorten catalyst life below that mandated. Contamination also can come from sulfur dioxide in the exhaust.

Catalyst life might be extended by the development of low-ash lubricating oils and by modifying carrier properties such as surface chemistry, pore structure, and surface area, to create contamination-resistant catalysts.

DOCs also convert sulfur dioxide to sulfur trioxide, which forms sulfuric acid droplets or solid sulfate particles. These add to the amount of particulates emitted and can put an engine out of compliance. One approach to this problem is to lower fuel sulfur from the 1994 level to 0.01 wt% or less. A second approach is to develop a catalyst that oxidizes HC and CO but not sulfur dioxide. Newly developed catalysts that make almost no sulfate at temperatures as high as 400°C have brought this approach a step closer to reality. A third approach concerns catalyst placement. Sulfur dioxide oxidizes above 350–400°C over a DOC, while hydrocarbons do so below this temperature. A DOC could be located to have an inlet temperature favoring HC conversion.

The physical durability of DOCs depends heavily on mechanical and thermal properties of catalyzed substrates and their operating conditions; see Table 19. Since the diesel exhaust is substantially colder than that from a gasoline engine, and since the conversion temperature for diesel emissions is considerably lower, the thermal stresses associated with radial and axial temperature gradients in an LFA converter are well below the threshold strength, thereby eliminating the thermal fatigue potential and ensuring crack-free operation over the required 290K vehicle miles.

With thermal durability under control, the mechanical durability takes on a major focus to ensure total durability. To this end, it is necessary to ascertain high mechanical strength of the coated monolith, build in a resilient packaging system, and ensure positive and moderately high mounting pressure to guard against vibrational and impact loads. Much like automotive catalysts, the DOCs can continue to function catalytically even in

Table 19 Operating Conditions for Gasoline vs. Diesel Catalysts

	Gasoline	Diesel
Temperature range	300–1100°C	100–550°C
Temperature grad.	100–300°C	100–200°C
RH	<100%	100%
Space velocity	30,000–100,000 hr^{-1}	60,000–150,000 hr^{-1}
Vibration acceleration	28 g's	10–20 g's

the fractured state as long as there is a sufficient mounting pressure to keep the cracks shut and adequate catalytic activity to oxidize organic particulates over the required 290K vehicle miles; i.e., the packaging design is just as important as the catalyst formulation to meet the 290K vehicle mile durability.

The circular contour of DOC is ideal from a packaging point of view because it experiences a uniform mounting pressure and an axisymmetric temperature distribution, both of which are beneficial to long-term durability. However, since DOC is both larger and heavier and may experience different vibrational loads than the automotive catalyst, its mounting design requires special considerations. Moreover, since its operating temperature is lower than the intumescent temperature of ceramic mat, the converter assembly may have to be preheated to remove the organic binder and ensure an adequate mounting pressure before installation [33].

Finally, since packaging plays a key role in preserving the mechanical integrity of a diesel converter, both the mat thickness and its mounting density must be carefully tailored to provide substantial mounting pressure on the monolith to meet the 290K-mile durability. Guided by the successful packaging designs for European automotive converters and North American heavy-duty gasoline truck converters, both subjected to harsher driving conditions, a 6200-g/m^2 mat with a mount density of 1 g/cm^3 would be a good starting point. Such a design would result in a nominal mounting pressure of 50 psi, which is 10 times the minimum required value. It would also enhance the initial tangential strength of DOC and help contain any partial fragments over the required 290K vehicle miles should the converter experience any cracking. Additional mounting support, if necessary, may be provided in the axial direction by welding two end rings to the can at the entry and exit faces. The loss of active catalytic area in the peripheral cells, however, should be weighed against the improved mounting design afforded by the additional axial constraint. Other robust catalytic systems, suitable for practical applications, have also been discussed in a recent publication [34].

VII. SUMMARY

Since the enactment of the Clean Air Act, diesel engine manufacturers have been challenged to cut exhaust emission components in several stages, starting in 1988. The exhaust aftertreatment must include solid, liquid, particulate, and gaseous components that are hard to characterize. Diesel filters of cordierite ceramic composition have proven effective in trapping solid particulates and regenerating them at present intervals without imposing a high back pressure penalty.

Following the characterization of diesel exhaust and a brief review of performance requirements, this chapter provided design guidelines for selecting cell configuration, plugging pattern, contour, and overall size of the filter for a specific application. The role of microstructure in meeting filtration efficiency and back pressure requirements has also been emphasized. An approximate pressure drop model was included here to help define the maximum interval for soot loading and regeneration so as to preserve engine power and filter durability. This was followed by a brief discussion of four different regeneration techniques. Both electrical and catalytic regeneration techniques offer simplicity and reliability for the filter system.

The importance of physical properties on filter performance and durability was illustrated by comparing different filter compositions and microstructures. Advances in raw materials and manufacturing process have recently led to an optimum filter composition, EX-80, which yields the highest filtration efficiency, the lowest back pressure, and excellent long-term durability. This particular filter is an ideal candidate not only for urban buses and trucks, but also for diesel-powered passenger cars. Its low CTE permits soot regeneration, even under severe operating conditions, without compromising filter durability.

The importance of a systems approach to filter design was illustrated by three different examples where the appropriate choice of filter composition, packaging design, soot loading, and regeneration technique helps meet performance and durability requirements under specific operating conditions. The increasing use of diesel oxidation catalysts in view of advanced engine technologies was briefly touched on to provide an alternate and complimentary approach to the abatement of diesel emissions. Indeed, over 700,000 DOCs are in use in various diesel-powered vehicles and helping meet both HC and CO emissions regulations worldwide.

NOTATION

BPI	back pressure index
CFM	cubic feet per minute
CO	carbon monoxide
CTE	coefficient of thermal expansion
DOC	diesel oxidation catalyst
EPA	Environmental Protection Agency
HC	hydrocarbons
MIF	mechanical integrity factor
MOE	modulus of elasticity
MOR	modulus of rupture
MPH	miles per hour
NO_x	nitrous oxides
OFA	open frontal area
PM	particulate matter
SCFM	standard cubic feet per minute
SFA	specific filtration area
SOF	soluble organic fraction
TFA	total filtration area

REFERENCES

1. Springer, K.J., VII SIMEA, Symp. for Auto. Engineers, Sao Paulo (1993).
2. Lox, E.S., Engler, B.H., and Koberstein, E, CAPoC-II, Brussels (1990).
3. Kulkarni, N.S., *Automotive Eng. 100*:1 (1992).
4. Howitt, J.S., and Montierth, M.R., SAE Paper No. 810114 (1981).
5. Howitt, J.S., Elliott, W.T., Morgan, J.P., and Dainty, E. D., SAE Paper No. 830181 (1983).
6. Wade, W.R., White, J.E., and Florek, J.J., SAE Paper No. 810118 (1981).
7. Murtagh, M. J., Sherwood, D.L., and Socha, L.S. Jr., SAE Paper No. 940235 (1994).
8. Amann, C.A., Stivender, D.L., Plee, S.L., and MacDonald, J.S., SAE Paper No. 800251 (1980).
9. Weaver, C.S., SAE Paper No. 840174 (1984).
10. Gulati, S.T., and Helfinstine, J.D., SAE Paper No. 850010 (1985).
11. Gulati, S.T., and Sherwood, D.L., SAE Paper No. 910135 (1991).
12. Gulati, S.T., in *Structured Catalysts and Reactors*, A. Cybulski and J.A. Moulijn, eds., Dekker, New York (1997), Chap. 2.
13. Kitagawa, J., Asami. S., Ushara, K., and Hijikata, T., SAE Paper No. 920144 (1992).
14. Kitagawa, J., Hijikata, T., and Makino, M, SAE Paper No. 900113 (1990).
15. Brown, G.G., *Unit Operations*, Wiley, New York (1955).
16. Carman, P.C., *Flow of Gases Through Porous Media*, Butterworth, London (1956).
17. Wade, W.R., et al., SAE Paper No. 810118 (1981).
18. Ludecke, O. A., and Dimick, D. L., SAE Paper No. 830085 (1983).
19. Rao, V.D., et al., SAE Paper No. 850014 (1985).
20. Wade, W.R., et al., SAE Paper No. 830083 (1983).
21. Gulati, S.T., SAE Paper No. 830079 (1983).
22. Vergeer H.C., Gulati, S.T., Morgan J.P., and Dainty, E.D., SAE Paper No. 850152 (1985).
23. Gulati, S.T., SAE Paper No. 860008 (1986).
24. Gulati, S.T., and Lambert, D.W., ENVICERAM '91; Saarbücken, Germany (1991).
25. Gulati, S.T., Lambert, D.W., Hoffman, M.B., and Tuteja, A.D., SAE Paper No. 920143 (1992).
26. Widerhorn, S.M., in *Frac. Mech. Ceramics*, Vol. 2, Bradt et al., ed. Plenum Press, New York (1974).
27. Ritter, J.E., in *Frac. Mech. Ceramics*, Vol. 4, Bradt et al., ed. Plenum Press; New York (1978).
28. Gulati, S.T., SAE Paper No. 960471 (1996).
29. ANSYS, Swanson Systems, Elizabeth, PA.
30. Shinozaki, O., et al., SAE Paper No. 900107 (1990).
31. Gulati, S.T., SAE Paper No. 920145 (1992).
32. Farrauto, R., et al., *Automotive Eng.* February (1992).
33. Stroom, P.D., et al., SAE Paper No. 900500 (1990).
34. Neeft, J.P.A., et al., *Fuel Processing Technology*, Elsevier, Amsterdam (1995).

19

Zeolitic Membranes

Jolinde M. van de Graaf, Freek Kapteijn, and Jacob A. Moulijn
Delft University of Technology, Delft, The Netherlands

I. INTRODUCTION

Various types of membranes exist nowadays, applied mainly to separation processes. They consist of porous inorganic materials (silica, alumina, carbon), mixed solid oxides, metals, polymers, and liquid systems. For operation at elevated temperatures, for reasons of thermal efficiency in separation systems or membrane reactors (see Chapter 17), only the former three are in principle applicable. Metals and solid oxides can be used only for hydrogen and oxygen transport, respectively, leaving the porous inorganic materials for general use. The pores in these membranes have to be at least small enough so that Knudsen diffusion takes place, which discriminates on the basis of molecular mass. If the pore size approaches that of the molecule, other transport mechanisms, like surface diffusion and configurational diffusion, apply. Also, molecules that are too large will be unable to permeate, and molecular sieving takes place.

This is the area of zeolitic materials, which can be characterized as crystalline, porous materials built predominantly of oxygen, silica, and alumina (or phosphor) [1,2]. These crystals contain straight or sinusoidal channels that can be interconnected, resulting in a one-, two-, or three-dimensional pore network. At the intersections, small or large cavities may exist. Due to the crystallinity, the pores are uniform and can be prepared with great reproducibility, which is the biggest advantage over attempts to produce amorphous membranes with pores of molecular dimensions. Therefore, in the last decade much research effort has been put into the development of zeolitic membranes, and with success [3–29].

The use of zeolitic membranes in separation or combined reaction and separation processes is very appealing. Advantages of using this type of membrane include not only their ability to discriminate between molecules based on molecular size but also their thermal stability. The large variety of zeolite types could provide a tailor-made separation medium for specific processes. Moreover, the properties of zeolites are often easily adjustable (ion exchange, Si/Al ratio, etc.). This makes zeolitic membranes also very promising for use as catalytic membranes.

In this chapter examples will be given of permeation and separation characteristics of reported membranes. Accurate models for the description of permeation through zeolitic membranes are indispensable for the engineering implementation of these membranes in separation or reactor units. Therefore, the second section is devoted to models that are developed to describe these phenomena.

II. PERMEATION CHARACTERISTICS

A. Introduction

The first reported zeolite-based membranes were composed of zeolite-filled polymers [3–9]. The incorporation of zeolite crystals into these polymers resulted in a change of both permeation behavior and selectivity, due to the alteration of the affinity of the membrane for the components studied. Up to now, most known inorganic, zeolitic membranes have consisted of supported or unsupported ZSM-5 or silicalite [10–27]. Other reported membranes are prepared from zeolite-X [21], zeolite-A [21,28], or $AlPO_4$-5 [29]. The materials used as support are metals, glass, or alumina. The membrane configurations employed are flat sheet modules and annular tubes.

The above-mentioned studies reveal several features that determine the permeation through zeolitic membranes as well as their selectivity. Apart from size exclusion due to molecular sieving, both the affinity of the membranes for a given component and the mobility of that component in the pore network of the zeolite play a major role. In this section the importance of these features is shown on the basis of several examples. The emphasis will be on inorganic zeolitic membranes.

B. Size Exclusion

The pores of zeolites are of molecular dimensions (Fig. 1). In principle molecules can be excluded from the pores if their diameter is larger than the pore apertures. The zeolite framework is, however, not a rigid structure. Especially at higher temperatures, the pores become more flexible. As a result, molecules with larger diameters than the dimensions of the pores will be able to penetrate the channels. The maximum diameter of molecules that are able to adsorb within the zeolite is called the adsorption cutoff diameter. This diameter is about 0.95 nm for zeolite X and Y [30], 0.65 nm for ZSM-5 [30], and 0.4 nm for zeolite-4A [1] at 300 K. This means that, in principle, all molecules listed in Fig. 1 can permeate through a zeolitic membrane made from zeolite X, Y or ZSM-5, except for the large amines. Complete exclusion of molecules from the pores will therefore occur only when using a zeolite-A membrane.

Gas separations based on size exclusion thus require membranes made from small-pore zeolites. Although larger molecules penetrate with more difficulty through the zeolite channels than do smaller ones, true molecular sieving will occur only when using zeolites with maximum pore apertures of eight-membered T-O-T rings (T = Si or Al). Apart from zeolite A there are several other zeolites that could meet this criterion, for example, DOH or DD3R [31]. DOH is an all-silica zeolite that has pore apertures containing six oxygen atoms. Its pore diameter of only 0.28 nm is very promising for application in, for example, hydrogen recovery.

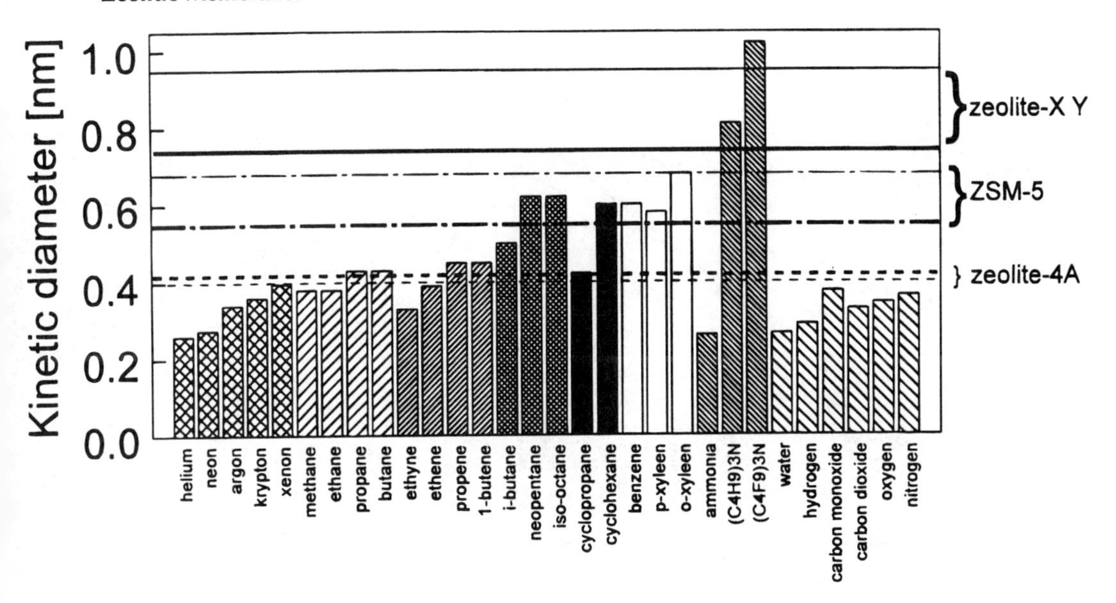

Figure 1 Kinetic diameters of several molecules compared to the free apertures of the pore (bold lines) and the adsorption cutoff diameter (thin lines) of zeolite 4A, ZSM-5, and zeolite X or Y. (From Refs. 1, 2, 30.)

C. Affinity and Mobility

1. *Single-Component Permeation*

Upon comparing the permeation of linear and branched paraffins through a silicalite membrane, the shape selectivity of the zeolite is demonstrated. Several authors report an iso-butane permeation rate that is a factor of up to 2–1000 lower than the permeation rate of *n*-butane, depending on the quality of the membrane and on the temperature [14,18,19,24].

In Fig. 2 the fluxes of several light hydrocarbons through a silicalite-1 membrane are shown as a function of their partial pressure on the feed side. The trend that can be deduced from this figure is that as the molecules get larger, their flux becomes lower. This decrease in flux is, however, smaller than expected on the basis of differences in diffusion coefficients [14]. The increase in the size of the molecule results in a lower mobility in the pores, but this effect is partly compensated by the higher concentration in the membrane, due to better adsorption of the larger molecules. This compensation effect is also the reason that at low partial pressures, ethane permeates faster through the membrane than does methane.

The nonlinear trend in the flux as a function of partial pressure in the feed also reflects the importance of adsorption in the membrane. The higher the adsorption of a compound, the sooner the concentration in the membrane reaches saturation, and permeation will no longer increase with pressure.

Figure 3 shows the temperature dependence of the permeation of some light hydrocarbons. The permeability is the permeation flux divided by the partial pressure gradient over the membrane:

$$P_i = \frac{N_i l}{\Delta p_i} \tag{1}$$

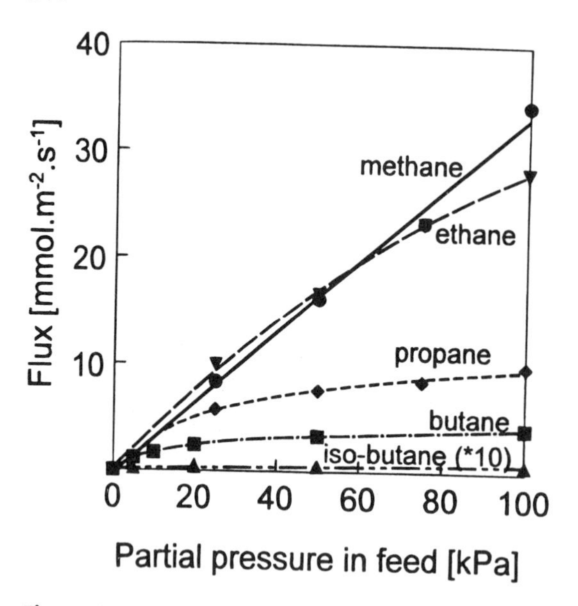

Figure 2 Flux of methane (●), ethane, (▼), propane (♦), *n*-butane (■), and iso-butane (▲) through a silicalite-1 membrane as a function of partial pressure on the feed side (T = 298 K, P_{tot} = 100 kPa). Feed was composed of hydrocarbon and balance helium; sweep gas used was helium. There was no absolute pressure difference across the membrane. (Adapted from Ref. 14.)

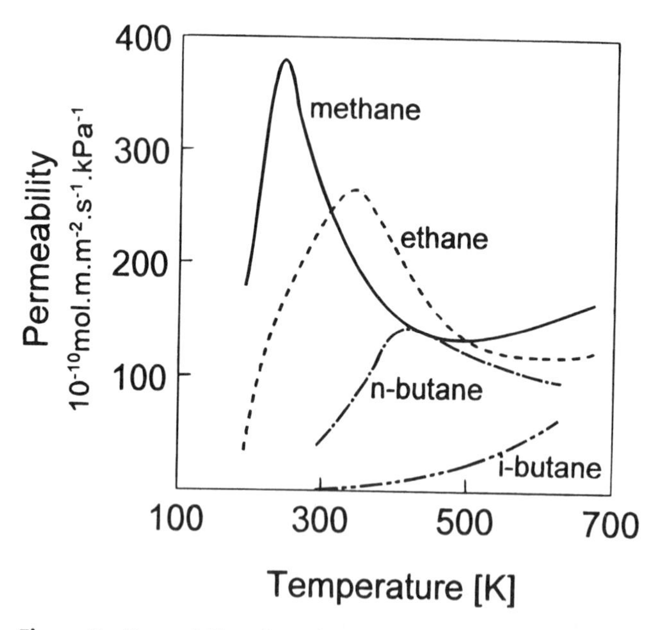

Figure 3 Permeability of methane (—), ethane (---), *n*-butane (-·-·-·), and iso-butane (-··-··) through a silicalite-1 membrane as a function of temperature. (Adapted from Ref. 14.)

A rise in temperature can have both a positive and a negative effect on the permeability. Transport through the membrane is an activated process, so it will benefit from a rise in temperature. However, as the temperature increases, the concentration in the membrane diminishes. These effects counteract each other, and one of them will dominate, depending on the difference in activation energy for diffusion and the heat of adsorption and the temperature. This will be discussed further in Section III.

2. *Multicomponent Permeation*

For binary systems, the separation selectivity of the membrane is defined as the ratio of the molar fractions of the two components in the permeate divided by that of the feed:

$$\alpha_{i,j} = \frac{x_{i,l}/x_{j,l}}{x_{i,0}/x_{j,0}} \tag{2}$$

This selectivity can differ significantly from the permeation ratio of the single components. In Fig. 4 transient permeation data of hydrogen, *n*-butane, and a hydrogen/*n*-butane mixture are shown. Clearly, hydrogen is the faster-permeating component on comparing the unary systems. Looking at the mixture permeation flux, it can be seen that initially hydrogen permeates faster than *n*-butane. When *n*-butane starts to permeate, the hydrogen permeation drops considerably and *n*-butane is selectively removed from the mixture (selectivity is greater than 100). The reason for this is that at this temperature the adsorption of *n*-butane is much stronger than that of hydrogen. Therefore, the pores of the zeolite become blocked, and hydrogen is not able to permeate around the sorbed *n*-butane molecules. In this case separation is based on differences in adsorption between the components. As the tempera-

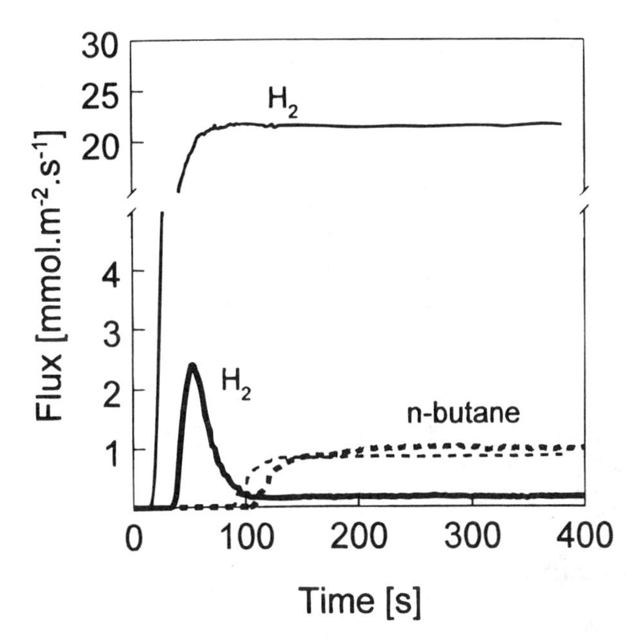

Figure 4 Breakthrough curves of hydrogen ($P_{i,0}$ 95 kPa, balance helium, thin line), *n*-butane ($P_{i,0}$ 5 kPa, balance helium, thin line), and a hydrogen/*n*-butane mixture ($P_{i,0}$ 95 kPa/5 kPa, bold lines) through a silicalite-1 membrane (T = 295 K, P_{tot} = 100 kPa). (Adapted from Ref. 14.)

ture increases, adsorption becomes less and less important, resulting, eventually, in the selective separation of hydrogen from the mixture (Fig. 5, [14]). At high temperatures the separation selectivity reflects the differences in mobility and approaches the flux ratio from unary experiments.

The separation selectivity is also found to depend on concentration, as is shown in Figs. 6a and 6b for permeation of an ethane/ethene mixture through a silicalite membrane [32]. Both unary (Fig. 6a) and binary (Fig. 6b) data are given. Again, it can be seen that the faster-moving component in the unary system is decelerated in the mixture by the stronger-adsorbing species. The separation selectivity found for the mixture ranges from 1.2 to 1.8 in favor of ethane. The separation selectivity based on unary fluxes is totally different, leading to the prediction that under certain conditions the separation of ethene from the mixture is preferred. This clearly demonstrates the importance of the affinity of the membrane for ethane in the ethane/ethene system and the need for a model that takes this feature into account.

A good example of separation on the basis of affinity is the separation of alcohol/water mixtures using a hydrophobic, silicalite membrane. Pervaporation of an ethanol/water mixture through such a membrane resulted the removal of the alcohol from the mixture [16]. The separation selectivities achieved are between 10 and 60, depending on temperature and the alcohol content in the feed. In this way azeotropes can be broken. The reason for this is that the principle of separation, namely, differences in adsorptive behavior, is different from separation based on vapor pressure differences, used in distillation. Another example of such a separation is the pervaporation of an acetic acid/water mixture through a silicalite membrane, resulting in the removal of acetic acid [17].

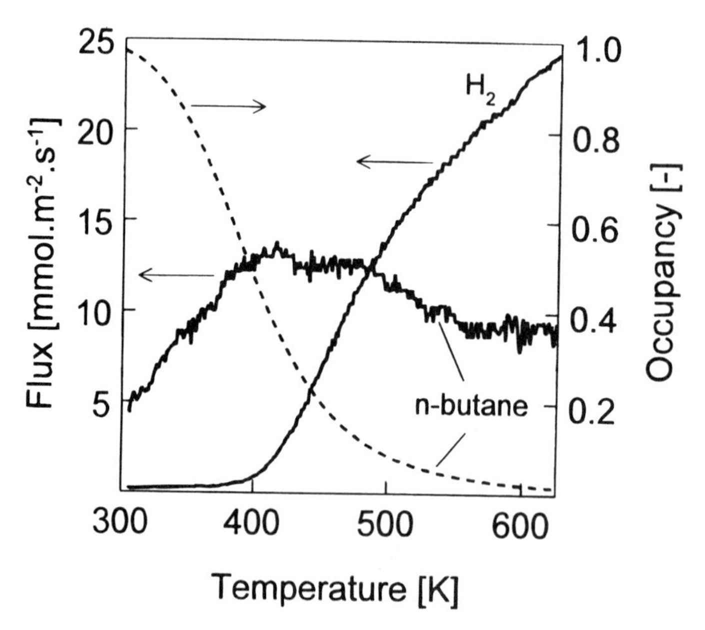

Figure 5 Fluxes of a hydrogen/n-butane ($P_{i,0}$ 50 kPa/50 kPa) mixture through a silicalite-1 membrane and n-butane occupancy in the membrane as a function of temperature (P_{tot} = 100 kPa). (Adapted from Ref. 14.)

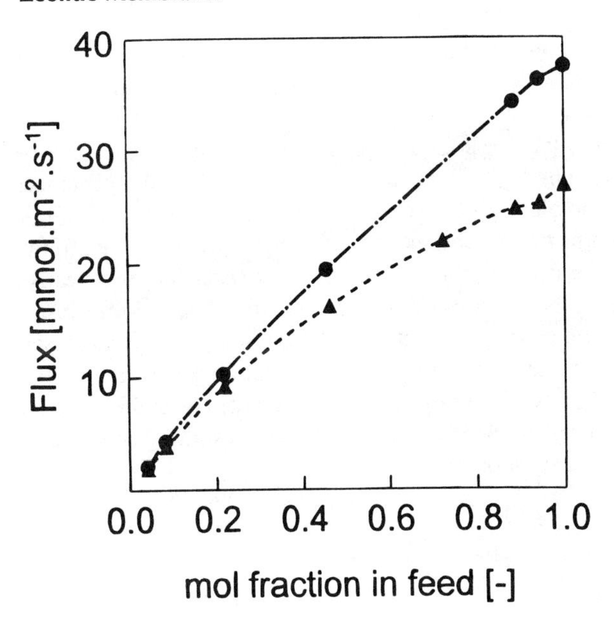

Figure 6a One-component fluxes of ethane (▲) and ethene (●) through a silicalite-1 membrane as a function of mol fraction in the feed (P_{tot} = 101.3 kPa, T = 297 K). Feed was composed of hydrocarbon and balance helium; sweep gas used was helium.

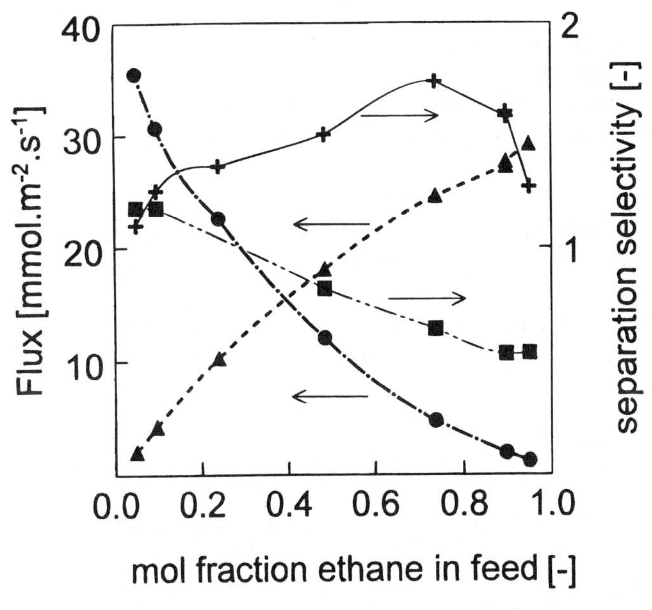

Figure 6b Fluxes of mixtures of ethane (▲) and ethene (●) through a silicalite-1 membrane as a function of mol fraction in the feed (P_{tot} = 101.3 kPa, T = 297 K). Feed was composed of 100% hydrocarbon (mol fraction of ethene = 1 − mol fraction of ethane); sweep gas used was helium. The measured separation selectivity toward ethane is also given (+) together with the separation selectivity predicted from single-component fluxes for identical partial pressures ratios (■).

Shape selectivity was demonstrated by the separation of *n*-paraffins from a naphtha distillate using a zeolite–CaA membrane [21]. The remaining fluid is enriched in branched parrafins and has a high octane number.

These examples demonstrate that separation by using zeolitic membranes can be done on the basis of differences in mobility/molecular size and on the basis of differences in affinity/adsorption. A good example for separation on the basis of molecular size is the separation of linear hydrocarbons from branched hydrocarbons. Differences in affinity between water and alcohols for the hydrophobic zeolite are the reason for the selective removal of alcohols from alcohol/water mixtures using a silicalite membrane. Temperature and concentration have a large effect on the selectivity. As the temperature increases, adsorption in the membrane becomes less important, and the separation selectivity reflects the differences in mobility between two molecules. At low temperatures the strongly adsorbing component blocks the pores for the more mobile but less adsorbing component, resulting in a separation based on adsorption differences.

D. Permeation and Reaction

The combination of reaction and separation in one multifunctional membrane reactor is an interesting option. In such a reactor the membrane could be catalytically active itself, or it could serve only as a separation medium. There are several types of operation for such a reactor [33]. It could be used to separate the formed products from the reaction mixture. In this way it is possible to overcome equilibrium limitations or to improve the selectivity of the reaction. Another possibility is the controlled addition of reactant via the membrane, which might be of use in, for example, oxidation reactions or sequential reactions. The advantage of using zeolitic membranes in a membrane reactor is that they have a high thermal stability and exhibit a good selectivity. Moreover, they can be made catalytically active.

Despite the attractive options of zeolitic membranes in these combined processes, there are only few reports of such applications. An improvement of selectivity by the removal of reaction products was found for the decomposition of cumene to propene and benzene over a H-ZSM-5 catalyst. In the membrane reactor, consisting of a packed bed of H-ZSM-5 surrounded by a silicalite membrane, a high selectivity toward propene and benzene was reported [22]. In the conventional reactor the main products were benzene and di- or triisopropylbenzene. A similar configuration, with a silicalite membrane, was used by Dalmon et al. [34] in order to enhance the yield in the equilibrium-limited dehydrogenation of iso-butane. The application of the membrane reactor resulted in an increase in the iso-butene yield of 70% compared to the conventional reactor. An example of the controlled addition of reactants through a ZSM-5 membrane to a packed bed of catalyst is the oxidation of cyclohexane to cyclohexanone and cyclohexanol. In a conventional packed-bed reactor the formation of CO_2 during this reaction was considerably higher than in the membrane reactor [22]. Suzuki [21] reported on the application of catalytically active zeolitic membranes. The membranes were composed of zeolite-A, zeolite-X, or ZSM-5 in the pores of macroporous supports like alumina and glass. The zeolitic membranes were made catalytically active by ion exchange. Several (de)hydrogenation and cracking reactions were carried out. The conversion of the less bulky reactants was always favored using this catalytic membrane.

III. THE MODELING OF PERMEATION THROUGH ZEOLITIC MEMBRANES

A. Introduction

A general model for transport through porous crystal membranes was described by Barrer [35]. The model involves five steps (Fig. 7):

1. Adsorption on the external surface
2. Transport form the external surface into the pores
3. Intracrystalline transport
4. Transport out of the pores to the external surface
5. Desorption from the external surface

These are all activated steps, which can be modeled assuming that molecules jump between low-energy sites. Each jump can be correlated to an activation energy, and the net flow is calculated from the forward and reverse jumps. Obstructions in the pores can be modeled as occasional intracrystalline energy barriers. The rate-determining step in this model is dependent on operating conditions (temperature, partial pressure) and the characteristics of the molecule and the crystalline material. Step 1, 2, 4, and 5 are referred to as interfacial processes. The importance of these processes will be discussed in Section III.B.

There are several models to describe intracrystalline diffusion (step 3) in microporous media. Diffusion in zeolites is extensively described in Ref. 30. For the modeling of permeation through zeolitic membranes, such a model should take the concentration dependence of zeolitic diffusion into account. Moreover, it should be easy applicable to multicomponent systems. In Section III.C, several models will be discussed.

Adsorption plays an important role in permeation through microporous membranes. First of all, steps 1 and 5 involve adsorption and desorption processes. Second, the concentration dependence of the diffusion coefficient is often described by the adsorption isotherm. Some data on adsorption in zeolites will be presented in Section III.D.

Finally, examples regarding the modeling of experimental data with some of the models discussed are given in Section III.E.

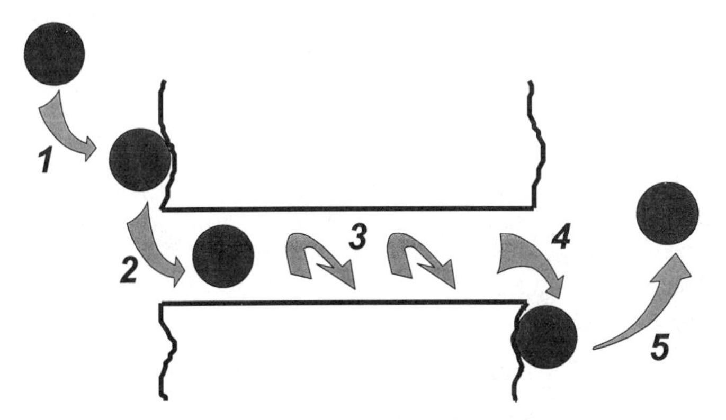

Figure 7 Five-step model for mass transfer through crystal membranes. *Step 1*: adsorption from the gas phase to the external surface. *Step 2*: transport from the external surface into the pores. *Step 3*: intracrystalline diffusion. *Step 4*: transport out of the pores to the external surface. *Step 5*: desorption from the external surface into the gas phase. (From Ref. 35.)

B. Interfacial Processes

It is easy to imagine that for weakly adsorbing molecules or at high temperatures, adsorption on the external surface might become the rate-limiting step for permeation through the membrane. However, if the activation energy for exit from the pores is larger than the activation energy for intracrystalline diffusion, a rise in temperature would diminish the influence of interfacial effects on the flux. For bulky molecules, entrance in the pores or intracrystalline diffusion is likely to be rate controlling as a result of a higher activation energy for these steps. Interfacial effects will be more important when the membrane becomes thinner.

Barrer [35] derived a criterion for the importance of interfacial effects:

$$\frac{N_i}{N_{i,\mathrm{id}}} = \frac{1}{1 + \left(\frac{1 - K'p_{i,0}}{1 - Kp_{i,0}} + \frac{1 - K'p_{i,l}}{1 - Kp_{i,l}}\right)\frac{\alpha}{l}\exp\left\{\frac{\delta\ \Delta E}{RT}\right\}} \tag{3}$$

In Eq. (3), $N_{i,id}$ is the flux when intracrystalline transport is rate controlling, N_i is the real flux. $\delta\ \Delta E$ is the difference between the activation energy for escape from within the crystal to the externally adsorbed layer and the activation energy for diffusion. It is generally a positive quantity [35]. K' and K represent the Langmuir parameters for adsorption on the external and internal surfaces, respectively. When internal and external adsorption isotherms are taken to be identical or are within Henry's law range, Eq. (3) becomes

$$\frac{N_i}{N_{i,id}} = \frac{1}{1 + 2\frac{\alpha}{l}\exp\left\{\frac{\delta\ \Delta E}{RT}\right\}} \tag{4}$$

In Figs. 8a and 8b the effect of the activation energy and the membrane thickness on the importance of interfacial processes, calculated from Eq. (4), are presented. $\delta\ \Delta E$ is taken to be a positive quantity, since it is likely that the activation energy for intracrystalline diffusion is smaller than the activation energy for exit from the pores [35]. It is clear from these figures that at high temperatures, for small values of $\delta\ \Delta E$ or for large crystal thicknesses, surface processes are less important.

If a difference in adsorption on the external and internal surfaces of the membrane is taken into account, the effects become more complicated. There are two possibilities. First, adsorption on the external surface is weaker than adsorption within the crystals. In this case interfacial processes become less important the higher is the adsorption strength on the outer surface. This is because the concentration in the pores is always higher than on the external layer. On the other hand, when adsorption on the external surface is strong compared to adsorption in the pores, the flux becomes restricted by the interfacial processes. The reason for this is that escape from the crystal becomes more difficult when the external surface is highly occupied. The effects of adsorption on the ratio between the ideal and the measured flux are summarized in Table 1.

C. Diffusion in Zeolites

1. *Fick's First Law of Diffusion*

A general form for the description of diffusion processes is Fick's First Law of Diffusion:

$$N_i = -D_i^{\mathrm{Fick}}\ \nabla c_i \tag{5}$$

Equation (5) implies that the driving force for diffusion is the concentration gradient.

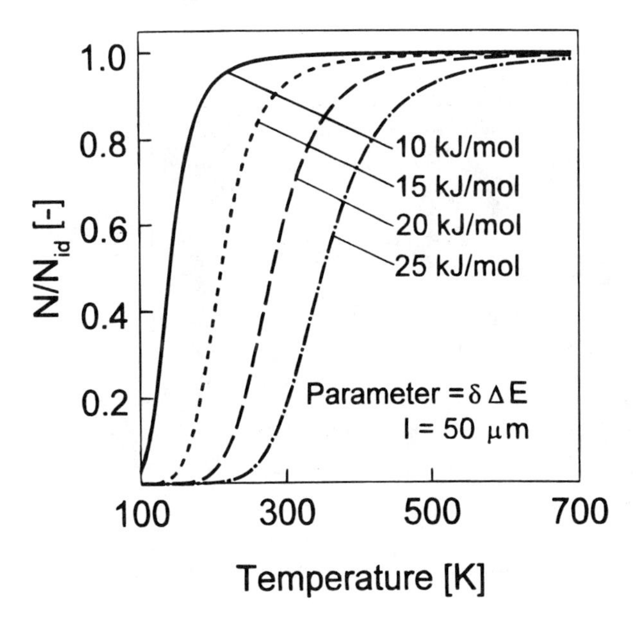

Figure 8a Importance of interfacial effects according to Eq. (3) as a function of temperature for a membrane thickness of 50 μm. Activation energy difference between exit from the pores to the externally adsorbed layer and intracrystalline diffusion is varied. If $N/N_{id} = 1$, interfacial effects are not important.

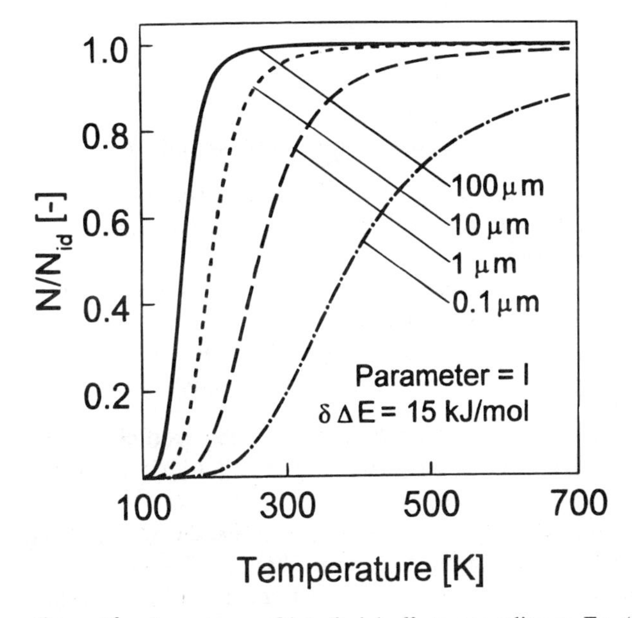

Figure 8b Importance of interfacial effects according to Eq. (3) as a function of temperature for an activation energy difference between exit from the pores to the externally adsorbed layer and intracrystalline diffusion of 15 kJ/mol. Membrane thickness is varied. If $N/N_{id} = 1$, interfacial effects are not important.

Table 1 Effect of Differences in Adsorption Constants on Interfacial Effects During Permeation of Gases Through Porous Crystal Membranes

$\frac{K'}{K} < 1$	$K'\uparrow$	$\frac{N}{N_{id}} \rightarrow 1$
$\frac{K'}{K} > 1$	$K'\downarrow$	$\frac{N}{N_{id}} \rightarrow 1$
K_s constant	$\frac{K'}{K}\downarrow$	$\frac{N}{N_{id}} \rightarrow 1$

Note: K' and K are the Langmuir adsorption parameters on the external and internal surfaces, respectively. (From Ref. 35.)

However, since the true driving force for diffusion is the chemical potential gradient, the expression for the flux should be

$$N_i = -B_i c_i \nabla \mu_i \tag{6}$$

For an ideal gas the chemical potential gradient and the partial pressure are related in the following way:

$$\mu_i = \mu_i^0 + RT \ln p_i \tag{7}$$

Combining Eqs. (5)–(7) leads to the definition of the corrected or self-diffusivity, the Darken equation:

$$D_i^{\text{Fick}} = D_i^0 \frac{\partial \ln p_i}{\partial \ln c_i} \qquad \text{with} \qquad D_i^0 = B_i RT \tag{8}$$

Three well-known types of diffusion can be distinguished in pore diffusion:

1. Bulk or molecular diffusion, dominated by molecule–molecule collisions in the gas phase. This type of diffusion becomes important for relatively large pore diameters or at high system pressures.
2. Knudsen diffusion, dominated by molecule–wall collisions. This mechanism is prevailing at low pressures or high temperatures.
3. Surface diffusion, which represents activated transport of adsorbed species along the pore wall.

In Fig. 9a and 9b the orders of magnitude of the different diffusion coefficients and the activation energy for diffusion are given as a function of pore size [36]. For diffusion in micropores (<2 nm in diameter), the diffusivity can vary over several orders of magnitude, depending on the size and nature of the diffusing species and the microporous media. Diffusion in micropores is often referred to as configurational diffusion.

In the case of surface diffusion, which is likely to occur in micropores, the partial pressure p_i is related to the adsorbed-phase concentration by the adsorption isotherm. Consequently, the choice of the adsorption isotherm influences the diffusion coefficient according to Eq. (8).

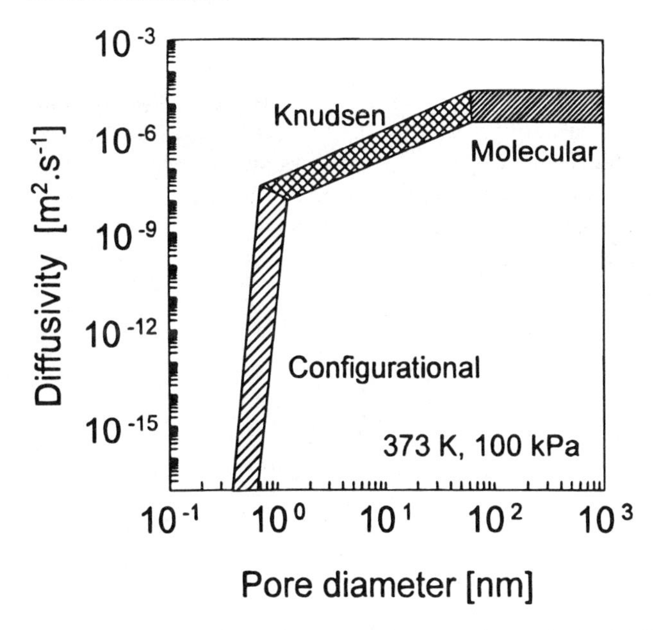

Figure 9a Effect of pore diameter on the order of magnitude of diffusivities. (Adapted from Ref. 36.)

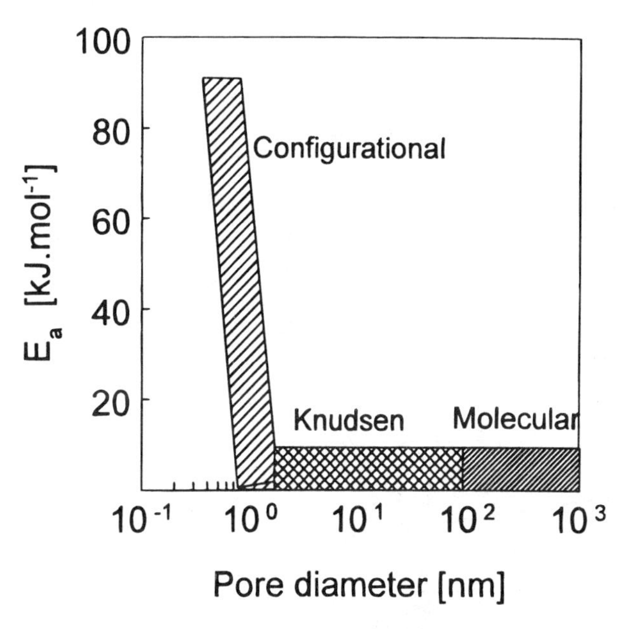

Figure 9b Effect of pore diameter on the order of magnitude of the activation energy for diffusion. (Adapted from Ref. 36.)

2. *Combined Gaseous and Surface Diffusion*

One way of looking at transport through microporous media is that at the starting point molecules can either retain their gaseous character in the pores or are adsorbed on the surface. Figure 10 presents the contribution of the gas-phase concentration and the surface concentration of a number of gases to the total concentration in silicalite. The contribution of the gas-phase concentration becomes more and more important with increasing temperature. The net flow through the pores is a combination of gaseous and surface flow, which are both assumed to be activated processes:

$$N_i^{\text{tot}} = N_i^s + N_i^g \tag{9}$$

where

$$N_i^s = -D_i^s \nabla c_{i,\text{ads}} \qquad \text{and} \qquad N_i^g = -\frac{D_i^g}{RT} \nabla p_i \tag{10}$$

This type of approach was developed by Wei et al. [37], Ma et al [38], and Inoue et al. [39] for gaseous systems. Yoshida et al. [40,41] studied the use of combined liquid and surface flow for the description of dye permeation through a cellulose membrane.

A general expression for diffusion coefficients is [37]:

$$D = gdv \exp\left\{\frac{-E_a}{RT}\right\} \tag{11}$$

In this, g is a geometrical factor, d is a diffusional length, v is the velocity of a molecule, and E_a is the activation energy for diffusion. In ordinary gas diffusion the activation

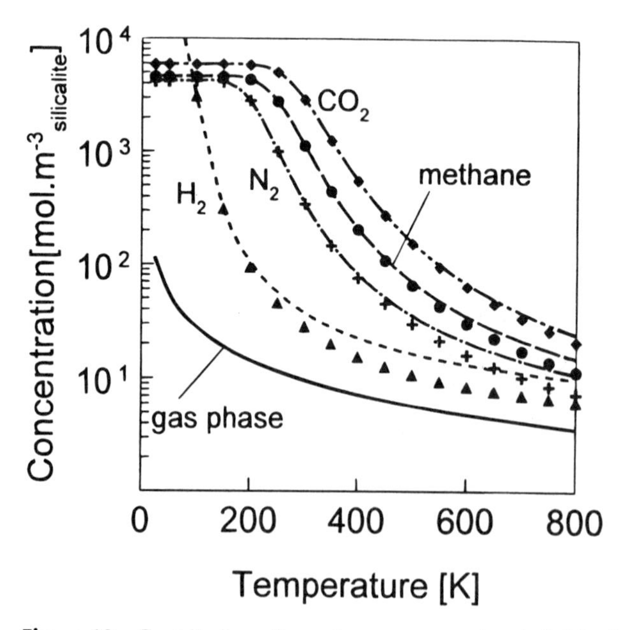

Figure 10 Contribution of gas-phase concentration (solid line) and adsorbed-phase concentration (symbols: ▲ = H_2, + = N_2, ● = CH_4, ◆ = CO_2) to the total concentration in silicalite-1 (dotted lines) as a function of temperature (p_i = 1 atm). (Adsorption data are taken from Ref. 61.)

energy is zero. For gaseous diffusion in microporous media, also called gas translational diffusion, the diffusivity has been correlated to the Knudsen diffusivity by the introduction of an activation energy for diffusion [37,38]:

$$D_i^g = \frac{1}{z}\, d\sqrt{\frac{8RT}{\pi M_i}}\exp\left\{\frac{-E_a}{RT}\right\} \tag{12}$$

For Knudsen diffusion, d is the pore diameter and the activation energy for diffusion is zero. For activated gas translational diffusion d is the diffusional length between adjacent low-energy sites [37] or the pore diameter [38], and the activation energy is between 10 and 60 kJ/mol [37]. The parameter z, the lattice coordination number, depends on the zeolite type. z is equal to 4 for ZSM-5 and to 6 for 5A, compared to 3 for Knudsen diffusion.

The activation energy for gas translational diffusion represents the energy barrier that a molecule has to overcome upon moving from one channel intersection to the other. The order of magnitude can be calculated from the difference in potential field in a channel and in an intersection. The transition from Knudsen to configurational diffusion depends on the size of the molecule with respect to the zeolite channels and the proportions of the molecule (ratio of length to diameter), as is shown in Fig. 11 for ZSM-5 and zeolite-A.

The surface diffusion coefficient can be related to a jump frequency, ν, and a jump distance, α [42–44]:

$$D^s = \frac{1}{z}\,\alpha^2\nu(\theta)\exp\left\{\frac{-E_a}{RT}\right\} \tag{13}$$

Compared to Eq. (11), $\alpha\nu$ represents the velocity of the molecule, α represents the

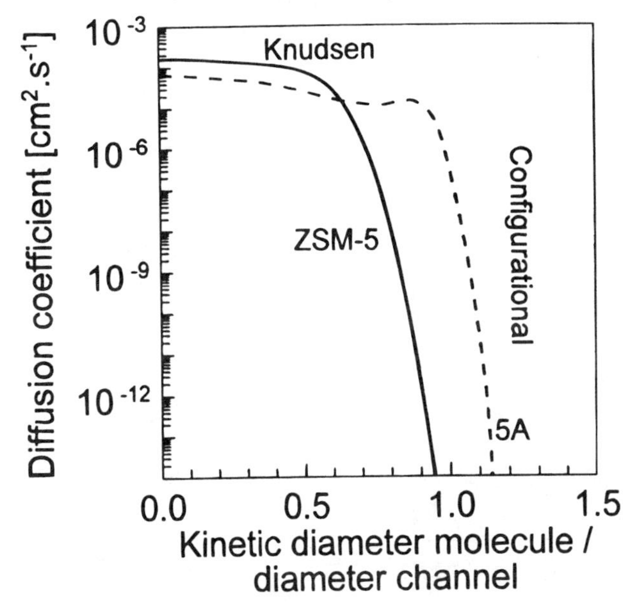

Figure 11 Transition from Knudsen diffusion to configurational diffusion in ZSM-5 and 5A as a function of the ratio between the minimum kinetic diameter of the molecule and the maximum channel diameter of the zeolite at 300 K. The ratio between the length and the kinetic diameter of the diffusing molecule was taken to be 1.25. (Adapted from Ref. 37.)

diffusional length, and $1/z$ is the geometrical factor. The jump frequency ν can be dependent on occupancy.

Xiao et al. [37] included interaction between molecules in the diffusion coefficient by allowing each site to be occupied by a maximum of two molecules. If molecules are on the same site, they interact with each other. The diffusion coefficient increases with increasing interaction between molecules. The apparent diffusivity becomes

$$D_{\text{app}} = D\left[1 + w\theta\,(1 - \theta)\,\frac{\partial\theta(2)}{\partial\theta}\right] \exp\{\theta(2)w\} \tag{14}$$

in which $\theta(2)$ is the probability of double occupancy and w is the dimensionless energy change due to the molecule–molecule interaction. Figure 12a represents the normalized diffusion coefficient (D_{app}/D) as a function of the total occupancy and w. Experimental evidence showed indeed that for benzene in ZSM-5 at occupancies below 0.5, the diffusivity is independent of occupancy, whereas at higher loadings the diffusivity increases with occupancy [45]. Figure 12b gives experimental data on the normalized diffusivity of benzene in ZSM-5 as a function of occupancy, together with the prediction according to Eq. (14). A physical picture of the increasing diffusion coefficient with occupancy is that when a molecule jumps to a site that is already occupied, the energy well becomes less deep due to the repulsive force between the molecules. As a result, displacement becomes easier.

3. *The Generalized Maxwell–Stefan Equations*

The principle of the Maxwell–Stefan diffusion equations is that the force acting on a species is balanced by the friction that is exerted on that species. The driving force for

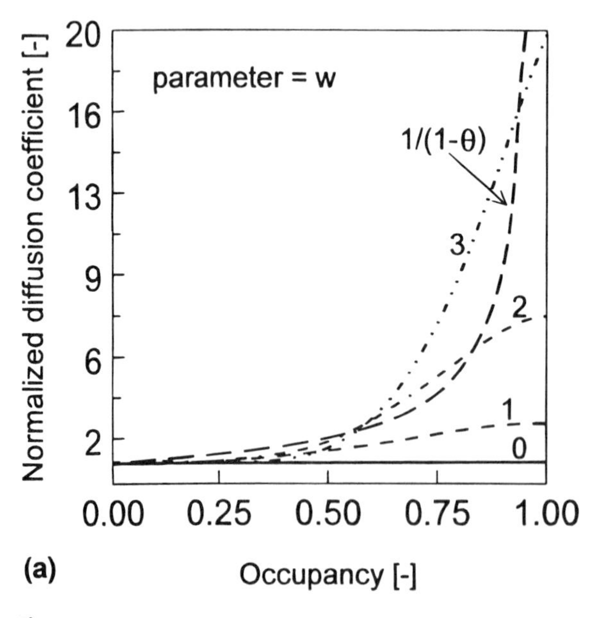

Figure 12a Occupancy dependence of the normalized diffusivity (D_{app}/D) as a function of the dimensionless activation energy, w, according to Eq. (14). (Adapted from Ref. 37.)

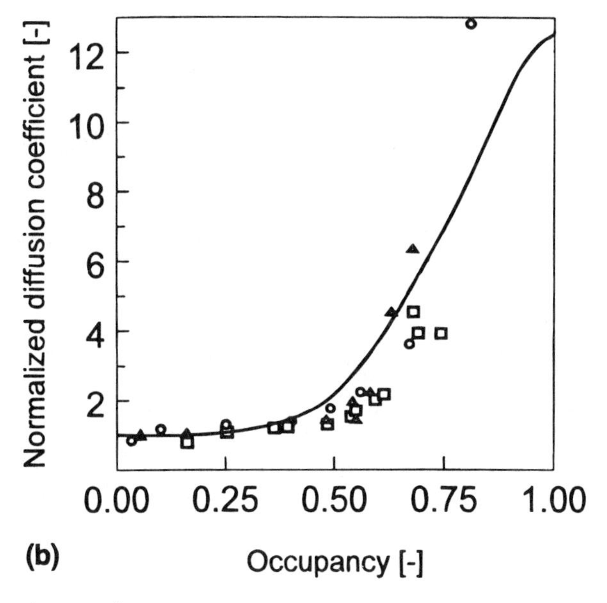

Figure 12b Occupancy dependence of the normalized diffusivity (D_{app}/D) of benzene in ZSM-5. Experimental data (○ = 296 K, □ = 308 K, △ = 318 K) and model prediction (w = 2.5, ΔE = 6.5 kJ/mol) according to Eq. (14). (Adapted from Ref. 45.)

diffusion is the chemical potential gradient. The Maxwell–Stefan equations were applied to surface diffusion in microporous media by Krishna [46]. During surface diffusion, a molecule experiences friction from other molecules and from the surface, which is included in the model as a pseudospecies, $n + 1$ (dusty-gas model). The balance between force and friction in a multicomponent system can thus be written as [46]

$$-\frac{1}{RT}\nabla\mu_i = \sum_{j=1}^{n}\theta_j\frac{v_i - v_j}{D_{i,j}^{MS}} + \theta_{n+1}\frac{v_i - v_{n+1}}{D_{i,n+1}^{MS}} \qquad i, j = 1, 2, \ldots, n \tag{15}$$

The first term on the right-hand side of the equation denotes the friction between species; the second term represents friction between a species and the surface. $D_{i,j}^{MS}$ and $D_{i,n+1}^{MS}$ are the Maxwell–Stefan diffusivities. The first term on the right side is often referred to as an exchange term that represents the probability of molecules exchanging places on the surface. Since this exchange is not likely to occur in narrow zeolite channels, it is commonly neglected. This is called single-file diffusion.

The Maxwell–Stefan surface diffusivity is defined, by analogy to the definition of the Knudsen diffusivity [46], as

$$D_i^{MS} = \frac{D_{i,n+1}^{MS}}{\theta_{n+1}} \tag{16}$$

The chemical potential gradient can be related to a matrix of thermodynamic factors [46]:

$$\frac{\theta_i}{RT}\nabla\mu_i = \sum_{j=1}^{n}\Gamma_{ij}\nabla\theta_i \qquad i,j = 1, 2, \ldots, n \tag{17}$$

where

$$\Gamma_{ij} \equiv \theta_i \frac{\partial \ln p_i}{\partial \theta_i}$$

The occupancies can be derived from the adsorption isotherms. For Langmuir adsorption this relation is

$$\theta_i = \frac{c_i}{c_i^{\text{sat}}} = \frac{K_i p_i}{1 + \sum K_j p_j} \tag{18}$$

Combining Eqs. (16), (17), and (18) and taking

$$N_i = \rho c_i^{\text{sat}} \theta_i v_i \tag{19}$$

gives Eqs. (20) and (21) for one- and two-component systems, respectively:

$$N_1 = -c_1^{\text{sat}} \rho \cdot \frac{D_i^{MS}}{1 - \theta_1} \cdot \nabla \theta_1 \tag{20}$$

$$N_1 = -c_1^{\text{sat}} \rho \cdot \frac{D^{MS}}{1 - \theta_1 - \theta_2} \cdot [(1 - \theta_2) \cdot \nabla \theta_1 + \theta_1 \cdot \nabla \theta_2] \tag{21}$$

(single-file diffusion)

On comparing Eqs. (20) and (5) and (8), it can be seen that for Langmuir adsorption, the Maxwell–Stefan diffusivity in a one-component system is identical to corrected diffusivity in the Darken equation.

From Eqs. (20) and (21) it follows that if D_i^{MS} is independent of concentration, binary fluxes can be predicted on the basis of single-component permeation data once the multicomponent adsorption isotherm is known.

Diffusion through zeolites, and consequently permeation through zeolitic membranes, is concentration dependent [10–14,16,45]. It is therefore worthwhile to compare how this concentration dependence is taken into account in different models. Yang and Chen [47] proposed a model for diffusion in micropores in which interaction between molecules on the surface was taken into account by introducing a sticking probability, λ. This parameter represents the chance of a molecule to move to an occupied site compared to the chance of moving to a vacant site on the surface. λ is related to the activation energy of diffusion. For a single-component system this results in the following relationship for the flux [47]:

$$N_i = -c_i^{\text{sat}} \rho \frac{D_i^0}{1 - (1 - \lambda_i)\theta_i} \nabla \theta_i \quad 0 \leq \lambda_i \leq 1 \tag{22}$$

In fact, λ is a correction parameter for the Fick diffusion coefficient. This correction has a similar effect on the apparent diffusivity as the correction given in Eq. (14). When λ is less then 1, the diffusivity increases with occupancy. This correction can also be applied to the Maxwell–Stefan diffusivity, which results in an even larger effect of concentration on the flux. The concentration dependence of the flux in the Maxwell–Stefan equations depends largely on the adsorption isotherm chosen, since this isotherm determines the thermodynamic factor. For Langmuir adsorption the concentration dependence of the flux increases in the following order using different models:

$$N_i \propto$$

$$\underset{\text{Fick}}{\nabla\theta_i} < \underset{\text{Fick + Yang}}{\frac{1}{1-(1-\lambda_i)\theta_i} \cdot \nabla\theta_i} < \underset{\text{Maxwell-Stefan}}{\frac{1}{1-\theta_i} \cdot \nabla\theta_i} < \underset{\text{Maxwell-Stefan + Yang}}{\frac{1}{1-(1-\lambda_i)\theta_i} \cdot \frac{1}{1-\theta_i} \nabla\theta_i} \tag{23}$$

Figure 13 gives the dependence of the flux on concentration, normalized to the flux according to the Fick model, assuming that the diffusion coefficients are constant in all models.

The temperature dependence of surface diffusion is expressed in the adsorption constant and the diffusivity. For the adsorption constant the following relationship holds:

$$K_i = \exp\left\{\frac{\Delta S_i}{R} + \frac{Q_i}{RT}\right\} \tag{24}$$

If the diffusivity is independent of occupancy, its temperature dependence satisfies an Ahrenius-type relation:

$$D_i = D^0 \exp\left\{\frac{-E_{D,i}}{RT}\right\} \tag{25}$$

Figure 14 presents the effect of varying the activation energy for diffusion on the temperature dependence of the flux. From this it can be seen that, when using the Maxwell–Stefan description (assuming that intracrystalline transport is rate controlling and adsorption follows a Langmuir isotherm), a maximum in the flux is to be expected when the activation energy for diffusion is smaller than the heat of adsorption. When the activation energy for diffusion is higher than the heat of adsorption, the flux increases with temperature.

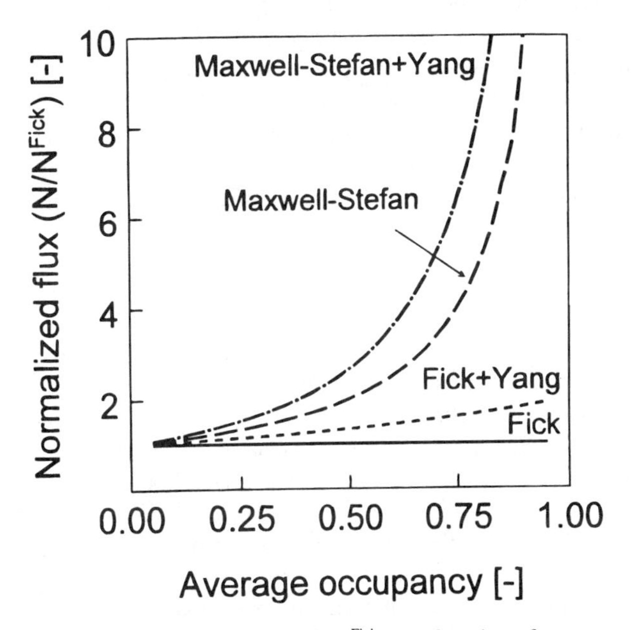

Figure 13 Normalized flux (N/N^{Fick}) as a function of occupancy according to Eq. (23). Diffusivity is taken constant, $\lambda = 0.5$.

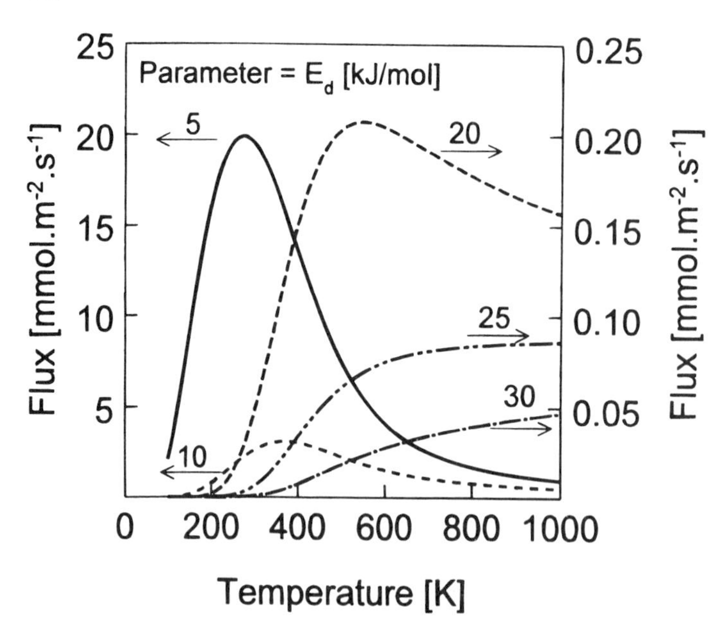

Figure 14 Temperature dependence of the flux according to the Maxwell–Stefan model for a one-component system. Activation energy for diffusion was varied, heat of adsorption was taken as 25 $kJ \cdot mol^{-1}$. Other parameters are: $\Delta S = -75\ J \cdot mol^{-1} \cdot K^{-1}$, $c^{sat} = 1\ mmol \cdot g^{-1}$, $D^0 = 1 \times 10^{-6}\ m^2 \cdot sec^{-1}$, $E_d^s = 15\ kJ \cdot mol^{-1}$, $l = 50\ \mu m$, $\rho = 1.8 \times 10^6\ g \cdot m^{-3}$

The application of the Maxwell–Stefan theory for diffusion in microporous media to permeation through zeolitic membranes implies that transport is assumed to occur only via the adsorbed phase (surface diffusion). Upon combination of surface diffusion according to the Maxwell–Stefan model (Eq. 20) with activated-gas translational diffusion (Eq. 12) for a one-component system, the temperature dependence of the flux shows a maximum and a minimum for a given set of parameters (Fig. 15). At low temperatures, surface diffusion is the most important diffusion mechanism. This type of diffusion is highly dependent on the concentration of adsorbed species in the membrane, which is calculated from the adsorption isotherm. At high temperatures, activated-gas translational diffusion takes over, causing an increase in the flux until it levels off at still-higher temperatures.

D. Adsorption on Zeolites

Adsorption plays an important role in permeation through microporous media. The selectivity of a zeolitic membrane at low temperatures is largely determined by differences in adsorption between species, as was shown in Section II. Moreover, the surface concentration, which is related to the partial pressure by the adsorption isotherm, plays an important role in the models for zeolitic diffusion. Finally, the thermodynamic factor from Eq. (17) is related to the adsorption isotherm.

Adsorption in zeolites depends highly on the characteristics (type of cation, Si/Al ratio) of the zeolite [2]. The adsorption isotherms in zeolites are mostly of type I, which means that monolayer adsorption takes place and a saturation concentration is observed. Extensive reviews on adsorption on microporous media are given in several textbooks [48,49].

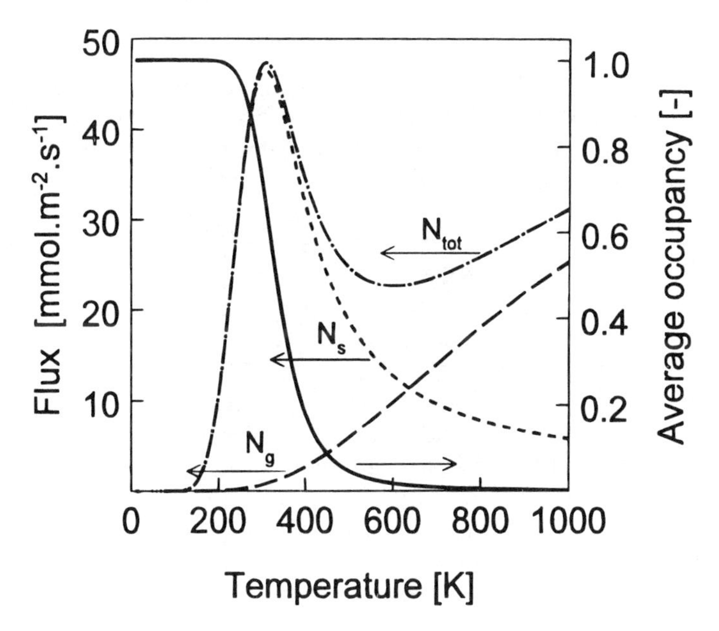

Figure 15 Combined gas translational diffusion and surface diffusion as a function of temperature. The surface coverage on the feed side is also included. Parameters are: $\Delta S = -50\ \mathrm{J\cdot mol^{-1}\cdot K^{-1}}$, $c^{\mathrm{sat}} = 1\ \mathrm{mmol\cdot g^{-1}}$, $Q = 25\ \mathrm{kJ\cdot mol^{-1}}$, $D^0 = 1 \times 10^{-6}\ \mathrm{m^2\cdot sec^{-1}}$, $E_d^s = 15\ \mathrm{kJ\cdot mol^{-1}}$, $l = 50\ \mu\mathrm{m}$, $\rho = 1.8 \times 10^6\ \mathrm{g\cdot m^{-3}}$, $E_d^g = 15\ \mathrm{kJ\cdot mol^{-1}}$, $d = 5.5 \times 10^{-10}$, $M = 16\ \mathrm{g\cdot mol^{-1}}$.

Adsorption of single components in zeolites can often be described by a simple Langmuir isotherm, Eq. (18), or the Langmuir–Freundlich isotherm, Eq. (26) [50–52]:

$$\theta_i = \frac{c_i}{c_i^{\mathrm{sat}}} = \frac{(K_i p_i)^n}{(1 + K_i p_i)^n} \tag{26}$$

For faujasite-type zeolites and zeolite-A, where several molecules can occupy one cage, a special type of isotherm has been derived, the statistical model isotherm [49]. This isotherm treats each cage in the zeolite as a subsystem. Each cage can contain a fixed maximum number of molecules, and the molecules can interchange between the cages. Within a cage, interaction between the molecules is taken into account, which is not accounted for in the Langmuir description.

For multicomponent adsorption the most commonly used isotherm is the extended Langmuir isotherm (Eq. 18). Another, frequently used approach is the Ideal Adsorption Solution theory (IAS theory), which was developed by Prausnitz [53] and applied to mixtures of gases by, for example, Kaul [54] and Rees [52,55].

E. Examples

Application of the Maxwell–Stefan equations to permeation through zeolitic membranes was done by Kapteijn et al. [50,56] and Krishna and van den Broeke [57]. Kapteijn showed that both the temperature and occupancy dependence of the steady-state n-butane flux can accurately be described by Eqs. (20), (24), and (25) [56]. The advantage of using the Maxwell–Stefan description is that it is able to describe both occupancy and temperature dependence by using a constant diffusivity. This is not the case if one were to use Fick's first law of diffusion. Figures 16a and 16b [56] show the occupancy and temperature dependence of the Fickian and Maxwell–Stefan diffusivity, calculated from the data in Figs. 2 and 3. The calculated activation energy for diffusion is higher than the heat of adsorption. The observed maximum in the permeation is thus in agreement with the theory, as shown in Fig. 14.

The same model was applied to permeation of lighter hydrocarbons (C_1–C_3) through the silicalite-1 membrane [50]. In the case of methane, ethane, and ethene, some concentration dependence of the Maxwell–Stefan diffusivity was observed. This can be caused either by the importance of interfacial effects, which are not taken into account, or by the contribution of activated-gas translational diffusion to the net flux. The diffusivities calculated from these permeation experiments were, however, in rather good agreement with diffusivity values from the literature, which implies that these zeolitic membranes could also be a valuable tool for the determination of diffusion coefficients in zeolites.

The transient permeation results from Refs. 12 and 14 were modeled qualitatively by Krishna et al. [57] by using the Maxwell–Stefan formulation for two components. Figure 17 compares the modeling results for a two-component system to the Fick formulation with constant diffusivity. A maximum, like that observed for hydrogen in the hydrogen/*n*-butane system (Fig. 4), can be predicted when using the Maxwell–Stefan description but is absent when using the Fick formulation with constant diffusivity. From these results it can be concluded that a concentration-dependent model is indispensable for the description of multicomponent permeation through zeolitic membranes.

The Maxwell–Stefan diffusivities defined in Eqs. (20) and (21) for one- and two-component systems should be identical if they are independent of occupancy. Diffusivities

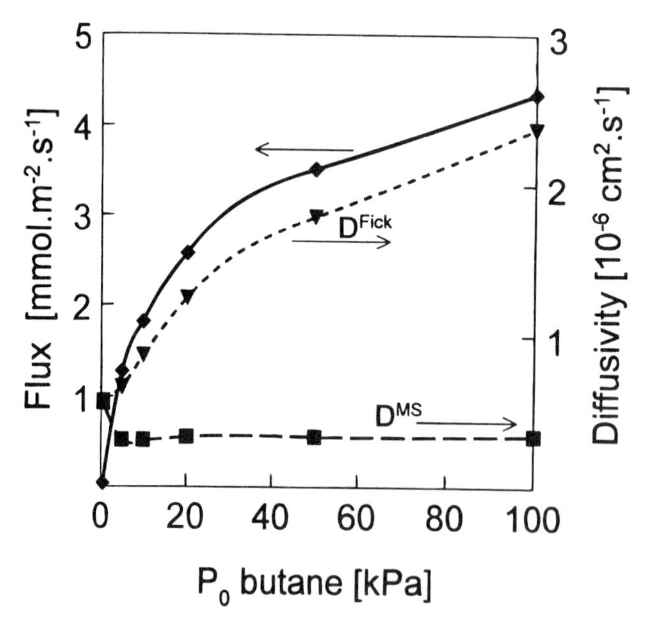

Figure 16a Permeation flux of *n*-butane (◆) through a silicalite-1 membrane as a function of the feed partial pressure of *n*-butane (T = 300 K, P_{tot} = 100 kPa). Included are the calculated Fickian (▼) (Eq. 5) and Maxwell–Stefan (■) (Eq. 20) diffusivities. (Adapted from Ref. 56.)

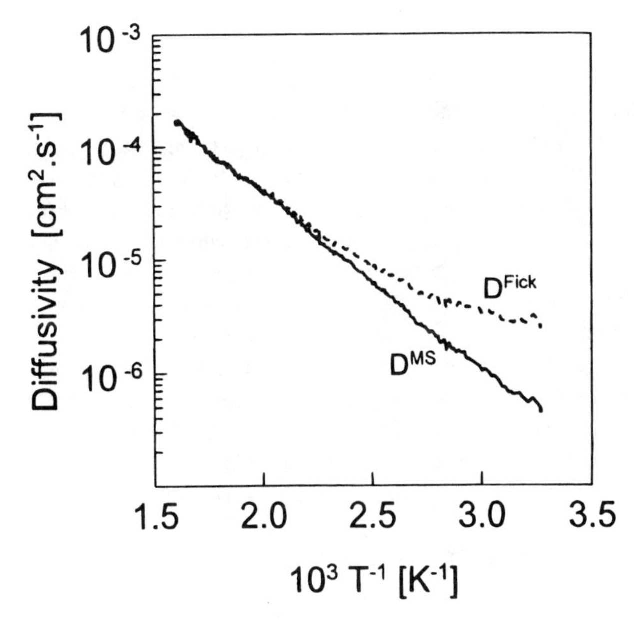

Figure 16b Fickian and Maxwell–Stefan diffusivities of *n*-butane as a function of $1/T$, calculated from the *n*-butane permeation results in Fig. 3. (Adapted from Ref. 56.)

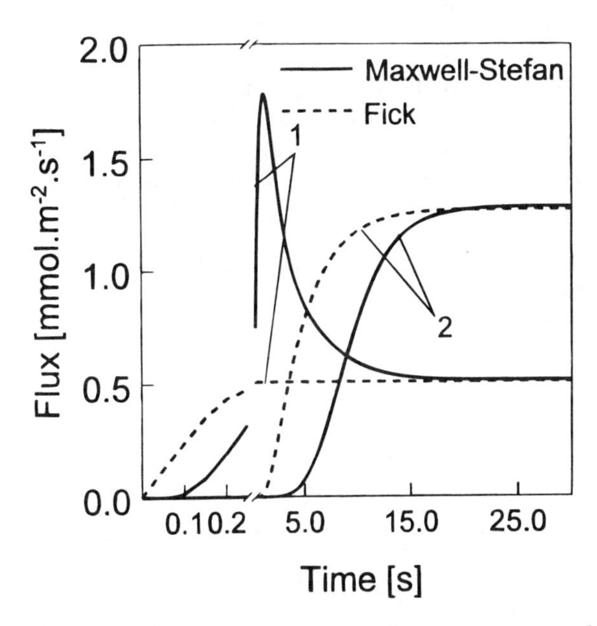

Figure 17 Simulations of transport of a two-component mixture ($p_1, p_2 = 50$ kPa) across a zeolitic membrane using the Fickian and Maxwell–Stefan descriptions (Eqs. 5 and 21, respectively). Permeate partial pressures are taken to be zero. The following parameters were used: $K_1 = 0.01\ \text{kPa}^{-1}$, $K_2/K_1 = 20$, $c_1^{sat} = c_2^{sat} = 1\ \text{mmol}\cdot\text{g}^{-1}$, $D_1^{MS} = 1 \times 10^{-9}\ \text{m}^2\cdot\text{sec}^{-1}$, $D_1^{Fick} = 4 \times 10^{-9}\ \text{m}^2\cdot\text{sec}^{-1}$, $D_1/D_2 = 40$.

of ethane and ethene were calculated from steady-state ethane, ethene, and ethane/ethene permeation experiments (Figs. 6a and 6b [32]). Some results are given in Table 2. It can be seen that diffusivities calculated from one- and two-component experiments at a given concentration in the membrane deviate by up to 50%. This was also observed when using corrected diffusivities according to the model proposed by Yang [47]. Yang, however, successfully predicted the uptake rate of nitrogen and oxygen in a mixture in carbon molecular sieve pellets on the basis of single-component uptake experiments, using this model [47]. On the other hand, the model did not work for a methane/ethane system through a carbon molecular sieve membrane [58]. Apparently there is some influence of the presence of other molecules that is not taken into account in these models.

The temperature dependence of the methane permeation through a silicalite membrane, showing a maximum and a minimum as a function of temperature (Fig. 3 [14]), can not be predicted by using the Maxwell–Stefan description for surface diffusion only. Such a maximum and minimum in the permeation as a function of temperature can be predicted only when the total flux is described by a combination of surface diffusion and activated-gas translational diffusion (Fig. 15).

Shelekhin et al. [38] used this combined model for the description of permeation of permanent gases through a microporous glass membrane. Adsorption data were calculated from separate experiments and fitted with a Dubinin–Radushkevich isotherm. Figure 18 shows the permeability of CO_2 as a function of pressure at two temperatures. When combining surface and gaseous flow, only the permeability resulting from surface flow is dependent on pressure, which can easily be seen on combining Eqs. (1), (9), and (10). The pressure dependence shown in Fig. 18 thus points to an important contribution of surface flow to the overall permeability. For less adsorbing gases (He, H_2), a permeability independent of partial pressure was observed, indicating that for these gases the contribution of surface flow was negligible.

Evaluating the not-abundant results on zeolitic membrane modeling, it becomes apparent that the most successful direction is the approach based on two contributions: surface diffusion and gas translational diffusion, with the former being dominant at low, the latter at high temperatures. The description of the surface diffusion based on the

Table 2 Single- and Multicomponent Diffusivities of Ethane and Ethene, Calculated Using the Maxwell–Stefan Model for Zeolitic Diffusion

Single component						Multicomponent				
Ethane			Ethene			Ethane/Ethene				
p_0 (kPa)	θ^*_{av}	D^{MS} (10^{-10} $m^2 \cdot sec^{-1}$)	p_0 (kPa)	θ^*_{av}	D^{MS} (10^{-10} $m^2 \cdot sec^{-1}$)	$p_{0,c2}$ (kPa)	$p_{0,c_2^=}$ (kPa)	θ^*_{av}	D^{MS}_{c2} ($10^{-10} m^2 \cdot sec^{-1}$)	$D^{MS}_{c_2^=}$ ($10^{-10} m^2 \cdot sec^{-1}$)
101.3	0.86	8.4	101.3	0.75	14.7					
76.0	0.81	7.9				76.0	25.3	0.83	9.7	8.5
50.7	0.75	7.1	50.7	0.61	11.2	50.6	50.7	0.81	9.8	10.2
25.3	0.61	6.0	25.3	0.45	9.3	25.3	76.0	0.78	10.3	11.9
10.1	0.25	4.4	10.1	0.25	6.2					

*Average occupancy in the membrane.
From Ref. 32.

thermodynamic potential gradient yields the best results, although the permeation results are not yet completely predictable. The generalized Maxwell–Stefan equations offer good prospects to extend this to multicomponent systems. In all models the knowledge about the concentrations of the components in the zeolite is indispensable, so accurate single- and multicomponent adsorption data are essential to the success of modeling permeation through zeolitic membranes.

IV. CHALLENGES FOR FURTHER DEVELOPMENT

The potentially most attractive applications of zeolitic membranes are in the high-temperature region. At high temperature, only inorganic materials, metals, and solid mixed oxides are stable, whereas polymeric membranes are thermally highly unstable. This does not rule out application at lower temperatures. But in the lower-temperature range, the competition with polymeric membranes turns up. Polymeric membranes can be produced with higher surface-area-to-volume ratios, in bundles of small tubes, a mode not easily envisaged for zeolitic systems. For the latter, high separation factors and permeation fluxes must be determining factors.

Application at high temperature requires robust and thermostable systems. Both for ceramically and stainless-steel-supported systems the thermostability has been demonstrated. So, in spite of the different thermal expansion coefficients, the asymmetric membrane remains intact. However, there are no data available on the resistance of zeolitic membranes to thermal stresses, as a result of, for example, large sudden changes in temperature. The stainless-steel-supported system seems the most promising configuration

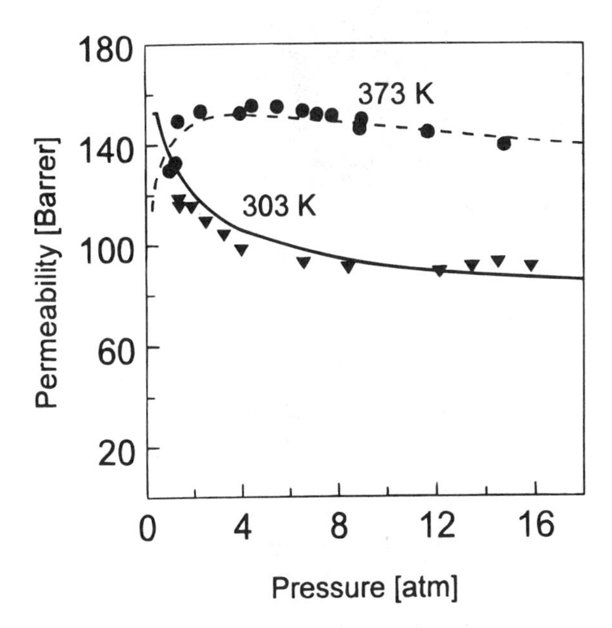

Figure 18 Pressure dependence of the CO_2 permeability through a microporous glass membrane. Experimental results are compared to theory of combined gaseous and surface flow. (1 Barrer = 3.35×10^{-14} mol·m·m^{-2}·Pa^{-1}·sec^{-1}.) (Adapted from Ref. 38.)

due to the easier mounting opportunities compared to ceramic materials, which are brittle too, with a danger of crack formation and sealing failures. The experimental results apply mainly to silicalite-1 systems, and further experience with other zeolites has to be acquired in order to verify if these systems also exhibit this thermal stability.

There are no experimental data regarding the resistance of zeolitic membranes to fouling. Regeneration of zeolites after, for example, coke deposition is known to be feasible from industrial applications of zeolites. However, the influence of regeneration conditions on zeolitic, composite membranes is not known.

The next step for application is the development of handsome modules that can be combined in larger units for separation or catalytic membrane reactors. In separation one strives to high surface-area-to-volume ratios; in catalytic membrane reactors this ratio will depend on the volumetric catalytic rates compared to the required permeation fluxes.

To open up more applications, the preparation of membranes from other zeolites is required. Silicalite-1 has pores consisting of 10 oxygen ions, but 12-, 8-, and 6-ring systems also exist [1,2]. Figure 19 shows the pore apertures of different zeolites, together with a methane molecule, in order to give an idea of the size of these different pores. The 6- or 8-ring pores seem especially suited for separation based on size exclusion, for example, hydrogen recovery. An alternative approach is the modification of the pore entrance or of the adsorption characteristics of the external and/or internal surface, thus affecting the selectivity and permeation properties. Some interesting results have been obtained in this way with zeolites for application in pressure swing adsorption [59,60].

Zeolitic membranes is still a young research area with a lot of promise for future applications. A long road, however, has to be traveled before they will become a commod-

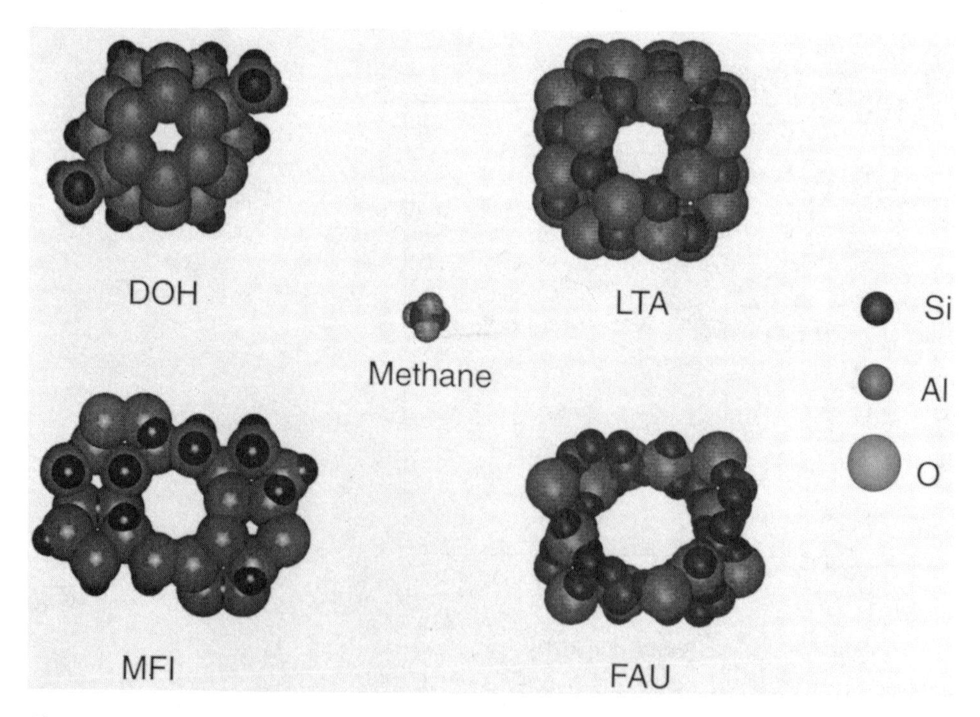

Figure 19 Pore apertures of DOH (6-membered ring, 0.28 nm), zeolite-A (LTA, 8-membered ring, 0.41 nm), silicalite-1 (MFI, 10-membered ring, 0.52×0.55 nm), and a zeolite-X or -Y (FAU, 12-membered ring, 0.74 nm) and a methane molecule (kinetic diameter = 0.38 nm).

ity, a road with a lot of opportunities for engineers, theoreticians, chemists, and materials scientists for further developments and breakthroughs.

LIST OF SYMBOLS

B	mobility of a component	$m^2 \cdot mol \cdot J^{-1} \cdot sec^{-1}$
c	concentration	$mol \cdot m^{-3}$ or $mol \cdot g^{-1}$
D	diffusivity	$m^2 \cdot sec^{-1}$
d	diffusional length	m
E_a	activation energy	$J \cdot mol^{-1}$
E_d	activation energy for diffusion	$J \cdot mol^{-1}$
g	geometrical factor	—
K	adsorption constant	kPa^{-1}
l	membrane thickness	m
M	molecular mass	$kg \cdot mol^{-1}$
N	flux	$mol \cdot m^{-2} \cdot sec^{-1}$
P	permeability	$mol \cdot m \cdot m^{-2} \cdot sec^{-1} \cdot kPa^{-1}$
p	(partial) pressure	kPa
Q	heat of adsorption	$J \cdot mol^{-1}$
R	gas constant	$J \cdot mol^{-1} \cdot K^{-1}$
ΔS	adsorption entropy	$J \cdot mol^{-1} \cdot K^{-1}$
T	temperature	K
v	velocity	$m \cdot sec^{-1}$
w	dimensionless energy change	—
x	mol fraction	—
z	coordination number	—

Greek

α	jump distance or selectivity	m or —
Γ	thermodynamic correction factor for binary mixture	—
λ	sticking probability	—
μ	chemical potential	$J \cdot mol^{-1}$
ν	jump frequency	s^{-1}
θ	occupancy	—
ρ	density	$g \cdot m^{-3}$

Subscripts

ads	adsorbed
app	apparent
av	average
id	ideal
i,j	component i, j
l	permeate side
$n + 1$	pseudospecies
0	feed side

Superscripts

Fick	Fick
g	gas phase
MS	Maxwell–Stefan
s	surface
sat	saturation
tot	total
0	standard or corrected

REFERENCES

1. R. Szostak, *Handbook of Molecular Sieves*, Van Nostrand-Reinold, New York, 1992.
2. D.W. Breck, *Zeolite Molecular Sieves, Structure, Chemistry, and Use*, Wiley, New York, 1974.
3. F. Wolf, W. Hentschel, and E. Krell, Zur Trennung von Methan und niederen Olefinen durch Molekularsiebmembranen, *Z. Chem. 16*:107 (1976).
4. H.J.C. te Hennepe, D. Bargeman, M.H.V. Mulder, and C.A. Smolders, Zeolite-filled silicone rubber membranes. Part 1. Membrane preparation and pervaporation results, *J. Membrane Sci. 35*:39 (1987).
5. H.J.C. te Hennepe, C.A. Smolders, D. Bargeman, and M.H.V. Mulder, Exclusion and tortuosity effects for alcohol/water separation by zeolite filled PDMS membranes, *Sep. Sci. Technol. 26(4)*:585 (1991).
6. J.M. Duval, B. Folkers, M.H.V. Mulder, G. Desgrandchamps, and C.A. Smolders, Adsorbent filled membranes for gas separation. Part 1. Improvement of the gas separation properties of polymeric membranes by incorporation of microporous adsorbents, *J. Membrane Sci. 80*:189 (1983).
7. J.P. Boom, D. Bargeman, and H. Strathmann, Zeolite filled membranes for gas separation and pervaporation, *Zeolites and related microporous materials: State of the art 1994 Part B*. Proc. 10th Int. Zeol. Conf., Garmisch-Partenkirchen (J. Weitkamp, H.G. Karge, H. Pfeifer, and W. Holderich, eds.), Elsevier, Amsterdam, 1994, p. 1167.
8. M.-D. Jia, K.-V. Peinemann, and R.-D. Behling, Molecular sieving effect of the zeolite-filled silicone rubber membranes in gas permeation, *J. Membrane Sci. 57*:289 (1991).
9. Y.-C. Long, X. Chen, Z.-H. Ping, S.-K. Fu, and Y.-J. Sun, MFI-type zeolite filled silicone rubber membranes: Preparation, composition, and performance, *Zeolites and related microporous materials: State of the art 1994 Part B*. Proc. 10th Int. Zeol. Conf., Garmisch-Partenkirchen (J. Weitkamp, H.G. Karge, H. Pfeifer, and W. Holderich, eds.), Elsevier, Amsterdam, 1994, p. 1083.
10. E.R. Geus, W.J.W. Bakker, J.A. Moulijn, and H. van Bekkum, High-temperature stainless steel supported zeolite (MFI) membranes: Preparation, module construction and permeation experiments, *Microporous Mater. 1*:131 (1993).
11. E.R. Geus, W.J.W. Bakker, P.J.T. Verheijen, M.J. den Exter, J.A. Moulijn, and H. van Bekkum, Permeation experiments on in situ grown ceramic MFI type films, Proc. 9th Int. Zeol. Conf., Stoneham, Masachusetts (J.B. Higgins, R. von Ballmoos, and M.M.J. Treacy eds.), Butterworth-Heinemann, 1993, pp. 371–378.
12. W.J.W. Bakker, G. Zheng, M. Makkee, F. Kapteijn, J.A. Moulijn, E.R. Geus, and H. van Bekkum, Single- and Multi-component transport through metal-supported MFI zeolite membranes, in *Precision Process Technology* (M.P.C. Weijnen and A.A.H. Drinkenburg, eds.), Kluwer Academic Publishers, Amsterdam, 1993, p. 425.
13. W.J.W. Bakker, F. Kapteijn, J.C. Janssen, H. van Bekkum, and J.A. Moulijn, Doorbraak in ontwikkeling zeoliet membranen (in Dutch), *Procestechnologie 3(12)*:7 (1993).

14. W.J.W. Bakker, F. Kapteijn, J. Poppe, and J.A. Moulijn, Permeation of a metal-supported Silicalite-1 zeolite membrane, *J. Membrane Sci. 177*:57 (1996).
15. V. Valtchev and S. Mintova, The effect of the metal substrate composition on the crystallization of zeolite coatings, *Zeolites 15*:171 (1995).
16. T. Sano, H. Yangishita, Y. Kiyozumi, F. Mizukami, and K. Haraya, Separation of ethanol/water mixture by silicalite membrane on pervaporation, *J. Membrane Sci. 95*:221 (1994).
17. T. Sano, S. Ejiri. M. Hasegawa, Y. Kawakami, N. Enomoto, Y. Tamai, and H. Yangishita, Silicalite membrane for separation of acetic acid/water mixture, *Chem. Lett*:153 (1995).
18. M.-D. Jia, K.V. Peinemann, and R.D. Behling, Ceramic zeolite composite membranes. Preparation, characterization and gas permeation, *J. Membrane Sci. 82*:15 (1993).
19. M.-D. Jia, B. Chen, R.D. Noble, and J.L. Falconer, Ceramic–zeolite composite membranes and their application for separation of vapor/gas mixtures, *J. Membrane Sci. 90*:1 (1994).
20. C. Bai, M.-D. Jia, J.L. Falconer, and R.D. Noble, Preparation and separation properties of silcalite composite membranes, *J. Membrane Sci. (105)*:79 (1995).
21. H. Suzuki, Composite having a zeolite, a layered compound or a crystalline-lattice material in the pores of a porous support and processes for production thereof, European Pat. 0,180,200 (1986).
22. W.O. Haag and J.G. Tsikoyiannis, Membrane composed of a pure molecular sieve, U.S. Pat. 5,019,263 (1991).
23. A. Ishikawa, T.H. Chiang, and F. Toda, Separation of water-alcohol mixtures by permeation through a zeolite membrane on porous glass, *J. Chem. Soc. Chem. Commun.*:764 (1989).
24. Z.A.E.P. Vroon, K. Keizer, M.J. Gilde, H. Verweij, and A.J. Burggraaf, Transport properties of alkanes through ceramic thin zeolite MFI membranes, *J. Membrane Sci.*: *113(2)*:296 (1996).
25. Y. Yan, M. Tsapatsis, G.R. Gavalas, and M.E. Davis, Zeolite ZSM-5 membranes grown on porous α-Al_2O_3, *J. Chem. Soc., Chem. Commun.*:227 (1995).
26. P. Kölsch, D. Venzke, M. Noack, P. Toussaint, and J. Caro, Zeolite-in-metal-membranes: Preparation and testing, *J. Chem. Soc., Chem. Commun. (21)*:2491 (1994).
27. M. Matsukata, N. Nishiyama, and K. Ueyama, Preparation of a thin zeolitic membrane, *Zeolites and related microporous materials: State of the art 1994 Part B.* Proc. 10th Int. Zeol. Conf., Garmisch-Partenkirchen (J. Weitkamp, H.G. Karge, H. Pfeifer, and W. Holderich, eds.), Elsevier, Amsterdam, 1994, p. 1183.
28. S. Yamazaki and K. Tsutsumi, Synthesis of an A-type zeolite membrane on silicon oxide film–silicon, quartz plate and quartz fiber filter, *Microporous Materials 4*:205 (1995).
29. M. Noack, P. Kölsch, D. Venzke, P. Toussaint, and J. Caro, New one-dimensional membrane: Aligned $AlPO_4$-5 molecular sieve crystals in a nickel foil, *Microporous Mater. 3*:201 (1994).
30. J. Karger and D.M. Ruthven, *Diffusion in zeolites and other microporous solids*, Wiley, New York, 1992.
31. M.J. Den Exter, J.C. Jansen, and H. van Bekkum, Separation of permanent gases on the all-silica 8-ring clathrasil DD3R, *Zeolites and related microporous materials: State of the art 1994 Part B.* Proc. 10th Int. Zeol. Conf., Garmisch-Partenkirchen (J. Weitkamp, H.G. Karge, H. Pfeifer, and W. Holderich, eds.), Elsevier, Amsterdam, 1994, p. 1159.
32. J.M. van de Graaf, to be published, presented at the 7th annual meeting of the North American Membrane Society, Portland (Ore.), May 20–24 (1995).
33. F. Kapteijn, W.J.W. Bakker, J.M. van de Graaf, G. Zheng, J. Poppe, and J.A. Moulijn, Permeation and separation behavior of a silicalite-1 membrane, *Cat. Today 25*:213 (1995).
34. D. Casanave, A. Giroir-Fendler, J. Sanchez, R. Loutaty, and J.A. Dalmon, Control of transport properties with a microporous membrane reactor to enhance yields in dehydrogenation reactions, *Cat. Today 25*:309 (1995).
35. R.M. Barrer, Porous crystal membranes, *J. Chem. Soc. Faraday Trans. 86(7)*:1123 (1990).

36. M.F.M. Post, Diffusion in zeolite molecular sieves, in *Introduction to Zeolite Science and Practice* (H. van Bekkum, E.M. Flanigen, and J.C. Jansen, eds.), Elsevier, Amsterdam, 1991, p. 391.
37. J. Xiao and J. Wei, Diffusion mechanism of hydrocarbons in zeolites—I. Theory, *Chem. Eng. Sci. 47(5)*:1123 (1992).
38. A.B. Shelekhin, A.G. Dixon, and Y.H. Ma, Theory of gas diffusion and permeation in inorganic molecular-sieve membranes, *AIChE-J. 41(1)*:58 (1995).
39. Y. Shindo, T. Hakuta, H. Yoshitome, and H. Inoue, Gas diffusion in microporous media in Knudsen's regime, *J. Chem. Eng. Japan 16(2)*:120 (1983).
40. H. Yoshida, M. Maekawa, and M. Nango, Parallel transport by surface and pore diffusion in a porous membrane, *Chem. Eng. Sci. 46(2)*:429 (1991).
41. R. Gutsche and H. Yoshida, Solid Diffusion in the pores of cellulose membrane, *Chem. Eng. Sci. 49(2)*:179 (1994).
42. L. Riekert, Rates of sorption and diffusion of hydrocarbons in zeolites, *AIChE-J. 17(2)*:446 (1971).
43. D.A. Reed and G. Ehrlich, Surface diffusion, atomic jump rates and thermodynamics, *Surface Sci. 102*:588 (1981).
44. V.P. Zhdanov, General equations for description of surface diffusion in the framework of the lattice-gas model, *Surface Sci 149*:L13 (1985).
45. J. Xiao and J. Wei, Diffusion mechanism of hydrocarbons in zeolites—II. Analysis of experimental observations, *Chem. Eng. Sci. 47(5)*:1143 (1992).
46. R. Krishna, A unified approach to the modeling of intraparticle diffusion in adsorption processes, *Gas Sep. Purif. 7(2)*:91 (1993).
47. Y.D. Chen, R.T. Yang, and P. Uawithya, Diffusion of oxygen, nitrogen and their mixtures in carbon molecular sieve, *AIChE Journal. 40(4)*:577 (1994).
48. R.M. Barrer, *Zeolites and Clay Minerals as Sorbents and Molecular Sieves*, Academic Press, London, 1978.
49. D.M. Ruthven, *Principles of adsorption and adsorption processes*, Wiley, New York, 1984.
50. F. Kapteijn, W.J.W. Bakker, G. Zheng, J. Poppe, and J.A. Moulijn, Permeation and separation of light hydrocarbons through a silicalite-1 membrane; application of the generalized Maxwell–Stefan equations, *Chem. Eng. J. 57*:145 (1995).
51. D.M. Ruthven and K.F. Loughlin, Sorption of light paraffins in type-A zeolites. Analysis and interpretation of equilibrium isotherms, *J. Chem., Soc. Faraday Trans. I 68*:696 (1972).
52. L.V.C. Rees, J. Hampson, and P. Bruckner, Sorption of single gases and their binary mixtures in zeolites, in *Zeolites Mocroporous Solids: Synthesis, Structure and Reactivity* (E.G. Derouane and F. Lemos, eds.), Kluwer Academic Publishers, Dordrecht, The Netherlands, 1992, p. 133.
53. A.L. Meyers and J.M. Prausznitz, Thermodynamics of mixed gas adsorption, *AIChE-J. 11*:121 (1965).
54. B.K. Kaul, Correlation and prediction of adsorption isotherm data for pure and mixed gases, *Ind. Eng. Chem. Process. Des. Dev. 23*:711 (1984).
55. P. Graham, A.D. Hughes, and L.V.C. Rees, Sorption of binary gas mixtures in zeolites I. Sorption of nitrogen and carbon dioxide mixtures in silicalite, *Gas Sep. & Purif. 3*:56 (1989).
56. F. Kapteijn, W.J.W. Bakker, G. Zheng, and J.A. Moulijn, The temperature and occupancy dependent diffusion of *n*-butane through a silicalite membrane, *Microporous Mater. 3(3)*:227 (1994).
57. R. Krishna and L.P.J. van den Broeke, The Maxwell–Stefan description of mass transport across zeolite membranes, *Chem. Eng. J. 57*:155 (1995).
58. Y.D. Chen and R.T. Yang, Preparation of carbon molecular sieve membrane and diffusion of binary mixtures in the membrane, *Ind. Eng. Chem. Res. 33*:3146 (1994).

59. R.T. Yang and E.S. Kikkenides, New sorbents for olefin/paraffin separations by adsorption via π-complexation, *AlChE-J. 41(3)*:509 (1995).
60. E.F. Vansant, *Pore size engineering in zeolites*, Wiley, Chichester, 1990.
61. C. Goldon and S. Sircar, Gas adsorption on silicalite, *J. Colloid and Interface Sci. 162*:182 (1994).

20

Cross-Flow Reactors with Permeable Walls

Nils-Herman Schöön
Chalmers University of Technology, Göteborg, Sweden

I. SOME INTRODUCTORY NOTES ABOUT THE CROSS-FLOW TERM AND ITS DEFINITION

The cross-flow principle is defined and applied in different ways in chemical engineering theory and practice. Before describing the more narrow application of the cross-flow principle in chemical reaction engineering, it may be informative to illustrate an extended interpretation and application of this concept. Most often the cross-flow principle implies that two different fluids are flowing perpendicularly to or from each other in a process apparatus. These fluids can be combined into a common flow while separated from each other by a dense or a porous wall. The communication and exchange between these separated flows can proceed via different mechanisms, depending on the type of separating wall and the type of exchange process.

A. Cross Flow Without a Separating Wall Between the Fluids

Cross flow without a separating wall between two streams of liquids, gases, solid particles, or combinations of these can be illustrated with the following four examples.

- Distributed inflow or outflow to favor the yield of one product in competing reactions
- Distributed inflow to an annular fixed-catalyst bed to decrease the temperature rise and the pressure drop
- Distributed inflow to a moving catalyst bed of continuously deactivated catalyst accompanied by continuous regeneration
- Distributed inflow to a cascade of fluidized-bed reactors to minimize coke formation

1. Distributed Inflow or Outflow to Favor the Yield

The cross-flow principle, where the two fluids are successively brought together in one and the same flow, is applied if one wants to maximize one of two parallel or consecutive reactions [1]. Figure 1 illustrates this type of cross flow for the reaction system

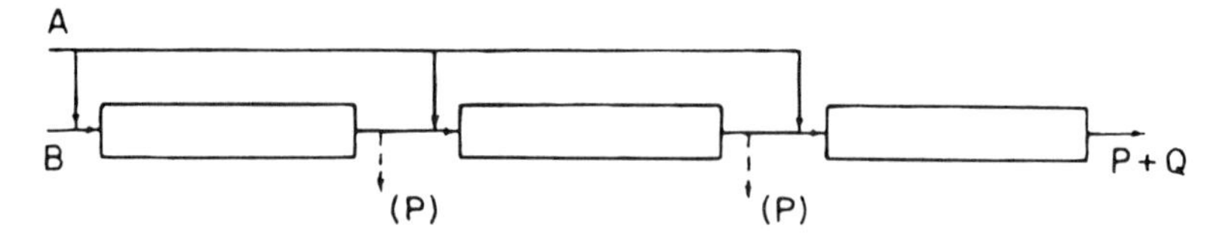

Figure 1 Possible cross-flow reactor systems consisting of tubular reactors. (From Ref. 1.)

$$A + B \begin{matrix} \nearrow P \\ \searrow Q \end{matrix}$$

when the tubular reactor alternative is the most favorable choice.

This principle may also be illustrated by some real cases. In the codimerization of propene and hexene it is important primarily to minimize the dimerization of the reactive propene. In order to favor the codimerization, a stage injection of propene according to the principle in Fig. 1 was therefore performed [2]. A similar process design with distributed additions of chlorine was applied in the chlorination of propene to allyl chloride in order to suppress different side reactions [3]. For liquid-phase processes, a distributed feed to the cascade of stirred reactors was a more natural variant. This was applied in the sulfuric acid alkylation of *iso*butane, where the olefin feed has to be subdivided due to selectivity reasons and the goal was to reach a desired octane number of the product [4].

2. *Distributed Inflow to an Annular Fixed Catalyst Bed*

Annular fixed beds with radial cross flow have been described in a number of papers [5–10], where, among others, a design criterion [9] and the effect of flow maldistribution on conversion and selectivity [10] have been dealt with.

3. *Distributed Inflow to a Moving Catalyst Bed*

Until recently only a few papers were available on moving beds in cross flow [11–18]. This type of reactor is sometimes a favorable process solution for a selective catalytic process with a moderate catalyst residence time and with a short gas residence time, especially when the process is accompanied by a continuous catalyst regeneration. The use of conventional short-contact-time reactors like fluidized-bed reactors, risers, and fixed-bed reactors does not always yield satisfactory results. This may be explained by problems connected with gas back-mixing, channeling of gas, low catalyst holdup, attrition of the solid catalyst, or difficulties in temperature control.

Cross-flow moving beds are found in the petroleum industry [19,20] (catalytic cracking, reforming) and in flue gas desulfurization plants [21].

The reaction or regeneration cross-flow gas in the moving bed is imposed by the phenomenon of "pinning" [14,16,17], which means that the drag exerted on the catalyst bed by the gas flowing through this bed forces the bed against the downstream wall. The consequence will be that the downward motion ceases and the bed is said to be pinned.

4. *Distributed Inflow to Fluidized Beds*

The cross-flow principle has also been applied in fluidized-bed reactor technology. In order to minimize the coke formation in continuous decomposition of kerogen to oil and gas, a favorable reactor solution may consist of a cross-current multistage fluidized-bed reactor [3]. It should also be noted here that in mathematical modeling of the flow and reaction in a fluidized bed, an important cross flow between the bubble phase and the emulsion phase is often considered [22].

B. Cross Flow with a Separating Wall Between the Fluids

The cross-flow principle was probably first applied to the design of compact heat exchangers. This may explain why the nomenclature in cross-flow technology derives from the heat exchange field. Cross-flow heat exchangers are commonly used in air or gas heating and cooling. In cross-flow heat exchangers, the gas flowing across the tubes is said to be a mixed stream, while the fluid inside the tubes is said to be unmixed [23] (see Fig. 2). The gas is mixed because it can move almost freely in the exchanger as it exchanges heat. The other fluid is confined in separate tubular channels and cannot mix with itself during the heat transfer process. Two types of cross-flow heat exchangers, with two unmixed fluids, are shown in Fig. 3. In the chemical reactor application of the cross-flow principle, the alternative with two unmixed fluids is used most often. In these reactors, sinusoidal-shaped ducts in the two directions of flow are also frequent. The application of the principle of one unmixed flow and one mixed flow in cross-flow reactors corresponds to the annulus reactor with a separating cylinder wall between the two fluids. One example of this type of reactor is the hollow-fiber reactor, which has found its major application in biotechnology. The parallel-channel cross-flow structure and the annulus unit may be combined into a so-called rotary cross-flow apparatus (Fig. 4) with a multiple application of the cross-flow principle.

In the multitubular cross-flow reactor, one or more baffles are used to force the coolant to flow across the tubes, and some parallel flow arises where the coolant flow direction reverses.

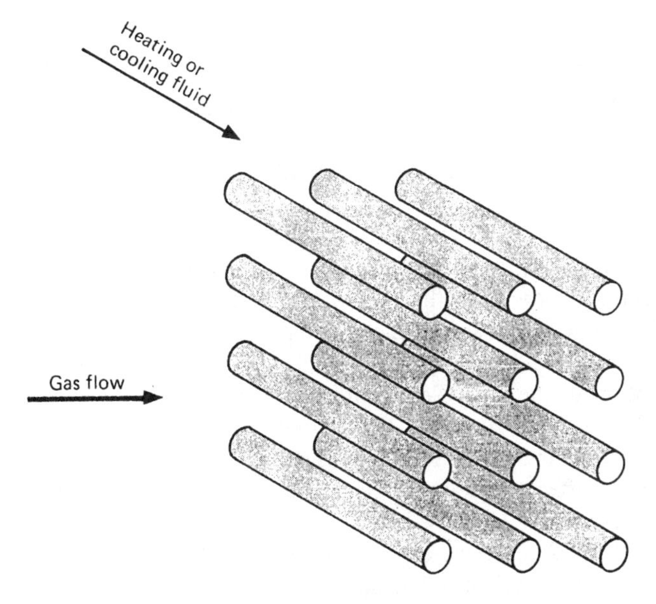

Figure 2 Cross-flow heat exchanger, with one fluid mixed and one unmixed. (From Ref. 23.)

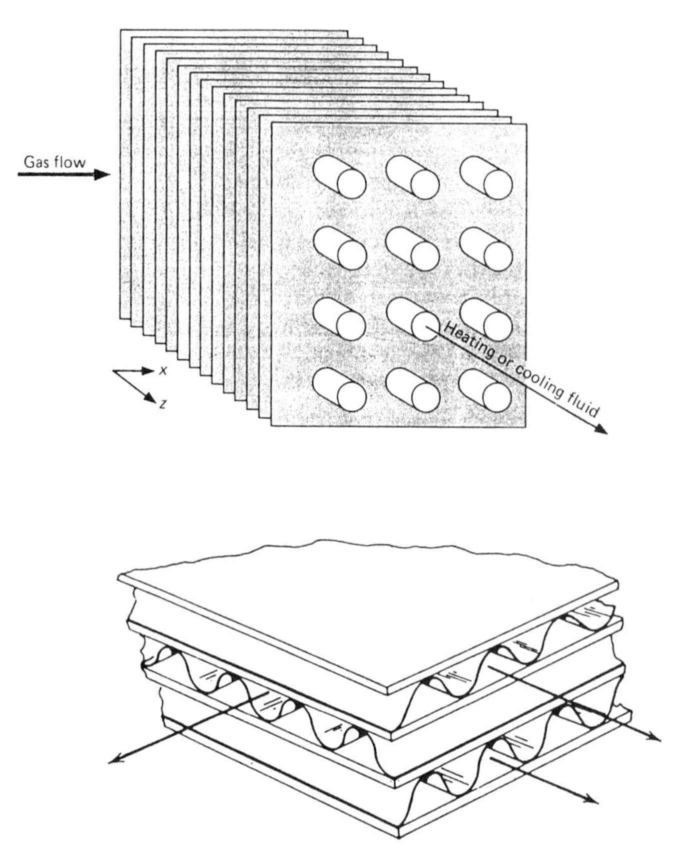

Figure 3 Cross-flow heat exchangers, with both fluids unmixed. (From Refs. 23 and 24.)

C. An Extended Cross-Flow Definition

In filtration unit operation, especially in microfiltration, one usually differentiates between dead-end filtration (with cake formation) and cross-flow filtration [25] (Fig. 5). The cross-flow filter can have different geometries (Fig. 6): phase membranes, tubular membranes, or pleated membranes, of which the tubular and pleated ones are already accepted as cross-flow geometries in reactor technology, as mentioned above. In filtration engineering the cross-flow term means that the filtrate flows perpendicularly to the suspension stream. Cross-flow may not be considered a sufficiently illustrative term here [25]. A better term would be *parallel filtration*, but the term *cross-flow filtration* has been accepted generally and may be difficult to change at present.

A transfer of this definition of the cross-flow term to the chemical reactor field implies that the so-called cell reactor [26], consisting of thin, porous catalyst plates mounted in a rack like a filter press, should also be of interest to describe here. The same may apply to the great number of different electrochemical filter-press cell reactors [27]. It may be noted that the cell reactor principle is, however, not valid for the so-called parallel-passage reactor [28,29]. In this case the same fluid flows on both sides of the catalyst plates without any need for communication and exchange between the fluids through the plates. The advantage of this reactor is its being dust-proof, since dust present

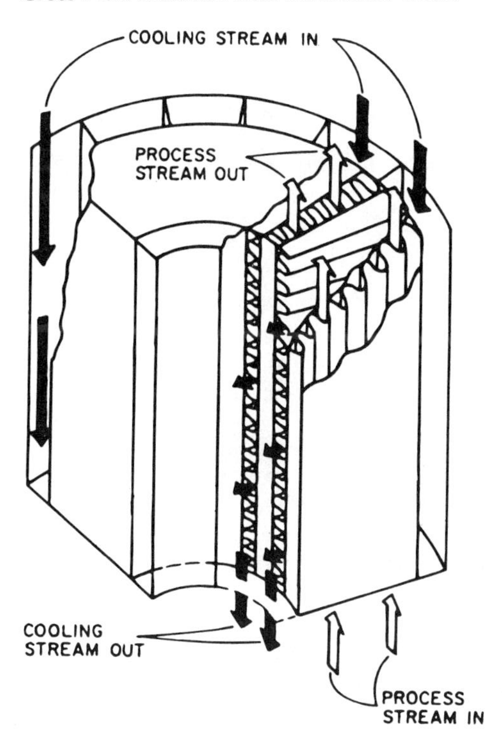

Figure 4 Schematic view of rotary cross-flow equipment. (From Ref. 37.)

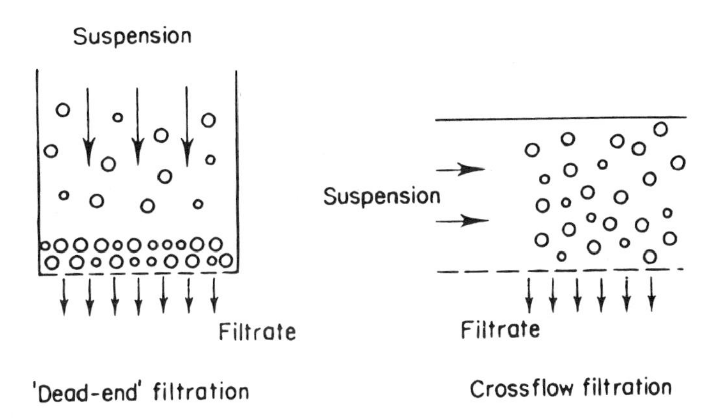

Figure 5 Dead-end filtration with cake formation and cross-flow cake-free filtration. (From Ref. 25.)

in the gas can flow straight through the reactor, without impinging upon obstacles on its way. The parallel-passage reactor should, rather, be compared to the honeycomb reactor with catalytic channel walls.

In addition to the above cross-flow reactors of extended definition, the block of thin-walled spiral-tubular-membrane catalyst reactor and the double-spiral coiled-plate-membrane reactor may be included [30–33].

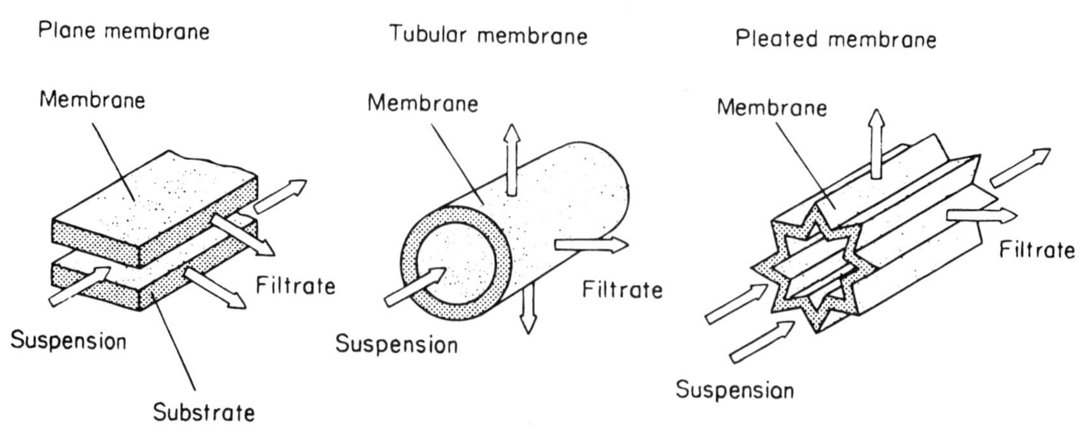

Figure 6 Different geometries of cross-flow filter membranes. (From Ref. 25.)

II. SURVEY OF CROSS-FLOW REACTORS WITH SEPARATING WALLS IN REGARD TO THEIR FUNCTION AND APPLICABILITY

Examples of different cross-flow reactors are shown in Table 1, where different characteristics such as separating wall material, type of transport through the wall, and model reactions studied are given. As seen from Table 1 the reactors may be divided into five groups with respect to their fields of application:

Heat exchange coupled with chemical reactions
Air-drying and regeneration of the hygroscopic salt in the dryer
Power cells and electrochemical-synthesis reactors
Trickle-bed alternatives
Bioreactors

In all these fields the traditional cross-flow reactor is represented. This reactor consists of a great number of parallel porous plates separated from each other by corrugated planes or by a similar regular structure, giving rise to a compact system of parallel closed channels, with only entrance and exit openings perpendicular to the main flow direction (cf. Fig. 3). The parallel-channel system of this traditional cross-flow reactor is arranged so that the inflows of the two fluids are separated 90°. This means that the two fluids will not be mixed in the same channels and the fluids will penetrate the plane plates from different sides. The fluids can meet only inside the porous catalytic plates where the reactions proceed.

A. Heat Exchange Coupled with Chemical Reactions

In an early study of cross-flow heat exchangers with an intended use as automobile exhaust converters and noncatalytic afterburners, Johnson et al. [34] made a comprehensive determination of the pressure drop of the flow at different wavelengths of the corrugated planes of the cross-flow structure, in order to find a reasonable choice between a tolerable pressure drop (long wavelength) and sufficient heat transfer area per unit volume (short wavelength). The optimum height of the heat exchanger and the duct wavelength was given in a diagram. It was established that the cross-flow heat exchanger had a low

Table 1 Examples of Cross-Flow Reactors with Separating Walls Between the Fluids

Reactor	Separating wall material	Type of transport through the wall	Model reaction	Comments	Refs.
Noncatalytic afterburner	Ceramics	Heat conduction	Oxidation of hydrocarbons		34
Cocurrent reactor–heat exchanger	Cordierite	Heat conduction	Oxidation of CO	Pellet-filled or coated monolith wall	35, 36
Cross-flow dryer and regenerator	Calcium aluminum silicate	Heat conduction and water vapor diffusion	Drying of air	Hygroscopic salt-coated monolith wall	37
Solid-state electrocatalytic reactor	Yttria-doped zirconia	Heat, electronic, and O^{2-} conduction	Oxidation of CO	Pt-coated monolith wall	48
Fuel cell	Graphite and phenolic resin	Ionic conduction	Oxidation of CH_3OH	Pt electrodes	39
Salinity power cell	Ion-permeable membranes	Ionic conduction		Ag electrodes	50
Electrochemical filter-press cell reactor	Cation-selective membrane	Diffusion of cations	Synthesis of inorganic and organic chemicals		27
Double-spiral coiled-plate membrane	Pd alloy	Diffusion of H	Hydrogenation of unsaturated alcohols and aromatic hydrocarbons		30–32
Spiral tubular membrane	Pd alloy	Diffusion of H			30–33
Tubular multiphase catalyst reactor	α-Al_2O_3	H_2 diffusion in liquid-filled pores	Hydrogenation of α-methylstyrene	Pd/γ-Al_2O_3 coated tube	56
Hollow-fiber trickle-bed reactor	Polypropylene	O_2 and SO_2 diffusion in liquid-filled pores	Oxidation of SO_2	Pellet-filled outer shell side	39

Table 1 *Continued*

Reactor	Separating wall material	Type of transport through the wall	Model reaction	Comments	Refs.
Catalytic cross-flow reactor	Calcium aluminum silicate	H_2 diffusion in liquid-filled pores	Hydrogenation of *p*-nitrobenzoic acid	Pd-coated monolith wall	60
Catalytic cross-flow reactor	Calcium aluminum silicate	H_2 diffusion in liquid-filled pores	Hydrogenation of alkylbenzenes	Rh-coated monolith wall	61
Cell reactor	PTFE and Pd on an inorganic carrier	H_2 diffusion in liquid-filled pores	Hydrogenation of *p*-nitrobenzoic acid	The PTFE layer on the gas side	26
Membrane bioreactor	Asymmetric polysulfone membrane	Diffusion of carbohydrates in liquid-filled pores	Enzymatic hydrolysis of lactose	Whole cells of *Sulfolobus solfataricus* entrapped in the tube wall	65

pressure drop, short heat-up time, and no attrition losses. A series of tests was run on a prototype noncatalytic afterburner with a 1957 Chevrolet six-cylinder stationary test engine. Unlike the usual way to utilize the cross-flow reactor, one and the same gas was here flowing through all the channels. Figure 7 illustrates an efficient way to control the gas flow.

Degnan and Wei [35,36] used a block of four cross-flow monoliths (Fig. 8) as a combined catalytic reactor and heat exchanger because of its large heat transfer area per volume. The reaction channels were filled with solid Cu–Cr catalyst pellets, or their walls were coated with this catalytic material. The transverse channels functioned as cooling channels. Five different flow combinations were studied from a heat-exchange point of view, when using the strongly exothermic oxidation of CO as a model reaction. It was found from these runs that, by feeding the coolant and reactant streams parallel into the same cross-flow block, a good approximation to a cocurrent reactor–heat exchanger was obtained, and nearly isothermal temperature profiles were attained. It was also demonstrated that the coolant of the wall-coated monolith was more effective on the reaction side than in the pellet-filled monolith. This is easy to understand since the deposited metal reduces the thermal mass of the monolith and increases the heat transfer because the thermal gas film resistance close to the channel wall no longer exists.

It should be noted in these two reactor examples that where the cross-flow structure functioned as both a reactor and a heat exchanger, the channel walls separating the flows were not permeable to mass transfer through the walls.

B. Air-Drying and Regeneration of the Hygroscopic Salt in the Dryer

In an air-dryer the channel walls are partially made of a sorbent such as silica-gel, lithium chloride, or alumina. These walls exchange moisture and heat with a stream of air flowing

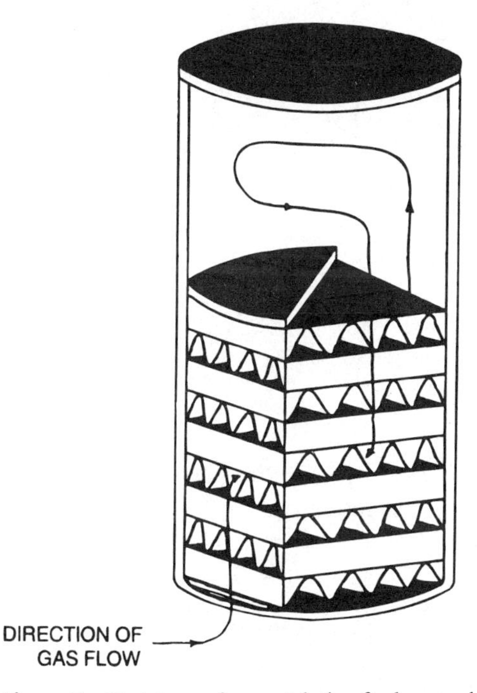

Figure 7 Prototype of noncatalytic afterburner design. (From Ref. 34.)

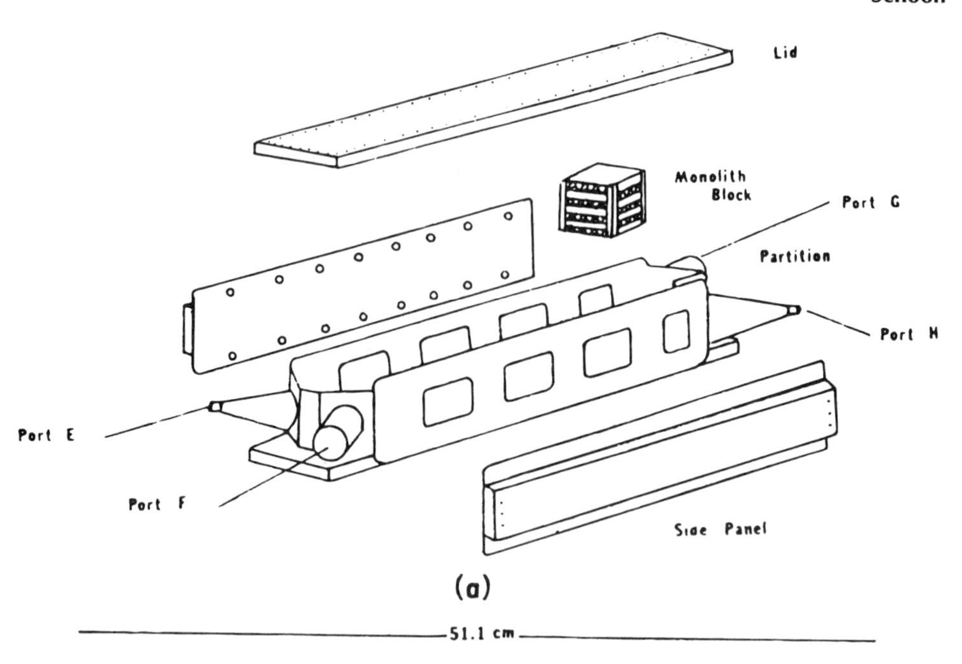

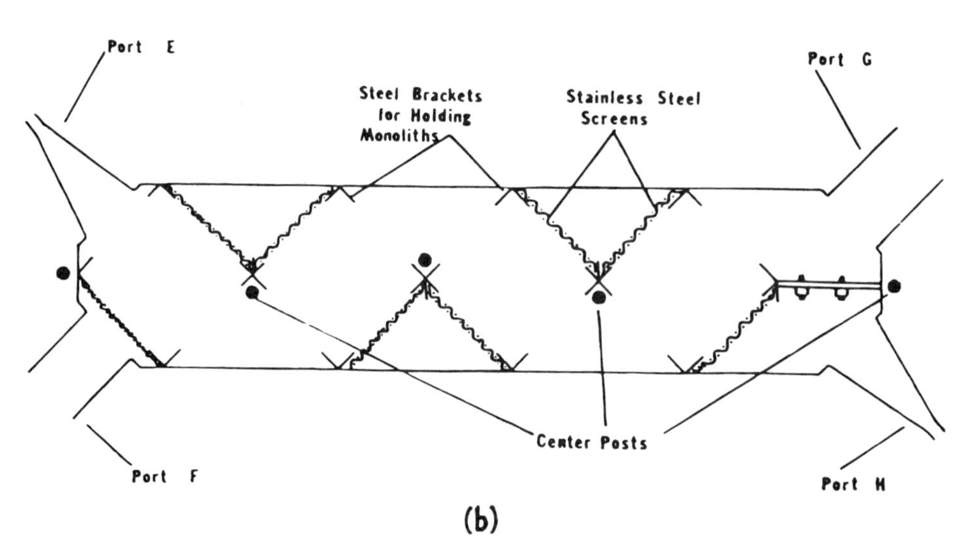

Figure 8 A reactor consisting of four cross-flow monoliths in a steel manifold. (a) Reactor before assembly; (b) top view of reactor. (From Ref. 36.)

through the dryer. The sorption capacity of this dryer is greatly reduced due to large heat effects associated with the sorption of water and due to the high temperature left in the matrix after the regeneration. These two disadvantages were easily eliminated by using a dryer of the conventional cross-flow type [37]. An extended variant of the cross-flow dryer is the rotary type of dryer [37]. One stream is arranged in this dryer to flow in the radial direction and the other stream in the axial direction (Fig. 4). Such a cross-cooling may also be effective when the cross-flow structure is used as a catalytic reactor.

C. Power Cells and Electrochemical-Synthesis Reactors

1. Compact Fuel Cells

The cross-flow geometry soon became a natural structural solution when developing compact fuel cells and solid-state electrolytic reactors. Fuel cell technology is an extensive part of chemical engineering science [38] and would require a chapter or book of its own to be described.

From a cross-flow point of view it may be of interest to mention the phosphoric acid fuel cell with the so-called DiGas system (Fig. 9), which is an air-cooled cross-flow configuration for use in utility-power stations [39]. The process air stream is diverted into two types of channels: into individual cells with relatively small cross-sectional area, and into cooling plates (approximately one for every five cells) with a large cross-section. Bipolar plates were molded from a mixture of graphite and phenolic resin, with a Pt-on-carbon cathode and a Pt anode combined with colloidal PTFE on a graphite-paper backing.

2. Solid-State Electrolytic Reactors

The goal of using solid-state electrolytic reactors is not only to generate electrical power, but also to combine this with an industrially important catalytic reaction, such as dissociation of oxygen-containing compounds like NO [40,41], quantitative oxidation of NH_3 to NO [42–44], oxidation of SO_2 [45], and methanol [46], ethylene epoxidation [46], or Fischer–Tropsch synthesis [47]. The cross-flow reactor used in this type of study (Fig. 10) [48,49] has a solid electrolyte consisting of yttria-doped zirconia. The plates are electrically connected in series, with a varying number of plates in parallel. The oxidant flow channels

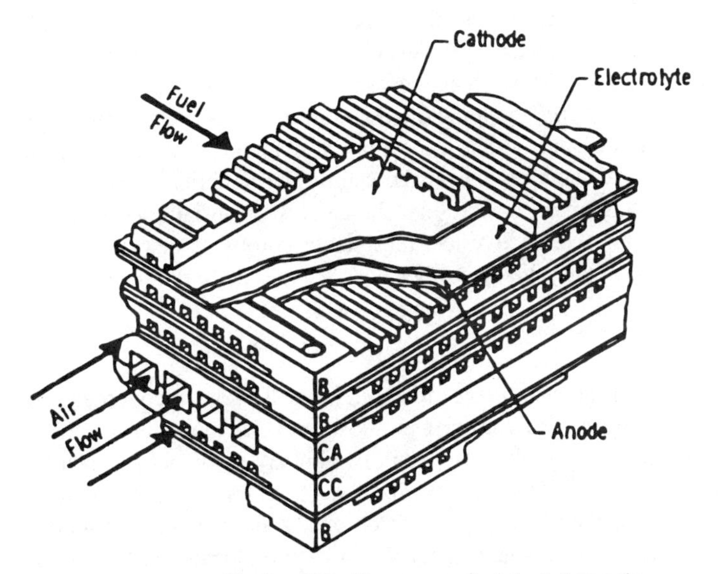

B - Bipolar plate with process air & fuel channels
CA - Anode DIGAS cooling plate
CC - Cathode DIGAS cooling plate

Figure 9 The ERC-WE DiGas system. (From Ref. 38.)

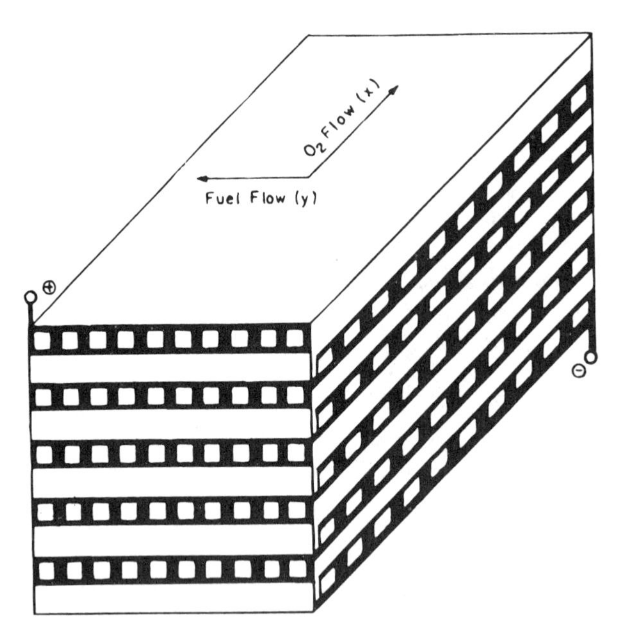

Figure 10 A cross-flow monolith fuel cell consisting of five unit batteries connected in series. Each unit battery consists of 100 unit cells electrically connected in parallel. Dark areas are covered with conductive metal film. (From Ref. 48.)

are coated with a metal such as Pt or Ag to catalyze the reduction of O_2 to O^{2-}. Since yttria-doped zirconia is a good O^{2-} conductor, these oxygen ions will migrate through the solid electrolyte wall to the fuel flow channels, which are perpendicular to the oxidant flow channels. The walls of these fuel flow channels are coated with a suitable metal or a conductive metal oxide that catalyzes the anodic oxidation of the fuel. One advantage of solid electrolyte cells is the possibility that the operation temperature can be chosen high enough to fulfill a catalytic function and at the same time to diminish activation polarization phenomena. It is important to note that the catalyst material coated on the channel walls must be electrically conductive and reasonably porous so that they allow fuel or oxygen molecules access to the solid electrolyte–catalyst–gas three-phase boundary.

3. *Salinity Power Cells*

The oil crisis during the 1970s stimulated intense research in most Western countries and Japan, effectively to save energy and to find proper substitutes for the suddenly more expensive use of oil. At a sufficiently high price of oil, even many risky and speculative projects were economically justified. One of these projects was based on the idea of obtaining electrical energy from the electromotive force of a concentration cell with a flow of seawater and fresh river water. This salinity cell was to be located in the mouth of a river, where fresh water meets seawater. The plates in the cross-flow salinity cell consist of alternate anion-selective membranes and cation-selective membranes. In a seawater channel of the cross-flow cell, the anions and cations migrate through respective membranes to the nearest channels above and below the seawater channel. These neighboring channels are filled with flowing fresh river water, and the concentration difference over the membranes between the channels will give rise to an electromotive force. If a great number of such channels are piled together in a cross-flow structure and connected electrically, it was found possible in laboratory experiments [50] to attain a voltage difference somewhat lower than the theoretical one. The discrepancy was explained by water and electricity leakages. Every cross-flow segment of about 5-cm height consisted of 400 parallel membranes, and 20–40 such segments constituted a complete pile, provided with one electrode at each end (Fig. 11). Twenty-four of these piles formed a unit with an estimated output of 20 kW at a salinity content of 3% in the seawater. The whole power plant was planned to contain 200 rows of 50 units.

It should be noted that the cross-flow salinity power cell idea was first proposed by Pattle [51], who already had a cell running in 1953. For diverse reasons his experiments were interrupted and forgotten until 1974. Since then the method has been proposed by several authors. Weinstein and Leitz [52] used a desalination stack for energy production. Loeb and Block [53], however, established that membranes and membrane separations used in desalination stacks are economically impractical, but Emrén [54,55] and Emrén and Bergström [50] indicated that economic and environmental problems are solvable.

4. *Electrochemical Filter-Press Cell Reactors*

The three kinds of reactors already described in this section are all traditional cross-flow reactors with permeable plates or membranes. The electrochemical filter-press cell reactors used, e.g., for electrosynthesis, are equipped with cation-selective membranes to prevent mixing of the anolyte and the catholyte. These cell reactors are therefore good examples of the extended type of cross-flow reactors according to the definition transferred from the filtration field. The application of the electrochemical filter-press cell reactor technique

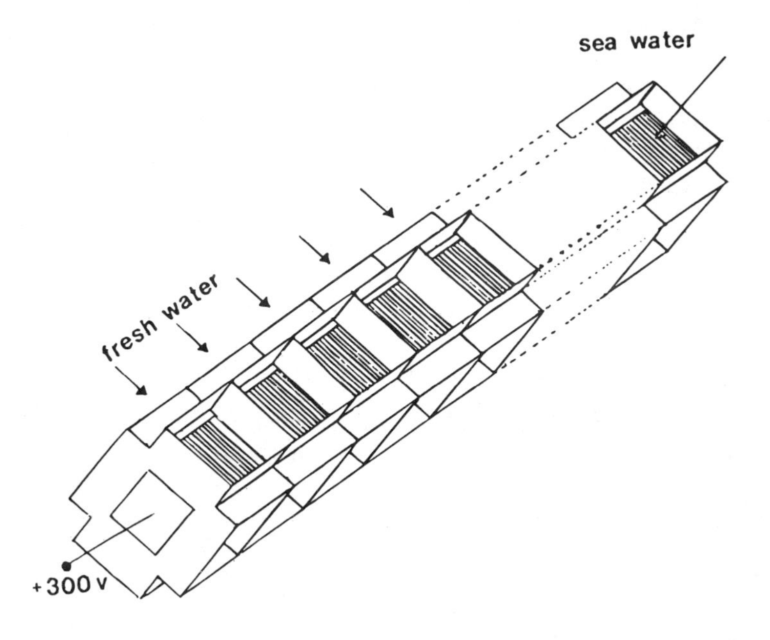

Figure 11 Membrane pile in a salinity power station. (From Ref. 50.)

is very extensive, but may be considered to lie outside the core of this chapter. However, it is of interest to note a very educational and informative paper recently published in this field [27] that mentions that five companies are suppliers of electrochemical filter-press cells and related equipment.

It should also be observed that the catalytic cell reactor (described in Section II.D), which is intended to be an alternative reactor to trickle beds for liquid-phase hydrogenations, is a further-developed electrochemical filter-press cell based on the firm Electro Cell AB's concept with respect to the preparation of thin, porous electrodes.

D. Trickle-Bed Alternatives

The world production capacity of liquid-phase hydrogenation processes and similar hydrotreating processes is very extensive and important. The dominating reactor in these processes is the trickle bed. In this type of reactor, the gas and liquid phases flow cocurrently downward over a fixed bed of catalyst pellets. Most often, both intraparticle and external mass transport resistances are significant. Moreover, the complex multiphase flows often give rise to many operational difficulties. Liquid maldistribution and channeling lead to poor contacting of the catalyst in the so-called trickling flow regime, and the catalyst pellets become partially wetted at low liquid flow rates. In processing a volatile liquid taking part in an exothermic reaction, partial wetting can lead to hot zones in the reactor, initiating unwanted side reactions and finally a catalyst sintering and deactivation.

1. *Monolithic Catalyst Reactors*

The disadvantages of processing with the trickle bed have long since provoked a search for alternative reactor solutions. Hitherto only monolithic catalyst reactors seem to have been industrially successful, as may be described in detail in Part I of this book. The monolithic catalyst means here a honeycomb catalyst, and the gas and liquid streams are not separated, but flow in the same channels. The advantages of the monolithic catalyst for three-phase processes are explained by the fact that, at Taylor flow conditions (i.e., a flow of well-separated plugs of liquid and gas), the liquid film between the gas plug and the channel wall is very thin, and, moreover, the liquid flow rate is locally very high in this film. This will give a high mass transfer rate across the liquid film, from the gas plug to the catalyst channel wall. The high liquid flow rate in this thin film also results in a jetlike liquid flow into the liquid plug just when the gas plug passes. This jetlike flow causes a secondary mixing effect in the following liquid plug and disturbs the laminar layer close to the channel wall, with a consequently increased mass transfer to the channel wall.

2. *Tubular Membrane and Coiled-Plate Membrane Reactors*

Even if the monolithic catalyst reactor is a well-accepted alternative to the trickle bed, it cannot be the ideal final option. In such an alternative, the two phases have to be separated, and therefore the cross-flow reactor may have all the qualifications of an ideal option.

The key problem of the cross-flow reactor is not how to construct an effective separation of the two flowing phases. It is instead connected with how to design the porosity and location of the catalytic active zones of the separating walls so that the transport resistance across the wall does not limit the conversion and the selectivity of the chemical reactions. Palladium-alloy membranes, or thin films of these alloys on porous ceramic tubs, seem to have the potential to be good solutions of the separating-wall problem for cross-flow reactors used for hydrogenation reactions.

Since Chapter 16 deals with metal-containing membrane reactors and Chapter 17 with inorganic membrane reactors, the design, preparation, activity, and selectivity aspects concerning these types of reactors will not be commented on here. Only aspects relevant to chemical reaction engineering will be discussed.

Gryaznov et al. [30–32] have done pioneering work in the study of differently designed palladium-alloy membrane reactors for reactions in both gas phase and liquid phase. Most interest was directed to the composition and properties of the membranes, which were decisive questions at this stage of the development, and still are. Selectivity problems in various organic chemical reactions were also of importance to study.

Deactivation of the catalyst is always an industrially important problem. For fixed-bed reactors, to which class the cross-flow reactors also belong, catalyst poisoning is a particularly delicate matter, since the reactivation is often complicated and expensive. Some poisoning effects may be difficult to explain and understand, and this of course causes extra uncertainty. One example of such poisoning was the observation by Amor and Farris [33] that a special deactivation effect appeared in liquid-phase hydrogenation of toluene using a spiral tubular membrane reactor. Toluene was not hydrogenated at all over the palladium foil used. This phenomenon and reactivation of the catalyst have recently been studied by Ali et al. [56].

Liquid-phase hydrogenation using an α-Al_2O_3 tubular reactor coated on its inside wall with palladium on γ-Al_2O_3 was studied by Cini and Harold [57] and Harold et al. [58]. The liquid flowed on the outside of the tube, and hydrogen flowed in its core. The whole reactor was envisioned to consist of a vessel containing a bank of such tubes. In hydrogenation of α-methylstyrene it was found that the surface reaction was the rate-determining step of the process up to 40°C. A strong decrease of the activation energy of the global rate process could be observed above the reaction temperature, 45°C. This high temperature in combination with heat effects of the exothermic reaction may have resulted in a depletion of the volatile liquid reactant and a pore emptying. This may have changed the process from a liquid-phase hydrogenation to a gas-phase hydrogenation in some parts of the reactor. Under these conditions the transport within the now-gas-filled pores may be the rate-limiting step of the process. The authors verified that this reactor with permeable tubular walls gave a reaction rate that was 15–20 times greater than for catalyst pellets exposed to the unsaturated liquid reactant. This effect was shown to be a result of the partial wetting of the hollow tubular catalyst caused by the partial pore emptying. In the case of fully wetted pellets, the performance of the pellets depends strongly on the degree of saturation of the liquid phase with hydrogen, since the gas is not in direct contact with the pellet surface.

The hollow-fiber trickle-bed reactor, according to Yang and Cussler [59], is another variant of the hollow-tube theme. In this case the porous tube is not coated with a catalytic material. The outer shell surrounding the fibers is instead filled with catalyst pellets. The liquid is added to this outer shell, and the gas reactant is added to the inside of the fibers. Since no catalyst is present in the gas–liquid contact, this type of reactor functions merely as an effective gas-absorber. In comparison with the trickle bed, no flooding occurs with the hollow-fiber trickle-bed reactor at high liquid loads, which means a much higher reaction rate at high liquid flow rates than obtained with the traditional trickle bed.

3. *The Traditional Cross-Flow Reactor*

The above-cited studies reflect the most recent alternatives to the trickle-bed reactor using different types of permeable tubes as reactors. The hitherto-only-published studies using the traditional cross-flow reactor with permeable catalytic plates as an alternative to the trickle bed are some years older. De Vos et al. [60,61] studied both the effectiveness and the selectivity obtained in this type of reactor and made comparisons with the trickle bed. Despite the fact that the quality of the permeable plates with respect to porosity was not as high as in corresponding permeable materials today, the authors could demonstrate the advantages of the cross-flow catalytic reactor compared with the trickle bed.

The reactor capacity was determined using the hydrogenation of *p*-nitrobenzoic acid as a model reaction. This reaction is a first-order process in hydrogen and zero-order in the liquid reactant.

In the study of the effectiveness factor, the intrinsic activity of the catalyst was determined from slurry experiments with the ground cross-flow catalyst. Since the effective diffusivity of hydrogen in the liquid-filled permeable plates was also determined by means of a special technique, it was possible to calculate the effectiveness factor of the catalytic reaction based on a mathematical model for the process in a channel. The calculated and experimental values for the effectiveness factor were found to agree well, so it was also possible to determine the coordinate, in the porous plate, from the liquid channel side, where the *p*-nitrobenzoic acid vanishes because of the reaction. Since the flow in the channels is laminar, one could also calculate the ratio between the gradients of *p*-nitrobenzoic acid concentration in the pores and in the channels close to the pore mouth, i.e., the Biot number (Bi_m) of mass transfer. Low values of Bi_m were obtained at high reaction rates, indicating that large gradients were obtained even in the channels. At these high reaction rates, the *p*-nitrobenzoic acid already vanished after half the length of the pores. A comparison was also made with hydrogenations using a laboratory trickle bed with rather small particles ($d_p = 3 \times 10^{-3}$ m) but of course much larger than the thickness of the permeable plates (2×10^{-4} m). The effectiveness factor values in these trickle-bed runs were on the order of magnitude 0.1–1%, i.e., somewhat greater than commonly reported in the literature. These values may be compared with the values 0.6–7.3% obtained in the cross-flow reactor experiments, which clearly demonstrate the larger capacity of the traditional cross-flow reactor.

Selectivity properties were also important to compare. Such a comparison was performed by De Vos et al. [61] concerning the complicated hydrogenation of 1,4 di-*tert*-butylbenzene in cyclohexane in the presence of a Rh catalyst. Siegel et al. [62,63] showed that this reaction has very special selectivity properties. The yield of di-*tert*-butylcyclohexenes was found to increase with increasing hydrogen pressure between 0.34 and 0.98 bar, but decreased with increasing pressure between 7.8 and 152 bar.

The yield of *cis*- (equatorial) and *trans*- (axial) 1,4-di-*tert*-butylcyclohexane was of interest to compare in slurry hydrogenations using ground cross-flow catalyst with the cross-flow reactor hydrogenation. It was found from this comparison that the selectivity of cis to trans at high conversion was equal to about 10 in slurry experiments, but increased to between 28 and 322 in comparing experiments with the cross-flow reactor.

The hydrogenation of 1,4-di-*tert*-butylbenzene may be described with the complex reaction scheme in Fig. 12. Due to this complexity, an attempt to explain the different results of the two hydrogenation series was performed only by mathematical simulation, based on different sets of parameter values of the rate equations describing the reaction network and of the effective diffusivities. The result of this simulation showed that it was quite possible to find a reasonable set of parameter values that can explain the great selectivity differences between slurry and cross-flow experiments. The mathematical simulation also included a calculation of the concentration profiles of the different compounds

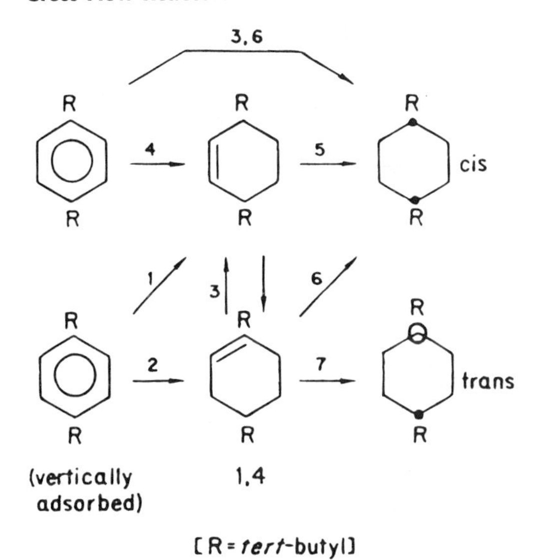

Figure 12 Reaction networks for the hydrogenation of 1,4-di-*tert*-butylbenzene. (From Ref. 61.)

in the porous catalyst. From these profiles it was easy to understand why the selectivity can be so different. The explanation is the combination of pore transport effects and the fact that hydrogen and the arene are added from opposite sides of the catalyst plate in the cross-flow case, whereas transport effects are lacking in the slurry case.

A simulation of trickle-bed hydrogenations was also performed, giving almost the same high selectivity as in the cross-flow case. This shows that the high selectivity is primarily a pore transport effect and not a result of the addition of the arene and hydrogen to opposite sides of the porous plate.

4. *The Cell Reactor*

The catalytic plates are not completely accessible in the monolithic cross-flow structure, since a certain part of the plates is lost by contact with the corrugated interstitial planes of this structure. Moreover, the successful development in the preparation of thin, porous plates for electrochemical purposes created the idea that this preparation technique should also be used to produce permeable catalyst plates for cross-flow catalyst reactors. Instead of using these plates for the complicated cross-flow structure, the whole step was taken to using them in an electrochemical cell-like design of the reactor. The catalytic plates were thus mounted in a special rack like a filter press (Fig. 13), giving the so-called cell reactor [26].

The key problem here is the preparation of these plates. The plates contained one inert hydrophobic part close to the hydrogen gas side, and another part consisting of a catalytically active metal on various types of carrier powder. The hydrophobic layer was made of 30–50-μm nonporous PTFE particles. The catalyst carrier particles were porous (mean pore diameter of 10 nm) with a particle size of about 5 μm. The catalytic material was of three different types: 10% Pd on alumina, 10% Pd on carbon, and 1.9% Pd on NiO/SiO_2. In addition to these powder materials, the plates contained nets of nickel wire (0.16 mm) or glass fibers (0.2 mm) as reinforcement. The catalytic plates were prepared

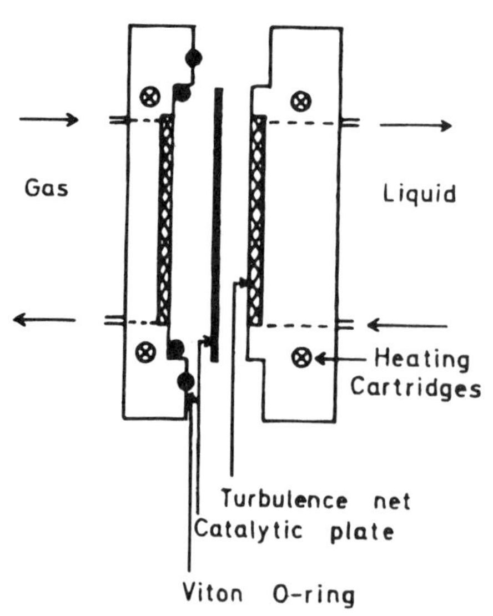

Figure 13 Model unit of the laboratory cell reactor. (From Ref. 26.)

in a hydraulic press at a pressure from 250 to 750 bar. The interparticle pores of the plates were dependent on the pressure during preparation and were found to be between 1 and 5 μm. In plates tested in hydrogenation of *p*-nitrobenzoic acid, the catalytic layer was between 0.5 and 0.005 mm and the PTFE layer between 0.12 and 0.4 mm. Moreover, interparticle porosity in the catalytic layer was between 0.37 and 0.86, and the BET area was between 250 and 900 m^2/g.

As seen from Fig. 13, which gives a model unit of the laboratory cell reactor, this cell contains specially designed nets, to provide turbulence of the liquid flow. From the experimental result it was shown that the effectiveness factor was very high (up to 0.84) in experiments with plates with a very thin catalytic layer located close to the liquid side. In experiments with plates with a thick catalytic layer located in the middle of the plate, the effectiveness factor was very low (down to 0.02), and only a small part of the catalyst layer was utilized, since *p*-nitrobenzoic acid vanished in the plate after only 10% of the thickness of the catalytic layer had been passed in this particular run.

E. Bioreactors

Membrane fixed-enzyme reactors or fixed-whole-cell reactors designed as permeable tube reactors are well-known types of reactors in bioengineering [64] and should be mentioned here only as a complement to the chemical-engineering-oriented description above. One typical example is the tube reactor consisting of asymmetric capillary polysulphone membranes loaded with whole thermophilic bacteria *Sulfolobus solfataricus*, which is catalytically active via a cytoplasmic β-galactosidase [65]. The enzymatic activity was determined by studying the rate of lactose hydrolysis to glucose and galactose. The lactose solution was added to the lumen of the capillary membranes. The goal of this study was to find the optimal operating conditions and the optimal design of the reactor on the basis of a mathematical model of the process consisting of a set of partial differential equations.

III. MATHEMATICAL MODELING OF CROSS-FLOW REACTORS

While membrane-material problems have been the stumbling block in the development of cross-flow reactors with permeable walls, many mathematical modeling studies have been published concerning the flow in the channels coupled with reactions and transport in the porous walls. Hamrin, Jr., et al. [66–69] made a thorough mathematical analysis of the reaction between gaseous and liquid reactants that approach each other from opposite sides of a porous wall in a traditional cross-flow reactor and in a tubular reactor. Derived equations were solved numerically and analytically using a simplified model. The study was restricted to reactions of first order with respect to the gaseous reactant and zero order with respect to the liquid reactant. This analysis may be representative for many liquid-phase hydrogenation processes. The effect of different process parameters was clearly demonstrated with these mathematical studies.

In mathematical modeling of cross-flow solid-state electrochemical reactors, the dimension of the mathematical model increases with two additional variables compared to gas-liquid processes, since both the heat balance and the electron balance have to be considered. Introduction of an integral electron conservation balance results in an integro-differential problem. A comprehensive study of this kind was performed by Vayenas et al. [48] and by Debenedetti and Vayenas [49].

Mathematical models of cross-flow air-dryers and regenerators differ from those of the cross-flow reactors, since one must consider the fact that the sides of the channel walls are partly made of a reactive solid substance that is capable of exchanging a species with the process streams. These mass and heat balances result in nonlinear Volterra-type integral equations, which have been studied by Roy and Gidaspow [37,70].

Stability aspects of the cross-flow reactor–heat exchanger were the aim of a comprehensive study by Degnan and Wei [35,71], where the theoretical results were also verified experimentally. Of particular interest was the experimental demonstration of the multiplicity of the steady states for the autothermal countercurrent process case.

A special stability problem was studied by Yakhnin et al. [12,13] for cross-flow moving-bed reactors. The instability here is caused by the differential flow between heat acting as autocatalyst and the reacting matter at elevated Lewis numbers. The cross flow removes the system from equilibrium all along the reactor and thereby furthers the occurrence of instabilities. The study is purely theoretical without any experimental verification for the present.

IV. SCALE-UP OF CROSS-FLOW REACTORS WITH PERMEABLE WALLS

Scale-up based on the often-used principle of magnification of all the length dimensions is not applicable in the scale-up of cross-flow reactors with permeable walls. The best scale-up rule here may be the so-called "multiplication of units."

The scale-up of hollow-fiber reactors and tubular reactors is easily performed by the multiplication of the number of fibers and tubes.

The scale-up of the traditional cross-flow reactor with corrugated interstitial planes has been solved by Degnan and Wei [35,36], who showed that, for example, four cross-flow monoliths can be positioned in a clever way inside a steel manifold (Fig. 8). This technique is not limited to only four monoliths but may be scaled-up to a much larger number. Emrén and Bergström [50] have even proposed a scale-up including 10,000 such monolith units.

The easiest cross-flow reactor to scale-up may be the cell reactor. Here the scale-up of the filter press serves as a good model.

The questions of scale-up take us closer to industrial realities. A common problem for fixed-bed reactors is how to decrease the tendency of catalyst poisoning and deactivation. Since this seems to be unavoidable, one has to find methods to restore the catalytic activity. This problem affects all fixed-bed reactors, and consequently it will not degrade the cross-flow reactors compared to other fixed-bed reactors.

REFERENCES

1. K.R. Westerterp, W.P.M. Van Swaaij, and A.A.C. Beenackers, *Chemical Reactor Design and Operation*, Wiley, Chichester, 1984, p. 101.
2. P. Trambouze, H. Van Landeghem, and J.P. Wanquier, *Chemical Reactors. Design, Engineering and Operation*, Editions Technip, Paris, 1987, p. 511.
3. R. Krishna and S.T. Sie, Strategies for multiphase reactor selection, *Chem. Eng. Sci. 49*:4029 (1994).
4. H. Lerner and V.A. Citarella, Improve alkylation efficiency, *Hydrocarbon Processing, November*:89 (1991).
5. Anonymous, Radial flow ammonia converter—New ammonia synthesis design, *Nitrogen 31*:22 (1962).
6. O.J. Quartulli and G.A. Wagner, Why horizontal NH_3-converters?, *Hydrocarbon Processing 57*:115 (1978).
7. A. Strauss and K. Budde, Die Anwendung von Radialstromreaktoren für heterogen-katalytische Processe, *Chem. Tech. 30*:73 (1978).
8. V. Balakotaiah and D. Luss, Effect of flow direction on conversion in isothermal radial flow fixed-bed reactors, *AIChE J. 27*:442 (1981).
9. H.-C. Chang, M. Saucier, and J.M. Calo, Design criterion for radial flow fixed-bed reactors, *AIChE J. 29*:1039 (1983).
10. P.R. Ponzi and L.A. Kay, Effect of flow maldistribution on conversion and selectivity in radial flow fixed-bed reactors, *AIChE J. 25*:100 (1979).
11. E.H.P. Wolff, P. Veenstra, and L.A.: Chewter, A novel circulating cross-flow moving bed reactor system for gas-solids contacting, *Chem. Eng. Sci. 49*:5427 (1994).
12. V.Z. Yakhnin, A.B. Rovinsky, and M. Menzinger, Differential flow instability of the exothermic standard reaction in a tubular cross-flow reactor, *Chem. Eng. Sci. 49*:3257 (1994).
13. V.Z. Yakhnin, A.B. Rovinsky, and M. Menzinger, Differential-flow-induced pattern formation in the exothermic $A \rightarrow B$ reaction, *J. Phys. Chem. 98*:2116 (1994).
14. K.A. Pilcher and J. Bridgwater, Pinning in a regular moving bed reactor with gas cross-flow, *Chem. Eng. Sci. 45*:2535 (1990).
15. C.M. Marb and D. Vortmeyer, Multiple steady states of a crossflow moving bed reactor: Theory and experiment, *Chem. Eng. Sci. 43*:811 (1988).
16. F.J. Doyle, III, R. Jackson, and J.C. Ginestra, The phenomenon of pinning in an annular moving bed reactor with crossflow of gas, *Chem. Eng. Sci. 41*:1485 (1986).
17. J.C. Ginestra and R. Jackson, Pinning of a bed of particles in a vertical channel by a cross flow of gas, *Ind. Eng. Chem. Fundam. 24*:121 (1985).
18. H. Jüntgen and C. Peuckert, Wanderbettreaktoren, *Fortschritte der Verfahrenstechnik. Abt. D. Reaktionstechnik 16*:367 (1978).

19. M. Sittig, Catalytic cracking techniques in review, *Petrol, Refiner 31*:263 (1952).
20. B.J. Cha, R. Huin, Van Landeghem, and A. Vidal, Regenerate reformers continuously, *Hydrocarbon Processing 52*:98 (1973).
21. K. Knoblauch, E. Richter, and H. Jüntgen, Simultane SO_2- und NO_x-Entfernung aus Rauchgasen durch Adsorptionskatalyse on Aktivkoksen, *Chem. Ing. Tech. 57*:239 (1985).
22. D. Kunii and O. Levenspiel, *Fluidization Engineering*, Wiley, New York, 1969, p. 179.
23. J.P. Holman, *Heat Transfer*, 4th edition, McGraw-Hill, New York, 1976, p. 393.
24. J.R. Welty, C.E. Wicks, and R.E. Wilson, *Fundamentals of Momentum, Heat, and Mass Transfer*, 3rd edition, Wiley, New York, 1984, p. 403.
25. J. Murkes and C.G. Carlsson, *Crossflow Filtration*, Wiley, Chichester, 1988, p. 3.
26. V. Hatziantoniou, B. Andersson, T. Larsson, N.H. Schöön, L. Carlsson, S. Schwarz, and K.B. Widéen, Preparation, characterization, and testing of a new type of porous catalytic plates usable for liquid-phase hydrogenations at enhanced mass-transfer conditions, *Ind. Eng. Chem. Process Des. Dev. 25*:143 (1986).
27. F. Walsh and D. Robinson, Electrochemical synthesis and processing in modern filterpress reactors, *Chem. Techn. Europe 2*(3):16 (1995).
28. F.M. Dautzenberg, J.E. Naber, and A.J.J. Van Ginneken, Shell's flue gas desulfurization process, *Chem. Eng. Prog. 67*(8):86 (1971).
29. H.P. Calis, T.S. Everwijn, A.W. Gerritsen, C.M. Van Den Bleek, F. Goudriaan, and F.G. Van Dongen, Mass transfer characteristics of parallel passage reactors, *Chem. Eng. Sci. 49*:4289 (1994).
30. V.M. Gryaznov, Hydrogen permeable palladium membrane catalysts. An aid to the efficient production of ultra pure chemicals and pharmaceuticals, *Platinum Metals Rev. 30*(2): 68 (1986).
31. V.M. Gryaznov, Platinum metals as components of catalyst-membrane systems, *Platinum Metals Rev. 36*(2):70 (1992).
32. V.M. Gryaznov, Surface catalytic properties and hydrogen diffusion in palladium alloy membranes, *Z. Phys. Chem. Neue Folge 174*:761 (1986).
33. J.N. Armor and T.S. Farris, Membrane catalysis over palladium and its alloys, *New Frontiers in Catalysis* (L. Guczi et al., eds.), Proceedings of the 10th International Congress on Catalysis, 1992, Budapest, Elsevier, Amsterdam, 1993, p. 1363.
34. L.L. Johnson, W.C. Johnson, and D.L. O'Brien, The use of structural ceramics in automobile exhaust converters, *AIChE Symp. Ser. 57*(35):55 (1961).
35. T.F. Degnan, Jr. and J. Wei, Monolithic reactor heat exchanger, *ACS Symp. Ser. 65*:83 (1978).
36. T.F. Degnan and J. Wei, The co-current reactor–heat exchanger, *AIChE J. 26*:60 (1980).
37. D. Roy and D. Gidaspow, Nonlinear coupled heat and mass exchange in a cross-flow regenerator, *Chem. Eng. Sci. 29*:2101 (1974).
38. K. Kordesh and M. Reindl, Fuel cells, in *Electrochemical Reactors, Their Science and Technology* (I.M. Ismail, ed.), Elsevier, Amsterdam, 1989, p. 450.
39. B.S. Baker et al., as Ref. 22 in *Electrochemical Reactors, Their Science and Technology* (I.M. Ismail, ed.), Elsevier, Amsterdam, 1989, p. 501.
40. S. Pancharatnam, R.A. Huggins, and D.M. Mason, Catalytic decomposition of nitric oxide on zirconia by electrolytic removal of oxygen, *J. Electrochem. Soc. 122*:869 (1975).
41. T.M. Gür and R.A. Huggins, Decomposition of nitric oxide on zirconia in a solid-state electrochemical cell, *J. Electrochem. Soc. 126*:1067 (1979).
42. C.G. Vayenas and R.D. Farr, Cogeneration of electric energy and nitric oxide, *Science 208*:593 (1980).
43. R.D. Farr and C.G. Vayenas, Ammonia high temperature solid electrolyte fuel cell, *J. Electrochem. Soc. 127*:1478 (1980).
44. C.T. Sigal and C.G. Vayenas, Ammonia oxidation to nitric oxide in a solid electrolyte fuel cell, *Solid St. Ionics 5*:567 (1981).

45. Y. Yang, P.G. Debenedetti, H. Britt, C.G. Vayenas, and L.B. Evans, Proceedings of International Symposium on Process Systems Engineering, Tokyo, 1982.
46. M. Stoukides and C.G. Vayenas, Transient and steady-state vapor phase electrocatalytic ethylene epoxidation, *ACS Symp. Ser. 178*:181 (1982).
47. T.M. Gür and R.A. Huggins, Electrocatalytic synthesis of methane on stabilized zirconia from molecular hydrogen/carbon dioxide mixtures, *Solid St. Ionics 5*:563 (1981).
48. C.G. Vayenas, P.G. Debenedetti, I. Yentekakis, and L.L. Hegedus, Cross-flow, solid-state electrochemical reactors: A steady-state analysis, *Ind. Eng. Chem. Fundam. 24*:316 (1985).
49. P.G. Debenedetti and C.G. Vayenas, Steady-state analysis of high temperature fuel cells, *Chem. Eng. Sci. 38*:1817 (1983).
50. A.T. Emrén and S.B. Bergström, Salinity power station at the Swedish west-coast. Possibilities and energy price for a 200 MW plant, Proceedings of Miami International Conference on Alternative Energy Sources, 1977, p. 2909.
51. R.E. Pattle, Experiments on the production of electricity without fuel using the first model of the hydroelectric pile, Report to National Research Development Corporation, 1954 (see Ref. 50).
52. J.N. Weinstein and F.B. Leitz, Electric power from differences in salinity: The dialytic battery, *Science 191*:557 (1976).
53. S. Loeb and M.R. Block, Salinity power, potential and processes, especially membrane processes, The Joint Oceanographic Assembly, Edinburgh, Scotland, Sept. (1976).
54. A.T. Emrén, Concentration cell for salinity power production. Economic potential of the concentration cell, Proceedings of the 3rd Miami International Conference on Alternative Energy Sources, 1980.
55. A.T. Emrén, Concentration cell for salinity power production. Economic potential of the concentration cell, *Energy Newslett. 2*:41 (1981).
56. J.K. Ali, E.J. Newson, and D.W.T. Rippin, Deactivation and regenation of Pd–Ag membranes for dehydrogenation reactions, *J. Membrane Sci. 89*:171 (1994).
57. P. Cini and M.P. Harold, Experimental study of the tubular multiphase catalyst, *AIChE J. 37*:997 (1991).
58. M.P. Harold, P. Cini, B. Patenaude, and K. Venkataraman, The catalytically impregnated ceramic tube: An alternative multiphase reactor, *AIChE Symp. Ser. 85*:26 (1989).
59. M.-C. Yang and E.L. Cussler, A hollow-fiber trickle-bed reactor, *AIChE J. 33*:1754 (1987).
60. R. De Vos, V. Hatziantoniou, and N.-H. Schöön, The cross-flow catalyst reactor. An alternative for liquid phase hydrogenations, *Chem. Eng. Sci. 37*:1719 (1982).
61. R. De Vos, G. Smedler, and N.-H. Schöön, Selectivity aspects of using the cross-flow catalyst reactor for liquid-phase hydrogenations, *Ind. Eng. Chem. Process Des. Dev. 25*:197 (1986).
62. S. Siegel and N. Garti, The effect of pressure on the catalytic hydrogenation of aromatic hydrocarbons on rhodium, in *Catalysis in Organic Syntheses 1977* (G.V. Smith, ed.), Academic Press, New York, 1977, p. 9.
63. S. Siegel, J. Outlaw Jr., and N. Garti, The kinetics, stereochemistry, and mechanism of hydrogenation of some *tert* butylbenzenes on a rhodium catalyst, *J. Catal. 58*:370 (1979).
64. M. Cheryan and M.A. Mehaia, Membrane bioreactors, in *Membrane Separations in Biotechnology* (W.C. Mc Gregor, ed.), Dekker, New York, 1986, p. 255.
65. G. Catapano, G. Iorio, E. Drioli, and M. Filosa, Experimental analysis of a cross-flow membrane bioreactor with entrapped whole cells: Influence of trans-membrane pressure and substrate feed concentration on reactor performance, *J. Membrane Sci 35*:325 (1988).
66. R. De Vos and C.E. Hamrin, Jr., A cross-flow reactor: Theoretical model for first order kinetics, *Chem. Eng. Sci. 37*:1711 (1982).
67. A. Akyurtlu, J.F. Akyurtlu, and C.E. Hamrin, Jr., Theoretical evaluation of a catalytic porous wall gas–liquid reactor, *Chem. Eng. Sci. 40*:1785 (1985).

68. A. Akyurtlu, J.F. Akyurtlu, K.S. Denison, and C.E. Hamrin, Jr., Application of the general purpose, collocation software, PDECOL to the Graetz problem, *Comp. Chem. Eng. 10*:213 (1986).
69. J.F. Akyurtlu, A. Akyurtlu, and C.E. Hamrin, Jr., A study of the performance of the catalytic porous-wall three-phase reactor, *Chem. Eng. Comm. 66*:169 (1988).
70. D. Roy and D. Gidaspow, A cross flow regenerator—A Green's matrix representation, *Chem. Eng. Sci. 27*:779 (1972).
71. T.F. Degnan and J. Wei, The cocurrent reactor–heat exchanger, Part I—Theory, *AIChE J. 25*:338 (1979).

21

Transformation of a Structured Carrier into Structured Catalyst

Xiaoding Xu and Jacob A. Moulijn
Delft University of Technology, Delft, The Netherlands

I. INTRODUCTION

Despite many applications of monolithic catalysts, articles describing their preparation in some detail are scarce [1–5 and the references therein]. The main reason probably is that most preparation work is done in industry and there is no commercial interest in publishing the information gained. In this chapter, methods and techniques used in the preparation of monolithic catalysts are described, partly based on our own research. It will be shown that conventional preparation methods are successfully applied, though more special precautions are advisable.

Ceramic monoliths can be manufactured either by extrusion [6–31] or by corrugation [32–44], the former being the technique mainly used. By extrusion, ceramic monoliths of various materials can be produced, though cordierite or mullite monoliths are most used, especially as catalyst carriers in exhaust gas treatment [6–31].

Metallic monoliths are produced exclusively by corrugation, followed by rolling up or folding into monoliths of the shape and size required [1,4,5]. We will discuss the production of monolith supports in the following sections.

A. Ceramic Monoliths

1. Ceramic Monoliths by Extrusion

Merkel and Murtagh et al. [6–8] described a process for producing cordierite monoliths. The process comprises preparing a mixture of talc, clay, an alumina-yielding component, and silica, mixing the mixture to form a moldable composition, molding the mixture, drying the greenware, and heating it at a temperature of 1473–1773 K to form a ceramic containing mainly cordierite and having a low coefficient of thermal expansion. It is also possible to use other materials.

Ito et al. [13] described the preparation of a batch consisting of talc, kaolin, calcined kaolin, and alumina that collectively provide a chemical compound of SiO_2 45–55, Al_2O_3 32–40, and MgO 12–15 wt%. This mixture is used to produce cordierite monoliths, as just described. Talc is a material consisting mainly of hydrous magnesium silicate, $Mg_3Si_4O_{10}(OH)_2$ [29]. Depending on the source and purity of talc, it may also be associated with other minerals, such as tremolite [$CaMg_3(SiO_3)_4$], serpentine ($3MgO \cdot 2SiO_2 \cdot 2H_2O$), anthophyllite [$Mg_7(OH)_2(Si_4O_{11})_2$], magnesite ($MgCO_3$), mica, and chlorite [29].

Not only can cordierite monoliths be produced by extrusion, but the technique can be used to produce monoliths of other materials, such as SiC, B_4C, Si_3N_4, BN, AlN, Al_2O_3, ZrO_2, mullite, Al titanate, ZrB_2, and sialon [15,16].

In extrusion, in addition to the nature and the properties of the materials used to make the moldable mixture, the additives used, the pH, the water content, and the force used in extrusion are also of importance with respect to the properties of the monolith products [31]. The additives applied in extrusion are, e.g., celluloses, $CaCl_2$, ethylene, glycols, diethylene glycols, alcohols, wax, paraffin, acids [15,16,24,30], and heat-resistant inorganic fibers [24]. Besides water, other solvents can also be used, such as ketones, alcohols, and ethers [15,16]. The use of additives may lead to improved properties of the monoliths, such as the production of microcracks that enhance the resistance to thermal shock [4,12], better porosity and adsorbability [15,16], and enhanced mechanical strength or a low thermal expansion [24].

2. Ceramic Monoliths by Corrugation

Corrugation was the first method used to prepare monoliths [1,2,4]. Both ceramic [32–44] and metallic monoliths [36–50] can be made by this method.

Han et al. [32] described the production of high-temperature ceramic monoliths by corrugation. Thus, fibers of silicates, aluminosilicates, fiberglass, and SiC are used in forming sheets. The sheets are impregnated with silicon-containing ceramic precursor solution or suspension. Subsequently, the impregnated sheets are corrugated while the impregnant is in liquid form. The sheets are heated and the impregnant is converted to a solid ceramic material. As a result the sheets are bonded together in an open cellular arrangement, and, provided everything goes well, a monolithic structure is obtained.

3. Wall-Flow Monoliths

Wall-flow monoliths have porous walls that trap the solid particulates in the gas flow, whereas gases are permitted to flow through [1,4,45,46,51,52]. This type of monolith is widely used in the treatment of exhaust gases from diesel engines. Usually, half of the channels are regularly blocked from one end, and the other channels are blocked at the opposite end, so that solid particulates can be trapped in the channels. When the accumulation of solid particulates is built up to a certain degree, it can be regenerated by burning off the soot [52].

Those monoliths can be produced from a piece of structured foam polymer with macropores. The piece of polymer is soaked in a sol that will form a ceramic of the desired material after heat treatment. The sol-soaked structure is dried, and it is burned at a suitable temperature to remove the polymer. The remaining structure will be a ceramic one with macropores, permitting the wall-flow of gases. This technique is also used to produce heat plates and pipes with macroporous walls for gas separation purposes. The polymers used are often derived from polyurethanes [45–46].

B. Metallic Monoliths

Metallic monoliths are produced almost exclusively by corrugation techniques. Kamimura [37] gives an example in which the metallic monoliths are prepared by laminating flat and corrugated metal plates, winding the laminates, pressing the coiled stacks from the periphery, followed by heat treating at 873–1173 K to form oxide on the surface of the plates and heat treating at more than 1273 K to join the plates.

Thin metallic sheets or strips can be corrugated or rolled to form metallic monoliths. Often the sheets or strips are made of ferric alloys, e.g., stainless steel, and/or contain a small amount of aluminum [39,44], which after oxidation forms a layer of alumina that is helpful in bonding to an extra oxidic layer later on, for supporting the active phase when the sheets or strips are used to prepare monolithic catalysts [43,48–54].

Similar methods can be used to produce wall-flow monoliths, as described in Section I.A.3 [1,4].

II. COATING OF A SUPPORT LAYER ON CERAMIC STRUCTURED CARRIERS

A. Washcoating

In principle, the surface area of monolith substrates is low. Of course, it is possible to extrude porous materials, leading to high-surface-area monolithic structures. However, in the production process, calcination is essential and, as a consequence, sintering takes place, destroying the high surface area. When, in order to preserve a high surface area, the calcination temperature is chosen very low, the result is a mechanically weak structure. A layer of oxide(s) is coated onto it in order to increase its surface area for applications as a support for catalysts. Various techniques can be used to coat an oxidic layer on a monolith [55–59]. Washcoating is by far the most used technique for both metallic and ceramic monoliths.

1. *Preparation of Sols*

In order to washcoat a monolith, a suitable sol has to be made. The sol can be prepared via a hydrolytic route [60–70] or via a nonhydrolytic route [71 and the references therein].

a. Hydrolytic Route. Many articles describe the preparation of sols [48–84]. The most used method for the preparation of a sol is hydrolysis of an appropriate alkoxide [60,61]. The general reaction is as follows:

$$M(OR)_n + nH_2O \rightarrow M(OH)_n + nROH \qquad M = Al, Si, Ti, Zr, \ldots \tag{1}$$

The hydrolysis of metal alkoxide is usually accelerated by the presence of an acid or a base. Therefore, the relative amounts of acid or base, water, and the alkoxide and the temperature used are important parameters in sol preparation [60]. During aging of the sol, a polycondensation process proceeds, leading to cross-linking and the formation of polymerlike compounds:

$$-M-OH + HO-M'- \rightarrow -M-O-M'- + H_2O \tag{2}$$

This type of condensation proceeds three dimensionally. The polycondensation process also modifies the viscosity of the sol as well as the properties of the coated oxidic washcoat, such as the porosity, after heat treatment. Therefore, aging time is an important parameter [55].

Alumina is by far the most used material for washcoating monolithic catalysts. In addition to alkoxide hydrolysis mentioned above, Al-sol can be prepared from other aluminum precursors, e.g., from pseudo-boehmite or from hydrolysis of $AlCl_3$ [55 and the references therein], which will be discussed later.

Si-sol is often prepared from hydrolysis of tetraalkoxysilicate (TAOS) [61,72–75], e.g., tetramethoxysilicate (TMOS), tetraethoxysilicate (TEOS), and tetrapropoxysilicate (TPOS). Due to the fact that the TAOS's are usually immiscible with water, alcohols are often added as a cosolvent to obtain a homogeneous sol. Other oxides can be washcoated similarly [14,60,61,72–78].

When mixed sols are used in washcoating, a mixed oxidic layer may be formed on the monolith surface [79–84], so a large flexibility exists.

Mixed oxide powders or films can also be obtained via reaction of alkoxide with metal or metal hydroxide. The reactions for mixed oxides of BaO or MgO and TiO_2 are as follows:

$$Ba(OH)_2 + Ti(OBu)_4 + H_2O \rightarrow BaTiO_3 + 4BuOH \tag{3}$$

$$Mg + Ti(OBu)_4 + BuOH\ (\text{absolute}) \rightarrow MgTiO_x(OBu)_{6-2x} + H_2 \tag{4}$$

If other suitable reactants are used, other mixed oxides are formed. The number of possibilities is essentially endless, comparable with the production of ceramic materials.

b. Nonhydrolytic Route. Besides the conventional hydrolysis route, sols can also be prepared using a nonhydrolytic sol-gel route [71 and the references therein].

Acosta et al. [71] described the nonhydrolytic sol-gel route for the preparation of oxides, in particular, for silica, alumina, silica-alumina, and titania. In a classical hydrolytic route, the M—O bond of the alkoxide is cleaved [Eq. (1)], whereas in the nonhydrolytic route, the O—C bond is cleaved.

$$MOR + MX \rightarrow \text{intermediates} \rightarrow M{-}O{-}M + RX \tag{5}$$

Thus, silica and alumina can be prepared as follows.

Silica. An oxygen donor, ROH, can react with silicium tetrachloride, producing the corresponding hydroxychloride:

$$ROH + SiCl_4 \rightarrow Cl_3SiOH + RCl \tag{6}$$

The oxygen donor can be ter-BuOH, $PhCH_2OH$, and $Si(OCH_2Ph)_4$.

The hydroxychloride formed can react further with another tetrachloride [Eq. (7)] or another hydroxychloride [Eq. (8)], forming Cl_3Si—O—$SiCl_3$.

$$Cl_3SiOH + SiCl_4 \rightarrow Cl_3Si{-}O{-}SiCl_3 + HCl \tag{7}$$

$$2Cl_3SiOH \rightarrow Cl_3Si{-}O{-}SiCl_3 + H_2O \tag{8}$$

The gel obtained is referred to as nonhydrolytic gel (NHG). After calcination, silica is formed.

Alumina. Similarly, alumina can be produced [71]. The corresponding reactions for alumina with (i—PrO)$_3$Al and i—Pr_2O as the oxygen donor are shown in Eqs. (9) and (10):

$$(i{-}PrO)_3Al + AlCl_3 \xrightarrow{Et_2O/CCl_4} \text{gel} \xrightarrow{\text{calcination}} \text{NHG alumina} \tag{9}$$

$$1.5i{-}Pr_2O + AlCl_3 \xrightarrow{CH_2Cl_2} \text{gel} \xrightarrow{\text{calcination}} \text{NHG alumina} \tag{10}$$

Equation (9) proceeds in the presence of Et_2O/CCl_4 and Eq. (10) in the presence of CH_2Cl_2. Similarly, Ti-sol can be prepared using $(i—PrO)_4Ti$ or $i—Pr_2O$ as the oxygen donor and $TiCl_4$ as the Ti precursor.

It appears that this method may lead to high-surface-area gels. It was reported that the surface areas of alumina made via this route after calcination at 923 K were 370–400 m^2/g and those of silica were up to 850 m^2/g [71]. A disadvantage of this route is that chloride is formed during the process, which is environmentally unfriendly.

2. *Washcoating Procedure*

Washcoating is a method to coat a thin oxidic layer onto a solid surface. Often it is closely linked with the sol-gel method [1,4,5,48,50,53,55]. A sol is a colloidal suspension containing precursor(s) of oxide(s) [60,61].

Usually, a predried and evacuated piece of monolith is dipped into a suitable sol. After a certain time, the monolith is withdrawn from the sol. It is drained and the sol blown off, in order to remove the remaining sol; subsequently, it is dried and calcined, forming a thin layer of oxide on the surface of the monolith.

The hydroxide deposited onto the surface from an aqueous sol before calcination is called a *hydrogel*; the corresponding gel from a sol using alcohol as a solvent is called *alcogel*. Due to a smaller surface tension of alcohols compared to that of water, the washcoat layer from an alcogel is less easy to crack and often is superior to that from a hydrogel [60].

Compared to washcoating of ceramic structure, washcoating of metallic monolith is more difficult. It is possible to washcoat metallic monoliths with or without a prior oxidation. In the former case, the adhesion of the washcoat layer is better [48–51,53].

Alumina washcoating is taken as an example to illustrate the washcoating of an alumina layer onto a monolith surface.

a. Alumina Washcoating. As the material of washcoat, alumina is the most used, due to its resistance to high temperature and other advantages over other oxidic materials. This situation is completely analogous to "normal" heterogeneous catalysis. The preparation of sol has been described earlier.

The most popular method in washcoating is the so-called dipping technique.

Al-sol can be prepared using various aluminum precursors, such as pseudo-boehmite, $AlO(OH)\cdot xH_2O$, and aluminum alkoxide [48–50,55,60]. Xu et al. [55] described several possibilities. Additives, e.g., urea, or organic amines, e.g., hexamethylenetetramine (HMT), can be added to the sol in order to improve the quality of alumina obtained. The decomposition of these additives in heat treatment (calcination) may lead to a better porosity of the alumina. The reaction for urea decomposition is as follows:

$$NH_2CONH_2 + H_2O \rightarrow 2NH_3 + CO_2 \quad (11)$$

Moreover, the additives may influence the stability of the sols, which is of importance in their applications.

Optionally, cations, e.g., La, Mg, Zr, Si, which inhibit the transition of active alumina into inert α-phase, can be incorporated in the sol in order to stabilize the washcoated alumina against sintering upon heat treatment.

Figure 1 shows the block scheme of washcoating. A dry monolith is dipped in an Al-sol. Afterwards, it is drained or blown with air to remove the remaining sol. After drying and calcination at appropriate temperatures, the alumina washcoating is completed.

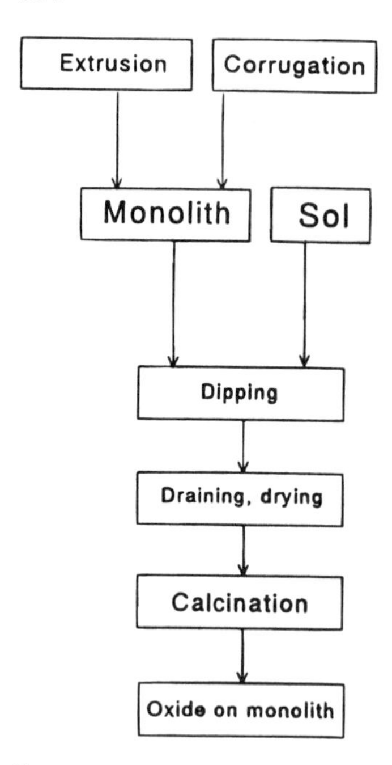

Figure 1 Washcoating of a monolith by the sol-gel method.

b. Washcoating of Other Materials. Washcoating is not restricted to alumina; various oxidic phases, e.g., zirconia, titania, and silica, can also be deposited by this technique [14,53,57,60,61,72–78]. The washcoating can be carried out analogously to the procedures described above for alumina washcoating. Sometimes, ceramic fibers or metallic fibers are incorporated, resulting in better properties of the monoliths, e.g., improved mechanical strength [21,41].

B. Carbon Coating

Sometimes, other support materials besides oxides, e.g., carbon, are attractive and advantageous for various reasons. In fine chemicals production a lot of experience exists with carbon-supported catalysts. Often they show good performance, and a high intrinsic activity is observed. Carbon support is pronounced for a weak interaction with the active phase and has a high surface area. An often-cited disadvantage of conventional carbon support is its mechanical weakness. Carbon-coated monolith can overcome this disadvantage due to the strength of the monolithic substrate.

Carbon coating can be achieved using pyrolysis of hydrocarbons at elevated temperatures [69]. Figure 2 shows a device used for carbon coating via hydrocarbon pyrolysis. In the example described here, an alumina-washcoated monolith is covered with carbon by pyrolysis of cyclohexene. A gas mixture of cyclohexene in nitrogen is passing the reactor at a certain flow rate. The monolith block to be coated is placed in the middle of the heated tubular reactor. The reaction takes place at 873–973 K, and the amount of carbon deposited can be controlled by the temperature and the time on stream. Up to 3–10 wt% carbon can be homogeneously coated onto the monolith in this way. It appears that the surface area of the carbon-coated alumina-washcoated cordierite monolith is of

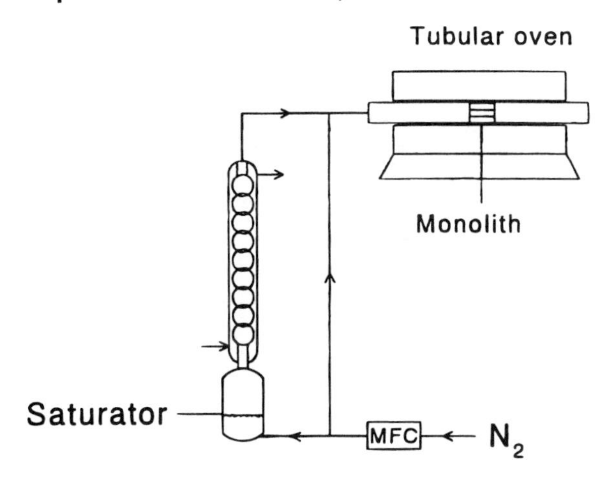

Figure 2 An apparatus for carbon coating of a monolith.

the same order of magnitude as that of the original alumina-washcoated substrate [69], indicating that the carbon is deposited over the alumina but that it has no microporosity.

Our preliminary results of nickel on carbon-coated monolithic catalysts show that in a hydrogenation reaction it is five times more active than the corresponding nickel on alumina-washcoated monolithic catalyst without carbon coating.

C. Deposition of an Oxide Layer on Metallic Structured Carriers

In principle, metallic monoliths have attractive features, e.g., fast warm-up, high mechanical strength, flexibility in shaping (see also Chapter 3 of this book). However, they are usually less porous and have, in general, small surface areas. Often, deposition of an oxidic layer is necessary to increase the surface area or improve the surface properties of the monoliths, but adhesion of an oxide layer on a metal surface can lead to weak structure.

Many metallic monoliths are made of alloys that often contain more than one metal. It is possible first to oxidize the surface of a metallic monolith. The existence of an oxidic layer on a metallic monolith is favorable for the adhesion of a subsequent washcoat layer. It appears that this oxidic layer by oxidation is hardly porous. Often, deposition of another porous metal oxide layer is needed to give a sufficiently large surface area of the support in catalysis. Coating of a thin layer of oxide(s) is carried out mainly by washcoating using the sol-gel method explained earlier.

III. INCORPORATION OF CATALYTICALLY ACTIVE SPECIES

Various methods are possible to incorporate a catalytically active phase to the monolith [48–59,85–95]. Figure 3 shows the general scheme for preparing a monolithic catalyst structure from a washcoated monolith. In fact, no fundamental differences exist between incorporation of an active phase in a conventional support (beads, extrudates, spheres) and in monoliths. In practice, precautions are needed because, besides concentration profile on a particle scale, such profile over the length of the monolith also can easily arise.

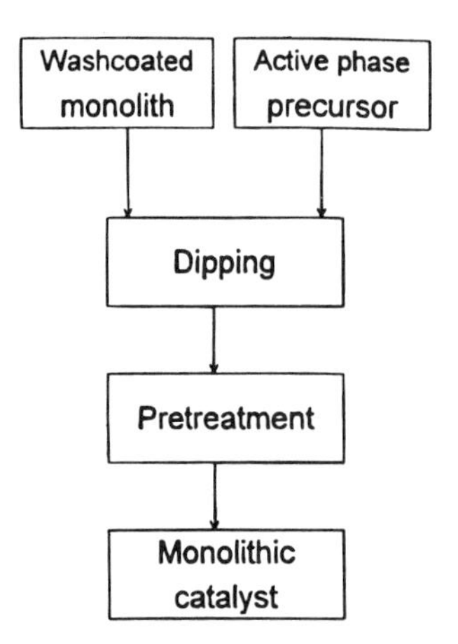

Figure 3 A general scheme for preparation of a monolithic catalyst.

A. Impregnation

Impregnation is one of the most used techniques to incorporate an active phase in a support. It can also be used to deposit active phase to a monolith [85]. Usually, a high-surface-area monolith is dried, evacuated, and dipped in a solution containing a precursor of the active phase. After drying and calcination a monolithic catalyst is obtained. Often, an activation step is necessary to convert the precursor of the active phase into the active phase, e.g., the transformation of a metal oxide in the corresponding metal or metal sulfide. Monolithic catalysts with complex compositions of active phases can be prepared by sequential impregnations with suitable solutions or with a common solution containing various precursors of the components.

This method is simple. Often, however, due to the difficulties involved in the drying step, a homogeneous distribution of the active phase is difficult to obtain. Moreover, when the salts have a low melting point, a redistribution of the active phase may occur during heat treatments, leading to an inhomogeneous distribution of the active phases, such as in the case of nickel deposition using nickel acetate or nitrate [55]. This can be circumvented by using another precursor of the active phase or by using a more sophisticated method, e.g., deposition precipitation.

B. Adsorption and Ion Exchange

Similar to the impregnation method, the adsorption and ion exchange method [84–86] is often used in preparing monolithic catalysts from washcoated monoliths. This method is similar to that of impregnation, except after dipping in a salt solution, a draining step is introduced and, as a consequence, only the species adsorbed on the monolith or ion exchanged with the surface groups remain on the (internal) surface of the washcoat layer. This implies that the amount of precursor of the active phase may be lower than that in impregnation. However, the interaction of the adsorbed or ion-exchanged species with the support is very strong. It is not self-evident that a homogeneous distribution is obtained.

In order to increase the homogeneity of the active-phase distribution, a circulating system is helpful (cf. Fig. 4).

In ion exchange, the pH should be measured. Dependent on the pH, the surface is negatively or positively charged. The pH when the surface is just neutral is called the point of zero charge (PZC). An example will show the relevance of PZC.

In the preparation of palladium on alumina-washcoated monolithic catalysts, ion exchange with a palladium complex has been applied. As palladium complex, either $PdCl_4^{2-}$ or $Pd(NH_3)_n^{2+}$ complex in a solvent can be used. Note that in the former case, a palladium anionic complex is used, and in the latter, a palladium cationic one. When the negative complex is used, the surface should be positively charged and, as a consequence, the pH should be below the PZC. The PZC of alumina is ~8 (depending on the alumina in case), so the pH should below 8. In the opposite case, $Pd(NH_3)_n^{2+}$ is used, so the pH should be higher than 8. The pH should, of course, be chosen so that the support does not dissolve. For alumina this means that the window is about $4 < pH < 13$. In this respect, applying $PdCl_4^{-2}$ leads to unsatisfactory results. When $PdCl_4^{2-}$ complex is prepared from $PdCl_2$, HCl is usually added to enhance the solubility of the salt in water. As a result, the pH of the solution is very low and alumina is partially dissolved, resulting in a low palladium distribution. On the contrary, when $Pd(NH_3)_n^{2+}$ complex solution is used, the pH is above the PZC of alumina yet not high enough to dissolve alumina, and a high dispersion of palladium is the result.

It appears that ion exchange may lead to a homogeneous distribution of the metal deposited, provided a suitable complex solution is used. Of course, the amount of active phase that can be deposited is limited by the number of ionic sites at the surface. For

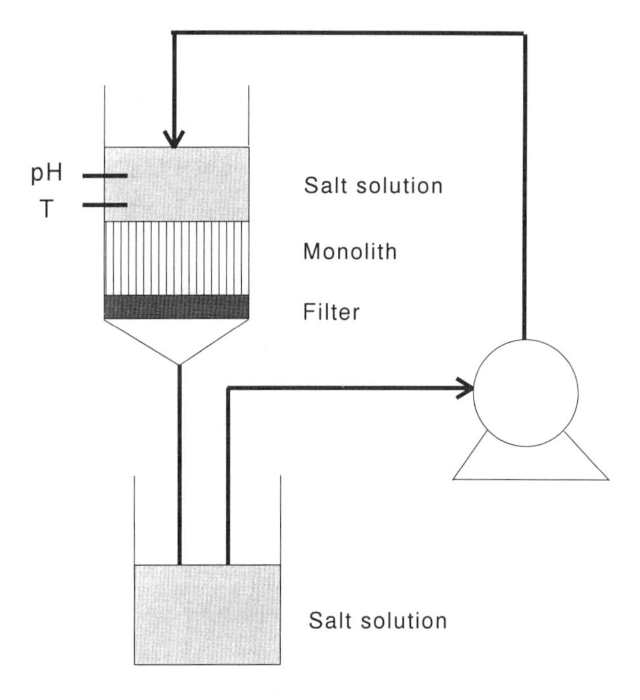

Figure 4 Scheme showing the loading of active phase by adsorption.

noble metals, which are both expensive and very active, ion exchange methods are extensively used [55,86–88].

C. Precipitation or Coprecipitation

An alternative method to deposit the oxidic layer or precursors of the active phase is precipitation or coprecipitation. This is widely used in conventional catalyst manufacture. An advantage is that a high loading of the active phase can be reached. As in monolithic reactions, catalyst loading is a point of concern. It is not surprising that precipitation methods are often applied in monolithic catalyst synthesis.

Aoki et al. [89,90] reported the deposition of iron oxide on a monolith by the precipitation method. Thus, iron-based monolithic catalysts for wastewater treatment were manufactured by immersing a monolithic ceramic or metallic support in an aqueous solution of Fe(II) salts, followed by immersing in aqueous solutions of alkali hydroxide or carbonates; precipitated particles of hydrous ferric hydroxide on the supports are formed by subsequently blowing oxygen-containing gas, washing, drying, and calcination at a high temperature. The reactions involved are given in Eqs. (12–14). Ferrous ions are precipitated as hydroxide [Eq. (12)], and with oxygen, transformation to ferric hydroxide takes place [Eq. (13)], which during calcination forms iron oxide [Eq. (14)].

$$Fe^{2+} + 2OH^- \rightarrow Fe(OH)_2 \tag{12}$$

$$2Fe(OH)_2 + {}^1\!/_2O_2 + H_2O \rightarrow 2Fe(OH)_3 \tag{13}$$

$$2Fe(OH)_3 \rightarrow Fe_2O_3 + 3H_2O \tag{14}$$

Of course, a high flexibility exists with respect to the active phase to be deposited.

The precipitation method is not only applicable to one hydroxide; in case a solution of mixed salts is used, coprecipitation takes place [54]. As in conventional coprecipitation, the solubilities of various hydroxides may not be the same, leading to a molecular ratio of the two metals in the deposited layer being different from that in the solution. Due to difficulties with respect to stirring, concentration gradients may exist that will in turn influence the homogeneity of the precipitated phase. The pore volume of the substrate may restrict the amount deposited.

As the deposited oxide layer is well mixed, strong interaction between the oxides is expected, leading often to mechanically strong materials, but pretreatment procedures can be hindered. For instance, in the preparation of a metal-based catalyst, a reduced reducibility of the precursor is often encountered, and, as a result, a reduced availability of the catalytically active phase is encountered. Moreover, under strongly acidic or basic conditions, some support materials, e.g., alumina, may be dissolved, as mentioned before. Furthermore, the adhesion of the precipitated layer with the monolith substrate is often a point of concern, especially during drying and heat treatment.

D. Deposition Precipitation

Sometimes, due to a low melting point of a salt, e.g., when a nitrate is chosen as the precursor of the active phase, homogeneous distribution is difficult to obtain. Then, the deposition precipitation method might be appreciated [91–93]. This method is illustrated for the synthesis of a nickel monolithic catalyst.

A dry monolith is dipped in an aqueous solution containing both a nickel salt and urea at suitable concentrations. Upon heating of this solution (above 363 K), urea decomposes:

$$CO(NH_2)_2 + 3H_2O \rightarrow CO_2 + 2NH_4^+ + 2OH^- \tag{15}$$

Due to the reaction, the pH increases and the hydroxide is deposited:

$$Ni(OH)^+ + AlOH + OH^- \rightarrow Al{-}O{-}Ni(OH) + H_2O \tag{16}$$

An attractive point is that throughout the porous structure the pH is the same, leading to a high homogeneity. Moreover, due to a much higher melting point of nickel hydroxide, compared to those of the nickel nitrate and acetate, redistribution of nickel species during calcination is prevented [55]. A highly homogeneous distribution of nickel species can be achieved.

The hydrolysis of urea is a slow step [93]; a relatively long heating period (5–20 hr) is necessary to precipitate the nickel ions completely. Compared to impregnation, the amount of nickel deposited by deposition precipitation is relatively low. Nevertheless, due to a more homogeneous distribution of the active phase or its precursor, this method is often preferred [55,91,93,94]. In a recent optimization study, it was found that a higher nickel or urea concentration in the solution increases the amount of nickel deposited and shortens the time needed for the deposition. Nickel loadings up to 10.5 wt% were realized in a time frame of 5 hr [69,91]. This method is, of course, not restricted to nickel deposition; other species, e.g., manganese, can be deposited similarly [93].

E. Sol-Gel Method

By the sol-gel method, an oxidic layer together with the precursor of the active phase sometimes can be deposited simultaneously [58,59,70]. The resulting monolithic catalyst will contain both the washcoated oxide layer and the precursor of the active phase.

An example is washcoating of platinum and silica using a sol containing both a silica and a platinum precursor [58,59]. After calcination a catalyst containing platinum on silica-washcoated monolith is obtained. Since the platinum precursor is homogeneously mixed with silica, encapsulation of platinum may occur, which reduces the effectiveness of platinum as compared to that of a two-step preparation. In this respect, the method may be more suitable for less expensive active phases.

Alternatively, a sol can be used as a binder to glue powder of active phases to the surface of a monolith. For example, a Si-sol mixed with zeolite crystallites was used to make a zeolite on monolith catalyst [16].

F. Slurry Dip-Coating

Coating of a porous oxidic layer or incorporation of an active phase can be performed by slurry dip-coating [1,43,54,88,94,95]. Various active phases or porous supports, e.g., SiO_2, Cr_2O_3, ZSM-5, $LaCrO_3$, and other perovskites, have been deposited on metallic monoliths using this method. This method is successfully used to prepare commercial palladium-only monolithic three way catalysts on a cordierite substrate [95]. For example, a SiO_2 of a suitable size with known textural properties was mixed with water and vigorously stirred for 2 hr. Afterwards it was ball-milled overnight. This slurry was used in dip-coating of silica on a metallic monolith [94].

The technique is quite suitable for manufacturing bimodal pore systems. For example, a ZSM-5/mesoporous silica layer was deposited as follows: porous silica powder with a suitable-size fraction (<40 μm) and ZSM-5 powder (<63 μm) were added slowly to a mixture consisting of potassium water glass and colloidal Si-sol under stirring. The mixture was kept at room temperature under vigorous stirring during at least 30 min. A preoxidized metallic monolith was dipped in this slurry and withdrawn at a certain speed. Afterwards, the slurry present in the channels of the monolith was removed by centrifuging or air-blowing. Subsequently, the product was dried at 348 K and calcined at 773–873 K. In order to prevent sudden gelation, ammonia was added to maintain a pH greater than 5 in the slurry. A porous coating layer with ZSM-5 and a dual pore-size distribution were the result.

In application of the method, care should be taken not to change the properties of the active phase (or its precursor) in the process of slurry dip-coating, and attention should be paid to the adhesion of the coated layer. Encapsulation of the active phase may occur, which reduces the availability of the active phase. An attractive feature of slurry dip-coating is that it enables depositing of a synthesized active phase or a support material with known properties on a monolith surface. When the active phase or its precursor is difficult to synthesize or is prepared under conditions that may damage the support (e.g., perovskite) [88,94], the method is very useful.

G. In Situ Crystallization

In the preparation of zeolite membranes, in situ crystallization in an autoclave is used to deposit a zeolite layer on a macroporous metallic membrane substrate. For example, Geus [96] reported the preparation of ZSM-5 zeolite on a macroporous metal substrate, which can be used in separations at high temperatures.

In principle, the same method can be applied to a monolith substrate to obtain a zeolite-on-monolith catalyst. Of course, this method is not limited to zeolite as the active phase. It is not difficult to envisage that many other active phases can be deposited in this way.

H. Addition of Catalytic Species to the Mixture for Extrusion

It is also possible to incorporate the active phase into the moldable mixture used in the extrusion process to manufacture the monoliths [14,97]. At first sight, this might seem the most convenient method. However, extrusion is a difficult technique, in which the rheological properties of the extrusion mixture are critical [97]. Nevertheless, promising results have been obtained [1,4 and the literature therein; 14,25,26,97,98].

Isopova et al. described the preparation of various perovskite-based monolithic catalysts for fuel combustion by extrusion of synthesized perovskite powders [14]. Blanco et al. [25], Lachman and Williams [97], and del Valle et al. [98] reported titania-based and other monolithic catalysts by extrusion. The titania catalysts were tested in a coal-fired power pilot plant for electrostatic separation of fly-ash [98]. Lyakhova et al. studied the WO_3-doped titania–vanadia monolithic catalysts for selective catalytic reduction (SCR: NO_x conversion) by extrusion [26]. The rheological properties of the paste for extrusion and the effect of various organic plastisizers on catalytic activity in SCR were discussed.

Of course, care should be taken to avoid encapsulation of active species in the substrate. Moreover, during the heat treatment of the extruded greenware, the active phase may be sintered. For example, when a γ-Al_2O_3 structure is aimed at, sintering will probably destroy the micro- and mesoporosity. The method is not recommended when an expensive active phase is used.

I. Other Coating Techniques

Nowadays, many advanced techniques are available in the ceramic industry to coat a solid layer onto a solid surface or to make ceramic materials with special properties [99–115], such as spin-coating [99], chemical vapor deposition [100–106], and chemical vapor infiltration [106–109], thermal spray [110–112], plastic spray [113], and spray-coating [114]. The deposition can be caused by conventional heating, by laser beam, or by microwave heating.

Some of these techniques may be applied to coat a solid layer on monoliths, provided suitable reactions and appropriate reaction conditions are chosen. Moene et al. [106,107] described the coating of a SiC layer on a microporous active carbon surface using the CVD or CVI techniques. Their results can easily be translated to monolithic catalyst support synthesis.

Microwave Heating. Since many ceramics are isolators, during manufacture of catalytic monoliths, isothermal conditions are difficult to maintain in all the channels of a monolith. In this respect, it appears that microwave heating might be quite promising [91,115]. Heating by microwave takes place simultaneously in all the channels of the monolith. Moreover, due to high polarity, many reactants are heated more efficiently by microwave, compared to ceramics, viz., the reactants are selectively heated. In this way overheating of the ceramic support can be prevented. This can be advantageous, e.g., when a porous structure is the designed product.

IV. CONCLUDING REMARKS

A number of methods can be used to prepare monoliths. Extrusion is widely used for the manufacture of ceramic monoliths, whereas corrugation is used mostly for those of metallic monoliths. As to the techniques used to coat a thin layer of oxide(s) on a monolith, the sol-gel method is used the most. Other methods, e.g., surface oxidation, slurry dip-coating, or CVD and CVI techniques, can also be used.

For the deposition of active phase(s), impregnation, adsorption and ion exchange, (co)precipitation, deposition precipitation, and in situ crystallization methods can be used. Moreover, it is possible to mix the active phase in the mixture for extrusion or to deposit the active phase by using a mixed sol containing both the precursors of the oxidic species and the active phase or a slurry with the precursor powder of the active phase. Other coating techniques, e.g., CVD or CVI techniques, can also be used. The dispersion of the active phase depends strongly on the method and conditions used, its precursors form as well as on the history of the active phase.

In some preparation steps of monolithic catalysts, microwave heating appears to be promising. This heating is faster and more selective than the conventional ones. Moreover, it may lead to a more homogeneous distribution of the active phase in the catalysts.

A wide variety of methods has been reported for preparing monolithic catalysts. In fact, the methods stem from preparation methods developed in heterogeneous catalysts and methods for the manufacture of thin-layer materials. The latter are developed in the fields of material science and are directly focused on product development, such as the production of integrated circuits and composites. It is expected that monolithic catalyst manufacture will benefit from development in these areas.

REFERENCES

1. J.P. DeLuca and L.E. Campbell. In: J.J. Burton and R.L. Garten, eds., *Advanced Materials in Catalysis*. Academic Press, London, 1977, p. 293.
2. I.M. Lachman and R.N. McNally. *Chem. Eng. Prog. Jan.*:29 (1985).
3. A.B. Stiles. *Catalyst Supports and Supported Catalysts. Theoretical and Applied Concepts.* Butterworths, Boston, 1987, p. 6.
4. S. Irandoust and B. Andersson. *Catal. Rev.-Sci. Eng. 30(3)*:341 (1988).
5. A. Cybulski and J.A. Moulijn. *Catal. Rev.-Sci. Eng. 36(3)*:179 (1994).
6. G.A. Merkel and M.J. Murtagh. *E.P. 545008* (1993).
7. M.J. Murtagh. *U.S. 5141686* (1992).
8. D.M. Beall, E.M. DeLiso, D.L. Guile, and M.J. Murtagh. *U.S. 5114644* (1992).
9. Y. Kasai, K. Kumazawa, T. Hamanaka, and T. Itoh. *E.P. 506301* (1992).
10. G.M. Bustamante. *Braz. Pedido PI BR8906554* (1991).
11. H. Kanazawa, Y. Fujimoto, and Y. Ogura. *J.P. 03242213* (1991).
12. G.D. Forsoythe. *E.P. 455451* (1991).
13. T. Ito, K. Kumazawa, T. Hamanaka, and Y. Kasai. *E.P. 514205* (1992).
14. L.A. Isupova, V.A. Sadykov, L.P. Solovyova, M.P. Andrianova, V.P. Ivanov, G.N. Kryukova, V.N. Kolomiichuk, E.G. Avvakumov, I.A. Pauli, O.V. Andryushkova, V.A. Poluboyarov, A. Ya. Rozovskii, and V.F. Tretyakov. In: G. Poncelet, J. Martens, B. Delmon, P.A. Jacobs, and P. Grange, eds. *Preparation of Catalysts VI*. Elsevier, Amsterdam, 1995, p. 637.
15. H. Yamanchi and Y. Ohashi. *J.P. 03271151* (1991).
16. H. Yamanchi and Y. Ohashi. *J.P. 03271152* (1991).
17. K. Rennebecks. *Ger. Offen. DE 4033227* (1991).
18. K. Arai. *J.P. 03242227* (1991).
19. T. Kusuda and M. Yonemura. *J.P. 03258347* (1991).
20. D. Scholl, J.P. Gabachuler, and K.C. Eckert. *Patentschrift (Switz.) CH 679394* (1992).
21. R. Ostertag, T. Hang, R. Renz, and W. Zankl. *E.P. 477505* (1992).
22. H. Tsunola, M. Araoka, and T. Kobayashi. *J.P. 0436645* (1992).
23. Y. Murano, K. Hasegawa, S. Wade, Y. Ikeda, M. Ogawa, K. Sasaki, and H. Tagi. *J.P. 04357179* (1992).
24. E. Przeradzki. *Pol. PL 154356* (1991).
25. J. Blanco, P. Avila, M. Yates, and A. Bahamonde. In: G. Poncelet, J. Martens, B. Delmon, P.A. Jacobs, and P. Grange, eds. *Preparation of Catalysts VI*. Elsevier, Amsterdam, 1995, p. 755.
26. V. Lyakhova, G. Barannyk, and Z. Ismagilov. In: G. Poncelet, J. Martens, B. Delmon, P.A. Jacobs, and P. Grange, eds. *Preparation of Catalysts VI*. Elsevier, Amsterdam, 1995, p. 775.
27. O. Shinozaki, E. Shinoyama, and K. Saito. *Soc. Automot. Eng., SP-816*:1 (1990).
28. H. Katsuki, A. Kawahara, H. Ichinose, S. Furuta, and H. Nakao. *J.P. 0465372* (1992).
29. Kirk-Othmev. *Encyclopedia of Chemical Technology 23*, 3rd ed. Wiley, New York, 1983, p. 523.
30. Y. Matsuo, T. Takenchi, and M. Tamura. *J.P. 0450157* (1992).
31. G.J. Piderit, P.F. Toro, and E. Cordova. An.-Congr. Bras. Ceram. III Iberroam. Ceram., Vidrios Refract., 35th, Assoc. Bras. Ceram. 2, 1991, p. 631 (Span).

32. J.H. Han and A.G. Hegedus. *PCT Int. Appl. WO 9116277* (1991).
33. T. Kuma. *PCT Int. Appl. WO 9116971* (1991).
34. H. Shino and A. Nara. *J.P. 0475278* (1992).
35. D. Myamoto, M. Oota, K. Ishikawa, Y. Suketa, and T. Kaji. *J.P. 0523529* (1993).
36. M. Nishizawa and T. Yamada. *J.P. 03245851* (1991).
37. F. Kamimura. *J.P. 04166234* (1992).
38. T. Matsumoto. *J.P. 0478447* (1992).
39. M. Fukaya, K. Omura, and M. Yamanaka. *J.P. 04156945* (1992).
40. I. Yamaguchi and M. Murotani. *J.P. 03285079* (1991).
41. N. Ito. *J.P. 04260446* (1992).
42. S. Matsuda, K. Nehashi, K. Isemura, H. Endo, and H. Kobayashi. *J.P. 04308075* (1992).
43. M.F.M. Zwinkels, S.G. Järås, and P.G. Menon. In: G. Poncelet, J. Martens, B. Delmon, P.A. Jacobs, and P. Grange, eds. *Preparation of Catalysts VI*. Elsevier, Amsterdam, 1995, p. 85.
44. C.M.T. Tsang, and R.E. Bedford. *U.S. 5114901* (1992).
45. M.V. Twigg and J.T. Richardson. In: G. Poncelet, J. Martens, B. Delmon, P.A. Jacobs, and P. Grange, eds. *Preparation of Catalysts VI*. Elsevier, Amsterdam, 1995, p. 345.
46. D. Scholl, J.P. Gabathuler, K.L. Eckert, and T. Mizrah. *CH680788 (Switz)* (1992).
47. R.L. Goldsmith and B.A. Bishop. *U.S. 5114581* (1992).
48. M. Nishizawa and T. Yamada. *J.P. 03245851* (1991).
49. C.J. Pereira and K.W. Plumlee. *Catal. Today 13*:23 (1992).
50. A. Talo, J. Lahtinen, and P. Hautojärvi. *Appl. Catal., B: Environmental 5*:221 (1995).
51. T. Igarashi and T. Otani. *J.P. 0499184* (1992).
52. J.P.A. Neeft. *Catalytic Oxidation of Soot: Potential for the Reduction of Diesel Particulate Emission*. Ph.D. thesis, Delft University of Technology, 1995, The Netherlands.
53. J.-F. Quinson, C. Chino, A.-M. de Becdelievre, and C. Guizard. In: A.K. Cheetham, C.J. Brinker, M.L. Mecartney, and C. Sanchez, eds. *Better Ceramic Through Chemistry VI*. Material Research Soc., Pittsburgh, 1994, p. 703.
54. M.F.M. Zwinkels, S.G. Järås, and P.G. Menon, *Catal. Rev.-Sci. Eng. 35(3)*:319–358 (1993).
55. Xiaoding Xu, H. Vonk, A. Cybulski, and J.A. Moulijn. In: G. Poncelet, J. Martens, B. Delmon, P.A. Jacobs, and P. Grange, eds. *Preparation of Catalysts VI*. Elsevier, Amsterdam, 1995, p. 1069.
56. W.B. Kolb, A.A. Papadimitriou, R.L. Cerro, D.D. Leavitt, and J.C. Summers. *Chem. Eng. Prog. Feb.*:61 (1993).
57. Y. Mizushima and M. Hori. *J.P. 0543344* (1993).
58. R.D. Gonzalez. Proc. AIChE 1992 Annual Meeting, paper 46f, 1992, Miami Beach.
59. K. Balakrishnan and R.D. Gonzalez. Proc. AIChE 1992 Annual Meeting, paper 49a, 1992, Miami Beach.
60. C.J. Brinker and G.W. Scherer. *Sol-Gel Science: The Physics and Chemistry of Sol-Gel Processing*. Academic Press, Boston, 1990.
61. R.K. Iler. *The Chemistry of Silica*. Wiley, New York, 1979.
62. R.W. Chorley and P.W. Lednor. *Adv. Mater. 3(10)*:474 (1991).
63. I.S. Sakka and T. Yoko. *Ceram. Int. 17(4)*:217 (1991).
64. I.S. Sakka and T. Yoko. *Ceramurgia 21(1)*:24 (1991) (Ital).
65. Y. Mizushima and M. Hori. *J.P. 0406180* (1992).
66. L. Spiccia, B.O. West, J. Cullen, D. De Villiers, I. Watkins, J.M. Bell, B. Ben-Nissan, M. Anast, and G. Johnston. *Key Eng. Mater. 53–55*:445 (1991).
67. T. Akiba. *Shinsozai 3(4)*:45 (1992) (Japan).
68. L.C. Klein, C. Yu, R. Woodman, and R. Pavilik. *Catal. Today 14(2)*:165 (1992).

69. H. Vonk. *Preparation and Characterization of Monolithic Catalysts.* Masters thesis, Delft University of Technology, 1994, The Netherlands.
70. V.G. Kessler, E.P. Turevskaya, S.I. Kucheiko, N.I. Kozlova, E.P. Turevskaya, I.E. Obvintseva, and M.I. Yanovskaya. In: A.K. Cheetham, C.J. Brinker, M.L. Mecartney, and C. Sanchez, eds. *Better Ceramic Through Chemistry VI.* Material Research Soc., Pittsburgh, 1994, p. 3.
71. S. Acosta, P. Arnal, R.J.P. Corriu, D. Leclercq, P.H. Mutin, and A. Vioux. In: A.K. Cheetham, C.J. Brinker, M.L. Mecartney, and C. Sanchez, eds. *Better Ceramic Through Chemistry VI.* Material Research Soc., Pittsburgh, 1994, p. 43.
72. D. Collina, G. Fornasari, A. Rinaldo, F. Trifiro, G. Leofanti, G. Paparatto, and G. Petrini. In: G. Poncelet, J. Martens, B. Delmon, P.A. Jacobs, and P. Grange, eds. *Preparation of Catalysts VI.* Elsevier, Amsterdam, 1995, p. 401.
73. Shengcheng Luo, Linlin Gui, Xiaozhi Fu, and Youqi Tang. In: A.K. Cheetham, C.J. Brinker, M.L. Mecartney, and C. Sanchez, eds. *Better Ceramic Through Chemistry VI.* Material Research Soc., Pittsburgh, 1994, p. 445.
74. L. Chu, M.I. Tejedor-Tejedor, and M.A. Anderson. In: A.K. Cheetham, C.J. Brinker, M.L. Mecartney, and C. Sanchez, eds. *Better Ceramic Through Chemistry VI.* Material Research Soc., Pittsburgh, 1994, p. 855.
75. M.N. Logan, S. Prabakar, and C.J. Brinker. (A.K. Cheetham, C.J. Brinker, M.L. Mecartney, and C. Sanchez, eds.) *Better Ceramic Through Chemistry VI.* Material Research Soc., Pittsburgh, 1994, p. 115.
76. H.C. Zeng. In: A.K. Cheetham, C.J. Brinker, M.L. Mecartney, and C. Sanchez, eds. *Better Ceramic Through Chemistry VI* Material Research Soc., Pittsburgh, 1994, p. 715.
77. M. Marella, M. Tomaselli, L. Meregalli, M. Battagliarin, P. Gerontopoulos, F. Pinna, M. Signoretto, and G. Strukul. In: G. Poncelet, J. Martens, B. Delmon, P.A. Jacobs, and P. Grange, eds. *Preparation of Catalysts VI.* Elsevier, Amsterdam, 1995, p. 327.
78. M.A. Anderson and Xu Qunyin. *PCT Int. Appl. WO9219369* (1992).
79. Y. Abe and H. Hosono. *J.P. 0426573* (1992).
80. S. Bernal, J.J. Calvino, M.A. Cauqui, J.M. Rodriguez-Izquierdo, and H. Vidal. In: G. Poncelet, J. Martens, B. Delmon, P.A. Jacobs, and P. Grange, eds. *Preparation of Catalysts VI.* Elsevier, Amsterdam, 1995, p. 461.
81. Y. Sun and P.A. Sermon. In: G. Poncelet, J. Martens, B. Delmon, P.A. Jacobs, and P. Grange, eds. *Preparation of Catalysts VI.* Elsevier, Amsterdam, 1995, p. 471.
82. M.H. Han and K.C. Park. *Yop Hakhoechi 27(5)*:625 (1990) (Korean).
83. J.J. Calvino, M.A. Cauqui, J.M. Gatica, J.A. Perez, and J.M. Rodriguez-Izquierdo. In: A.K. Cheetham, C.J. Brinker, M.L. Mecartney, and C. Sanchez, eds. *Better Ceramic Through Chemistry VI.* Material Research Soc., Pittsburgh, 1994, p. 685.
84. M.I. Yanovskaya, N.M. Kotova, I.E. Obvintseva, E.P. Turevskaya, E.P. Turevskaya, K.A. Vorotilov, L.I. Solov'yova, and E.P. Kovsman. In: A.K. Cheetham, C.J. Brinker, M.L. Mecartney, and C. Sanchez, eds. *Better Ceramic Through Chemistry VI.* Material Research Soc., Pittsburgh, 1994, p. 15.
85. S.J. Schmieg and D.N. Belton. *Appl. Catal., B: Environmental 6*:127 (1995).
86. L. Evaldsson, L. Löwendahl, and J.-E. Otterstedt. *Appl. Catal. 55*:123 (1989).
87. Xiuren Zhao and Junhang Jing. In: G. Poncelet, J. Martens, B. Delmon, P.A. Jacobs, and P. Grange, eds. *Preparation of Catalysts VI.* Elsevier, Amsterdam, 1995, p. 949.
88. P.-Y. Lin, M. Skoglundh, L. Löwendahl, J.-E. Otterstedt, L. Dahl, K. Janssen, and M. Nygren. *Appl. Catal., B: Environmental 6*:237 (1995).
89. I. Aoki, T. Matsui, T. Imai, and N. Horiishi. *J.P. 05177138* (1993).
90. I. Aoki, T. Matsui, T. Imai, and N. Horiishi. *J.P. 05177139* (1993).
91. Xu Xiaoding, H. Vonk, A.C.J.M. van de Riet, A. Cybulski, A. Stankiewicz, and J.A. Moulijn. *Catal. Today 30*:91 (1996).

92. L.M. Knijff, R.H. Bolt, R. van Yperen, A.J. van Dillen, and J.W. Geus. In: B. Delmon, P. Grange, P. Jacobs, and G. Poncelet, eds. *Scientific Bases for the Preparation of Heterogeneous Catalysts*. Elsevier, Amsterdam, 1991, p. 166.
93. K.P. de Jong. In: B. Delmon, P. Grange, P. Jacobs, and G. Poncelet, eds. *Scientific Bases for the Preparation of Heterogeneous Catalysts*. Elsevier, Amsterdam, 1991, p. 19.
94. M. Zwinkels. *High-Temperature Catalytic Combustion*. Ph.D. thesis, Royal Institute of Technology, Stockholm, 1996.
95. Z. Hu, C.Z. Wan, Y.K. Lui, J. Dettling, and J.J. Steger, *Catal. Today 30*:83 (1996).
96. E.R. Geus. *Inorganic Zeolite membranes*. Ph.D. thesis, Delft University of Technology, 1993, The Netherlands.
97. I.M. Lachman and J.L. Williams. *Catal. Today 14*:317 (1992).
98. J.O. del Valle, L.S. Matrinez, B.M. Baum, and V.C. Galeano. *Proc. of the 2nd. Int. Conf. on "Catalysis and adsorption in fuel processing and environmental protection,"* Sep. 18–21, 1996, Szklarska Poreba, Poland, pp. 43–55.
99. R. Ochoa and R. Miranda. In: A.K. Cheetham, C.J. Brinker, M.L. Mecartney, and C. Sanchez, eds. *Better Ceramic Through Chemistry VI*. Material Research Soc., Pittsburgh, 1994, p. 553.
100. M.A. Petrich. *CHEMTECH Dec.*:740 (1989).
101. J. Chin. In: K.S. Mazdiyasni, ed. *Fiber Reinf. Ceram. Compos.* 1990, p. 342.
102. A.W. Moore. *Mater. Res. Soc. Symp. Proc., 1992, 250* (Chemical Vapor Deposition of Refractory Metals and Ceramics II):269 (1992).
103. S. Köhler, M. Reiche, C. Frobel, and M. Baerns. In: G. Poncelet, J. Martens, B. Delmon, P.A. Jacobs, and P. Grange, eds. *Preparation of Catalysts VI*. Elsevier, Amsterdam, 1995, p. 1009.
104. Y.S. Lin and A.J. Burggraaf. *Chem. Eng. Sci. 46(12)*:3067 (1991).
105. R.K. Shiba, V.I. Srdanov, M. Hay, and H. Eckert. In: A.K. Cheetham, C.J. Brinker, M.L. Mecartney, C. Sanchez, eds. *Better Ceramic Through Chemistry VI*. Material Research Soc., Pittsburgh, 1994, p. 587.
106. R. Moene. *Application of Chemical Vapor Deposition in Catalyst Design*. Ph.D. thesis, Delft University of Technology, 1995, The Netherlands.
107. R. Moene, L.F. Kramer, J. Schoonman, M. Makkee, and J.A. Moulijn. In: G. Poncelet, J. Martens, B. Delmon, P.A. Jacobs, and P. Grange, eds. *Preparation of Catalysts VI*. Elsevier, Amsterdam, 1995, p. 371.
108. R. Naslain. In: R. Warren, ed. *Ceram.—Matrix. Compos.* 1992, p. 199.
109. J.W. Evans and D. Gupta. *Mater. Res. Soc. Symp. Proc., 1991, 189*, 1991, pp. 101–107.
110. T.L. Schohest. *IEEE Trans. Plasma Sci. 19(5)*:725 (1991).
111. Y. Harada. *Nippon Kinzoku Gakkai Kaiho 31(5)*:413 (1992) (Japan).
112. M. Kurita and K. Toyama. *J.P. 04255569* (1992).
113. O. Kubo and M. Kawai. *Jidosha Gijutsu 47(5)*:70 (1993) (Japan).
114. S. Kagawa. *J.P. 0419541* (1992).
115. A.C. Metaxas and J.G.P. Binner. In: J.G.P. Binner, ed. *Adv. Ceram. Process. Technol.* 1990, p. 285.

22

Computer-Aided Characterization and Design of Catalyst Pore Structure

Reginald Mann
University of Manchester, Institute of Science and Technology (UMIST), Manchester, England

I. INTRODUCTION

Catalytic reactions take place by individual molecular interactions at active sites. These sites, specifically favorable to reaction, exist only at the interface between solid catalyst and some fluid phase that conveys reagent species and exports products of catalytic reaction. Usually, to achieve practical reaction rates, any catalytic agent should be deployed over a large surface area. This is facilitated by dispersing the catalyst within some support solid. Other components, like binders and promoters, may also be incorporated to enhance reactivity.

A good catalyst is then one that has an optimized assembly of the catalyst, the support, and any promoters. The resultant geometry and architecture of the porosity must give good accessibility, large surface area, and local surface properties that induce high reactivity, good chemical selectivity, and resistance to deactivation.

The requirement for surface areas on the order of hundreds of square meters per gram means the porosity must be composed mainly of nanometer (10^{-9}) scale pore voids (or pores). Catalyst particles, even at a few microns' scale, will probably contain more than 10^4 such pore elements. Particles at millimeter and centimeter scale will have pore numbers exceeding 10^7. The fabrication of optimized pore geometries with well-configured components demands a degree of micromanipulation and control that lie in the realms of nanotechnology but that are far from present practicalities of catalyst preparation.

Indeed it has been said that catalysts are not so much designed as thrown together. Processing techniques of co-precitation and solution impregnation, the washcoating of porous monoliths, as well as the fabrication of porous membranes all tend to furnish pore spaces that are typically a mass of randomly tortuous interconnected pore voids. The catalytic performance of such empirically random porous solids is readily measurable. However, the empiricism of the pore structure conceals the fundamentals of diffusion, catalytic reaction, and any induced or forced permeation that together govern the catalytic performance at an overall macroscopic scale.

Furthermore, the deployment of catalytic particles inside a reactor requires an appreciation of the integration of the microscopic intraparticle processes with the macroscale extraparticle reaction flows. Many fixed-bed reactors comprise a randomized mass of particles, with consequently randomized extraparticle voids. The overall reactor behavior then ensues from the interactions between randomly porous particles and random reactor voids. From this seemingly intractable randomness, present from a scale of nanometers to meters, has begun to emerge the notion that perhaps, in the face of such empirical ignorance, some suitably intelligent combination of voids and surfaces in a structured catalyst coupled with an appropriately structured reactor should produce improved performance. Evidence is accumulating that intraparticle randomness is suboptimal, and it seems likely that the randomly packed bed is similarly less than ideal—hence the appearance of this volume on structured catalysts and reactors.

In a recent wide-ranging survey, Reyes and Iglesia [1] have described the use of pore structural models based on Bethe lattices [2] and aggregates of spheres [3]. Bethe lattices, which are based on the repeated branching of pore segments, have the benefit of providing analytical expressions for transport but suffer the disadvantage of a physically quite unrealistic interconnection topology. Aggregates of spheres, when randomly assembled, provide probably the most realistic physical pore structure model. However, even when the spheres forming the aggregate are all of the same size, the pore space geometries are so individually complicated and varied that the tractability of diffusional transport calculations is severely compromised.

Current pore structure models all have some inadequacy or other. In this chapter, treatments will be put forward that use stochastic networks of simple cylindrical pores [4–6]. Such networks confer the benefit of computational tractability at the expense of some shortcomings in representing the physical reality of random pore spaces within typical catalyst particles. A quantitative understanding of the consequences of randomness at a microscopic scale permits prediction of the impact of structural reordering. Resulting changes and improvements in catalytic performance should provide a new impetus for practical implementation of more precisely engineered pore structure designs.

A. Nature of Porosity

Some of the arguments and issues can be examined with reference to Fig. 1. This shows an SEM view at ×500 magnification of the porous interior of a ring-shaped particle of a large-pore reforming catalyst [7]. The qualitative nature of the porosity is clear in Fig. 1, and some quantitative features can be extracted by simple inspection. In this figure, pore spaces show as dark voids penetrating into the alumina support. Their sizes appear to be predominantly in the range 5–15 μm, although the existence of many smaller pores in the submicron range can be discerned. The overall porosity in Fig. 1 is 24%. The pore voids within the particle evidently comprise a jumbled tangle. Related to porosity are the properties of pore volume and surface area within which and upon which intraparticle processes take place.

In terms of catalysis, important equilibrium processes include low-temperature gas adsorption (capillary condensation) and nonwetting fluid invasion, both of which are routinely used to characterize pore size distribution. Static diffusion in a Wicke–Kallenbach cell characterizes effective diffusivity. The simultaneous rate processes of diffusion and reaction determine catalyst effectiveness, which is the single most significant measure of practical catalytic reactor performance.

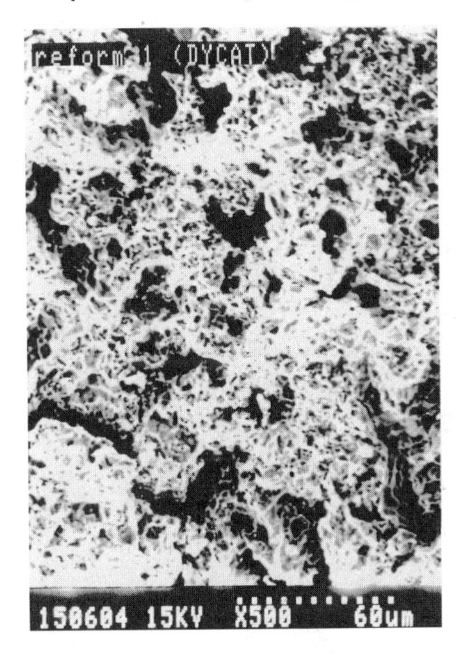

Figure 1 SEM view (×500) of porous catalyst interior. (From Ref. 7.)

B. Simple 1-D Parallel-Bundle Models

Simple single pores as uniform diameter cylinders date back some 50 years [8]. As indicated in Fig. 2, a simple diffusion reaction balance describes surface catalysis in such pores. This analysis is the basis of almost all textbook treatments [9]. The approximation of typically complex labyrinthal pore spaces by such a simplified model is certain to introduce inadequacies into process simulation.

One of the simplest ways of extending this single-pore model is to assume that variations in the size of pore spaces can be represented by a variable-diameter assembly of such pores, referred to as a parallel bundle of pores. An example is shown in Fig. 3, for a variation of the model applied to a supported zeolite cracking catalyst. In this example [10] the zeolite pores are simply configured along the pore walls, so that the parallel bundle represents the Si/Alumina-support pore spaces.

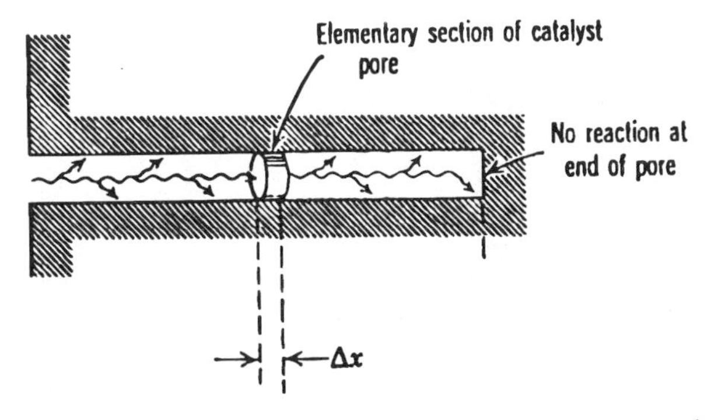

Figure 2 Simple-cylindrical-pore model for a catalyst. (From Ref. 9.)

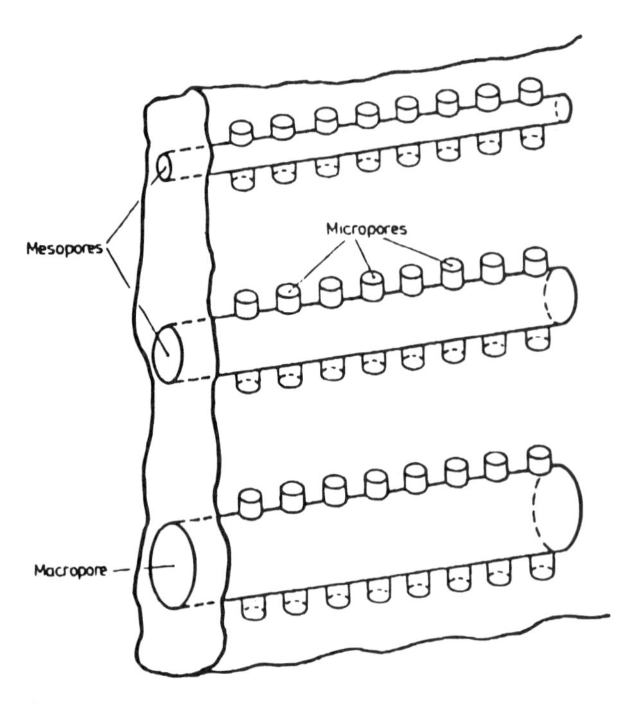

Figure 3 Parallel-bundle model for a supported zeolite FCC catalyst. (From Ref. 12.)

C. Inadequacies of 1-D Models

However, the greatest deficiency with this parallel-bundle approach is that it cannot account for the interconnectedness amongst pores, which is intuitively expected to affect the diffusion/reaction process in the pore spaces.

Some authors have purported to incorporate interconnectivity by randomly overlapping parallel bundles in 2 D. This configuration is depicted in Fig. 4. However, the random interconnectedness in this case is more apparent than real, because this arrangement leaves pores of all sizes just as accessible as they would have been in a nonintersecting parallel bundle (as in Fig. 3). As Fig. 1 has already suggested, random interconnectedness amongst pores would be expected to result in large and small pores being intermingled, an effect not produced in Fig. 4. However, this figure does have the merit of rendering the parallel bundle more diffusively isotropic in 2 D. This contrasts with the obvious diffusive anisotropy of the more conventional parallel bundle in Fig. 3.

II. RANDOM PORE NETWORKS IN 2 D

A. Constructing Simple 2-D Examples

If a large number of open-ended cylindrical pore segments (like the one in Fig. 2) are interconnected such that the diameter of any pore is independent of the size of neighbor pores, a so-called randomized, or stochastic, pore network is formed. Such a set can be assembled from a cohort obeying any stipulated pore diameter distribution function. If all the pore segments are of equal length with a connectivity of 4, a square network

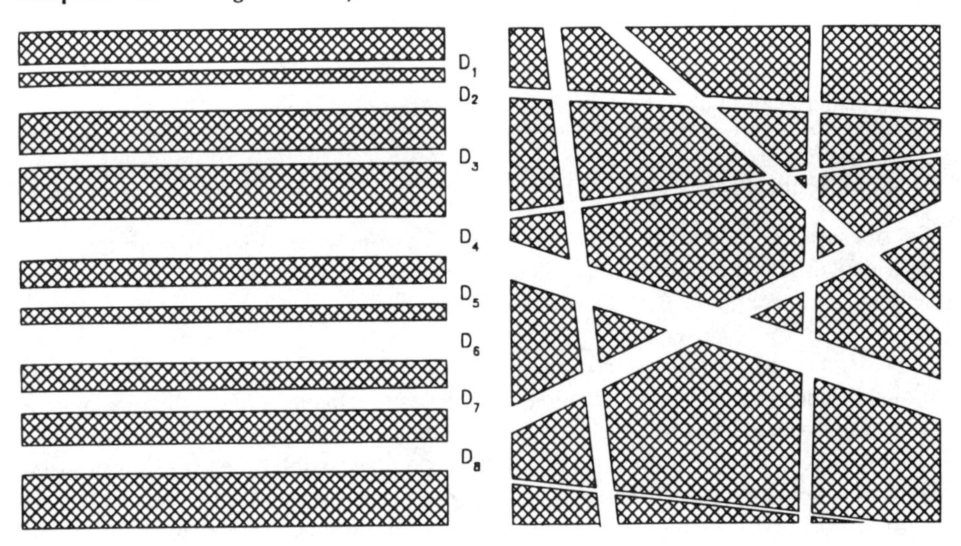

Figure 4 Randomly overlapping parallel-bundle model (in 2 D).

results. Such networks obviously allow for the random juxtaposition of large and small pores. As a consequence, many larger pores will be hidden amongst and shielded by small pores. This has an effect on diffusive accessibility, since the potentially greater reactivity of large pores (with large surface areas) is restricted by the bottleneck effect of smaller neighbor pores.

The qualitative nature of the variable porosity of pore size is best appreciated by computer-generated pictorializations. An example is shown in Fig. 5, which conveys the impression of large pores hidden amongst smaller ones. The pores in this illustrative 30×30 network obey a uniform distribution between 10 Å and 2000 Å and comprise a

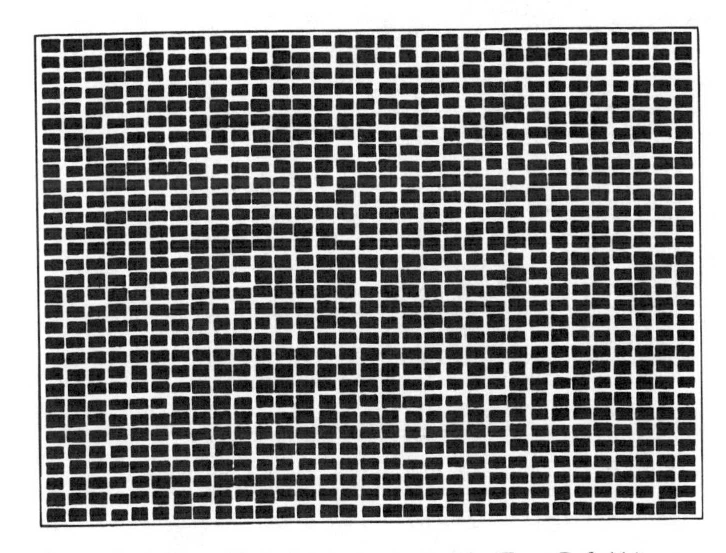

Figure 5 A 30×30 stochastic pore network. (From Ref. 11.)

total of 1860 cylindrical pore segments (shown in section). Once such "virtual" networks have been realized by computer generation of random numbers, computer "experiments" can be undertaken to analyze processes taking place within them.

B. Equilibrium Processes in 2-D Networks

The stochastic pore network visualized in Fig. 5 can now be examined with respect to the widely used characterization techniques of mercury porosimetry and low-temperature capillary gas adsorption. Both these equilibrium processes can be computed on a pore-by-pore basis.

Mercury porosimetry is governed in each pore by an equilibrium force/surface tension balance (the Washburn equation) that relates the diameter of a cylindrical pore to the pressure needed to force mercury into it. The pressured step-by-step invasion of a pore network is then controlled by a pattern of pore accessibly at each given pressure. Systematic penetration, starting from an empty network surrounded by mercury, can be readily performed. Results for the network in Fig. 5 are given in Fig. 6, showing both the penetration curve and the retraction curve. Stochastic pore networks implicitly predict hysteresis between penetration and retraction as well as a residual final entrapment of mercury. In Fig. 6, the final entrapment is about 45%, with much of the retained mercury entrapped in the larger pores [11]. More details of the pore-by-pore calculation have been published [4].

The parallel bundle, by its nature, cannot give hysteresis or entrapment in porosimetry, and the stochastic pore network indicates that both these phenomena arise from randomness and connections amongst pores. Applications of this simple model to oil-

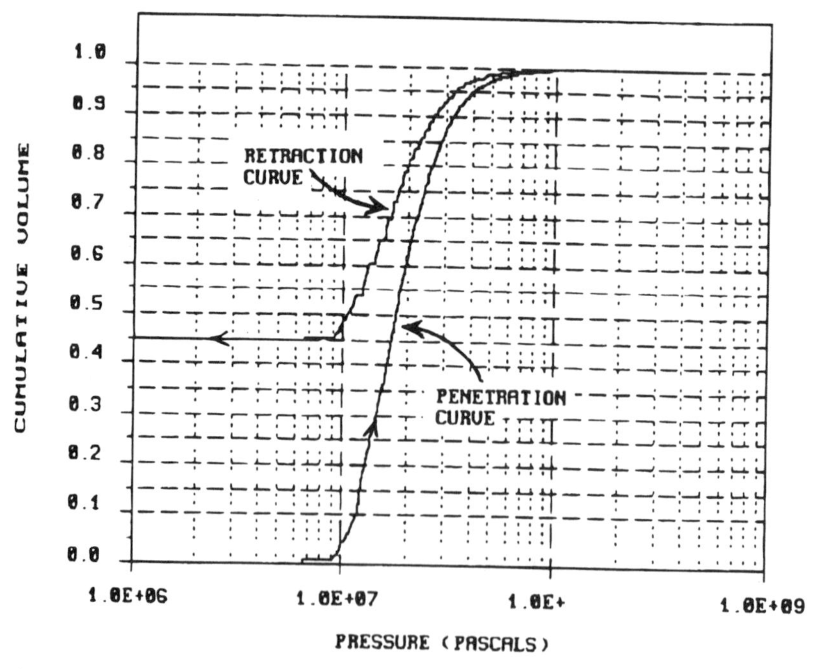

Figure 6 Mercury porosimeter curves for the network in Fig. 5. (From Ref. 11.)

bearing rock [5] have indicated that these simple random pore networks can provide a realistic model for understanding processes such as oil recovery and catalyst deactivation by coke laydown.

The filling and emptying of pore networks during the process of low-temperature capillary gas adsorption can also be analyzed for stochastic pore networks [12]. The adsorption/desorption isotherms for water for the network of Fig. 5 are given in Fig. 7. In contrast to porosimetry, in adsorption a network fills in accordance with the ascending order of pore diameters. Therefore, the adsorption branch gives the intrinsic pore diameter distribution directly. However, the emptying process relies on accessibility to the network (particle) exterior. As Fig. 7 indicates, these access limitations, caused by larger pore's being "hidden" amongst smaller ones, result in a delayed emptying producing a hysteresis effect. In adsorption, however, there is no entrapment effect and the pore network empties completely. A direct application of this model to an activated charcoal has recently been reported [13].

C. Rate Processes in 2-D Networks

The rate processes of diffusion and catalytic reaction in simple square stochastic pore networks have also been subject to analysis. The usual second-order diffusion and reaction equation within individual pore segments (as in Fig. 2) is combined with a balance for each node in the network, to yield a square matrix of individual node concentrations. Inversion of this $2N^2$ matrix gives (subject to the limitation of equimolar counterdiffusion) the concentration profiles throughout the entire network [14]. Figure 8 shows an illustrative result for a 20×20 network at an intermediate value of the Thiele modulus. The same approach has been applied to diffusion (without reaction) in a Wicke–Kallenbach configuration. As a result of large and small pores' being randomly juxtaposed inside a network, there is a 2-D distribution of the frequency of pore fluxes with pore diameter,

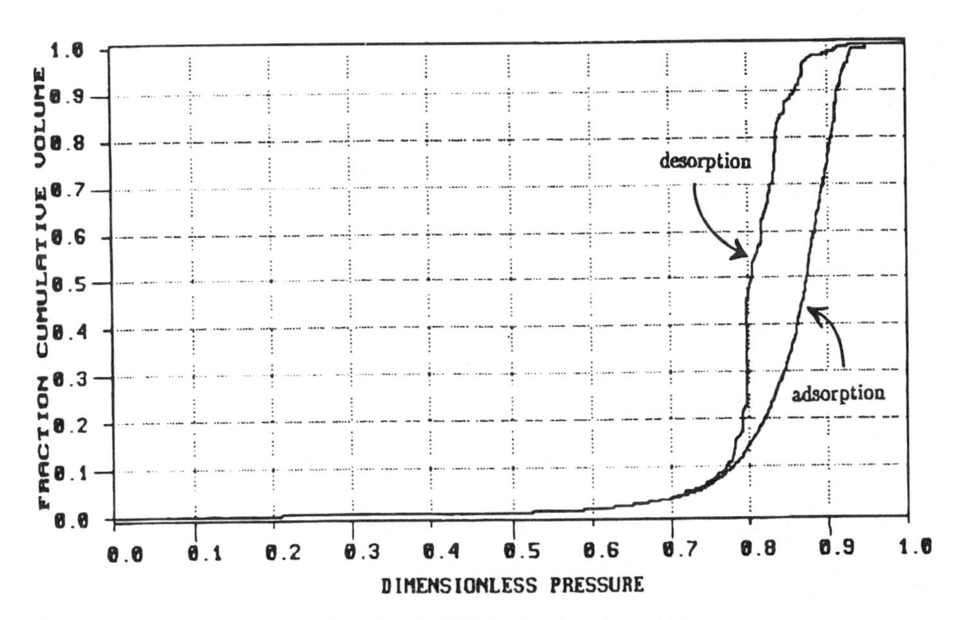

Figure 7 Water isotherm for the network in Fig. 5. (From Ref. 11.)

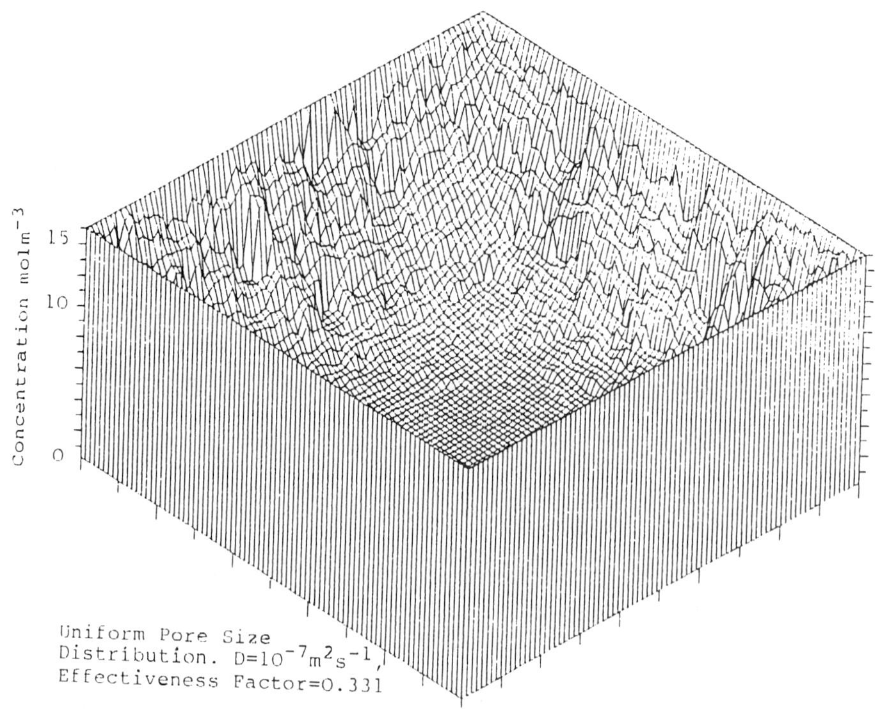

Figure 8 Concentration profile in a 2-D stochastic pore network. (From Ref. 11.)

as shown in Fig. 9. Some large pores have small fluxes, and, similarly, small pores have relatively larger diffusion flows. It has been shown that such stochastic pore networks can be applied to give direct interpretation of a tortuosity value (subject, of course, to the validity of the model) and that, moreover, anomalously high tortuosities can arise from bimodal pore size distributions [14].

The ability to solve the diffusion and reaction processes inside stochastic pore networks also led to analyses of coke laydown in catalysis, from which the process of catalyst deactivation by surface coverage and pore blocking could be quantitatively assessed. Figure 10 shows illustrative calculations [15] for a 30 $\times$ 30 pore network in which significant pore blocking is predicted to occur even in the face of uniform coke laydown (in the absence of any diffusional resistance). Once again, the random early blocking of the smaller pores progressively isolates more and more of the larger pores and causes faster deactivation. Random pore spaces are thus less efficient then uniform-sized ones. The diffusion and reaction analysis has also been extended to a pair of consecutive reactions [16].

D. Extended Degrees of Randomness

The regular square network of Fig. 5 can be modified in a number of simple ways in order to depart from the perfectly random assumption. A simple and elegant way of allowing pore lengths to vary randomly, with or without any correlation with pore diameter, can be achieved by permitting the nodes in the square network to undergo random

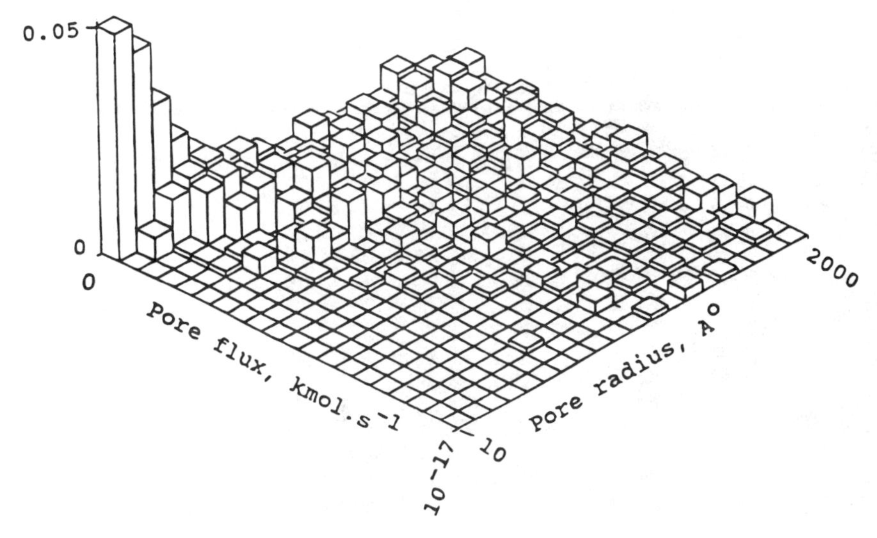

Figure 9 Distribution of pore fluxes in a 2-D network. (From Ref. 14.)

Figure 10 Coke laydown and pore blocking in a 2-D network. (From Ref. 15.)

Figure 11 Random-pattern stochastic pore network. (From Ref. 17.)

displacements. An example network is shown in Fig. 11, in which diameter and length are uncorrelated. An alternative situation might be expected for pores spaces found between approximately spherical particles (such as sand grains in a consolidated sandstone), when length and diameter are correlated, as examined for a North Sea reservoir sandstone [17].

There are many possible simple topological variations for a regular square 2-D network. Some examples are shown in Fig. 12. Allowing pore diameter to vary between nodes gives rise to randomly corrugated individual pores. A trivial subordering of segment sizes gives larger pores congregating preferentially close to nodes, a behavior that is possibly appropriate to the pore spaces arising from assemblies of approximately spherical particles.

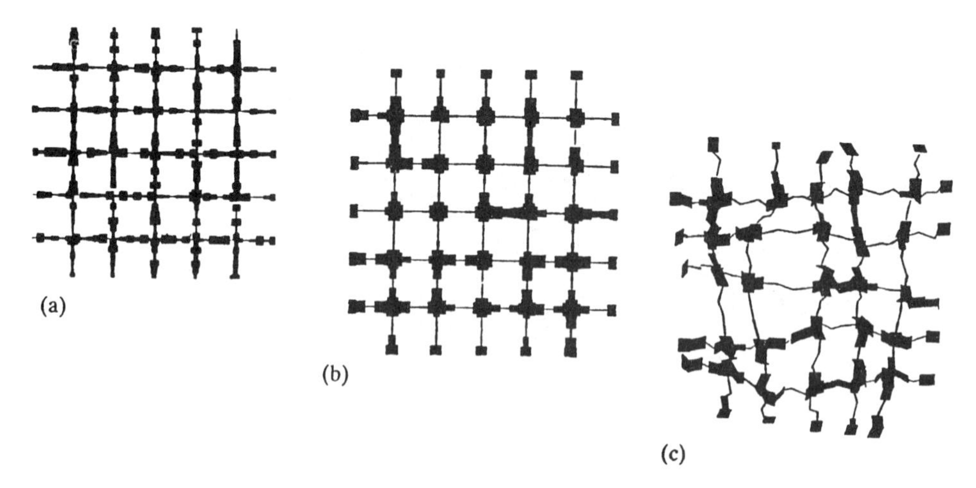

Figure 12 Variations on a regular 2-D stochastic pore network. (a) Random configuration; (b) subordering of corrugations between nodes; (c) random node displacement with subordering. (From Ref. 11.)

III. RANDOM PORE NETWORKS IN 3 D

A. Visualising 3-D Networks

The construction of pore networks in 3 D is a simple extension of the 2-D arrangement depicted in Fig. 5. A regular cubic network with a connectivity of 10 (each pore has 10 neighbors) contains $2N(N^2 + 1)$ cylindrical pore segments. Such networks can be pictorialized in a perspective view as shown in Fig. 13 for a $15 \times 15 \times 15$ example that contains almost 7000 pores. In this particular illustration, pores shown range in diameter from 10 to 10^5 nm. They are drawn to scale by using up to 10 multiple lines between the nodes to be indicative of pore diameter. They correspond to the pore size distribution fitted to a catalyst particle porosimeter curve; see Section IV.C. As with the 2-D case in Fig. 5, this 3-D network shows clearly how a small number of large pores may be hidden amongst the more predominant smaller ones.

B. Extending to Five Degrees of Randomness

The regular cubic 3-D network in Fig. 13 exhibits one degree of randomness, namely, random variation of pore diameter. In a manner similar to the 2-D case in Fig. 11, pores can take random lengths by random relocation of the nodes (junctions of pores) in 3 D.

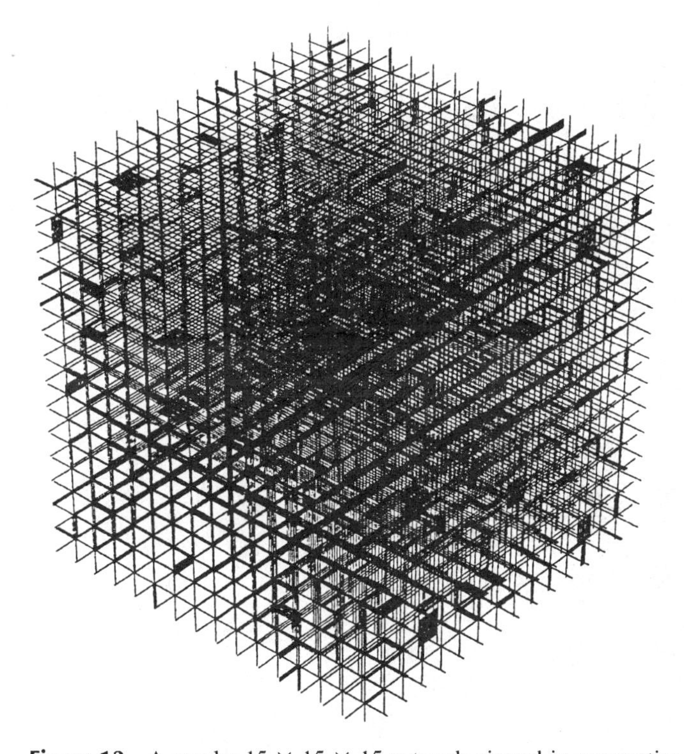

Figure 13 A regular $15 \times 15 \times 15$ network viewed in perspective.

The network of Fig. 13 when randomized in this way appears as in Fig. 14. It now comprises two degrees of randomness, diameter and length. In Fig. 14, pore lengths have been varied randomly without changes to diameter. In practice, any form of correlation between length and diameter may be applied. If length and diameter are not correlated, then the mercury porosimeter behavior has been shown to be identical to the case of regular cubic equal pore length for both 2-D and 3-D networks [17,18].

Three further possible elements of random variability are:

Tortuous paths between nodes (junctions)
Random surface roughness
Random distorted cross section

These three further randomness elements are depicted schematically in Fig. 15. Figure 15(a) shows a standard circular cross section void (black) for a pore segment intersected at right angles by a sectioning plane. A random angle is imposed on the section to imitate the allowance of a tortuous path between nodes. This would distort the section on an ellipse, whose major and minor axes depend on the "random" angle to the straight line path between nodes, as shown in Fig. 15(b). Surface roughness is then introduced by slicing the void section into a number of parallel thin strips and displacing each strip randomly up to some maximum fraction of the basic pore diameter. This provides a realistic-looking perimeter while conserving the pore cross-sectional area. Figure 15(c) shows a roughness factor of 0.2, representative of a 20% uniformly random displacement of each "strip" of the pore cross section. This value creates a roughened pore perimeter

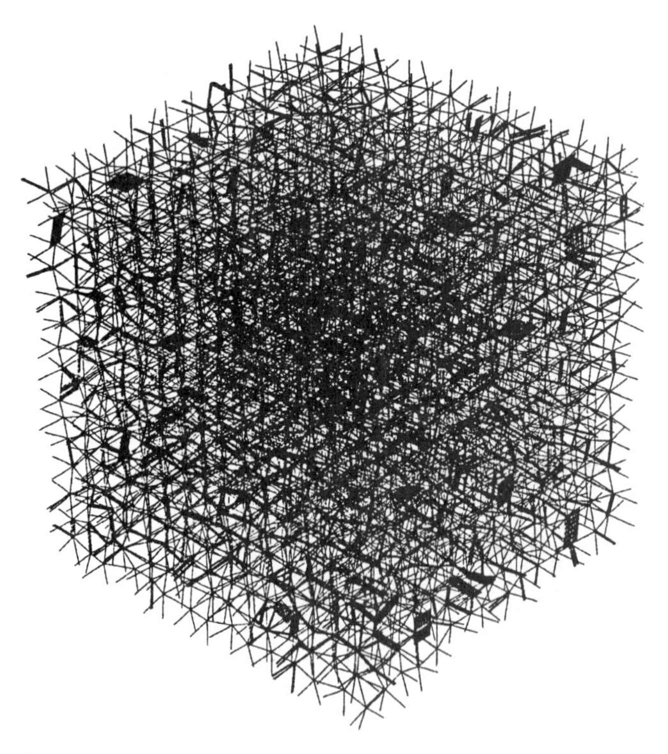

Figure 14 An irregular $15 \times 15 \times 15$ network (perspective view).

(a)

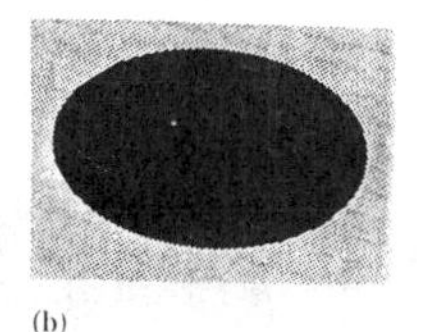
(b)

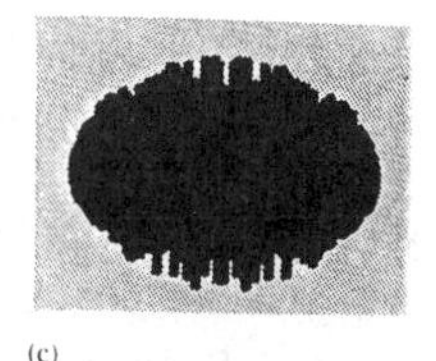
(c)

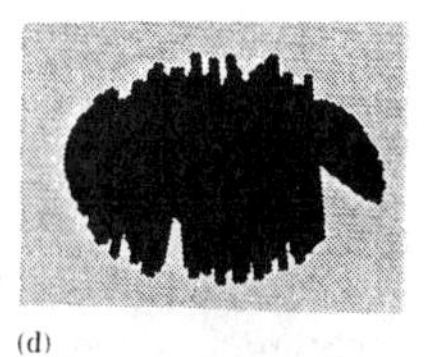
(d)

Figure 15 Additional randomness elements for individual pore segments. (a) Regular pore sectioned at right angles; (b) tortuous pore section; (c) randomly roughened tortuous pore section; (d) random adjustment to shape of cross section. (From Ref. 7.)

that seems qualitatively comparable to what is observed in typical experimental sections. This surface roughness is assumed to have no significant impact on the average "cylindrical" pore diameter of a basic pore segment. Therefore, such roughness does not affect the porosimeter result or consequently the deduced pore size distribution. The final element of randomness involves breaking the resultant elliptical section at two random interior points and introducing a random angular orientation to produce the effect shown schematically in Fig. 15(d).

C. Imaging "Virtual" Cross-Sectional Slices

If a theoretical random slice is cut through a network like that in Fig. 14, the cross section produced will show pore spaces exposed amongst the solid support. Figure 16 illustrates how such a sliced section appears for five degrees of randomness (with a surface roughness factor of 0.2) for the instance where the pores are relatively uniform in diameter, obeying

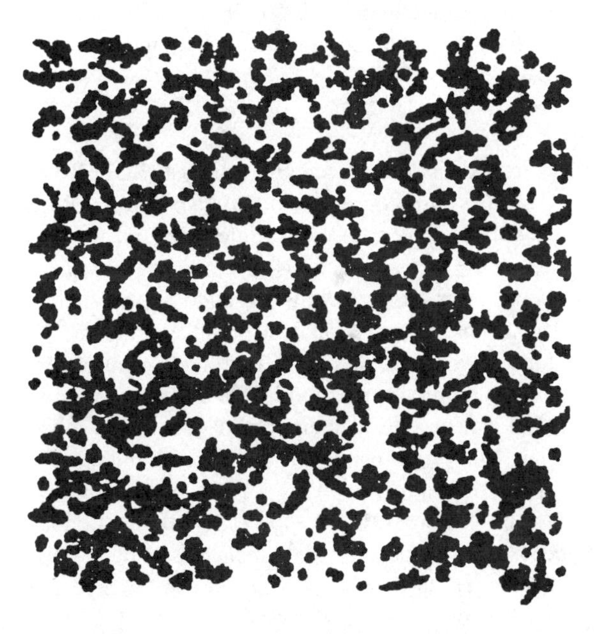

Figure 16 Virtual slice through a network of five degrees of randomness. (From Ref. 7.)

a normal distribution with a mean of 1 μm and variance of 0.3 μm. If the sectioning slice happens to pass close by a node, a pore feature containing areas from up to six neighbor pores can be created. This is the explanation for the many large features in Fig. 16. It is also evident from this figure that even though the basic pore elements are mostly very close to 1 μm in size, there is a much greater variation in the size of exposed overlapping features in the vicinity of the nodes. Thus, sectional slices seem likely to exaggerate the intrinsic variation in pore element diameters that are used to build the pore network.

IV. APPLICATIONS OF 3-D NETWORKS

A. Porosity Visualization by Alloy Impregnation

In this section, two examples are presented for the application of a technique of low-melting-point alloy (LMPA) impregnation that provides for a visualization of the invasion of a nonwetting fluid into the pore spaces in a typical porous article. The visualization can be linked to the modeling of mercury porosimeter curves using 3-D stochastic pore networks. This makes the quantification of pore structure more direct. Quantified structures can be visually examined against sample particle sections. The visual comparison can be made more precise by image analysis of the accessible porosity made visible by metal penetration over a series of pressures.

Experimental details of the technique have previously been reported [7]. The low-melting-point alloy (MP47, Mixing and Chemical Products Ltd) melts at 47°C. Impregnation tests were undertaken at 60°C in a pressurized bomb immersed in a thermostat bath. The alloy has the same surface tension as mercury, and similar penetration to mercury porosimetry is assumed. Pressuring is preceded by prolonged evacuation to below 0.05 mm Hg. On raising the pressure, up to 1 hour is allowed for equilibrium to be established. The penetrated alloy sets solid on cooling to room temperature (20°C). Solidified samples are carefully sectioned and polished to reveal a visualization of the penetrated porosity.

B. Application to FCC Catalyst

This has been done for a sample of an (unused) industrial cracking catalyst (SUPER-D); the results are shown in Fig. 17 [19]. This section is of size 500 μm × 600 μm, which exposed cross sections of some 200 (or so) cracking catalyst particles. Several features of this result are worth noting. First, although the exposed sections appear to show particles of variable diameter, this is because the sections pass randomly through approximately spherical particles of roughly equal size. In this case the catalyst particles are some 70 μm in diameter. Second, and perhaps most significant, there is a startling nonuniformity of penetrated internal matrix/support pore structure, as indicated by the extent of invasion of the low-melting-point alloy. The individual particle's internal pore spaces are much more widely variable than they are similar. This variability ranges from the total nonpenetration of some 50 particles (about 25%) to the largest exposed section, which contains giant-sized pore voids. In between these two extremes, there appears to be a continuous variation in the indicated porosity and associated pore space dimensions. Third, most particles seem to show a significant skin effect, so that interior access can be expected to be restricted by smaller exterior pores. This outer skin is also probably the explanation for the large fraction of totally unpenetrated particles, which appear completely black in Fig. 17.

Figure 17 Low-melting-point alloy impregnation of FCC catalyst particles.

As a postscript to Fig. 17, it ought to be possible to draw some inferences as to the average particle structure. This should serve as some guide to improvements, since it is reasonable to assume that the disparate set of particles in Fig. 17 on average delivers reasonable performance. Thus, Fig. 18 shows the particle that is subjectively judged to present a pore structure average. This detailed view of an individual particle clearly shows an outer skin, which has been penetrated by alloy through three closely adjacent large peripheral pores (upper right particle periphery). Figure 19 then shows a "virtual" computer-generated random section through a 3-D stochastic pore network by a low-melting-point alloy (LMPA) impregnation.

The computer-generated reconstruction in Fig. 19 is consistent with the mercury porosimetry result. The random 3-D network of Fig. 19 contains pore elements up to 3 μm in diameter. The computer image, however, shows pore features up to 9 μm in (equivalent) diameter. The imaged representation exhibits these large pore features due to overlapping of pores as the plane section passes close by pore junctions (nodes). A feature may thus contain elements of several pores and thereby appear enlarged in the plane of the section. This aggregation of pore sections is readily discernible in Fig. 19, just as it was in the previous illustrative example in Fig. 16.

C. Application to Ni/Al_2O_3 Reforming Catalyst [7]

Figure 20 shows a 45° plane section (apex to apex) through a single stochastic realization of a $30 \times 30 \times 30$ random network with a psd providing for an exact replication of the experimental porosimeter curve. This section shows void "features" that are aggregations

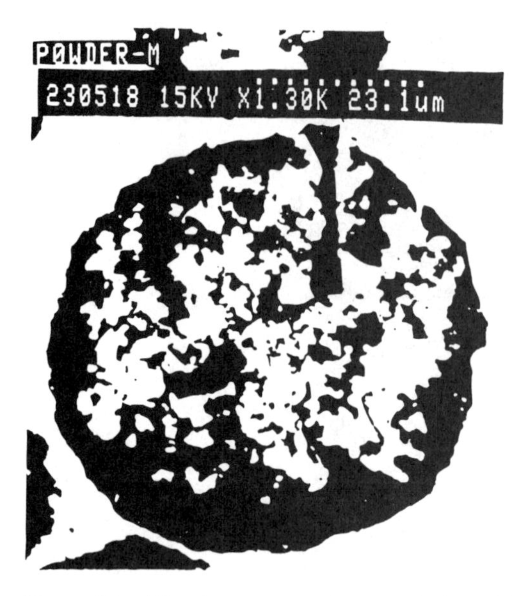

Figure 18 Alloy impregnation for a typical particle of FCC catalyst. (From Ref. 19.)

of individual pore-segment sections close to nodes, where the plane can in principle slice through up to a maximum of six pores. These pore features can obviously be much larger in area than the largest individual pore segments. Hence, the deduction of the intrinsic pore size distribution of the individual pore segments forming the network is complicated by this overlapping effect near junctions. This potential difficulty is, however, avoided by using virtual reality sections, because comparison can be effected directly with real sections, which inherently exhibit the same aggregation phenomenon. The underlying psd of the pore segments, because it is the starting point of the construction of the theoretical irregular network in 3-D, is of course always known.

The results for alloy impregnation of the pore network used to create the section in Fig. 20 are shown in Fig. 21 for three pressure levels: 1, 10, and 50 atm. In Fig. 21(a), at 1 atm the penetration of the alloy (shown white) is restricted to a few peripheral pores corresponding to accessible segments on the outermost part of the network. As the pressure is increased to 10 atm, the penetration is increased significantly, as indicated by Fig. 21(b), although the innermost pores remain substantially unpenetrated. However, on increasing the pressure to 50 atm, the majority of pores are penetrated, although a number of larger pore features (shown black) are still not penetrated, even though ostensibly they would have been had they not been rendered inaccessible by shielding behind small pores.

In Figure 22(a), an approximately 2-mm catalyst fragment shows an evidently similar peripheral penetration of alloy at ×40 magnification. In Fig. 22(b), the penetration is increased toward the interior. At the highest pressure, 50 atm, there is substantial penetration throughout the whole interior, and Fig. 22(c) shows behavior exactly similar to the virtual section in Fig. 21(c). Also clearly evident in Fig. 22(c) are dark spaces identifiable with unpenetrated pore voids. These have an apparently similar appearance and spatial distribution to the dark unpenetrated hypothetical voids shown black in Fig. 21(c).

Each of the six images in Figs. 21 and 22 have been subjected to quantitative analysis on a Magiscan 840 Image Analyzer. The results are summarized in Table 1. The

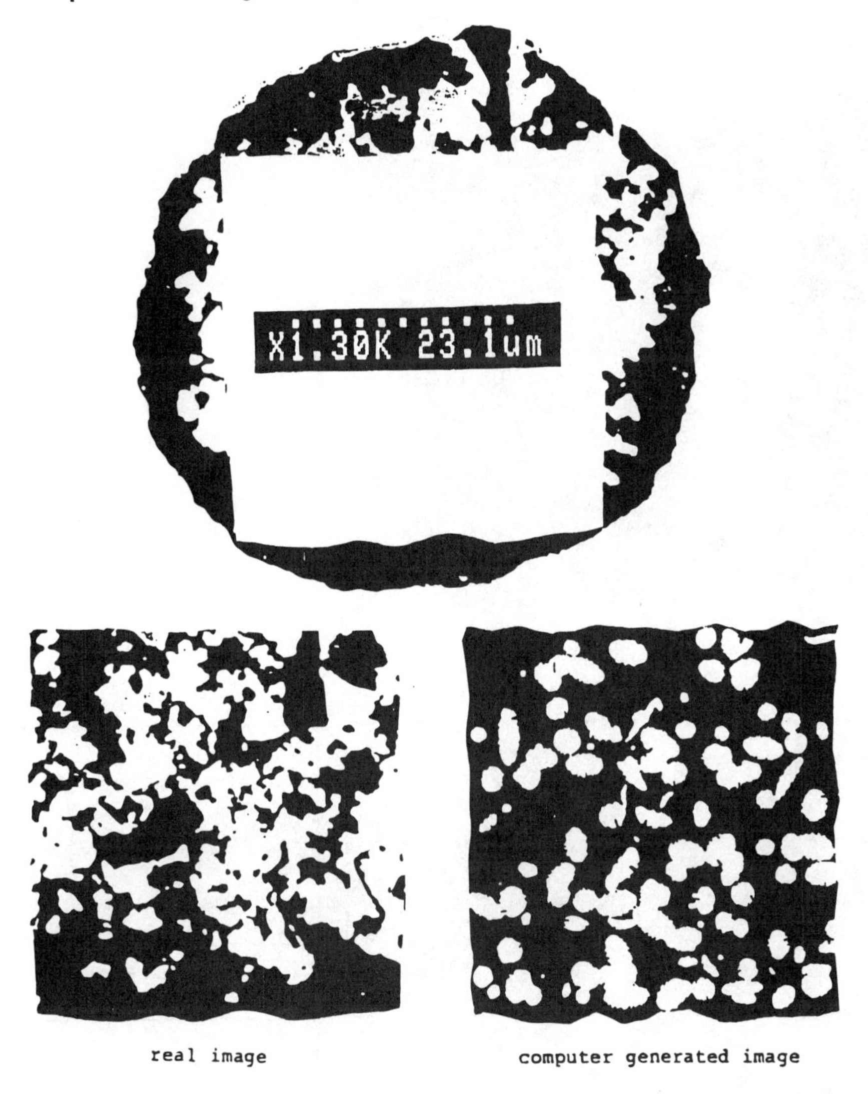

Figure 19 "Virtual" section through a 3-D stochastic pore network. (From Ref. 19.)

experimental and hypothetical section results are presented side by side for comparison. As Table 1 shows, for each criterion of comparison, the real and virtual sections are closely self-consistent, showing only a small amount of variation. This degree of noise is statistically reasonable, since the samples are one-off arbitrarily random choices from an infinity of possibilities. Thus, the number of penetrated voids (shown as white) increases from around 20 at 1 atm, to around 100 at 10 atm, increasing to 200 at 50 atm. Likewise, the extent of porosity exposed by alloy impregnation increases from 2% to 10% and then to 20% for the three experimental pressures. Since the number frequency of penetrated features and the penetrated porosity are integrated values, they show less statistical variation (noise) than the actual values of the largest and smallest penetrated features/pores detected in the sectioned fields subjected to pixel counting by quantitative image analysis.

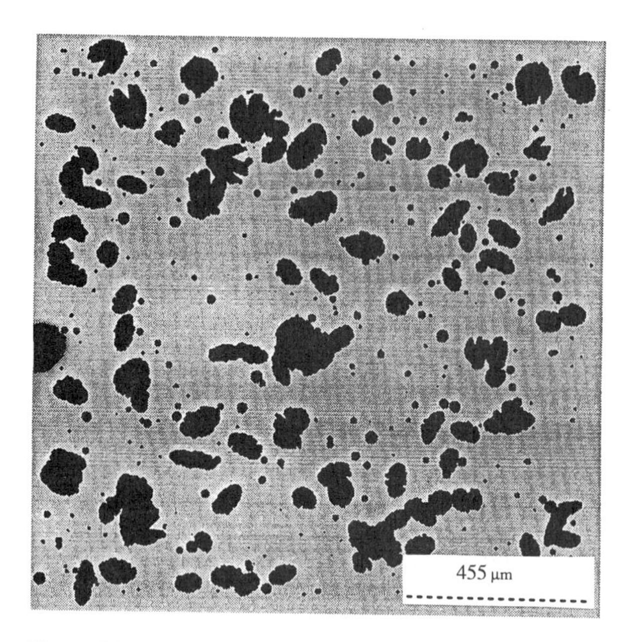

Figure 20 Catalyst particle "virtual" section for Ni/Al_2O_3 reforming catalyst. (From Ref. 7.)

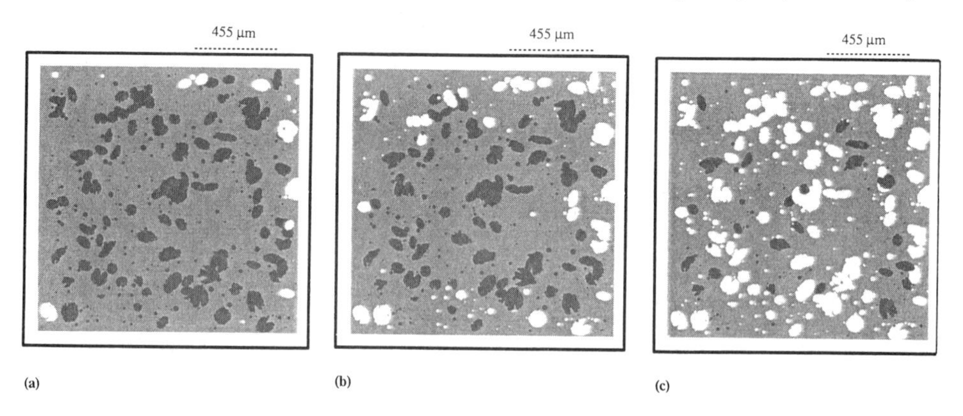

Figure 21 "Virtual" sections for LMPA impregnation. (a) At 1 atm; (b) at 10 atm; (c) at 50 atm.

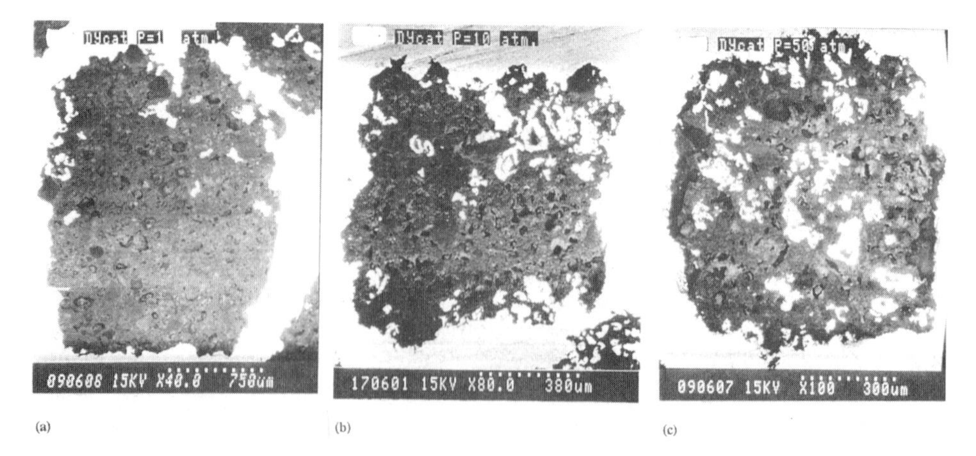

Figure 22 Experimental "real" sections for LMPA impregnation. (a) At 1 atm; (b) at 10 atm; (c) at 50 atm.

Table 1 Image Analysis of Real Experimental and Virtual Theoretical Particle Sections

	Penetration pressure atmos	Number of penetrated features	Penetrated porosity	Minimum diameter (μm)	Maximum diameter (μm)
Real SEM image	1	20	0.022	20.5	140.4
Virtual network image	1	16	0.015	11.9	104.0
Real SEM image	10	99	0.119	3.9	115.3
Virtual network image	10	97	0.108	5.9	123.4
Real SEM image	50	193	0.207	3.1	114.2
Virtual network image	50	201	0.211	4.2	174.3

The nature of the inherent statistical variation is most easily examined for separate stochastic realizations of additional "virtual" network sections (using the same parameters, but varying the value of the "seed" initiation for generation of sequences of random numbers). Three results are presented in Table 2, using nominal seed values of 1, 2, and 3. Seed 1 was already used for the theoretical sections in Table 1. The results for the additional seeds, 2 and 3, are consistent with the expected degree of variation amongst the repeat stochastic realizations.

V. SOME CHALLENGES FOR PORE STRUCTURE DESIGN

A. FCC Catalyst: Design for Staged Cracking

The importance of pore architecture in determining cracking performance was initially limited to the internal pore structure of zeolite crystallites. This shape-selecting concept has now been recognized as being important for the larger pores, exterior to the zeolite, that arise from the amorphous support matrix and possibly also from the exterior surfaces of zeolite crystallites. As an idealized simplification, three possible stages of the cracking process can be envisaged [19]. First, large asphaltenes crack, and metals are deposited.

Table 2 Variation Between Virtual Section Stochastic Repeats

Penetration pressure	Seed	Number of penetrated features	Penetrated porosity	Minimum diameter (μm)	Maximum diameter (μm)
1	1	16	0.015	11.9	104.0
1	2	20	0.018	12.9	110.1
1	3	7	0.011	23.0	113.3
10	1	99	0.108	5.9	123.4
10	2	97	0.103	5.1	148.5
10	3	94	0.100	5.1	131.7
50	1	201	0.211	4.2	174.3
50	2	208	0.215	4.2	166.6
50	3	187	0.195	4.2	158.2

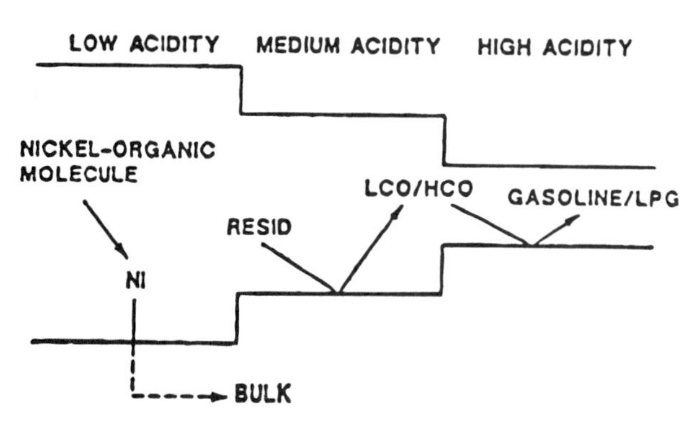

Figure 23 Idealized pore structure for staged cracking. (From Ref. 19.)

Second, products of the primary cracking precrack further in mesopores. Third, final cracking of the smaller molecules takes place inside zeolite cages. This concept of an architecture for the catalyst that promotes staged cracking is clearly important for the design of FCC resid catalyst that will be simultaneously efficient/effective/selective and resistant to deactivation. Moreover, as Fig. 23 shows schematically, size, shape, acidity, and accessibility have to be designed for, and particles have to be fabricated and assembled accordingly.

In practice, the pore size distribution ought to be manipulated into a continuous form similar to that shown in Fig. 24. The micropores (<20) promote the cracking of the smallest hydrocarbon fragments; the mesopores (between 20 and 1000) accommodate the intermediate second stage of cracking. The large pores (>1000) serve as "liquid catching" pores and give preliminary cracking of asphaltenes as well as accommodating metals deposition.

The major conclusion from Fig. 24 is that whatever the optimal pore structure of the matrix/support should be in terms of mesopores linked to precracking pores as staged

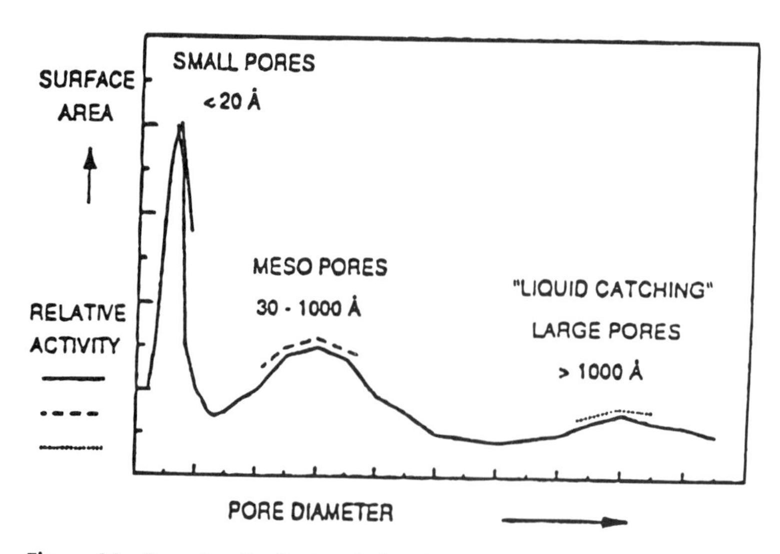

Figure 24 Pore size distribution designed for staged cracking. (From Ref. 19.)

pathways to the zeolite micropores, the assembly of particles in Fig. 17 could hardly be expected to begin to approximate it. Because of the intrinsic wide variation from particle to particle, each one would have different flux/permeation/diffusivity characteristics. Moreover, each one would behave quite differently with respect to coke laydown and buildup. Finally, each one would then have quite different burn-off behavior at the regeneration stage. The subsidiary conclusion from Fig. 17 is that the procedures used to fabricate cracking catalyst particles need to be examined and revised so as to be able to control individual pore structures much more precisely. Only in this way will it become possible to fabricate particles that have optimally tailored pore structural characteristics.

The foregoing assessment of the pore structures of actual cracking catalyst particles has indicated the need for catalyst manufacturing and preparation techniques to recognize the importance of dovetailed pore structures that could optimize architectures so as to deliver staged cracking of the heavier resid feedstocks. Techniques developed for imaging stochastic pore networks can be readily adapted to construct pore assemblies with any geometric properties. In this respect, Fig. 25 shows a "concept" network in which pore sizes are controlled from the interior (smallest) to the exterior (largest). The diffusional and reaction performance of such networks can be readily calculated, so that candidate pore structures can be screened by a modeling stage. The spiraled network of Fig. 25 would appear to have a desirable geometry for a staged cracking reaction that requires precracking pores capable of accommodating metal deposits.

B. Reforming Catalyst: Design for Intrapellet Convection

Reactions that are strongly diffusion influenced benefit from intraparticle convection. This is because the supply of primary reagents can now be driven by bulk diffusion. Furthermore, reversible reactions of the type $A + B \rightleftharpoons C + D$ will show an additional benefit because the reaction products will be swept from the particle interior by convection, when otherwise their greater accumulation would reverse (and therefore hinder) the local rate of reaction.

However, creating an optimum balance of surface reaction and intraparticle convection requires the pore structure to be appropriately designed and fabricated. This has so far not been achieved, although the basic requirements to do this are clear.

Some insights are available from some preliminary work on this problem [20] although this has so far not been achieved for the full 3-D case outlined in Section IV.C. Calculations have so far only been based on a 2-D structure, and these will now be outlined.

The appropriate pore structure in 2 D (for a 30×30 network) is depicted in Fig. 26. This psd matches the porosimeter curve for a commercial (Dycat) reforming catalyst, and the number frequency distribution is shown in Fig. 27 [20]. In this particular case, the pore elements have a bimodal distribution, with the highest frequencies for largest and smallest pores, although the small pores predominate numerically. However, the less frequent, larger pores thoroughly dominate the pore volume distribution, and a small number of large pores dominate the visual appearance of the pore structure in Fig. 26, especially the two largest pores.

To evaluate the impact of intraparticle convection it is necessary to impose a pressure gradient across the network. Such pressure gradients arise naturally in fixed-bed operation, though the pressure difference across a particle is usually only about 1 cm H_2O. By solving the Hagen–Poisenille equation across every pore in the network, the overall flow through the particle (network) is known.

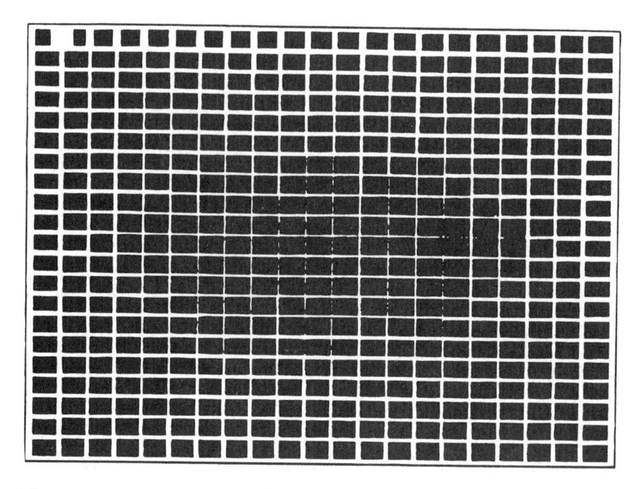

Figure 25 Concept pore network for FCC stage cracking. (From Ref. 19.)

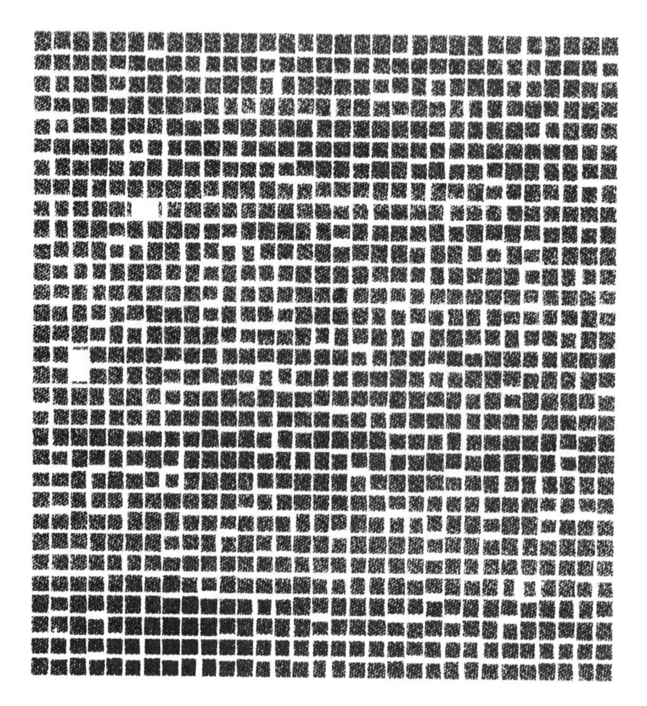

Figure 26 Stochastic pore network for reforming catalyst. (From Ref. 20.)

Predictions of the benefits then show up in Figs. 28 and 29. Without any convection (zero imposed pressure difference), at a Thiele modulus of 5, the concentration profile inside the network is as shown in Fig. 28. There is a severe depletion of primary reagent into the network, and only the ends see sufficient reagent. With imposed convection (at a λ value of 750) for the same Thiele modulus, the effectiveness factor ratio becomes 2.84. The impact of the imposed convection is shown in Fig. 29. The flow generated within the network is quite nonuniform, as the fingers of high concentration show. This

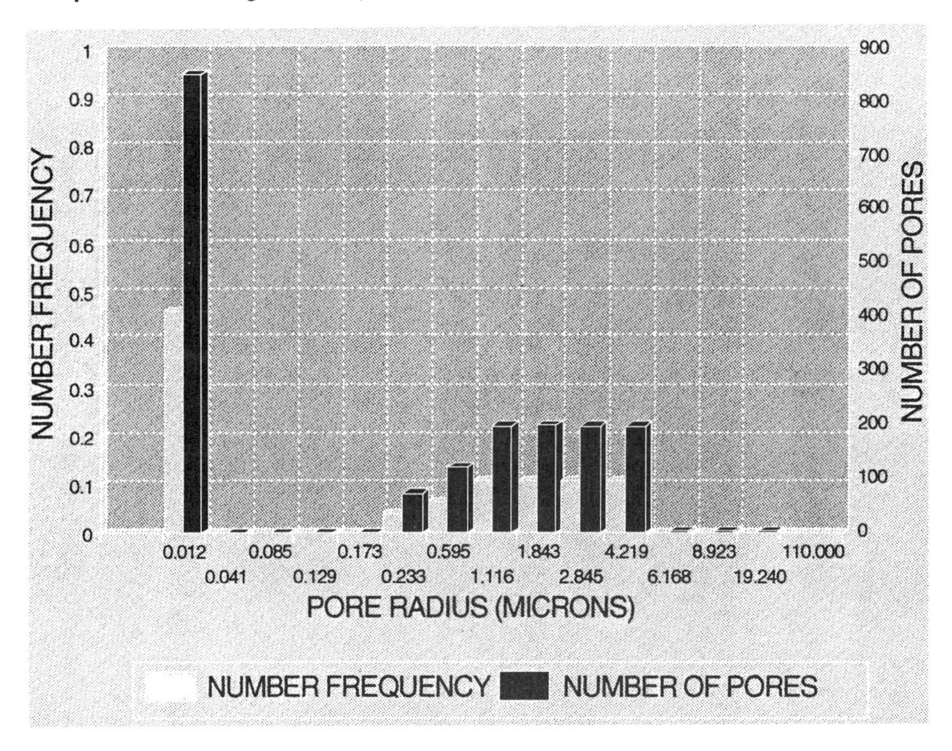

Figure 27 Pore size distribution for reforming catalyst. (From Ref. 20.)

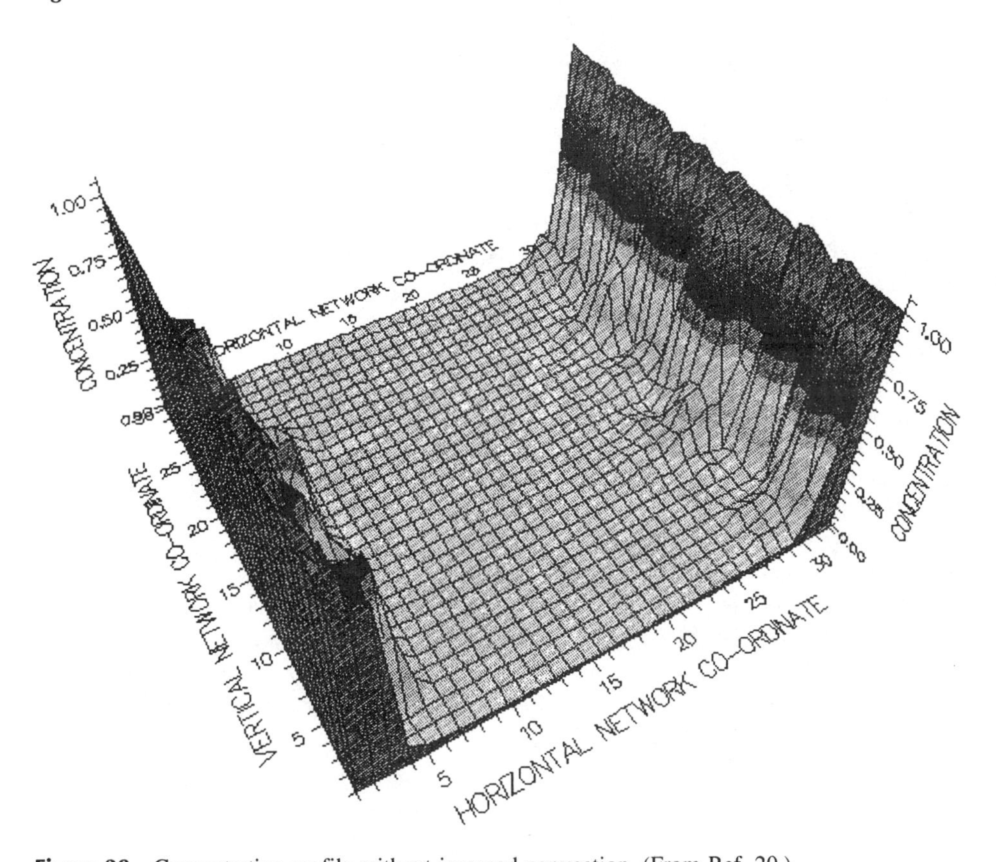

Figure 28 Concentration profile without imposed convection. (From Ref. 20.)

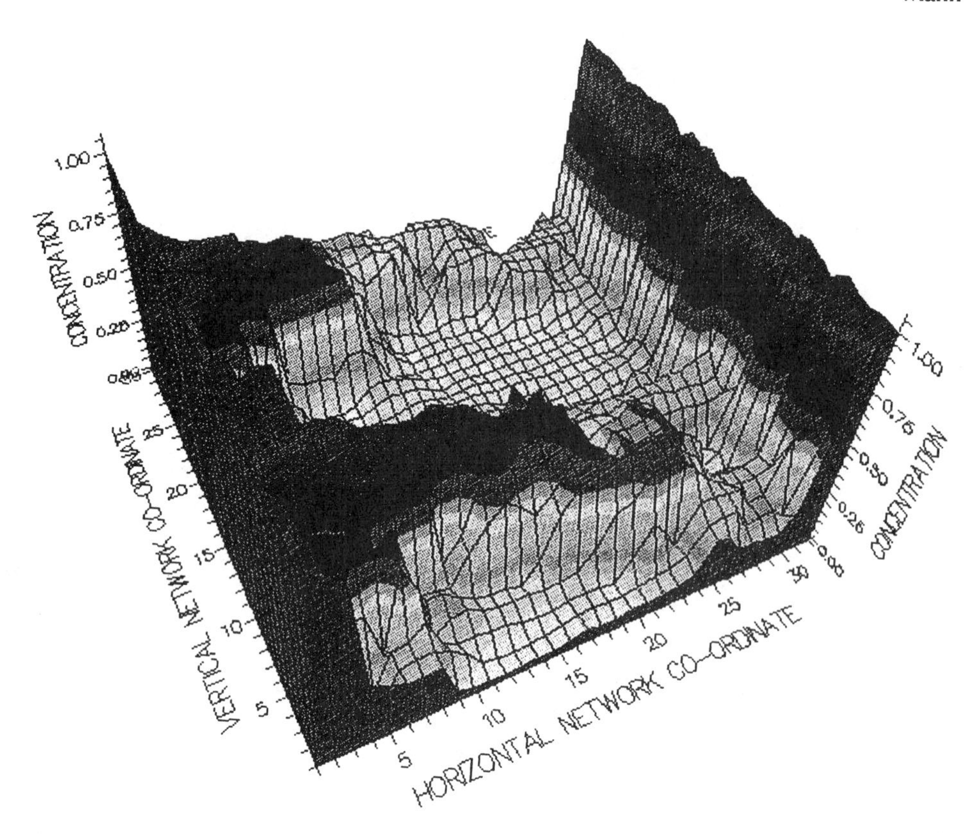

Figure 29 Concentration profile with imposed convection. (From Ref. 20.)

"noisy" pattern of intrapellet convection arises from the chance creation of preferred flow pathways due to the chance accretion of larger-than-average pores, which in turn provide opportunities for enhanced permeation (percolation) pathways. The network in Fig. 29 thus has a 284% increase in reaction rate due to the convection.

Unfortunately, however, the value of "λ" arising from the pore network in Fig. 26 is negligible at typical pressure gradients. If, on the other hand, the pore network were assembled by design, so as to give convection by having smaller pores nested amongst networks of larger ones, as depicted in Fig. 30, there would be a potentially huge improvement in particle (network) effectiveness. Indeed, the bimodal psd in Fig. 27 could readily be rearranged to promote such convection-enhanced reactivity. Thus it is clear that in this case, the randomized spatial mixing of large and small pores, dominated as it is by the numerically preponderant small pores, is very inefficient because the permeation potential of the larger pores is nullified by being shielded by the smaller pores.

Nominating a beneficial conceptual pore structure like that in Fig. 30 does not immediately tell us how it could be achieved. However, this nested "pore networks in networks" concept gives a strong incentive to seek fabrication procedures that will deliver nonrandom structured pore designs, tailored to combine diffusion and permeation to improve catalyst particle reactivity.

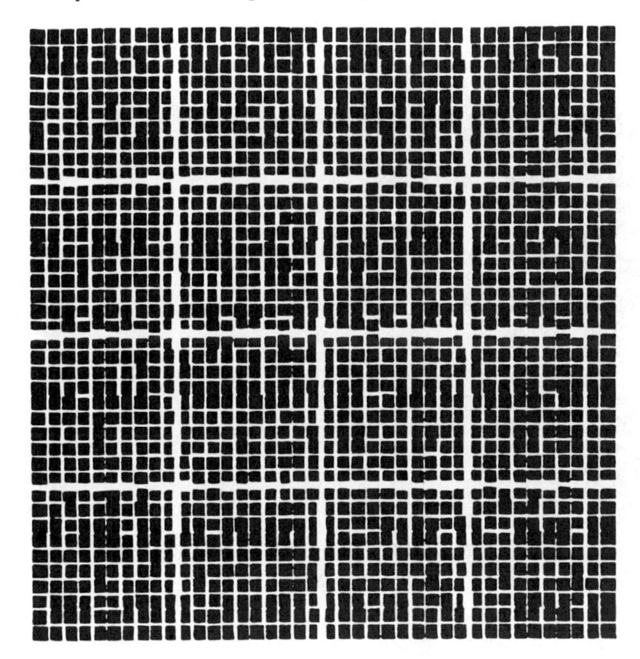

Figure 30 Networks in networks, designed to promote convection.

C. Macro- and Mesoporous Monoliths

The above considerations on the design of "particles" of catalyst can be extended to the "geometrical engineering" of a monolith. For traditional automobile applications these have been made from nonporous ceramic, with the catalyst washcoated onto the walls. However, for more demanding chemical process applications needing higher reactivity, it would be useful to provide additional surface area by having a porous/permeable monolith. The major channels of the monolith would promote bulk transport and throughflow. If the monolith walls are porous, it would then be potentially beneficial to structure the macro- and microporosity as depicted (schematically) in Fig. 31.

The monolith channels in Fig. 31 are of illustrative size $2L \times 2L$, where L is the (equal) pore length (node spacing) for a stochastic regular network of pores. The dimensions of the monolith channels should be balanced against the diffusive flux induced by the intervening stochastic pore network, which in this case has the same pore size distribution (psd) as the network in Fig. 26. Such quantitative evaluations have yet to be undertaken, but the intelligent combination of monolith porosity (size and spacing of monolith channels) and the wall microporosity, so as to optimize the balance of convective flow, pressure-driven permeation, and concentration-driven diffusion within the whole assembly, is a challenging problem in structured catalytic reactor design.

The above ideas can be readily extended to a three-level nesting of monolith porosity and meso + microporosity, as shown in Fig. 32, where the pore network has been relaxed to a more realistic random-pattern (irregular) configuration. Nesting the porosity as networks in networks should provide improved access and possibly additional convection through the larger macropore channels, thereby enhancing the reactivity of the mesoporous network and any associated micropore structure.

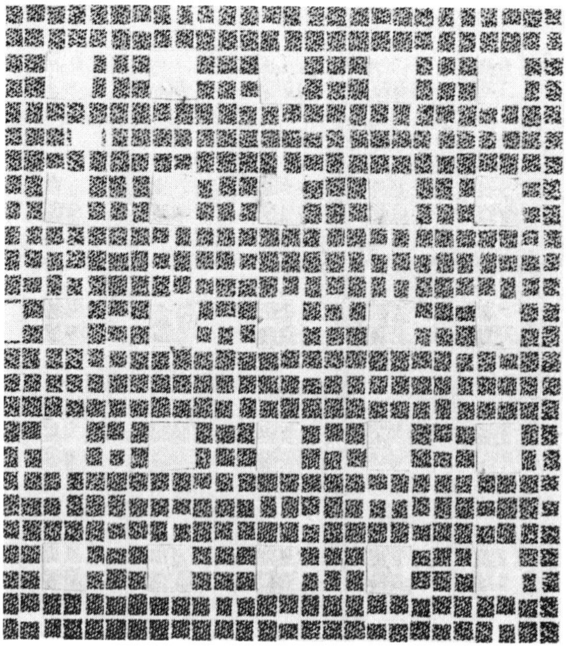

Figure 31 Combination of monolith and wall pore structure.

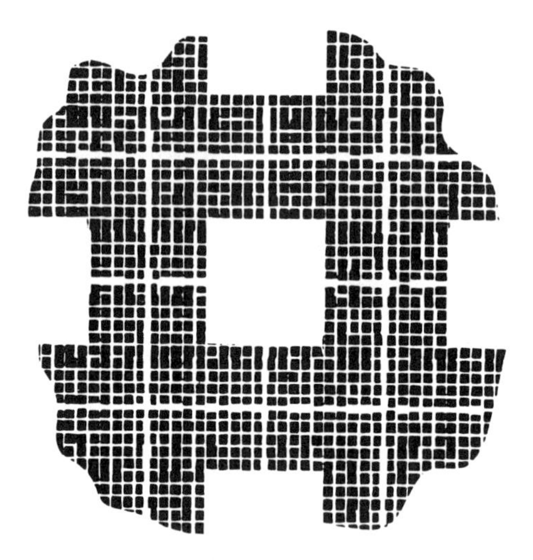

Figure 32 Three integrated layers of porosity.

The optimization of the conceptualized geometry in Fig. 32 raises many novel questions on designing for the correct balance of reaction, diffusion, permeation, and bulk flow. There is no doubt that these questions should now be addressed. The answers should spawn new generations of compact, low-pressure-drop, highly reactive monolithic catalysts.

REFERENCES

1. S.C. Reyes and E. Iglesia, Simulation techniques for the characterization of structural and transport properties of catalyst pellets, in *Computer-Aided Design of Catalysts*, E.R. Becker, and C.J. Pereira, eds., Dekker, New York, 1993.
2. S.C. Reyes and K.F. Jensen, Estimates of effective transport coefficients in porous solids based on percolation concepts. *Chem. Eng. Sci. 40*:1723 (1985).
3. S.C. Reyes and E. Iglesia, Effective diffusivities in catalyst pellets: New model porous structures and transport simulation techniques, *J. Catal. 129*:457 (1991).
4. G.P. Androutsopolos and R. Mann, Interpretation of mercury porosimeter experiments using network pore models, *Chem. Eng. Sci. 34*:1203 (1979).
5. R. Mann, G.P. Androutsopoulos, and H. Golshan, Application of stochastic pore networks to oil bearing rock with observations relevant to oil recovery, *Chem. Eng. Sci. 36*:337 (1981).
6. R. Mann and H. Golshan, Application of stochastic pore networks to a catalyst pellet, *Chem. Eng. Commun. 12*:377 (1981).
7. R .Mann, A. Al-Lamy, and A. Holt, Visualized porosimetry for pore structure characterization of a Nickel/Alumina Reforming Catalyst, *Trans. I. Chem. E. 73(A)*:147 (1995).
8. A. Wheeler, *Advances in Catalysis 3*:250 (1951).
9. O. Levenspiel, *Chemical Reaction Engineering*, 2nd ed., Wiley, New York, 1972, p. 470.
10. R. Mann and G. Thomson, Deactivation of a supported zeolite catalyst: Diffusion, reaction and coke deposition in a parallel bundle, *Chem. Eng. Sci. 42*:555 (1987).
11. R. Mann, Developments in chemical reaction engineering: Issues relating to particle pore structures and porous materials, *Trans. I. Chem. E. 71(A)*:551 (1993).
12. G. Thomson and R. Mann, Application of stochastic pore networks to interpret adsorption isotherms, in *Adsorption: Science and Technology*, E.A. Rodrigues, ed., Kluwer Academic Publishers, Dordrecht, The Netherlands, 1989, p. 63.
13. R. Mann, H.N.S. Yousef, D.K. Friday, and J.J. Mahle, Interpretation of water isotherm hysteresis for an activated charcoal using stochastic pore networks, *Adsorption 1*:256 (1995).
14. A.N. Patwardhan and R. Mann, Effective diffusivity and tortuosity in Wicke–Kallenbach experiments: Direct interpretation using stochastic pore networks, *Trans. I. Chem. E. 69(A)*:205 (1991).
15. R. Mann, P.N. Sharratt, and G. Thomson, Catalyst deactivation by fouling: Diffusion, reaction and coke deposition in stochastic pore networks, *Chem. Eng. Sci. 41*:711 (1986).
16. R. Mann and P.N. Sharratt, Diffusion and consecutive reaction in stochastic pore networks, *Chem. Eng. Sci. 43*:1875 (1988).
17. R. Mann, J. Almeida, and M. Mugerwa, A random pattern extension to the stochastic network pore model, *Chem. Eng. Sci. 41*:2663 (1986).
18. A. Al-Lamy, Characterization of catalyst pore structure by image reconstruction from 3-D stochastic pore networks, Ph.D. dissertation, UMIST (1995).
19. R. Mann, Fluid catalytic cracking: Some recent developments in catalyst particle design and unit hardware, *Catalysis Today 18*:509 (1993).
20. A.E. Forrest, Stochastic network modelling of convection and diffusion in porous catalysts, M.Sc. thesis, UMIST (1994).

Index